Applied Mathematical Sciences

Volume 69

Applied Mathematical Sciences

1. John: **Partial Differential Equations,** 4th ed.
2. Sirovich: **Techniques of Asymptotic Analysis.**
3. Hale: **Theory of Functional Differential Equations,** 2nd ed.
4. Percus: **Combinatorial methods.**
5. von Mises/Friedrichs: **Flid Dynamics.**
6. Freiberger/Grenander: **A Short Course in Computational Probability and Statistics.**
7. Pipkin: **Lectures on Viscoelasticity Theory.**
9. Friedrichs: **Spectral Theory of Operators in Hilbert Space.**
11. Wolovich: **Linear Multivariable Systems.**
12. Berkovitz: **Optimal Control Theory.**
13. Bluman/Cole: **Similarity Methods for Differential Equations.**
14. Yoshizawa: **Stability Theory and the Existence of Periodic Solution and Almost Periodic Solutions.**
15. Braun: **Differential Equations and Their Applications,** 3rd ed.
16. Lefschetz: **Applications of Algebraic Topology.**
17. Collatz/Wetterling: **Optimization Problems.**
18. Grenander: **Pattern Synthesis: Lectures in Pattern Theory, Vol I.**
20. Driver: **Ordinary and Delay Differential Equations.**
21. Courant/Friedrichs: **Supersonic Flow and Shock Waves.**
22. Rouche/Habets/Laloy: **Stability Theory by Liapunov's Direct Method.**
23. Lamperti: **Stochastic Processes: A Survey of the Mathematical Theory.**
24. Grenander: **Pattern Analysis: Lectures in Pattern Theory, Vol. II.**
25. Davies: **Integral Transforms and Their Applications,** 2nd ed.
26. Kushner/Clark: **Stochastic Approximation Methods for Constrained and Unconstrained Systems**
27. de Boor: **A Practical Guide to Splines.**
28. Keilson: **Markov Chain Models—Rarity and Exponentiality.**
29. de Veubeke: **A Course in Elasticity.**
30. Sniatycki: **Geometric Quantization and Quantum Mechanics.**
31. Reid: **Sturmian Theory for Ordinary Differential Equations.**
32. Meis/Markowitz? **Numerical Solution of Partial Differential Equations.**
33. Grenander: **Regular Structures: Lectures in Pattern Theory, Vol. III.**
34. Kevorkian/Cole: **Perturbation methods in Applied Mathematics.**
35. Carr: **Applications of Centre Manifold Theory.**
36. Bengtsson/Ghil/Källén: **Dynamic Meterology: Data Assimilation Methods.**
37. Saperstone: **Semidynamical Systems in Infinite Dimensional Spaces.**
38. Lichtenberg/Lieberman: **Regular and Stochastic Motion.**
39. Piccini/Stampacchia/Vidossich: **Ordinary Differential Equations in $\mathbf{R}^n$.**
40. Naylor/Sell: **Linear Operator Theory in Engineering and Science.**
41. Sparrow: **The Lorenz Equations: Bifurcations, Chaos, and Strange Attractors.**
42. Guckenheimer/Holmes: **Nonlinear Oscillations, Dynamical Systems and Bifurcations of Vector Fields.**
43. Ockendon/Tayler: **Inviscid Fluid Flows.**

(continued following index)

Martin Golubitsky
Ian Stewart
David G. Schaeffer

Singularities and Groups in Bifurcation Theory

Volume II

With 96 Illustrations

Springer-Verlag
New York Berlin Heidelberg
London Paris Tokyo

Martin Golubitsky
Department of Mathematics
University of Houston
Houston, TX 77004
USA

Ian Stewart
Mathematics Institute
University of Warwick
Coventry CV4 7AL
England

David G. Schaeffer
Department of Mathematics
Duke University
Durham, NC 27706
USA

Mathematics Subject Classification (1980): 35B32, 34D05, 22E70, 58F22, 76E30, 58F14, 34D20

Library of Congress Cataloging-in-Publication Data
(Revised for vol. 2)
Golubitsky, Martin.
Singularities and groups in bifurcation theory.
(Applied mathematical sciences; 51, 69)
Bibliography: v. 1, p. [455]-459.
Includes index.
1. Bifurcation theory. 2. Groups, Theory of.
I. Schaeffer, David G. II. Title. III. Series:
Applied mathematical sciences (Springer-Verlag
New York Inc.); 51, 69.
QA374.G59 1985 515.3′53 84-1414

Typeset by Asco Trade Typesetting Ltd., Hong Kong.
Printed and bound by R.R. Donnelley & Sons, Harrisonburg, Virginia.
Printed in the United States of America.

9 8 7 6 5 4 3 2 1

ISBN 0-387-96652-8 Springer-Verlag New York Berlin Heidelberg
ISBN 3-540-96652-8 Springer-Verlag Berlin Heidelberg New York

To the memory
of
Walter Kaufmann-Bühler

Preface

Bifurcation theory studies how the structure of solutions to equations changes as parameters are varied. The nature of these changes depends both on the number of parameters and on the symmetries of the equations. Volume I discusses how singularity-theoretic techniques aid the understanding of transitions in multiparameter systems. In this volume we focus on bifurcation problems with symmetry and show how group-theoretic techniques aid the understanding of transitions in symmetric systems.

Four broad topics are covered: group theory and steady-state bifurcation, equivariant singularity theory, Hopf bifurcation with symmetry, and mode interactions. The opening chapter provides an introduction to these subjects and motivates the study of systems with symmetry. As in Volume I, detailed Case Studies which illustrate how group-theoretic methods can be used to analyze specific problems arising in applications are included. In particular, the intriguing subject of pattern-formation in fluid systems and its relation to symmetry is discussed in Case Study 4 on Bénard convection and Case Study 6 on the Taylor–Couette system. All three Case Studies demonstrate the importance of spontaneous symmetry-breaking. The deformation of an incompressible elastic cube under traction, analyzed in Case Study 5, illustrates this topic so clearly that in the introductory chapter we have also used it to motivate the idea of spontaneous symmetry-breaking.

This volume may be used as a basis for two somewhat distinct one-semester courses, which may be summarized as "Equivariant Singularity Theory" and "Equivariant Dynamics." Only the first of these depends heavily on material from Volume I. More specifically, if the singularity-theoretic material in Volume I, Chapters I–V, IX, X, can be assumed, then equivariant singularity theory and its applications to bifurcation problems can be covered using Chapters XI–XV. If basics of steady-state and Hopf bifurcation can be assumed, as in Volume I, Chapters II and VIII, then a course on equivariant dynamics (as it relates to bifurcation theory) can be taught by omitting Chapters XIV and XV.

It has not escaped our notice that this volume contains much material. One reason is that we have worked out, in detail, the bifurcation theory for a number of symmetry groups and, where possible, have described specific applications based on those groups. To understand the theory of bifurcations with symmetry it is not necessary to work through the details of all of these examples. For this reason we have marked with a dagger (†) those sections that deal with examples of specific groups which are not needed later. In the same vein, sections which either develop aspects of the theory that are not needed elsewhere or contain proofs of theorems whose statements are all that is needed are marked with an asterisk (*). We suggest that these marked sections should be skimmed briefly on a first reading.

This volume has benefited greatly from discussions with many individuals. These include Giles Auchmuty, Pascal Chossat, Andrew Cliffe, John David Crawford, Jim Damon, Benoit Dionne, Gerhard Dangelmayr, Bill Farr, Bernold Fiedler, Terry Gaffney, Stephan van Gils, John Guckenheimer, Ed Ihrig, Gérard Iooss, Barbara Keyfitz, Edgar Knobloch, Martin Krupa, Bill Langford, Reiner Lauterbach, Jerry Marsden, Jan-Cees van der Meer, Ian Melbourne, James Montaldi, Mark Roberts, David Sattinger, Pat Sethna, Mary Silber, Jim Swift, Harry Swinney, Randy Tagg, and Andre Vanderbauwhede. We are grateful to them all and, indeed, to others too numerous to mention. We thank Wendy Aldwyn for drawing the figures. Our research in bifurcation theory has been generously supported by the National Science Foundation, NASA-Ames, the Applied Computational Mathematics Program of DARPA, the Energy Laboratory of the University of Houston, and the Science and Engineering Research Council of the United Kingdom.

Walter Kaufmann-Bühler provided us with constant encouragement and editorial expertise. We acknowledge our debt to him here and dedicate this volume to his memory.

Houston, Warwick, and Durham
September 1987

Martin Golubitsky
Ian Stewart
David G. Schaeffer

Contents of Volume II

Contents of Volume I

CHAPTER XI

Introduction

§0. Introduction

In Volume I we showed how techniques from singularity theory may be applied to bifurcation problems, and how complicated arrangements of bifurcations may be studied by unfolding degenerate singularities. Both steady-state and Hopf bifurcations proved amenable to these methods.

In this volume we extend the methods to systems with symmetry. There are many reasons for wishing to make such an extension. Many natural phenomena possess more or less exact symmetries, which are likely to be reflected in any sensible mathematical model. Idealizations such as periodic boundary conditions can produce additional symmetries. Certain mathematical contexts reveal unanticipated symmetry: for example, we saw in Chapter VIII that Hopf bifurcation can be treated as steady-state bifurcation with circle group symmetry $\mathbf{S}^1$.

In this chapter we attempt to explain why and how the occurrence of symmetries in systems of differential equations affects the types and multiplicity of solutions that bifurcate from an invariant steady state.

The chapter divides into three main sections, together with a final overview. In the first we explain what we mean by symmetries of a differential equation, and symmetries of a solution, either steady or time-periodic. We use these symmetries to define the problem of spontaneous symmetry-breaking. The discussion is motivated by two physical examples (suitable for "thought experiments"): the traction problem for deformation of an elastic cube and the oscillation of a circular hosepipe.

In §2 we briefly describe three techniques, which together form the basis of the theory that we present for analyzing symmetric systems of differential equations. They are:

Restriction to fixed-point subspaces,
Invariant theory,
Equivariant singularity theory.

We apply each technique to the traction problem, to illustrate the kind of information that it can provide.

In §3 we describe what is meant by mode interactions and indicate some of the bifurcation phenomena that they can lead to. We again illustrate the ideas with two examples. The first, exemplifying steady-state mode interaction, is the buckling of a rectangular plate (originally presented in Case Study 3). The second, illustrating the interaction between steady and periodic states, or between distinct periodic states, is the Taylor–Couette problem. This concerns the flow of fluid between two concentric rotating cylinders.

Finally in §4 we provide a brief overview of the structure of the main part of the book.

The theory of bifurcations with symmetry is very rich, combining methods from several areas of mathematics. It will take some time to draw all of these threads together. The aim of this chapter is to sketch, with a very broad brush, the main features of the overall framework; to describe a small number of key examples which motivate the point of view to be adopted; and to hint at some of the applications. We hope that such a preview will make the main part of the book easier to follow. Of course, this approach has a cost: we cannot expect the reader to appreciate the fine details of many of the arguments—though we hope that their spirit will be comprehensible.

§1. Equations with Symmetry

Bifurcation of steady-state and periodic solutions for systems of ODEs with symmetry differs from bifurcation in systems without symmetry. When studying specific model equations it is often possible to ignore any symmetries that may be present and to determine the bifurcation behavior directly. However, it is then impossible to disentangle those aspects of the analysis that depend on the specifics of the model from those that are *model-independent*, that is, due to symmetry alone. In addition, explicit use of symmetry-based principles may make the analysis easier, or at least more coherent.

This section is divided into four subsections. In the first we introduce symmetries of ODEs and discuss some simple implications for steady-state and periodic solutions. We illustrate this discussion in the next two subsections using two physically motivated examples: the deformation of an elastic cube under dead-load traction and the oscillation of a hosepipe induced by internal fluid flow. Finally we abstract some general principles from these examples and phrase them in the language of group theory. This last step allows us to introduce the fundamental notion of *spontaneous symmetry-breaking*—that equations may have more symmetry than their solutions—and state the basic problem of spontaneous symmetry-breaking in a general framework.

(a) Symmetries of Equations and of Solutions

We begin with some definitions and observations which are simple but fundamental. Let

$$dx/dt = f(x) \tag{1.1}$$

be a system of ODEs, where $f: \mathbb{R}^n \to \mathbb{R}^n$ is smooth, and let γ be an invertible $n \times n$ matrix. We say that γ is a *symmetry* of (1.1) if

$$f(\gamma x) = \gamma f(x) \tag{1.2}$$

for all $x \in \mathbb{R}^n$. An easy consequence of (1.2) is that if $x(t)$ is a solution to (1.1), then so is $\gamma x(t)$. In particular if $x(t) \equiv x_0$ is a steady state of (1.1), then so is γx_0. Either $\gamma x_0 \neq x_0$, in which case we have found a "new" steady state; or $\gamma x_0 = x_0$, in which case we say that γ is a *symmetry of the solution* x_0. When enumerating steady-state solutions of (1.1), it makes sense to enumerate only those that are *not* related by symmetries of f, since the remaining equilibria may be found by applying these symmetries.

There is a similar consequence for periodic solutions: if $x(t)$ is a T-periodic solution of (1.1), then so is $\gamma x(t)$. However, in the periodic case it is natural to widen the definition of a symmetry of a solution. Uniqueness of solutions to the initial value problem for (1.1) implies that the trajectories of $x(t)$ and $\gamma x(t)$ are either disjoint, in which case we have a "new" periodic solution, or identical, in which case $x(t)$ and $\gamma x(t)$ differ only by a phase shift. That is,

$$x(t) = \gamma x(t - t_0) \tag{1.3}$$

for some t_0. In this case we say that the pair (γ, t_0) is a *symmetry of the periodic solution* $x(t)$. Thus symmetries of periodic solutions have both a spatial component γ and a temporal component t_0.

(b) Deformation of an Elastic Cube

The first example is an incompressible elastic body in the shape of a unit cube, subjected to a uniform tension λ normal to each face, as in Figure 1.1. For small values of λ the undeformed cube is a stable equilibrium configuration for the body. But for large λ this shape, though still an equilibrium configuration, is unstable. What new equilibrium shapes should we expect to occur as a result of this loss of stability? Our discussion of this question is based on results of Ball and Schaeffer [1983].

First we must prescribe which deformations of the cube are permitted in

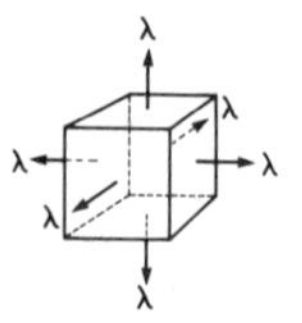

Figure 1.1. An elastic cube under uniform traction.

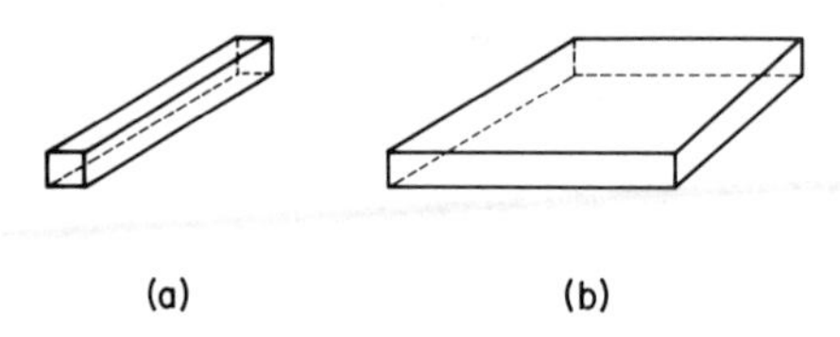

Figure 1.2. Possible equilibrium configurations of a deformed cube with two sides of equal length: (a) rod-like; (b) plate-like.

our model. For pedagogical purposes we consider only deformations whereby the cube becomes a rectangular parallelepiped. The consequences of symmetry already manifest themselves in this very restricted class of deformations. See Ball and Schaeffer [1983] for a more general setting.

Equilibrium configurations of the body may then be described completely by the (positive) side lengths l_1, l_2, l_3 subject to the incompressibility constraint

$$l_1 l_2 l_3 = 1. \tag{1.4}$$

Clearly any sensible mathematical model for the deformations of the cube should remain unchanged by permutations of the lengths of the sides. Thus if (l_1, l_2, l_3) represents an equilibrium configuration for a given load λ, then so do (l_2, l_1, l_3), (l_2, l_3, l_1), and so on. We can use this observation to classify the possible types of equilibria. After permuting the l_j we may assume that

$$0 < l_1 \le l_2 \le l_3. \tag{1.5}$$

Should all the l_j be equal, then by (1.4) we must have

$$l_1 = l_2 = l_3 = 1, \tag{1.6}$$

the trivial undeformed cube. There are two possible ways in which two of the l_j might be equal:

$$\begin{aligned} &\text{(a)} \quad l_1 = l_2 < l_3 \\ &\text{(b)} \quad l_1 < l_2 = l_3. \end{aligned} \tag{1.7}$$

These equilibria are pictured in Figure 1.2. In (1.7a) one side is longer than the other two, yielding a "rodlike" deformation; in (1.7b) one side is shorter than the other two, yielding a "platelike" deformation.

The remaining possibility is that all sides are of unequal length:

$$l_1 < l_2 < l_3. \tag{1.8}$$

The symmetries of the equilibria (1.6–1.8) are easily found. There are six permutations of the l_j, of which five are nontrivial. All six are symmetries of the trivial solution (1.6). The rodlike and platelike equilibria (1.7) have only one nontrivial permutational symmetry (interchanging (12) for rods and (23) for plates). Equilibria (1.8) have no nontrivial symmetries. These symmetries imply that solutions of the type (1.7) occur in threes whereas solutions of type (1.8) come six at a time.

We can now refine the question posed earlier. *At a loss of stability of the fully symmetric cubic shape, do we expect to find deformations having two equal sides, or deformations in which all sides are of unequal length*? This question is by its nature model-independent. If we knew the exact model then we could in principle solve the model and answer the question without any speculation. What we are asking is, given a typical f in (1.1) having as symmetries all permutations on three symbols, what kinds of equilibria do we expect to see?

There is an *incorrect* approach to this question which illustrates some of the difficulties. In some sense, in the "typical" case all of the l_j are unequal: imagine selecting them "at random." It is thus tempting to imagine that a typical branch of solutions bifurcating from the cubic state will consist of rectangular parallelepipeds whose sides have three different lengths. However, this argument applies "typicality" in the wrong context, because we are not asking the question for arbitrary f, but for functions *f with certain specified symmetries.* And in fact we show in §2 that generically bifurcating solutions will correspond to either rodlike or platelike equilibria. This is what we mean by a model-independent result depending only on the symmetries of the problem.

(c) Oscillation of a Hosepipe

Our second example concerns bifurcation to periodic solutions rather than steady states as previously. Consider a flexible hose of circular cross section suspended vertically, with water flowing through it at a rate λ. See Figure 1.3.

Figure 1.3. A circular hose suspended vertically.

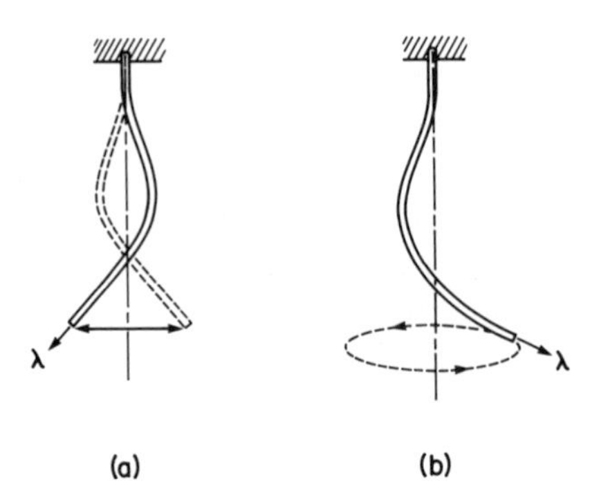

Figure 1.4. Standing waves (a) and rotating waves (b) in the motion of a hosepipe.

Assume that the stiffness is large compared to the gravitational restoring force, so that the effect of gravitation may be ignored. When the flow rate is small the hose remains steady, but when the rate is large the hose begins to oscillate. The hose is circular, and we assume a circularly symmetric model. That is, the ODE will have as symmetries all rotations and reflections about the origin in the horizontal plane.

As shown in Bajaj and Sethna [1982] the occurrence of circular symmetry does affect the expected types of oscillation. They show that there are two types of periodic solution that bifurcate from the invariant steady state (at the same value of λ). These are a *standing wave*, where the end of the hose swings like a pendulum in a fixed plane containing the vertical axis, and a *rotating wave*, where the end of the hose traverses a circle in a horizontal plane. See Figure 1.4. In both cases the shape of the hose is approximately that of a cantilever beam in its second mode.

These two types of oscillation may be distinguished by their symmetries. The standing wave has a spatial symmetry: reflection across the plane in which the hose oscillates. There is also a spatiotemporal symmetry: reflection in the vertical axis of the plane of oscillation coupled with a phase shift of half a period. As remarked in subsection (a), the overall symmetry of the equations (circular) implies that if one standing wave solution is found, then there must exist a family of such solutions (related by applying the circular symmetries of the system). This yields one standing wave solution for each plane containing the vertical axis. The set of all such solutions glues together in phase space to form an invariant 2-torus.

The rotating waves exhibit a mixture of spatial and temporal symmetry. The oscillation of the end of the hose takes a very special form. We can imagine—as is indeed the case—that the time t evolution of the hose is identical to rotation of the hose through an angle kt in the horizontal plane, where k is some constant. Now the overall circular symmetry implies that if a counterclockwise rotating wave solution occurs, then a clockwise rotating wave must also be present.

Bajaj and Sethna [1982] show that either of these periodic solutions can be stable. Which solution is stable depends on the system parameters, in particular the *mass ratio*—the ratio of the masses per unit length of the fluid and of the solid pipe.

At this stage we shall not attempt to explain why these two modes of oscillation appear, or why no other mode of oscillation is expected. What we stress at this point is that there is a theory of Hopf bifurcation with symmetry which gives model-independent information about periodic solutions, depending only on the symmetry assumed in the system. Chapter XVI contains the theoretical development, and Chapter XVII applies it in particular to the circularly symmetric case.

(d) Abstraction to the Language of Groups

The purpose of group theory is to provide a language for making rigorous statements about symmetry, so it is not surprising that groups play a prominent role in the sequel. In particular we note that for any given f the invertible matrices γ satisfying (1.2) always form a group. To see this we must show that if γ and δ are invertible matrices that satisfy (1.2), then

$$\begin{array}{ll}\text{(a)} & \gamma^{-1} \text{ satisfies (1.2), and} \\ \text{(b)} & \gamma\delta \text{ satisfies (1.2).}\end{array} \tag{1.9}$$

It is easy to check (1.9b), because

$$f(\gamma\delta x) = \gamma f(\delta x) = \gamma\delta f(x).$$

To check (1.9a) set $y = \gamma x$ in (1.2) to obtain

$$f(y) = \gamma f(\gamma^{-1} y) \tag{1.10}$$

and multiply (1.10) by γ^{-1}.

We are thus led to define the *group of symmetries* Γ of the ODE (1.1). The symmetries of any particular steady-state solution x_0 form a subgroup of Γ called the *isotropy subgroup* of x_0, defined by

$$\Sigma_{x_0} = \{\gamma \in \Gamma: \gamma x_0 = x_0\}. \tag{1.11}$$

Observe that in the traction problem of subsection (b) earlier the group of symmetries is $\mathbf{S}_3$, the six-element group of all permutations on three symbols. The isotropy subgroup of the trivial undeformed solution (1.6) is $\mathbf{S}_3$ itself. The isotropy subgroup of rod- or platelike solutions (1.7) is a two-element subgroup. The isotropy subgroup of solutions (1.8) is the trivial subgroup $\mathbb{1}$ consisting only of the identity permutation.

The distinct solutions to (1.1) forced by symmetry may also be described group-theoretically. For $x_0 \in \mathbb{R}^n$ we define the *group orbit* through x_0 to be

$$\Gamma x_0 = \{\gamma x_0: \gamma \in \Gamma\}. \tag{1.12}$$

If x_0 is an equilibrium solution to (1.1) then so is every point in its group orbit. In other words, a mapping f satisfying (1.2) for all $\gamma \in \Gamma$ must vanish on (unions of) group orbits. Elements γx_0 and δx_0 of a group orbit are equal if and only if $\delta^{-1}\gamma x_0 = x_0$, that is, $\delta^{-1}\gamma \in \Sigma_{x_0}$. In particular, if Γ is a finite group, then the number of distinct equilibria forced on (1.1) by the existence of one equilibrium solution x_0 is

$$|\Gamma|/|\Sigma_{x_0}|$$

where $|\ |$ indicates the number of elements, or *order*, of the group. The reader may verify this formula for the different types of equilibrium in the traction problem.

Next we rephrase, in general terms, the question concerning which types of solution to the traction problem can be expected to occur. Suppose that the ODE (1.1) depends on a bifurcation parameter λ, and that the equation has a symmetry group Γ for all λ. Further assume that there is a Γ-invariant steady state (that is, one whose isotropy subgroup is Γ), which for convenience we take to be $x_0 = 0$. Thus we have an ODE

$$dx/dt = f(x, \lambda)$$

where

$$f(\gamma x, \lambda) = \gamma f(x, \lambda)$$

for all $\gamma \in \Gamma$, and

$$f(0, \lambda) \equiv 0.$$

Finally assume that f has a singularity at $\lambda = 0$, so that

$$\det(df)_{0,0} = 0.$$

The fundamental question is:

> Generically, for which isotropy subgroups Σ should we expect to find bifurcating branches of steady states, having Σ as their group of symmetries?

We think of the full symmetry group Γ of the Γ-invariant steady state breaking to the symmetry group Σ of the bifurcating solution branch. This process is called *spontaneous symmetry-breaking.*

For the traction problem the answer will turn out to be the following: we expect symmetry to break from the permutation group $\mathbf{S}_3$ to the two-element subgroup $\mathbf{Z}_2$, rather than to the trivial subgroup $\mathbb{1}$.

The whole discussion may be repeated for Hopf bifurcation of a Γ-invariant steady state to periodic solutions. However, we must enlarge Γ to accommodate phase shifts. Two phase shifts $t_1, t_2 \in \mathbb{R}$ will have the identical effect if they differ by an integer multiple of the period T, so the "correct" group of phase shifts is $\mathbb{R}$ modulo $T\mathbb{Z}$, the *circle group* $\mathbf{S}^1$ for period T. The pairs (γ, t_0) of symmetry operations for T-periodic solutions thus lie in the group $\Gamma \times \mathbf{S}^1$ rather than Γ.

For the hosepipe the group of symmetries is $\mathbf{O}(2)$, the rotations and reflections of the plane that keep the origin fixed. Thus the symmetries of periodic solutions obtained by Hopf bifurcation are isotropy subgroups inside $\mathbf{O}(2) \times \mathbf{S}^1$. The standing waves have a $\mathbf{Z}_2$ reflectional symmetry inside $\mathbf{O}(2)$, and also a $\mathbf{Z}_2$ spatiotemporal reflection inside $\mathbf{O}(2) \times \mathbf{S}^1$. The rotating waves have the symmetry group

$$\widetilde{\mathbf{SO}}(2) = \{(\theta, \theta): \theta \in \mathbf{S}^1\}$$

where we identify $\mathbf{S}^1$ with the rotations in $\mathbf{O}(2)$. For further details and fine points on $\mathbf{O}(2)$ Hopf bifurcation see Chapter XVII.

It is worth remarking that this subsection contains the germ of an important idea, worth bearing in mind throughout. *There is an analogy between steady-state and Hopf bifurcation, in which* Γ *is replaced by* $\Gamma \times \mathbf{S}^1$.

§2. Techniques

The study of bifurcation problems with symmetry is complicated because symmetry often forces eigenvalues of high multiplicity. Thus, even after a Liapunov–Schmidt reduction (see Chapter VII), it is necessary to study bifurcation problems with several state variables. The main theme of this volume is that techniques exist to simplify the analysis of symmetric bifurcation problems, and that these techniques exploit the very same symmetries that cause the initial complication.

The three basic techniques are:

(a) Restriction to fixed-point subspaces,
(b) Invariant theory,
(c) Equivariant singularity theory.

In this section we illustrate these techniques by applying them to the traction problem described in §1b. We begin by discussing why symmetry forces multiple eigenvalues and then proceed to the three techniques.

Consider a bifurcation problem

$$\Phi(y, \lambda) = 0, \qquad \Phi(0, \lambda) \equiv 0 \tag{2.1}$$

with symmetry group Γ, so that

$$\Phi(\gamma y, \lambda) = \gamma \Phi(y, \lambda) \quad \text{for all } \gamma \in \Gamma. \tag{2.2}$$

Without loss of generality we may assume that a bifurcation occurs along the trivial solution $y = 0$ at $\lambda = 0$; that is, we let $L = (d\Phi)_{0,0}$ and assume that

$$\ker L \neq \{0\}.$$

By definition, we have a *multiple eigenvalue* whenever

$$\dim \ker L \geq 2. \tag{2.3}$$

Recall two facts from VII, §3:

(i) Matrices in Γ map $\ker L$ into itself (see Lemma VII, 3.2).
(ii) The Liapunov–Schmidt reduced bifurcation equation $g\colon \ker L \times \mathbb{R} \to \ker L$ may be chosen to commute with the matrices in Γ, so that

$$g(\gamma x, \lambda) = \gamma g(x, \lambda) \quad \text{for all } \gamma \in \Gamma. \tag{2.4}$$

(See Proposition VII, 3.3.)

The generalities of the Liapunov–Schmidt reduction imply that

$$(dg)_{0,0} = 0. \tag{2.5}$$

Thus, in the abstract study of bifurcation with symmetry, we may assume that the bifurcation problem $g(x, \lambda)$ satisfies (2.4, 2.5).

The occurrence of multiple eigenvalues is intimately related to the notion of irreducible group actions. We say that a subspace $V \subset \ker L$ is Γ-*invariant* if each matrix $\gamma \in \Gamma$ maps V to itself. The group Γ acts *irreducibly* on $\ker L$ if there are no nonzero proper Γ-invariant subspaces of $\ker L$. It is not difficult to prove (see XIII, §3 later) that generically, for bifurcation problems with symmetry group Γ, the action of Γ on $\ker L$ is irreducible.

When the group is trivial, that is, $\Gamma = \mathbb{1}$, every subspace is Γ-invariant, and irreducibility implies that $\dim \ker L = 1$. Thus, when no symmetry is present, generically we expect simple eigenvalues. That is, we expect bifurcation problems with only one state variable (after Liapunov–Schmidt reduction). Much of Volume 1 was devoted to a study of this important "special case." In fact we showed in Chapter IX that the simplest multiple eigenvalue problems for bifurcation without symmetry occur in codimension three.

All (compact Lie) groups except $\mathbf{Z}_2$ and $\mathbb{1}$ have irreducible actions on vector spaces of dimension greater than one. Thus, when symmetries are present, we may expect to find bifurcation problems where $\ker L$ has dimension greater than one. Similar remarks apply in the case of Hopf bifurcation: again multiple eigenvalues are commonplace.

In the remainder of this section we outline the three basic techniques mentioned earlier.

(a) Restriction to Fixed-point Subspaces

It is possible to reduce the effective dimension of $\ker L$ by prescribing in advance the symmetries of solutions being sought. Suppose we want to find steady states with symmetry Σ, a subgroup of Γ. Such solutions must lie in

$$\mathrm{Fix}(\Sigma) = \{y \in \ker L\colon \sigma y = y \text{ for all } \sigma \in \Sigma\}, \tag{2.6}$$

the *fixed-point subspace* for Σ. (Note that $\mathrm{Fix}(\Sigma)$ is a vector subspace of $\ker L$ since it is defined by a system of linear equations.) To find such steady states it suffices to solve the restricted system of equations

$$g|\text{Fix}(\Sigma) \times \mathbb{R} = 0.$$

This task is made easier by the fact that fixed-point subspaces are invariant for g; that is,

$$g\colon \text{Fix}(\Sigma) \times \mathbb{R} \to \text{Fix}(\Sigma). \tag{2.7}$$

To verify (2.7), suppose that $y \in \text{Fix}(\Sigma)$ and $\sigma \in \Sigma$. Then

$$\sigma g(y, \lambda) = g(\sigma y, \lambda) = g(y, \lambda)$$

so σ fixes $g(y, \lambda)$, and $g(y, \lambda) \in \text{Fix}(\Sigma)$. Thus, even though g may be a highly nonlinear mapping, symmetry forces g to have invariant linear subspaces.

By (2.7) the system of equations

$$g(y, \lambda) = 0, \qquad y \in \text{Fix}(\Sigma)$$

consists of $m = \dim \text{Fix}(\Sigma)$ equations, and m may be considerably smaller than $\dim \ker L$. The extreme case is when

$$\dim \text{Fix}(\Sigma) = 1. \tag{2.8}$$

Then (2.7) is a bifurcation problem in one state variable.

The fundamental observation about one-dimensional fixed-point subspaces is the *equivariant branching lemma*, which asserts that generically, for each Σ satisfying (2.8), there exists a unique branch of nontrivial steady-state solutions to $g = 0$ lying in $\text{Fix}(\Sigma) \times \mathbb{R}$. The proof of this lemma, first stated in this abstract form by Vanderbauwhede [1980], is elementary (see Theorem XIII, 3.2, and also Cicogna [1981]). The lemma's usefulness is based on the fact that an analytic statement (the existence of a branch of solutions with certain symmetry properties) is replaced by an algebraic one (the existence of subgroups with one-dimensional fixed-point subspace). The equivariant branching lemma provides one instance where symmetry complicates a bifurcation analysis (by forcing eigenvalues of high multiplicity), yet these same symmetries help simplify the search for solutions.

We consider here an example which enables us to show that generically rod- and platelike solutions occur in the traction problem of §1. We first discuss this example and then return to the traction problem later.

Identify $\mathbb{R}^2$ with $\mathbb{C}$ and consider the dihedral group $\mathbf{D}_3$ of all symmetries of an equilateral triangle. More precisely, $\mathbf{D}_3$ is a six-element group generated by an element of order 2 and an element of order 3:

$$\begin{aligned} &\text{(a)} \quad z \mapsto \bar{z} \\ &\text{(b)} \quad z \mapsto e^{2\pi i/3} z. \end{aligned} \tag{2.9}$$

The action of $\mathbf{D}_3$ on $\mathbb{C}$ is irreducible. For if V is a nonzero proper $\mathbf{D}_3$-invariant subspace it must be a line through the origin, but no line can be fixed by a $2\pi/3$ rotation.

It is easy to find a subgroup of $\mathbf{D}_3$ that has a one-dimensional fixed-point space. Let Σ be the two-element group generated by (2.9a). Then $\text{Fix}(\Sigma)$

consists of all z for which $z = \bar{z}$; that is, $\text{Fix}(\Sigma) = \mathbb{R}$. The equivariant branching lemma implies that generically there exist branches of solutions, to bifurcation problems with $\mathbf{D}_3$ symmetry, that have at least a reflectional symmetry.

We now relate this result to the traction problem. Recall from §1b that the symmetry group of the traction problem is the permutation group $\mathbf{S}_3$, and that equilibria are determined by a bifurcation problem defined on the space of side lengths (l_1, l_2, l_3) of the body, subject to the incompressibility constraint

$$l_1 l_2 l_3 = 1. \tag{2.10}$$

We seek solutions that bifurcate from the "trivial" cubic shape

$$l_1 = l_2 = l_3 = 1.$$

We claim that such a problem is abstractly isomorphic to the $\mathbf{D}_3$ bifurcation problem discussed previously. To see this observe that $\mathbf{S}_3$ and $\mathbf{D}_3$ are isomorphic groups. The isomorphism maps $z \to \bar{z}$ to the permutation (l_2, l_1, l_3) and $z \to e^{2\pi i/3}z$ to the permutation (l_2, l_3, l_1). Next, observe that the traction problem is posed on a two-dimensional submanifold of $\mathbb{R}^3$ defined by (2.10). Locally near $(1, 1, 1)$ we can pose this bifurcation problem on the tangent plane

$$l_1 + l_2 + l_3 = 3 \tag{2.11}$$

to the manifold (2.10). It is a simple exercise to show that $\mathbf{S}_3$ acts on the plane (2.11) as symmetries of an equilateral triangle (whose sides lie on the intersection of the plane (2.11) with the coordinate planes $l_j = 0$).

These identifications, coupled with the $\mathbf{D}_3$ result stated previously, show that generically we expect a unique branch of solutions to the traction problem whose solutions are symmetric under the permutation (l_2, l_1, l_3). That is, rod- and platelike solutions are to be expected. (The original convention $l_1 \leq l_2 \leq l_3$ is best abandoned at this stage: the two cases should be thought of as $l_1 = l_2 < l_3$ and $l_1 = l_2 > l_3$.) In the next subsection we use invariant theory to show that generically no other solutions occur near bifurcation.

We summarize the discussion. There is an algebraic criterion which gives a partial answer to the problem of spontaneous symmetry-breaking. Given a group of matrices Γ acting on $\mathbb{R}^n$, find all isotropy subgroups Σ whose fixed-point subspace is one-dimensional. Then (generically) the equivariant branching lemma lets us associate to each such isotropy subgroup a unique branch of steady-state solutions. The general problem of spontaneous symmetry-breaking—for which subgroups Σ do we generically have solutions?—remains unsolved; nevertheless this technique provides us with much information. In Chapter XIII we present several examples, the most interesting being bifurcation problems with spherical symmetry.

The same complications of high multiplicity occur in Hopf bifurcation for symmetric systems, since the purely imaginary eigenspace of dg is again invariant under Γ. There is a result, analogous to the equivariant branching lemma, which guarantees the existence of branches of periodic solutions corresponding to isotropy subgroups of $\Gamma \times \mathbf{S}^1$ with two-dimensional fixed-point subspace. See Theorem XVI, 3.1. Indeed the rotating and standing waves describing oscillations of the hosepipe in §1b may be understood from this point of view; see XVII, §2.

(b) Invariant Theory

It is well known that every even function in x is a function of x^2 and that every odd function is the product of an even function with x. These statements are trivial instances of two important theorems from the invariant theory of compact groups, theorems that provide a theoretical basis for many of the calculations presented in this volume. In particular, invariant theory lets us organize, in a rational way, the Taylor expansion of mappings satisfying the commutativity condition (2.4).

We begin our discussion with a definition. A function $f\colon \ker L \to \mathbb{R}$ is Γ-*invariant* if

$$f(\gamma x) = f(x) \quad \text{for all } \gamma \in \Gamma.$$

For example, if $\Gamma = \mathbf{Z}_2 = \{\pm 1\}$ acting on $\mathbb{R}$ by multiplication, then the invariant functions are just the even functions. Moreover, if f is a $\mathbf{Z}_2$-invariant polynomial, then there exists a polynomial p such that

$$f(x) = p(x^2).$$

In general, the Hilbert–Weyl theorem (XII, 4.2) implies that for compact groups Γ there always exist finitely many (homogeneous) Γ-invariant polynomials $u_1, \ldots, u_s$ such that every Γ-invariant polynomial f has the form

$$f(x) = p(u_1(x), \ldots, u_s(x)) \tag{2.12}$$

for some polynomial p.

Similarly, $g\colon \ker L \to \ker L$ is Γ-*equivariant* (or *commutes with* Γ) if

$$g(\gamma x) = \gamma g(x) \quad \text{for all } \gamma \in \Gamma.$$

The $\mathbf{Z}_2$-equivariant mapping are just odd functions, having the form

$$g(x) = p(x^2)x.$$

The theorem on invariant functions (2.12) may also be used to show that there exists a finite set of Γ-equivariant polynomial mappings $X_1, \ldots, X_t$ such that every Γ-equivariant polynomial mapping has the form

$$g = f_1 X_1 + \cdots + f_t X_t \tag{2.13}$$

where each f_j is a Γ-invariant polynomial. See Theorem XII, 5.2. Both theorems (2.12) and (2.13) may be generalized from polynomials to smooth germs (without changing the u_j or the X_k): see Theorems XII, 4.3, and 5.3.

Having stated these general facts, we consider $\mathbf{D}_3$ acting on $\mathbb{C}$ as in (2.9). It is easy to check that

$$u = z\bar{z} \quad \text{and} \quad v = \mathrm{Re}(z^3) \tag{2.14}$$

are $\mathbf{D}_3$-invariant, and that

$$X(z) = z \quad \text{and} \quad Y(z) = \bar{z}^2 \tag{2.15}$$

are $\mathbf{D}_3$-equivariant. A longer but still elementary calculation shows that every $\mathbf{D}_3$-invariant polynomial f has the form

$$f(z) = p(u, v) \tag{2.16}$$

and that every $\mathbf{D}_3$-equivariant polynomial mapping g has the form

$$g(z) = p(u, v)X(z) + q(u, v)Y(z). \tag{2.17}$$

See Examples XII, 4.1c and 5.4c.

Observe that (2.17) lets us enumerate the terms in the Taylor series of g. In particular there is

one linear term z,
one quadratic term $\bar{z}^2$,
one cubic term uz.

Their coefficients are $p(0,0)$, $q(0,0)$, and $p_u(0,0)$, respectively.

We discuss the implications of (2.17) for the traction problem. In particular we show that generically *only* rod- or platelike solutions bifurcate from the cubic shape. To establish this, observe that (2.17) implies that the traction problem leads to a bifurcation problem of the form

$$g(z, \lambda) = p(u, v, \lambda)z + q(u, v, \lambda)\bar{z}^2 = 0 \tag{2.18}$$

where the trivial cubic shape corresponds to $z = 0$. At a bifurcation point (which we take to be $\lambda = 0$) the linear term in (2.18) vanishes, so

$$p(0, 0, 0) = 0.$$

Generically in (2.18) the coefficient of the quadratic term is nonzero; that is,

$$q(0, 0, 0) \neq 0. \tag{2.19}$$

Using the special form (2.18) we now attempt to solve $g = 0$. If z and $\bar{z}^2$ are linearly independent (as vectors in $\mathbb{R}^2$) then $g = 0$ has a solution only if $p = q = 0$. The genericity assumption (2.19) precludes such solutions near $z = 0$. Thus all solutions bifurcating from the trivial solution must have z a real multiple of $\bar{z}^2$; that is, $\mathrm{Im}(z^3) = 0$. It is easy to check that points z with $\mathrm{Im}(z^3) = 0$ are those with a $\mathbf{Z}_2$ symmetry. (Note that there are *three* $\mathbf{Z}_2$ subgroups of $\mathbf{D}_3$, corresponding to the three reflectional axes of an equilateral triangle: our previous analysis refers to a particular one of these.) These points correspond to rod- or platelike solutions, which perforce are the generic possibilities.

Up till now we have not discussed whether, in our theory, the rod- and platelike solutions can be asymptotically stable. In the next subsection we show that generically they are unstable near bifurcation. Thus stable solutions can be found only globally, and we use singularity-theoretic methods to achieve this.

For general groups Γ it is a difficult task to find explicitly the generators $u_1, \ldots, u_s$ and $X_1, \ldots, X_t$ whose existence is asserted in (2.12, 2.13). There are, however, many examples of group actions where such generators can be found explicitly. In these cases the technique of writing the general Γ-equivariant mapping g in the form (2.13), and then using this form to analyze the zero-set $g = 0$, is a useful one. Even when explicit generators cannot be found, it may be possible to analyze the form of a Γ-equivariant mapping up to some particular order (say cubic or quintic) in the Taylor series, and to proceed from that description.

(c) Equivariant Singularity Theory

Until this point we have discussed only issues concerning the generic behavior of bifurcation problems with symmetry. In Volume I we described situations where the study of degenerate bifurcations proved useful, and we showed that singularity theory is an appropriate tool for such studies. As might be expected, the study of degenerate bifurcation problems with symmetry can be equally profitable, and here equivariant singularity theory is appropriate. We motivate our discussion by returning to the traction problem.

The rod- and platelike solutions to the traction problem which we have found previously are unstable (Exercise 2.1). In particular the genericity assumption (2.19) is precisely what is needed to prove these solutions unstable. Hence, to find asymptotically stable solutions with local methods, we must consider degenerate bifurcation problems and their universal unfoldings. (We note that the instability of the rod- and platelike solutions can be obtained by a general group-theoretic argument; see Theorem XII, 4.3.) We now provide a short summary of equivariant singularity theory, apply it to the least degenerate bifurcation problems with $\mathbf{D}_3$ symmetry, and interpret the results for the traction problem.

A singularity theory analysis depends on having a class of mappings, and a notion of equivalence sufficiently robust for the determinacy and unfolding theorems to hold. Such a setting exists for symmetric bifurcation problems. The class of mappings is the space of Γ-equivariant mappings. The notion of equivalence is Γ-*equivalence*, defined as follows: two Γ-equivariant bifurcation problems g and h are Γ-*equivalent* if

$$g(x, \lambda) = S(x, \lambda)h(X(x, \lambda), \Lambda(\lambda)) \tag{2.20}$$

where S is an invertible matrix, X is a diffeomorphism depending on λ, and Λ is a diffeomorphism. Certain equivariance conditions are imposed on S and X in order to preserve Γ-equivariance. See Definition XIV, 1.1, for details. Chapters XIV and XV present equivariant singularity theory, including examples.

We now preview the discussion of $\mathbf{D}_3$ symmetry in XV, §4. Suppose that

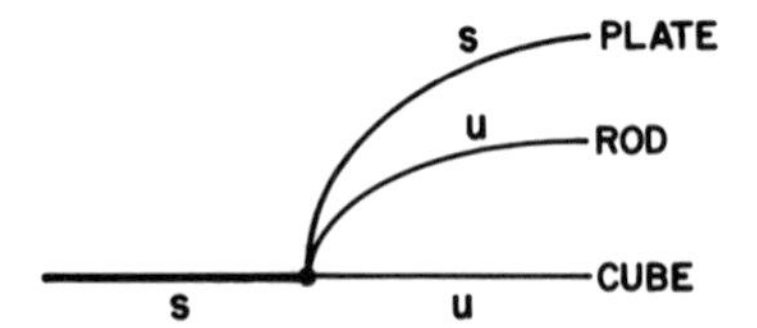

Figure 2.1. Least degenerate $\mathbf{D}_3$ bifurcation diagram.

(2.19) fails, so that

$$q(0, 0, 0) = 0.$$

Then a normal form for the least degenerate $\mathbf{D}_3$-equivariant bifurcation problem with $q = 0$ is

$$h(z, \lambda) = (\varepsilon u + \delta\lambda)z + (\sigma u + mv)\bar{z}^2 \tag{2.21}$$

where $\varepsilon, \delta, \sigma = \pm 1$ and $m \neq 0$ is a modal parameter (see Chapters V and X). This normal form is obtained subject to certain nondegeneracy conditions on terms of order 3, 4, and 5 in the general $\mathbf{D}_3$-equivariant bifurcation problem g (see Theorem XIV, 4.4). If we assume that the trivial solution is asymptotically stable for $\lambda < 0$ and that the $\mathbf{Z}_2$-symmetric solutions bifurcate supercritically, then $\varepsilon = 1$ and $\delta = -1$. A schematic bifurcation diagram is shown in Figure 2.1.

Two branches of $\mathbf{Z}_2$-symmetric solutions bifurcate from the trivial solution. These correspond to rod- and platelike solutions, respectively. The sign of the fourth order coefficient σ determines which is stable. We assume, for argument's sake, that the platelike solutions are asymptotically stable.

Of course our real interest lies not in the degenerate bifurcation problem itself, but in its perturbations. The universal unfolding for (2.21) in the world of $\mathbf{D}_3$ symmetry is obtained by adding one perturbation term $\alpha\bar{z}^2$. The notable qualitative features of the perturbed bifurcation diagrams are the changes in the stability of the $\mathbf{Z}_2$-symmetric solutions and the introduction of solutions with no nontrivial symmetry via secondary bifurcation. Moreover, depending on the sign of the fifth order coefficient m, these solutions with trivial symmetry may be asymptotically stable. Figure 2.2 shows the schematic bifurcation diagrams when $\sigma = 1$.

Finally, we interpret these diagrams for the traction problem. Suppose that by varying a parameter we find a degenerate $\mathbf{D}_3$ bifurcation problem in which the fifth order coefficient m happens to be positive (Figure 2.2b). Then, when the trivial cubic shape loses stability as the tension λ on the faces is increased, we see a jump to a platelike shape. As λ is increased further, the platelike shape loses stability and the body deforms into a rectangular parallelepiped with three unequal sides. Eventually, as λ is increased still further, the body deforms into a rodlike shape. In Case Study 5 we show, following Ball and Schaeffer [1983], that this scenario does occur in equations modeling an elastic body made of "Mooney–Rivlin material."

We end this section by noting that degeneracies can also be important for

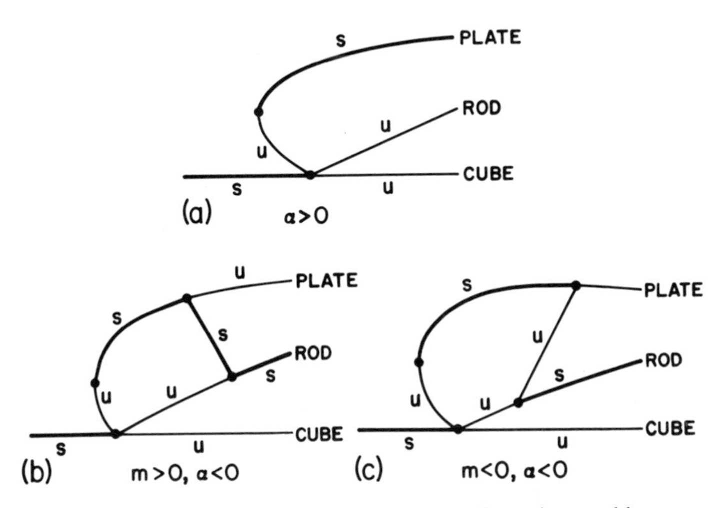

Figure 2.2. Universal unfolding of a $\mathbf{D}_3$ bifurcation problem.

Hopf bifurcation with symmetry, just as in the asymmetric case (VIII, §5). For example, recall the oscillations of a hosepipe from §1b. There are two possible dominant motions: rotating and standing waves. Either mode can be asymptotically stable, depending on the mass ratio β. Bajaj and Sethna [1982] show that there are critical values of β at which a transition of stability (or a degeneracy) occurs. This type of degeneracy will be studied in XVII, §5, using singularity-theoretic methods. The main result is the existence of an invariant 2-torus in the dynamics.

EXERCISES

2.1. Show that generically both rod- and platelike solutions to the traction problem must be unstable. *Hint*: Write the g of (2.18) in real coordinates (x, y) where $z = x + iy$. Compute the Jacobian matrix (dg) at $(x, 0)$, a typical point with $\mathbf{Z}_2$ symmetry, and show that the eigenvalues are $p \pm 2xq$. Finally observe that solutions $(x, 0)$ to $g = 0$ satisfy $p + xq = 0$. Hence show that the eigenvalues $p \pm 2xq$ are of opposite sign when $q \neq 0$.

2.2. Verify that u and v in (2.14) are $\mathbf{D}_3$-invariant and that X and Y in (2.15) are $\mathbf{D}_3$-equivariant.

2.3. Consider the action of $\mathbf{D}_3$ on $\mathbb{C}$ defined in (2.9). Show that points $z \in \mathbb{C}$ with isotropy subgroup isomorphic to $\mathbf{Z}_2$ satisfy $\mathrm{Im}(z^3) = 0$, $z \neq 0$.

2.4. Let $\mathbf{S}_3$ act on (l_1, l_2, l_3)-space by permuting coordinates. Show that the $\mathbf{S}_3$-invariant polynomials are generated by the elementary symmetric polynomials

$$u_1 = l_1 + l_2 + l_3$$

$$u_2 = l_1 l_2 + l_1 l_3 + l_2 l_3$$

$$u_3 = l_1 l_2 l_3.$$

§3. Mode Interactions

The simplest kinds of degeneracy occur in the linear terms of bifurcation equations. A linear degeneracy occurs when, at an equilibrium, the Jacobian matrix has eigenvalues on the imaginary axis. Steady-state bifurcation corresponds to zero eigenvalues, and Hopf bifurcation to purely imaginary eigenvalues. Of course, nondegeneracy conditions involving higher order terms must be satisfied in order to have the simplest steady-state bifurcation (the limit point $x^2 + \lambda$) and the simplest Hopf bifurcation. See Table IV, 2.3, and Theorem VIII, 3.2.

The standard dynamical systems theory of local bifurcation (see Guckenheimer and Holmes [1983]) counts codimension by the number of degeneracies that must be satisfied in order for a given singularity to occur. From this point of view (where no distinguished bifurcation parameter is present) limit points (or saddle-nodes) and Hopf bifurcations have codimension one. The codimension two singularities occur either through a degeneracy in the higher order terms or through a second linear degeneracy.

In Volume I we gave an account of a theory for organizing *nonlinear* degeneracies at simple eigenvalues. In the last two chapters of this volume we discuss phenomena associated with double *linear* degeneracies. Chapter XIX treats aspects of these double degeneracies in systems without symmetry, and Chapter XX deals with the same issues when circular $\mathbf{O}(2)$ symmetry is present. In this section we preview some of the points raised in those chapters.

The eigenspaces associated with eigenvalues of the Jacobian matrix are often called *modes*. In this language we may speak of "modes going unstable" as the corresponding eigenvalues cross the imaginary axis. Of course, the codimension two linear degeneracies then correspond to two modes going unstable simultaneously. Such degeneracies are interesting because nonlinear terms often couple the modes to create behavior more complicated than might be expected from them individually. These complications are said to be produced by nonlinear interactions of the two modes; more briefly, by *mode interaction.*

We divide mode interactions into three types:

(i) Steady-state/steady-state,
(ii) Steady-state/Hopf,
(iii) Hopf/Hopf.

We discuss briefly some notable features of these interactions. In (i) we see immediately a difference between the steady-state and dynamic theories of ODEs. A double zero eigenvalue can occur in a steady-state equation only in codimension three (see Chapter IX); however, the steady-state/steady-state mode interaction (two zero eigenvalues) occurs in codimension two. This apparent inconsistency is resolved by recalling that a 2×2 matrix with a double zero eigenvalue need not be zero. Consider the system of ODEs

$$\dot{x} = \begin{bmatrix} 0 & 1 \\ 0 & 0 \end{bmatrix} x + \cdots \qquad (x \in \mathbb{R}^2). \tag{3.1}$$

Locally, the steady states of this system may be found by Liapunov–Schmidt reduction. Since the Jacobian matrix is nonzero the system reduces to a bifurcation problem in one state variable where only limit points are expected. However, under perturbation the singularity (3.1) produces interesting dynamic behavior, namely periodic and homoclinic orbits. See Bogdanov [1975], Takens [1974], and Guckenheimer and Holmes [1983].

The most notable feature of steady-state/Hopf mode interactions is the occurrence of an invariant 2-torus in the dynamics. For Hopf/Hopf interactions it is the occurrence of an invariant 3-torus. In our subsequent discussion we emphasize those aspects of the dynamics that can be obtained by "steady-state" techniques. For example, the existence (but not the stability) of the 2-torus can be deduced from a change of stability along a branch of periodic solutions, and these periodic solutions can be found by static bifurcation techniques as in Chapter VIII. Thus "steady-state" techniques can find many states that are not steady.

Our discussion in §2 shows that when symmetries are present, multiple eigenvalues may be expected. This happens because the symmetries leave the associated eigenspace invariant. The multiplicity of eigenvalues is least when this action is irreducible, and this is the generic situation. It is then convenient to speak of the entire (irreducible) eigenspace as a single mode.

As we might expect, the high multiplicity causes complicated behavior in mode interactions. For example, when symmetry is present, just the enumeration of the possible mode interactions is complicated, because each irreducible representation of a group produces its own type of steady-state and Hopf bifurcation. Thus the mode interactions must be indexed according to the irreducible representations of the group. See Chapter XX, where details of mode interaction with $\mathbf{O}(2)$ symmetry are presented.

Mode interactions with symmetry usually lead to large numbers of eigenvalues on the imaginary axis. For example, when $\mathbf{O}(2)$ symmetry is present we expect to find, among other things, Hopf/Hopf interactions of codimension two, having eight purely imaginary eigenvalues. Thus in systems with circular symmetry and two parameters, we may expect to find parameter values where eight eigenvalues of the Jacobian lie on the imaginary axis. Despite this, the complications generated by these large numbers of critical eigenvalues often remain amenable to the techniques inspired by symmetry, as described in the previous section.

We mention two applications of mode interactions with symmetry: the buckling of a rectangular plate and the Couette–Taylor system. We showed in Case Study 3 that a steady-state/steady-state mode interaction occurs in the buckling of a rectangular plate, when only two parameters are varied. In the experiment, the plate is subjected to an end load λ. As λ is increased the undeformed state loses stability and the plate buckles. The buckled state has a well-defined wave number, the number of maxima of the deformed plate along its midline. The value of this wave number depends on the aspect ratio α, the ratio of width to length. At certain critical values of α the plate loses stability simultaneously to buckling patterns with two consecutive wave numbers. The $\mathbf{Z}_2 \oplus \mathbf{Z}_2$ symmetry permits this double eigenvalue to occur in codimension two. One product of this nonlinear interaction is the existence of mixed mode buckled states.

One of the most famous experiments in fluid mechanics, the Couette–Taylor system, provides another physical example where mode interactions with symmetry occur. We summarize some of the results here; details are in Case Study 6. The experimental apparatus consists of a fluid contained between two independently rotating coaxial cylinders. The flow is described by the Navier–Stokes equations, in which the velocity field is the principal unknown. A solution of these equations is time-independent (or a steady state) if the velocity field does not vary in time (even though the fluid moves). For example, Couette flow, in which fluid particles move in circles around the axis at a speed depending only on the radius, is time-independent. Although, strictly speaking, Couette flow is only a mathematical idealization which treats the apparatus as infinitely long, nonetheless it is the basis of the bifurcation analysis.

Both time-independent and time-dependent transitions from Couette flow have been observed experimentally. G.I. Taylor [1923] noticed that when the outer cylinder is held fixed and the speed λ of the inner cylinder is increased, Couette flow loses stability to a time-independent state now known as *Taylor vortices*; see Figure 1.1(b) of Case Study 6. In his experiments Taylor also found that in the counterrotating case, when the outer cylinder is set in motion at a fixed and sufficiently large speed $\alpha < 0$, then as $\lambda > 0$ is increased, Couette flow loses stability to a time-dependent state called *spirals*. See Figure 1.1(c) of Case Study 6 and Andereck, Liu, and Swinney [1986]. It is not surprising, then, that there is a critical speed of counterrotation for which Couette flow loses stability simultaneously to two modes as λ is increased. That is, there is a codimension two steady-state/Hopf interaction. Moreover, because of various symmetries in the mathematical models for this experiment, the eigenvalues are double.

Many other states have been observed for different speeds of the cylinders. We mention two here: wavy vortices and twisted vortices, illustrated in Figures 1.1(d, e) of Case Study 6. These states can be produced mathematically from this mode interaction, as we show in Case Study 6.

Hopf/Hopf mode interactions also occur in the Couette–Taylor system. The spiral state has an associated azimuthal wave number m defined by the number of distinct spirals that intertwine in the apparatus. When Couette flow loses stability to spirals, the actual wave number depends on the speed α of counterrotation. In fact, there are critical speeds α where Couette flow loses stability simultaneously to spirals with wave numbers m and $m + 1$. Since the bifurcation to spirals is a Hopf bifurcation, we have a Hopf/Hopf interaction (with eight eigenvalues on the imaginary axis). These interactions have been studied by Chossat [1985a] and Chossat *et al.* [1987].

§4. Overview

Bifurcations of systems without symmetry occur generically at a simple eigenvalue, and may be of steady-state or Hopf type, corresponding respectively to zero or imaginary eigenvalues. Degeneracies are of two basic types: *linear* degeneracies (or multiple eigenvalues) leading to mode interaction and *nonlinear* degeneracies in higher order terms leading to complicated branching patterns.

When symmetry is present, a similar picture applies, although multiple eigenvalues are commonplace even in the generic case. Instead, the basic "unit" of which bifurcations are built is an *irreducible* representation of the group Γ (or of $\Gamma \times \mathbf{S}^1$ in the Hopf case). Again it is necessary to understand:

(a) The generic case (steady-state or Hopf),
(b) Nonlinear degeneracies (steady-state or Hopf),
(c) Linear degeneracies (steady-state/steady-state, steady-state/Hopf, and Hopf/Hopf mode interaction).

The behavior depends on the choice of the representations involved, so that many more individual cases arise, even for a fixed group Γ.

The mathematical methods employed may be described as being either *geometric* or *algebraic*. Geometric methods center around the group-theoretic features of the analysis: representation theory and the yoga of fixed-point subspaces and isotropy subgroups. In a sense, these methods attempt to extract as much information as possible from *linear* data. They apply especially to the generic case, and to finding the correct setting for mode interactions.

The algebraic methods are invariant theory and equivariant singularity theory. They are designed to deal with the effects of higher order terms—"genuine nonlinearities." They occur in their purest form in the study of nonlinear degeneracies. In practice the boundaries are not so clear-cut, and both methods tend to become intertwined.

The pattern of development of the remainder of this volume is as follows.

Geometry (Chapters XII–XIII). Chapter XII is a relatively concrete introduction to Lie groups and their representations, and to invariant theory, concentrating on those results (mainly the simpler ones) that prove useful in the sequel. Chapter XIII describes the geometry of group actions and proves the equivariant branching lemma, a fundamental result in steady-state bifurcation. The groups $\mathbf{D}_n$ (symmetries of a regular n-gon), $\mathbf{SO}(3)$ (rotations in $\mathbb{R}^3$), and $\mathbf{O}(3)$ (rotations and reflections in $\mathbb{R}^3$) are discussed as examples, the latter two at some length since they have an infinite family of irreducible representations. These geometric ideas are applied in Case Study 4 to the problem of Bénard convection in the plane.

Algebra (*Chapters XIV–XV*). Chapter XIV sets up equivariant singularity theory, concentrating on the recognition problem, tangent spaces, and intrinsic ideals, by analogy with Chapter II. Similarly Chapter XV develops unfolding theory, by analogy with Chapter III, and includes proofs of the main theorems, promised from Volume I. The ideas are illustrated using the dihedral group $\mathbf{D}_3$ (the symmetry group of an equilateral triangle) and its relation to the spherical Bénard problem via spherical harmonics of order 2. Case Study 5 shows how to apply the algebraic methods to the traction problem for an elastic cube, continuing the analysis outlined previously in §2.

Hopf Bifurcation (*Chapters XVI–XVIII*). At this stage the theory moves away from static bifurcation and begins to acquire dynamic aspects. Chapter XVI develops a general theory of equivariant Hopf bifurcation, concentrating on existence and stability results. Chapter XVII applies this methodology to Hopf bifurcation with circular symmetry (the group $\mathbf{O}(2)$), considering both the generic case and nonlinear degeneracies (by a trick: reduction to amplitude equations). Quasi-periodic motion on a torus occurs here. The closely related groups $\mathbf{SO}(2)$, $\mathbf{Z}_n$, and $\mathbf{O}(n)$ (acting on $\mathbb{R}^n$) are also discussed. More complicated examples are dealt with in Chapter XVIII. Hopf bifurcation with dihedral group symmetry $\mathbf{D}_n$ is studied in detail and applied to oscillators coupled in a ring. Hopf bifurcation with spherical symmetry and Hopf bifurcation on the hexagonal lattice (relevant to doubly diffusive systems) are sketched.

Mode Interactions (*Chapters XIX–XX*). Chapter XIX discusses mode interactions without any prescribed symmetry, concentrating on the steady-state/Hopf and Hopf/Hopf cases. Because of the natural $\mathbf{S}^1$ symmetry of Hopf bifurcation, these problems acquire $\mathbf{Z}_2$ and $\mathbf{Z}_2 \oplus \mathbf{Z}_2$ symmetry during the analysis (rendering results from Volume I applicable). Finally Chapter XX considers mode interactions with $\mathbf{O}(2)$ symmetry.

The results are applied to Taylor–Couette flow in long cylinders (i.e., subject to periodic boundary conditions) in Case Study 6, which brings together virtually all of the ideas developed in this volume. The outcome is a coherent description, in symmetry terms, of some of the prechaotic behavior observed in this much-studied experiment.

CHAPTER XII

Group-Theoretic Preliminaries

§0. Introduction

The basic theme of this volume is that the symmetries of bifurcating systems impose strong restrictions on the form of their solutions and the way in which the bifurcation may take place. There are two major subthemes, which we might term "geometric" and "algebraic." These lead us to introduce two pieces of mathematical machinery: group representation theory and equivariant singularity theory. The aim of this chapter is to describe, in a fairly concrete fashion, the requisite mathematical background. In this manner we hope to make the methods accessible to a wide audience.

A symmetry of a system $\mathscr{X}$ is a transformation of $\mathscr{X}$ that preserves some particular structure. The set Γ of all such transformations has seveal pleasant properties, which can be summarized by saying that Γ is a group. In this book $\mathscr{X}$ is a real vector space $\mathbb{R}^n$, the transformations are linear mappings $\gamma: \mathbb{R}^n \to \mathbb{R}^n$, and the structure to be preserved is a particular bifurcation problem. For example, consider the static bifurcation problem

$$g(x, \lambda) = 0 \tag{0.1}$$

where $g: \mathbb{R}^n \times \mathbb{R} \to \mathbb{R}^n$ is a smooth (C^∞) mapping, λ being the bifurcation parameter. By "preserved" we mean that for all $\gamma \in \Gamma$

$$g(\gamma x, \lambda) = \gamma g(x, \lambda) \tag{0.2}$$

so that every $\gamma \in \Gamma$ commutes with g. It follows that x is a solution if and only if γx is, so the *solution set* to $g = 0$ is preserved by the symmetries γ. The "geometric" subtheme is the study of how Γ transforms $\mathbb{R}^n$; the "algebraic" subtheme deals with the use of (0.2) to restrict the form of g.

We therefore begin in §1a with some group theory. We introduce the idea of a Lie group Γ acting on a space $\mathbb{R}^n$ and describe fundamental examples including the orthogonal group $\mathbf{O}(n)$, the circle group $\mathbf{S}^1$, the dihedral group $\mathbf{D}_n$ and the n-torus $\mathbf{T}^n$. A given abstract group Γ can act as transformations of a space in many ways; this is discussed in §1b and leads to the ideas of an *action* and a *representation* of Γ. These are two slightly different ways of looking at the same basic idea: a group of $n \times n$ matrices that is isomorphic, as an abstract group, to Γ.

A Lie group has topological as well as algebraic properties, and the important ones for this book are compactness and, to a lesser extent, connectedness. The representation theory of a *compact* Lie group is especially well understood, and we shall confine attention throughout to the compact case. (Every finite group is compact, and so are $\mathbf{O}(n)$, $\mathbf{S}^1$, and $\mathbf{T}^n$.) In §1c we discuss the existence on a compact Lie group of an invariant (Haar) integral, which is important in a number of situations because it allows us to average over the group. For example, it permits us to assume that Γ acts by orthogonal transformations of $\mathbb{R}^n$.

In §2 we describe the decomposition of a given representation into simpler ones, called *irreducible* representations. In fact, if Γ is a compact Lie group acting on $V = \mathbb{R}^n$, then we can write V as a direct sum

$$V = V_1 \oplus V_2 \oplus \cdots \oplus V_s$$

of subspaces V_j, each invariant under Γ, such that V_j has no Γ-invariant subspaces other than $\{0\}$ and V_j. These "irreducible components" of V are the fundamental building blocks of representation theory. The process of decomposing V is in a sense analogous to that of diagonalizing a matrix and is done for the same purpose—to simplify the mathematics.

In §3 we discuss *linear* maps $\mathbb{R}^n \to \mathbb{R}^n$ that commute with an action of Γ. This discussion has important implications for bifurcation problems (0.1) that satisfy (0.2), because the linearization $(dg)_0$ must commute with Γ. Two main points are made. The first is that there is a notion stronger than irreducibility, *absolute irreducibility*, which ensures that only scalar multiples of the identity commute with Γ. The second is that certain uniquely defined subspaces must be invariant under any mapping that commutes with Γ. We shall use these ideas to restrict the form of $(dg)_0$.

§§4–6 develop the "algebraic" subtheme. In §4 we consider *invariant* functions $f\colon \mathbb{R}^n \to \mathbb{R}$, that is, functions such that

$$f(\gamma x) = f(x), \qquad (x \in \mathbb{R}^n, \gamma \in \Gamma).$$

There are two main results. The first, due to Hilbert and Weyl, states that (when Γ is compact) the *polynomial* invariants are generated by a finite set of polynomials $u_1, \ldots, u_s$. The second, due to Schwarz, states that every *smooth* invariant f is of the form $h(u_1, \ldots, u_s)$ for a smooth function h. We give examples for the main groups of interest: the proofs are postponed until §6.

In §5 we describe analogous results, due to Poénaru, for *equivariant* mappings, that is, mappings $g\colon \mathbb{R}^n \to \mathbb{R}^n$ that commute with Γ as in (0.2). We emphasize the simple but crucial fact that if $f(x)$ is invariant and $k(x)$ is equivariant, then $f(x)k(x)$ is also equivariant. In more abstract language, the space $\vec{\mathscr{E}}(\Gamma)$ of equivariant mappings is a module over the ring $\mathscr{E}(\Gamma)$ of invariant functions. These results are needed in Chapters XIV–XV to set up equivariant singularity theory.

In §6 we discuss the proofs of four theorems from §§4–5: the Hilbert–Weyl theorem, Schwarz's theorem, and their equivariant analogues. This section may be omitted if desired.

In §7 we return to group theory and present three results about torus groups which will be needed in Chapters XIX–XX on mode interactions. This section may be omitted on first reading.

§1. Group Theory

In order to make precise statements about symmetries, the language and point of view of group theory are indispensable. In this section and the next we present some basic facts about Lie groups. We assume that the reader is familiar with elementary group-theoretic concepts such as subgroups, normal subgroups, conjugacy, homomorphisms, and quotient (or factor) groups. We also assume familiarity with elementary topological concepts in $\mathbb{R}^n$ such as open, compact, and connected sets. See Richtmeyer [1978]. Fortunately we do not require the deeper results from the theory of Lie groups, so the material presented here should prove reasonably tractable. We have adopted a fairly concrete point of view in the hope that this will make the ideas more accessible to readers having only a nodding acquaintance with modern algebra.

We treat three main topics in this section. The first consists of basic definitions and examples. The second is the beginnings of representation theory. The third is the existence of an invariant integral, allowing us to employ averaging arguments which in particular let us identify any representation of a compact Lie group with a group of orthogonal transformations.

(a) Lie Groups

Let $\mathbf{GL}(n)$ denote the group of all invertible linear transformations of the vector space $\mathbb{R}^n$ into itself, or equivalently the group of nonsingular $n \times n$ matrices over $\mathbb{R}$. For our purposes we shall define a *Lie group* to be a closed subgroup Γ of $\mathbf{GL}(n)$. In the literature these are called *linear Lie groups*, and the term *Lie group* is given a more general definition. However, it is a theorem that every compact Lie group in this more general sense is topologically isomorphic to a linear Lie group; see Bourbaki [1960]. By *closed* we mean the following. The space of all $n \times n$ matrices may be identified with $\mathbb{R}^{n^2}$, which contains $\mathbf{GL}(n)$ as an open subset. Then Γ is a closed subgroup if it is a closed subset of $\mathbf{GL}(n)$ as well as a subgroup of $\mathbf{GL}(n)$. A *Lie subgroup* of Γ is just a closed subgroup in the same sense.

By defining Lie groups as closed groups of matrices we avoid discussing some of their topological and differentiable structure. However, we often wish to refer to a Lie group by the name of its associated abstract group, a practice that is potentially confusing. For example, the two-element group $\mathbf{Z}_2 = \{\pm 1\}$ is isomorphic as an abstract group to the subgroup $\{I_n, -I_n\}$ of $\mathbf{GL}(n)$ for any n, where I_n is the $n \times n$ identity matrix. Usually, the precise group of matrices in question will be specified by the context. We often use a phrase such as "$\mathbf{Z}_2$ is the Lie group $\{\pm I_2\}$" rather than the more precise but cumbersome phrase "$\mathbf{Z}_2$ is isomorphic to the Lie group $\{\pm I_2\}$." This practice should not cause confusion.

We now give some examples of Lie groups which will prove useful throughout the book.

EXAMPLES 1.1.

(a) The n-dimensional *orthogonal group* $\mathbf{O}(n)$ consists of all $n \times n$ matrices A satisfying

$$AA^t = I_n.$$

Here A^t is the transpose of A.

(b) The *special orthogonal group* $\mathbf{SO}(n)$ consists of all $A \in \mathbf{O}(n)$ such that $\det A = 1$. The group $\mathbf{SO}(n)$ is often called the n-dimensional *rotation group.* In particular $\mathbf{SO}(2)$ consists precisely of the planar rotations

$$R_\theta = \begin{bmatrix} \cos\theta & -\sin\theta \\ \sin\theta & \cos\theta \end{bmatrix}. \tag{1.1}$$

In this way, $\mathbf{SO}(2)$ may be identified with the *circle group* $\mathbf{S}^1$, the identification being $R_\theta \mapsto \theta$. The group $\mathbf{O}(2)$ is generated by $\mathbf{SO}(2)$ together with the *flip*

$$\kappa = \begin{bmatrix} 1 & 0 \\ 0 & -1 \end{bmatrix}. \tag{1.2}$$

(c) Let $\mathbf{Z}_n$ denote the *cyclic group* of order n. (Recall that the *order* of a finite group is the number of elements that it contains). We may identify $\mathbf{Z}_n$ with the group of 2×2 matrices generated by $R_{2\pi/n}$; thus $\mathbf{Z}_n$ is a Lie group.

(d) The *dihedral group* $\mathbf{D}_n$ of order $2n$ is generated by $\mathbf{Z}_n$, together with an element of order 2 that does not commute with $\mathbf{Z}_n$. For definiteness, we identify $\mathbf{D}_n$ with the group of 2×2 matrices generated by $R_{2\pi/n}$ and the flip κ (1.2). This clearly exhibits $\mathbf{D}_n$ as a Lie group. Geometrically $\mathbf{D}_n$ is the symmetry group of the regular n-gon, whereas $\mathbf{Z}_n$ is the subgroup of rotational symmetries.

(e) All finite groups are isomorphic to Lie groups; see Exercise 1.2.

(f) The n-dimensional torus $\mathbf{T}^n = \mathbf{S}^1 \times \cdots \times \mathbf{S}^1$ (n times) is isomorphic to a

Lie group. To show this, identify $\theta \in \mathbf{T}^n$ with the matrix

$$\begin{bmatrix} R_{\theta_1} & 0 & 0 & \dots & 0 \\ 0 & R_{\theta_2} & 0 & \dots & 0 \\ 0 & 0 & R_{\theta_3} & \dots & 0 \\ \cdot & \cdot & \cdot & \dots & \cdot \\ 0 & 0 & 0 & \dots & R_{\theta_n} \end{bmatrix}$$

in $\mathbf{GL}(2n)$.

(g) $\mathbb{R}^n$ is isomorphic to the group of matrices of the form

$$\begin{bmatrix} 1 & a_1 & a_2 & \dots & a_n \\ 0 & 1 & 0 & \dots & 0 \\ & & \dots & & \\ 0 & 0 & \dots & & 1 \end{bmatrix} \in \mathbf{GL}(n+1)$$

where $a_j \in \mathbb{R}, j = 1, \dots, n$.

It is important at the outset to eliminate one potential source of confusion. We have already seen that it is possible for a single abstract group to occur in more than one way as a group of matrices. The question that must be addressed is, when should two matrix groups which are isomorphic as abstract groups be considered as essentially the same? This question leads directly into representation theory and is dealt with in subsection (b). To illustrate what is involved, observe on the one hand that changing the basis in $\mathbb{R}^n$ will change the actual matrices that appear in a given Lie group—surely just a cosmetic change. On the other hand, consider the following two groups of matrices isomorphic to $\mathbf{Z}_2$:

$$\{I_2, -I_2\} \tag{1.3}$$

and

$$\left\{\begin{bmatrix} 1 & 0 \\ 0 & 1 \end{bmatrix}, \begin{bmatrix} -1 & 0 \\ 0 & 1 \end{bmatrix}\right\}. \tag{1.4}$$

There is a definite geometric distinction between (1.3), where the element of order 2 in $\mathbf{Z}_2$ is a rotation, and (1.4), where it is a reflection. Such a distinction is often important in the theory.

Because $\mathbb{R}^{n^2}$ is a topological space, we can talk about topological properties of Lie groups as well as algebraic ones. In particular we say that a Lie group Γ is *compact* or *connected* if it is compact or connected as a subset of $\mathbb{R}^{n^2}$. Equivalently, Γ is compact if and only if the entries in the matrices defining Γ are bounded. It follows that $\mathbf{O}(n)$, $\mathbf{SO}(n)$, $\mathbf{T}^n$, and all finite groups are compact; but $\mathbb{R}^n$ and $\mathbf{GL}(n)$ are not. Compactness is crucial for much of the theory we develop here. It would be of great significance for applications to modify the theory so that it extends to suitable noncompact groups. For an example, see Case Study 4.

The identity element of Γ is denoted by I. In fact, if $\Gamma \subset \mathbf{GL}(n)$ we must have $I = I_n$, the $n \times n$ identity matrix. The trivial group $\{I\} = \{I_n\}$ is denoted by $\mathbb{1}_n$, or more commonly by $\mathbb{1}$ when the size of matrix is clear from the context.

As a subset of $\mathbb{R}^{n^2}$, the group Γ splits into connected components. The connected component that contains I is denoted Γ^0. For example

$$\mathbf{O}(n)^0 = \mathbf{SO}(n).$$

Being a connected component, Γ^0 is a closed subset of Γ. Since Γ is closed in $\mathbf{GL}(n)$, so is Γ^0. Thus Γ^0 is a Lie subgroup of Γ and is compact if Γ is. Moreover, Γ^0 is a normal subgroup of Γ. To see why, recall that $\Sigma \subset \Gamma$ is *normal* if for each $\gamma \in \Gamma$ we have $\Sigma = \gamma\Sigma\gamma^{-1}$ as a set of matrices. Now $\gamma\Gamma^0\gamma^{-1}$ is a connected component of Γ since matrix multiplication is continuous, and it contains $\gamma I \gamma^{-1} = I$. Therefore, $\gamma\Gamma^0\gamma^{-1} = \Gamma^0$, so Γ^0 is normal.

It is not difficult (Exercise 1.3) to show that a compact Lie group Γ has a finite number of connected components, and hence that Γ/Γ^0 is finite.

(b) Representations and Actions

Let Γ be a Lie group and let V be a finite-dimensional real vector space. We say that Γ *acts* (*linearly*) *on* V if there is a continuous mapping (the *action*)

$$\begin{aligned} \Gamma \times V &\to V \\ (\gamma, v) &\mapsto \gamma \cdot v \end{aligned} \tag{1.5}$$

such that:

(a) For each $\gamma \in \Gamma$ the mapping $\rho_\gamma \colon V \to V$ defined by

$$\rho_\gamma(v) = \gamma \cdot v \tag{1.6}$$

is linear.

(b) If $\gamma_1, \gamma_2 \in \Gamma$ then

$$\gamma_1 \cdot (\gamma_2 \cdot v) = (\gamma_1 \gamma_2) \cdot v. \tag{1.7}$$

The mapping ρ that sends γ to $\rho_\gamma \in \mathbf{GL}(V)$ is then called a *representation of* Γ on V. Here $\mathbf{GL}(V)$ is the group of invertible linear transformations $V \to V$. By abuse of language we will also talk of "the representation V." In the sequel we shall often omit the dot and write γv for $\gamma \cdot v$, but for the remainder of this section we retain the dot for clarity. As illustrated shortly, linear actions and representations are essentially identical concepts, differing only in viewpoint. In fact, both (1.5) and ρ must be analytic; see Montgomery and Zippin [1955].

For example, there is an action of the circle group $\mathbf{S}^1$ on $\mathbb{C} \equiv \mathbb{R}^2$ given by

$$\theta \cdot z = e^{i\theta} z \qquad (\theta \in \mathbf{S}^1, z \in \mathbb{C}).$$

We verify that this is an action. Clearly (a) holds. To check (b), calculate

$$\theta_1 \cdot (\theta_2 \cdot z) = \theta_1 \cdot (e^{i\theta_2} z) = e^{i\theta_1} e^{i\theta_2} z = e^{i(\theta_1 + \theta_2)} z = (\theta_1 + \theta_2) \cdot z,$$

where by an accident of notation $\theta_1 + \theta_2$ is the "product" in the group $\mathbf{S}^1$. This action gives rise to a representation ρ of $\mathbf{S}^1$ for which ρ_θ is the rotation matrix

$$\begin{bmatrix} \cos\theta & -\sin\theta \\ \sin\theta & \cos\theta \end{bmatrix}$$

on $\mathbb{R}^2 \equiv \mathbb{C}$. The difference in viewpoint is that an *action* tells us how a group element γ transforms a given *element* $v \in V$, whereas a *representation* tells us how γ transforms the entire *space* V. More technically, ρ defines a homomorphism of Γ into $\mathbf{GL}(V)$; see Exercise 1.4. An action of Γ on V may be defined by specifying (1.5) only on generators of Γ, as long as this action is consistent in the sense that (1.7) is satisfied.

Examples 1.2.

(a) Every linear Lie group Γ is a group of matrices in $\mathbf{GL}(n)$ for some n. As such, Γ has a natural action on $V = \mathbb{R}^n$ given by matrix multiplication.

(b) Every group Γ has a *trivial* action on $V = \mathbb{R}^n$ defined by $\gamma \cdot x = x$ for all $x \in \mathbb{R}^n$, $\gamma \in \Gamma$.

(c) For every integer k the circle group $\mathbf{S}^1$ has an action on $V = \mathbb{C} \equiv \mathbb{R}^2$ defined by

$$\theta \cdot z = e^{ik\theta} z. \tag{1.8}$$

Notice that $k = 0$ corresponds to the trivial action of example (b). The action for $k = 1$ is the one discussed previously in the text.

(d) Each action of $\mathbf{S}^1 = \mathbf{SO}(2)$ defined in (c) extends to an action of $\mathbf{O}(2)$ on $\mathbb{C}$ by letting

$$\kappa \cdot z = \bar{z} \tag{1.9}$$

where κ is the flip (1.2).

(e) Each Lie group $\Gamma \subset \mathbf{GL}(n)$ acts on the space of $n \times n$ matrices A by similarity: $\gamma \cdot A = \gamma A \gamma^{-1}$.

It is often possible to give two different descriptions of "the same" action. More precisely, the two actions may be isomorphic in the following sense. Let V and W be n-dimensional vector spaces and assume that the Lie group Γ acts on both V and W. Say that these actions are *isomorphic*, or that the spaces V and W are Γ-*isomorphic*, if there exists a (linear) isomorphism $A: V \to W$ such that

$$A(\gamma \cdot v) = \gamma \cdot (Av) \tag{1.10}$$

for all $v \in V$, $\gamma \in \Gamma$. Note that the action of γ on the left-hand side of (1.10) is that on V, whereas on the right it is on W. Another way to say this is that we get the *same* group of matrices if we identify the *spaces* V and W (via the linear isomorphism A). To avoid cumbersome terminology we say that V and W are Γ-*isomorphic*. It is easy to extend these ideas to the case where Γ acts on V and a group Δ, isomorphic to Γ, acts on W.

For example, the actions (1.8, 1.9) of $\mathbf{O}(2)$ for k and $-k$ are isomorphic. To see why, denote the two actions by the symbols $\cdot$ and $*$. Define A by $A(z) = \bar{z}$. Then for $\gamma \in \mathbf{SO}(2)$ we have

$$A(\gamma \cdot z) = \overline{e^{ik\theta}z} = e^{-ik\theta}\bar{z} = e^{-ik\theta}(Az) = \gamma * (Az),$$

and further

$$A(\kappa \cdot z) = \bar{\bar{z}} = z = \kappa * \bar{z} = \kappa * (Az),$$

so (1.10) holds.

In the same way, the groups $\mathbf{SO}(2)$ and $\mathbf{S}^1$ are isomorphic, and the action (1.8) of $\mathbf{S}^1$ on $\mathbb{C}$ with $k = 1$ is isomorphic to the standard action of $\mathbf{SO}(2)$ defined in Example 1.2a.

(c) Invariant Integration

Every compact Lie group Γ in $\mathbf{GL}(n)$ can be identified with a subgroup of the orthogonal group $\mathbf{O}(n)$. Since it is often useful to assume this, we sketch the proof. The identification is made using Haar integration, a form of integration that is invariant under translation by elements of Γ. In this subsection we define Haar integration, show how its existence leads to the identification of Γ with a subgroup of $\mathbf{O}(n)$, and give explicit examples of Haar integration.

Haar integration may be defined abstractly as an operation that satisfies three properties. Let $f: \Gamma \to \mathbb{R}$ be a continuous real-valued function. The operation

$$\int_{\gamma \in \Gamma} f(\gamma) \quad \text{or} \quad \int_\Gamma f \quad \text{or} \quad \int_\Gamma f\, d\gamma \in \mathbb{R}$$

is an integral on Γ if it satisfies the following two conditions:

(a) *Linearity.* $\int_\Gamma (\lambda f + \mu g) = \lambda \int_\Gamma f + \mu \int_\Gamma g$
where $f, g: \Gamma \to \mathbb{R}$ are continuous and $\lambda, \mu \in \mathbb{R}$. (1.11)
(b) *Positivity.* If $f(\gamma) \geq 0$ for all $\gamma \in \Gamma$ then $\int_\Gamma f \geq 0$.
It is a *Haar integral* if it also has the property
(c) *Translation-Invariance.* $\int_{\gamma \in \Gamma} f(\delta\gamma) = \int_{\gamma \in \Gamma} f(\gamma)$
for any fixed $\delta \in \Gamma$. (1.12)

The Haar integral can be proved to be unique. Because Γ is compact, $\int_\Gamma 1$ is finite. We may therefore scale the Haar integral so that $\int_\Gamma 1 = 1$. This yields the *normalized Haar integral.* For compact groups the Haar integral is also invariant under right translations; i.e.,

$$\int_{\gamma \in \Gamma} f(\gamma\delta) = \int_{\gamma \in \Gamma} f(\gamma) \quad \text{for all } \delta \in \Gamma. \tag{1.13}$$

The proof of existence and uniqueness of the Haar integral is in Hochschild [1965], p. 9. Vector-valued mappings may also be integrated, by performing the integration separately on each component.

Proposition 1.3. *Let Γ be a compact Lie group acting on a vector space V and let ρ_γ be the matrix associated with $\gamma \in \Gamma$. Then there exists an inner product on V such that for all $\gamma \in \Gamma$, ρ_γ is orthogonal.*

Remark. Proposition 1.3 implies that we may identify compact Lie groups in $\mathbf{GL}(n)$ with closed subgroups of $\mathbf{O}(n)$.

PROOF. The idea is to use the Haar integral to construct an *invariant* inner product $\langle\ ,\ \rangle_\Gamma$ on V, that is, one that satisfies

$$\langle \rho_\delta v, \rho_\delta w\rangle_\Gamma = \langle v, w\rangle_\Gamma \tag{1.14}$$

for all $\delta \in \Gamma$. The construction proceeds as follows. Let $\langle\ ,\ \rangle$ be any inner product on V and define

$$\langle v, w\rangle_\Gamma = \int_\Gamma \langle \rho_\gamma v, \rho_\gamma w\rangle. \tag{1.15}$$

This is also an inner product by (1.11). Invariance of the Haar integral (1.12) shows that the inner product (1.15) satisfies (1.14). □

EXAMPLES 1.4.

(a) Let Γ be a finite Lie group of order $|\Gamma|$. Then the normalized Haar integral on Γ is

$$\int_\Gamma f \equiv \frac{1}{|\Gamma|} \sum_{\gamma \in \Gamma} f(\gamma). \tag{1.16}$$

(b) Let $\Gamma = \mathbf{SO}(2)$. Every continuous function $f: \mathbf{SO}(2) \to \mathbb{R}$ uniquely determines a continuous 2π-periodic function $\tilde{f}: \mathbb{R} \to \mathbb{R}$ such that

$$\tilde{f}(\theta) = f(R_\theta).$$

The normalized Haar integral on $\mathbf{SO}(2)$ is

$$\int_\Gamma f \equiv \frac{1}{2\pi} \int_0^{2\pi} \tilde{f}(\theta)\, d\theta. \tag{1.17}$$

The abstract definition of the Haar integral that we have given is sufficient for our purposes, because we use it only as a tool to prove abstract results such as Proposition 1.3. There is, however, an explicit definition of the Haar integral that uses the manifold structure of Lie groups; see Exercise 1.8.

Exercises

1.1. Two elements α, β of a Lie group Γ are *conjugate* in Γ if $\alpha = \gamma^{-1}\beta\gamma$ for some $\gamma \in \Gamma$. Show that all elements of $\mathbf{O}(2) \sim \mathbf{SO}(2)$ are conjugate in $\mathbf{O}(2)$.

1.2. Two subgroups H, K of a Lie group Γ are *conjugate* in Γ if $H = \gamma^{-1}K\gamma$ for some $\gamma \in \Gamma$.
(a) Show that the closed subgroups of $\mathbf{O}(2)$ are conjugate to $\mathbf{SO}(2)$, $\mathbf{D}_n$, or $\mathbf{Z}_n$.
(b) Find up to conjugacy all subgroups of $\mathbf{D}_n$, $n \geq 3$. (*Hint*: consider separately the cases n even, n odd.)

1.3. Show that every finite group G is isomorphic to a Lie group. (*Hint*: if $\gamma \in G$ then the map $\delta \mapsto \gamma\delta$ is a permutation of G. Now consider the corresponding permutation matrix.)

1.4. Show that every compact Lie group has a finite number of connected components.

1.5. Let $\rho\colon \Gamma \to \mathbf{GL}(V)$ be a representation of the group Γ as defined by (1.6, 1.7).
(a) Show that ρ is a group homomorphism.
(b) Show that $\ker \rho$ is a normal subgroup of Γ.

1.6. Let ρ and σ be representations of the Lie group Γ on the same space V. Show that if ρ and σ are isomorphic then $\ker \rho = \ker \sigma$. Conclude that if $\ker \rho \neq \ker \sigma$ then ρ and σ are distinct.

1.7. Let $f\colon V \to \mathbb{R}$ be continuous, and let a compact Lie group Γ act on V. Show that

$$\hat{f}(x) = \int_{\gamma \in \Gamma} f(\gamma x)$$

has the property that $\hat{f}(\gamma x) = \hat{f}(x)$ for all $\gamma \in \Gamma$.

1.8. (*Warning*: This exercise requires knowledge of the elementary theory of manifolds.) To define the Haar integral explicitly we must use the fact that every Lie group Γ is a smooth manifold. Let U be an open neighborhood of 0 in $\mathbb{R}^k$ where $k = \dim \Gamma$ and let $\mathscr{X}\colon U \to \Gamma$ be a smooth parametrization satisfying $\mathscr{X}(0) = 1$. Let $f\colon \Gamma \to \mathbb{R}$ be continuous with the support $\operatorname{supp}(f)$ of f contained in $\mathscr{X}(U)$. Define $\int_\Gamma f$ as follows.

Let $L_\delta\colon \Gamma \to \Gamma$ be left translation by δ; that is, $L_\delta(\gamma) = \delta\gamma$. For $\delta \in \mathscr{X}(U)$ the composition

$$\tilde{L}_\delta = \mathscr{X}^{-1} \circ L_\delta \circ \mathscr{X}$$

is a smooth mapping on a neighborhood of 0 in $\mathbb{R}^k$. Let

$$J(\delta) = \det(d\tilde{L}_\delta)_0.$$

Now define

$$\int_\Gamma f = \int_U f[\mathscr{X}(u)]J(u)^{-1}\,du. \tag{1.18}$$

Suppose that $\sigma \in \Gamma$ and $\sigma(\operatorname{supp}(f)) \subset \mathscr{X}(U)$. Show that

$$\int_U f[\sigma\mathscr{X}(u)]J(\sigma u)^{-1}\,du = \int_\Gamma f.$$

(*Comment*: When the Lie group Γ has a single parametrization $\mathscr{X}$ such that

$$\overline{\mathscr{X}(U)} = \Gamma$$

then (1.18) defines a Haar integral on Γ since $\Gamma \sim \mathscr{X}(U)$ has "measure zero" and $\int_\Gamma = \int_{\mathscr{X}(U)}$.)

§2. Irreducibility

The study of a representation of a compact Lie group is often made easier by observing that it decomposes into a direct sum of simpler representations, which are said to be irreducible. We describe the basic properties of this decomposition in this section. The main result, Theorem 2.5, states that the decomposition always exists. In general it is not unique, but the sources of nonuniqueness can be described and controlled.

Let Γ be a Lie group acting linearly on the vector space V. A subspace $W \subset V$ is called Γ-*invariant* if $\gamma w \in W$ for all $w \in W$, $\gamma \in \Gamma$. A representation or action of Γ on V is *irreducible* if the only Γ-invariant subspaces of V are $\{0\}$ and V. A subspace $W \subset V$ is said to be Γ-*irreducible* (or *irreducible* if it is clear which group Γ is intended) if W is Γ-invariant and the action of Γ on W is irreducible. For example, the actions of $\mathbf{SO}(2)$ and $\mathbf{O}(2)$ on $\mathbb{R}^2$ defined in Example 1.2c, d are irreducible when $k \neq 0$.

One of the fundamental features of actions of compact Lie groups is that invariant subspaces always have invariant complements. More precisely:

Proposition 2.1. *Let Γ be a compact Lie group acting on V. Let $W \subset V$ be a Γ-invariant subspace. Then there exists a Γ-invariant complementary subspace $Z \subset V$ such that*

$$V = W \oplus Z.$$

Proof. By Proposition 1.3 there exists a Γ-invariant inner product $\langle\ ,\ \rangle_\Gamma$ on V. Let $Z = W^\perp$ where

$$W^\perp = \{v \in V: \langle w, v\rangle_\Gamma = 0 \text{ for all } w \in W\}.$$

The Γ-invariance of the inner product implies that $W^\perp$ is a Γ-invariant complement to W. □

It follows directly from this proposition that every representation of a compact Lie group may be written as a direct sum of irreducible subspaces:

Corollary 2.2 (Theorem of Complete Reducibility). *Let Γ be a compact Lie group acting on V. Then there exist Γ-irreducible subspaces $V_1, \ldots, V_s$ of V such that*

$$V = V_1 \oplus \cdots \oplus V_s. \tag{2.1}$$

PROOF. We may assume V nonzero. Then there exists a nonzero Γ-irreducible subspace $V_1 \subset V$ (take V_1 to be of minimal dimension among the nonzero Γ-invariant subspaces). By Proposition 2.1 there is a Γ-invariant complement Z to V_1 in V. Now repeat the process on Z, choosing a nonzero Γ-invariant subspace $V_2 \subset V$. Since V is finite-dimensional this process must terminate, yielding the desired decomposition (2.1). □

Some specific examples may help to clarify the implications of this result.

EXAMPLES 2.3.
(a) Define an action of $\mathbf{O}(2)$ on $\mathbb{R}^3$ as follows. Let the rotations $R_\theta \in \mathbf{SO}(2)$ act by rotating the (x, y)-plane through angle 2θ and leaving the z-axis fixed: that is, define

$$\theta \cdot (x, y, z) = (x \cos 2\theta - y \sin 2\theta, x \sin 2\theta + y \cos 2\theta, z).$$

Let the flip $\kappa \in \mathbf{O}(2)$ act by

$$\kappa \cdot (x, y, z) = (x, -y, -z).$$

Observe that

$$V_1 = \mathbb{R}^2 \times \{0\} = \{(x, y, 0)\}$$
$$V_2 = \{0\} \times \mathbb{R} = \{(0, 0, z)\}$$

are $\mathbf{O}(2)$-invariant subspaces and that $\mathbf{O}(2)$ acts irreducibly on each.

(b) There is a standard irreducible action of $\mathbf{O}(3)$ on $\mathbb{R}^5$. Let V be the vector space of symmetric 3×3 matrices of trace zero. Such matrices have the form

$$\begin{bmatrix} a & b & c \\ b & d & e \\ c & e & -(a+d) \end{bmatrix}$$

so $\dim V = 5$. Define

$$\gamma \cdot A = \gamma^t A \gamma$$

for $\gamma \in \mathbf{O}(3)$ and $A \in V$. Thus $\mathbf{O}(3)$ acts on V by similarity.

Next view $\mathbf{O}(2) \subset \mathbf{O}(3)$ as follows. Identify the matrix $\delta \in \mathbf{O}(2)$ with

$$\left[\begin{array}{cc|c} \multicolumn{2}{c|}{\delta} & 0 \\ & & 0 \\ \hline 0 & 0 & 1 \end{array}\right]$$

in $\mathbf{O}(3)$. In this way, we can view $\mathbf{O}(2)$ as acting on V. It is a straightforward calculation to show that

$$V_1 = \begin{bmatrix} a & b & 0 \\ b & -a & 0 \\ 0 & 0 & 0 \end{bmatrix},$$

$$V_2 = \begin{bmatrix} 0 & 0 & c \\ 0 & 0 & d \\ c & d & 0 \end{bmatrix}, \text{ and}$$

$$V_3 = \begin{bmatrix} a & 0 & 0 \\ 0 & a & 0 \\ 0 & 0 & -2a \end{bmatrix}$$

are invariant irreducible subspaces of V under the action of $\mathbf{O}(2)$. Since $V_1 \oplus V_2 \oplus V_3 = V$ we have decomposed V as in Corollary 2.2.

In general, the decomposition of V in (2.1) is not unique. It will be useful in later sections to understand the sources of nonuniqueness and to find conditions under which the decomposition (2.1) *is* unique. In particular, such a discussion will simplify the computation of linearized asymptotic stability for solutions of differential equations. The remainder of this section is devoted to the issue of nonuniqueness, beginning with an example.

EXAMPLE 2.4. Let V be the four-dimensional space of 2×2 matrices and let $\mathbf{SO}(2)$ act on V by matrix multiplication on the left. That is,

$$\theta \cdot A = R_\theta A$$

where $\theta \in \mathbf{SO}(2)$ and $A \in V$. Observe that $V = V_1 \oplus V_2$ where

$$V_1 = \begin{bmatrix} a & 0 \\ b & 0 \end{bmatrix}, \qquad V_2 = \begin{bmatrix} 0 & c \\ 0 & d \end{bmatrix},$$

and that $\mathbf{SO}(2)$ acts irreducibly on V_1 and V_2.

However, we also have $V = V_1 \oplus V_2'$, where (say)

$$V_2' = \begin{bmatrix} 8c & c \\ 8d & d \end{bmatrix}$$

and $\mathbf{SO}(2)$ acts irreducibly on V_2'.

It will turn out that the reason for nonuniqueness in the decomposition of Corollary 2.2 is the occurrence in V of two isomorphic irreducible representations. Recall the definition (1.10) of Γ-isomorphism. We state this more precisely in Corollary 2.6 later. The main result of this section is as follows:

Theorem 2.5. *Let Γ be a compact Lie group acting on V.*
(a) *Up to Γ-isomorphism there are a finite number of distinct Γ-irreducible subspaces of V. Call these $U_1, \ldots, U_t$.*
(b) *Define W_k to be the sum of all Γ-irreducible subspaces W of V such that W is Γ-isomorphic to U_k. Then*

$$V = W_1 \oplus \cdots \oplus W_t. \tag{2.2}$$

Remark. The subspaces W_k are called the *isotypic components* of V, of type U_k, for the action of Γ. The name is chosen to reflect that fact that all irreducible subspaces of W_k have the same isomorphism type. By construction the *isotypic decomposition* (2.2) is unique.

Before proving Theorem 2.5 we show how it implies that the nonuniqueness in the choice of irreducible summands in Proposition 2.1 is directly related to the repetition of irreducible representations among the V_j in (2.1).

Corollary 2.6.
(a) *If $W \subset V$ is Γ-irreducible then $W \subset W_k$ for a unique k, namely, that k for which W is Γ-isomorphic to U_k.*
(b) *Let Γ be a compact Lie group acting on V. Let $V = V_1 \oplus \cdots \oplus V_s$ be a decomposition of V into a direct sum of Γ-invariant irreducible subspaces. If the representations of Γ on the V_j are all distinct (not Γ-isomorphic) then the only nonzero Γ-irreducible subspaces of V are $V_1, \ldots, V_s$.*

Proof. Part (a) follows directly from Theorem 2.5 since if W is Γ-irreducible then it is Γ-isomorphic to some unique U_k, and then by definition $W \subset W_k$. It is useful, however, to have (a) stated explicitly.

For (b), consider the isotypic components W_k of V. Each V_j is isomorphic to some U_k, hence lies in W_k for some k. It follows that the W_k are just the V_j, perhaps written in a different order. If $W \neq 0$ is a Γ-irreducible subspace of V then by part (a) we have $W \subset W_k$ for some k. But $W_k = V_j$ for suitable j, and irreducibility of V_j implies that $W = V_j$. □

The proof of Theorem 2.5 depends on two lemmas, which we deal with first.

Lemma 2.7. *Let Γ be a compact Lie group acting on W. Suppose that*

$$W = \sum_\alpha U_\alpha$$

where each U_α is a Γ-invariant subspace that is Γ-isomorphic to some fixed irreducible representation U of Γ. Then every Γ-irreducible subspace of W is Γ-isomorphic to U.

Remark. Because of nonuniqueness, a Γ-irreducible subspace of W may not be one of the U_α. The lemma says that, provided all the U_α are Γ-isomorphic, every Γ-irreducible subspace of W is Γ-isomorphic to any one of the U_α.

Proof. Because of our intended application, the theorem is stated in a way that allows the index α to run over an infinite set. In fact this represents no real increase in generality, since we first show that

$$W = U_{\alpha_1} \oplus \cdots \oplus U_{\alpha_t}, \tag{2.3}$$

a *direct* sum of a finite subset of the U_α. The proof is by induction. Suppose we have found a subspace

$$W' = U_{\alpha_1} \oplus \cdots \oplus U_{\alpha_{t-1}} \subset W.$$

If $W' = W$ we are done. If not, some U_{α_t} is not contained in W'. Then $U_{\alpha_t} \cap W' \subset U_{\alpha_t}$ must be $\{0\}$ by irreducibility. Therefore the sum $W' + U_{\alpha_t}$ is direct and we have a subspace

$$W'' = U_{\alpha_1} \oplus \cdots \oplus U_{\alpha_t}.$$

By finiteness of dimension, (2.3) must hold for large enough s.

Now let X be a Γ-irreducible subspace of W. There exists $t \leq s$ such that

$$X \not\subset U_{\alpha_1} \oplus \cdots \oplus U_{\alpha_{t-1}} \tag{2.4}$$

$$X \subset U_{\alpha_1} \oplus \cdots \oplus U_{\alpha_t}. \tag{2.5}$$

There is only one such t. By irreducibility of X,

$$X \cap (U_{\alpha_1} \oplus \cdots \oplus U_{\alpha_{t-1}}) = 0. \tag{2.6}$$

Let π be the projection

$$\pi\colon U_{\alpha_1} \oplus \cdots \oplus U_{\alpha_t} \to U_{\alpha_t}.$$

Then (2.6) implies that $\pi|X$ is a Γ-isomorphism of X onto $\pi(X)$; and $\pi(X) \subset U_{\alpha_t}$ implies that $\pi(X) = U_{\alpha_t}$ by irreducibility of U_{α_t}. Therefore X is Γ-isomorphic to U_{α_t}, hence to U. □

Lemma 2.8. *Let Γ be a compact Lie group acting on V. Let X, Y be Γ-invariant subspaces of V such that no two Γ-irreducible subspaces $W \subset X$, $Z \subset Y$ are Γ-isomorphic. Then*:

(a) $X \cap Y = \{0\}$,
(b) *If $W \subset X \oplus Y$ is Γ-irreducible, then $W \subset X$ or $W \subset Y$.*

PROOF.
(a) Since $X \cap Y$ is Γ-invariant, any Γ-irreducible subspace of $X \cap Y$ would be in both X and Y, contrary to the assumptions on X and Y. Thus $X \cap Y$ has no nonzero Γ-irreducible subspaces, and this is possible only when $X \cap Y = \{0\}$; see Corollary 2.2.
(b) The subspaces $W \cap X$ and $W \cap Y$ of W are Γ-invariant. By the irreducibility of W, either $W \cap X = \{0\}$ or $W \subset X$; and similarly for Y. If $W \not\subset X$ and $W \not\subset Y$ then $W \cap X = \{0\} = W \cap Y$. Let π_X and π_Y denote the projections of $X \oplus Y$ onto X and Y, respectively. Then W is Γ-isomorphic to $\pi_X(W)$ and to $\pi_Y(W)$ as in the proof of Lemma 2.7. But this contradicts the hypotheses on X and Y. □

PROOF OF THEOREM 2.5. Choose a Γ-irreducible subspace $U_1 \subset V$. Let W_1' be the sum of all Γ-invariant subspaces of V that are Γ-isomorphic to U_1. If $W_1' \neq V$, then choose a Γ-invariant complement Z to W_1' and repeat the process on Z to obtain W_2'. By finiteness of dimension this process terminates with

$$V = W_1' \oplus W_2' \oplus \cdots \oplus W_s' \tag{2.7}$$

where each W_k' is the sum of a set of Γ-isomorphic Γ-irreducible subspaces of V, say isomorphic to $U_k \subset V$; and if $i \neq j$ then U_i is not Γ-isomorphic to U_j. We do not yet know that W_k' is the sum W_k of *all* Γ-invariant subspaces of V that are Γ-isomorphic to U_k, because we defined W_k' in Z, not in V. We shall quickly see that in fact $W_k' = W_k$.

Suppose that U is a Γ-irreducible subspace of V. By Lemma 2.8(b) and a simple inductive argument, it follows that

$$U \subset W_k' \tag{2.8}$$

for some k. By Lemma 2.7, U is Γ-isomorphic to U_k. This proves part (a). But now we see that

$$W_k' = W_k, \tag{2.9}$$

as defined in the statement of Theorem 2.5b, and (2.7) implies (2.2), proving part (b). □

EXERCISES

2.1. (a) Show that every two-dimensional irreducible representation of $\mathbf{S}^1$ is isomorphic to

$$\rho_\theta^k(z) = e^{ki\theta}z \tag{2.10}$$

for some integer $k > 0$.

(b) Show that the representations ρ^k and ρ^l in (2.10) are not isomorphic if $k > l > 0$. (*Hint*: Use Exercise XII, 1.6.)

(c) Show that the only one-dimensional irreducible representation of $\mathbf{S}^1$ is the trivial representation.

2.2. Let $\mathbf{O}(2)$ act on the four-dimensional space V of 2×2 matrices by similarity:

$$\gamma \cdot A = \gamma^{-1} A \gamma \qquad (\gamma \in \mathbf{O}(2), A \in V).$$

Show that $V = V_1 \oplus V_2 \oplus V_3$ where

$$V_1 = \left\{\begin{bmatrix} a & 0 \\ 0 & a \end{bmatrix}\right\}$$

$$V_2 = \left\{\begin{bmatrix} 0 & -b \\ b & 0 \end{bmatrix}\right\}$$

$$V_3 = \left\{\begin{bmatrix} c & d \\ d & -c \end{bmatrix}\right\}.$$

Show that

(a) The $\mathbf{O}(2)$-action on V_1 is trivial.

(b) The $\mathbf{O}(2)$-action on V_2 is the nontrivial one-dimensional representation, in which $\gamma \in \mathbf{O}(2) \sim \mathbf{SO}(2)$ acts as $-I$ and $\gamma \in \mathbf{SO}(2)$ acts as I.

(c) The $\mathbf{O}(2)$-action on V_3 is isomorphic to the standard action on $\mathbb{R}^2 \equiv \mathbb{C}$.

2.3. In the notation of Exercise 2.2, let $\mathbf{O}(2)$ act on V by matrix multiplication:

$$\gamma \cdot A = \gamma A.$$

Show that $V = V_1 \oplus V_2$, where

$$V_1 = \left\{\begin{bmatrix} a & 0 \\ c & 0 \end{bmatrix}\right\}$$

$$V_2 = \left\{\begin{bmatrix} 0 & b \\ 0 & d \end{bmatrix}\right\},$$

and that the $\mathbf{O}(2)$-action on each of V_1, V_2 is isomorphic to the standard action. Hence show that V has only one isotypic component, namely V itself. Find an irreducible subspace of V that is not equal to V_1 or V_2.

§3. Commuting Linear Mappings and Absolute Irreducibility

In later sections when we compute linearized asymptotic stability of steady-state solutions to ODEs we will need to understand the structure of linear mappings that commute with the action of a compact Lie group. We explore this issue here. The main result is Theorem 3.5, which lets us put commuting linear mappings into a certain block diagonal form.

Let Γ be a compact Lie group acting linearly on V. A mapping $F: V \to V$ *commutes with* Γ or is Γ-*equivariant* if

$$F(\gamma v) = \gamma F(v) \tag{3.1}$$

for all $\gamma \in \Gamma$, $v \in V$.

EXAMPLES 3.1.
(a) Consider the standard action of $\Gamma = \mathbf{SO}(2)$ on $V = \mathbb{R}^2$ defined by rotation through angle θ. That is,

$$R_\theta = \begin{bmatrix} \cos\theta & -\sin\theta \\ \sin\theta & \cos\theta \end{bmatrix}$$

acts on

$$\mathbb{R}^2 = \left\{\begin{bmatrix} x \\ y \end{bmatrix}\right\}$$

by matrix multiplication.

We claim that the linear mappings that commute with this action of $\mathbf{SO}(2)$ all have the form cR_θ where $c \in \mathbb{R}$ is a scalar; that is, such linear maps have the matrix form

$$\begin{bmatrix} a & -b \\ b & a \end{bmatrix}. \tag{3.2}$$

Certainly such matrices commute with $\mathbf{SO}(2)$ because $\mathbf{SO}(2)$ is a commutative group, that is $\mathbf{SO}(2)$ satisfies

$$R_\theta R_\varphi = R_\varphi R_\theta.$$

The proof of the converse is a straightforward calculation. Suppose that

$$R_\theta \begin{bmatrix} a & b \\ c & d \end{bmatrix} = \begin{bmatrix} a & b \\ c & d \end{bmatrix} R_\theta \tag{3.3}$$

for all θ. Equate matrix entries on the first row of (3.3) to obtain

$$\begin{aligned} &\text{(a)} \quad a\cos\theta - c\sin\theta = a\cos\theta + b\sin\theta \\ &\text{(b)} \quad b\cos\theta - d\sin\theta = -a\sin\theta + b\cos\theta. \end{aligned} \tag{3.4}$$

Since (3.4) holds for all θ it follows that $b = -c$ and $a = d$. Therefore, the matrix has the desired form.

(b) Now consider the standard action of $\mathbf{O}(2)$ on $\mathbb{R}^2$. We claim that the only linear mappings that commute with $\mathbf{O}(2)$ are cI, $c \in \mathbb{R}$. Note that scalar multiples of the identity commute with any group representation since they commute with any matrix. To prove the claim let M be a matrix commuting with $\mathbf{O}(2)$. Since it commutes with $\mathbf{SO}(2)$ it must have the form (3.2). It is now a simple matter to show that if M commutes with

$$\begin{bmatrix} 1 & 0 \\ 0 & -1 \end{bmatrix}$$

then $b = 0$.

Definition 3.2. A representation of a group Γ on a vector space V is *absolutely irreducible* if the only linear mappings on V that commute with Γ are scalar multiples of the identity.

To justify the terminology we prove:

Lemma 3.3. *Let Γ be a compact Lie group acting on V. If the action of Γ is absolutely irreducible then it is irreducible.*

Proof. Suppose the action of Γ is not irreducible. Then there is a proper Γ-invariant subspace $W \neq \{0\}$ having a Γ-invariant complement $W^\perp$, by Proposition 2.1. Define $\pi\colon W \oplus W^\perp \to V$ to be projection onto W with $\ker \pi = W^\perp$. It is easy to check that π commutes with Γ and is not a scalar multiple of the identity. Hence V is not absolutely irreducible. □

Remark. We hasten to point out that if we work with *complex* representations of compact Lie groups then Schur's lemma (Adams [1969], 3.22, p. 40) implies that the complex versions of irreducibility and absolute irreducibility are equivalent concepts. However, this is not true for real representations, as Example 3.1(a) shows. We provide further discussion at the end of this section.

We now discuss several points about linear maps that commute with nonirreducible representations. The following observation is quite useful.

Lemma 3.4. *Let Γ be a compact Lie group acting on V, let $A\colon V \to V$ be a linear mapping that commutes with Γ, and let $W \subset V$ be a Γ-irreducible subspace. Then $A(W)$ is Γ-invariant, and either $A(W) = \{0\}$ or the representations of Γ on W and $A(W)$ are isomorphic.*

Proof. To show that $A(W)$ is Γ-invariant let $z \in A(W)$, so that $z = A(w)$ for $w \in W$. Since A commutes with Γ we have

$$\gamma z = \gamma A(w) = A(\gamma w)$$

so $\gamma z \in A(W)$.

Similarly, $\ker A$ is Γ-invariant since $A(v) = 0$ implies that $A(\gamma v) = \gamma A(v) = \gamma 0 = 0$. Then $\ker A \cap W$ is a Γ-invariant subspace of W, and irreducibility implies that either $W \subset \ker A$ or $\ker A \cap W = \{0\}$. In the first case $A(W) = \{0\}$. In the second, $A(W)$ is isomorphic to W as a vector space, the isomorphism being A; but Γ commutes with A so A is a Γ-isomorphism between A and $A(W)$. □

Lemma 3.4 implies:

Theorem 3.5. *Let Γ be a compact Lie group acting on the vector space V. Decompose V into isotypic components*

$$V = W_1 \oplus \cdots \oplus W_s.$$

Let $A\colon V \to V$ be a linear mapping commuting with Γ. Then

$$A(W_k) \subset W_k \tag{3.5}$$

for $k = 1, \ldots, s$.

Proof. Write $W_k = V_1 \oplus \cdots \oplus V_r$, where all V_j are Γ-isomorphic to an irreducible U_k. By Lemma 3.4 either $A(V_j) = \{0\}$ or $A(V_j)$ is also Γ-isomorphic to U_k. In either case $A(V_j) \subset W_k$. By linearity, $A(W_k) \subset W_k$. □

Finally we return to the question of irreducible but not absolutely irreducible representations. Suppose Γ acts irreducibly on V and let

$$\mathscr{D} = \{A\colon V \to V | A \text{ linear, } A\gamma = \gamma A \text{ for all } \gamma \in \Gamma\}$$

be the set of all commuting mappings. The real version of Schur's lemma (Kirillov [1976], Theorem 2, p. 119) states that $\mathscr{D}$ is an associative algebra over $\mathbb{R}$ and is isomorphic to one of $\mathbb{R}$, $\mathbb{C}$, or $\mathbb{H}$, where $\mathbb{H}$ is the four-dimensional algebra of quaternions. The reason is that by Lemma 3.4 $\mathscr{D}$ is a skew field, and skew fields may be classified into the preceding three types. The case $\mathscr{D} \cong \mathbb{R}$ occurs if and only if V is absolutely irreducible. The example of $\mathbf{SO}(2)$ acting on $\mathbb{R}^2$ is a case where $\mathscr{D} \cong \mathbb{C}$. To verify this, recall that the commuting mappings are the matrices

$$\begin{bmatrix} a & -b \\ b & a \end{bmatrix}.$$

The isomorphism $\mathscr{D} \cong \mathbb{C}$ identifies such a matrix with $a + ib \in \mathbb{C}$. Note that

$$\begin{bmatrix} a & b \\ -b & a \end{bmatrix}\begin{bmatrix} c & d \\ -d & c \end{bmatrix} = \begin{bmatrix} ac - bd & ad + bc \\ -(ad + bc) & ac - bd \end{bmatrix}$$

and $(a + ib)(c + id) = (ac - bd) + i(ad + bc)$, so this map *is* an isomorphism. The case $\mathscr{D} \cong \mathbb{H}$ can also occur; see Exercise 3.1. The distinction between $\mathbb{C}$ and $\mathbb{H}$ is a basic one when considering nonabsolutely irreducible representations. Most representations discussed in this book will in fact be absolutely irreducible, $\mathscr{D} \cong \mathbb{R}$; but the case $\mathscr{D} \cong \mathbb{C}$ arises repeatedly in the context of Hopf bifurcation. We have found no such natural context for representations with $\mathscr{D} \cong \mathbb{H}$.

Exercises

3.1. Let Γ be the group $\mathbf{SU}(2)$ of unit quaternions

$$\{a + bi + cj + dk: a^2 + b^2 + c^2 + d^2 = 1\}.$$

Show that Γ is a compact Lie group. Let Γ act on $\mathbb{R}^4 \equiv \mathbb{H}$ by left multiplication,

$$\gamma \cdot x = \gamma x.$$

Prove that $\mathscr{D}$ consists of mappings δq, $q \in \mathbb{H}$, acting as right multiplication,

$$\delta q(x) = xq.$$

Hence show that $\mathscr{D} \cong \mathbb{H}$.

3.2. Let Γ be a Lie group acting irreducibly on a space V. Let $A: V \to V$ be a nonzero linear map commuting with Γ. Show that A is invertible and that A^{-1} commutes with Γ.

3.3. Let A, B be commuting matrices. Let E be an eigenspace, or a generalized eigenspace, of A. Show that B leaves E invariant.

3.4. If a 2×2 matrix A commutes with $\kappa = \left[\begin{smallmatrix} 1 & 0 \\ 0 & -1 \end{smallmatrix}\right]$ then show that A is diagonal. If a diagonal matrix commutes with a rotation matrix R_θ, where θ is not an integer multiple of π, show that it is a scalar multiple of the identity. Hence show that the standard action of $\mathbf{D}_n$, $n \geq 3$, is absolutely irreducible.

3.5. Let Γ act on $V = V_1 \oplus V_2$ where V_1 and V_2 are absolutely irreducible and nonisomorphic. Let $A: V \to V$ commute with Γ. Prove that A has real eigenvalues and that at most two distinct eigenvalues occur.

3.6. Let $\mathbf{O}(3)$ act on the space

$$V = \{3 \times 3 \text{ symmetric trace 0 matrices}\}$$

by similarity:

$$\gamma \cdot A = \gamma^{-1} A \gamma.$$

Show that V is absolutely irreducible. (*Hint*: Let D be the set of diagonal matrices in V.) Observe that

$$D = \{A: \sigma_1 A = A, \sigma_2 A = A\} \tag{3.6}$$

where

$$\sigma_1 = \begin{bmatrix} -1 & & \\ & 1 & \\ & & 1 \end{bmatrix}, \qquad \sigma_2 = \begin{bmatrix} 1 & & \\ & -1 & \\ & & 1 \end{bmatrix}.$$

Let $\alpha: V \to V$ commute with Γ. Use (3.6) to show that $\alpha(D) \subset D$. Since every symmetric matrix can be diagonalized, show that α is uniquely determined by its effect on D. Let $\beta = \alpha | D$. Show that β commutes with the permutation matrices $\mathbf{S}_3$, and that the $\mathbf{S}_3$-action on D is absolutely irreducible. Deduce that the action of $\mathbf{O}(3)$ on V is absolutely irreducible.

3.7. Let Γ act on V and let H be a subgroup of Γ. If V is absolutely irreducible for H, prove that it is absolutely irreducible for Γ.

§4. Invariant Functions

The goal of this section and the next is to present an efficient way of describing *nonlinear* mappings that commute with a group action. We begin with a discussion of invariant functions. There are two main results: the Hilbert–Weyl theorem, which gives a theoretical foundation for describing invariant polynomials, and Schwarz's theorem (Schwarz [1975]), which builds on Hilbert and Weyl's result, yielding a description of invariant C^∞ germs. See II, §1, for a definition and discussion of germs.

Let Γ be a (compact) Lie group acting on a vector space V. Recall that a real-valued function $f: V \to \mathbb{R}$ is *invariant* under Γ if

$$f(\gamma x) = f(x) \tag{4.1}$$

for all $\gamma \in \Gamma$, $x \in V$. An *invariant polynomial* is defined in the obvious way by taking f to be polynomial. Note that it suffices to verify (4.1) for a set of generators of Γ.

EXAMPLES 4.1.
(a) Let $\Gamma = \mathbf{Z}_2$ act nontrivially on $V = \mathbb{R}$. That is, $-1 \cdot x = -x$, where $\mathbf{Z}_2 = \{\pm 1\}$. For this example the invariant functions are just the *even* functions since (4.1) becomes $f(-x) = f(x)$. It is easy to see that if f is an invariant

polynomial then there exists another polynomial h such that

$$f(x) = h(x^2). \tag{4.2}$$

(b) Let $\mathbf{S}^1$ act on $\mathbb{R}^2 \equiv \mathbb{C}$ in the standard way; that is, $\theta z = e^{i\theta}z$ for $\theta \in \mathbf{S}^1$. Equation (4.1) states that $f(e^{i\theta}z) = f(z)$ for every $\theta \in \mathbf{S}^1$. Since $\theta \mapsto e^{i\theta}$ traces out a circle centered at 0 with radius $|z|$ we see that $\mathbf{S}^1$-invariants are functions that are constant on circles. We now show (as is already plausible) that if f is an $\mathbf{S}^1$-invariant polynomial on $\mathbb{C}$ then there exists a polynomial $h: \mathbb{R} \to \mathbb{R}$ such that

$$f(z) = h(z\bar{z}). \tag{4.3}$$

(This observation is contained in the proof of Proposition VIII, 2.3; we give a different proof here.) The proof of (4.3) will be carried out using complex notation, a trick that is often useful. Write f as a polynomial in the "real" coordinates $z, \bar{z}$ on $\mathbb{C}$ in the form

$$f(z) = \sum a_{\alpha\beta} z^\alpha \bar{z}^\beta \tag{4.4}$$

where $a_{\alpha\beta} \in \mathbb{C}$. (They are "real" coordinates in the sense that they coordinatize $\mathbb{C}$ as a real vector space. However, for $z = x + iy$ we have $x = (z + \bar{z})/2$, $y = -i(z - \bar{z})/2$, so the coefficients required may be complex. Thus we have to impose on all polynomials a reality condition: their *values* must be in $\mathbb{R}$.) Here the reality condition is that f is real-valued; that is, $\bar{f} = f$. So the coefficients $a_{\alpha\beta}$ satisfy

$$\bar{a}_{\alpha\beta} = a_{\beta\alpha}. \tag{4.5}$$

Direct computation from (4.4) shows that

$$f(e^{i\theta}z) = \sum a_{\alpha\beta} e^{i\theta(\alpha-\beta)} z^\alpha \bar{z}^\beta. \tag{4.6}$$

Since $f(e^{i\theta}z) \equiv f(z)$ as polynomials, they have identical coefficients. From (4.4, 4.6) we obtain the identity

$$a_{\alpha\beta} = e^{i\theta(\alpha-\beta)} a_{\alpha\beta}. \tag{4.7}$$

Now (4.7) holds for *all* $\theta \in \mathbf{S}^1$ only if $\alpha = \beta$ or $a_{\alpha\beta} = 0$. Thus $\mathbf{S}^1$-invariance implies that

$$f(z) = \sum a_{\alpha\alpha}(z\bar{z})^\alpha$$

where, by (4.5), $a_{\alpha\alpha} \in \mathbb{R}$. If

$$h(x) = \sum a_{\alpha\alpha} x^\alpha$$

then (4.3) is satisfied.

(c) Let $\Gamma = \mathbf{D}_n$ in its standard action on $\mathbb{C}$. We claim that for every $\mathbf{D}_n$-invariant polynomial $f(z)$ there exists a polynomial $g: \mathbb{R}^2 \to \mathbb{R}$ such that

$$f(z) = g(z\bar{z}, z^n + \bar{z}^n). \tag{4.8}$$

We verify (4.8) in a similar way to (4.3). We may again assume f has the form (4.4) and satisfies the reality condition (4.5). Since the action of $\mathbf{D}_n$ is generated by

$$\theta z = e^{i\theta} z \quad (\theta = 2\pi/n) \quad \text{and} \quad \kappa z = \bar{z}$$

we need verify (4.1) only for these elements. The restriction placed on f by the first generator is (4.7) when $\theta = 2\pi/n$. The restriction placed by κ is

$$a_{\alpha\beta} = a_{\beta\alpha}, \tag{4.9}$$

and from (4.5, 4.9) we conclude that $a_{\alpha\beta} \in \mathbb{R}$. In summary, we require

$$\begin{aligned} &\text{(a)} \quad a_{\alpha\beta} \in \mathbb{R} \\ &\text{(b)} \quad a_{\alpha\beta} = a_{\beta\alpha} \\ &\text{(c)} \quad a_{\alpha\beta} = 0 \quad \text{unless} \quad \alpha \equiv \beta (\operatorname{mod} n). \end{aligned} \tag{4.10}$$

Using (4.10) we may rewrite (4.4) as

$$f(z) = \sum_{\alpha \leq \beta} A_{\alpha\beta}(z^\alpha \bar{z}^\beta + \bar{z}^\alpha z^\beta)$$

where

$$A_{\alpha\beta} = \begin{cases} a_{\alpha\beta} & \text{if} \quad \alpha \neq \beta, \\ a_{\alpha\beta}/2 & \text{if} \quad \alpha = \beta. \end{cases}$$

Next, we factor out the largest powers of $z\bar{z}$ and use (4.10c) to arrive at the form

$$f(z) = \sum_{j,k} B_{jk}(z\bar{z})^j(z^{kn} + \bar{z}^{kn}) \tag{4.11}$$

for certain coefficients B_{jk}. Finally we use the identity

$$z^{kn} + \bar{z}^{kn} = (z^n + \bar{z}^n)(z^{(k-1)n} + \bar{z}^{(k-1)n}) - z\bar{z}(z^{(k-2)n} + \bar{z}^{(k-2)n})$$

inductively, to write the polynomial in the form

$$f(z) = \sum C_{lm}(z\bar{z})^l(z^n + \bar{z}^n)^m$$

for certain real coefficients C_{lm}. Now define

$$h(x, y) = \sum C_{lm} x^l y^m.$$

We make one very important observation about the invariant polynomials in Examples 4.1. There is a *finite* subset of invariant polynomials $u_1, \ldots, u_s$ such that *every* invariant polynomial may be written as a polynomial function of $u_1, \ldots, u_s$. This finite set of invariants (which is not unique) is said to *generate* the set of invariants, or to form a *Hilbert basis*. We denote the set of invariant polynomials by $\mathscr{P}(\Gamma)$. Note that $\mathscr{P}(\Gamma)$ is a ring since sums and products of Γ-invariant polynomials are again Γ-invariant. The existence of this finite set of generators is a general phenomenon. The main theoretical result, initiated by Hilbert and proved by Weyl [1946], is as follows:

Theorem 4.2 (Hilbert–Weyl Theorem). *Let Γ be a compact Lie group acting on V. Then there exists a finite Hilbert basis for the ring $\mathscr{P}(\Gamma)$.*

Remarks.

(a) The actual computation of a generating set for $\mathscr{P}(\Gamma)$ can be extremely difficult. In many cases, such as those in Examples 4.1, a set of invariant generators may be obtained by a combination of tricks and direct calculation.

(b) Since Γ is a compact Lie group, we may assume it is a subgroup of the orthogonal group $\mathbf{O}(n)$ by Proposition 1.3. In this case, the norm

$$\|x\|^2 = x_1^2 + \cdots + x_n^2$$

is always Γ-invariant.

We prove this theorem in §6; similar proofs are given in Weyl [1946] and Poénaru [1976]. In individual examples, such as those of Examples 4.1, we may verify Theorem 4.2 explicitly by exhibiting a finite set of invariant generators.

It is not surprising that a similar result to Theorem 4.2 holds for real analytic functions. It is perhaps more surprising, however, that this sort of result remains true for C^∞ germs, and it is in this category that we wish to work. Although a finitude theorem for C^∞ germs was known in special cases (see Whitney [1943] for $\mathbf{Z}_2$ acting on $\mathbb{R}$, and Glaeser [1963] for the symmetric group $\mathbf{S}_n$ acting as permutations on $\mathbb{R}^n$) it was not until Schwarz [1975] that the C^∞ germ result was proved for general compact Lie groups. We state Schwarz's theorem here and sketch its proof in §6. We use the notation $\mathscr{E}(\Gamma)$ for the ring of Γ-invariant germs $V \to \mathbb{R}$.

Theorem 4.3 (Schwarz [1975]). *Let Γ be a compact Lie group acting on V. Let $u_1, \ldots, u_s$ be a Hilbert basis for the Γ-invariant polynomials $\mathscr{P}(\Gamma)$. Let $f \in \mathscr{E}(\Gamma)$. Then there exists a smooth germ $h \in \mathscr{E}_s$ such that*

$$f(x) = h(u_1(x), \ldots, u_s(x)). \tag{4.12}$$

Here $\mathscr{E}_s$ is the ring of C^∞ germs $\mathbb{R}^s \to \mathbb{R}$.

We conclude this section with a discussion of some special structure often found in the ring $\mathscr{P}(\Gamma)$, which is quite useful when making explicit calculations. It implies in particular that when f in (4.12) is polynomial then there is a unique choice of the polynomial h.

More precisely, say that a set of Γ-invariant polynomials has a *relation* if there exists a nonzero polynomial $r(y_1, \ldots, y_s)$ such that

$$r(u_1(x), \ldots, u_s(x)) \equiv 0. \tag{4.13}$$

The ring $\mathscr{P}(\Gamma)$ is a *polynomial ring* if it has a Hilbert basis without relations. (*Warning*: A polynomial ring is *not* just a ring of polynomials.)

An example of a group action for which $\mathscr{P}(\Gamma)$ is not a polynomial ring is given by $\Gamma = \mathbf{Z}_2$ acting on $\mathbb{R}^2$, where the action of $-1 \in \mathbf{Z}_2$ is defined by $x \mapsto -x$. It is easy to see that $\mathscr{P}(\mathbf{Z}_2)$ is generated by all monomials of even total degree. The polynomials

$$u_1 = x_1^2, \qquad u_2 = x_1 x_2, \qquad u_3 = x_2^2$$

form a Hilbert basis for $\mathscr{P}(\mathbf{Z}_2)$, but there is a relation

$$u_1 u_3 - u_2^2 \equiv 0.$$

Indeed it can be shown that no choice of Hilbert basis can eliminate all relations, so $\mathscr{P}(\mathbf{Z}_2)$ is *not* a polynomial ring.

There is a simple test to determine whether a given Hilbert basis $u_1, \ldots, u_s$ for $\mathscr{P}(\Gamma)$ makes it into a polynomial ring. Define the mapping $\rho\colon V \to \mathbb{R}^2$, called the *discriminant* of Γ, by

$$\rho(x) = (u_1(x), \ldots, u_s(x)). \tag{4.14}$$

Lemma 4.4. *If the Jacobian $(d\rho)_x$ is onto for some x, then $\mathscr{P}(\Gamma)$ is a polynomial ring.*

Proof. If $(d\rho)_x$ is onto, then by the implicit function theorem $\rho(V)$ contains an open subset of $\mathbb{R}^s$. Hence any polynomial mapping $r\colon \mathbb{R}^s \to \mathbb{R}$ is uniquely determined by $r|\rho(V)$. Now suppose r satisfies (4.13); that is, $r|\rho(V) \equiv 0$. It follows that $r \equiv 0$ and that there are no nontrivial relations. □

Note that in the preceding example of $\mathbf{Z}_2$ $\rho(x_1, x_2) = (x_1^2, x_1 x_2, x_2^2)$ and $(d\rho)_x\colon \mathbb{R}^2 \to \mathbb{R}^3$. Hence it is impossible for $(d\rho)_x$ to be onto. (However, the converse of Lemma 4.4 has not been proved, so this does not show that $\mathscr{P}(\Gamma)$ is not a polynomial ring.)

We may use Lemma 4.4 to check that for Examples 4.1, $\mathscr{P}(\Gamma)$ is a polynomial ring. For instance, consider Example 4.1c, where $\Gamma = \mathbf{D}_n$ acts on $\mathbb{C}$. Recall from (4.8) that

$$u_1(z, \bar{z}) = z\bar{z}, \qquad u_2(z, \bar{z}) = z^n + \bar{z}^n$$

is a Hilbert basis. Then

$$\rho(z, \bar{z}) = (z\bar{z}, z^n + \bar{z}^n)$$

so that

$$d\rho = \begin{bmatrix} z & \bar{z} \\ nz^{n-1} & n\bar{z}^{n-1} \end{bmatrix}.$$

It follows that $\det d\rho = n(z^n - \bar{z}^n)$, which is (often) nonzero. By Lemma 4.4, $\mathscr{P}(\mathbf{D}_n)$ is a polynomial ring.

Remarks.

(a) When $\mathscr{P}(\Gamma)$ is a polynomial ring in the Hilbert basis $u_1, \ldots, u_s$, then every invariant polynomial f has *uniquely* the form

$$f(x) = h(u_1(x), \ldots, u_s(x)).$$

To prove this, suppose not. Then also $f = k(u_1(x), \ldots, u_s(x))$. If $r = h - k$ then $r(u_1(x), \ldots, u_s(x)) \equiv 0$, so r is a relation. This is a contradiction.

(b) Even when $\mathscr{P}(\Gamma)$ is a polynomial ring, uniqueness need not hold in (4.12) for C^∞ germs. For example, let $\mathbf{Z}_2$ act on $\mathbb{R}$ in the standard way. Then $u_1(x) = x^2$ is a Hilbert basis. By Theorem 4.3 every invariant germ $f \in \mathscr{E}(\Gamma)$ has the form $f(x) = h(x^2)$ for some $h \in \mathscr{E}_x$. However, define

$$k(x) = \begin{cases} e^{-1/x} & \text{if} \quad x < 0 \\ 0 & \text{if} \quad x \geq 0. \end{cases}$$

Then k is smooth, and

$$f(x) = h(x^2) + k(x^2)$$

so uniqueness fails. More generally, if $\operatorname{Im} \rho$ in Lemma 4.4 does not contain a neighborhood of the origin in $\mathbb{R}^s$ then uniqueness in (4.12) fails in $\mathscr{E}(\Gamma)$.

(c) It is, however, true that if $\mathscr{P}(\Gamma)$ is a polynomial ring then the Taylor expansion of h in (4.12) at the origin is uniquely defined. Since in our analysis of bifurcation problems we consider only finitely determined situations (that is, those in which the problem may be reduced to a finite part of the Taylor expansion of f), it follows that to all intents and purposes uniqueness in $\mathscr{E}(\Gamma)$ does hold.

(d) Another test to show that $\mathscr{P}(\Gamma)$ is a polynomial ring, even simpler than Lemma 4.4, is give in XIII, §1.

Exercises

4.1. Let $\mathbf{S}^1$ act on $\mathbb{C}^n$ by $(z_1, \ldots, z_n) \mapsto (e^{i\theta} z_1, \ldots, e^{i\theta} z_n)$. Show that a Hilbert basis is $\{\operatorname{Re}(z_j \bar{z}_k), \operatorname{Im}(z_j \bar{z}_k)\}$.

4.2. Let $\mathbf{S}^1$ act on $\mathbb{C}^2$ by $(z_1, z_2) \mapsto (e^{ki\theta} z_1, e^{li\theta} z_2)$ where k, l are coprime. Show that a Hilbert basis is $\{\operatorname{Re}(z_1^l \bar{z}_2^k), \operatorname{Im}(z_1^l \bar{z}_2^k), |z_1|^2, |z_2|^2\}$.

4.3. Let $(\theta, \varphi) \in \mathbf{T}^2$ act on $\mathbb{C}^2$ by $(z_1, z_2) \mapsto (e^{k_1 i\theta + k_2 i\varphi} z_1, e^{l_1 i\theta + l_2 i\varphi} z_2)$ where k_1, l_1 and k_2, l_2 are coprime. Find a Hilbert basis. (*Hint*: Apply Exercise 4.2 to the θ-action and observe the action of φ on a Hilbert basis. Or use brute force on monomials $z_1^\alpha \bar{z}_1^\beta z_2^\gamma \bar{z}_2^\delta$.)

4.4. Which of the preceding rings of invariants are polynomial rings?

4.5. Let the symmetric group $\mathbf{S}_3$, consisting of all permutations of $\{1, 2, 3\}$, act on $\mathbb{R}^3$ by permuting a basis. Show that the invariant functions are generated by $s_1 = x_1 + x_2 + x_3$, $s_2 = x_1 x_2 + x_2 x_3 + x_1 x_3$, and $s_3 = x_1 x_2 x_3$. Prove that the ring of invariants $\mathscr{E}(\mathbf{S}_3)$ is a polynomial ring.

4.6. Prove results analogous to the preceding for $\mathbf{S}_n$ acting on $\mathbb{R}^n$.

4.7. Let Γ be the group of all symmetries, including reflections, of a cube centered at the origin of $\mathbb{R}^3 = \{(x, y, z)\}$ with edges parallel to the axes. (In the notation of

XIII, §9, this is the group $\mathbb{O} \oplus \mathbf{Z}_2^c$.) Prove that the ring of Γ-invariants is generated by

$$u = x^2 + y^2 + z^2$$
$$v = x^2y^2 + y^2z^2 + x^2z^2$$
$$w = x^2y^2z^2$$

and that it is a polynomial ring.

The next group of exercises investigates conditions under which a function $\mathrm{Fix}(\Sigma) \to \mathbb{R}$ extends to a Γ-invariant function.

4.8. Let Γ act on V and let Σ be an isotropy subgroup. Let $f: V \to \mathbb{R}$ be Γ-invariant and let $\varphi = f|\mathrm{Fix}(\Sigma)$. Let $N = N_\Gamma(\Sigma)$.
   (a) Show that φ is N-invariant. Hence a *necessary* condition for a function $\psi: \mathrm{Fix}(\Sigma) \to \mathbb{R}$ to extend to a Γ-invariant function on V is that ψ be N-invariant.
   (b) φ has the following more general *hidden symmetry property*: If there exist $\gamma \in \Gamma \sim N$ and $v \in V$ such that $v, \gamma v \in \mathrm{Fix}(\Sigma)$ then $\varphi(v) = \varphi(\gamma v)$.
   (c) If $\psi: \mathrm{Fix}(\Sigma) \to \mathbb{R}$ is N-invariant, then a necessary condition that ψ should extend to a Γ-invariant function $\hat{\psi}: V \to \mathbb{R}$ is that ψ satisfies the hidden symmetry condition.

4.9. Find an example where the hidden symmetry condition is violated, showing that the condition in Exercise 4.8(a) is not sufficient for an extension to exist.

4.10. If $\psi: \mathrm{Fix}(\Sigma) \to \mathbb{R}$ is N-invariant and satisfies the hidden symmetry condition prove that there exists a *continuous* Γ-invariant extension $\hat{\psi}: V \to \mathbb{R}$. (*Hint*: Work inside a suitable closed ball center 0. Define $\mathscr{X} = \bigcup_{\gamma \in \Gamma} \gamma\, \mathrm{Fix}(\Sigma)$. Prove that $\mathscr{X}$ is a closed subset of V and that ψ extends uniquely to a Γ-invariant function $\tilde{\psi}$ on $\mathscr{X}$. Use the Tietze extension theorem to extend $\tilde{\psi}$ from $\mathscr{X}$ to V, and average over Γ by Haar integration.)

4.11. Let $\Gamma = \mathbf{D}_5$ acting on $\mathbb{C}$, $\Sigma = \mathbf{Z}_2(\kappa)$, and $\psi(x) = x^3$ $(x \in \mathrm{Fix}(\Sigma) = \mathbb{R})$. Observe that ψ trivially satisfies the hidden symmetry condition and is N-invariant. By considering 3-jets (Taylor expansions to degree 3) show that ψ has no *smooth* extension to a Γ-invariant function $\mathbb{C} \to \mathbb{R}$.

4.12. Investigate analogous results to Exercises 4.8–4.11 for the extension of N-equivariant mappings on $\mathrm{Fix}(\Sigma)$ to Γ-equivariant mappings on V.

§5. Nonlinear Commuting Mappings

As usual we let Γ be a compact Lie group acting on a vector space V. Recall that a mapping $g: V \to V$ *commutes with* Γ or is Γ-*equivariant* if

$$g(\gamma x) = \gamma g(x) \tag{5.1}$$

for all $\gamma \in \Gamma$, $x \in V$. In §3 we discussed some of the restrictions placed on linear mappings g by (5.1). In this section we describe the restrictions placed on nonlinear g.

The main observation is that the product of an equivariant mapping and an invariant function is another equivariant mapping.

Lemma 5.1. *Let $f: V \to \mathbb{R}$ be a Γ-invariant function and let $g: V \to V$ be a Γ-equivariant mapping. Then $fg: V \to V$ is Γ-equivariant.*

PROOF. This follows from an easy calculation. For all $\gamma \in \Gamma$ and $x \in V$ we have:

$$\begin{aligned}(fg)(\gamma x) &= f(\gamma x)g(\gamma x)\\ &= f(x)\cdot \gamma g(x)\\ &= \gamma f(x)g(x)\\ &= \gamma fg(x).\end{aligned} \tag{5.2}$$

The first and fourth equalities in (5.2) use the definition of fg; the second equality follows by Γ-invariance and Γ-equivariance; and the third follows because γ acts linearly on V and $f(x)$ is a scalar. □

For example, when $\Gamma = \mathbf{Z}_2$ acts on $\mathbb{R}$ by $-1\cdot x = -x$, then the $\mathbf{Z}_2$-equivariant mappings are just the odd functions; that is, they satisfy $g(-x) = -g(x)$. It is well known that every odd function may be written as an even function times x. This was proved in Corollary VI, 2.2; nevertheless we reproduce the argument here. Since $g(0) = 0$ we use Taylor's theorem to write $g(x) = f(x)x$. Since g is odd,

$$f(-x)x = f(x)x,$$

so f is even. Moreover, we know that $f(x) = h(x^2)$ for a suitably chosen smooth h, by (4.3) and Theorem 4.2 (or by Lemma VI.2.1). Hence

$$g(x) = h(x^2)x. \tag{5.3}$$

We now abstract some general principles from the preceding observations. Let $\vec{\mathscr{P}}(\Gamma)$ be the space of Γ-equivariant polynomial mappings of V into V, and let $\vec{\mathscr{E}}(\Gamma)$ be the space of Γ-equivariant germs (at the origin) of C^∞ mappings of V into V. Lemma 5.1 implies that $\vec{\mathscr{P}}(\Gamma)$ is a module over the ring of invariant polynomials $\mathscr{P}(\Gamma)$, and equally that $\vec{\mathscr{E}}(\Gamma)$ is a module over the ring of invariant function germs $\mathscr{E}(\Gamma)$. This means that if $f \in \mathscr{P}(\Gamma)$ and $g \in \vec{\mathscr{P}}(\Gamma)$ then $fg \in \vec{\mathscr{P}}(\Gamma)$, with a similar statement for $\mathscr{E}$, and this is the content of Lemma 5.1.

The results for $\Gamma = \mathbf{Z}_2$ can be stated in symbols:

$$\begin{aligned}&\text{(a)}\quad \vec{\mathscr{P}}(\mathbf{Z}_2) = \mathscr{P}(\mathbf{Z}_2)\{x\},\\ &\text{(b)}\quad \vec{\mathscr{E}}(\mathbf{Z}_2) = \mathscr{E}(\mathbf{Z}_2)\{x\}.\end{aligned} \tag{5.4}$$

In words, the module $\vec{\mathscr{E}}(\mathbf{Z}_2)$ (or $\vec{\mathscr{P}}(\mathbf{Z}_2)$) is generated over the ring $\mathscr{E}(\mathbf{Z}_2)$ (or $\mathscr{P}(\mathbf{Z}_2)$) by the single $\mathbf{Z}_2$-equivariant mapping x. In general, we say that the equivariant polynomial mapping $g_1, \ldots, g_r$ *generate* the module $\vec{\mathscr{P}}(\Gamma)$ over the ring $\mathscr{P}(\Gamma)$ if every Γ-equivariant g may be written as

$$g = f_1 g_1 + \cdots + f_r g_r \tag{5.5}$$

for invariant polynomials $f_1, \ldots, f_r$. A similar definition may be made for $\vec{\mathscr{E}}(\Gamma)$. The next theorem follows from, and is similar in spirit to, the Hilbert–Weyl theorem. A proof is given in §6.

Theorem 5.2. *Let Γ be a compact Lie group acting on V. Then there exists a finite set of Γ-equivariant polynomials $g_1, \ldots, g_r$ that generates the module $\vec{\mathscr{P}}(\Gamma)$.*

The Γ-equivariant version of Schwarz's theorem (Theorem 4.3) is proved in Poénaru [1976]. We present this proof in §6 too.

Theorem 5.3 (Poénaru [1976]). *Let Γ be a compact Lie group and let $g_1, \ldots, g_r$ generate the module $\vec{\mathscr{P}}(\Gamma)$ of Γ-equivariant polynomials over the ring $\mathscr{P}(\Gamma)$. Then $g_1, \ldots, g_r$ generate the module $\vec{\mathscr{E}}(\Gamma)$ over the ring $\mathscr{E}(\Gamma)$.*

The implications of Theorems 5.2 and 5.3 are illustrated by the following examples.

Examples 5.4.
(a) Let $\Gamma = \mathbf{S}^1$ in its standard action on $V = \mathbb{C}$. We claim that every $\mathbf{S}^1$-equivariant mapping $g \in \vec{\mathscr{E}}(\mathbf{S}^1)$ has the form

$$g(z) = p(z\bar{z})z + q(z\bar{z})iz \tag{5.6}$$

where p and q are real-valued $\mathbf{S}^1$-invariant functions. This has already been proved in Proposition VIII, 2.5 in slightly different notation; we give a different proof here.

Let $g\colon \mathbb{C} \to \mathbb{C}$ be an $\mathbf{S}^1$-equivariant polynomial. In the coordinates $z, \bar{z}$ it has the form

$$g = \sum b_{jk} z^j \bar{z}^k \tag{5.7}$$

where $b_{jk} \in \mathbb{C}$. The equivariance condition (5.1) can be restated as an invariance condition

$$g(x) = \gamma^{-1} g(\gamma x), \tag{5.8}$$

which is often more convenient to use. In the case $\Gamma = \mathbf{S}^1$ we have

$$g = e^{-i\theta} \sum b_{jk} e^{(j-k)i\theta} z^j \bar{z}^k = \sum b_{jk} e^{(j-k-1)i\theta} z^j \bar{z}^k. \tag{5.9}$$

Hence $b_{jk} = 0$ unless $j = k + 1$. Thus

$$g(z) = \sum b_{k+1,k} (z\bar{z})^k z$$

and g has the form (5.6), where

$$p(y) = \sum \operatorname{Re}(b_{k+1,k}) y^k,$$

$$q(y) = \sum \operatorname{Im}(b_{k+1,k}) y^k.$$

Now apply Theorem 5.3.

(b) Let $\Gamma = \mathbf{O}(2)$ in its standard action on $\mathbb{C}$. We claim that every $\mathbf{O}(2)$-equivariant mapping $g \in \vec{\mathscr{E}}(\mathbf{O}(2))$ has the form

$$g(z) = p(z\bar{z})z. \tag{5.10}$$

To prove this, observe that g is in particular $\mathbf{S}^1$-equivariant, hence has the form (5.6). But $\mathbf{O}(2)$ is generated by $\mathbf{S}^1 \cong \mathbf{SO}(2)$ and the flip κ, which acts by $\kappa z = \bar{z}$. Now compute

$$\overline{g(\bar{z})} = p(z\bar{z})z - q(z\bar{z})iz. \tag{5.11}$$

The only way that $\overline{g(\bar{z})}$ can equal $g(z)$ is if $q(z\bar{z}) = 0$, thus proving the claim.

(c) Let $\Gamma = \mathbf{D}_n$ in its standard action on $\mathbb{C}$. We claim that every $\mathbf{D}_n$-equivariant germ $g \in \vec{\mathscr{E}}(\mathbf{D}_n)$ has the form

$$g(z) = p(u, v)z + q(u, v)\bar{z}^{n-1} \tag{5.12}$$

where $u = z\bar{z}$ and $v = z^n + \bar{z}^n$.

We begin again with a $\mathbf{D}_n$-equivariant polynomial g of the form

$$g(z) = \sum b_{jk} z^j \bar{z}^k \tag{5.13}$$

where $b_{jk} \in \mathbb{C}$. We first obtain restrictions on the b_{jk} by using the equivariance of g with respect to κ, where $\kappa z = \bar{z}$. Now

$$\overline{g(\bar{z})} = \sum \bar{b}_{jk} z^j \bar{z}^k.$$

Hence $\overline{g(\bar{z})} = g(z)$ implies that b_{jk} is real.

Recall that $\mathbf{D}_n$ is generated by κ and $\zeta = 2\pi/n$, which acts as multiplication by $e^{i\zeta}$. Now equivariance with respect to ζ implies that

$$\begin{aligned} g(z) &= e^{-i\zeta} g(e^{i\zeta} z) \\ &= \sum b_{jk} e^{(j-k-1)i\zeta} z^j \bar{z}^k. \end{aligned} \tag{5.14}$$

Hence $b_{jk} = 0$ unless $j \equiv k + 1 \pmod{n}$. (It is here that the analysis begins to differ from the case $\Gamma = \mathbf{S}^1$.)

We now show that z and $\bar{z}^{n-1}$ generate the module $\vec{\mathscr{P}}(\mathbf{D}_n)$ over $\mathscr{P}(\mathbf{D}_n)$. In individual terms in (5.13) we can factor out powers of $z\bar{z}$, which are $\mathbf{D}_n$-invariants, until we are left either with $j = 0$ or $k = 0$. Since $j \equiv k + 1 \pmod{n}$ the terms z^{ln+1} and $\bar{z}^{(l+1)n-1}$, $l = 0, 1, 2, \ldots,$ generate the module $\vec{\mathscr{P}}(\mathbf{D}_n)$. However, the identities

$$\begin{aligned} &\text{(a)} \quad z^{(l+2)n+1} = (z^n + \bar{z}^n)z^{(l+1)n+1} - (z\bar{z})^n z^{ln+1} \\ &\text{(b)} \quad \bar{z}^{(l+3)n-1} = (z^n + \bar{z}^n)\bar{z}^{(l+2)n-1} - (z\bar{z})^n \bar{z}^{(l+1)n-1} \end{aligned} \tag{5.15}$$

show that the generators z^{ln+1}, $\bar{z}^{(l+1)n-1}$ are redundant for $l \geq 2$. Similarly

$$\begin{aligned} &\text{(c)} \quad z^{n+1} = (z^n + \bar{z}^n)z - (z\bar{z})\bar{z}^{n-1}, \\ &\text{(d)} \quad \bar{z}^{2n-1} = (z^n + \bar{z}^n)\bar{z}^{n-1} - (z\bar{z})^{n-1} z. \end{aligned}$$

Hence the generators z^{n+1} and $\bar{z}^{2n-1}$ are redundant. This proves the claim.

To end this section we discuss when the representation of a Γ-equivariant g in (5.5) in terms of given generators $g_1, \ldots, g_r$ is unique. We say that $g_1, \ldots, g_r$ *freely generate* the module $\vec{\mathscr{E}}(\Gamma)$ over $\mathscr{E}(\Gamma)$ if the relation

$$f_1 g_1 + \cdots + f_r g_r \equiv 0, \tag{5.16}$$

where $f_j \in \mathscr{E}(\Gamma)$, implies that

$$f_1 \equiv \cdots \equiv f_r \equiv 0. \tag{5.17}$$

We also say that $\vec{\mathscr{E}}(\Gamma)$ is a *free* module over $\mathscr{E}(\Gamma)$. (This definition is the module version of linear independence in vector spaces.) It is clear that if $g_1, \ldots, g_r$ freely generate $\vec{\mathscr{E}}(\Gamma)$ then every $g \in \vec{\mathscr{E}}(\Gamma)$ may be written uniquely as $g = f_1 g_1 + \cdots + f_r g_r$ where $f_j \in \mathscr{E}(\Gamma)$.

Each module discussed in the preceding examples is free. We show this for Example 5.3(c), where $\Gamma = \mathbf{D}_n$. Suppose that

$$p(z\bar{z}, z^n + \bar{z}^n)z + q(z\bar{z}, z^n + \bar{z}^n)\bar{z}^{n-1} \equiv 0. \tag{5.18}$$

Suppose there exists $z \in \mathbb{C}$ at which $q(z\bar{z}, z^n + \bar{z}^n) \neq 0$. By continuity $q \neq 0$ in a neighbourhood of z. Multiply (5.18) by $\bar{z}$ and solve for

$$\bar{z}^n = p(z\bar{z}, z^n + \bar{z}^n)z\bar{z}/q(z\bar{z}, z^n + \bar{z}^n). \tag{5.19}$$

The right-hand side of (5.18) is real, but $\bar{z}^n$ is never real-valued on an open set (or else it would be everywhere real) so we have a contradiction. Hence $q \equiv 0$. But (5.18) now implies $p \equiv 0$. Hence $\vec{\mathscr{E}}(\mathbf{D}_n)$ is a free module over the ring $\mathscr{E}(\mathbf{D}_n)$ with free generators z and $\bar{z}^{n-1}$.

Exercises

5.1. Let $\mathbf{S}^1$ act on $\mathbb{C}^n$ as in Exercise 4.1. Prove that the equivariants are generated as a module over the invariants by the mappings $z \mapsto z_k$, $z \mapsto iz_k$, for $k = 1, \ldots, n$.

5.2. Let $\mathbf{S}^1$ act on $\mathbb{C}^2$ as in Exercise 4.2. Prove that the equivariants are generated as a module over the invariants by the mappings $(z_1, z_2) \mapsto$

$$(z_1, 0), (iz_1, 0), (\bar{z}_1^{l-1} z_2^k, 0), (i\bar{z}_1^{l-1} z_2^k, 0),$$

$$(0, z_2), (0, iz_2), (0, z_1^l \bar{z}_2^{k-1}), (0, iz_1^l \bar{z}_2^{k-1}).$$

5.3. Let $\mathbf{T}^2$ act on $\mathbb{C}^2$ as in Exercise 4.3. Find generators for the equivariants.

5.4. Which of the preceding modules of equivariants are free?

5.5. Let $\Gamma = \mathbb{O} \oplus \mathbf{Z}_2^c$ be the symmetry group of a cube acting on $\mathbb{R}^3$ as in Exercise 4.7. Prove that the module of Γ-equivariants is generated by the mappings

$$X_1 = \begin{bmatrix} x \\ y \\ z \end{bmatrix} \quad X_2 = \begin{bmatrix} x^3 \\ y^3 \\ z^3 \end{bmatrix} \quad X_3 = \begin{bmatrix} y^2 z^2 x \\ z^2 x^2 y \\ x^2 y^2 z \end{bmatrix},$$

and is a free module over the ring of invariants.

§6.* Proofs of Theorems in §§4 and 5

In this section we present the promised proofs of four theorems from the previous two sections, namely:

(a) The Hilbert–Weyl theorem, Theorem 4.2,
(b) Schwarz's theorem, Theorem 4.3,
(c) $\vec{\mathscr{P}}(\Gamma)$ is finitely generated, Theorem 5.2,
(d) Poénaru's theorem, Theorem 5.3.

However, the proof of Theorem 4.3 is only sketched, because a complete proof would involve too much extra machinery.

(a) Proof of Theorem 4.2

The proof of the Hilbert–Weyl theorem follows by an induction argument from the Hilbert basis theorem, which we state later. First we recall some standard facts about polynomials. Let R be a commutative ring. An expression of the form

$$r_n x^n + r_{n-1} x^{n-1} + \cdots + r_0 \tag{6.1}$$

with $r_0, \ldots, r_n \in R$ is a *polynomial* in the *indeterminate* x with *coefficients* in R. If $r_n \neq 0$ then the polynomial (6.1) has *degree n*. The set of all polynomials in x with coefficients in R is also a ring, with addition and multiplication of polynomials being defined in the obvious way. Denote this ring by $R[x]$. Inductively define

$$R[x_1, \ldots, x_n] = R[x_1, \ldots, x_{n-1}][x_n], \tag{6.2}$$

the ring of polynomials over R in n indeterminates $x_1, \ldots, x_n$.

Theorem 6.1 (Hilbert Basis Theorem). *Let R be a commutative ring such that every ideal in R is finitely generated. Then every ideal in the ring $R[x]$ is finitely generated.*

Before proving this we derive the Hilbert–Weyl theorem. We need:

Corollary 6.2. *Every ideal of $\mathbb{R}[x_1, \ldots, x_n]$ is finitely generated.*

Proof. The corollary is proved by induction on n. When $n = 0$ the only ideals of $\mathbb{R}$ are $\{0\}$ and $\mathbb{R}$, generated by $\{0\}$ and $\{1\}$, respectively. When $n > 0$ the induction step follows from Theorem 6.1, setting

$$R = \mathbb{R}[x_1, \ldots, x_{n-1}].$$

□

It is this choice of R that requires us to prove a sufficiently general version of the Hilbert basis theorem: it would not suffice to state it just for $R = \mathbb{R}$.

The proof of the Hilbert–Weyl theorem requires a minor variant of Corollary 6.2.

Proposition 6.3. *Let U be a nonempty subset of $\mathbb{R}[x_1, \ldots, x_n]$. Then there exists a finite set of elements $\{u_1, \ldots, u_s\}$ of U such that every $u \in U$ may be written in the form*

$$u = f_1 u_1 + \cdots + f_s u_s \tag{6.3}$$

where $f_1, \ldots, f_s \in \mathbb{R}[x_1, \ldots, x_n]$.

PROOF. Let $\mathscr{I}$ be the ideal in $\mathbb{R}[x_1, \ldots, x_n]$ generated by U. Corollary 6.2 states that $\mathscr{I}$ is finitely generated, say by $p_1, \ldots, p_l$. Since $\mathscr{I}$ is generated by U we may write each p_j in the form

$$p_j = f_{j,1} u_{j,1} + \cdots + f_{j,m(j)} u_{j,m(j)}$$

for $j = 1, \ldots, l$, where $u_{j,k} \in U$. Therefore, the $u_{j,k}$ generate $\mathscr{I}$ and give the desired subset of U. □

Now we can prove the Hilbert–Weyl theorem.

PROOF OF THEOREM 4.2. Let Γ be a compact Lie group acting on V. Identify V with $\mathbb{R}^n$ and let $x_1, \ldots, x_n$ be coordinates. Recall that $\mathscr{P}(\Gamma)$ denotes the ring of invariant polynomials. We must show that there exists a Hilbert basis for $\mathscr{P}(\Gamma)$, that is, that there is a finite set $u_1, \ldots, u_s \in \mathscr{P}(\Gamma)$ such that every $u \in \mathscr{P}(\Gamma)$ may be written in the form

$$u = f(u_1, \ldots, u_s) \tag{6.4}$$

where f is a polynomial function.

If $v \in \mathscr{P}(\Gamma)$ is of degree m and we write

$$v = v_0 + v_1 + \cdots + v_m \tag{6.5}$$

where each v_j is a homogeneous polynomial of degree j, then each $v_j \in \mathscr{P}(\Gamma)$. This is valid since Γ acts linearly on V; hence $v_j(\gamma x)$ is a homogeneous polynomial of degree j for every $\gamma \in \Gamma$. Thus the polynomials $v(\gamma x)$ and $v(x)$ are equal only if $v_j(\gamma x) = v_j(x)$ for each j.

By (6.5) we need verify (6.4) only for homogeneous u. We claim, moreover, that we can choose $u_1, \ldots, u_s$ to be homogeneous polynomials. Let U be the set of nonconstant homogeneous polynomials in $\mathscr{P}(\Gamma)$, and apply Proposition 6.3 to U, obtaining $u_1, \ldots, u_s$, satisfying (6.3).

We now verify (6.4) for homogeneous $u \in \mathscr{P}(\Gamma)$, using induction on the degree $\deg u$ of u. If $\deg u = 0$ then u is constant, and (6.4) is obvious: just define $f = u$. For the induction step, assume (6.4) is valid for all $v \in U$ with $\deg v \leq k$, and let $\deg u = k + 1$. Since $u \in U$ there exist polynomials $f_1, \ldots, f_s \in \mathbb{R}[x_1, \ldots, x_n]$ such that

$$u = f_1 u_1 + \cdots + f_s u_s. \tag{6.6}$$

We may assume each f_j is homogeneous, with

$$\deg f_j = \deg u - \deg u_j. \tag{6.7}$$

In particular, if $\deg u < \deg u_j$ then $f_j = 0$. This assumption follows directly since u and the u_j are homogeneous.

We claim that we can replace the homogeneous polynomials f_j in (6.6) by homogeneous Γ-invariant polynomials F_j, with

$$\deg F_j = \deg f_j \tag{6.8}$$

for each $j = 1, \ldots, s$. To establish this claim, integrate (6.6) over Γ. The Γ-invariance of u and u_j leads to

$$F_j(x) = \int_{\Gamma} f_j(\gamma x)\, d\gamma.$$

Averaging a homogeneous polynomial of degree d produces another (Γ-invariant) homogeneous polynomial of degree d since Γ acts linearly, hence leaves the space of homogeneous polynomials of degree d invariant.

Finally, we use the induction hypothesis to write each f_j in the form

$$f_j = g_j(u_1, \ldots, u_s).$$

This is possible since each u_j has degree ≥ 1, so by (6.7)

$$\deg f_j \leq \deg u - 1 \leq k.$$

Now we set

$$f(u_1, \ldots, u_s) = \sum_{j=1}^{s} g_j(u_1, \ldots, u_s) u_j,$$

and (6.4) holds as required. □

The remainder of this subsection is devoted to a proof of the Hilbert basis theorem, Theorem 6.1, which is required to complete the preceding proofs. We begin this task by introducing some notation and proving a preliminary lemma.

Let $f(x)$ be a polynomial in $R[x]$ of degree m. Of course, f has the form (6.1), where $r_m \neq 0$. We call r_m the *leading coefficient* of f and denote r_m by $\hat{f}$. By convention $\hat{0} = 0$.

Suppose that $\mathscr{I} \subset R[x]$ is an ideal. We define

$$\hat{\mathscr{I}} = \{\hat{f} \in R \mid f \in \mathscr{I}\}. \tag{6.9}$$

We claim that $\hat{\mathscr{I}}$ is an ideal, the *ideal of leading terms* in $\mathscr{I}$. To verify this we must show that

$$\begin{aligned} &\text{(a)} \quad \text{If } f_1, f_2 \in \mathscr{I} \quad \text{then} \quad \hat{f}_1 + \hat{f}_2 \in \hat{\mathscr{I}}, \\ &\text{(b)} \quad \text{If } r \in R, f \in \mathscr{I} \quad \text{then} \quad r\hat{f} \in \hat{\mathscr{I}}. \end{aligned} \tag{6.10}$$

To prove (6.10(a)) we assume

$$d_1 \equiv \deg f_1 \leq \deg f_2 \equiv d_2.$$

Let $f = x^{d_2-d_1}f_1 + f_2$; then $f \in \mathscr{I}$ since $\mathscr{I}$ is an ideal. Moreover, $\hat{f} = \hat{f}_1 + \hat{f}_2$ as desired. The proof of (6.10(b)) is simpler since clearly $\widehat{(rf)} = r\hat{f}$.

Given an ideal $\mathscr{I} \in R[x]$ we define

$$\mathscr{I}_k = \{f \in \mathscr{I} | \deg f \leq k\}. \tag{6.11}$$

That is, $\mathscr{I}_k$ consists of all polynomials in $\mathscr{I}$ of degree $\leq k$. Observe that $\mathscr{I}_k$ is an R-module, since $\deg(rf) \leq \deg f$.

Lemma 6.4. *Suppose that all ideals in R are finitely generated. Then for each k, $\mathscr{I}_k$ is a finitely generated R-module.*

Remark. Recall that $\mathscr{I}_k$ is a finitely generated R-module if there exist finitely many generators $q_1, \ldots, q_s \in \mathscr{I}_k$ such that every $q \in \mathscr{I}_k$ has the form

$$q = r_1q_1 + \cdots + r_sq_s$$

where $r_1, \ldots, r_s \in R$.

Proof. We use induction on k. The result is trivially true if $k = 0$ since $\mathscr{I}_0 \subset R$ is an ideal of R, thus finitely generated (as an ideal, hence as an R-module).

Inductively, suppose that $\mathscr{I}_{k-1}$ is a finitely generated R-module, with generators $f_1, \ldots, f_s$. As we showed earlier, $\hat{\mathscr{I}}_k$ is an ideal in R and hence is finitely generated, say by $\hat{g}_1, \ldots, \hat{g}_t$. Moreover, we may assume $\deg g_i = k$ for all i. (If $\deg g_i < k$ then replace g_i by $x^{k-\deg g_i}g_i$.) We claim that

$$\{f_1, \ldots, f_s, g_1, \ldots, g_t\}$$

is a set of generators for $\mathscr{I}_k$ as an R-module. To verify the claim, suppose that $g(x) \in \mathscr{I}_k$ has the form

$$g(x) = r_kx^k + r_{k-1}x^{k-1} + \cdots + r_0 \tag{6.12}$$

where $r_0, \ldots, r_k \in R$. If $r_k = 0$, then $g \in \mathscr{I}_{k-1}$ and by induction is a linear combination of the f_j. If $r_k \neq 0$ then, by definition, $r_k \in \hat{\mathscr{I}}$. It follows that

$$r_k = a_1\hat{g}_1 + \cdots + a_t\hat{g}_t \tag{6.13}$$

where $a_1, \ldots, a_t \in R$. We may use (6.12) ad (6.13) to conclude that

$$g - (a_1g_1 + \cdots + a_tg_t)$$

has the form (6.12) with $r_k = 0$ (since $\deg g_i = k$ for all i) and is a linear combination of the f_j as previously. This construction proves the claim. □

Proof of Theorem 6.1. Let $\mathscr{I} \subset R[x]$ be an ideal. We must show that $\mathscr{I}$ is finitely generated. We construct a set of generators for $\mathscr{I}$ as follows. By assumption every ideal in R is finitely generated; hence $\hat{\mathscr{I}}$ is finitely generated. Let $\hat{p}_1, \ldots, \hat{p}_s$ be generators for $\hat{\mathscr{I}}$ and let

$$k = \max_{1 \leq i \leq s} \deg p_i.$$

Let $\{q_1, \ldots, q_t\}$ be a set of generators for the R-module $\mathscr{I}_k$, whose existence is guaranteed by Lemma 6.4. We claim that

$$\{p_1, \ldots, p_s, q_1, \ldots, q_t\}$$

is a set of generators for the ideal $\mathscr{I}$.

To prove this we must show that for every $f \in \mathscr{I}$ there exist $a_1, \ldots, a_s$, $b_1, \ldots, b_t \in R[x]$ such that

$$f = a_1 p_1 + \cdots + a_s p_s + b_1 q_1 + \cdots + b_t q_t. \tag{6.14}$$

This is trivially true if $\deg f \leq k$, since then f is a linear combination of the q_j with constant polynomials $b_j \in R$ as coefficients. To prove (6.14) holds in general, we use induction on $\deg f$. Assume that whenever $\deg f \leq k + l$, (6.14) holds. Now suppose f has degree $k + l + 1$. We can write the leading coefficient of f as

$$\hat{f} = r_1 \hat{p}_1 + \cdots + r_s \hat{p}_s$$

since the $\hat{p}_j$ generate the ideal $\hat{\mathscr{I}}$. Now observe that

$$g = f - \sum_{j=1}^{s} r_j x^{k+l+1-\deg p_j} p_j$$

has degree $\leq k + l$ since the leading term of f has been cancelled away. It follows by induction that g has the form (6.14), hence so does f. □

(b) Proof of Theorem 4.3

Here we sketch (with a broad brush) the proof of Schwarz's theorem, Theorem 4.3, on smooth invariants. Complete proofs may be found in Schwarz [1975], Mather [1977], and Bierstone [1980].

Recall the setting: Γ is a compact Lie group acting (orthogonally) on $V = \mathbb{R}^n$, and $\{u_1, \ldots, u_s\}$ is a Hilbert basis for the ring $\mathscr{P}(\Gamma)$ of Γ-invariant polynomials. We wish to show that every germ $g \in \mathscr{E}_x(\Gamma)$ is of the form

$$g(x) = f(u_1(x), \ldots, u_s(x)) \tag{6.15}$$

for some germ $f \in \mathscr{E}_y$, where $y = (y_1, \ldots, y_s)$. It is sufficient to verify (6.15) for one Hilbert basis. For suppose that $v_1, \ldots, v_t$ is another Hilbert basis. By the Hilbert–Weyl theorem

$$u_i = w_i(v_1, \ldots, v_t) \qquad (i = 1, \ldots, s),$$

so

$$\begin{aligned} g(x) &= f(w_1(v_1, \ldots, v_t), \ldots, w_s(v_1, \ldots, v_t)) \\ &= F(v_1, \ldots, v_t) \end{aligned}$$

for suitable smooth F. This verifies (6.15) for the Hilbert basis $v_1, \ldots, v_t$.

In particular, we may assume for the remainder of this section that the u_j are homogeneous polynomials.

We begin our discussion by recalling a result of Borel. Details of the proof may be found in Theorem 4.10 of Bröcker [1975].

Lemma 6.5 (E. Borel). *Let $\varphi(x)$ be any formal power series in $x = (x_1, \ldots, x_n)$, with real coefficients. Then there exists a smooth germ $f \in \mathscr{E}_x$ such that*

$$jf(x) = \varphi(x), \tag{6.16}$$

where jf is the infinite Taylor series of f.

Remark. Recall from Chapter II, §3, that the Taylor series of f may be written using multi-index notation as

$$jf(x) = \sum_\alpha \frac{1}{\alpha!}\left(\frac{\partial}{\partial x}\right)^\alpha f(0)\cdot x^\alpha.$$

Uniqueness is not asserted in Lemma 6.5. The germs $f \in \mathscr{E}_x$ with $jf \equiv 0$ are said to be *flat*. We show later that the difficulties in proving Schwarz's theorem all reside in the flat germs.

Lemma 6.6. *If $\varphi(x)$ is a Γ-invariant formal power series, then there is a formal power series ψ in s variables such that*

$$\varphi(x) = \psi(u_1(x), \ldots, u_s(x)). \tag{6.17}$$

Proof. Write $\varphi(x) = \sum_0^\infty \varphi_i(x)$ where φ_i consists of those terms in φ which are homogeneous of degree i. By the Hilbert–Weyl theorem, Theorem 4.2, there exist polynomials $\psi_i(y_1, \ldots, y_s)$ such that

$$\varphi_i(x) = \psi_i(u_1(x), \ldots, u_s(x)). \tag{6.18}$$

Let $l = \max \deg u_j(x)$ and recall that we are assuming the u_j to be homogeneous. Observe that to satisfy (6.18) we may assume that the smallest degree of a nonzero term in ψ_i is $[i/l]$. It follows that

$$\psi(y) = \sum_0^\infty \psi_i(y)$$

is a well-defined formal power series, since in any fixed degree there are only a finite number of ψ_i contributing nonzero terms to the sum. Now ψ satisfies (6.17). □

Corollary 6.7. *If every flat germ $g \in \mathscr{E}_x(\Gamma)$ has the form* (6.15), *then every germ $g \in \mathscr{E}_x(\Gamma)$ has the form* (6.15).

Proof. Let $g(x) \in \mathscr{E}_x(\Gamma)$. Then jg is a Γ-invariant power series. Hence Lemma 6.6 implies that there exists a formal power series $\psi(y_1, \dots, y_s)$ such that

$$jg(x) = \psi(u_1(x), \dots, u_s(x)).$$

By Lemma 6.5, there is a smooth germ $f \in \mathscr{E}_y$ such that $jf = \psi$. It follows that

$$j(g(x) - f(u_1(x), \dots, u_s(x)) \equiv 0.$$

Hence $g(x) - f(u_1(x), \dots, u_s(x))$ is flat. By assumption there exists a germ $h \in \mathscr{E}_y$ such that

$$g(x) - f(u_1(x), \dots, u_s(x)) = h(u_1(x), \dots, u_s(x)).$$

Therefore, g satisfies (6.15). □

Of course, the flat case is the heart of the problem. Nevertheless, the reduction to the flat case is important, as we can see by considering the example of $\mathbf{Z}_2$ acting on $\mathbb{R}$ by reflection. Suppose that g is flat and satifies $g(x) = g(-x)$. We claim there is a smooth germ f such that $g(x) = f(x^2)$. Moreover, we can define f easily since g is flat. Namely, let $f(y) = g(\sqrt{|y|})$. It is clear that $f(x^2) = g(x)$, and that f is smooth away from the origin. However, since g vanishes to infinite order at the origin, f is also smooth at 0. Thus we have proved Schwarz's theorem when $\Gamma = \mathbf{Z}_2$ and $V = \mathbb{R}$ (and we have given an alternative proof to Lemma VI, 2.1, since the bifurcation parameter λ can easily be introduced without affecting the argument).

Let us try to generalize this approach. Let g be a Γ-invariant germ and let

$$\begin{aligned} &\rho\colon \mathbb{R}^n \to \mathbb{R}^s \\ &\rho(x) = (u_1(x), .., u_s(x)) \end{aligned} \tag{6.19}$$

be the discriminant of Γ, with $u_1, \dots, u_s$ a Hilbert basis for $\mathscr{E}_x(\Gamma)$. We wish to find a smooth germ $f \in \mathscr{E}_y$ such that

$$g(x) = f(\rho(x)). \tag{6.20}$$

(Of course, (6.20) is just another way to write (6.15).) Equation (6.20) defines f uniquely on the image Δ of ρ in $\mathbb{R}^s$, known as the (*real*) *discriminant variety* of Γ. We indicate here why f is well defined on Δ. Suppose that $\rho(x) = \rho(x')$. By standard results on invariants we see that x and x' must lie on the same orbit under the group Γ. So $x' = \gamma x$ for some $\gamma \in \Gamma$, whence $g(x') = g(x)$ and f is well defined on Δ.

The main difficulty lies in showing that f extends from Δ to a smooth function on $\mathbb{R}^s$. Very roughly, this is relatively easy to achieve except at the origin. However, at the origin we may use the flatness of g to construct the desired extension. The central idea in finding this extension is the fact that ρ is a polynomial mapping. Hence the singularity of ρ at 0 is no worse than algebraic, much as in the case $\Gamma = \mathbf{Z}_2$ where it is $\sqrt{\ }$. The flatness of g then swamps this singularity of ρ and allows the extension.

The actual proof involves numerous technical details. Again we refer the reader to Schwarz [1975], Mather [1977], and Bierstone [1980].

(c) Proof of Theorems 5.2 and 5.3

In this subsection we discuss the structure of the modules $\vec{\mathscr{P}}(\Gamma)$ and $\vec{\mathscr{E}}(\Gamma)$ over the rings $\mathscr{P}(\Gamma)$ and $\mathscr{E}(\Gamma)$, respectively. We prove the basis theorems here in somewhat greater generality than is stated in §5. Let us introduce this generality now.

Let the compact Lie group Γ act on two different spaces V and W. We can still speak of Γ-equivariant mappings of V into W as those maps g satisfying

$$g(\gamma x) = \gamma g(x) \tag{6.21}$$

where the action of γ on the left-hand side of (6.21) is its action on V, and the action on the right-hand side is that on W. We denote the Γ-equivariant polynomials by $\vec{\mathscr{P}}(\Gamma; V, W)$ and the Γ-equivariant germs by $\vec{\mathscr{E}}(\Gamma; V, W)$. Both these spaces are modules over the rings of Γ-invariant functions on V, that is, $\mathscr{P}_x(\Gamma)$ and $\mathscr{E}_x(\Gamma)$, respectively.

Following Poénaru [1976], who also credits Malgrange, we shall prove:

Theorem 6.8.
(a) *The module $\vec{\mathscr{P}}(\Gamma; V, W)$ is finitely generated over the ring $\mathscr{P}(\Gamma)$.*
(b) *Let $g_1, \ldots, g_l$ be generators for the module $\vec{\mathscr{P}}(\Gamma; V, W)$. Then $\{g_1, \ldots, g_l\}$ is a set of generators for the module $\vec{\mathscr{E}}(\Gamma; V, W)$ over the ring $\mathscr{E}_x(\Gamma)$.*

Remark. Theorems 5.2 and 5.3 follow immediately by setting $V = W$ and assuming that the actions of Γ on V and W are identical.

PROOF.
(a) The basic idea is to convert the equivariant situation to the invariant case. Here we must use the fact that Γ is compact and may be assumed to act orthogonally on both V and W. Let $\langle\ ,\ \rangle$ denote a Γ-invariant inner product on W. Suppose that $g: V \to W$ is Γ-equivariant and let $y \in W$. Then

$$f(x, y) = \langle g(x), y \rangle \tag{6.22}$$

is a Γ-invariant function $V \times W \to \mathbb{R}$, where the action of Γ on $V \times W$ is $\gamma(x, y) = (\gamma x, \gamma y)$. To check Γ-invariance in (6.22) we compute

$$f(\gamma x, \gamma y) = \langle g(\gamma x), \gamma y \rangle = \langle \gamma g(x), \gamma y \rangle = \langle g(x), y \rangle = f(x, y)$$

where the penultimate equality follows from orthogonality of the action of Γ on W.

Conversely, we can recover Γ-equivariant mappings g from Γ-invariant functions f by the relation

$$g(x) = (d_y f)^t_{x,0}, \tag{6.23}$$

where t indicates the transpose. It is easy to see that (6.23) is a consequence of (6.22). More generally, however, we claim that the mapping g defined by (6.23) is Γ-equivariant for any Γ-invariant $f: V \times W \to \mathbb{R}$. To prove this, differentiate the relation

$$f(\gamma x, \gamma y) = f(x, y)$$

with respect to the y-variables and evaluate at $y = 0$, obtaining

$$(d_y f)_{\gamma x, 0}\gamma = (d_y f)_{x, 0}. \tag{6.24}$$

It follows from (6.23) and (6.24) that

$$\gamma^t g(\gamma x) = g(x).$$

However, since Γ acts orthogonally on W, $\gamma^t = \gamma^{-1}$ and we have

$$g(\gamma x) = \gamma g(x)$$

as claimed.

These calculations show that the Γ-equivariant polynomial mappings in $\vec{\mathscr{P}}(\Gamma; V, W)$ (respectively, germs in $\vec{\mathscr{E}}_x(\Gamma; V, W)$) may be obtained by the construction (6.23) from the Γ-invariant polynomial functions in $\mathscr{P}(\Gamma; V \times W)$ (respectively, germs in $\mathscr{E}_{x,y}(\Gamma, V \times W)$) in a natural way. We now claim that generators for the module $\vec{\mathscr{P}}(\Gamma, V, W)$ can be obtained from a Hilbert basis for the Γ-invariant functions $\mathscr{P}(\Gamma, V \times W)$, whose existence is guaranteed by the Hilbert–Weyl theorem. The general Γ-invariant function in $\mathscr{P}(\Gamma, V \times W)$ has the form

$$f(u_1(x, y), \ldots, u_s(x, y)), \tag{6.25}$$

where f is a polynomial in s variables. Using the construction (6.23) we can write the general Γ-equivariant mapping in $\vec{\mathscr{P}}(\Gamma; V, W)$ as

$$g(x) = \sum_{j=1}^{s} \frac{\partial f}{\partial u_j}(u_1, \ldots, u_s)|_{y=0}(d_y u_j)^t_{x,0}. \tag{6.26}$$

Since $\frac{\partial f}{\partial u_j}(u_1(x, 0), \ldots, u_j(x, 0))$ is a Γ-invariant function in $\mathscr{P}(\Gamma, V)$ we have shown that the s equivariants

$$(d_y u_1)_{x,0}, \ldots, (d_y u_s)_{x,0}$$

generate the module $\vec{\mathscr{P}}(\Gamma; V, W)$, thus proving part (a) of this theorem.

(b) This can be proved by slightly adapting the preceding argument, which in particular shows that all Γ-equivariant smooth germs in $\vec{\mathscr{E}}_x(\Gamma; V, W)$ may be obtained from the Γ-invariant functions in $\mathscr{E}_x(\Gamma, V \times W)$ by the construction (6.23). We can of course represent the general Γ-invariant smooth germ in $\mathscr{E}_x(\Gamma, V \times W)$ in the form (6.25) using Schwarz's theorem in place of the Hilbert–Weyl theorem. The only difference is that f is now a smooth germ in s variables. The remainder of the argument is identical to that in part (a). □

§7.* Tori

In §1 we introduced the n-dimensional torus $\mathbf{T}^n = \mathbf{S}^1 \times \cdots \times \mathbf{S}^1$ (n times). Tori are important in Lie group theory, and we shall need some of their properties in Chapters XIX and XX on mode interactions. These properties are collected here for reference; this section may be omitted on first reading.

We realize $\mathbf{T}^n$ as a Lie group by its *standard* representation on $\mathbb{R}^{2n}$, in which $\theta = (\theta_1, \ldots, \theta_n) \in \mathbf{T}^n$ acts as the matrix

$$\begin{bmatrix} R_{\theta_1} & & & \\ & R_{\theta_2} & & \\ & & \ddots & \\ & & & R_{\theta_n} \end{bmatrix}.$$

As a topological manifold, $\mathbf{T}^n$ is n-dimensional. The main results of this section are as follows:

Theorem 7.1. *Every irreducible representation of a compact abelian Lie group (in particular of a torus) is of dimension at most* 2.

Theorem 7.2. *A Lie group is compact connected abelian if and only if it is isomorphic to a torus.*

Theorem 7.3. *Every torus* $\mathbf{T}^n$ *contains a dense subgroup* Δ *isomorphic to the additive group of real numbers.*

The reader willing to take these results on trust may omit the remainder of this section.

For the less trusting we begin with the following:

PROOF OF THEOREM 7.1. Let Γ be compact abelian acting irreducibly on V. We may suppose that the action on V is not trivial, otherwise $\dim V = 1$. Identify V with $\mathbb{R}^n$ and complexify to get $\mathbb{C}^n$. We can make Γ act on $\mathbb{C}^n$ by extending the action on V via linearity over $\mathbb{C}$. To do this let $z \in \mathbb{C}^n$ and write it as $z = x + iy$ where $x, y, \in \mathbb{R}^n$. Define a Γ-action on $\mathbb{C}^n$ by

$$\gamma z = \gamma x + i\gamma y.$$

An easy calculation shows that each transformation

$$\rho_\gamma \colon \mathbb{C}^n \to \mathbb{C}^n$$

$$\rho_\gamma(z) = \gamma z$$

is $\mathbb{C}$-linear. Now any commuting set of $\mathbb{C}$-linear transformations (finite or infinite) on $\mathbb{C}^n$ has a simultaneous eigenvector, so we may let $w = u + iv$ be a simultaneous eigenvector for all the ρ_γ, $\gamma \in \Gamma$. Let $\lambda(\gamma) = \mu(\gamma) + i\nu(\gamma)$ be the corresponding eigenvalue. Now $\mathbb{C}\{w\}$ is a one-dimensional (over $\mathbb{C}$) $\mathbb{C}$-linear subspace of $\mathbb{C}^n$ invariant under Γ. We construct a "real form" of this as follows. We have

$$\gamma w = \lambda(\gamma) w$$

whence

$$\gamma u + i\gamma v = (\mu(\gamma) + i\nu(\gamma))(u + iv)$$

so that

$$\gamma u = \mu(\gamma) u - \nu(\gamma) v$$

$$\gamma v = \mu(\gamma) v + \nu(\gamma) u.$$

Hence the *real* vector space $W \subset V$ spanned by u and v is Γ-invariant. By irreducibility, $W = V$. But $\dim_{\mathbb{R}} W \leq 2$ as claimed. □

Corollary 7.4. *If Γ is a compact connected abelian Lie group then every nontrivial irreducible representation has dimenion 2 and is isomorphic to a representation on $\mathbb{R}^2 \equiv \mathbb{C}$ with action*

$$\gamma z = e^{i\theta(\gamma)} z$$

where $z \in \mathbb{C}$ and $\theta\colon \Gamma \to \mathbf{S}^1$ is a homomorphism.

PROOF. Without loss of generality the action of Γ is orthogonal. Suppose $\dim V = 1$. The only orthogonal transformations of V are then $\pm I$. Since the action is nontrivial, $\rho_\gamma = -I$ for some $\gamma \in \Gamma$. Since $\{\pm I\}$ is discrete, this contradicts connectedness of Γ.

Thus $\dim V = 2$, and Γ acts via elements of $\mathbf{O}(2)$. Connectedness implies that Γ acts via $\mathbf{SO}(2) \cong \mathbf{S}^1$. Hence we may identify V with $\mathbb{C}$, and γ acts by $z \to e^{i\theta(\gamma)} z$. For this to be an action we require

$$\theta(\gamma_1 + \gamma_2) = \theta(\gamma_1) + \theta(\gamma_2)$$

so θ is a group homomorphism. □

An immediate consequence is the following:

Proposition 7.5. *Every compact connected abelian Lie group is isomorphic to a subgroup of a standard torus.*

PROOF. Let $\Gamma \subset \mathbf{GL}(V)$ be compact connected abelian. Decompose V into irreducible subspaces, $V = V_1 \oplus \cdots \oplus V_k$. By Corollary 7.4, $\dim V_j = 2$, and in relation to an appropriate basis Γ acts on V by

$$\rho_\gamma = \begin{bmatrix} R_{\theta_1(\gamma)} & & & \\ & R_{\theta_2(\gamma)} & & \\ & & \ddots & \\ & & & R_{\theta_k(\gamma)} \end{bmatrix},$$

for suitable homomorphisms $\theta_j\colon \Gamma \to \mathbf{S}^1$. This exhibits $\rho(\Gamma)$, which is isomorphic to Γ, as a subgroup of $\mathbf{T}^k$. □

In fact, a stronger result, namely Theorem 7.2, is true. The proof can be completed by showing that every connected closed subgroup of a torus is a torus or by exploiting additional machinery from the theory of Lie groups (in particular the "exponential map"). See Adams [1969], p. 16, Corollary 2.20.

Finally we turn to Theorem 7.3, the existence in any torus $\mathbf{T}^n$ of a dense subgroup Δ isomorphic to $\mathbb{R}^+$, the group of reals under addition. By *dense* we mean that the closure of Δ in $\mathbf{T}^n$ is the whole of $\mathbf{T}^n$. To see how such a subgroup can arise, consider the 2-torus $\mathbf{T}^2$. For any $\alpha \in \mathbb{R}$, define a map $\varphi_\alpha\colon \mathbb{R}^+ \to \mathbf{T}^2$ by

$$\varphi_\alpha(\theta) = \left[\begin{array}{c|c} R_\theta & \\ \hline & R_{\alpha\theta} \end{array}\right].$$

We ask when φ_α is an isomorphism onto its image, that is, $\ker \varphi_\alpha = 0$. Now $\theta \in \ker \varphi_\alpha$ if and only if

$$\theta \equiv 0 \pmod{2\pi}$$
$$\alpha\theta \equiv 0 \pmod{2\pi}.$$

Therefore, $\theta = 2q\pi$, $\alpha\theta = 2p\pi$ where $p, q \in \mathbb{Z}$. Hence either $\theta = 0$ or $\alpha = p/q$ is rational. Therefore, φ_α is an isomorphism onto its image if and only if α is irrational. It is well known in this case that the image Δ of φ_α is dense in $\mathbf{T}^2$. See Abraham and Marsden [1978], p. 259, Proposition 4.1.11.

More generally, we have a strengthening of Theorem 7.3:

Proposition 7.6. *Let $\alpha_1, \ldots, \alpha_n \in \mathbb{R}$ be linearly independent over the rationals. Define $\varphi_\alpha\colon \mathbb{R}^+ \to \mathbf{T}^n$ by*

$$\varphi_\alpha(\theta) = \begin{bmatrix} R_{\alpha_1\theta} & & & \\ & R_{\alpha_2\theta} & & \\ & & \ddots & \\ & & & R_{\alpha_n\theta} \end{bmatrix}.$$

Then the image of φ_α is dense in $\mathbf{T}^n$ and isomorphic to $\mathbb{R}^+$.

Proof. See Adams [1969], p. 79, Proposition 4.3, or Palis and DeMelo [1982], p. 35, Exercise 11.13. □

Exercises

7.1. Classify all irreducible representations of $\mathbf{S}^1$. (*Hint*: Use Corollary 7.4, and compare with Exercise 2.1.)

7.2. Classify all irreducible representations of $\mathbf{O}(2)$. There are:

(a) Two one-dimensional irreducibles, the trivial representation and the representation in which $\gamma \in \mathbf{O}(2)$ acts as multiplication by $\det \gamma$.

(b) A countably infinite family of two-dimensional irreducibles defined by

(i) $z \mapsto e^{ki\theta} z$

(ii) $z \mapsto \bar{z}$

where $z \in \mathbb{C} \equiv \mathbb{R}^2$ and $k = 1, 2, 3, \ldots$.

(*Hint*: If $\mathbf{O}(2)$ acts irreducibly on a vector space V, show that the subgroup $\mathbf{SO}(2)$ also acts irreducibly on V.)

7.3. Show that all irreducibles for $\mathbf{Z}_n$ and $\mathbf{D}_n$ are of dimension 1 or 2.

CHAPTER XIII

Symmetry-Breaking in Steady-State Bifurcation

§0. Introduction

In this chapter we begin to study the structure of bifurcations of steady-state solutions to systems of ODEs

$$\frac{dx}{dt} + g(x, \lambda) = 0 \tag{0.1}$$

where $g: \mathbb{R}^n \times \mathbb{R} \to \mathbb{R}^n$ commutes with the action of a compact Lie group Γ on $V = \mathbb{R}^n$. Steady-state solutions satisfy $dx/dt = 0$; that is,

$$g(x, \lambda) = 0. \tag{0.2}$$

We focus here on the symmetries that a solution x may possess and in particular define some simple "geometric" notions that will prove to be of central importance.

In §1 we note that since Γ commutes with g, if x is a solution then so is γx for all $\gamma \in \Gamma$. The set of all γx for $\gamma \in \Gamma$ is the *orbit* of x under Γ. The amount of symmetry present in a solution x is measured by its *isotropy subgroup*

$$\Sigma = \Sigma_x = \{\sigma \in \Gamma: \sigma x = x\}.$$

The smaller Σ is, the larger is the orbit of x.

In §2 we introduce the *fixed-point subspace*

$$\mathrm{Fix}(\Sigma) = \{v \in V | \sigma v = v \text{ for all } \sigma \in \Sigma\}.$$

It is a linear subspace of V and, remarkably, is invariant under g (even when g is *nonlinear*). This leads to a strategy for finding solutions to (0.2) with preassigned isotropy subgroups Σ: restrict g to $\mathrm{Fix}(\Sigma)$ and solve there. This strategy will be used repeatedly in the sequel. It is important to be able to compute $\dim \mathrm{Fix}(\Sigma)$, and we prove a trace formula for this.

The main result of this chapter, proved in §3, is the equivariant branching lemma (Theorem 3.2) due to Vanderbauwhede [1980] and Cicogna [1981]. This states that, with certain conditions on Σ, a unique branch of solutions to (0.2) with isotropy subgroup Σ exists. The main hypothesis is that the fixed-point subspace $\mathrm{Fix}(\Sigma)$ is one-dimensional. Thus the point of view is to prescribe in advance the symmetries required of x and to reduce the problem to a study of $g|\mathrm{Fix}(\Sigma)$.

The restriction $\dim \mathrm{Fix}(\Sigma) = 1$ is not as arbitrary as it may appear, and this condition is often satisfied. The problem (0.2) is connected with "spontaneous symmetry-breaking" as follows. Suppose that (0.1) has for each λ a trivial solution $x = 0$ (which manifestly has isotropy subgroup Γ). Suppose it to be asymptotically stable for $\lambda < 0$ and to lose stability at $\lambda = 0$. Usually such a loss of stability is associated with the occurrence of new branches of solutions $x \neq 0$ to (0.2), emanating from the trivial branch at $\lambda = 0$. Such solutions often have isotropy subgroups Σ smaller than Γ. We may ask, Which Σ typically arise in this way? In the language of symmetry-breaking, one says that the solution spontaneously breaks symmetry from Γ to Σ. "Spontaneously" here means that the *equation* $g = 0$ still commutes with all of Γ. Instead of a unique solution $x = 0$ with all of Γ as its symmetries, we see a set of symmetrically related solutions (orbits under Γ modulo Σ) each with symmetry group (conjugate to) Σ. In many examples it turns out that the subgroups Σ are maximal isotropy subgroups—not contained in any larger isotropy subgroup other than Γ. (Exceptions to this statement do occur; see §10.) If $\dim \mathrm{Fix}(\Sigma) = 1$ then Σ is maximal, and such Σ are the most tractable maximal isotropy subgroups.

Thus the equivariant branching lemma yields a set of solution branches in a relatively simple way. It is important to decide whether the solutions associated with any of these branches can be asymptotically stable. In §4 we show that for some group actions Γ on $\mathbb{R}^n$, *all* such branches are unstable. This means that in some problems it is essential to consider degeneracies; this leads to problems that can be solved using singularity theory. See Chapters XIV and XV.

In §5 we discuss in more detail how to represent Γ-equivariant bifurcations by a (schematic) bifurcation diagram. Such diagrams are very convenient, but we make their schematic nature explicit to avoid misunderstandings.

§§6–9 apply the theory thus developed to two classes of examples: the groups $\mathbf{SO}(3)$ and $\mathbf{O}(3)$ acting in *any* irreducible representation. The proofs may be omitted if so desired. These representations are obtained in §7, which links them to the classical idea of "spherical harmonics."

Finally in §10 we discuss to what extent we may expect spontaneous symmetry-breaking to occur to maximal isotropy subgroups. This section is optional. Although many questions remain unanswered, it is possible to establish a number of facts. In particular there are three distinct types of maximal isotropy subgroup, which we call real, complex, and quaternionic. A theorem due to Dancer [1980a] effectively rules out all but the real maximal isotropy subgroups. On the other hand, Chossat [1983] and Lauterbach [1986] give examples in which submaximal isotropy subgroups arise generically, and we outline their results. We also describe two contexts in which solutions occur for *all* maximal isotropy subgroups. These contexts are variational equations (Michel [1972]) and periodic solutions near equilibria of Hamiltonian systems (Montaldi, Roberts, and Stewart [1986]).

§1. Orbits and Isotropy Subgroups

Let Γ be a Lie group acting on the vector space V. There are two simple notions used in describing aspects of a group action, which are intimately related to the way we think of bifurcation problems with symmetry. We explain these ideas and relations in the following discussion.

The *orbit* of the action of Γ on $x \in V$ is the set

$$\Gamma x = \{\gamma x \colon \gamma \in \Gamma\}. \tag{1.1}$$

Suppose that $f\colon V \to V$ is Γ-equivariant; then when f vanishes, it vanishes on orbits of Γ. For if $f(x) = 0$, then

$$f(\gamma x) = \gamma f(x) = \gamma 0 = 0.$$

In other words, this calculation shows that symmetric equations (Γ-equivariants) cannot distinguish between points (solutions) on the same orbit.

The *isotropy subgroup* of $x \in V$ is

$$\Sigma_x = \{\gamma \in \Gamma \colon \gamma x = x\}. \tag{1.2}$$

See the following for an example. We think of isotropy subgroups as giving the symmetries of the point x (under the action of Γ). In later sections we shall attempt to find solutions to $f = 0$, for some unspecified Γ-equivariant mapping f, by specifying required symmetries for the solution x, that is, by specifying the isotropy subgroup of x.

It is natural to ask how the isotropy subgroups of two points on the same orbit compare. The answer is as follows:

Lemma 1.1. *Points on the same orbit of* Γ *have conjugate isotropy subgroups. More precisely,*

$$\Sigma_{\gamma x} = \gamma \Sigma_x \gamma^{-1}. \tag{1.3}$$

Remarks.

(a) Let $\Sigma \subset \Gamma$ be a subgroup and let $\gamma \in \Gamma$. Then

$$\gamma \Sigma \gamma^{-1} = \{\gamma \sigma \gamma^{-1} \colon \sigma \in \Sigma\}$$

is a subgroup of Γ, said to be *conjugate* to Σ.

(b) The *conjugacy class* of Σ consists of all subgroups of Γ that are conjugate to Σ.

Proof. Let $x \in V$ and $\gamma \in \Gamma$. Suppose that $\sigma \in \Sigma_x$. We claim that $\gamma\sigma\gamma^{-1} \in \Sigma_{\gamma x}$. We may check this directly:

$$\gamma\sigma\gamma^{-1}(\gamma x) = \gamma\sigma(\gamma^{-1}\gamma)x = \gamma\sigma x = \gamma x,$$

the last equality holding since $\sigma \in \Sigma_x$. It follows that

$$\Sigma_{\gamma x} \supset \gamma\Sigma_x\gamma^{-1}.$$

Replacing x by γx and γ by γ^{-1} yields $\Sigma_x \supset \gamma^{-1}\Sigma_{\gamma x}\gamma$, which proves the lemma. □

A convenient method for describing geometrically the group action of Γ on V is to lump together in a set W all points of V that have conjugate isotropy subgroups. We say that W is an *orbit type* of the action.

We illustrate these ideas by considering the action of the dihedral group $\mathbf{D}_n$ on $\mathbb{C}$ generated by

$$\kappa: z \mapsto \bar{z} \quad \text{and} \quad \zeta: z \mapsto e^{2\pi i/n}z.$$

Geometrically we picture the action of $\mathbf{D}_n$ as the symmetries of a regular n-gon centered at the origin in the plane. This n-gon is shown in Figure 1.1 by dashed lines, when $n = 5$. We derive in the following the orbit types of this group action. The result depends on whether n is odd or even, and for simplicity we consider only the case when n is odd. The complete results may be found in §5. The vertices on the n-gon, shown as □ in Figure 1.1, are mapped into each other by Γ. More precisely, these vertices constitute a single orbit of the action of Γ. The isotropy subgroup of a vertex on the real axis (not at the origin) is the group $\mathbf{Z}_2$ generated by κ. The other vertices have isotropy subgroups conjugate to $\mathbf{Z}_2$, by Lemma 1.1. Finally, if $t \neq 0$ then linearity of the action implies that $\Sigma_{tz} = \Sigma_z$. So all points on the lines joining the origin to a vertex have conjugate isotropy subgroups and belong to the same orbit type.

Next we consider a point near, but not on, the real axis, indicated by a • in Figure 1.1. By reflection and rotation we see that its orbit contains $2n$ points,

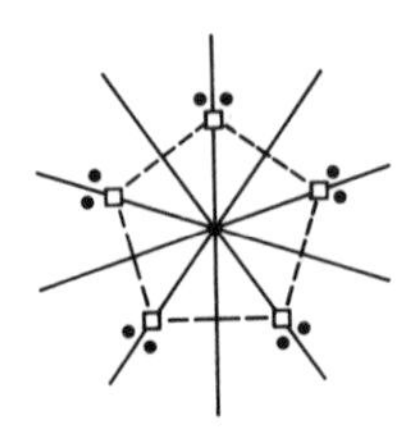

Figure 1.1. Orbits of the action of $\mathbf{D}_5$ on $\mathbb{C}$.

Table 1.1. Orbit Types and Isotropy Subgroups for $\mathbf{D}_n$ on $\mathbb{C}$, n odd

Orbit Type	Isotropy Subgroup	Size of Orbit
$\{0\}$	$\mathbf{D}_n$	1
$\{z \in \mathbb{C} \mid \mathrm{Im}(z^n) = 0, z \neq 0\}$	$\mathbf{Z}_2$	n
$\{z \in \mathbb{C} \mid \mathrm{Im}(z^n) \neq 0\}$	$\mathbb{1}$	$2n$

and the only group element that fixes one of these points is the identity in $\mathbf{D}_n$. Hence all points in the wedges between the vertex-origin lines belong to the same orbit type.

Finally, of course, the origin forms an orbit on its own and is fixed by the whole group $\mathbf{D}_n$. Thus there are three orbit types. We list these, along with their (conjugacy class of) isotropy subgroups, in Table 1.1. (In the case n even, points on the lines joining the origin to midpoints of edges of the n-gon have nontrivial isotropy subgroups not conjugate to those listed in Table 1.1; see §5.)

In this example "almost all" points—an open dense set—have trivial isotropy subgroup. It is a general theorem (Bredon [1972], p. 179) that there exists a unique minimal isotropy subgroup $\Sigma_{\min}$ for any linear action of a Lie group Γ on a vector space V and that points with this isotropy form an open dense subset of V. Since $\mathrm{Fix}(\Sigma_{\min})$ contains an open dense subset of V and is a vector space, it must be the whole of V; therefore, $\Sigma_{\min}$ is the *kernel* of the action—the subgroup of all elements of Γ that act on V as the identity. The points with isotropy group $\Sigma_{\min}$ are said to have *principal orbit type.*

We see that in this example, the larger the orbit, the smaller the isotropy subgroup. We formalize this observation as follows:

Proposition 1.2. *Let Γ be a compact Lie group acting on V. Then*
(a) *If $|\Gamma| < \infty$, then $|\Gamma| = |\Sigma_x||\Gamma x|$.*
(b) $\dim \Gamma = \dim \Sigma_x + \dim \Gamma x$.

Remarks.
(a) Proposition 1.2(a) states that the order of the group Γ is the product of the order of Σ_x and the size of the orbit of x. This formula may be checked for $\Gamma = \mathbf{D}_n$ from Table 1.1, using the fact that $|\mathbf{D}_n| = 2n$.
(b) Lie groups are always smooth manifolds and have well-defined dimensions. Since isotropy subgroups are always Lie subgroups both $\dim \Gamma$ and $\dim \Sigma_x$ make sense. Similarly orbits of Lie groups are always submanifolds and have well-defined dimensions. Thus $\dim \Gamma x$ makes sense.

Sketch of Proof. There is a natural map $\varphi\colon \Gamma \to \Gamma x$ defined by

$$\varphi(\gamma) = \gamma x. \tag{1.4}$$

By definition φ maps onto the orbit Γx, and $\varphi^{-1}(x) = \Sigma_x$. Define the *coset space* of a subgroup Σ of Γ to be

$$\Gamma/\Sigma = \{\gamma\Sigma | \gamma \in \Gamma\}$$

where we recall that the cosets of Σ in Γ are the sets

$$\gamma\Sigma = \{\gamma\sigma | \sigma \in \Sigma\}.$$

Then φ induces a map

$$\begin{aligned} \psi &: \Gamma/\Sigma_x \to \Gamma x \\ \psi(\gamma) &= \gamma x \end{aligned} \tag{1.5}$$

which is both one-to-one and onto. In the case that Γ is finite, a simple counting of the cosets in Γ/Σ_x verifies part (a). In general, both φ and ψ are smooth mappings and $(d\psi)_0$ is invertible. It follows from the inverse function theorem that

$$\dim \Gamma x = \dim(\Gamma/\Sigma_x)$$

from which part (b) is immediate. □

Remark (d) of XII, §4, promised a simple criterion for $\mathscr{P}(\Gamma)$ to be a polynomial ring. We have now defined the concepts needed to state this; we omit the proof. Suppose that Γ acts on V with minimal isotropy subgroup $\Sigma_{\min}$. Let $\{u_1(x), \ldots, u_s(x)\}$ be a Hilbert basis for $\mathscr{P}(\Gamma)$. If

$$s = \dim V - \dim \Gamma + \dim \Sigma_{\min} \tag{1.6}$$

then $\mathscr{P}(\Gamma)$ is a polynomial ring.

In particular, if Γ is finite then (1.6) reduces to

$$s = \dim V. \tag{1.7}$$

For example, (1.7) trivially implies that $\mathscr{P}(\mathbf{D}_n)$ is a polynomial ring whenever $\mathbf{D}_n$ acts irreducibly on $\mathbb{C}$.

Exercises

1.1. Let $\mathbf{O}(n)$ act on $\mathbb{R}^n$ in its standard representation. Find the orbits and the corresponding isotropy subgroups.

1.2. Let Γ be the group of all symmetries, including reflections, of a cube center the origin of $\mathbb{R}^3$ with edges parallel to the axes. (In the notation of XIII, §9, Γ is the group $\mathbb{O} \oplus \mathbf{Z}_2^c$.) Show that
(a) $|\Gamma| = 48$ and Γ is generated by

$$\kappa_x = \begin{bmatrix} -1 & 0 & 0 \\ 0 & 1 & 0 \\ 0 & 0 & 1 \end{bmatrix} \quad R_x = \begin{bmatrix} 1 & 0 & 0 \\ 0 & 0 & -1 \\ 0 & 1 & 0 \end{bmatrix} \quad R_y = \begin{bmatrix} 0 & 0 & 1 \\ 0 & 1 & 0 \\ -1 & 0 & 0 \end{bmatrix}.$$

(b) Show that the orbit data for Γ are as follows:

Orbit Representative	Isotropy Subgroup
$(0,0,0)$	Γ
$(x,0,0)$	$\mathbf{D}_4$
$(x,x,0)$	$\mathbf{Z}_2^r \oplus \mathbf{Z}_2^t$
(x,x,x)	$\mathbf{S}_3$
$(x,y,0)$	$\mathbf{Z}_2^r$
(x,x,z)	$\mathbf{Z}_2^t$
(x,y,z) $\lvert x\rvert, \lvert y\rvert, \lvert z\rvert$ distinct and $\neq 0$	$\mathbb{1}$

where $\mathbf{D}_4$ is generated by R_x and κ_z (in obvious notation, compare (a)), $\mathbf{Z}_2^r$ by κ_z, and $\mathbf{Z}_2^t$ by $(x, y, z) \mapsto (x, z, y)$.

(c) Verify Proposition 1.2(a) directly for this example.

1.3. Find the orbits and isotropy subgroups for $\mathbf{O}(3)$ in its five-dimensional representation (as in Exercise XII, 3.6). Verify Proposition 1.2(b) for this example. (*Hint*: Every symmetric matrix can be diagonalized.)

1.4. (a) Show that in $\mathbf{D}_{2n+1}$ all reflections are conjugate.
(b) In $\mathbf{D}_{2n}$ show that there are two geometrically distinct types of reflection: those through lines joining the origin to a vertex and those through lines joining the origin to the midpoint of an edge. Prove that all reflections of the same type are conjugate in $\mathbf{D}_{2n}$, but that different types of reflection are not conjugate.
(c) Prove that all reflections in $\mathbf{D}_{2n}$ are conjugate in $\mathbf{D}_{4n}$.

1.5. Let $\mathbf{Z}_2$ act on $\mathbb{R}^2$ so that $-1 \in \mathbf{Z}_2$ acts as $(x, y) \mapsto (-x, y)$. Prove that $\mathscr{P}(\mathbf{Z}_2)$ is not a polynomial ring.

§2. Fixed-Point Subspaces and the Trace Formula

This section divides into three subsections, devoted to the following topics:

(a) The existence of invariant subspaces for *nonlinear* equivariant mappings: the fixed-point subspaces,
(b) A method for computing the dimensions of fixed-point subspaces: the trace formula,
(c) Ways to use the dimensions of fixed-point subspaces to find an important class of isotropy subgroups: the maximal isotropy subgroups.

(a) Fixed-Point Subspaces

One of the most remarkable as well as one of the simplest features of nonlinear Γ-equivariant mappings is that their equivariance forces them to have invariant linear subspaces. Moreover, these invariant subspaces correspond naturally to certain subgroups of Γ.

Let $\Sigma \subset \Gamma$ be a subgroup. The *fixed-point subspace* of Σ is

$$\mathrm{Fix}(\Sigma) = \{x \in V: \sigma x = x \text{ for all } \sigma \in \Sigma\}. \tag{2.1}$$

If it is important to display the space V explicitly we write $\mathrm{Fix}_V(\Sigma)$. Observe that $\mathrm{Fix}(\Sigma)$ is always a linear subspace of V since

$$\mathrm{Fix}(\Sigma) = \bigcap_{\sigma \in \Sigma} \ker(\sigma - Id)$$

and each kernel is a linear subspace.

Note that the simplest fixed-point subspaces are $\mathrm{Fix}(\mathbb{1})$ and $\mathrm{Fix}(\Gamma)$. Since the identity subgroup $\mathbb{1}$ fixed every point, we have $\mathrm{Fix}(\mathbb{1}) = V$. At the other extreme, $\mathrm{Fix}(\Gamma)$ consists of all vectors in V that are fixed by every element in Γ. Thus $\mathrm{Fix}(\Gamma)$ is the subspace of V on which Γ acts trivially. We shall often adopt the hypothesis that $\mathrm{Fix}(\Gamma) = \{0\}$.

We now show that the fixed-point subspaces have the invariance property asserted earlier.

Lemma 2.1. *Let $f: V \to V$ be Γ-equivariant. Let $\Sigma \subset \Gamma$ be a subgroup. Then*

$$f(\mathrm{Fix}(\Sigma)) \subset \mathrm{Fix}(\Sigma). \tag{2.2}$$

Proof. Let $\sigma \in \Sigma$, $x \in \mathrm{Fix}(\Sigma)$. Then

$$f(x) = f(\sigma x) = \sigma f(x) \tag{2.3}$$

where the first equality follows from the definition of $\mathrm{Fix}(\Sigma)$, and the second from equivariance. From (2.3) we see that σ fixed $f(x)$. Therefore, $f(x) \in \mathrm{Fix}(\Sigma)$. □

Remark. In Lemma 2.1 we do not require Σ to be an isotropy subgroup. However, for any subgroup Σ, $\mathrm{Fix}(\Sigma)$ is equal to the sum W of all subspaces $\mathrm{Fix}(\Delta)$ where $\Delta \supset \Sigma$ is an isotropy subgroup. To prove this, first let $v \in \mathrm{Fix}(\Sigma)$. Then $\Sigma_v \supset \Sigma$ and $v \in \mathrm{Fix}(\Sigma_v)$. Hence we may take $\Delta = \Sigma_v$ to show that $v \in W$, so $\mathrm{Fix}(\Sigma) \subset W$. On the other hand, if $w \in W$ the $w = w_1 + \cdots + w_k$ where $w_j \in \mathrm{Fix}(\Delta_j)$, for an isotropy subgroup $\Delta_j \supset \Sigma$. But this means that $\sigma w_j = w_j$ for all $\sigma \in \Sigma$, so $w_j \in \mathrm{Fix}(\Sigma)$; therefore, $w \in \mathrm{Fix}(\Sigma)$ and so $W \subset \mathrm{Fix}(\Sigma)$. Hence $W = \mathrm{Fix}(\Sigma)$.

Thus in theory there is no real loss of generality if we let Σ run through just the isotropy subgroups of Γ. However, it may sometimes be convenient *not* to require Σ to be an isotropy subgroup, since this condition may not be easy to check.

For an example where we can check Lemma 2.1 directly, consider once more $\Gamma = \mathbf{D}_n$ in its standard action on $\mathbb{C}$. We find $\mathrm{Fix}(\Sigma)$ for the isotropy subgroups Σ. Obviously if $\Sigma = \mathbb{1}$ then $\mathrm{Fix}(\Sigma) = V$; and if $\Sigma = \mathbf{D}_n$ then $\mathrm{Fix}(\Sigma) = \{0\}$. If $\Sigma = \mathbf{Z}_2$ then $\mathrm{Fix}(\Sigma)$ is the real axis; and if Σ is a conjugate of $\mathbf{Z}_2$ then $\mathrm{Fix}(\Sigma)$ is the image of the real axis under an element of $\mathbf{D}_n$, that is, one of the lines through the origin and a vertex.

Taking $\Sigma = \mathbf{Z}_2$ in Lemma 2.1 it follows that every $\mathbf{D}_n$-equivariant mapping f must leave the real axis invariant. By (XII, 5.12) the general f has the form

$$f(z) = p(u, v)z + q(u, v)\bar{z}^{n-1}$$

where $u = z\bar{z}$, $v = z^n + \bar{z}^n$. If $z = x$ is real, then

$$f(x) = p(x^2, 2x^n)x + q(x^2, 2x^n)x^{n-1}$$

is also real. So $\mathrm{Fix}(\mathbf{Z}_2)$ is invariant under f as predicted.

An immediate consequence of Lemma 2.1 is the existence of trivial solutions for Γ-equivariant mappings f. More precisely, if $\mathrm{Fix}(\Gamma) = \{0\}$ then $\{0\}$ must be invariant under f, so that $f(0) = 0$. In fact, we have three equivalent properties:

Proposition 2.2. *Let Γ be a compact Lie group acting on V. The following are equivalent*:
(a) $\mathrm{Fix}(\Gamma) = \{0\}$.
(b) *Every Γ-equivariant map $f: V \to V$ satisfies $f(0) = 0$ (there always exist trivial solutions).*
(c) *The only Γ-invariant linear function is the zero function.*

Remark. The most important implication (a) $\Rightarrow$ (b) we showed previously, using Lemma 2.1.

Proof. The converse (b) $\Rightarrow$ (a) is proved easily as follows. We claim that for every $v \in \mathrm{Fix}(\Gamma)$, the constant mapping $f(x) = v$ is Γ-equivariant. If so, (b) will imply that $v = f(0) = 0$, proving (a). To verify the claim, compute

$$\gamma f(x) = \gamma v = v = f(\gamma x).$$

The first equality is by definition of $f(x)$, the second follows since $v \in \mathrm{Fix}(\Gamma)$, and the third holds since f is constant.

Next we show that (a) implies (c). Let $L: V \to \mathbb{R}$ be linear and invariant. We may write L in the form

$$L(x) = \langle v, x \rangle$$

for some $v \in V$. We claim that $v \in \mathrm{Fix}(\Gamma)$, whence (a) implies (c). Since L is Γ-invariant, $L(x) = L(\gamma^{-1}x)$ for all $\gamma \in \Gamma$. Since Γ acts orthogonally $\gamma^{-1} = \gamma^t$. Thus

$$\langle v, x \rangle = \langle v, \gamma^{-1}x \rangle = \langle v, \gamma^t x \rangle = \langle \gamma v, x \rangle$$

for all x. Hence $\gamma v = v$ for all γ and $v \in \mathrm{Fix}(\Gamma)$, as claimed.

Finally we prove that (c) $\Rightarrow$ (b). Let $f: V \to V$ be Γ-equivariant. We must show that $f(0) = 0$. To do this, define

$$L(x) = \langle f(0), x \rangle$$

where $\langle \ , \ \rangle$ is a Γ-invariant inner product on V. We claim that the linear function L is Γ-invariant. If so, then $L \equiv 0$ and $f(0) = 0$. To verify the claim, compute

$$L(\gamma x) = \langle f(0), \gamma x \rangle = \langle \gamma^{-1} f(0), x \rangle = \langle f(0), x \rangle = L(x). \qquad \square$$

(b) The Trace Formula

In later sections we shall want to compute the dimension of $\mathrm{Fix}(\Sigma)$. There is an elegant formula for this, which depends only on the trace $\mathrm{tr}(\sigma)$ for $\sigma \in \Sigma$. Because Γ acts linearly on V we may think of $\gamma \in \Gamma$ as acting by the linear mapping $\rho_\gamma: x \mapsto \gamma x$. By $\mathrm{tr}(\sigma)$ we mean the trace of ρ_σ on V.

Theorem 2.3 (Trace Formula). *Let Γ be a compact Lie group acting on V and let $\Sigma \subset \Gamma$ be a Lie subgroup. Then*

$$\dim \mathrm{Fix}(\Sigma) = \int_\Sigma \mathrm{tr}(\sigma) \tag{2.4}$$

where $\int$ denotes the normalized Haar integral on Σ.

Remark. If Σ is finite then (2.4) can be rephrased as

$$\dim \mathrm{Fix}(\Sigma) = \frac{1}{|\Sigma|} \sum_{\sigma \in \Sigma} \mathrm{tr}(\sigma). \tag{2.5}$$

See Example XIII, 1.4.

Proof. Define the linear transformation $A: V \to V$ by

$$A = \int_\Sigma \sigma. \tag{2.6}$$

Because the Haar integral is Σ-invariant, we see that

$$A = \int_\Sigma \sigma' \sigma$$

where σ' is any fixed element of Σ. It follows that

$$A^2 = A; \tag{2.7}$$

that is, A is a linear projection. To check (2.7), compute

$$\begin{aligned}A^2 = A \circ A = A\left(\int_{\sigma \in \Sigma} \sigma\right) &= \int_{\sigma' \in \Sigma} \sigma' \left(\int_{\sigma \in \Sigma} \sigma\right) \\ &= \int_{\sigma' \in \Sigma} \left(\int_{\sigma \in \Sigma} \sigma'\sigma\right) \\ &= \int_{\sigma' \in \Sigma} A \\ &= A.\end{aligned}$$

By (2.7)

$$\begin{aligned}&\text{(a)} \quad V = \ker A \oplus \operatorname{Im} A \\ &\text{(b)} \quad A|\operatorname{Im} A = Id.\end{aligned} \tag{2.8}$$

We verify (2.8b) first. Suppose $x \in \operatorname{Im} A$, so $x = Ay$. Using (2.7) we have

$$Ax = A^2y = Ay = x,$$

proving (2.8b). To verify (2.8a) observe that $\dim \ker A + \dim \operatorname{Im} A = \dim V$, since A is linear. Thus it suffices to show that $\ker A \cap \operatorname{Im} A = \{0\}$. However, if $x \in \ker A \cap \operatorname{Im} A$ then $x = Ax$ by (2.8b), and $Ax = 0$.

It follows directly from (2.8) that

$$\operatorname{tr}(A) = \dim \operatorname{Im} A. \tag{2.9}$$

We claim that $\operatorname{Im} A = \operatorname{Fix}(\Sigma)$. The theorem will then follow since $\dim \operatorname{Im} A = \dim \operatorname{Fix}(\Sigma)$ and

$$\operatorname{tr}(A) = \int_{\sigma \in \Sigma} \operatorname{tr}(\sigma).$$

To prove the claim, observe that $\operatorname{Fix}(\Sigma) \supset \operatorname{Im} A$ by (2.8(b)). Conversely, $\operatorname{Fix}(\Sigma) \subset \operatorname{Im} A$ by (2.8(a)). More precisely, suppose $x \in \operatorname{Fix}(\Sigma)$. Write $x = k + y$ where $k \in \ker A$ and $y \in \operatorname{Im} A$. Then $x = Ax = Ak + Ay = y$. This can happen only if $k = 0$ and $x \in \operatorname{Im} A$. □

In certain cases it is possible to use the trace formula to reduce the calculation of $\dim \operatorname{Fix}(\Sigma)$ to finding the dimensions of fixed-point spaces $\operatorname{Fix}(\Delta)$ for certain subgroups Δ of Σ. This reduction, stated in Lemma 2.5 later, will be of particular use when we discuss the fixed-point subspaces for subgroups of **SO**(3) and **O**(3) in §§6–9.

Definition 2.4. Let $H_1, \ldots, H_k$ be subgroups of a group Σ. We say that Σ is the *disjoint union* of $H_1, \ldots, H_k$ if

$$\text{(a)} \quad \Sigma = H_1 \cup \cdots \cup H_k$$

$$\text{(b)} \quad H_i \cap H_j = \mathbb{1} \quad \text{for all } i \neq j.$$

We use the notation $\Sigma = H_1 \dot{\cup} \cdots \dot{\cup} H_k$ to denote disjoint unions.

When Σ has a disjoint union decomposition then we can compute $\dim \operatorname{Fix}(\Sigma)$ in terms of the numbers $\dim \operatorname{Fix}(H_j)$:

Lemma 2.5. *Let* $\Sigma = H_1 \dot{\cup} \cdots \dot{\cup} H_k$ *be a finite subgroup of* Γ, *with* Γ *acting on* V. *Then*

$$\dim \operatorname{Fix}(\Sigma) = \frac{1}{|\Sigma|}\left[\sum_{i=1}^{k} |H_i| \dim \operatorname{Fix}(H_i) - (k-1)\dim V\right]. \tag{2.10}$$

Proof. From (2.5) we see that

$$\begin{aligned} \dim \operatorname{Fix}(\Sigma) &= \frac{1}{|\Sigma|} \sum_{\sigma \in \Sigma} \operatorname{tr}(\sigma) \\ &= \frac{1}{|\Sigma|}\left[\sum_{i=1}^{k} \sum_{h \in H_i} \operatorname{tr}(h) - (k-1)\operatorname{tr}(I)\right] \end{aligned} \tag{2.11}$$

where the second equality is obtained by splitting the sum over Σ into a sum over the H_i. Since Σ is a disjoint union of the H_i we must add $\operatorname{tr}(I)$ (k times) for the overlap on the identity element. Since we want to count $\operatorname{tr}(I)$ only once we subtract the overenumeration, obtaining (2.11).

To derive (2.10) from (2.11) we make two observations. First, $\operatorname{tr}(I) = \dim V$. Second, we apply the trace formula (2.5) directly to each H_i, obtaining

$$\dim \operatorname{Fix}(H_i) = \frac{1}{|H_i|} \sum_{h \in H_i} \operatorname{tr}(h).$$

Substitute this in (2.11) to yield the desired result. □

(c) Maximal Isotropy Subgroups

It is important to be able to determine, in as simple a manner as possible, whether a given closed subgroup is an isotropy subgroup. That is, we wish to do this without knowing the orbit structure of Γ. We now consider a distinguished class of isotropy subgroups for which this question may be answered using the dimensions of fixed-point subspaces.

Definition 2.6. Let Γ be a Lie group acting on V. An isotropy subgroup $\Sigma \subseteq \Gamma$ is *maximal* if there does not exist an isotropy subgroup Δ of Γ satisfying $\Sigma \subsetneq \Delta \subsetneq \Gamma$.

Lemma 2.7. *Let* $\operatorname{Fix}(\Gamma) = \{0\}$, *and let* Σ *be a subgroup of* Γ. *Then* Σ *is a maximal isotropy subgroup of* Γ *if and only if*:

$$\begin{aligned} &\text{(a)} \quad \dim \operatorname{Fix}(\Sigma) > 0 \\ &\text{(b)} \quad \dim \operatorname{Fix}(\Delta) = 0 \textit{ for every closed subgroup } \Delta \supsetneq \Sigma. \end{aligned} \tag{2.12}$$

Proof. Suppose Σ is a maximal isotropy subgroup of Γ. Then $\dim \operatorname{Fix}(\Sigma) > 0$ since Σ must fix some nonzero vector, by the definition of an isotropy subgroup. Suppose $\Delta \supsetneq \Sigma$ and suppose there is a vector $x \in V$ fixed by Δ. Then the isotropy subgroup Σ_x of x satisfies $\Sigma_x \supset \Delta \supseteq \Sigma$. Since Σ is a maximal isotropy subgroup we must have $\Sigma_x = \Gamma$. But $\operatorname{Fix}(\Gamma) = \{0\}$, so $x = 0$. Therefore, $\dim \operatorname{Fix}(\Delta) = 0$.

Conversely, suppose that Σ satisfies (2.12). Then some nonzero vector $x \in V$ is fixed by Σ, so Σ_x contains Σ. Since Σ_x is an isotropy subgroup, it is closed. If $\Sigma_x \neq \Sigma$ then (2.12(b)) implies that $\dim \operatorname{Fix}(\Sigma_x) = 0$, contrary to Σ_x being an isotropy subgroup. Therefore $\Sigma = \Sigma_x$, so Σ is an isotropy subgroup. The same argument now proves that Σ is maximal. □

Lemma 2.7 provides a strategy for finding the maximal isotropy subgroups of Γ if we know enough about the dimensions of fixed-point spaces of subgroups of Γ. Namely, we find the largest closed subgroups with nonzero fixed-point subspaces. We use this strategy in §§6–9 to compute the maximal isotropy subgroups of $\mathbf{SO}(3)$ and $\mathbf{O}(3)$.

Exercises

2.1. Find the fixed-point subspaces for the isotropy subgroups of Exercises 1.1 and 1.3.

2.2. Let Σ be an isotropy subgroup of Γ. Show that the largest subgroup of Γ that leaves $\operatorname{Fix}(\Sigma)$ setwise invariant is $N = N_\Gamma(\Sigma)$. If $\dim \operatorname{Fix}(\Sigma) = 1$ show that N/Σ is either $\mathbb{1}$ or $\mathbf{Z}_2$. If it is $\mathbf{Z}_2$ show that the corresponding bifurcation is of pitchfork type.

2.3. Show that for the group $\mathbb{O} \oplus \mathbf{Z}_2^c$ of Exercise 1.2, the fixed-point subspaces are as follows:

Isotropy Subgroup	Fixed-Point Subspace	Dimension
Γ	$\{(0,0,0)\}$	0
$\mathbf{D}_4$	$\{(x,0,0)\}$	1
$\mathbf{Z}_2^r \oplus \mathbf{Z}_2^t$	$\{(x,x,0)\}$	1
$\mathbf{S}_3$	$\{(x,x,x)\}$	1
$\mathbf{Z}_2^r$	$\{(x,y,0)\}$	2
$\mathbf{Z}_2^t$	$\{(x,x,z)\}$	2
$\mathbb{1}$	$\mathbb{R}^3$	3

2.4. Let $\Gamma = \mathbf{Z}_2 \oplus \mathbf{Z}_2$ act on $\mathbb{R}^2$ by $(x, y) \mapsto (\pm x, \pm y)$ as in X, §1(a). Show that the action of Γ is not irreducible, but that $\text{Fix}(\Gamma) = \{0\}$; that is, Γ-equivariant bifurcation problems have a trivial solution.

2.5. Show that the group $\mathbb{O}$ of rotational symmetries of a cube has a disjoint union decomposition into cyclic subgroups.

2.6. Verify Theorem 2.3 directly for the three maximal isotropy subgroups of $\mathbb{O} \oplus \mathbf{Z}_2^c$ listed in Exercise 2.2.

2.7. Let Γ act on V and let Σ be an isotropy subgroup. It is clear that $\dim \text{Fix}(\Sigma)$ is the dimension of the trivial part of the isotypic decomposition of V for Σ, that is, the multiplicity with which the trivial representation of Σ occurs on V. If instead we ask the multiplicity of some other representation, then there is an analogous formula to Theorem 2.3 which may be deduced from the orthogonality relations for characters (see XIII, §7(f)). This exercise asks for a bare-hands proof of a special case.

Let $\Sigma = \mathbf{O}(2)$, and let ρ be the representation on $\mathbb{R}$ in which $\mathbf{SO}(2)$ acts trivially and κ acts as -1. Show that the dimension of the isotypic component corresponding to ρ is

$$\int_{\sigma \in \mathbf{SO}(2)} \operatorname{tr} \sigma - \int_{\sigma \in \mathbf{O}(2) \sim \mathbf{SO}(2)} \operatorname{tr} \sigma.$$

(*Hint*: Let

$$A = \int_{\sigma \in \mathbf{SO}(2)} \sigma - \int_{\sigma \in \mathbf{O}(2) \sim \mathbf{SO}(2)} \sigma$$

and mimic the proof of Theorem 2.3.)

§3. The Equivariant Branching Lemma

In this section we prove a simple but useful theorem of Vanderbauwhede [1980] and Cicogna [1981] to the effect that isotropy subgroups with one-dimensional fixed-point subspaces lead to solutions of bifurcation problems with symmetry.

Definition 3.1. Let Γ be a Lie group acting on a vector space V. A *bifurcation problem with symmetry group* Γ is a germ $g \in \vec{\mathscr{E}}_{x,\lambda}(\Gamma)$ satisfying $g(0,0) = 0$ and $(dg)_{0,0} = 0$.

Here we recall notation used earlier in this volume as well as in Volume I. A germ $g \in \vec{\mathscr{E}}_{x,\lambda}(\Gamma)$ is the germ of a Γ-equivariant mapping, which by abuse of notation we also denote by g. Here $g\colon V \times \mathbb{R} \to V$ satisfies

$$g(\gamma x, \lambda) = \gamma g(x, \lambda) \tag{3.1}$$

for all $\gamma \in \Gamma$. By convention our germs are based at the origin $(x, \lambda) = (0, 0)$.

In Definition 3.1 we require that $g(0,0) = 0$ to avoid trivial complications.

If $\text{Fix}(\Gamma) = \{0\}$ then Proposition 2.2 implies that $g(0, \lambda) \equiv 0$, and hence $g(0, 0) = 0$. However, in general $g(0, 0)$ need not vanish.

We also require that $(dg)_{0,0} = 0$. Recall that dg is the $n \times n$ Jacobian matrix obtained by differentiating g *in the* V-directions. Here $n = \dim V$. If $(dg)_{0,0}$ is nonzero, then we can use the Liapunov–Schmidt reduction with symmetries (see VIII, §3) to reduce g to the case where the Jacobian vanishes. Of course, this process will change n to a smaller value n' and will also change the representation of Γ. Nevertheless, we assume that this reduction *has already been performed* and we therefore assume $(dg)_{0,0} = 0$.

We claim that generically we may assume the action of Γ on $V = \mathbb{R}^n$ to be absolutely irreducible. Before stating the result more precisely, we must discuss the term *generic*. A rigorous definition is somewhat technical, and we try instead to convey the underlying idea.

Recall from Chapter II that a bifurcation problem $g(x, \lambda)$ is equivalent to a limit point singularity $\pm x^2 \pm \lambda$ precisely when the defining conditions

$$g(0, 0) = 0, \qquad g_x(0, 0) = 0 \tag{3.2}$$

and the nondegeneracy conditions

$$g_{xx}(0, 0) \neq 0, \qquad g_\lambda(0, 0) \neq 0 \tag{3.3}$$

are satisfied. We say that among those bifurcation problems g in one state variable having a singularity at the origin (i.e., those g satisfying (3.2)) it is generic for the singularity to be a limit point. More succinctly, we say that the "generic singularity" is a limit point.

We abstract this process as follows. Let g be a germ satisfying some property $\mathscr{P}$, where the defining conditions for $\mathscr{P}$ consist of a finite number of equalities involving a finite number of derivatives of g evaluated at the origin. The equalities in (3.2) provide an example, with $\mathscr{P}$ being the property "g has a singularity at the origin." A set S of germs is *generic* for property $\mathscr{P}$ if there exists a finite number of inequalities Q involving a finite number of derivatives of g at the origin, such that $g \in S$ if and only if g has property $\mathscr{P}$ and g satisfies the inequalities in Q. Thus, in the example, Q is given by (3.3) and limit points—those germs satisfying (3.2, 3.3)—are generic singularities.

Actually, even this definition must be qualified. The inequalities Q must not *contradict* any of the defining equalities of $\mathscr{P}$. For example, if $\mathscr{P}$ is defined by $g_x(0, 0) = 0$ then Q should not include the inequality $g_x(0, 0) \neq 0$. We do not intend that the empty set S be considered generic.

We find it convenient to use the word *generic* when we do not wish to specify the inequalities Q explicitly. The important point is that a "typical" germ with property $\mathscr{P}$ will be generic, where by *typical* we mean "not satisfying any additional constraints" (e.g., on derivatives). This follows since an atypical germ must violate an inequality in Q, that is, satisfy a further *equality*.

For example, in applications one expects to see only limit point singularities in steady-state bifurcation problems $g(x, \lambda)$, *unless* some other constraint such as symmetry is placed on g. (The effect of symmetry is to constrain certain terms of the Taylor series of g, so symmetry effectively imposes conditions on derivatives of g at the origin.) In Volume I we focused on nongeneric or degenerate singularities, since these are expected to occur "generically" in *multiparameter* systems. A major theme of this volume is to identify a "generic" class of one-parameter bifurcation problems with symmetry.

The following proposition, whose proof will be sketched at the end of this section, is a first step in that direction.

Proposition 3.2. *Let $G: \mathbb{R}^N \times \mathbb{R} \to \mathbb{R}^N$ be a one-parameter family of Γ-equivariant mappings with $G(0,0) = 0$. Let $V = \ker(dG)_{0,0}$. Then generically the action of Γ on V is absolutely irreducible.*

Remark. When one is interpreting this proposition in the preceding framework, $\mathscr{P}$ is defined as follows. A germ G has property $\mathscr{P}$ if it is a germ of a one-parameter family of Γ-equivariant mappings, and $G(0,0) = 0$. The inequalities Q which imply that the action of Γ on $\ker(dG)_{0,0}$ is absolutely irreducible are left unstated.

Proposition 3.2 supports our assumption later that Γ acts absolutely irreducibly on $\mathbb{R}^n$ and that $g: \mathbb{R}^n \times \mathbb{R} \to \mathbb{R}^n$ is a Γ-equivariant bifurcation problem. We use the assumption of absolute irreducibility as follows. Apply the chain rule to the identity $g(\gamma x, \lambda) = \gamma g(x, \lambda)$ to obtain

$$(dg)_{0,\lambda}\gamma = \gamma(dg)_{0,\lambda}. \tag{3.4}$$

Absolute irreducibility states that the only matrices commuting with all $\gamma \in \Gamma$ are scalar multiples of the identity. Therefore $(dg)_{0,\lambda} = c(\lambda)I$. Since $(dg)_{0,0} = 0$ by Definition 3.1, we have $c(0) = 0$. We now assume the hypothesis

$$c'(0) \neq 0, \tag{3.5}$$

which is valid generically.

We next state the result of Vanderbauwhede and Cicogna, which—despite the simplicity of its proof—forms the basis of many bifurcation results for symmetric problems.

Theorem 3.3 (Equivariant Branching Lemma). *Let Γ be a Lie group acting absolutely irreducibly on V and let $g \in \vec{\mathscr{E}}_{x,\lambda}(\Gamma)$ be a Γ-equivariant bifurcation problem satisfying* (3.5). *Let Σ be an isotropy subgroup satisfying*

$$\dim \operatorname{Fix}(\Sigma) = 1. \tag{3.6}$$

Then there exists a unique smooth solution branch to $g = 0$ such that the isotropy subgroup of each solution is Σ.

Remarks 3.4.

(a) We may restate the equivariant branching lemma as follows: Generically, bifurcation problems with symmetry group Γ have solutions corresponding

to all isotropy subgroups with one-dimensional fixed-point subspaces. Since Σ is an isotropy subgroup satisfying (3.6) it follows that Σ is a maximal isotropy subgroup. Thus the equivariant branching lemma gives us a method for finding solutions corresponding to a special class of maximal isotropy subgroups. To see that Σ is maximal, suppose $\Delta \supsetneqq \Sigma$ is an isotropy subgroup. Then $\mathrm{Fix}(\Delta) \subsetneqq \mathrm{Fix}(\Sigma)$, whence $\mathrm{Fix}(\Delta) = \{0\}$, which is impossible.
(b) Cicogna [1981] generalizes Theorem 3.3 to the case in which $\dim \mathrm{Fix}(\Sigma)$ is odd, using a topological degree argument. However, to obtain effective information in this case we must also assume that Σ is a maximal isotropy subgroup. Otherwise, the solutions in $\mathrm{Fix}(\Sigma)$ whose existence is being asserted might actually have a larger isotropy subgroup than Σ.

In fact, we prove a slightly more general result than Theorem 3.3:

Theorem 3.5. *Let Γ be a Lie group acting on V. Assume*
(a) $\mathrm{Fix}(\Gamma) = \{0\}$,
(b) $\Sigma \subset \Gamma$ *is an isotropy subgroup satisfying* (3.6),
(c) $g\colon V \times \mathbb{R} \to V$ *is a Γ-equivariant bifurcation problem satisfying*

$$(dg_\lambda)_{0,0}(v_0) \neq 0 \tag{3.7}$$

where $v_0 \in \mathrm{Fix}(\Sigma)$ is nonzero.

Then there exists a smooth branch of solutions $(tv_0, \lambda(t))$ to the equation $g(t, \lambda) = 0$.

Two remarks make it clear why Theorem 3.3 follows from Theorem 3.5. First, it is easy to show that nontrivial irreducible actions satisfy $\mathrm{Fix}(\Gamma) = \{0\}$, since by Lemma 2.1 $\mathrm{Fix}(\Gamma)$ is an invariant subspace. Second, when Γ acts absolutely irreducibly,

$$(dg_\lambda)_{0,0}(v_0) = Kc'(0)$$

for some nonzero constant K. Hence (3.5) is equivalent to (3.7).

Remarks.
(a) The advantage of hypothesis (3.5) over (3.7) is that it holds simultaneously for all subgroups Σ of Γ.
(b) The advantage of Theorem 3.5 is that it does not require that Γ act irreducibly on V. However, a separate nondegeneracy condition (3.7) is required for each subgroup Σ satisfying (3.6).
(c) Since the solution branch $(tv_0, \lambda(t))$ lies in $\mathrm{Fix}(\Sigma) \times \mathbb{R}$, each solution for $t \neq 0$ has as its symmetries the isotropy subgroup Σ.

Proof of Theorem 3.5. It follows from Lemma 2.1 that

$$g\colon \mathrm{Fix}(\Sigma) \times \mathbb{R} \to \mathrm{Fix}(\Sigma).$$

Since $\dim \mathrm{Fix}(\Sigma) = 1$ we have

$$g(tv_0, \lambda) = h(t, \lambda)v_0.$$

Moreover, the assumption that $\text{Fix}(\Gamma) = \{0\}$ implies by Corollary 2.2 that g has a trivial solution. So $h(0, \lambda) = 0$. Applying Taylor's theorem to h yields

$$g(tv_0, \lambda) = k(t, \lambda)tv_0.$$

By Definition 3.1

$$k(0,0)v_0 = (dg)_{0,0}(v_0) = 0$$

and further

$$k_\lambda(0,0)v_0 = (dg_\lambda)_{0,0}(v_0) \neq 0,$$

by assumption. Apply the implicit function theorem to solve $k(t, \lambda) = 0$ for $\lambda = \lambda(t)$ as required. □

EXAMPLE 3.6. $\Gamma = \mathbf{D}_n$ acting on $V = \mathbb{C}$. We know that the isotropy subgroup of every point on the real axis is a two-element subgroup $\mathbf{Z}_2$ generated by the reflection $\kappa: z \mapsto \bar{z}$. See Table 1.1. Moreover, the only complex numbers fixed by κ are the reals. Thus $\text{Fix}(\mathbf{Z}_2) = \mathbb{R}$ and $\dim \text{Fix}(\mathbf{Z}_2) = 1$. We conclude, using the equivariant branching lemma, that generically $\mathbf{D}_n$-equivariant bifurcation problems have solution branches consisting of solutions with $\mathbf{Z}_2$ symmetry.

We end this section with the following, as promised.

SKETCH OF PROOF OF PROPOSITION 3.2. In this sketch we show only that there exist small perturbations G_ε of G such that Γ acts absolutely irreducibly on $\ker(dG_\varepsilon)_{0,0}$. This argument can be expanded, with some effort, to give a proof of genericity.

We begin by claiming that the action of Γ on V may be assumed irreducible. Write $\mathbb{R}^N = V \oplus W$ where W is Γ-invariant and write

$$V = V_1 \oplus \cdots \oplus V_k$$

where each V_j is irreducible. In fact we can take W to be the sum of the generalized eigenspaces corresponding to nonzero eigenvalues of $(dG)_{0,0}$. Define $M: \mathbb{R}^N \to \mathbb{R}^N$ to be the unique linear mapping such that

$$M|W = 0$$

$$M|V_1 = 0$$

$$M|V_j = Id_{V_j}.$$

Let $\varepsilon \in \mathbb{R}$ and consider the Γ-equivariant perturbation

$$G_\varepsilon(x, \lambda) = G(x, \lambda) + \varepsilon Mx.$$

The eigenvalues of $(dG_\varepsilon)_{0,0}$ are 0 on V_1, and nonzero on W. Apply a Liapunov–Schmidt reduction to G_ε near $(0,0)$ to obtain a bifurcation problem on V_1. Since Γ acts irreducibly on V_1, we have verified the claim.

We now assume that $g: V \times \mathbb{R} \to V$ is a bifurcation problem with symmetry group Γ and that Γ acts irreducibly but not absolutely irreducibly on V. We claim that in these circumstances there exist small perturbations of g which have *no* steady-state bifurcations near the origin. Let $\mathcal{D}$ be the vector space of linear mappings on V that commute with Γ. Recall from XII, §3, that $\mathcal{D}$ is isomorphic to one of $\mathbb{R}$, $\mathbb{C}$, or $\mathbb{H}$, and that $\mathcal{D} \cong \mathbb{R}$ means that Γ acts absolutely irreducibly on V. Now Γ acts irreducibly on V, so $g(0, \lambda) \equiv 0$. The linear maps $L_\lambda = (dg)_{0,\lambda}$ commute with Γ and form a curve in $\mathcal{D}$. Since g is a bifurcation problem, $L_0 = 0$, so the curve passes through the origin. Generically we may assume that $\rho = (d/d\lambda)L_\lambda|_{\lambda=0} \neq 0$; that is, the curve L_λ has a nonzero tangent vector at $\lambda = 0$.

Assume that $\dim_{\mathbb{R}} \mathcal{D} > 1$, so that Γ does not act absolutely irreducibly on V. We can choose $0 \neq \delta \in \mathcal{D}$ such that ρ and δ are linearly independent. For $\varepsilon \in \mathbb{R}$ define the Γ-equivariant perturbation

$$g_\varepsilon(x) = g(x, \lambda) + \varepsilon\delta x.$$

When $\varepsilon = 0$, the curve

$$(dg_\varepsilon)_{0,\lambda} = (dg)_{0,\lambda} + \varepsilon\delta = L_\lambda + \varepsilon\delta$$

in $\mathcal{D}$ misses the origin entirely for λ near 0.

Thus $L_\lambda + \varepsilon\delta$ is not zero. A general argument now shows that it has no zero eigenvalues. Indeed, if $\alpha \in \mathcal{D}$ has a zero eigenvalue then $\alpha = 0$. To see this, suppose that $\alpha v = 0$ where $\alpha \neq 0$, $v \neq 0$. Since $\mathcal{D}$ is a division algebra, α^{-1} exists, and $v = 1v = \alpha^{-1}\alpha v = \alpha^{-1}0 = 0$. This contradiction forces $\alpha = 0$ as claimed.

Thus when g is a bifurcation problem whose symmetry group Γ acts irreducibly but not absolutely irreducibly, small perturbations of g have no steady-state bifurcation whatsoever. □

Remark. Proposition 3.2 does not exclude the possibility that Γ-invariant equilibria can lose stability by having center subspaces with irreducible but not absolutely irreducible representations of Γ. This can happen generically with Hopf bifurcation, but *not* with steady-state bifurcation. See the definition of Γ-simple in XVI, §1.

EXERCISES

3.1. Use the results of Exercises 1.2 and 2.3 to investigate steady-state bifurcation with the symmetry $\mathbb{O} \oplus \mathbf{Z}_2^c$ of the cube (see Melbourne [1987a]). Prove that
   (a) Generically three branches of solutions bifurcate, with isotropy subgroups $\mathbf{D}_4$, $\mathbf{S}_3$, and $\mathbf{Z}_2^r \oplus \mathbf{Z}_2^t$.
   (b) Generically there are no solution branches corresponding to the isotropy subgroups $\mathbf{Z}_2^r$, $\mathbf{Z}_2^t$, and $\mathbb{1}$.

3.2. Let $\mathbf{Z}_2 \oplus \mathbf{Z}_2$ act on $\mathbb{R}^2$ by $(\pm x, \pm y)$ as in Chapter X. Show that the existence of the pure mode solutions (X, 1.11(b), (c)) can be obtained by applying the equivariant branching lemma. Note that a separate nondegeneracy condition is needed for each branch.

3.3. Let **O**(3) act in its five-dimensional representation. Using the results of Exercises 1.3 and 2.1 show that generically there exists a branch of axisymmetric solutions, where a solution is *axisymmetric* if its isotropy subgroup contains **SO**(2).

3.4. Let **SO**(2) act on $\mathbb{R}^2 \equiv \mathbb{C}$ in its standard representation. Recall from Lemma VIII, 2.2, that the invariants are generated by $|z|^2$ and the equivariants by z and iz. Let $g(z, \lambda)$ be an **SO**(2)-equivariant bifurcation problem.
   (a) Show that generically no steady-state bifurcation can occur.
   (b) Assume further that the vector field is a gradient and show that generically there now exist branches of steady states.
   (c) Observe that **SO**(2)-invariant functions are also **O**(2)-invariant, and deduce part (b) from the equivariant branching lemma.
   (d) If g depends on two bifurcation parameters (λ, μ) rather than just one, so that $g = g(z, \lambda, \mu)$, show that generically steady-state branching does occur, even in the nongradient case. (Compare Hopf bifurcation, XVI, §4, where a second parameter τ, the perturbed period, plays a similar role.)

§4. Orbital Asymptotic Stability

As discussed in Chapter VIII, an equilibrium solution x_0 to a system of ODEs

$$\frac{dx}{dt} + g(x) = 0 \tag{4.1}$$

is *asymptotically stable* if every trajectory $x(t)$ of the ODE which begins near x_0 stays near x_0 for all time $t > 0$, and also $\lim_{t\to\infty} x(t) = x_0$. The equilibrium is *neutrally stable* if the trajectory stays near x_0 for all $t > 0$. It is *unstable* if there always exist trajectories beginning near x_0 which do not stay near x_0 for all $t > 0$.

We repeat here the well-known condition for asymptotic stability known as *linear stability*: the eigenvalues of $(dg)_{x_0}$ all have positive real part. The standard theorem states that if x_0 is linearly stable then x_0 is asymptotically stable. Moreover, if some eigenvalue of $(dg)_{x_0}$ has negative real part, then x_0 is unstable. See Hirsch and Smale [1974], p. 187.

In this section we discuss the stability properties of equilibria for systems of ODEs (4.1) when the mapping g commutes with the action of a Lie group Γ. We address three issues:

(a) If the isotropy subgroup of an equilibrium has dimension less than that of Γ, then neither linear stability nor asymptotic stability is possible. The orbit of equilibria has positive dimension in this case, forcing dg to have zero eigenvalues. However, these concepts may be replaced by linear orbital stability and (asymptotic) orbital stability, respectively.

(b) The explicit computation of $(dg)_x$ is aided by knowledge of the representation of the isotropy subgroup Σ_x.
(c) For certain group actions, generically all of the solutions found using the equivariant branching lemma are unstable.

(a) Orbital Stability

Let Γ be a Lie group acting on V and let $g\colon V \to V$ be a Γ-equivariant map. Let x_0 be an equilibrium of the system (4.1), and let $\Sigma = \Sigma_{x_0}$ be the isotropy subgroup of x_0. We claim that if $\dim \Sigma < \dim \Gamma$ then x_0 cannot be asymptotically stable. To see this, recall from Proposition 1.2(b) that the orbit Γx_0 is a submanifold of V of positive codimension. It follows that there are steady states of the system (4.1) arbitrarily close to x_0. The trajectories starting at these equilibria are fixed for all time and so do not tend to x_0. Thus x_0 is not asymptotically stable. However, x_0 can be neutrally stable. In fact, x_0 can satisfy a specific kind of neutral stability, as follows.

The equilibrium x_0 is *orbitally stable* if x_0 is neutrally stable and if whenever $x(t)$ is a trajectory beginning near x_0, then $\lim_{t\to\infty} x(t)$ exists and lies in Γx_0.

There is a linear criterion for orbital stability. To show this, we first indicate why linearized stability fails. We claim that

$$\ker(dg)_{x_0} \supset T_{x_0}\Gamma x_0 \tag{4.2}$$

where $T_{x_0}\Gamma x_0$ denotes the tangent space of Γx_0 at x_0. It follows from (4.2) that $(dg)_{x_0}$ must, of necessity, have 0 as an eigenvalue; so linear stability is not possible at x_0.

To verify (4.2) let $y(t) = \gamma(t)x_0$ be a smooth curve in the orbit Γx_0 with $\gamma(t)$ a smooth curve in Γ and $\gamma(0) = 1$. Since x_0 is an equilibrium *and* since g is Γ-equivariant we see that

$$g(y(t)) \equiv 0. \tag{4.3}$$

Differentiate (4.3) with respect to t, to get

$$\frac{d}{dt} g(y(t))|_{t=0} = (dg)_{x_0}\left(\frac{d\gamma}{dt}(0)\cdot x_0\right) = 0. \tag{4.4}$$

Thus (4.4) shows that $(d\gamma/dt)(0)\cdot x_0$ is an eigenvector of $(dg)_{x_0}$ with eigenvalue zero.

Remark 4.1. Equation (4.4) provides a method for calculating null vectors of $(dg)_{x_0}$ by considering curves in the group Γ.

Definition 4.2. Let x_0 be an equilibrium of (4.1), where g commutes with the action of Γ. The steady state x_0 is *linearly orbitally stable* if the eigenvalues of $(dg)_{x_0}$ *other than those arising from* $T_{x_0}\Gamma x_0$ have positive real part.

In other words, x_0 is linearly orbitally stable if those eigenvalues of $(dg)_{x_0}$, not forced by the group action to be zero, have positive real part.

The basic result is as follows:

Theorem 4.3. *Linear orbital stability implies orbital (asymptotic) stability.*

SKETCH OF PROOF. We can motivate this by considering the linearized equation

$$\frac{dx}{dt} + (dg)_{x_0}x = 0. \tag{4.5}$$

If (4.1) is linearly orbitally stable then $\ker(dg)_{x_0} = T_{x_0}\Gamma_{x_0}$ and the remaining nonzero eigenvalues of $(dg)_{x_0}$ all have positive real part. Let W be the vector subspace generated by the generalized eigenspaces of these remaining eigenvalues, so that W is a complement to $T_{x_0}\Gamma_{x_0}$. Then it is easy to check that trajectories of the linear equation (4.5) lie in planes parallel to W and approach $T_{x_0}\Gamma_{x_0}$ exponentially. We may now relate this linearized flow to the original nonlinear flow by methods similar to those used in showing that linearized stability implies asymptotic stability; see Aulbach [1984], p. 2. The result is a proof that trajectories of (4.1) tend exponentially to some point on the orbit Γx_0, if x_0 is linearly orbitally stable. □

(b) Isotropy Restrictions on dg

As we have seen in (3.4), dg satisfies the commutativity constraint

$$(dg)_{\gamma x}\gamma = \gamma(dg)_x. \tag{4.6}$$

Let $\Sigma \subset \Gamma$ be the isotropy subgroup of x. Then for every $\sigma \in \Sigma$ (4.6) takes the form

$$(dg)_x\sigma = \sigma(dg)_x; \tag{4.7}$$

that is, $(dg)_x$ commutes with the isotropy subgroup Σ of x.

The commutativity relation (4.7) restricts the form of $(dg)_x$ as follows. Given Σ we can decompose V into isotypic components

$$V = W_1 \oplus \cdots \oplus W_k$$

as in Theorem XII, 2.5(b). By Theorem XII, 3.5,

$$(dg)_x(W_j) \subset W_j. \tag{4.8}$$

We can always take $W_1 = \mathrm{Fix}(\Sigma)$ since $\mathrm{Fix}(\Sigma)$ is the sum of all subspaces of V on which Σ acts trivially.

In summary, the group Γ affects the form of $(dg)_x$ in two ways.

(a) Γ/Σ forces null vectors of $(dg)_x$ as in (4.2). That is, $\dim \ker(dg)_x \geq \dim \Gamma/\Sigma$.
(b) $(dg)_x$ has invariant subspaces as in (4.8).

The restriction of $(dg)_x$ to W_j is often subject to extra conditions. For example, suppose that Σ acts absolutely irreducibly on W_j. Then $(dg)_x | W_j$ is a scalar multiple of the identity. Even when the action of Γ on W_j is not absolutely irreducible, the form of $(dg)_x | W_j$ may be constrained by the symmetry, but we shall not pursue this matter here.

We now consider two examples: $\mathbf{D}_n$ and $\mathbf{O}(2)$ in their standard representations on $\mathbb{C}$. In each case, we let $x \in \mathbb{C}$ be real and recall that the isotropy subgroup of x is the group $\mathbf{Z}_2$ generated by the reflection $\kappa: z \mapsto \bar{z}$. In real coordinates the matrix of κ is

$$L = \begin{bmatrix} 1 & 0 \\ 0 & -1 \end{bmatrix}.$$

Thus $W_1 = \mathbb{R}$ and $W_2 = i\mathbb{R}$. The action of L on W_1 is the identity, and on W_2 minus the identity. These representations are distinct and absolutely irreducible. Therefore,

$$(dg)_x = \begin{bmatrix} a & 0 \\ 0 & b \end{bmatrix} \tag{4.9}$$

for $a, b \in \mathbb{R}$; that is, $(dg)_x$ is diagonal.

In the case $\Gamma = \mathbf{D}_n$, the form (4.9) is all that we can say. But when $\Gamma = \mathbf{O}(2)$ a null vector is forced on $(dg)_x$ by the construction in (4.4). We perform this calculation in real coordinates:

$$\frac{d}{dt}\begin{bmatrix} \cos t & -\sin t \\ \sin t & \cos t \end{bmatrix}\Bigg|_{t=0} \begin{bmatrix} 1 \\ 0 \end{bmatrix} = \begin{bmatrix} 0 & -1 \\ 1 & 0 \end{bmatrix}\begin{bmatrix} 1 \\ 0 \end{bmatrix} = \begin{bmatrix} 0 \\ 1 \end{bmatrix}$$

is a null vector for $(dg)_x$. Thus

$$(dg)_x = \begin{bmatrix} a & 0 \\ 0 & 0 \end{bmatrix}.$$

Remark. We emphasize that we have arrived at this form for $(dg)_x$ without ever having to compute a derivative of g. In general, the isotropy subgroup will not reduce the form of $(dg)_x$ so substantially, but every little bit helps.

(c) Unstable Solutions in the Equivariant Branching Lemma

In this subsection we prove that generically, for certain group actions, the solutions obtained from the equivariant branching lemma are all unstable. We prove this theorem using the hypotheses of Theorem 3.3; we indicate in the text where weaker hypotheses are appropriate. The hypotheses we assume are:

(a) Γ acts absolutely irreducibly on V,
(b) $g: V \times \mathbb{R} \to V$ is Γ-equivariant with $(dg)_{0,\lambda} = c(\lambda)I$,
(c) $c(0) = 0$ and $c'(0) < 0$,
(d) Σ is an isotropy subgroup with $\dim \operatorname{Fix}(\Sigma) = 1$, (4.10)
(e) Some term in the Taylor expansion of $g|\operatorname{Fix}(\Sigma) \times \{0\}$ is nonzero,
(f) $(dq)_{x_0}$ has eigenvalues off the imaginary axis, where q is the quadratic part of g and $x_0 \in \operatorname{Fix}(\Sigma)$.

Remarks.
(a) We assume that $c'(0) < 0$ so that the trivial solution $x = 0$ is asymptotically stable for $\lambda < 0$ and unstable for $\lambda > 0$. Similar results, however, hold when $c'(0) > 0$.
(b) The action of Γ often forces all quadratic terms (in x) of g to be zero. This happens, for example, when $-I \in \Gamma$. Then symmetry requires that g be an odd function. In such circumstances $q(x) \equiv 0$ and (f) is never valid. However, when (f) holds for one q, it holds generically.
(c) All quadratic equivariants may be zero even if $-I \notin \Gamma$. For example, consider the standard action of $\mathbf{D}_5$ on $\mathbb{R}^2$.

Theorem 4.4. *Assume hypotheses* (4.10(a)–(f)). *Then the unique branch of solutions to* $g(x, \lambda) = 0$ *in* $\operatorname{Fix}(\Sigma)$ *whose existence is guaranteed by the equivariant branching lemma consists of unstable solutions.*

Remark. Suppose $\dim \operatorname{Fix}(\Sigma) = 1$. Then in suitable circumstances the bifurcation problem $g|\operatorname{Fix}(\Sigma) \times \mathbb{R} = 0$ must be a pitchfork. This happens when $N(\Sigma) \neq \Sigma$; see Exercise 4.1. If in particular (4.10(f)) holds, then the nontrivial branch of this pitchfork is unstable, even if it is supercritical. An example of this phenomenon may be found in Case Study 4.

Let $0 \neq v_0 \in \operatorname{Fix}(\Sigma)$. In the proof of the equivariant branching lemma we saw that the nontrivial branch of solutions to $g|\operatorname{Fix}(\Sigma) \times \mathbb{R} = 0$ has the form $(tv_0, \Lambda(t))$ where $\Lambda(0) = 0$. We call this nontrivial solution branch *transcritical* if $\Lambda'(0) \neq 0$ and *degenerate* if $\Lambda'(0) = 0$. The proof of Theorem 4.4 divides into two parts, depending on whether the nontrivial branch is transcritical or degenerate. Hypothesis (4.10(f)) is required only in the degenerate case.

We show below that

$$\operatorname{sgn}(\Lambda'(0)) = \operatorname{sgn}(c'(0))\operatorname{sgn}(d^2 g)_{0,0}(v_0, v_0). \tag{4.11}$$

Identity (4.11) provides a method to determine whether a given branch is transcritical. It also shows that transcriticality implies (4.10(e), (f)).

The first part of Theorem 4.4 is as follows:

Theorem 4.5. *Assume* (4.10(a)–(d)) *and suppose that the unique branch of solutions to* $g|\operatorname{Fix}(\Sigma) \times \mathbb{R} = 0$, *whose existence is guaranteed by the equivari-*

ant branching lemma, is transcritical. Then this branch consists of unstable solutions.

Remark. It is perhaps surprising that transcritical solutions are unstable in symmetric systems. Indeed for bifurcation problems in one state variable, the supercritical part (see following discussion) of a transcritical branch is stable; consider $x^2 - \lambda x = 0$. The fundamental difference between the two situations is stated in the next lemma.

Lemma 4.6. *Let Γ be a Lie group acting on V. Let $q: V \to V$ be a Γ-equivariant homogeneous quadratic polynomial. Then*

$$L(x) = \operatorname{tr}(dq)_x$$

is a Γ-invariant linear function.

Moreover, if $\operatorname{Fix}(\Gamma) = \{0\}$ *then* $\operatorname{tr}(dq) = 0$.

Proof. Differentiate the equivariance condition as usual to obtain $(dq)_{\gamma x}\gamma = \gamma(dq)_x$, and rewrite this as

$$(dq)_{\gamma x} = \gamma(dq)_x\gamma^{-1}.$$

Take traces to obtain

$$L(\gamma x) = L(x).$$

The entries of dq are linear in x since q is quadratic. If $\operatorname{Fix}(\Gamma) = \{0\}$ then by Proposition 2.2 every linear invariant function is zero, so $L \equiv 0$. □

There is another common classification of bifurcating branches. The branch $(tv_0, \Lambda(t))$ is *subcritical* if for all nonzero t near 0,

$$t\Lambda'(t) < 0. \tag{4.12}$$

It is *supercritical* if $t\Lambda'(t) > 0$. This definition makes sense for the two parts of the branch $t > 0$ and $t < 0$. Of course, when the branch is transcritical, it has one subcritical part and one supercritical part. It is convenient to prove Theorem 4.4 for these parts separately.

The instability of subcritical branches is well known; see Crandall and Rabinowicz [1973]. It is included here for completeness. We prove the following:

Proposition 4.7. *Assume* (4.10(a)–(e)) *and suppose that the branch* $(tv_0, \Lambda(t))$, $t > 0$, *is subcritical. Then this branch consists of unstable solutions.*

Remark. The same result holds when the branch $(tv_0, \Lambda(t))$, $t < 0$ is subcritical.

Proof. Since some derivative of $g|\operatorname{Fix}(\Sigma) \times \{0\}$ is nonzero, by (4.10(e)), the sign of $\Lambda'(t)$ is uniquely defined for small $t > 0$. The solution branch is subcritical precisely when $\Lambda'(t) < 0$. We claim that v_0 is an eigenvector for $(dg)_{tv_0,\Lambda(t)}$. To prove this, recall from (4.8) and the ensuing discussion that $(dg)_{tv_0,\Lambda(t)}(\mathrm{Fix}(\Sigma)) \subset \mathrm{Fix}(\Sigma)$. Since $\dim \mathrm{Fix}(\Sigma) = 1$ by (4.10(d)) and $v_0 \in \mathrm{Fix}(\Sigma)$ we see that v_0 is an eigenvector for $(dg)_{tv_0,\Lambda(t)}$.

Next, we claim that the sign of the corresponding eigenvalue is that of $t\Lambda'(t)$, which by assumption is negative. The theorem follows since, by our convention, negative eigenvalues imply instability.

To establish the claim, recall how the branch $(tv_0, \Lambda(t))$ is constructed in the proof of Theorem 3.5. Since g maps $\mathrm{Fix}(\Sigma) \times \mathbb{R}$ to $\mathrm{Fix}(\Sigma)$, we have

$$g(tv_0, \lambda) = h(t, \lambda)v_0.$$

Since g has a trivial solution,

$$h(t, \lambda) = tk(t, \lambda).$$

Finally (4.10(c)) implies that $k(0,0) = 0$, $k_\lambda(0,0) < 0$. The unique branch of solutions is found by applying the implicit function theorem to the equation $k = 0$. So

$$k(t, \Lambda(t)) = 0. \tag{4.13}$$

We compute $(dg)_{x,\lambda}v_0$ by evaluating

$$\frac{d}{ds} g(x + sv_0, \lambda)|_{s=0}.$$

Set $x = tv_0$; then

$$\begin{aligned}(dg)_{tv_0,\lambda}v_0 &= \frac{d}{ds} g((s + t)v_0, \lambda)|_{s=0}\\ &= \frac{d}{ds} h(s + t, \lambda)|_{s=0}v_0\\ &= h_t(t, \lambda)v_0.\end{aligned}$$

Thus the eigenvalue associated to v_0 is $h_t(t, \lambda)$. We now compute h_t along the branch of solutions, that is, where $k = 0$ as in (4.13). This yields

$$h_t(t, \Lambda(t)) = tk_t(t, \Lambda(t)). \tag{4.14}$$

Now differentiate (4.13) implicitly with respect to t, obtaining

$$k_t(t, \Lambda(t)) + k_\lambda(t, \Lambda(t))\Lambda'(t) \equiv 0. \tag{4.15}$$

Substitute (4.15) in (4.14), to yield

$$h_t(t, \Lambda(t)) = -t\Lambda'(t)k_\lambda(t, \Lambda(t)). \tag{4.16}$$

Since $k_\lambda(0,0) < 0$, (4.16) shows that for small t the sign of the eigenvalue associated with v_0 is given by $\mathrm{sgn}(t\Lambda'(t))$, as claimed. □

VERIFICATION OF (4.11). From (4.15)

$$\mathrm{sgn}(\Lambda'(0)) = \mathrm{sgn}\, k_\lambda(0,0) \cdot \mathrm{sgn}\, k_t(0,0).$$

Now $k_\lambda(0,0) = c(0)$, and by twice differentiating $g|\mathrm{Fix}(\Sigma) \times \mathbb{R}$ we have $k_t(0,0) = h_{tt}(0,0) = (d^2g)_{0,0}(v_0, v_0)$. Thus (4.11) holds. □

Proof of Theorem 4.4. In the transcritical case the solution branch in $\mathrm{Fix}(\Sigma) \times \mathbb{R}$ has both a supercritical and a subcritical part. By Proposition 4.7 we need consider only the supercritical branch. By (4.12), along the supercritical part of the transcritical branch we have

$$t\Lambda'(t) > 0. \tag{4.17}$$

We define

$$T(t) = \mathrm{tr}(dg)_{tv_0, \Lambda(t)}.$$

We claim that in the transcritical case

$$T(t) = nc'(0)\Lambda'(0)t + O(t^2). \tag{4.18}$$

It follows from (4.17) and (4.10c) that for small t

$$T(t) < 0.$$

Hence for small t at least one eigenvalue of $(dg)_{tv_0, \Lambda(t)}$ has negative real part, and these solutions are unstable as well.

We now prove (4.18). The assumptions on g, namely (4.10(b), (c), (f)), imply that the Taylor expansion of g has the form

$$g(x, \lambda) = c(\lambda)x + q(x, \lambda) + O(x^3). \tag{4.19}$$

Hence

$$\mathrm{tr}(dg)_{x,\lambda} = nc(\lambda) + \mathrm{tr}(dq)_{x,\lambda} + O(x^2)$$

where $n = \dim V$. Lemma 4.6 implies that for each λ, $\mathrm{tr}(dq)_{x,\lambda} = 0$. Thus

$$\begin{aligned} T(t) &= \mathrm{tr}(dg)_{tv_0, \Lambda(t)} \\ &= nc(\Lambda(t)) + O(t^2) \\ &= nc'(0)\Lambda'(0)t + O(t^2) \end{aligned}$$

as claimed. □

We complete the proof of Theorem 4.3 by proving the following:

Proposition 4.8. *Assume* (4.10(a)–(f)) *and suppose that the branch* $(tv_0, \Lambda(t))$ *is degenerate*; *i.e.*, $\Lambda'(0) = 0$. *Then the branch consists of unstable solutions.*

Proof. Here we compute the matrix $(dg)_{tv_0, \Lambda(t)}$ up to order 2, rather than just its trace. We show that

$$(dg)_{tv_0, \Lambda(t)} = t[(dq)_{v_0, 0} + O(t))]. \tag{4.20}$$

From (4.19)

$$(dg)_{tv_0, \Lambda(t)} = c(\Lambda(t))I + (dq)_{tv_0, \Lambda(t)} + O(t^2). \tag{4.21}$$

Now $c(\Lambda(t)) = O(t^2)$ since $c(0) = 0$ and $\Lambda(0) = \Lambda'(0) = 0$. Also

$$(dq)_{tv_0, \Lambda(t)} = t(dq)_{v_0, \Lambda(t)} \tag{4.22}$$

since $(dq)_{x, \lambda}$ is linear in x and

$$(dq)_{v_0, \Lambda(t)} = (dq)_{v_0, 0} + O(t) \tag{4.23}$$

since $\Lambda(0) = 0$. Of course, (4.21)–(4.23) together yield (4.20).

By (4.10f), $(dq)_{v_0, 0}$ has one eigenvalue off the imaginary axis. By Lemma 4.6 the trace of this matrix is zero; hence $(dq)_{v_0, 0}$ has at least one eigenvalue with positive real part and one eigenvalue with negative real part. Now write (4.20) as

$$(dg)_{tv_0, \Lambda(t)} = t[(dq)_{v_0, 0} + tK(t)] \tag{4.24}$$

for some matrix $K(t)$. Since the eigenvalues of a matrix depend continuously on parameters, it follows that for small t at least one eigenvalue of

$$(dq)_{v_0, 0} + tK(t)$$

has positive real part, and at least one has negative real part. Finally, from (4.24) we see that for small t the matrix $(dg)_{tv_0, \Lambda(t)}$ has at least one eigenvalue with negative real part, whatever the sign of t. Thus the solutions are unstable in this case too. □

Remark. We repeat that the assumptions on g in (4.10) are *generic* for any absolutely irreducible group action that admits a nonzero equivariant homogeneous quadratic.

Exercises

4.1. Let $\Sigma \subset \Gamma$ be an isotropy subgroup and let g be Γ-equivariant. Assume $\dim \mathrm{Fix}(\Sigma) = 1$ and $N(\Sigma) \neq \Sigma$. Show that g: $\mathrm{Fix}(\Sigma) \to \mathrm{Fix}(\Sigma)$ is an odd function.

4.2. This continues Exercise 3.1 and is based on Melbourne [1987a].

(a) Using the results and notation of Exercises XII, 4.7; XII, 5.5; XIII, 1.2; and XIII, 2.3 show that the general bifurcation problem on $\mathbb{R}^3$ with the symmetries $\mathbb{O} \oplus \mathbf{Z}_2^c$ of the cube has the form $g(x, y, z, \lambda) = PX_1 + QX_2 + RX_3$ where P, Q, R are functions of u, v, w, λ.

(b) Show that if the nondegeneracy conditions

$$Q(0) \neq 0, P_u(0) \neq -1, -\tfrac{1}{2}, -\tfrac{1}{3}$$

are satisfied, then the branches of solutions corresponding to the three maximal isotropy subgroups satisfy the following equations:

$$\mathbf{D}_4: \qquad \lambda = -(P_u(0) + Q(0))x^2/P_\lambda(0) + \cdots$$

$$\mathbf{Z}_2^r \oplus \mathbf{Z}_2^t: \qquad \lambda = -(2P_u(0) + Q(0))x^2/P_\lambda(0) + \cdots$$

$$\mathbf{S}_3: \qquad \lambda = -(3P_u(0) + Q(0))x^2/P_\lambda(0) + \cdots.$$

(c) Find the directions of branching and stability conditions for these solutions. Show that in the nondegenerate case the $\mathbf{Z}_2^r \oplus \mathbf{Z}_2^t$ branch is always unstable, that for a branch to be stable all three must be supercriticial, and that in this case either the $\mathbf{D}_4$ branch or the $\mathbf{S}_3$ branch (but not both) is stable.

§5. Bifurcation Diagrams and $\mathbf{D}_n$ Symmetry

Let $g: V \times \mathbb{R} \to V$ be a bifurcation problem with symmetry group Γ. In this section we describe what we mean by the bifurcation diagram associated with g and illustrate the notion by discussing the generic bifurcation problems with $\mathbf{D}_n$-symmetry. Bifurcation diagrams are important vehicles for summarizing analytic information efficiently. To accomplish this task some information must be suppressed. Our purpose here is to specify precisely what is to be suppressed and what included.

(a) General Description of Bifurcation Diagrams

The simplest view of a bifurcation diagram is the zero set of g,

$$\{(x, \lambda) \in V \times \mathbb{R}: g(x, \lambda) = 0\}.$$

We have two reasons for not wishing to picture this set.

(a) If $\dim V > 2$, we would be trying to draw a figure in a space of at least four dimensions, where visualization is at best tricky.
(b) If g is Γ-equivariant then the set $\{g = 0\}$ contains redundant information since g must vanish on entire orbits of the action.

Because of these observations, we prefer to draw *schematic* bifurcation diagrams where each point represents an orbit of solutions to $g = 0$. These schematics will always be drawn in the plane according to the following conventions:

(a) The horizontal axis is the λ-axis, and the vertical axis is, *loosely speaking*, the norm of the (orbit of) solution(s).
(b) Each solution branch is labeled with its (conjugacy class of) isotropy subgroup.
(c) Bifurcation points and limit points are indicated by bold dots.
(d) The asymptotic orbital stability of solutions, determined by eigenvalues of the Jacobian, is marked. Orbitally stable branches are indicated by heavy lines.
(e) Predictions of transitions under quasistatic variation of λ are provided.

We now discuss a hypothetical bifurcation diagram illustrating these five points. For the sake of argument assume that Γ is a four-dimensional group acting on a six-dimensional space V. Suppose that the bifurcation problem g

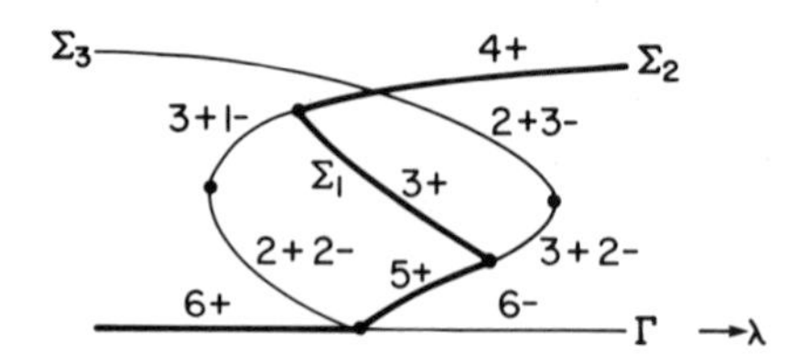

Figure 5.1. Fictitious bifurcation diagram with symmetry.

has branches of nontrivial solutions corresponding to isotropy subgroups Σ_j ($j = 1, 2, 3$) where $\dim \Sigma_j = j$. Consider the fictitious bifurcation diagram in Figure 5.1. Each point on branch Σ_j corresponds to a group orbit of dimension $\dim \Gamma - \dim \Sigma_j = 4 - j$. For example, a point on the Σ_2 branch actually corresponds to a two-dimensional manifold of solutions, and where the branches corresponding to Σ_2 and Σ_1 intersect we actually have a 3-manifold of solutions in $\mathbb{R}^6$ merging into a 2-manifold. Obviously such intersections can be very complicated, but fortunately, for most aspects of bifurcation theory, the detailed geometric picture of how such transitions take place is not particularly relevant.

Because our schematic bifurcation diagrams are "projections" into $\mathbb{R}^2$, branches of solutions may appear to intersect, even though they do not actually intersect in $V \times \mathbb{R}$. We "solve" this problem by placing bold dots at genuine intersection points. Thus Figure 5.1 illustrates a situation where branch Σ_1 does intersect branches Σ_2 and Σ_3, but branches Σ_2 and Σ_3 intersect only at the origin.

We now turn to the question of orbital stability. Recall that equivariance under Γ forces several eigenvalues of dg at $g = 0$ to zero. The number of these zero eigenvalues is equal to the dimension of the orbit of solutions. When making stability assignments we employ two conventions. First, we indicate eigenvalues of dg with positive real part by "$+$" and those with negative real part by "$-$". Thus, along the Σ_2 branch, the annotation $3 + 1 -$ indicates solutions where dg has three eigenvalues with positive real part and one with negative real part. Second, eigenvalues forced to zero by the group action are *not* included. Along each branch the total number of eigenvalues must equal $\dim V$, which is 6 in this case. Indeed along branch Σ_2 the number of eigenvalues forced by the group action to be zero is 2, so that $2 + 4 = 6$ gives the correct number of eigenvalues altogether. Note that at limit points the stabilities of solutions change. For this reason limit points are also indicated by bold dots.

To end the discussion of stabilities, note that we are seeking equilibrium solutions to a system of ODEs written in the form

$$\frac{dx}{dt} + g(x, \lambda) = 0.$$

Using this form, eigenvalues with positive real part indicate (linearized) stability, whereas those with negative real part indicate instability. Thus a solution is orbitally stable when no "$-$" signs appear in the stability assignments. Orbitally stable solutions are shown by heavy lines. Note that there are orbitally stable solutions on part of the branches Σ_1, Σ_2, Σ_3, and Γ (with 3, 4, 5, and 6 positive eigenvalues, respectively).

The most important information preserved in these schematic diagrams is the answers to the following two questions:

(a) For each λ, how many orbits of solutions are there to the equation $g = 0$, and which are stable?
(b) For which values of λ do transitions in the number of solutions, or their stability, occur?

The answers to these questions are preserved by projection onto the λ-axis, allowing us to keep track of smooth bifurcations, jump transitions (when solutions cease to exist or change stability as λ varies), and hysteretic phenomena.

We end this section by discussing the simplest bifurcation diagrams for problems with $\mathbf{D}_n$ symmetry. Not all features of the diagram in Figure 5.1 appear, but all of these features will be important in later sections. For example, see Case Study 4.

(b) Bifurcation Diagrams for $\mathbf{D}_n$ Symmetry

We begin by describing the isotropy subgroups of $\mathbf{D}_n$ in its standard action on $\mathbb{C} \equiv \mathbb{R}^2$, generated by

$$\begin{aligned} &\text{(a)} \quad \kappa z = \bar{z} \\ &\text{(b)} \quad \zeta z = e^{2\pi i/n} z. \end{aligned} \tag{5.1}$$

By computing the isotropy subgroups and applying the equivariant branching lemma we will be able to determine the expected number of solution branches. The actual bifurcation diagrams are given at the end of this section in Figures 5.3, and 5.4. In this way it should become apparent just how much information is contained in one of these pictures.

The lattice of isotropy subgroups, Figure 5.2, depends on whether n is odd or even. We compute the isotropy subgroups by choosing representative points on the group orbits. Recall (Lemma 1.1) that points on the same orbit have conjugate isotropy subgroups. Moreover, any two points on the same line through (but not including) the origin have the same isotropy subgroup. Thus it sufficies to compute Σ_z for points $z = e^{i\theta}$ on the unit circle.

We claim that z is on the same orbit as a point $e^{i\theta}$ with

$$0 \leq \theta \leq \pi/n. \tag{5.2}$$

It is easy to arrange for $0 \leq \theta \leq 2\pi/n$ by sending z to $(\zeta^l)z = e^{(\theta+(2\pi l/n))i}$ for

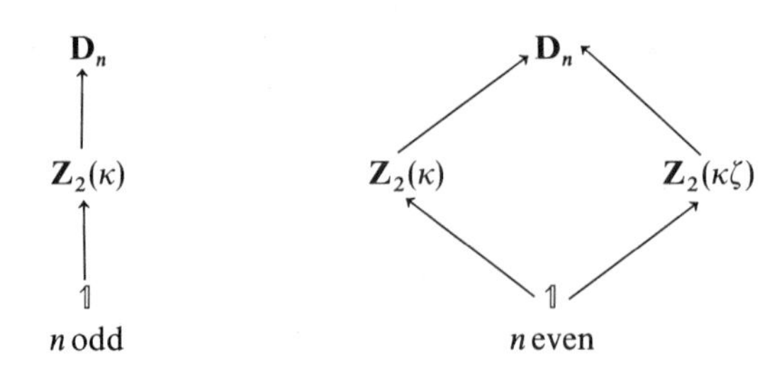

Figure 5.2. Lattice of isotropy subgroups of $\mathbf{D}_n$ ($n \geq 3$).

some appropriately chosen l. Now if $\pi/n < \theta \leq 2\pi/n$, we have

$$\zeta\kappa z = e^{((2\pi/n)-\theta)i} \tag{5.3}$$

and $0 < 2\pi/n - \theta \leq \pi/n$, as promised.

Next observe that if n is odd, we can assume that

$$0 \leq \theta < \pi/n. \tag{5.4}$$

To see this, let $n = 2m + 1$. Then $(\zeta^m)e^{i\pi/n} = e^{i\pi} = -1$, and this has the same isotropy subgroup as $z = 1$, i.e., $\theta = 0$.

We now leave it to the reader to check that the isotropy subgroup of $z = e^{i\theta}$ is $\mathbb{1}$ if $0 < \theta < \pi/n$; $\mathbf{Z}_2(\kappa)$ if $\theta = 0$; and $\mathbf{Z}_2(\zeta\kappa)$ if n is even and $\theta = \pi/n$ (see (5.3)). We emphasize that for n even the two $\mathbf{Z}_2$ isotropy subgroups, $\mathbf{Z}_2(\kappa)$ and $\mathbf{Z}_2(\zeta\kappa)$, are *not* conjugate in $\mathbf{D}_n$. This is clear on geometrical grounds, since vertices and midpoints of edges lie on different lines through the origin. Alternatively, a simple calculation shows that the conjugates of κ are the elements $\zeta^s\kappa$ where s is *even*. For even n, these do not include $\zeta\kappa$.

Since the fixed-point subspaces of both $\mathbf{Z}_2(\kappa)$ and $\mathbf{Z}_2(\zeta\kappa)$ are clearly one-dimensional, the equivariant branching lemma implies that generically there are unique branches of solutions to bifurcation problems with $\mathbf{D}_n$ symmetry, corresponding to these isotropy subgroups. In general a given "branch" of solutions, defined say for all $x \in \mathbb{R}$, may correspond to either one orbit, or two distinct orbits, of solutions, depending on whether or not x and $-x$ lie in the same orbit of Γ. See Exercise 4.1. In particular for $\mathbf{D}_n$ with n odd there is one branch, and when n is even there are two. We show later that when n is odd the one branch splits into two orbits of solutions, whereas when n is even each branch corresponds to a unique orbit.

We can determine more complete information about these branches by analyzing the general form of $\mathbf{D}_n$-equivariant mappings. Recall from Chapter XII, §5, that if $g: \mathbb{C} \times \mathbb{R} \to \mathbb{C}$ commutes with $\mathbf{D}_n$, then

$$g(z, \lambda) = p(u, v, \lambda)z + q(u, v, \lambda)\bar{z}^{n-1}, \tag{5.5}$$

where $u = z\bar{z}$ and $v = z^n + \bar{z}^n$. In order for g to be a bifurcation problem, the linear terms in (5.5) must vanish. Hence

$$p(0, 0, 0) = 0. \tag{5.6}$$

Table 5.1. Solution of $g = 0$ for $\mathbf{D}_n$-Equivariant g, $n \geq 3$

Isotropy Subgroup	Fixed-Point Subspace	Equations
$\mathbf{D}_n$	$\{0\}$	$z = 0$
$\mathbf{Z}_2(\kappa)$	$\mathbb{R}$	$p(x^2, 2x^n, \lambda) + x^{n-2}q(x^2, 2x^n, \lambda) = 0$
		$x \neq 0$ [n odd], $x > 0$ [n even]
$\mathbf{Z}_2(\zeta\kappa)$	$\mathbb{R}\{e^{i\pi/n}\}$	$p(x^2, -2x^n, \lambda) - x^{n-2}q(x^2, -2x^n, \lambda) = 0$
[n even]		$x > 0$
$\mathbb{1}$	$\mathbb{C}$	$p = q = 0$
		$\mathrm{Im}(z^n) \neq 0$

In addition, the genericity hypothesis of the equivariant branching lemma requires

$$p_\lambda(0, 0, 0) \neq 0. \tag{5.7}$$

We now prove that a second nondegeneracy hypothesis, namely

$$q(0, 0, 0) \neq 0, \tag{5.8}$$

implies that generically the *only* (local) solution branches to $g = 0$ are those obtained using the equivariant branching lemma.

Observe that z and $\bar{z}^{n-1}$ are collinear only when $\mathrm{Im}(z^n) = 0$. Thus when $\mathrm{Im}(z^n) \neq 0$, solving $g = 0$ is equivalent to solving

$$p = q = 0. \tag{5.9}$$

Thus, under the genericity hypothesis (5.8), it is not possible to find solutions to (5.9) near the origin. Now $\mathrm{Im}(z^n) \neq 0$ precisely when the isotropy subgroup of z is $\mathbb{1}$. Thus the only solutions to $g = 0$ are those corresponding to the maximal isotropy subgroups. The full solution to $g = 0$ is given in Table 5.1.

When n is even, $\zeta^{n/2} = -1$, so that the points z and $-z$ are on the same orbit. Thus when n is even we may assume $x > 0$ (not just $x \neq 0$) in Table 5.1.

In the remainder of this section we discuss the direction of branching and the asymptotic stability of the solutions we have found. In this discussion we restrict attention to $n \geq 5$ since $n = 3$ and $n = 4$ are exceptional. See Chapter XV, §4, for a discussion of the case $n = 3$ and Chapter XVII, §6, for $n = 4$.

We first explain why $n = 3$ and 4 are special. When $n = 3$ there is a nontrivial $\mathbf{D}_3$-equivariant quadratic $\bar{z}^2$. Then Theorem 4.4 implies that generically the branch of $\mathbf{Z}_2(\kappa)$ solutions is unstable. Therefore, in order to find asymptotically stable solutions to a $\mathbf{D}_3$-equivariant bifurcation problem by a local analysis, we must consider the degeneracy $q(0, 0, 0) = 0$ and apply unfolding theory. We return to this point in the discussion of the traction problem in Case Study 5 and the spherical Bénard problem in Chapter XV, §5.

In the case $n = 4$ the term $\bar{z}^{n-1}$ is cubic, and $q(0, 0, 0)$ enters nontrivially into the branching equations. See Table 5.1, Exercise 5.1, and Chapter XVII, §6.

We now restrict attention to $n \geq 5$. Observe from Table 5.1 that the lowest

Table 5.2. Data on Solutions of Generic $\mathbf{D}_n$-Equivariant Bifurcation Problems, $n \geq 5$

Isotropy	Branching Equation	Signs of Eigenvalues
$\mathbf{D}_n$	$z = 0$	$p_\lambda(0,0,0)\lambda$ (twice)
$\mathbf{Z}_2(\kappa)$	$\lambda = -\dfrac{p_u(0,0,0)}{p_\lambda(0,0,0)}x^2 + \cdots$	$p_u(0,0,0)$
	$x > 0$ [n even]	$-q(0,0,0)$ [n even]
	$x \neq 0$ [n odd]	$-q(0,0,0)x$ [n odd]
$\mathbf{Z}_2(\zeta\kappa)$	$\lambda = -\dfrac{p_u(0,0,0)}{p_\lambda(0,0,0)}x^2 + \cdots$	$p_u(0,0,0)$
[n even]	$x > 0$	$q(0,0,0)$

order terms in the equation for both $\mathbf{Z}_2$ solutions are

$$p_u(0,0,0)x^2 + p_\lambda(0,0,0)\lambda + \cdots. \tag{5.10}$$

Thus the $\mathbf{Z}_2$ branches are supercritical when $p_u(0,0,0)p_\lambda(0,0,0) < 0$ and subcritical when $p_u(0,0,0)p_\lambda(0,0,0) > 0$. Generically we may assume that

$$p_u(0,0,0) \neq 0 \tag{5.11}$$

so that the direction of branching is determined.

We now discuss stabilities. Both κ and $\zeta\kappa$ are reflections, having 1 and -1 as distinct eigenvalues. Therefore, from the restrictions imposed by isotropy (see (4.8)) dg leaves the corresponding one-dimensional eigenspaces invariant. Hence the eigenvalues of dg must be real.

A straightforward calculation shows that if we think of g as a function of real coordinates z, $\bar{z}$, then

$$(dg)(w) = g_z w + g_{\bar{z}}\bar{w}. \tag{5.12}$$

(The method for computing dg in (5.12) is typically the most efficient when g is defined using complex variables.) Compute (5.12) to obtain:

$$\begin{aligned} &\text{(a)} \quad g_z = p + p_u z\bar{z} + np_v z^n + (q_u\bar{z} + nq_v z^{n-1})\bar{z}^{n-1} \\ &\text{(b)} \quad g_{\bar{z}} = p_u z^2 + np_v z\bar{z}^{n-1} + (n-1)q\bar{z}^{n-2} + (q_u z + nq_v\bar{z}^{n-1})\bar{z}^{n-1}. \end{aligned} \tag{5.13}$$

We list the branching and eigenvalue information in Table 5.2 and now verify those data. Along the trivial solution $z = 0$ we have

$$(dg)_{0,\lambda}(w) = p(0,0,\lambda)w = (p_\lambda(0,0,0)\lambda + \cdots)w.$$

Thus dg is a multiple of the identity, having a repeated eigenvalue whose sign is $\operatorname{sgn}(p_\lambda(0,0,0)\lambda)$ since $p_\lambda(0,0,0)$ is assumed nonzero.

Next we consider the $\mathbf{Z}_2(\kappa)$ solutions. The fixed-point subspace is the real axis ($w = \bar{w}$) and the -1 eigenspace is the imaginary subspace ($w = -\bar{w}$). Since these subspaces are invariant under dg we can find the eigenvalues directly from (5.12). They are

$$g_z + g_{\bar{z}} \quad \text{and} \quad g_z - g_{\bar{z}}. \tag{5.14}$$

Using (5.13) and (5.14) we compute these eigenvalues to lowest order. They are

$$\begin{aligned} &\text{(a)} \quad g_z + g_{\bar{z}} = 2p_u(0,0,0)x^2 + \cdots \\ &\text{(b)} \quad g_z - g_{\bar{z}} = -(n-1)q(0,0,0)x^{n-2} + \cdots, \end{aligned} \tag{5.15}$$

assuming that $n \geq 5$. It follows that the signs of the eigenvalues are determined by $p_u(0,0,0)$ and $-q(0,0,0)x$, as recorded in Table 5.2.

Finally we consider the $\mathbf{Z}_2(\zeta\kappa)$ solutions which appear as a distinct orbit of solutions only when n is even. The eigenspaces of $\zeta\kappa$ corresponding to the eigenvalues 1 and -1 are, respectively, $\mathbb{R}\{e^{i\pi/n}\}$ and $\mathbb{R}\{ie^{i\pi/n}\}$. Using (5.13), the eigenvalues of $(dg)_z$ where $z = xe^{i\pi/n}$ are

$$\begin{aligned} &\text{(a)} \quad g_z|_{z=xe^{i\pi/n}} = p + x^2p_u - np_vx^n - q_ux^n + nq_vx^{2n-2} \\ &\text{(b)} \quad g_{\bar{z}}|_{z=xe^{i\pi/n}} = [p_ux^2 - np_vx^n - (n-1)qx^{n-2} - q_ux^n + nq_vx^{2n-2}]e^{2\pi i/n}. \end{aligned} \tag{5.16}$$

Using (5.12) we compute

$$\begin{aligned} &\text{(a)} \quad (dg)_z(e^{i\pi/n}) \\ &\qquad = [p - (n-1)qx^{n-2} + 2x^2p_u - 2np_vx^n - 2q_ux^n + 2nq_vx^{2n-2}]e^{i\pi/2}, \\ &\text{(b)} \quad (dg)_z(ie^{i\pi/n}) = [p + (n-1)qx^{n-2}]ie^{i\pi/n}. \end{aligned} \tag{5.17}$$

Since $p = x^{n-2}q$ along the $\mathbf{Z}_2(\zeta\kappa)$ solution branch (Table 5.1) and $n \geq 5$, the eigenvalues of $(dg)_z$ are

$$\begin{aligned} &\text{(a)} \quad 2p_u(0,0,0)x^2 + \cdots \\ &\text{(b)} \quad nq(0,0,0)x^{n-2} + \cdots, \end{aligned} \tag{5.18}$$

giving the last entry in Table 5.2.

Thus we have shown that for $n \geq 5$, if we assume

$$p_\lambda(0) \neq 0, \qquad p_u(0) \neq 0, \quad \text{and} \quad q(0) \neq 0 \tag{5.19}$$

then the bifurcation diagram of $g = 0$ is determined. For each n there are eight possible diagrams, depending on the signs of the terms in (5.19). To reduce the complexity we assume that the trivial solution $z = 0$ is stable subcritically, that is, that $p_\lambda(0) < 0$. The remaining possibilities for the bifurcation diagrams are drawn in Figures 5.3 [n odd] and 5.4 [n even]. These diagrams are constructed from the data in Table 5.2, along with a final observation. When n is even, $\zeta^{n/2} \in \mathbf{D}_n$ acts as -1 on $\mathbb{C}$. Hence solutions z and $-z$ to $g = 0$ lie on the same orbit of solutions. This fact accounts for the restriction $x > 0$ when n is even.

Remarks 5.1.

(a) There is a remarkable parallel between the generic bifurcation diagrams with $\mathbf{D}_n$ symmetry when n is odd and when n is even (as long as $n \geq 5$), despite

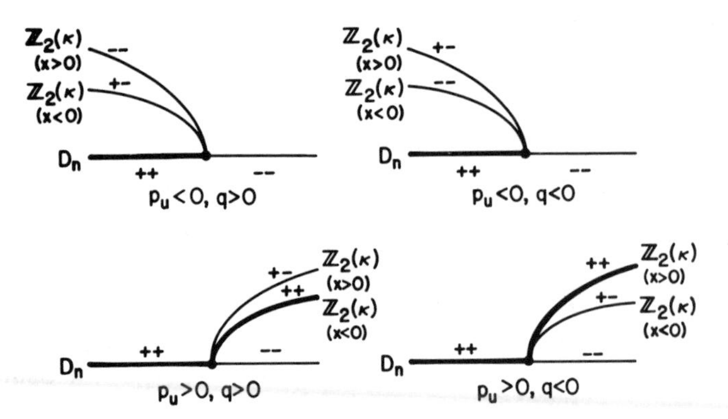
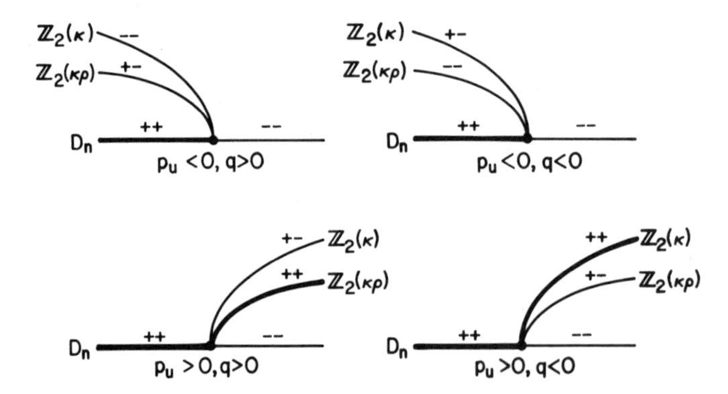

Figure 5.3. Bifurcation diagrams for $\mathbf{D}_n$ symmetry when $p_\lambda(0) < 0$, n odd, $n \geq 5$.

Figure 5.4. Bifurcation diagrams for $\mathbf{D}_n$ symmetry when $p_\lambda(0) < 0$, n even, $n \geq 6$.

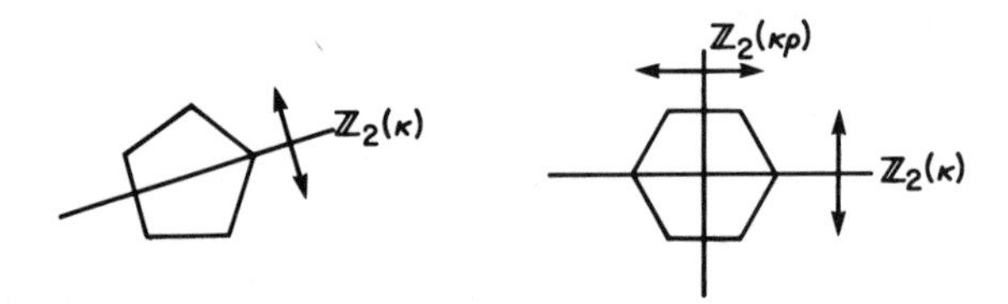

Figure 5.5. Geometric differences in solutions to a PDE with $\mathbf{D}_n$ symmetry.

the differences in the calculation. In particular, the diagrams for n even may be obtained from those for n odd by replacing $\mathbf{Z}_2(\kappa)$ $(x > 0)$ by $\mathbf{Z}_2(\kappa)$, and $\mathbf{Z}_2(\kappa)$ $(x < 0)$ by $\mathbf{Z}_2(\zeta\kappa)$. In spite of this apparent similarity, there is a subtle difference which is captured by the isotropy subgroups. Imagine a solution to a $\mathbf{D}_n$-equivariant PDE posed on the interior of a regular n-gon in $\mathbb{R}^2$. Suppose that there is a $\mathbf{D}_n$-invariant steady state which bifurcates to a solution which breaks the $\mathbf{D}_n$ symmetry. As we have shown, generically we expect exactly two different orbits of solutions to bifurcate, and both should be super- or subcritical together. However, when n is odd we expect both orbits of solutions to be invariant under a reflection in a line joining a vertex of the n-gon to the midpoint of the opposite edge; see Figure 5.5. On the other hand, when n is even, we expect one orbit of solutions to be invariant under reflection in a line joining opposite vertices, and another to be invariant under reflection in a line joining opposite midpoints of edges. See Figure 5.5.

EXERCISE

5.1. Compute the branching equations for generic $\mathbf{D}_4$ bifurcation.

§6.† Subgroups of **SO**(3)

We now embark upon an extended example, applying techniques developed previously to the orthogonal group $\mathbf{O}(3)$ and the special orthogonal group $\mathbf{SO}(3)$. Recall that $\mathbf{SO}(3)$ is the group of all orthogonal 3×3 matrices over $\mathbb{R}$ of determinant 1. In this section we discuss the closed subgroups of $\mathbf{SO}(3)$. We classify them, describe specific realizations, show that the finite ones have disjoint union decompositions, and list containment relations between them. In §7 we discuss the irreducible representations of $\mathbf{SO}(3)$. In §8 we find (for all irreducible representations) the dimensions of the fixed-point subspaces of closed subgroups of $\mathbf{SO}(3)$, and list the isotropy subgroups of $\mathbf{SO}(3)$ with one-dimensional fixed-point subspaces. These results may be found in Michel [1980], although we follow the presentation by Ihrig and Golubitsky [1984]. In §9 we extend the results to $\mathbf{O}(3)$. We do not prove everything in detail. In particular, results from Lie theory that would require substantial development—and yet are well known—are just stated along with appropriate references.

(a) Classification

We now describe the closed subgroups of $\mathbf{SO}(3)$. Geometrically, $\mathbf{SO}(3)$ is the group of orientation-preserving rigid motions of $\mathbb{R}^3$ that fix the origin. Its closed subgroups have nice geometric interpretations in terms of symmetries of subsets of $\mathbb{R}^3$. Choose a plane P in $\mathbb{R}^3$ and an axis A orthogonal to P. The subgroup of transformations leaving P invariant consists of rotations about A together with reflections through lines in P (combined with reversals in the sense of A to yield elements of $\mathbf{SO}(3)$; see the following discussion). This group is isomorphic to $\mathbf{O}(2)$. If we require the sense of A to be preserved this is reduced to the special orthogonal group in two dimensions, or the circle group, $\mathbf{SO}(2)$.

The symmetries of a regular n-gon lying in P yield a subgroup of $\mathbf{O}(2)$ isomorphic to the dihedral group $\mathbf{D}_n$. This consists of rotations through $2k\pi/n$

 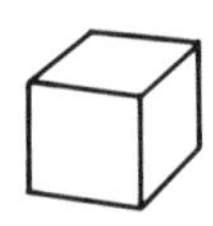 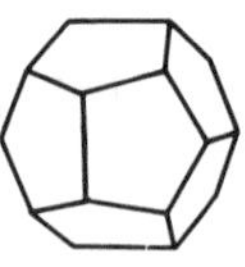

Figure 6.1. The regular polyhedra in $\mathbb{R}^3$, which give rise to exceptional subgroups of $\mathbf{SO}(3)$.

about A ($k = 0, 1, \ldots, n-1$) plus flips about symmetry axes of the n-gon (again combined with reversals of the sense of A). Preserving the sense of A restricts this to the cyclic group $\mathbf{Z}_n$, the rotations in $\mathbf{D}_n$. (We allow $n = 2$ in this description.) We call $\mathbf{O}(2)$, $\mathbf{SO}(2)$, $\mathbf{D}_n$, and $\mathbf{Z}_n$ the *planar* subgroups of $\mathbf{SO}(3)$.

In addition there are the "rotational" symmetry groups $\mathbb{T}, \mathbb{O}, \mathbb{I}$ of the regular tetrahedron, octahedron (or cube), and icosahedron (or dodecahedron); see Figure 6.1. By "rotational" we mean that reflections (in planes in $\mathbb{R}^3$) are excluded, since these do not lie in $\mathbf{SO}(3)$. We call these the *exceptional* groups and refer to them individually as the *tetrahedral*, *octahedral*, and *icosahedral* groups. (For purposes of visualization it is often convenient to replace the octahedron by the cube and the icosahedron by the dodecahedron—which does not change the group of symmetries—but the names *octahedral* and *icosahedral* are traditional.) Note that $\mathbf{O}(2)$ and $\mathbf{SO}(2)$ are infinite groups (of dimension 1), whereas $\mathbf{D}_n$, $\mathbf{Z}_n$, $\mathbb{T}$, $\mathbb{O}$, and $\mathbb{I}$ are finite, of orders $2n$, n, 12, 24, 60, respectively.

It is clear on geometric grounds that for the planar subgroups a different choice of the plane P and the sense of the axis A just yields a conjugate subgroup. Similarly, a different orientation of the relevant regular polyhedron yields a conjugate exceptional subgroup. These considerations motivate the following:

Theorem 6.1. *Every closed subgroup of* $\mathbf{SO}(3)$ *is conjugate to one of* $\mathbf{SO}(3)$, $\mathbf{O}(2)$, $\mathbf{SO}(2)$, $\mathbf{D}_n$ ($n \geq 2$), $\mathbf{Z}_n$ ($n \geq 2$), $\mathbb{T}$, $\mathbb{O}$, $\mathbb{I}$, *and* $\mathbb{1}$.

The main idea, which is classical, is to show that any closed subgroup acts as symmetries of a suitable geometric figure and then to classify these figures by exploiting certain arithmetic constraints given by counting arguments. A proof is given in Dubrovin, Fomenko, and Novikov [1984], p. 189. See also Exercises 6.1–6.4.

(b) Realizations

We now provide specific realizations of the planar subgroups of $\mathbf{SO}(3)$, in a form suitable for computations. Let (x, y, z) be coordinates in $\mathbb{R}^3$. The planar subgroups all leave a plane invariant, and after a suitable conjugacy we may assume this to be the (x, y)-plane. Then $\mathbf{SO}(2)$ is the group of rotations in this plane, leaving the z-axis fixed. $\mathbf{Z}_n$ is the unique cyclic subgroup of order n contained in $\mathbf{SO}(2)$.

As a group, $\mathbf{O}(2)$ is obtained from $\mathbf{SO}(2)$ by adjoining an element of order 2. It is tempting to adjoint the reflection $(x, y) \mapsto (-x, y)$ in the (x, y)-plane by letting this reflection also fix the z-axis. However, $(x, y, z) \mapsto (-x, y, z)$ has determinant -1 so does not lie in $\mathbf{SO}(3)$. We can avoid this problem by extending the reflection so that it sends z to $-z$. Thus $\mathbf{O}(2)$ is realized as a subgroup of $\mathbf{SO}(3)$ as follows. Let $\gamma \in \mathbf{O}(2)$ be written as a 2×2 matrix, and define

$$\gamma \cdot (x, y, z) = (\gamma \cdot (x, y), \det(\gamma) z).$$

Finally, the dihedral groups $\mathbf{D}_n$ occur uniquely as subgroups of $\mathbf{O}(2)$.

Coordinate forms of the exceptional subgroups may be found in Coxeter [1963]. We do not require them here.

(c) Disjoint Union Decompositions

The dihedral groups $\mathbf{D}_n$ and the exceptional subgroups all have disjoint union decompositions (Definition 2.4) into cyclic subgroups. These decompositions are most useful when determining the dimensions of fixed-point subspaces. They are rooted in the geometry of regular polyhedra.

Lemma 6.2.

(a) $\mathbf{D}_n = \dot{\cup}^n \mathbf{Z}_2 \dot{\cup} \mathbf{Z}_n$.
(b) $\mathbb{O} = \dot{\cup}^3 \mathbf{Z}_4 \dot{\cup}^4 \mathbf{Z}_3 \dot{\cup}^6 \mathbf{Z}_2$.
(c) $\mathbb{T} = \dot{\cup}^4 \mathbf{Z}_3 \dot{\cup}^3 \mathbf{Z}_2$.
(d) $\mathbb{I} = \dot{\cup}^6 \mathbf{Z}_5 \dot{\cup}^{10} \mathbf{Z}_3 \dot{\cup}^{15} \mathbf{Z}_2$.

Here $\dot{\cup}^k \mathbf{Z}_l$ *denotes a disjoint union of k copies of subgroups all conjugate* (*in* $\mathbf{SO}(3)$) *to* $\mathbf{Z}_l$.

PROOF.

(a) The dihedral group has $2n$ elements. Of these, n lie in the cyclic subgroup $\mathbf{Z}_n$. Each of the remaining n elements acts as a reflection of the (x, y)-plane and generates a subgroup conjugate (in $\mathbf{SO}(3)$) to $\mathbf{Z}_2$.

(b) The first step for the exceptional groups is to observe that every rotation in $\mathbf{SO}(3)$ has an axis of symmetry. Each axis intersects the invariant regular polyhedron in either the midpoint of a face, the midpoint of an edge, or a vertex. Think of the octahedral group $\mathbb{O}$ as the symmetries of a cube. The rotational symmetries whose axes meet a face form a subgroup conjugate to $\mathbf{Z}_4$. Each such axis meets two faces; since the cube has six faces we find three copies of $\mathbf{Z}_4$ in $\mathbb{O}$. The rotational symmetries of the cube whose axes meet the center of an edge generate a subgroup conjugate to $\mathbf{Z}_2$, and there are six such axes. Finally the rotational symmetries whose axes meet a vertex form a subgroup conjugate to $\mathbf{Z}_3$, and there are four such axes. This yields a total of

$$1 + 3\cdot(4 - 1) + 6\cdot(2 - 1) + 4(3 - 1) = 24$$

rotations, which equals the order of $\mathbb{O}$. Hence all elements of the octahedral group are accounted for.

(c, d) The proofs for $\mathbb{T}$ and $\mathbb{I}$ are similar and are left to the reader. □

(d) Containment Relations

Finally we discuss containment relations between conjugacy classes of subgroups of $\mathbf{SO}(3)$. We begin with a definition.

Definition 6.3. Let H_1, H_2 be subgroups of a group Γ. We say that the conjugacy class of (subgroups conjugate to) H_1 contains that of H_2 if there exists a subgroup of Γ, conjugate to H_1, which contains H_2. Then every subgroup conjugate to H_1 contains some subgroup conjugate to H_2. We write $H_1 > H_2$ or $H_2 < H_1$ to denote containment of conjugacy classes in this sense.

It is easy to determine all containment relations for the planar subgroups of $\mathbf{SO}(3)$. They are:

$$\begin{aligned}
&\text{(a)}\quad \mathbf{Z}_n < \mathbf{D}_n < \mathbf{O}(2) \qquad (n \geq 2).\\
&\text{(b)}\quad \mathbf{Z}_n < \mathbf{Z}_m \text{ and } \mathbf{D}_n < \mathbf{D}_m \qquad (\text{if } n \text{ divides } m).\\
&\text{(c)}\quad \mathbf{Z}_2 < \mathbf{D}_n \qquad (n \geq 2).\\
&\text{(d)}\quad \mathbf{Z}_n < \mathbf{SO}(2) < \mathbf{O}(2) \qquad (n \geq 2).
\end{aligned} \tag{6.1}$$

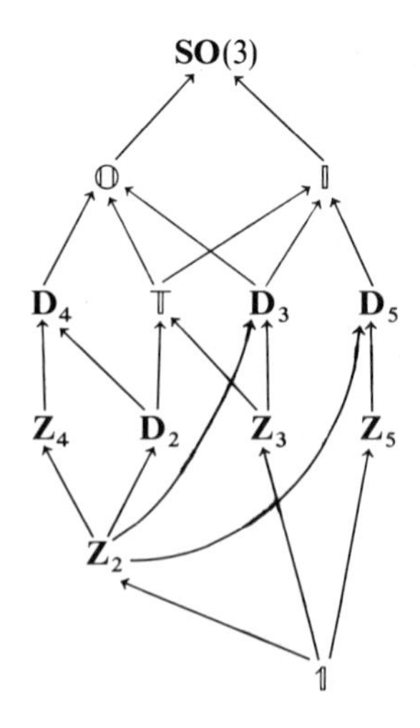

Figure 6.2. Containment relations for exceptional subgroups of $\mathbf{SO}(3)$.

The containment relations for the exceptional subgroups—with one exception—are determined directly from (6.1) and Lemma 6.2 and are summarized in Figure 6.2. The exception is that $\mathbb{T}$ is contained in $\mathbb{O}$ and $\mathbb{I}$. The easiest way to see this is to embed the regular tetrahedron in a cube or dodecahedron. This induces an embedding of the corresponding symmetry groups.

Exercises

These exercises lead to a proof of Theorem 6.1.

6.1. Let Γ be a proper closed subgroup of $\mathbf{SO}(3)$. Show that if Γ is infinite then there is a line L that is invariant under every $\gamma \in \Gamma$. If Γ fixes every point on L show that Γ is conjugate to $\mathbf{SO}(2)$. If some elements of Γ act nontrivially show that Γ is conjugate to $\mathbf{O}(2)$.

6.2. Let Γ be a finite subgroup of $\mathbf{SO}(3)$, of order g. For each $\gamma \in \Gamma$ define the *poles* of γ to be the two points in which the axis of γ meets the unit sphere. Show that Γ permutes (faithfully) the set of poles of the $g - 1$ nonidentity elements of Γ.

6.3. If P is a pole define ν_P to be the order of the isotropy subgroup Σ_P of P in Γ. Prove that Σ_P is cyclic and that the orbit of P under Γ has size g/ν_P. Calculate in two different ways the number of pairs (γ, P) where $1 \neq \gamma \in \Gamma$, P is a pole, and γ fixes P. Hence show that

$$2 - \frac{2}{g} = \sum_i \left(1 - \frac{1}{\nu_i}\right) \tag{6.2}$$

where i parametrizes the orbits of Γ on the set $\mathscr{P}$ of poles.

6.4. Using (6.2) show that there are either 2 or 3 orbits of Γ on $\mathscr{P}$. If there are two orbits show that Γ is conjugate to some $\mathbf{Z}_k$. If there are three orbits choose notation so that $\nu_1 \leq \nu_2 \leq \nu_3$. Prove that $\nu_1 = 2$ and $\nu_2 = 2$ or 3. If $\nu_2 = 2$ prove that Γ is conjugate to some $\mathbf{D}_k$. If $\nu_2 = 3$ prove that $\nu_3 = 3, 4$, or 5 and hence that $g = 12, 24$, or 60. Finally prove that in these cases Γ is conjugate to $\mathbb{T}$, $\mathbb{O}$, or $\mathbb{I}$, respectively.

§7.† Representations of SO(3) and O(3): Spherical Harmonics

It is a classical result that $\mathbf{SO}(3)$ has precisely one irreducible representation, up to isomorphism, in each odd dimension $2l + 1$, and no others. See Wigner [1959]. It easily follows that $\mathbf{O}(3)$ has precisely two distinct irreducible representations in each odd dimension, and no others. These representations may be realized in terms of special functions known as *spherical harmonics*. They may also be treated from a more abstract point of view, and much of the modern literature takes this approach. In this section we derive, by "bare hands" methods, the basic results on these irreducible representations. We try to steer a middle course between the classical and the modern treatments, to show how they are related.

(a) Classical Spherical Harmonics

Spherical harmonics often occur as linearized eigenfunctions in problems with spherical symmetry. They are normally defined in one specific setting: separation of variables for Laplace's equation in spherical polar coordinates. We sketch the theory here: see Duff and Naylor [1966] for further details.

Consider Laplace's equation in $\mathbb{R}^3$:

$$\Delta u \equiv u_{xx} + u_{yy} + u_{zz} = 0 \tag{7.1}$$

where $u = u(x, y, z)$ and subscripts denote partial derivatives. Transform to spherical polar coordinates (r, θ, φ), with $r \geq 0, 0 \leq \theta < 2\pi, 0 \leq \varphi \leq \pi$, where

$$x = r\cos\theta\sin\varphi$$

$$y = r\sin\theta\sin\varphi$$

$$z = r\cos\varphi.$$

Then (7.1) becomes

$$r^2 u_{rr} + 2ru_r + (\operatorname{cosec}^2\varphi)u_{\theta\theta} + u_{\varphi\varphi} + (\cot\varphi)u_\varphi = 0. \tag{7.2}$$

Separate the variables by assuming

$$u(r, \theta, \varphi) = R(r)\Theta(\theta)\Phi(\varphi).$$

Then R, Θ, Φ satisfy the ODEs

$$\begin{aligned} &\text{(a)} \quad r^2R'' + 2rR' - l(l+1)R = 0 \\ &\text{(b)} \quad \Theta'' + m^2\Theta = 0 \\ &\text{(c)} \quad (\sin^2\varphi)\Phi'' + (\sin\varphi\cos\varphi)\Phi' + (l(l+1)\sin^2\varphi - m^2) = 0. \end{aligned} \tag{7.3}$$

Here l and m are constants. We seek solutions $u(\theta, \varphi)$ that are independent of r. By (7.3b) we have

$$\Theta(\theta) = A\cos m\theta + B\sin m\theta$$

and for consistency at $\theta = 0, 2\pi$ we require $m \in \mathbb{Z}$. We may assume $m \geq 0$. Equation (7.3c) is the associated Legendre equation. It has a solution bounded on $0 \leq \varphi \leq \pi$ only when $l \in \mathbb{Z}$, and the solution is then $P_l^m(\cos\varphi)$ where P_l^m is the associated Legendre function. We may assume $m \leq l$, because P_l^m vanishes for $m > l$. Thus solutions are spanned by the functions

$$\begin{aligned} &\cos m\theta P_l^m(\cos\varphi) \\ &\sin m\theta P_l^m(\cos\varphi) \end{aligned} \tag{7.4}$$

where $0 \le m \le l$. Since $\sin 0 = 0$ there are $2l + 1$ such functions for a given l. They are known as *surface harmonics* of degree l. If multiplied by r^l they are called *spherical harmonics* of degree l, and they are then polynomials in x, y, z of degree l.

(b) Connection with Representations

To approach these polynomials from a representation-theoretic viewpoint, let W_l be the space of all homogeneous polynomials $p: \mathbb{R}^3 \to \mathbb{R}$ of degree l. Now **SO**(3) acts on W_l by

$$\gamma p(x) = p(\gamma^{-1}x). \tag{7.5}$$

For $l = 0$ this yields the trivial representation on $\mathbb{R}$; for $l = 1$ the natural representation on $\mathbb{R}^3$. For $l \ge 2$ it is not irreducible. To see why, let J_l be the subspace of polynomials in W_l that are multiples (by polynomials in W_{l-2}) of

$$\rho \equiv x^2 + y^2 + z^2.$$

That is, $J_l = \rho W_{l-2}$. Since ρ is invariant under **SO**(3), so is J_l. Let V_l be the space spanned by the spherical harmonics of degree l, that is, the elements $p \in W_l$ for which $\Delta p = 0$, where Δ is the Laplacian. V_l is also invariant under **SO**(3).

Proposition 7.1.
(a) $W_l = J_l \oplus V_l$, *that is,* V_l *is an* **SO**(3)-*invariant complement to* J_l.
(b) $\dim V_l = 2l + 1$.

PROOF. Clearly $\Delta: W_l \to W_{l-2}$ is linear, so

$$\dim \ker \Delta + \dim \operatorname{Im} \Delta = \dim W_l. \tag{7.6}$$

Now $\dim J_l = \dim W_{l-2} \ge \dim \operatorname{Im} \Delta$ since $\operatorname{Im} \Delta \subset W_{l-2}$. If we can show that $\ker \Delta \cap J_l = \{0\}$ then $\dim J_l \le \dim \operatorname{Im} \Delta$, since J_l is contained in a complement to $\ker \Delta$, whose dimension is $\le \dim \operatorname{Im} \Delta$ by (7.6). From (7.6) it follows that $W_l = J_l \oplus \ker \Delta$, which is part (a). Part (b) follows at once since $\dim W_l = \binom{l+2}{2} = \frac{1}{2}(l + 1)(l + 2)$, and $\dim V_l = \dim W_l - \dim J_l = \dim W_l - \dim W_{l-2}$.

It remains to prove that $J_l \cap \ker \Delta = \{0\}$. Assume that $0 \ne g \in J_l \cap \ker \Delta$. Write $g = \rho^k f$ where $k > 1$, $f \ne 0$, and ρ does not divide f. Calculate

$$\Delta(\rho^k f) = 2k(2k + 2m + 1)\rho^{k-1} f + \rho^k \Delta f, \tag{7.7}$$

where $m = l - 2k$ is the degree of f. Since $\Delta g = 0$ it follows that

$$f = -\rho \Delta f / 2k(2k + 2m + 1)$$

whence ρ divides f, a contradiction. Hence $g = 0$. □

(c) Cartan Decomposition

We next describe what in representation theory is called the *Cartan decomposition* of V_l. Let

$$\mathbf{SO}(2) = \left\{ \begin{bmatrix} \cos\theta & -\sin\theta & 0 \\ \sin\theta & \cos\theta & 0 \\ 0 & 0 & 1 \end{bmatrix} : \theta \in \mathbf{S}^1 \right\}$$

be contained in $\mathbf{SO}(3)$. We can decompose V_l into irreducible subspaces for the action of $\mathbf{SO}(2)$ as follows. Since $W_l = \rho W_{l-2} \oplus V_l$ we can think of V_l as the quotient space $W_l/\rho W_{l-2}$. Let $w = x + iy$, where $(x, y, z) \in \mathbb{R}^3$. For $k = 0, 1, \ldots, l$ define H_k to be the subspace of this quotient space spanned by the cosets of

$$\begin{aligned} s_k &= z^{l-k}\operatorname{Re}(w^k) \\ t_k &= z^{l-k}\operatorname{Im}(w^k), \end{aligned} \tag{7.8}$$

if $k \geq 1$. If $k = 0$ define H_0 to be the span of z^l. We claim that

$$\dim H_k = \begin{cases} 2 & (k > 0) \\ 1 & (k = 0). \end{cases} \tag{7.9}$$

Clearly this holds for $k = 0$. We must show s_k and t_k are linearly independent modulo ρW_{l-2}. So suppose that

$$as_k + bt_k \in \rho W_{l-2}$$

for $a, b, \in \mathbb{R}$. That is,

$$z^{l-k}(a\operatorname{Re}(w^k) + b\operatorname{Im}(w^k)) = \rho f$$

where $f \in W_{l-2}$. Then z^{l-k} divides f, so

$$a\operatorname{Re}(w^k) + b\operatorname{Im}(w^k) = \rho h$$

for some polynomial h. But the left-hand side is harmonic by the Cauchy–Riemann equations, whereas the right-hand side is not (by the proof of Proposition 7.1) unless $h = 0$. Therefore $\operatorname{Re}(w^k)$ and $\operatorname{Im}(w^k)$ are linearly dependent; this is absurd since they involve distinct monomials in x, y.

We now have the following:

Proposition 7.2. *$V_l = H_0 \oplus H_1 \oplus \cdots \oplus H_l$, where V_l is identified with $W_l/\rho W_{l-2}$. Moreover,* $\mathbf{SO}(2)$ *acts on H_k by rotation through $k\theta$.*

Proof. That $\theta \in \mathbf{SO}(2)$ acts on H_k by rotation through $k\theta$ is obvious, since $e^{i\theta}$ has this action on w^k and leaves z fixed. Since these representations are distinct, the sum $H_0 + \cdots + H_l$ is direct. By (7.7) this sum has dimension $2l + 1$, so is the whole of V_l. □

EXAMPLE. $l = 3$, $\dim V_l = 7$. To find a basis we seek harmonic cubics of the form

$$\begin{aligned}
k = 0:&\quad z^3 + \rho u_0\\
k = 1:&\quad z^2 x + \rho u_1\\
&\quad z^2 y + \rho v_1\\
k = 2:&\quad z(x^2 - y^2) + \rho u_2\\
&\quad z(2xy) + \rho v_2\\
k = 3:&\quad \left.\begin{matrix} x^3 - 3xy^2 \\ 3x^2 y - y^3 \end{matrix}\right\} \quad \text{already harmonic.}
\end{aligned}$$

Here the us and vs are linear. Now for linear u we have $\Delta(\rho u) = 10u$ by (7.7). So if g is cubic, then

$$0 = \Delta(g + \rho u) = \Delta g + 10u$$

leads to $u = -\Delta g/10$. Hence it is easy to obtain the basis

$$\begin{aligned}
k = 0:&\quad z^3 - \tfrac{3}{5}\rho z\\
k = 1:&\quad z^2 x - \tfrac{1}{5}\rho x\\
&\quad z^2 y - \tfrac{1}{5}\rho y\\
k = 2:&\quad z(x^2 - y^2)\\
&\quad xyz\\
k = 3:&\quad x^3 - 3xy^2\\
&\quad 3x^2 y - y^3.
\end{aligned}$$

Note that the **SO**(2)-action is obviously by rotations $k\theta$ when $k = 0, 2, 3$, and when $k = 1$ the two cubics are xq, yq where $q = z^2 - \frac{1}{5}\rho$ is **SO**(2)-invariant. So the action is the same as on the (x, y)-plane in $\mathbb{R}^3$, that is, the standard action.

(d) Absolute Irreducibility

We are now in a position to prove the key result:

Theorem 7.3. *The representation of* **SO**(3) *on* V_l *is absolutely irreducible.*

PROOF. Let $B: V_l \to V_l$ be linear, and commute with **SO**(3). Then B commutes with **SO**(2). Since the representations of **SO**(2) in Proposition 7.2 are distinct, it follows from XII, Theorem 3.5, that B leaves each subspace H_k invariant (and commutes with the **SO**(2) action on each H_k). Therefore B can be written, in relation to the decomposition of V_l as $H_0 \oplus \cdots \oplus H_l$, in the form

$$B = \begin{bmatrix} \lambda_0 & & & & & \\ & \lambda_1 & -\mu_1 & & 0 & \\ & \mu_1 & \lambda_1 & & & \\ & & & \ddots & & \\ & 0 & & & \lambda_l & -\mu_l \\ & & & & \mu_l & \lambda_l \end{bmatrix}$$

for $\lambda_k, \mu_k \in \mathbb{R}$. We must show that the λ_k are equal and the μ_k are zero.

Let $\gamma(x, y, z) = (x, -y, -z)$. Then γ acts as an element of $\mathbf{O}(2) \sim \mathbf{SO}(2)$ on H_k. In particular, $\gamma s_k = (-1)^{l-k} s_k$, $\gamma t_k = (-1)^{l-k+1} t_k$, where s_k and t_k are defined in (7.8). Since B commutes with γ, it follows that $\mu_k = 0$, and B is diagonal.

To prove that the λ_k are equal, form the subspace

$$\Lambda = \bigoplus_{\lambda_k = \lambda_0} H_k.$$

Clearly $\Lambda \supset H_0$. Since B commutes with $\mathbf{SO}(3)$ we know that Λ is invariant under $\mathbf{SO}(3)$. We claim, however, that the only $\mathbf{SO}(3)$-invariant subspace of V_l containing H_0 is the whole of V_l, leading to the desired conclusion that $\Lambda = V_l$.

Now the minimal $\mathbf{SO}(3)$-invariant subspace containing H_0 is

$$Z = \mathbb{R}\{\gamma z^l + \rho W_{l-2} : \gamma \in \mathbf{SO}(3)\}.$$

From now on we work modulo ρW_{l-2}. Thus Z is just $\mathbb{R}\{\gamma z^l\}$. We claim that in fact the span of all γz^l, as γ runs through $\mathbf{SO}(3)$, is the whole of W_l (modulo ρW_{l-2}). To verify this, note that the permutation σ which sends $x \mapsto y \mapsto z \mapsto x$ is in $\mathbf{SO}(3)$ and sends z^l to x^l. Then the rotation $R_\alpha \in \mathbf{SO}(2) \subset \mathbf{SO}(3)$ sends it to $(x \cos \alpha - y \sin \alpha)^l$. Letting $c = \tan \alpha$ we see that Z contains all $(x + cy)^l$ ($c \in \mathbb{R}$), that is, all polynomials $x^i y^{l-i}$ for $i = 0, 1, \ldots, l$.

Next fix some i, $0 \le i \le l$, and apply σ^{-1} to get the polynomial $z^i x^{l-i}$. Apply R_α to get all polynomials $z^i(x + cy)^{l-i}$, that is, all $z^i x^j y^{l-i-j}$ for $0 \le i \le l$, $0 \le j \le l - i$. But these span all of W_l. This completes the proof. □

In fact, *every* irreducible representation of $\mathbf{SO}(3)$ is isomorphic to some V_l. The proof of this uses extra machinery (orthogonality of characters) and is therefore placed at the end of this section.

(e) Irreducible Representations of $\mathbf{O}(3)$

We can now describe all irreducible representations of $\mathbf{O}(3) = \mathbf{SO}(3) \oplus \mathbf{Z}_2^c$, where $\mathbf{Z}_2^c = \{\pm I\}$.

Proposition 7.4. *The irreducible representations of* $\mathbf{O}(3)$ *are precisely the representations on* V_l *given by*

$$\begin{aligned} \gamma \cdot p(x) &= p(\gamma x) \qquad \gamma \in \mathbf{SO}(3) \\ -I \cdot p(x) &= p(x) \end{aligned} \tag{7.10}$$

and

$$\begin{aligned} \gamma \cdot p(x) &= p(\gamma x) \qquad \gamma \in \mathbf{SO}(3) \\ -I \cdot p(x) &= -p(x). \end{aligned} \tag{7.11}$$

PROOF. Let V be an irreducible representation of $\mathbf{O}(3)$. Decompose V into ± 1 eigenspaces for the action of $-I$,

$$V = V_1 \oplus V_{-1}.$$

Since $\mathbf{Z}_2^c$ commutes with $\mathbf{SO}(3)$, each $V_{\pm 1}$ is an invariant subspace. So either $V = V_1$ (when $-I$ acts as the identity) or $V = V_{-1}$ (when $-I$ acts as minus the identity). In either case, V is clearly irreducible as a representation of $\mathbf{SO}(3)$. By Proposition 7.5 later, $V \cong V_l$ for some l. Thus we obtain the two types of representation listed. □

We call (7.10 and 7.11), respectively, the *plus* and *minus* representations of $\mathbf{O}(3)$ on V_l.

In applications, the usual way that $\mathbf{O}(3)$ acts is induced from the natural action on $\mathbb{R}^3$, where $-I \cdot (x, y, z) = (-x, -y, -z)$. This leads to the representation on V_l whose sign is $(-1)^l$, since $p(-x, -y, -z) = (-1)^l p(x, y, z)$ when p is homogeneous of degree l.

(f)* Completeness

It remains to prove that every irreducible representation of $\mathbf{SO}(3)$ is isomorphic to some V_l. The most direct way is to appeal to properties of characters. The *character* of a representation V of a compact Lie group Γ is the function $\chi: \Gamma \to \mathbb{R}$ given by $\chi(\gamma) = \operatorname{tr} \rho_\gamma$, where $\rho_\gamma(x) = \gamma x$. Note that $\chi(\delta^{-1}\gamma\delta) = \operatorname{tr}(\rho_\delta^{-1}\rho_\gamma\rho_\delta) = \chi(\gamma)$ so χ is constant on conjugacy classes. There is an inner product on characters defined by

$$\langle \chi_1, \chi_2 \rangle = \int_\Gamma \chi_1(\gamma)\chi_2(\gamma)\, d\gamma$$

with the property that if χ_1 and χ_2 are characters of nonisomorphic irreducible representations, then $\langle \chi_1, \chi_2 \rangle = 0$, whereas $\langle \chi_1, \chi_1 \rangle > 0$. See Adams [1969], p. 49. Using this property, known as *orthogonality of characters*, we can prove the following:

Theorem 7.5. *Every irreducible representation of* $\mathbf{SO}(3)$ *is isomorphic to some* V_l.

PROOF. We first compute the character χ_l of V_l. This is constant on conjugacy classes. Every element of $\mathbf{SO}(3)$ is a rotation by θ about some axis, hence conjugate to $R_\theta \in \mathbf{SO}(2)$. Also R_θ and $R_{-\theta}$ are conjugate. By Proposition 7.2

$$\chi_l(R_\theta) = 1 + 2\cos\theta + \cdots + 2\cos l\theta.$$

Let χ be the character of an irreducible representation V, not isomorphic to any V_l. Then $\chi \neq \chi_l$, so $\langle \chi, \chi_l \rangle = 0$ for all l. Since χ is a smooth even function, in the sense that $\chi(R_\theta) = \chi(R_{-\theta})$, it can be expanded in a convergent cosine Fourier series

$$\chi(R_\theta) = \sum_{l=0}^{\infty} c_l \cos l\theta = \sum_{l=0}^{\infty} \tfrac{1}{2} c_l (\chi_l - \chi_{l-1}).$$

Since $\langle \chi, \chi_m \rangle = 0$ we have

$$0 = \langle \chi, \chi_m \rangle = \tfrac{1}{2} \sum c_l \langle \chi_l - \chi_{l-1}, \chi_m \rangle$$
$$= \begin{cases} \tfrac{1}{2}(c_m - c_{m+1}) \langle \chi_m, \chi_m \rangle & m > 0 \\ c_0 - \tfrac{1}{2} c_1 \langle \chi_0, \chi_0 \rangle & m = 0. \end{cases}$$

It follows that

$$\chi(R_\theta) = c_0 \left[\chi_0 + \sum_{l=1}^{\infty} (\chi_l - \chi_{l+1}) \right] \equiv 0,$$

which contradicts $\langle \chi, \chi \rangle \neq 0$. □

Exercises

7.1. This exercise relates our abstract definition of spherical harmonics to the usual classical coordinate system.

Let U_l ($l = 0, 1, \ldots$) denote the complexification of the space V_l of spherical harmonics of order l with its natural action of $\mathbf{O}(3)$. Let (r, θ, φ), where $0 \leq \theta \leq \pi$, $0 \leq \varphi < 2\pi$ denote spherical polar coordinates, related to cartesian coordinates (x, y, z) on $\mathbb{R}^3$ by

$$x = r\sin\theta\cos\varphi, \qquad y = r\sin\theta\sin\varphi, \qquad z = r\cos\theta.$$

Show that an arbitrary spherical harmonic of order l can be written as

$$\sum_{m=-l}^{l} z_m Y_{l,m}(\theta, \varphi)$$

where $z_m \in \mathbb{C}$, $z_{-m} = \bar{z}_m$, and

$$Y_{l,m}(\theta, \varphi) = P_{l,m}(\cos\theta) e^{im\varphi}, \qquad (-l \leq m \leq l),$$

the $P_{l,m}$ being polynomials. (These are the *associated Legendre polynomials*; see Whittaker and Watson [1948].)

7.2. Compute (up to a constant factor) the polynomials $Y_{l,m}$ for $\lambda = 0, 1, 2, 3$. (Classically, the constant factor is determined by a suitable normalization.)

7.3. Show that the action of $\mathbf{O}(2)$ on U_l is given by

$$\theta \cdot (z_{-l}, \ldots, z_l) = (e^{-il\theta} z_{-l}, \ldots, e^{ir\theta} z_r, \ldots, e^{il\theta} z_l),$$
$$\kappa \cdot (z_{-l}, \ldots, z_l) = (z_l, \ldots, (-1)^{l-r} z_{-r}, \ldots, z_{-l}).$$

§8.† Symmetry-Breaking from SO(3)

We now describe, for each irreducible representation V_l, the isotropy subgroups $\Sigma \subset \mathbf{SO}(3)$ for which $\dim \operatorname{Fix}(\Sigma) = 1$. The equivariant branching lemma implies that generically an $\mathbf{SO}(3)$-equivariant bifurcation problem will have branches that break symmetry to Σ. In fact, we calculate $\dim \operatorname{Fix}(\Sigma)$ for *all* closed subgroups Σ of $\mathbf{SO}(3)$, then use this information to find the maximal isotropy subgroups and those with $\dim \operatorname{Fix}(\Sigma) = 1$.

(a) Dimensions of Fixed-Point Subspaces

The first step is to compute $\dim \operatorname{Fix}(\Sigma)$ for all closed subgroups $\Sigma \subset \mathbf{SO}(3)$. Recall from Proposition 7.2 that we may write the space V_l of spherical harmonics as a direct sum of irreducible subspaces for $\mathbf{SO}(2) \subset \mathbf{SO}(3)$, via the Cartan decomposition

$$V_l = H_0 \oplus H_1 \cdots \oplus H_l, \tag{8.1}$$

where $\dim H_0 = 1$, and $\dim H_k = 2$ $(k \geq 1)$. The action of $\theta \in \mathbf{SO}(2)$ on H_k is by rotation through $k\theta$.

From (8.1) we obtain the following results:

$$\begin{aligned} &\text{(a)} \quad \dim \operatorname{Fix}_{V_l}(\mathbf{SO}(2)) = 1, \\ &\text{(b)} \quad \dim \operatorname{Fix}_{V_l}(\mathbf{Z}_m) = 2[l/m] + 1, \end{aligned} \tag{8.2}$$

where $[x]$ is the greatest integer less than or equal to x. The formula (8.2(a)) follows directly from (8.1) since the only vectors in V_l fixed by $\mathbf{SO}(2)$ are those in H_0. To verify (8.2(b)) recall that $\mathbf{Z}_m$ is generated by rotation through $2\pi/m$. Therefore $\mathbf{Z}_m$ fixes a nonzero vector in H_k if and only if m divides k. There are $[l/m]$ integers k between 1 and l such that m divides k. Since H_0 is also fixed by $\mathbf{Z}_m$, (7.2(b)) follows.

Let $d(\Sigma) = \dim \operatorname{Fix}(\Sigma)$. Then we may summarize our results as follows:

Theorem 8.1. *Let* $\mathbf{SO}(3)$ *act irreducibly on the space* V_l *of spherical harmonics of degree* l. *The dimensions of the fixed-point subspaces of closed subgroups are*:

$$\text{(a)} \quad d(\mathbf{Z}_m) = 2[l/m] + 1 \qquad (m \geq 1)$$

$$\text{(b)} \quad d(\mathbf{D}_m) = \begin{cases} [l/m] & (l \text{ odd}) \\ [l/m] + 1 & (l \text{ even}) \end{cases}$$

$$\text{(c)} \quad d(\mathbf{SO}(2)) = 1$$

$$\text{(d)} \quad d(\mathbf{O}(2)) = \begin{cases} 0 & (l \text{ odd}) \\ 1 & (l \text{ even}) \end{cases}$$

$$\text{(e)} \quad d(\mathbb{T}) = 2[l/3] + [l/2] - l + 1$$

(f) $d(\mathbb{O}) = [l/4] + [l/3] + [l/2] - l + 1$

(g) $d(\mathbb{I}) = [l/5] + [l/3] + [l/2] - l + 1.$

PROOF. Formulas (a) and (c) are restatements of (8.2). For the finite subgroups Σ of $\mathbf{SO}(3)$ we use the disjoint union decomposition in Lemma 6.2 together with Lemma 2.5, the corollary to the trace formula (2.5). For example, Lemma 6.2(a) states that

$$\mathbf{D}_m = \dot{\cup}^m \mathbf{Z}_2 \,\dot{\cup}\, \mathbf{Z}_m.$$

By (2.10)

$$\begin{aligned} d(D_m) &= \frac{1}{|\mathbf{D}_m|}\{m|\mathbf{Z}_2|d(\mathbf{Z}_2) + |\mathbf{Z}_m|d(\mathbf{Z}_m) - m \dim V_l\} \\ &= \frac{1}{2m}\{2m(2[l/2] + 1) + m(2[l/m] + 1) - m(2l + 1)\} \\ &= \tfrac{1}{2}\{4[l/2] + 2 + 2[l/m] + 1 - 2l - 1\} \\ &= 2[l/2] + [l/m] - l + 1 \\ &= \begin{cases} [l/m] & (l \text{ odd}) \\ [l/m] + 1 & (l \text{ even}). \end{cases} \end{aligned}$$

Similarly $d(\mathbb{T})$ is obtained using the disjoint union decomposition

$$\mathbb{T} = \dot{\cup}^4 \mathbf{Z}_3 \,\dot{\cup}^3 \mathbf{Z}_2$$

from Lemma 6.2(c). Now apply (2.10) to obtain

$$\begin{aligned} d(\mathbb{T}) &= \frac{1}{|\mathbb{T}|}\{4|\mathbf{Z}_3|d(\mathbf{Z}_3) + 3|\mathbf{Z}_2|d(\mathbf{Z}_2) - 6 \dim V_l\} \\ &= \tfrac{1}{12}\{12(2[l/3] + 1) + 6(2[l/2] + 1) - 6(2l + 1)\} \\ &= 2[l/3] + 1 + [l/2] + \tfrac{1}{2} - l - \tfrac{1}{2} \\ &= 2[l/3] + [l/2] - l + 1 \end{aligned}$$

as desired. The formulas (f) and (g) for $d(\mathbb{O})$ and $d(\mathbb{I})$ are obtained in identical fashion.

It remains only to derive the formula for $d(\mathbf{O}(2))$. To do this, observe that

$$\mathrm{Fix}_{V_l}(\mathbf{O}(2)) = \bigcap_{m=1}^{\infty} \mathrm{Fix}_{V_l}(\mathbf{D}_m).$$

A vector fixed by $\mathbf{O}(2)$ is certainly fixed by $\mathbf{D}_m$ for all m. But since the union of all $\mathbf{D}_m$ is dense in $\mathbf{O}(2)$, continuity implies that a vector fixed by every $\mathbf{D}_m$ is also fixed by $\mathbf{O}(2)$. Now formula (d) follows directly from (b) by considering m large. □

(b) Maximal Isotropy Subgroups of **SO**(3)

In the previous subsection we computed the dimensions of fixed-point subspaces for all closed subgroups of **SO**(3). We did not decide which of these subgroups are isotropy subgroups of **SO**(3), that is, exactly the symmetry groups of a given vector in V_l. The general question of when a subgroup is an isotropy subgroup is complicated; see Ihrig and Golubitsky [1984]. But by using the strategy developed in Lemma 2.7 we can obtain the *maximal* isotropy subgroups, as we now do.

Theorem 8.2. *Let* **SO**(3) *act irreducibly on* V_l. *Then the maximal isotropy subgroups are*:

$l = 2$:	$\mathbf{O}(2)$
$l = 4, 8, 14$:	$\mathbf{O}(2)$ *and* $\mathbb{O}$
all other even l:	$\mathbf{O}(2)$, $\mathbb{O}$, *and* $\mathbb{I}$.
$l = 1$:	$\mathbf{SO}(2)$
$l = 3$:	$\mathbf{SO}(2)$, $\mathbb{T}$, $\mathbf{D}_3$
$l = 5$:	$\mathbf{SO}(2)$, $\mathbf{D}_3$, $\mathbf{D}_4$, $\mathbf{D}_5$
$l = 7, 11$:	$\mathbf{SO}(2)$, $\mathbb{T}$, *and* $\mathbf{D}_m$ $(l/2 < m \leq l)$
$l = 9, 13, 17, 19, 23, 29$:	$\mathbf{SO}(2)$, $\mathbb{O}$, *and* $\mathbf{D}_m$ $(l/2 < m \leq l)$
all other odd l:	$\mathbf{SO}(2)$, $\mathbb{O}$, $\mathbb{I}$, *and* $\mathbf{D}_m (l/2 < m \leq l)$.

PROOF. First observe that $\mathbf{O}(2)$, $\mathbb{O}$, and $\mathbb{I}$ are maximal subgroups of $\mathbf{SO}(3)$. Hence these are maximal isotropy subgroups precisely when they are isotropy subgroups. By Lemma 2.7, this is the case precisely when the fixed-point subspace is nonzero.

$\mathbf{O}(2)$ is a maximal isotropy subgroup when l is even, since $d(\mathbf{O}(2)) = 1$ in this case. For $\mathbb{I}$ and $\mathbb{O}$ the argument is slightly different. Note that formulas (f) and (g) in Theorem 8.1 are "periodic" in l. More precisely, think of $d(\mathbb{O})$ and $d(\mathbb{I})$ as functions of l. Then

$$\begin{aligned} &\text{(a)} \quad d(\mathbb{O})(l + 12) = d(\mathbb{O})(l) + 1, \\ &\text{(b)} \quad d(\mathbb{I})(l + 30) = d(\mathbb{I})(l) + 1. \end{aligned} \tag{8.3}$$

It follows that $\mathbb{O}$ is a maximal isotropy subgroup when $l \geq 12$ and $\mathbb{I}$ is a maximal isotropy subgroup when $l \geq 30$. The other l for which $\mathbb{O}$ and $\mathbb{I}$ are maximal isotropy subgroups are obtained from Table 8.1. Just check in that table when $d(\mathbb{O})$ and $d(\mathbb{I})$ are positive.

Next observe that $\mathbf{SO}(2)$ is contained only in $\mathbf{O}(2)$ and that $d(\mathbf{SO}(2)) = 1$

Table 8.1. Fixed-Point Subspace Dimensions for the Exceptional Subgroups

l	$d(\mathbb{O})$	$d(\mathbb{I})$	$d(\mathbb{T})$	l	$d(\mathbb{I})$
1	0	0	0	16	1
2	0	0	0	17	0
3	0	0	1	18	1
4	1	0	1	19	0
5	0	0	0	20	1
6	1	1	2	21	1
7	0	0	1	22	1
8	1	0	1	23	0
9	1	0	2	24	1
10	1	1	2	25	1
11	0	0	1	26	1
12	2	1	3	27	1
13		0		28	1
14		0		29	0
15		1		30	2

for all l. Therefore for odd l, $\mathbf{SO}(2)$ is a maximal isotropy subgroup, since $d(\mathbf{O}(2)) = 0$ when l is odd.

The cyclic groups $\mathbf{Z}_m$ are contained in $\mathbf{SO}(2)$ and $\mathbf{O}(2)$. Hence they cannot be maximal isotropy subgroups of $\mathbf{SO}(3)$ for any l.

It remains to analyze the subgroups $\mathbb{T}$ and $\mathbf{D}_m$. Now $\mathbb{T}$ is contained only in $\mathbb{O}$ and $\mathbb{I}$. Since $\mathbb{O}$ is a maximal isotropy subgroup for all $l \geq 12$, $\mathbb{T}$ cannot be a maximal isotropy subgroup for $l \geq 12$. It is easy to check from Table 8.1 that $\mathbb{T}$ is a maximal isotropy subgroup only when $l = 3$, 7, or 11. (Note that $d(\mathbb{T})$ is also "periodic" of period 6.)

Since $\mathbf{D}_m \subset \mathbf{O}(2)$ for all m, and $\mathbf{O}(2)$ is a maximal isotropy subgroup when l is even, $\mathbf{D}_m$ cannot be a maximal isotropy subgroup for l even. When l is odd, $d(\mathbf{D}_m) = [l/m]$ by Theorem 8.1(b). When $m > l$, $d(\mathbf{D}_m) = 0$, so these cases are ruled out. On the other hand, when $m \leq l/2$, $\mathbf{D}_m \subset \mathbf{D}_{2m}$ and $d(\mathbf{D}_{2m}) > 0$. It follows from Lemma 2.7 that $\mathbf{D}_m$ cannot be a maximal isotropy subgroup. Therefore, we are left with the range of values

$$l/2 < m \leq l. \tag{8.4}$$

The only obstruction to $\mathbf{D}_m$'s being a maximal isotropy subgroup, when m is in this range, occurs when $\mathbf{D}_m$ is contained in an exceptional subgroup Σ with $d(\Sigma) > 0$. Now only $\mathbf{D}_2$, $\mathbf{D}_3$, $\mathbf{D}_4$ and $\mathbf{D}_5$ are contained in exceptional subgroups. Thus when l is odd and $l > 11$, $\mathbf{D}_m$ is a maximal isotropy subgroup for all m in the range (8.4). When $l \leq q$ and l is odd, $d(\mathbb{I}) = 0$. Therefore, $\mathbf{D}_5$ occurs whenever 5 is in the range (8.4). When $l \leq 7$ and l is odd, $d(\mathbb{O}) = 0$. So $\mathbf{D}_3$ and $\mathbf{D}_4$ occur whenever 3 and 4 are in the range (8.4). Finally $\mathbf{D}_2$ may be a maximal isotropy subgroup when $l = 3$. However, $d(\mathbb{T}) = 1$ when $l = 3$ so $\mathbf{D}_2$ is not a maximal isotropy subgroup. □

Finally, we list the isotropy subgroups $\Sigma \subset \mathbf{SO}(3)$ with $d(\Sigma) = 1$. These are maximal, so we can read them off from Theorems 8.2 and 8.1.

Theorem 8.3. *The (maximal) isotropy subgroups of* $\mathbf{SO}(3)$ *acting on* V_l, *which have one-dimensional fixed-point subspaces, are*:

$\mathbf{O}(2)$: *all even* l.

$\mathbf{SO}(2)$: *all odd* l.

$\mathbf{D}_m$: l *odd*, $l/2 < m \leq l$, $m \neq 2$.

$\mathbb{I}$: $l = 6, 10, 12, 15, 16, 18, 20–22, 24–28, 31–35, 37–39, 41, 43, 44, 47, 49, 53, 59$.

$\mathbb{O}$: $l = 4, 6, 8–10, 13–15, 17, 19, 23$.

$\mathbb{T}$: $l = 3, 7, 11$.

§9.† Symmetry-Breaking from $\mathbf{O}(3)$

Algebraically, the orthogonal group $\mathbf{O}(3)$ is just a direct sum

$$\mathbf{O}(3) = \mathbf{SO}(3) \oplus \mathbf{Z}_2^c \tag{9.1}$$

where $\mathbf{Z}_2^c = \{\pm 1\}$; and as we saw in §7 this leads to very close connections between the representations of $\mathbf{O}(3)$ and $\mathbf{SO}(3)$. We now address for $\mathbf{O}(3)$ the same questions that we addressed for $\mathbf{SO}(3)$ in §8, namely, the dimensions of fixed-point subspaces, the maximal isotropy subgroups, and the isotropy subgroups with one-dimensional fixed-point subspaces.

In order to understand the differences between the analyses of $\mathbf{SO}(3)$ and $\mathbf{O}(3)$, we first recall from §7 how (9.1) determines the irreducible representations of $\mathbf{O}(3)$. If $V = V_l$ and $\mathbf{SO}(3)$ acts on V as usual, then there are two irreducible actions of $\mathbf{O}(3)$ on V which extend that of $\mathbf{SO}(3)$. They are distinguished by whether $-I \in \mathbf{O}(3)$ acts as the identity or minus the identity. We claim that the *plus* representation, in which $-I$ acts as the identity, has essentially the same properties as the corresponding representation of $\mathbf{SO}(3)$. To make this statement precise we must discuss some properties of subgroups of $\mathbf{O}(3)$.

Subgroups of $\mathbf{O}(3)$ fall into three classes:

$$\begin{array}{ll} \text{(I)} & \text{Subgroups of } \mathbf{SO}(3), \\ \text{(II)} & \text{Subgroups containing } -I, \\ \text{(III)} & \text{Subgroups not in } \mathbf{SO}(3) \text{ and not containing } -I. \end{array} \tag{9.2}$$

Observe that subgroups of class II all have the form $\Sigma \oplus \mathbf{Z}_2^c$ where Σ is a subgroup of $\mathbf{SO}(3)$. Since we listed the closed subgroups of $\mathbf{SO}(3)$ in Theorem 6.1 we know the closed subgroups of classes I and II of $\mathbf{O}(3)$.

For the plus representation of $\mathbf{O}(3)$, $-I$ acts trivially and hence lies in every isotropy subgroup. Therefore, each isotropy subgroup is class II, of the form $\Sigma \oplus \mathbf{Z}_2^c$. Moreover, $\dim \mathrm{Fix}(\Sigma \oplus \mathbf{Z}_2^c) = \dim \mathrm{Fix}(\Sigma)$. Thus all results for $\mathbf{SO}(3)$ in §8 carry over to $\mathbf{O}(3)$ in this case, when Σ is replaced by $\Sigma \oplus \mathbf{Z}_2^c$.

Now we consider the minus representation, in which $-I$ acts as minus the identity on V. In this case $-I$ fixes only the origin, hence is not contained in any isotropy subgroup. Therefore, the isotropy subgroups are of class I or III. In order to determine the maximal isotropy subgroups for these actions of $\mathbf{O}(3)$ we must consider the class III subgroups.

We divide the remainder of this subsection into five parts:

(a) The description of class III subgroups,
(b) Containment of class III subgroups,
(c) Dimensions of fixed-point subspaces for class III subgroups,
(d) Maximal isotropy subgroups for irreducible representations of $\mathbf{O}(3)$,
(e) The natural representations on spherical harmonics.

We remark here on (e). Recall from §7(e) that the natural action of $\mathbf{O}(3)$ on spherical harmonics of degree l is by the plus representation when l is even and the minus representation when l is odd, since in the natural action

$$-I \cdot p(x) = p(-x) = (-1)^l p(x). \tag{9.3}$$

In subsection (e) we combine the results from §8(b) and subsection (d) of this chapter to discuss this natural action.

(a) The Class III Subgroups of $\mathbf{O}(3)$

We now show that each class III subgroup H is isomorphic to a subgroup of $\mathbf{SO}(3)$, though H is never conjugate to that subgroup. To see why, let π: $\mathbf{O}(3) \to \mathbf{SO}(3)$ be the homomorphism whose kernel is $\mathbf{Z}_2^c$. Since $-I \notin H$, it follows that $\pi | H$ is an isomorphism. Hence H is isomorphic to $\pi(H) \subset \mathbf{SO}(3)$. We claim that every class III subgroup H is uniquely determined by two data: the subgroups $\pi(H)$ and $H \cap \mathbf{SO}(3)$ of $\mathbf{SO}(3)$. Recall that if $L \subset K$ are groups, then the *index* of L in K is the number of cosets in the quotient K/L.

Lemma 9.1.
(a) *Let H be a class III subgroup of $\mathbf{O}(3)$. Then $H \cap \mathbf{SO}(3)$ is a subgroup of index 2 in $\pi(H)$.*
(b) *Let $L \subset K \subset \mathbf{SO}(3)$ be subgroups, where L has index 2 in K. Then there exists a unique class III subgroup $H \subset \mathbf{O}(3)$ such that $\pi(H) = K$ and $H \cap \mathbf{SO}(3) = L$.*

Remark. Lemma 9.1 allows us to determine the conjugacy classes of class III subgroups of $\mathbf{O}(3)$ by classifying pairs of subgroups $L \subset K \subset \mathbf{SO}(3)$ such that L has index 2 in K. Care must be taken, however, to analyze the conjugacy class of (K, L) considered as a pair of subgroups, and not just pairs of conjugacy classes of K and L. We amplify this remark in the following:

PROOF.
(a) Suppose $\gamma, \delta \in \mathbf{O}(3) \sim \mathbf{SO}(3)$. Then $\gamma\delta \in \mathbf{SO}(3)$ since $\mathbf{O}(3)/\mathbf{SO}(3) \cong \mathbf{Z}_2^c$. Now suppose $\gamma, \delta \in H \sim (H \cap \mathbf{SO}(3))$. Then $\gamma^{-1} \notin \mathbf{SO}(3)$ and $\gamma^{-1}\delta \in H \cap \mathbf{SO}(3)$. Therefore, $\delta \in \gamma(H \cap \mathbf{SO}(3))$, and δ and γ are in the same coset of $H \cap \mathbf{SO}(3)$ in H. It follows that the index of $H \cap \mathbf{SO}(3)$ in H is 2. Since $\pi(H \cap \mathbf{SO}(3)) = H \cap \mathbf{SO}(3)$, the index of $H \cap \mathbf{SO}(3)$ in $\pi(H)$ is also 2.
(b) Given $L \subset K \subset \mathbf{SO}(3)$, define

$$H = L \cup hL \tag{9.4}$$

where $h = (g, -I) \in \mathbf{SO}(3) \oplus \mathbf{Z}_2^c = \mathbf{O}(3)$ and $g \in K \sim L$. Since $h^2 = g^2 \in L$, it follows that H is a subgroup of $\mathbf{O}(3)$. Observe that $\pi(H) = L \cup gL = K$ and $H \cap \mathbf{SO}(3) = L$. Note that H has to be of class III. If $-I$ is in H then $K = \pi(H)$ equals $H \cap \mathbf{SO}(3) = L$. But $K \neq L$, so this is impossible.

Next, we claim that H is uniquely determined. Let H' be a subgroup of $\mathbf{O}(3)$ satisfying

$$\pi(H') = K \quad \text{and} \quad H' \cap \mathbf{SO}(3) = L.$$

We know that $H' \cap \mathbf{SO}(3) = H \cap \mathbf{SO}(3)$. Now suppose $h' \in H' \sim L$. Since $h' \notin \mathbf{SO}(3)$ it has the form $(g', -I) \in \mathbf{SO}(3) \oplus \mathbf{Z}_2^c$. Since $\pi(H') = K$ it follows that $g' \in K \sim L$. Therefore $g' = gl$ where g is the element in $K \sim L$ used to define H above, and $l \in L$. Thus $g'L = gL$ whence $H' = H$. □

We now use Lemma 9.1 to enumerate all conjugacy classes of class III subgroups. As a first step we enumerate up to conjugacy the subgroups $K \subset \mathbf{SO}(3)$ which have subgroups L of index 2. The pairs $K \supset L$ are:

$$\begin{array}{lll} \text{(a)} & \mathbf{O}(2) \supset \mathbf{SO}(2) & \\ \text{(b)} & \mathbb{O} \supset \mathbb{T} & \\ \text{(c)} & \mathbf{D}_m \supset \mathbf{Z}_m & (m \geq 2) \\ \text{(d)} & \mathbf{D}_{2m} \supset \mathbf{D}_m & (m \geq 2) \\ \text{(e)} & \mathbf{Z}_{2m} \supset \mathbf{Z}_m & (m \geq 2) \\ \text{(f)} & \mathbf{Z}_2 \supset \mathbb{1}. & \end{array} \tag{9.5}$$

Although we have enumerated all of the isomorphism classes of pairs of subgroups of index 2, we must check whether there are isomorphic but nonconjugate choices for L and K. This problem cannot arise in (9.5(a), (b), (c), (e), (f)) since L is the unique subgroup of K in its isomorphism class and of index 2.

Note, however, in (9.5(d)) there are two subgroups isomorphic to $\mathbf{D}_m$ which lie inside $\mathbf{D}_{2m}$. We specify them as follows. Let $\theta = \pi/m$ and let R_θ be rotation through θ. Observe that $\mathbf{Z}_m$ is generated by R_θ^2. Let κ be reflection in the plane across the x-axis. Then the two subgroups are

$$\mathbf{D}_m = \mathbf{Z}_m \cup \kappa\mathbf{Z}_m \quad \text{and} \quad \mathbf{D}'_m = \mathbf{Z}_m \cup \kappa R_\theta \mathbf{Z}_m.$$

These subgroups are not conjugate inside $\mathbf{D}_{2m}$ but they are conjugate inside $\mathbf{D}_{4m} \subset \mathbf{SO}(3)$. Since

$$R_{-\theta/2}\kappa R_{\theta/2} = \kappa R_\theta$$

it is easy to check that

$$R_{-\theta/2}\mathbf{D}_m R_{\theta/2} = \mathbf{D}'_m \quad \text{and} \quad R_{-\theta/2}\mathbf{D}_{2m} R_{\theta/2} = \mathbf{D}_{2m}.$$

Thus the pairs $\mathbf{D}_m \subset \mathbf{D}_{2m}$ and $\mathbf{D}'_m \subset \mathbf{D}_{2m}$ are conjugate in $\mathbf{SO}(3)$, so they generate conjugate class III subgroups of $\mathbf{O}(3)$.

Next we describe our notation for class III subgroups H. With one exception, we shall write them in the form $\pi(H)^-$, so that, for example, $\mathbb{O}^-$ indicates the type III subgroup that is isomorphic to $\mathbb{O}$. The exception is when $\pi(H) = \mathbf{D}_{2m}$. From (9.5(c), (d)) there are two nonisomorphic subgroups of index 2 in $\mathbf{D}_{2m}$, namely $\mathbf{D}_m$ and $\mathbf{Z}_{2m}$. We let $\mathbf{D}_m^z$ be the class III subgroup corresponding to the pair $\mathbf{D}_m \supset \mathbf{Z}_m$, and we let $\mathbf{D}_{2m}^d$ be that corresponding to $\mathbf{D}_{2m} \supset \mathbf{D}_m$.

We now list the conjugacy classes of closed subgroups of $\mathbf{O}(3)$.

Theorem 9.2. *Every closed subgroup of* $\mathbf{O}(3)$ *is conjugate to one of the following*:

(I) $\mathbf{SO}(3)$, $\mathbf{O}(2)$, $\mathbf{SO}(2)$, $\mathbb{T}$, $\mathbb{O}$, $\mathbb{I}$, $\mathbf{D}_{2m}$ $(m \geq 2)$, $\mathbf{Z}_m$ $(m \geq 2)$, $\mathbb{1}$.
(II) $K \oplus \mathbf{Z}_2^c$ *where* K *is a subgroup of* $\mathbf{SO}(3)$ *and* $\mathbf{Z}_2^c = \{\pm I\}$.
(III) $\mathbf{O}(2)^-$, $\mathbb{O}^-$, $\mathbf{D}_m^z$ $(m \geq 2)$, $\mathbf{D}_{2m}^d$ $(m \geq 2)$, $\mathbf{Z}_{2m}^-$ $(m \geq 1)$.

Remarks.
(a) All of these are nonconjugate. Note that $\mathbf{D}_2^d$ is omitted since it is conjugate to $\mathbf{D}_2^z$.
(b) There are three nonconjugate subgroups of $\mathbf{O}(3)$ of order 2, namely $\mathbf{Z}_2^c$, $\mathbf{Z}_2$, and $\mathbf{Z}_2^-$. They are generated, respectively, by

$$\begin{bmatrix} -1 & 0 & 0 \\ 0 & -1 & 0 \\ 0 & 0 & -1 \end{bmatrix}, \quad \begin{bmatrix} -1 & 0 & 0 \\ 0 & -1 & 0 \\ 0 & 0 & 1 \end{bmatrix}, \quad \text{and} \quad \begin{bmatrix} -1 & 0 & 0 \\ 0 & 1 & 0 \\ 0 & 0 & 1 \end{bmatrix}.$$

(c) There are three nonconjugate subgroups of $\mathbf{O}(3)$ isomorphic to $\mathbf{D}_{2m}$, namely $\mathbf{D}_{2m}$, $\mathbf{D}_{2m}^z$, and $\mathbf{D}_{2m}^d$. It is amusing to understand how each of these groups may be viewed as symmetries of the $2n$-gon in the (x, y)-plane. The cyclic subgroup $\mathbf{Z}_{2m}$, generated by rotation of the (x, y)-plane through π/m, is present in both $\mathbf{D}_{2m}$ and $\mathbf{D}^z_{2m}$. The group $\mathbf{D}_{2m} \subset \mathbf{SO}(3)$ is generated by $\mathbf{Z}_{2m}$ and a reflection across the x-axis in the (x, y)-plane, affected by

$$\begin{bmatrix} 1 & 0 & 0 \\ 0 & -1 & 0 \\ 0 & 0 & -1 \end{bmatrix}.$$

The group $\mathbf{D}^z_{2m} \not\subset \mathbf{SO}(3)$ is generated by $\mathbf{Z}_{2m}$ and a reflection across the x-axis in the (x, y)-plane affected by

$$\begin{bmatrix} 1 & 0 & 0 \\ 0 & -1 & 0 \\ 0 & 0 & 1 \end{bmatrix}.$$

The cyclic subgroup of order $2m$ in $\mathbf{D}^d_{2m}$ is not the standard cyclic subgroup $\mathbf{Z}_{2m} \subset \mathbf{SO}(3)$, but the class III subgroup $\mathbf{Z}^-_{2m}$, generated by the rotation

$$\left[\begin{array}{cc|c} \multicolumn{2}{c|}{R_{\pi/m}} & 0 \\ & & 0 \\ \hline 0 & 0 & -1 \end{array}\right].$$

The reflection across the x-axis in $\mathbf{D}^d_{2m}$ is the same as in $\mathbf{D}_{2m}$.

(b) Containments Involving Class III Subgroups

We begin our discussion by giving disjoint union decompositions into cyclic subgroups for the finite class III subgroups:

$$\begin{aligned} &\text{(a)} \quad \mathbb{O}^- = \dot\cup^3 \mathbf{Z}^-_4 \,\dot\cup^4 \mathbf{Z}_3 \,\dot\cup^6 \mathbf{Z}^-_2, \\ &\text{(b)} \quad \mathbf{D}^z_m = \mathbf{Z}_m \,\dot\cup^m \mathbf{Z}^-_2, \qquad (m \geq 2) \\ &\text{(c)} \quad \mathbf{D}^d_{2m} = \mathbf{Z}^-_{2m} \,\dot\cup^m \mathbf{Z}_2 \,\dot\cup^m \mathbf{Z}^-_2, \qquad (m \geq 2). \end{aligned} \tag{9.6}$$

The class III subgroups in (9.6) have disjoint union decompositions induced by their isomorphisms with subgroups of $\mathbf{SO}(3)$. The verification of (9.6) is based on a combinatorial argument using the disjoint unions for $\mathbb{O}$ and $\mathbf{D}_m$ in Lemma 6.2 and the fact that exactly half the elements in a class III subgroup are not in $\mathbf{SO}(3)$. We now give the details.

Since $\mathbf{D}^z_m$ is isomorphic to $\mathbf{D}_m$ it has a disjoint union decomposition with one subgroup isomorphic to $\mathbf{Z}_m$ and m subgroups isomorphic to $\mathbf{Z}_2$. Since $\mathbf{D}^z_m \cap \mathbf{SO}(3) = \mathbf{Z}_m$, we must have all other elements not in $\mathbf{SO}(3)$, whence (9.6(b)) holds. A similar argument applies to $\mathbf{D}^d_{2m}$, but now $\mathbf{Z}^-_{2m} \subset \mathbf{D}^d_{2m}$ so the decomposition must be as in (9.6(c)). For $\mathbb{O}^-$ the cyclic groups of order 3 must be $\mathbf{Z}_3$ since there is no such thing as $\mathbf{Z}^-_3$; the $\mathbf{Z}_4$ part is not contained in $\mathbf{SO}(3)$ so must be $\mathbf{Z}^-_4$, and a counting argument does the rest.

We now discuss containment relations involving class III subgroups.

Lemma 9.3. *Let H_1 and H_2 be class III subgroups. Then $H_2 \subset H_1$ if and only if*

$$\begin{aligned}&\text{(a)}\quad H_2 \cap \mathbf{SO}(3) \subset H_1 \cap \mathbf{SO}(3),\ \pi(H_2) \subset \pi(H_1), \quad \textit{and}\\ &\text{(b)}\quad \pi(H_2) \not\subset H_1 \cap \mathbf{SO}(3).\end{aligned} \tag{9.7}$$

PROOF. The necessity of condition (9.7(a)) is obvious. What is not so obvious is the need for condition (9.7(b)). To prove this, assume $H_2 \subset H_1$ and $\pi(H_2) \subset H_1 \cap \mathbf{SO}(3)$. We claim that $-I \in H_1$, contradicting H_1's being class III. To verify this, let $h \in H_2 \sim \mathbf{SO}(3)$. Then $\pi(h) \in \pi(H_2) \subset H_1 \cap \mathbf{SO}(3) \subset H_1$. Moreover, $h \in H_2 \subset H_1$. Therefore, $\pi(h)^{-1}h \in H_1$. However, since $h \notin \mathbf{SO}(3)$ it follows that $h = (\pi(h), -I) \in \mathbf{SO}(3) \oplus Z_2^c = \mathbf{O}(3)$, and hence $\pi(h)^{-1}h = (1, -I) = -I \in H_1$. This is what was claimed.

Conversely, suppose (9.7) holds. We show that $H_2 \subset H_1$. Assumption (9.7(a)) implies that $H_2 \cap \mathbf{SO}(3) \subset H_1$. We must show that $H_2 \sim \mathbf{SO}(3) \subset H_1$. Observe that

$$H_2 \sim \mathbf{SO}(3) = h(H_2 \cap \mathbf{SO}(3)) \tag{9.8}$$

for any $h \in H_2 \sim \mathbf{SO}(3)$, since the index of $H_2 \cap \mathbf{SO}(3)$ in H_2 is 2. Thus, if we can show that there exists $h \in H_2 \sim \mathbf{SO}(3)$ which is also in H_1, then the right-hand side of (9.8) is also in H_1, and $H_2 \subset H_1$ follows.

Using (9.7(b)) we choose $h \in H_2$ such that $\pi(h) \notin H_1 \cap \mathbf{SO}(3)$. Observe that $h \notin \mathbf{SO}(3)$. For if h is in $\mathbf{SO}(3)$, then h is in $H_2 \cap \mathbf{SO}(3) \subset H_1 \cap \mathbf{SO}(3)$ by (9.7(a)), and then $\pi(h) \in H_1 \cap \mathbf{SO}(3)$. We now show that $h \in H_1$, which proves the lemma. We know that $h = (\pi(h), -I) \in \mathbf{SO}(3) \oplus \mathbf{Z}_2^c$ since $h \notin \mathbf{SO}(3)$. Now $\pi(h) \in \pi(H_2) \subset \pi(H_1)$ by (9.7(a)). Thus there exists $h_1 \in H_1$ such that $\pi(h) = \pi(h_1)$. However, h_1 cannot be in $\mathbf{SO}(3)$ since $\pi(h) \notin H_1 \cap \mathbf{SO}(3)$. Therefore, $h_1 = (\pi(h_1), -I) = h$, so $h \in H_1$ as claimed. □

We next use Lemma 9.3 to prove the following:

Proposition 9.4. *The containments between conjugacy classes of subgroups of class III groups are as follows:*

(a) $\mathbb{O}^-$ *contains* $\mathbf{D}_4^d$, $\mathbf{Z}_4^-$, $\mathbf{D}_3^z$, $\mathbf{D}_2^z$, $\mathbf{Z}_2^-$, *and subgroups of* $\mathbb{T}$.
(b) $\mathbf{O}(2)^-$ *contains* $\mathbf{D}_m^z$ $(m \geq 2)$, $\mathbf{Z}_2^-$, *and subgroups of* $\mathbf{SO}(2)$.
(c) $\mathbf{Z}_{2m}^-$ *contains* $\mathbf{Z}_{2k}^-$ *where k divides m and 2k does not divide m, and subgroups of* $\mathbf{Z}_m$.
(d) $\mathbf{D}_{2m}^d$ *contains* $\mathbf{D}_{2k}^d$ *and* $\mathbf{Z}_{2k}^-$ *when k divides m and 2k does not divide m,* $\mathbf{D}_k^z$ *when k divides m,* $\mathbf{D}_2^z$, $\mathbf{Z}_2^-$, *and subgroups of* $\mathbf{D}_m$.
(e) $\mathbf{D}_m^z$ *contains* $\mathbf{D}_k^z$ *where k divides m,* $\mathbf{Z}_2^-$, *and subgroups of* $\mathbf{Z}_m$.

PROOF. We make two general comments. If H is a class III subgroup, then all subgroups of $H \cap \mathbf{SO}(3)$ are contained in H and all other subgroups of H are of class III. The class III subgroups of H may be obtained by using Lemma 9.3 and the pairs of subgroups $(\pi(H), H \cap \mathbf{SO}(3))$ of index 2 in (9.5). We prove (a, c, d), leaving (b, e) as exercises for the reader.

(a) Since $\mathbb{T} = \mathbb{O}^- \cap \mathbf{SO}(3)$ all subgroups of $\mathbb{T}$ are contained in $\mathbb{O}^-$, and the remaining subgroups are class III. From (9.7) the class III subgroups of $\mathbb{O}$ satisfy

$$\pi(H) \subset \mathbb{O}, \qquad H \cap \mathbf{SO}(3) \subset \mathbb{T}, \quad \text{and} \quad \pi(H) \not\subset \mathbb{T}.$$

From Figure 6.2 we see that $\mathbf{D}_4$, $\mathbf{D}_3$, and $\mathbf{Z}_4$ are possibilities for $\pi(H)$. In addition, there is a subtle possibility. There are subgroups $\mathbf{Z}_2$ and $\mathbf{D}_2$ which are in $\mathbb{O}$ but not in $\mathbb{T}$. Their existence follows from the disjoint union decompositions of $\mathbb{O}$ and $\mathbb{T}$ in Lemma 6.2(b, c). There are only three $\mathbf{Z}_2$ subgroups in $\mathbb{T}$ and six in $\mathbb{O}$. The possible class III subgroups are given by Theorem 9.2 (III), namely $\mathbf{D}_4^z$, $\mathbf{D}_4^d$, $\mathbf{Z}_4^-$, $\mathbf{D}_3^z$, $\mathbf{D}_2^z$, and $\mathbf{Z}_2^-$. The condition $H \cap \mathbf{SO}(3) \subset \mathbb{T}$ eliminates $\mathbf{D}_4^z$ and shows that the remainder occur as subgroups of $\mathbb{O}^-$.

(c) Since $\mathbf{Z}_m = \mathbf{Z}_{2m}^- \cap \mathbf{SO}(3)$, all subgroups of $\mathbf{Z}_m$ are contained in $\mathbf{Z}_{2m}^-$ and the remainder are class III. From (9.7) the class III subgroups H of $\mathbf{Z}_{2m}^-$ satisfy

$$\pi(H) \subset \mathbf{Z}_{2m}, \qquad H \cap \mathbf{SO}(3) \subset \mathbf{Z}_m, \quad \text{and} \quad \pi(H) \not\subset \mathbf{Z}_m.$$

Subgroups of $\mathbf{Z}_{2m}$ are cyclic, so $\pi(H)$ is cyclic. The only cyclic class III subgroups are $\mathbf{Z}_{2k}^-$. For $k > 1$, the only $\mathbf{Z}_{2k}^-$ that can occur are those for which k divides m (so that $\pi(H) \subset \mathbf{Z}_{2m}$ and $H \cap \mathbf{SO}(3) \subset \mathbf{Z}_m$) and $2k$ does not divide m (so that $\pi(H) \not\subset \mathbf{Z}_m$).

(d) Since $\mathbf{D}_{2m}^d \cap \mathbf{SO}(3) = \mathbf{D}_m$, all subgroups of $\mathbf{D}_m$ are contained in $\mathbf{D}_{2m}^d$ and the remainder are class III. By (9.7) the class III subgroups H of $\mathbf{D}_{2m}^d$ satisfy

$$\pi(H) \subset \mathbf{D}_{2m}, \quad H \cap \mathbf{SO}(3) \subset \mathbf{D}_m, \quad \text{and} \quad \pi(H) \not\subset \mathbf{D}_m.$$

Since $\pi(H) \subset \mathbf{D}_{2m}$ the only possibilities for H are $\mathbf{Z}_{2k}^-$, $\mathbf{D}_{2k}^d$, and $\mathbf{D}_k^z$, which we consider in order. Begin by observing that the containment of subgroups $\mathbf{Z}_{2k}^-$ in $\mathbf{D}_{2m}^d$ is analogous to that of $\mathbf{Z}_{2k}^-$ in (c) earlier.

Next let $H = \mathbf{D}_{2k}^d$ be contained in $\mathbf{D}_{2m}^d$. Since $\pi(H) = \mathbf{D}_{2k} \subset \mathbf{D}_{2m}$ we must have $k|m$. Similarly, $k|m$ implies that $H \cap \mathbf{SO}(3) = \mathbf{D}_k \subset \mathbf{D}_m$. If $2k$ does not divide m then $\pi(H) \not\subset \mathbf{D}_m$, and Lemma 9.3 guarantees that $\mathbf{D}_{2k}^d \subset \mathbf{D}_{2m}^d$.

Suppose now that H is conjugate to $\mathbf{D}_{2k}^d$ and that H is contained in $\mathbf{D}_{2m}^d$. As previously, k must divide m. Moreover, $H \cap \mathbf{SO}(3) \subset \mathbf{D}_{2m}^d \cap \mathbf{SO}(3) = \mathbf{D}_m$ and $H \cap \mathbf{SO}(3)$ is isomorphic to $\mathbf{D}_k$. We can always choose γ such that $\gamma^{-1}\mathbf{D}_{2m}\gamma = \mathbf{D}_{2m}$, $\gamma^{-1}\mathbf{D}_m\gamma = \mathbf{D}_m$, and $\gamma^{-1}(H \cap \mathbf{SO}(3))\gamma = \mathbf{D}_k$. Thus we may assume that $H \cap \mathbf{SO}(3) = \mathbf{D}_k$. Therefore, $\mathbf{D}_k \subset \pi(H) \subset \mathbf{D}_{2m}$ where $\pi(H)$ is isomorphic to $\mathbf{D}_{2k}$. We claim that in fact $\pi(H) = \mathbf{D}_{2k}$. It then follows from the preceding paragraph that $2k$ does not divide m. To verify the claim recall that the standard group $\mathbf{D}_l$ may be written as $\mathbf{Z}_l \cup \kappa\mathbf{Z}_l$, where κ is a fixed element of $\mathbf{SO}(3)$. Now $\pi(H)$ is isomorphic to $\mathbf{D}_{2k}$, so $\pi(H) \supset \mathbf{Z}_{2k}$. Since $\mathbf{D}_k \subset \pi(H)$ it follows that $\kappa \in \pi(H)$. Therefore, $\pi(H)$ contains the group generated by $\mathbf{Z}_{2k}$ and κ, which is $\mathbf{D}_{2k}$. Since $\pi(H)$ is isomorphic to $\mathbf{D}_{2k}$, we must have $\pi(H) = \mathbf{D}_{2k}$ as claimed.

Finally, let H be a subgroup conjugate to $\mathbf{D}_k^z$ and contained in $\mathbf{D}_{2m}^d$. Then $H \cap \mathbf{SO}(3)$ is cyclic and contained in $\mathbf{D}_m$. Thus $H \cap \mathbf{SO}(3) = \mathbf{Z}_k$. Since $\mathbf{Z}_k \subset \mathbf{D}_m$ we have $k|m$. Now assume that $k|m$. We claim that there is a subgroup $\mathbf{D}_k'$ isomorphic but not equal to $\mathbf{D}_k$ such that $\mathbf{D}_k' \subset \mathbf{D}_{2m}$ but $\mathbf{D}_k' \not\subset \mathbf{D}_m$. If so, then the class III subgroup H corresponding to $\mathbf{Z}_k \subset \mathbf{D}'_k$ is conjugate to $\mathbf{D}^z_k$ and is contained in $\mathbf{D}^d_{2m}$. To construct $\mathbf{D}'_k$ we use the idea in the proof that there is one conjugacy class of class III subgroups corresponding to (9.5(d)). Let

$$\mathbf{D}'_k = \mathbf{Z}_k \cup \kappa R_{\pi/m} \mathbf{Z}_k \tag{9.9}$$

where $R_{\pi/m}$ is rotation through π/m in the (x, y)-plane, and $\kappa(x, y, z) = (x, -y, z)$. It is easy to check that $\mathbf{D}'_k$ is a group isomorphic to $\mathbf{D}_k$. It is also easy to check that $\mathbf{D}'_k \subset \mathbf{D}_{2m}$ but $\mathbf{D}'_k \not\subset \mathbf{D}_m$. □

Note that $\mathbf{D}^z_2 \subset \mathbf{D}^d_{2m}$ for all m since D^z_2 is conjugate to $\mathbf{D}^d_2$.

(c) Dimensions of Fixed-Point Subspaces for Class III Subgroups

In this subsection and the next we consider the nontrivial irreducible representations of $\mathbf{O}(3)$, that is, the minus representations, where $-I$ acts as minus the identity. As noted previously, isotropy subgroups of $\mathbf{O}(3)$ in these representations are either class I or class III. The dimensions of the fixed-point subspaces of class I subgroups (subgroups of $\mathbf{SO}(3)$) are given in Theorem 8.1. In this section we derive the corresponding formulas for class III subgroups.

Theorem 9.5. *Let* $\mathbf{O}(3)$ *act irreducibly on* V_l *with* $-I$ *acting as minus the identity. Then the dimensions of the fixed-point subspaces for class III subgroups are*:

(a) $d(\mathbf{Z}^-_{2m}) = 2[(l + m)/2m]$.

(b) $d(\mathbf{D}^z_m) = \begin{cases} [l/m] & (l \text{ even}) \\ [l/m] + 1 & (l \text{ odd}) \end{cases}$

(c) $d(\mathbf{D}^d_{2m}) = [(l + m)/2m]$

(d) $d(\mathbb{O}^-) = [l/3] - [l/4]$

(e) $d(\mathbf{O}(2)^-) = \begin{cases} 0 & (l \text{ even}) \\ 1 & (l \text{ odd}). \end{cases}$

Proof.

(a) The group $\mathbf{Z}^-_{2m}$ is generated by $R_{\pi/m}$ followed by $-I$. The only way a vector $v \in V_l$ can be fixed by $\mathbf{Z}^-_{2m}$ is if

$$R_{\pi/m} v = -v. \tag{9.10}$$

Since $R_{\pi/m} \in \mathbf{SO}(2)$ we can use the Cartan decomposition, Proposition 7.2, of V_l to compute the number of independent vectors v satisfying (9.10). Recall that the Cartan decomposition is

$$V_l = H_0 \oplus H_1 \oplus \cdots \oplus H_l$$

where R_θ acts on H_k by rotation through $k\theta$. Thus vectors in H_k satisfy (9.10) if and only if k/m is an odd integer. There are $[(l+m)/2m]$ such integers k between 0 and l. Since $\dim H_k = 2$ when $k \neq 0$ we have $d(\mathbf{Z}_{2m}^-) = 2[(l+m)/2m]$ as claimed.

(b, c, d). Use the disjoint union decompositions in (9.6) and argue as in Theorem 8.1.

(e) Observe that

$$\mathrm{Fix}(\mathbf{O}(2)^-) = \bigcap_{m=2}^{\infty} \mathrm{Fix}(\mathbf{D}_m^z)$$

so that

$$d(\mathbf{D}(2)^-) = \lim_{m\to\infty} d(\mathbf{D}_m^z),$$

yielding the desired formula. □

Remark. The verification of b, c, d proceeds most easily using Exercise 9.1.

(d) Maximal Isotropy Subgroups for the Minus Representations

Theorem 9.6. *Let $\mathbf{O}(3)$ act irreducibly on V_l with $-I$ acting as minus the identity. Then the maximal isotropy subgroups are*:

$l = 1$:	$\mathbf{O}(2)^-$
$l = 3$:	$\mathbf{O}(2)^-, \mathbb{O}^-, \mathbf{D}_6^d$
$l = 5$:	$\mathbf{O}(2)^-, \mathbf{D}_{2m}^d \quad (2 \le m \le 5)$
$l = 7, 11$:	$\mathbf{O}(2)^-, \mathbb{O}^-, \mathbf{D}_{2m}^d \quad (l/3 < m \le l)$
$l = 9, 13, 17, 19, 23, 29$:	$\mathbf{O}(2)^-, \mathbb{O}^-, \mathbb{O}, \mathbf{D}_{2m}^d \quad (l/3 < m \le l)$
All other odd l:	$\mathbf{O}(2)^-, \mathbb{O}^-, \mathbb{O}, \mathbb{I}, \mathbf{D}_{2m}^d \quad (l/3 < m \le l)$
$l = 2$:	$\mathbf{O}(2), \mathbf{D}_4^d,$
$l = 4, 8$:	$\mathbf{O}(2), \mathbb{O}, \mathbf{D}_{2m}^d \quad (l/3 < m \le l)$
$l = 14$:	$\mathbf{O}(2), \mathbb{O}^-, \mathbb{O}, \mathbf{D}_{2m}^d \quad (5 \le m \le 14)$
All other even l:	$\mathbf{O}(2), \mathbb{O}^-, \mathbb{O}, \mathbb{I}, \mathbf{D}_{2m}^d \quad (l/3 < m \le l).$

PROOF. Observe that $\mathrm{Fix}(\mathbf{SO}(3)) = \{0\}$ since $\mathbf{SO}(3)$ acts irreducibly. Thus maximal subgroups of $\mathbf{SO}(3)$, namely $\mathbf{O}(2)$, $\mathbb{O}$, and $\mathbb{I}$, and maximal subgroups of $\mathbf{O}(3)$ which are not in $\mathbf{SO}(3)$, namely $\mathbf{O}(2)^-$ and $\mathbb{O}^-$, are maximal isotropy subgroups of $\mathbf{O}(3)$ provided their fixed-point subspaces are nonzero. From Theorem 8.1 and Theorem 9.5(e) we see that

$$d(\mathbf{O}(2)) = \begin{cases} 0 & (l \text{ odd}) \\ 1 & (l \text{ even}) \end{cases}$$

$$d(\mathbf{O}(2)^-) = \begin{cases} 1 & (l \text{ odd}) \\ 0 & (l \text{ even}). \end{cases}$$

Therefore, $\mathbf{O}(2)$ is a maximal isotropy subgroup when l is even, and $\mathbf{O}(2)^-$ is a maximal isotropy subgroup when l is odd.

The exceptional subgroups $\mathbb{O}$ and $\mathbb{I}$ are maximal isotropy subgroups for precisely the same l as for the irreducible representation of $\mathbf{SO}(3)$ on V_l.

The exceptional subgroup $\mathbb{O}^-$ is a maximal isotropy subgroup whenever $d(\mathbb{O}^-) > 0$. Now $d(\mathbb{O}^-)$ is "periodic" in l of period 12, by Theorem 9.5(d). That is,

$$d(\mathbb{O}^-)(l + 12) = d(\mathbb{O}^-)(l) + 1.$$

We enumerate $d(\mathbb{O}^-)(l)$ for $l < 12$ in Table 9.1. From the periodicity and Table 9.1 we see that $\mathbb{O}^-$ is a maximal isotropy subgroup for all l except 1, 2, 4, 5, and 8.

Next we decide which of the subgroups $\mathbf{D}^d_{2m}$ are maximal isotropy subgroups. Recall that

$$d(\mathbf{D}^d_{2m}) = [(l + m)/2m] \tag{9.11}$$

from Theorem 9.5(c). Certainly (9.11) implies that $m \leq l$, since otherwise $d(\mathbf{D}^d_{2m}) = 0$. Recall Lemma 2.7, where we showed that a subgroup Σ is a

Table 9.1 Dimensions of Fixed-Point Subspaces for $\mathbb{O}^-$

l	$d(\mathbb{O}^-)(l)$
1	0
2	0
3	1
4	0
5	0
6	1
7	1
8	0
9	1
10	1
11	1

maximal isotropy subgroup if $d(\Sigma) > 0$, and any subgroup $T \supsetneq \Sigma$ satisfies $d(T) = 0$. By Proposition 9.4, $\mathbf{D}^d_{2m}$ is contained in a subgroup T in only two ways:

$$\begin{aligned} &\text{(a)} \quad T = \mathbf{D}^d_{2s} && \text{where } m \text{ divides } s \text{ but } 2m \text{ does not divide } s. \\ &\text{(b)} \quad m = 2, && \text{when } \mathbf{D}^d_4 \subset \mathbb{O}^-. \end{aligned} \tag{9.12}$$

We consider when $\mathbf{D}^d_{2m}$ is not a maximal isotropy subgroup. This happens, for example, if $d(\mathbf{D}^d_{6m}) > 0$, since $\mathbf{D}^d_{2m} \subset \mathbf{D}^d_{6m}$. Now $d(\mathbf{D}^d_{6m}) > 0$ if $3m \leq l$, so $\mathbf{D}^d_{2m}$ is not a maximal isotropy subgroup if $m \leq l/3$.

It is easy to check that $\mathbf{D}^d_{2m} \not\subset \mathbf{D}^d_{4m}$ and that $\mathbf{D}^d_{2m}$ is a maximal isotropy subgroup when $l/3 < m \leq l$, with one possible exception, (9.12(b)). When $m = 2$, $\mathbf{D}^d_4 \subset \mathbb{O}^-$, so we need $d(\mathbb{O}^-) = 0$ for $\mathbf{D}^d_4$ to be a maximal isotropy subgroup. This can only occur when $l/3 < 2 \leq l$, that is, $l = 2, 3, 4, 5$. From Table 9.1 $d(\mathbb{O}^-) = 0$ unless $l = 3$. Thus $\mathbf{D}^d_4$ is a maximal isotropy subgroup when $l = 2, 4, 5$, but not 3.

Next we consider $\mathbf{D}^z_m$. Now $\mathbf{D}^z_m \subset \mathbf{O}(2)^-$, and $d(\mathbf{O}(2)^-) = 1$ when l is odd. So $\mathbf{D}^z_m$ can be a maximal isotropy subgroup only when l is even. When l is even, $d(\mathbf{D}^z_m) > 0$ precisely when $m \leq l$; see Theorem 9.5(b). However, when $m \leq l/2$, we have $d(\mathbf{D}^d_{2m}) > 0$ and Proposition 9.4(d) shows that $\mathbf{D}^z_m \subset \mathbf{D}^d_{2m}$. Hence $\mathbf{D}^z_m$ is never a maximal isotropy subgroup.

We have now shown that every subgroup listed in Theorem 9.6 is a maximal isotropy subgroup. To complete the proof we must rule out any others. If a subgroup of $\mathbf{SO}(3)$ is not a maximal isotropy subgroup for $\mathbf{SO}(3)$ then it cannot be one for $\mathbf{O}(3)$. Inspection of Theorem 8.2 shows that the only class I subgroups we need consider are

$$\begin{aligned} &\text{(a)} \quad \mathbf{SO}(2) \text{ when } l \text{ is odd}, \\ &\text{(b)} \quad \mathbf{D}_m \text{ when } l \text{ is odd and } l/2 < m \leq l,\ m \neq 2, \\ &\text{(c)} \quad \mathbb{T} \text{ when } l = 3, 7, 11. \end{aligned} \tag{9.13}$$

Further, the only class III subgroup not yet considered is $\mathbf{Z}^-_{2m}$.

We check the class I subgroups. We have $\mathbf{SO}(2) \subset \mathbf{O}(2)^-$, and $d(\mathbf{O}(2)^-) = 1$ when l is odd, eliminating (9.13(a)). From Proposition 9.4(d) we see that $\mathbf{D}_m \subset \mathbf{D}^d_{2m}$, which eliminates (9.13(b)). Using Proposition 9.4(a), $\mathbb{T} \subset \mathbb{O}^-$. We check in Table 9.1 to see that $d(\mathbb{O}^-) = 1$ when $l = 3, 7, 11$, eliminating (9.13(c)).

Finally we consider $\mathbf{Z}^-_{2m}$. Proposition 9.4(d) implies that $\mathbf{Z}^-_{2m} \subset \mathbf{D}^d_{2m}$. Moreover, Theorem 9.5(a, c) implies that $d(\mathbf{Z}^-_{2m}) = 2d(\mathbf{D}^d_{2m})$, so $\mathbf{Z}^-_{2m}$ is never a maximal isotropy subgroup. □

It is again easy to read off the isotropy subgroups Σ for which $d(\Sigma) = 1$.

Theorem 9.7. *Let $\mathbf{O}(3)$ act irreducibly on V_l with $-I$ acting as minus the identity. Then the isotropy subgroups with one-dimensional fixed-point subspaces are:*

$\mathbf{O}(2)^-$:	*all odd* l
$\mathbf{O}(2)$:	*all even* l
$\mathbb{O}^-$:	$l = 3, 6, 7, 9–14, 16, 17, 20$
$\mathbb{O}$:	$l = 4, 6, 8–10, 13–15, 17, 19, 23$
$\mathbb{I}$:	$l = 6, 10, 12, 15, 16, 18, 20–22, 24–28, 31–35, 37–39,$ $41, 43, 44, 47, 49, 53, 59$
$\mathbf{D}_{2m}^d$:	$l/3 < m \leq l$, *all* $l \geq 4$
$\mathbf{D}_6^d$:	$l = 3$
$\mathbf{D}_4^d$:	$l = 2.$

PROOF. For $\mathbb{O}$ and $\mathbb{I}$ the results are identical with those in Theorem 8.3. The remaining subgroups Σ are obtained by computing $d(\Sigma)$ from Theorem 9.5 and Table 9.1, for the maximal isotropy subgroups listed in Theorem 9.6. □

(e) The Natural Representation on Spherical Harmonics

Recall that the natural action of $\mathbf{O}(3)$ on the spherical harmonics V_l of order l is the representation of sign $(-1)^l$. It is easy to combine the preceding results to yield a list of maximal isotropy subgroups for the natural representation:

Theorem 9.8. *Let* $\mathbf{O}(3)$ *act on* V_l *in the natural representation. Then the maximal isotropy subgroups are*:

$l = 1$:	$\mathbf{O}(2)^-$
$l = 3$:	$\mathbf{O}(2)^-, \mathbb{O}^-, \mathbf{D}_6^d$
$l = 5$:	$\mathbf{O}(2)^-, \mathbf{D}_{2m}^d \quad (2 \leq m \leq 5)$
$l = 7, 11$:	$\mathbf{O}(2)^-, \mathbb{O}^-, \mathbf{D}_{2m}^d \quad (l/3 < m \leq l)$
$l = 9, 13, 17, 19, 23, 29$:	$\mathbf{O}(2)^-, \mathbb{O}^-, \mathbb{O}, \mathbf{D}_{2m}^d \quad (l/3 < m \leq l)$
all other odd l:	$\mathbf{O}(2)^-, \mathbb{O}^-, \mathbb{O}, \mathbb{I}, \mathbf{D}_{2m}^d \quad (l/3 < m \leq l)$
$l = 2$:	$\mathbf{O}(2) \oplus \mathbf{Z}_2^c$
$l = 4, 8, 14$:	$\mathbf{O}(2) \oplus \mathbf{Z}_2^c, \mathbb{O} \oplus \mathbf{Z}_2^c$
all other even l:	$\mathbf{O}(2) \oplus \mathbf{Z}_2^c, \mathbb{O} \oplus \mathbf{Z}_2^c, \mathbb{I} \oplus \mathbf{Z}_2^c.$

We can also deduce the isotropy subgroups Σ with $d(\Sigma) = 1$:

Theorem 9.9. *The (maximal) isotropy subgroups with one-dimensional fixed-point subspaces for the natural representation of* $\mathbf{O}(3)$ *on* V_l *are*:

$\mathbf{O}(2)^-$:	*all odd* l
$\mathbb{O}^-$:	3, 7, 9, 11, 13, 17
$\mathbb{O}$:	9, 13, 15, 17, 19, 23
$\mathbb{I}$:	15, 21, 25, 27, 31, 33, 35, 37, 39, 41, 43, 47, 49, 53, 59
$\mathbf{D}_{2m}^d$:	$l/3 < m \le l$, *all odd* $l \ge 5$
$\mathbf{D}_6^d$:	$l = 3$
$\mathbf{O}(2) \oplus \mathbf{Z}_2^c$:	*all even* l
$\mathbb{O} \oplus \mathbf{Z}_2^c$:	$l = 4, 6, 8, 10, 14$
$\mathbb{I} \oplus \mathbf{Z}_2^c$:	$l = 6, 10, 12, 16, 18, 20, 22, 24, 26, 28, 32, 34, 38, 44.$

We have enumerated the isotropy subgroups with one-dimensional fixed-point subspaces because these are the ones to which the equivariant branching lemma applies directly. However, recall Remark 3.6(a), that generically solutions with isotropy group Σ exist whenever $d(\Sigma)$ is odd. These subgroups could also be classified from our results with little extra work; we do not pursue this since we have no specific applications of such a classification in mind. When $d(\Sigma)$ is even we have no information on the existence or nonexistence of branches. Despite this we feel that it is important to determine the maximal isotropy subgroups, as part of a general program for understanding bifurcation with symmetry. We discuss this point further in the next section.

We have not discussed the asymptotic stability of the preceding solutions. By Theorem 4.3 we know that when l is even, generically the solutions found from the equivariant branching lemma are unstable. When l is odd, only partial results exist. Chossat and Lauterbach [1987] have shown that generically the axisymmetric solutions—those whose isotropy subgroup contains $\mathbf{SO}(2)$—are unstable.

Exercises

9.1. Consider the Lie group $\Gamma \oplus \mathbf{Z}_2$ where $\mathbf{Z}_2 = \{\pm 1\}$.
   (a) Show that subgroups $\Sigma \subset \Gamma \oplus \mathbf{Z}_2$ fall into three classes:
      (I) $\Sigma \subset \Gamma$,
      (II) $\Sigma = \Delta \oplus \mathbf{Z}_2$ where $\Delta \subset \Gamma$,
      (III) $\Sigma \not\subset \Gamma$ and $-I \notin \Sigma$.
   (b) Let $\pi\colon \Gamma \oplus \mathbf{Z}_2 \to \Gamma$ be projection and let Σ be a class III subgroup. Show that $H = \pi(\Sigma)$ is isomorphic to Σ and that $K = \Sigma \cap \Gamma$ has index two in H.
   (c) Let $\Gamma \oplus \mathbf{Z}_2$ act on the vector space V so that $\mathbf{Z}_2$ acts as $\pm I_V$. Use the trace formula to show that

$$\dim \mathrm{Fix}(\Sigma) = \dim \mathrm{Fix}(K) - \dim \mathrm{Fix}(H)$$

   for all class III subgroups Σ of $\Gamma \oplus \mathbf{Z}_2$.
   (d) Use (c) and Theorem 8.1 to verify Theorem 9.5.

§10.* Generic Spontaneous Symmetry-Breaking

In XI, §1, we posed the basic question of spontaneous symmetry-breaking. Let the compact Lie group Γ act on V and let Σ be an isotropy subgroup.

> Is there a generic set of conditions on Γ-equivariant bifurcation problems $g: V \times \mathbb{R} \to V$ which imply that there exists a branch of solutions to $g = 0$ with isotropy subgroup Σ? (10.1)

(See §3 for a discussion of the term *generic*.)

Proposition 3.2 states that generically we may assume the action of Γ on V to be absolutely irreducible, as we now do. By the equivariant branching lemma, Theorem 3.3, such conditions do exist when the fixed-point subspace of Σ is a one-dimensional, and the bifurcating solution branch is then unique.

The resolution of the problem of spontaneous symmetry-breaking also requires an answer to an equally important question:

> Is there a generic set of conditions on g which imply that there are no solutions to $g = 0$ with isotropy subgroup Σ? (10.2)

If (10.1 and 10.2) could be answered completely, then we could divide isotropy subgroups into three categories: those for which we expect solutions to bifurcate from a Γ-invariant equilibrium, those for which we do not expect solutions to bifurcate, and those for which solutions may or may not exist for open sets of bifurcation problems g.

In this section we discuss the limited results about these questions that are now known. In subsection (a) we define the isotropy lattice and show how the known results on (10.1) pertain to maximal isotropy subgroups in this lattice. In (b) we show that maximal isotropy subgroups are of three types—real, complex, and quaternionic—and that generically only the real ones generate solution branches. In (c) we discuss examples of submaximal isotropy subgroups for which solutions generically exist. In (d) we end by contrasting these results with those known for gradient systems.

(a) The Isotropy Lattice

Let Γ be a compact Lie group acting on V. Define the *isotropy lattice* of Γ to be the set $\mathscr{L}(\Gamma)$ of conjugacy classes $[\Sigma]$ of isotropy subgroups Σ of Γ and write $[\Sigma] < [T]$ if $\Sigma \subset T$ for suitable representatives. In other words, given two isotropy subgroups Σ and T, we have $[\Sigma] < [T]$ if and only if $\gamma^{-1}\Sigma\gamma \subset T$ for some $\gamma \in \Gamma$. We omit the square brackets in future. Note that $\mathscr{L}(\Gamma)$ depends on the representation V.

(In abstract algebra a lattice is a partially ordered set satisfying certain axioms. Strictly speaking, "isotropy lattice" is a misnomer: $\mathscr{L}(\Gamma)$ is in general just a finite partially ordered set. The finiteness is proved in Bredon [1972].)

EXAMPLES 10.1.
(a) $\Gamma = \mathbf{D}_n$ in its standard action on $\mathbb{C}$. The isotropy lattice is

$$\begin{array}{c} \mathbf{D}_n \\ \uparrow \\ \mathbf{Z}_2 \\ \uparrow \\ \mathbb{1} \end{array}$$

(b) $\Gamma = \mathbb{O} \oplus \mathbf{Z}_2^c \subset \mathbf{SO}(3) \oplus \mathbf{Z}_2^c = \mathbf{O}(3)$ acting on $\mathbb{R}^3$ by the restriction of the standard action of $\mathbf{O}(3)$. This is the full symmetry group of the cube, including reflectional symmetries. The isotropy lattice (See Melbourne [1986] and Exercise 10.1) is

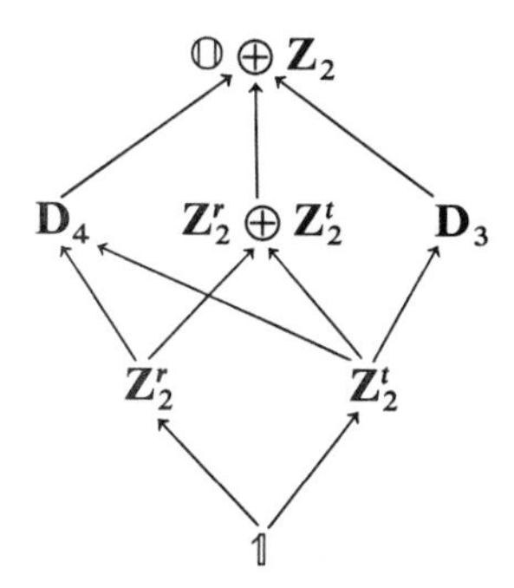

Here $\mathbf{D}_4$, $\mathbf{Z}_2^r \oplus \mathbf{Z}_2^t$, $\mathbf{D}_3$ are the subgroups that fix a line through midpoints of opposite faces, edges, and vertices of the cube, respectively. The two subgroups of order 2, $\mathbf{Z}_2^r$ and $\mathbf{Z}_2^t$, are generated, respectively, by a reflection and by a rotation leaving an edge invariant.

The evidence from many specific calculations may be summed up as follows. Isotropy subgroups "high up" in the lattice correspond to solution branches to the bifurcation equation. Those "low down" do not.

Define an isotropy subgroup Σ to be *maximal* if $\Sigma \neq \Gamma$ and Σ is contained in no other proper isotropy subgroup. (That is, $[\Sigma]$ is a maximal element of the lattice.) Otherwise say that Σ is *submaximal*. So in Example 10.1(a), $\mathbf{Z}_2$ is maximal and $\mathbb{1}$ is submaximal, and in Example 10.1(b) $\mathbf{D}_4$, $\mathbf{Z}_2^r \oplus \mathbf{Z}_2^t$, $\mathbf{D}_3$ are maximal and $\mathbf{Z}_2^r$, $\mathbf{Z}_2^t$, $\mathbb{1}$ are submaximal.

We summarize what is known about symmetry-breaking for these two examples. In (a) solution branches occur generically for $\mathbf{Z}_2$ but not for $\mathbb{1}$. In (b) (see Melbourne [1986] and Exercises 3.1, 10.1) branches occur generically for $\mathbf{D}_4$, $\mathbf{Z}_2^r \oplus \mathbf{Z}_2^t$, $\mathbf{D}_3$ but not for $\mathbf{Z}_2^r$, $\mathbf{Z}_2^t$, $\mathbb{1}$. The results are striking and suggest that the answer to (10.1) may be "the maximal isotropy subgroups ." In fact, this is *not* true in general, as we explain in subsections (b, c).

Nevertheless, we can make some remarks. For example, the equivariant branching lemma really involves a class of maximal isotropy subgroups—namely those with one-dimensional fixed-point subspace. To see this, let $\Sigma \subset \Delta$ be subgroups. Then $\mathrm{Fix}(\Delta) \subset \mathrm{Fix}(\Sigma)$. Therefore, any isotropy subgroup with one-dimensional fixed-point subspace must be maximal. On the other hand, the fixed-point subspaces of maximal isotropy subgroups can be of arbitrarily high dimension: just consider the octahedral group $\mathbb{O}$ in the various irreducible representations of $\mathbf{O}(3)$; see Theorems 8.2 and (8.3(a)).

Recall Cigogna's extension of the equivariant branching lemma, Remark 3.4(b). He uses the topological degree theory of Krasnoselskii [1964] to show that when Σ is a maximal isotropy subgroup with $\dim \mathrm{Fix}(\Sigma)$ odd, then there exist bifurcating solutions with isotropy Σ. His argument is as follows: Let g be a Γ-equivariant bifurcation problem and let Σ be an isotropy subgroup. Suppose that $\dim \mathrm{Fix}(\Sigma)$ is odd. By degree theory $g|\mathrm{Fix}(\Sigma)$ has nontrivial solutions. Let Δ be the isotropy subgroup of such a solution. Since the solution is nontrivial, Δ is a proper subgroup of Γ, and $\Sigma \subset \Delta$ since the solution lies in $\mathrm{Fix}(\Sigma)$. To conclude that there are solutions with isotropy exactly Σ we must know that Σ is a maximal isotropy subgroup. (This argument does not prove the existence of *branches* of solutions, neither does it assert uniqueness.)

In another direction, Field and Richardson [1987] show that for finite groups generated by reflections, all maximal isotropy subgroups have one-dimensional fixed-point subspaces. The octahedral group described earlier provides a special case of this result.

(b) Three Types of Maximal Isotropy Subgroups

In this section we show that maximal isotropy subgroups fall naturally into three types, which we call *real*, *complex*, and *quaternionic*, and we derive some related properties.

We begin with three simple observations whose proof is left to the reader:

Lemma 10.2. *Let $g: V \times \mathbb{R} \to \mathbb{R}$ commute with Γ and let Σ be an isotropy subgroup. Let $N = N_\Gamma(\Sigma)$ be the normalizer of Σ in Γ and let D be the quotient group N/Σ. Then*

(a) *N leaves* $\mathrm{Fix}(\Sigma)$ *invariant, whence D acts naturally on* $\mathrm{Fix}(\Sigma)$.
(b) *$g|\mathrm{Fix}(\Sigma) \times \mathbb{R}$ commutes with this action of D.*
(c) *Suppose that* $\mathrm{Fix}(\Gamma) = \{0\}$ *and that Σ is a maximal isotropy subgroup of Γ. Then the action of D on* $\mathrm{Fix}(\Sigma)$ *is fixed-point-free; that is, each nonidentity element of D fixes only the origin in* $\mathrm{Fix}(\Sigma)$.

Proof. For (a) see Exercise 2.2. Part (b) is obvious. See Exercise 10.2 for part (c). □

The groups that admit fixed-point-free actions have been classified. This lets us make a strong statement about D and its action:

Theorem 10.3. *Let* Γ *act on* V *with* $\mathrm{Fix}(\Gamma) = \{0\}$ *and let* Σ *be a maximal isotropy subgroup. Let* D^0 *be the connected component of the identity in* $D = N_\Gamma(\Sigma)/\Sigma$. *Then either*:

(a) $D^0 = \mathbb{1}$,
(b) $D^0 \cong \mathbf{S}^1$ *and* $\mathrm{Fix}(\Sigma)$ *is a direct sum of* D^0*-irreducible subspaces, on each of which the* D^0*-action is isomorphic to the natural action of* $\mathbf{S}^1$ *on* $\mathbb{C}$.
(c) $D^0 \cong \mathbf{SU}(2)$ *and* $\mathrm{Fix}(\Sigma)$ *is a direct sum of irreducible subspaces under* D^0, *an each of which the* D^0*-action is isomorphic to the natural action of* $\mathbf{SU}(2)$ *on the quaternions* $\mathbb{H}$.

Proof. Since the action of D is fixed-point-free, so is that of D^0. The result then follows from Theorem 8.5 of Bredon [1972]. A sketch proof, using Lie theory, is given in Golubitsky [1983]. □

Remark. $\mathbf{SU}(2)$, the special unitary group in two dimensions, can be identified with the group of unit quaternions

$$\{a + bi + cj + dk\colon a^2 + b^2 + c^2 + d^2 = 1\}$$

under quaternionic multiplication. This acts naturally on $\mathbb{H}$ by left multiplication

$$(q, x) \mapsto qx \qquad (q \in \mathbf{SU}(2), x \in \mathbb{H}).$$

In fact the possibilities for D, rather than just D^0, can be classified. When $D^0 = \mathbb{1}$, there is an extensive list of finite groups D; see Wolf [1967]. When $D^0 \cong \mathbf{S}^1$, D is either $\mathbf{S}^1$ or $\mathbf{O}(2)$. When $D^0 = \mathbf{SU}(2)$, the only possibility is $D = \mathbf{SU}(2)$. See Bredon [1972].

We say that Σ is *real* if $D^0 = \mathbb{1}$, *complex* if $D^0 \cong \mathbf{S}^1$, and *quaternionic* if $D^0 \cong \mathbf{SU}(2)$. In the complex case

$$\dim \mathrm{Fix}(\Sigma) \equiv 0 \pmod 2$$

and in the quaternionic case

$$\dim \mathrm{Fix}(\Sigma) \equiv 0 \pmod 4.$$

In particular, the maximal isotropy subgroups Σ with $\dim \mathrm{Fix}(\Sigma)$ odd are necessarily *real*. This also follows from Exercise 10.5.

This real/complex/quaternionic trichotomy is reminiscent of, but different from, the similar trichotomy that arises for the space $\mathscr{D}$ of commuting linear mappings of an irreducible representation, mentioned in XII, §3.

Examples 10.4.
(a) Complex or quaternionic maximal isotropy subgroups do occur. The simplest examples are very straightforward. For the complex case take $\mathbf{S}^1$

acting naturally on $\mathbb{C}$. Then $\mathbb{1}$ is a maximal isotropy subgroup (since the action is fixed-point-free) and obviously $D = \mathbf{S}^1$. Similarly for the quaternionic case let $\mathbf{SU}(2)$ act naturally on $\mathbb{H}$ and take $\Sigma = \mathbb{1}$.

(b) It is instructive to consider the generic situation in these two cases. For $\mathbf{S}^1$ we have (Lemma VIII, 2.5) the general equivariant bifurcation problem

$$p(u,\lambda)\begin{bmatrix} x \\ y \end{bmatrix} + q(u,\lambda)\begin{bmatrix} -y \\ x \end{bmatrix} = \begin{bmatrix} 0 \\ 0 \end{bmatrix}$$

where $u = x^2 + y^2$. Then for $(x, y) \neq (0,0)$ we obtain

$$p(u,\lambda) = 0,\ q(u,\lambda) = 0.$$

This system of equations is "overdetermined," and generically, for given λ, q will not vanish when p does. Thus there are *no* solution branches in the generic case, though there may be isolated solutions for certain λ. However, if p and q depend on an additional parameter τ, then we may be able to solve for τ and get a "branch" in (x, τ) space (intersecting x-space in isolated points). This actually occurs, and is important, in Hopf bifurcation—see VIII, §2b, and XVI, §3. Here τ is the perturbed period.

(c) The quaternionic case is similar. Let $x = a + bi + cj + dk \in \mathbb{H}$, and let $\mathbf{SU}(2)$ act on h by $q \cdot x = qx$. The invariants are generated by the norm

$$\|x\| = a^2 + b^2 + c^2 + d^2,$$

and the equivariants are generated by the maps α, β, γ, δ where $\alpha(x) = x$, $\beta(x) = xi$, $\gamma(x) = xj$, $\delta(x) = xk$; that is, by $\mathbb{H}$ under right multiplication. See Exercise 10.4. Thus the general $\mathbf{SU}(2)$-equivariant bifurcation problem has the form

$$A\alpha + B\beta + C\gamma + D\delta = 0$$

where A, B, C, D are functions of $\|x\|$ and λ. This is even more overdetermined: when $x \neq 0$, all four of A, B, C, D must vanish simultaneously. Even for isolated values of λ solutions do not occur generically; only if *three* additional parameters are added is a branch of solutions likely to occur.

Unlike the complex case, we know of no natural context where this type of bifurcation problem arises, and no interpretation for the three additional parameters.

Remarks 10.4(b, c) suggest that when considering steady-state bifurcation, generically we should not expect bifurcating solutions to occur with maximal isotropy subgroups that are complex or quaternionic. Indeed this is the case and can be proved using either a theorem of Dancer [1980a] or the equivariant transversality theorem of Bierstone [1977b] and Field [1976]. However, we currently know no example of an absolutely irreducible action having a complex or quaternionic maximal isotropy subgroup. Hence we know of no nontrivial example where this remark may be applied.

Indeed, on the available evidence it may in fact be the case that when Γ acts absolutely irreducibly on V then generically there exist bifurcating solutions corresponding to every maximal isotropy subgroup Σ. As we discuss in (d) later, results of Smoller and Wasserman [1986] and Chow and Lauterbach [1986] show that this possibility does occur for gradient systems. We suspect, however, that for general systems of ODEs this assertion may not always be valid when $\dim \mathrm{Fix}(\Sigma)$ is even.

(c) Submaximal Isotropy Subgroups

By considering $\mathbf{O}(3)$-equivariant bifurcation problems on the space V_l of spherical harmonics, Chossat [1983] and Lauterbach [1986] have found cases where branches with submaximal isotropy subgroups occur generically. Field and Richardson [1987] give examples of finite reflection groups in which branches with submaximal isotropy occur generically. Chossat's example is as follows:

Proposition 10.5. *Let* $\mathbf{O}(3)$ *act on* V_l *where either* $l = 4$ *or* $l \equiv 2 \pmod 4$, $l \neq 2$. *Let* g *be a generic* $\mathbf{O}(3)$*-equivariant bifurcation problem. Then there exists a branch of solutions to* $g = 0$ *whose isotropy subgroup is* $\mathbf{D}_l$.

Remark. In this case $\Delta = \mathbf{D}_l$ is submaximal (contained in $\Sigma = \mathbf{O}(2) \oplus \mathbf{Z}_2^c$) and $\dim \mathrm{Fix}(\Delta) = 2$, $\dim \mathrm{Fix}(\Sigma) = 1$.

Proof. We sketch this: see Chossat [1983] for further details. Restrict g to $\mathrm{Fix}(\Delta)$. Then $N_\Gamma(\Delta)/\Delta \cong \mathbf{Z}_2$ acts as $-I$ on $\mathrm{Fix}(\Delta)$, so $g|\mathrm{Fix}(\Delta)$ is $\mathbf{Z}_2$-equivariant. The linear and quadratic terms of $g|\mathrm{Fix}(\Delta)$ can be computed explicitly in terms of Wigner symbols and are

$$\lambda \begin{bmatrix} x \\ y \end{bmatrix} + \begin{bmatrix} q_1(x, y) \\ q_2(x, y) \end{bmatrix} \tag{10.3}$$

where the linear term can be predicted from absolute irreducibility. There are at least two solutions, the trivial one and the one derived from $\mathbf{O}(2) \oplus \mathbf{Z}_2^c$ by the equivariant branching lemma. The explicit forms for q_1, q_2 show that there are *four* solutions. The two extra ones must have Δ as isotropy subgroup (and are interchanged by $N_\Gamma(\Delta)$). Since (10.3) is 2-determined, or by the implicit function theorem, higher order terms do not destroy these solutions. □

Lauterbach's approach uses topological index theory, which we have not discussed: see Krasnoselskii [1964]. It leads to the following example:

Proposition 10.6. *For a generic* $\mathbf{O}(3)$*-equivariant bifurcation problem on* V_5, *there exists a solution branch with the submaximal isotropy subgroup* $\mathbf{D}_2^d$.

Proof. See Lauterbach [1986]. □

(d) The Variational Case

In this subsection we mention three contexts in which generically every maximal isotropy subgroup leads to solutions. These are:

(a) the bifurcation of steady states in gradient systems,
(b) the bifurcation of periodic solutions for general systems of ODEs,
(c) periodic solutions near equilibria of Hamiltonian systems.

We begin with gradient systems. Let Γ be a Lie subgroup of $\mathbf{O}(n)$ and let $f: \mathbb{R}^n \to \mathbb{R}$ be a Γ-invariant function. Let $g = \nabla f$, where ∇ indicates the gradient with respect to the x variables. Then the system of ODEs

$$\frac{dx}{dt} + g(x) = 0$$

is an *equivariant gradient system.* (In particular, g is here Γ-equivariant; see Exercise 10.6.)

Since we are interested in bifurcations, we may assume f vanishes to second order, whence ∇f vanishes to first order. We shall prove the following result due to Michel [1972]:

Proposition 10.7. *Suppose that Γ is a compact Lie group and g is a bifurcation problem of the form*

$$g(x, \lambda) = \nabla f(x) - \lambda x = 0 \tag{10.4}$$

where f is Γ-invariant and vanishes to second order in x. Let Σ be a maximal isotropy subgroup of Γ. Then for λ arbitrarily close to 0, (10.4) *has at least two distinct solutions in* $\text{Fix}(\Sigma)$.

PROOF. Note that g is Γ-equivariant. Let $h = g|\text{Fix}(\Sigma) \times \mathbb{R}$. Then h maps $\text{Fix}(\Sigma) \times \mathbb{R}$ into $\text{Fix}(\Sigma)$. For fixed $y \neq 0$ and arbitrary λ, $h(y, \lambda) = 0$ if and only if $\nabla f(y) = \lambda y$, which holds if and only if $\nabla f(y) \perp S$ where S is the sphere $\|x\| = \|y\|$ in $\text{Fix}(\Sigma)$. But this is equivalent to $f|S: S \to \mathbb{R}$ having a critical point at y. Since S is compact, $f|S$ has at least two critical points. □

Remarks 10.8.
(a) Smoller and Wasserman [1986] use the Conley index to improve substantially on Michel's result. They obtain generically the existence of solutions with maximal isotropy, to equations $g = \nabla f = 0$ where $f(x, \lambda)$ is a Γ-invariant function having a degenerate singularity at the origin. In particular, they do not assume that g has the restrictive form (10.4). Further, their method only requires g to be gradient-like. See also Chow and Lauterbach [1986].
(b) Smoller and Wasserman [1986a, b, 1987] consider steady-state bifurcation in reaction–diffusion equations in the n-ball with both Dirichlet and Neumann boundary conditions. These equations have a gradient structure and are $\mathbf{O}(n)$-invariant. When $n = 3$ the results of the last section and of their

general theorem couple to give the existence of a number of nontrivial steady states. For $n > 3$, Smoller and Wasserman prove the existence of a number of isotropy subgroups for irreducible actions of $\mathbf{O}(n)$ having one-dimensional fixed-point subspaces.

Hopf bifurcation of periodic solutions provides a second example in which maximal isotropy subgroups always lead to the existence of solutions; see Fiedler [1987]. We discuss this context in more detail in Chapter XVI, though we there confine ourselves to presenting a Hopf bifurcation analogue of the equivariant branching lemma.

Finally, the theory of periodic solutions of equivariant Hamiltonian systems near equilibria provides a third instance in which all maximal isotropy subgroups lead to solutions. See Montaldi, Roberts, and Stewart [1988].

Exercises

10.1. Let Σ be a maximal isotropy subgroup of Γ with $\dim \operatorname{Fix}(\Sigma)$ odd. Show that $D = N_\Gamma(\Sigma)/\Sigma$ has at most two elements.

10.2. Prove Lemma 10.2(c). (*Hint*: If $\delta \in N \sim \Sigma$ fixes an element x of $\operatorname{Fix}(\Sigma)$, then the isotropy subgroup Σ_x of x must be larger then Σ).

10.3. Let Γ act irreducibly on V and let $\Omega \subset \Gamma$ be any subgroup. Let $\mathscr{D}$ be the space of linear mappings on V that commute with Γ. (In XII, §3, we noted that $\mathscr{D}$ is isomorphic to one of $\mathbb{R}$, $\mathbb{C}$, or $\mathbb{H}$. Prove that if $\mathscr{D} \cong \mathbb{C}$ then $\dim \operatorname{Fix}(\Omega) \equiv 0 \pmod 2$, and if $\mathscr{D} \cong \mathbb{H}$ then $\dim \operatorname{Fix}(\Omega) \equiv 0 \pmod 4$. (*Hint*: Show that $\operatorname{Fix}(\Omega)$ is invariant under $\mathscr{D}$ and that the action of $\mathscr{D}$ on $\operatorname{Fix}(\Omega)$ is fixed-point-free, so $\operatorname{Fix}(\Omega)$ is a vector space over $\mathscr{D}$.)

10.4. Let $\mathbf{SU}(2)$ act on $\mathbb{H}$ by left multiplication. Prove that the invariants are generated by the norm $\|x\|$, and the equivariants are generated by the maps α, β, γ, δ where $\alpha(x) = x$, $\beta(x) = xi$, $\gamma(x) = xj$, $\delta(x) = xk$, that is, by $\mathbb{H}$ under right multiplication.

10.5. Let Γ act irreducibly on V. Use Exercise 10.3 to show that if any subgroup Ω of Γ has an odd-dimensional fixed-point subspace, then Γ acts absolutely irreducibly.

10.6. Let $\Gamma \subset \mathbf{O}(n)$ be a Lie subgroup and let $f\colon \mathbb{R}^n \to \mathbb{R}$ be a Γ-invariant function. Show that $g = \nabla f$ is Γ-equivariant.

10.7. This exercise establishes a version of Hopf bifurcation for certain equivariant systems and prepares the way for a quaternionic analogue in Exercise 10.8.

Let Γ act orthogonally on V and let $f\colon V \times \mathbb{R} \to V$ be Γ-equivariant. Consider the ODE $\dot{x} + f(x, \lambda) = 0$. Let Σ be a maximal isotropy subgroup of Γ with $W = \operatorname{Fix}(\Sigma)$. Assume that $\dim W = 2$ and $D = N_\Gamma(\Sigma)/\Sigma \cong \mathbf{S}^1$. Assume that there is a bifurcation "on W," in the sense that $df|W$ is singular at $(0, 0)$.

(a) Show that $df|W$ has either a double zero eigenvalue or a complex conjugate pair of eigenvalues $\pm\omega i$ ($\omega \neq 0$).

(b) Show that $g = f|W$ is $\mathbf{S}^1$-equivariant and hence has the form $g(z, \lambda) = (p + iq)z$ where $p = p(u, \lambda)$, $q = q(u, \lambda)$, $u = z\bar{z}$, and W is identified with $\mathbb{C}$.

(c) Show that for a steady-state branch we require $p = q = 0$, which generically does not occur.

(d) Show that an $\mathbf{S}^1$-equivariant vector field has rotational symmetry. Thus if $g(z, \lambda)$ is tangent to the circle through z at one point, then it is tangent everywhere. Deduce that this circle is invariant under the dynamics, and hence that there is a periodic solution.

(e) Show that the condition for tangency is $\mathrm{Re}(g(z, \lambda)\bar{z}) = 0$; that is, $p(u, \lambda) = 0$. Apply the implicit function theorem to show that this has a solution provided $p_\lambda(0, 0) \neq 0$, and deduce the generic occurrence of a branch of periodic solutions with isotropy Σ.

(f) Interpret the condition $p_\lambda(0, 0) \neq 0$ as the "eigenvalue crossing condition": the eigenvalues of dg cross the imaginary axis with nonzero speed as λ passes through 0.

10.8. This exercise establishes a quaternionic analogue of the results of Exercise 10.7, generalizing results of Cicogna and Gaeta [1987].

Assume the same hypotheses, except that that $\dim W = 4$ and $D \cong \mathbf{SU}(2)$.

(a) Show that at a bifurcation "on W" $df|W$ has either a quadruple zero eigenvalue or a double conjugate pair $\pm\omega i$ $(\omega \neq 0)$.

(b) Show that $g = f|W$ has the form $g(z, \lambda) = (p + iq + jr + ks)z$ $(z \in \mathbb{H})$ where p, q, r, s are functions of $z\bar{z}$ and λ, the bar denoting quaternionic conjugation.

(c) Show that for a steady-state branch we require $p = q = r = s = 0$, which generically does not occur.

(d) By using the symmetry of an $\mathbf{SU}(2)$-equivariant vector field, show that if $g(z, \lambda)$ is tangent to a 3-sphere $\mathbf{S}^3$ through z at one point, then it is tangent everywhere; hence $\mathbf{S}^3$ is invariant under the dynamics.

(e) Show that the condition for tangency is $\mathrm{Re}(g(z, \lambda)\bar{z}) = 0$; that is, $p(u, \lambda) = 0$. Apply the implicit function theorem to show that this has a solution provided $p_\lambda(0, 0) \neq 0$. (Again this is the "eigenvalue crossing condition.")

(f) Show that the invariant 3-sphere $\mathbf{S}^3$ is fibered by invariant circles corresponding to periodic solutions of identical period. (This is the *Hopf fibration*—Heinz Hopf, not Eberhard.) Hence deduce the generic occurrence of a branch of invariant 3-spheres, Hopf-fibered by periodic solutions, with isotropy Σ.

CASE STUDY 4

The Planar Bénard Problem

§0. Introduction

This case study focuses on Rayleigh–Bénard convection. The term *convection* refers to fluid motion caused by the interaction of temperature gradients with a gravitational field; motion occurs because hotter fluid is less dense and therefore tends to rise. In this case study we consider only carefully controlled laboratory experiments in which a horizontal layer of fluid is heated from below and the ensuing motion is observed. Of course, such experiments are intended to shed light on more dramatic geophysical occurrences of convection, such as in the atmosphere and in the interior of the earth (plate tectonics). See Koschmieder [1974], Schluter *et al.* [1965], Sattinger [1978].

In this section we (a) describe the experiments, (b) discuss how symmetry enters the problem, (c) show how to use Fourier analysis to obtain the group action, and (d) combine these ideas to obtain the appropriate symmetry in a suitably reduced model bifurcation problem.

(a) Description of the Experiments

A viscous fluid is contained in a rectangular box whose side walls are insulated and whose lower and upper faces are held at constant temperatures T_l and T_u, respectively. Convection occurs if T_l is sufficiently larger than T_u. In terms of the *Rayleigh number*

$$R = k(T_l - T_u), \tag{0.1}$$

where the constant k is chosen to make R nondimensional, convection occurs when R exceeds the critical Rayleigh number R_c. If R is only slightly greater

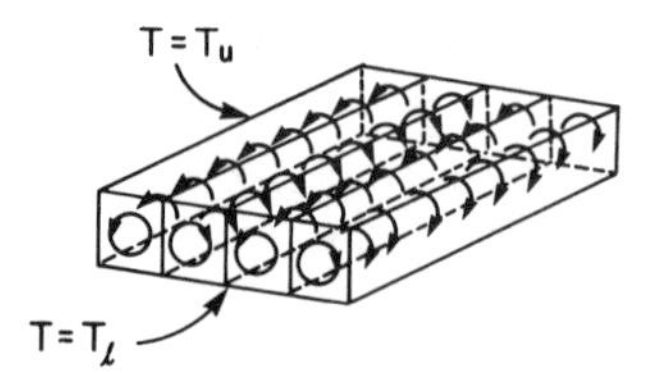

Figure 0.1. Schematic representation of steady rolls.

Figure 0.2. Schematic representation of steady hexagons.

than R_c, then convection ordinarily takes the form of steady *rolls* as indicated schematically in Figure 0.1.

In this case study we focus exclusively on the bifurcation from the conductive state (no motion) to such steady convection. As R is further increased, a bewildering variety of other bifurcations occur. One of these, involving Hopf bifurcation, is discussed in Chapter XVIII, §6.

In his original experiments, Bénard observed a hexagonal convection pattern (indicated schematically in Figure 0.2) rather than rolls. However, in his experiments the top surface of fluid was free, and subsequent analysis (Koschmieder [1974]) has shown that in his experiments surface tension had a larger effect than density variations. Most modern experiments are performed in closed containers, so that they may be more carefully controlled. It is now understood both theoretically and experimentally, (see Koschmieder [1974]) that in a closed container with fixed temperatures at top and bottom, rolls are generally the preferred motion. (We qualify this statement with "generally" since, because of various small effects such as temperature-dependent viscosity, the hexagonal pattern may be preferred over a short range of Rayleigh numbers.) In analyzing this problem we choose a mathematical framework in which it is possible to discuss the competition between roll and hexagonal patterns.

The roll pattern of Figure 0.1 involves only a few rolls, but in this case study we are most interested in experiments involving many rolls. Since the rolls have approximately a square cross section, there will be many rolls if the container is much wider than it is high. Indeed, we consider a mathematical idealization of the experiment in which the fluid is unbounded in the horizontal directions. Since an infinite container is invariant under translations, rotations, and reflections in the horizontal plane, symmetry plays a decisive role in the analysis of the model. Unfortunately the conclusions derived from such symmetry are only approximately applicable to any real experiment. It is a central, but very difficult, problem to understand how to modify these symmetry predictions to include finite size effects, and we shall not address it here. However, see Case study 6, §3, where finite size effects are discussed in a simpler situation.

(b) Symmetry in the Problem

We consider a mathematical idealization of a convection experiment in which the fluid is confined to a region

$$\Omega = \{(x, y, z)\colon 0 < z < 1\}. \tag{0.2}$$

This domain is invariant under $\mathbf{E}_2$, the group of Euclidean motions of the plane, i.e., translations, rotations, and reflections. As we shall see later, the PDE governing the fluid motion commutes with $\mathbf{E}_2$ (acting on functions on Ω by the obvious composition). We are considering bifurcations from the conduction state in which there is no motion and the temperature depends only on z; the convective solutions that bifurcate from this state have lower symmetry. We explore the symmetry issues implicit in this bifurcation before studying details of the PDE.

The domain Ω in (0.2) is not compact, leading to technical problems when one attempts to perform a Liapunov–Schmidt reduction. Specifically, the linearized equations have infinite-dimensional kernel for *every value* of R above a certain threshold. (Later we see this explicitly when discussing the PDE. The fact that irreducible representations of $\mathbf{E}_2$ may be infinite-dimensional serves as a warning of likely trouble.) The standard mathematical way of dealing with these difficulties calls upon the experimental fact that some observed flows seem to be spatially periodic. Thus one restricts attention by fiat to periodic solutions and in this way achieves a compact domain. This approach ignores the effect of other modes and leaves open several important questions. We adopt it with reluctance, but viable alternatives are as yet lacking.

To explain this more fully in simple terms, it is convenient to restrict attention to scalar functions $f\colon \mathbb{R}^2 \to \mathbb{R}$. Such functions can be obtained from a solution to the convection problem by (for example) restricting the vertical component of velocity to the midplane $z = 1/2$. Let e_1 and e_2 be two linearly independent vectors in $\mathbb{R}^2$, let

$$\mathscr{L} = \{n_1 e_1 + n_2 e_2\colon n_i \in \mathbb{Z}\} \tag{0.3}$$

be the lattice spanned by e_1 and e_2, and let $C^\infty_{\mathscr{L}}(\mathbb{R}^2)$ be the space of smooth functions that are periodic with respect to $\mathscr{L}$, i.e., functions f such that

$$f(x + e_1) = f(x + e_2) = f(x) \qquad (x \in \mathbb{R}^2).$$

When PDEs are posed on $C^\infty_{\mathscr{L}}(\mathbb{R}^2)$, rather than $C^\infty(\mathbb{R}^2)$, the symmetry properties of the problem are changed in that

(i) $C^\infty_{\mathscr{L}}(\mathbb{R}^2)$ is not invariant under most rotations in $\mathbf{E}_2$,
(ii) Translations belonging to $\mathscr{L}$ act trivially on $C^\infty_{\mathscr{L}}(\mathbb{R}^2)$.

Thus, given the restriction to periodic functions, the underlying symmetry group in the problem is modified from $\mathbf{E}_2$ to the semidirect sum

$$\Sigma \dotplus \mathbf{T}^2 \tag{0.4}$$

where Σ is the subgroup of $\mathbf{O}(2)$ that leaves $\mathscr{L}$ invariant and $\mathbf{T}^2$ is the quotient $\mathbb{R}^2/\mathscr{L}$.

Remark. The semidirect sum (or semidirect product) is defined as follows: Since Σ leaves $\mathscr{L}$ invariant, there is an action of Σ on the group $\mathbf{T}^2$ induced from its natural action on $\mathbb{R}^2$. Specifically, if $\sigma \in \Sigma$ and $p' \in \mathbf{T}^2$ is the image of $p \in \mathbb{R}^2$ under the natural map $\mathbb{R}^2 \to \mathbb{R}^2/\mathscr{L} = \mathbf{T}^2$, we define

$$\sigma \cdot p' = (\sigma p)'.$$

The semidirect sum of Σ and $\mathbf{T}^2$ is the group denoted by $\Sigma \dotplus \mathbf{T}^2$ and defined to be the cartesian product $\Sigma \times \mathbf{T}^2$ with the group operation

$$(\sigma_1, p_1')(\sigma_2, p_2') = (\sigma_1\sigma_2, \sigma_1 \cdot p_2' + p_1').$$

Then $\mathbf{T}^2$ is a normal subgroup with Σ as quotient group. Unlike the direct sum (or direct product) Σ is not a normal subgroup. In the same way, the Euclidean group $\mathbf{E}_2$ can be thought of as a semidirect sum $\mathbf{O}(2) \dotplus \mathbb{R}^2$.

Since later we study $\mathbf{D}_6 \dotplus \mathbf{T}^2$ in terms of a specific representation, with an explicitly stated action, we do not need this abstract description of its structure. However, it explains the notation used.

As mentioned previously, the competition between roll and hexagonal patterns is of great interest. Therefore, in choosing a space of periodic functions it is desirable that this space contain both types of pattern. Any choice of lattice permits rolls, since such functions depend on only one coordinate. We therefore choose a lattice with hexagonal symmetry, as follows: let

$$e_1 = c(1, 0), \qquad e_2 = c\left(\frac{1}{2}, \frac{\sqrt{3}}{2}\right) \tag{0.5}$$

where c is a unit of length to be chosen later. Then $\mathscr{L}$ is a hexagonal lattice, dual to that in Figure 0.3. For this lattice, the group Σ in (0.4) is the dihedral

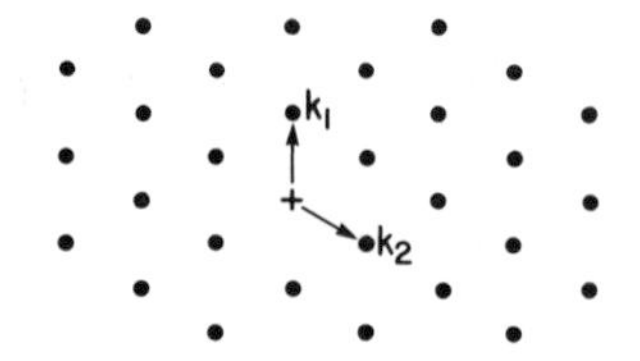

Figure 0.3. Dual of hexagonal lattice $\mathscr{L}$.

group $\mathbf{D}_6$. In the next subsection we discuss the representations of $\mathbf{D}_6 \dot{+} \mathbf{T}^2$ using Fourier analysis.

For other lattices, see Sattinger [1978] and Swift [1984a].

(c) Conclusions from Fourier Analysis

Any function in $C^\infty_{\mathscr{L}}(\mathbb{R}^2)$ can be expanded in a double Fourier series

$$f(x, y) = \sum_{j \in \mathbb{Z}^2} C_j e^{i(j_1 k_1 + j_2 k_2)\cdot(x, y)} \tag{0.6}$$

where k_1, k_2 are basis vectors for the dual lattice to $\mathscr{L}$, namely

$$k_1 = \frac{4\pi}{\sqrt{3}} c(0, 1), \qquad k_2 = \frac{4\pi}{\sqrt{3}} c\left(\frac{\sqrt{3}}{2}, -\tfrac{1}{2}\right). \tag{0.7}$$

If f is real-valued then in (0.6)

$$C_{-j} = \bar{C}_j. \tag{0.8}$$

In a moment we shall discuss how the group $\mathbf{D}_6 \dot{+} \mathbf{T}^2$ acts on the Fourier series (0.6). Before we do this, however, we discuss how a bifurcation problem would lead to an interest in (0.6). Suppose we want to solve a $\mathbf{D}_6 \dot{+} \mathbf{T}^2$-invariant PDE, which operates on functions in $C^\infty_{\mathscr{L}}(\mathbb{R}^2)$. Suppose in addition that there is a trivial ($\mathbf{D}_6 \dot{+} \mathbf{T}^2$-invariant) equilibrium existing independently of a parameter λ. Finally, suppose that as λ is varied a steady-state bifurcation from the trivial solution occurs at λ_0. Then 0 is an eigenvalue of the linear PDE (linearized about the trivial solution at $\lambda = \lambda_0$) and the associated space of eigenfunctions K is invariant under $\mathbf{D}_6 \dot{+} \mathbf{T}^2$. Generically, we expect the action of $\mathbf{D}_6 \dot{+} \mathbf{T}^2$ on K to be irreducible, by Proposition XIII, 3.2. The eigenfunctions in K can be written in double Fourier series (0.6), so we expect K to consist precisely of one of the irreducible representations that occur in the action of $\mathbf{D}_6 \dot{+} \mathbf{T}^2$ on Fourier series (0.6).

We claim that, aside from the trivial representation, the irreducible representations of $\mathbf{D}_6 \dot{+} \mathbf{T}^2$ that arise in this way are either six- or twelve-dimensional. To verify this claim we temporarily work with complex-valued functions; this enables us to discuss the action of $\mathbf{D}_6 \dot{+} \mathbf{T}^2$ on a single term in (0.6), say

$$e^{i(j_1 k_1 + j_2 k_2)\cdot(x, y)}. \tag{0.9}$$

Observe that any exponential (0.9) is mapped into a complex multiple of itself under the action of $\mathbf{T}^2$. Therefore (since the actions of $\mathbf{T}^2$ on distinct such exponentials are nonisomorphic), the $\mathbf{D}_6 \dot{+} \mathbf{T}^2$-invariant subspaces are generated by such terms. On the other hand, the action of $\mathbf{D}_6$ on (0.9) generates six linearly independent exponentials if $j_1 k_1 + j_2 k_2$ is invariant under one of the reflections in $\mathbf{D}_6$, and twelve exponentials if not. The spaces spanned by these exponentials yield six- or twelve-dimensional irreducible representations

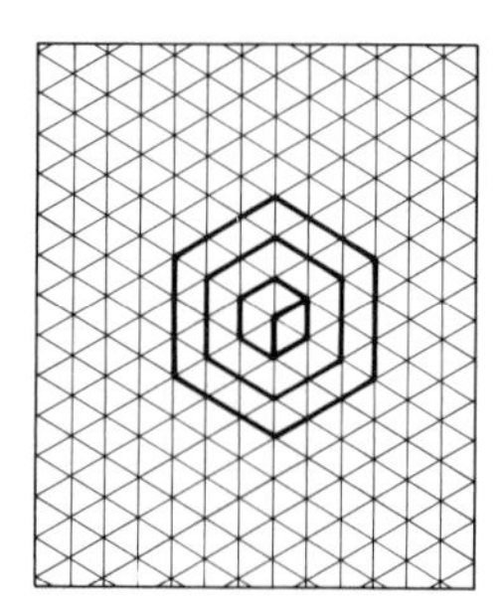

Figure 0.4. Dual hexagonal lattice to $\mathscr{L}$ indicating irreducible representations.

of $\mathbf{D}_6 \dotplus \mathbf{T}^2$. Formula (0.6) then shows that $C^\infty_{\mathscr{L}}(\mathbb{R}^2)$ decomposes as a (topological) direct sum of such irreducible representations. In fact, all of these irreducible representations are distinct, though we shall not need this fact.

There is a simple geometric way to enumerate the various irreducible representations of $\mathbf{D}_6 \dotplus \mathbf{T}^2$ constructed in this way. Consider the dual lattice to $\mathscr{L}$, pictured in Figure 0.4. Each lattice point represents the corresponding exponential (invariant under $\mathbf{T}^2$). The group $\mathbf{D}_6$ acts as the symmetries of a hexagon. The innermost hexagon marked contains exactly six points and corresponds to the *fundamental representation* of $\mathbf{D}_6 \dotplus \mathbf{T}^2$, spanned by the corresponding exponentials.

The next hexagon contains twelve lattice points which correspond to two six-dimensional irreducible representations. One, spanned by the six exponentials corresponding to vertices, is similar to the fundamental representation. (More precisely, $\mathbf{D}_6$ acts on it in exactly the same way, but the action of $p \in \mathbf{T}^2$ in this new representation is obtained by the action of $2p$ in the fundamental representation.) The six midpoints correspond to a "new" six-dimensional representation.

In the third hexagon, a twelve-dimensional irreducible representation appears for the first time, along with a $p \mapsto 3p$ rescaling of the fundamental representation. This analysis can easily be continued to larger hexagons.

In this case study we focus on the fundamental irreducible six-dimensional representation of $\mathbf{D}_6 \dotplus \mathbf{T}^2$. There are three reasons. It occurs first in the hierarchy, it is the simplest representation to consider, and—most importantly—it is a representation that must occur in such bifurcation problems. When we constructed the preceding hexagonal lattice by choosing in (0.5) the generating vectors e_1, e_2, we left in this definition a scale factor c which appears also in the basis vectors for the dual lattice k_1, k_2 as in (0.7). As we shall see in §1, it is possible to choose c so that the first eigenvalue (in λ) has the fundamental six-dimensional irreducible representation as its kernel.

We now give an explicit presentation of this fundamental representation. The generating exponentials are

$$e^{ik_j \cdot (x,y)} \qquad (j = i, \ldots, 6) \tag{0.10}$$

where $k_3 = -(k_1 + k_2)$ and $k_{j+3} = -k_j (j = 1, 2, 3)$. The subspace of $C^\infty_{\mathscr{L}}(\mathbb{R}^2)$ spanned by (0.10) is

$$\left\{ \sum_{j=1}^{6} z_j e^{ik_j \cdot (x,y)} : z_j \in \mathbb{C}, z_4 = \bar{z}_1, z_5 = \bar{z}_2, z_6 = \bar{z}_3 \right\}. \tag{0.11}$$

The conditions on the z_j arise from the restriction to real-valued functions.

Because of these conditions, an element of (0.11) is specified by the first three coefficients z_1, z_2, and z_3; we may thus identify (0.11) with $\mathbb{C}^3$. The action of $\mathbf{D}_6$ on $\mathbb{C}^3$ is as follows:

(a) The permutations $\mathbf{D}_3$ act on the coordinates (z_1, z_2, z_3),

(b) $(z_1, z_2, z_3) \mapsto (\bar{z}_1, \bar{z}_2, \bar{z}_3)$, and (0.12)

(c) for any $p \in \mathbf{T}^2$ the action of p on $\mathbb{C}^3$ is given by $p \cdot (z_1, z_2, z_3) = (e^{ik_1 \cdot p} z_1, e^{ik_2 \cdot p} z_2, e^{ik_3 \cdot p} z_3)$.

(On the right-hand side we as usual think of p as a vector in $\mathbb{R}^2$ representing the corresponding element of $\mathbf{T}^2 = \mathbb{R}^2/\mathscr{L}$.) Alternatively, if we choose p such that $k_1 \cdot p = s$ and $k_3 \cdot p = t$, then the action in (0.12c) may be rewritten as $p \cdot (z_1, z_2, z_3) = (e^{is} z_1, e^{-i(s+t)} z_2, e^{it} z_3)$. For further details see Buzano and Golubitsky [1983]. For a partial study of the twelve-dimensional representations see Kirchgässner [1979].

(d) Symmetry in Rayleigh–Bénard Convection

The preceding discussion of the action of $\mathbf{D}_6 \dotplus \mathbf{T}^2$ on periodic scalar-valued functions of two variables conveys the essentials of the way symmetry acts in the convection problem in an infinite layer, but it is incomplete in one important detail. The domain Ω is invariant (in a sense to be described later), and the governing PDE equivariant, with respect to reflection through the midplane

$$z \mapsto 1 - z. \tag{0.13}$$

The precise definition of this symmetry involves changes of the dependent variables in the PDE; see (1.19). The transformation (0.13) commutes with $\mathbf{D}_6 \dotplus \mathbf{T}^2$. Thus the full group of symmetries of the PDE is

$$\Gamma = (\mathbf{D}_6 \dotplus \mathbf{T}^2) \oplus \mathbf{Z}_2. \tag{0.14}$$

When Liapunov–Schmidt reduction is applied to the PDE, one obtains a bifurcation problem

$$g: \mathbb{C}^3 \times \mathbb{R} \to \mathbb{C}^3 \tag{0.15}$$

commuting with Γ. The action of $\mathbf{D}_6 \dotplus \mathbf{T}^2$ on $\mathbb{C}^3$ is that of (0.12), and it turns out that the midplane reflection in $\mathbf{Z}_2$ acts as minus the identity:

$$(z_1, z_2, z_3) \mapsto (-z_1, -z_2, -z_3). \tag{0.16}$$

In §1 we shall derive the action of this reflection as part of an analysis of the PDE, and in §2 we study bifurcation problems g commuting with Γ. The analysis of the PDE may be omitted without loss of continuity.

Exercises

0.1. Verity that for any integers j_1, j_2

$$f(x, y) = e^{i(j_1 k_1 + j_2 k_2)\cdot(x, y)}$$

is invariant under translation by e_1 and e_2. That is,

$$f((x, y) + e_1) = f(x, y) = f((x, y) + e_2).$$

0.2. Develop the representation theory of

$$\mathbf{D}_4 \dot{+} \mathbf{T}^2$$

for bifurcation problems posed on the planar square lattice; see Swift [1984a].

§1. Discussion of the PDE

We divide the discussion into two subsections. The first is an analytical treatment of the Navier–Stokes equations for Rayleigh–Bénard convection. The second continues the analysis of symmetries.

(a) The Boussinesq Equations

We consider Rayleigh–Bénard convection via the Navier–Stokes equations in the Boussinesq approximation. This means that the fluid is treated as incompressible except for buoyancy; that is, warmer fluid expands, becoming lighter, and therefore experiences an upward force. The state of the fluid is characterized by a velocity field $v(x, y, z, t)$ and a temperature $T(x, y, z, t)$. The fluid is heated from below, so even in the absence of motion there is a temperature gradient. Thus we write

$$T(x, y, z, t) = -\tilde{R}z + \theta(x, y, z, t) \tag{1.1}$$

where $\tilde{R}$ is proportional to the Rayleigh number (0.1) and θ measures the deviation of the temperature from the pure conduction state, $-\tilde{R}z$. The nondimensionalized Boussinesq equations are

$$\text{(a)} \quad \frac{1}{P}\left\{\frac{\partial v}{\partial t} + (v \cdot \nabla)v\right\} = -\nabla p + \theta \mathbf{g} + \Delta v$$

$$\text{(b)} \quad \operatorname{div} v = 0 \tag{1.2}$$

$$\text{(c)} \quad \frac{\partial \theta}{\partial t} + (v \cdot \nabla)\theta = Rv_3 + \Delta\theta.$$

Here R is the Rayleigh number, P the Prandtl number (the dimensionless ratio of viscosity to thermal conductivity), p is the pressure, $\mathbf{g}$ a unit vector in the z-direction representing gravity, v_3 the third component of v, and Δ the Laplace operator.

Apart from the term $\theta\mathbf{g}$ in (1.2a), equations (1.2a,b) are the Navier–Stokes equations for an incompressible fluid. The extra term $\theta\mathbf{g}$ represents the buoyancy force acting on warmer fluid. Similarly, apart from terms $(v \cdot \nabla)\theta$ and Rv_3, (1.2c) is the heat equation. The other two terms represent convection effects; that is, the temperature in a given region of space may change because fluid at a different temperature is flowing into that region. Note that there are two such terms because the temperature (1.1) is the sum of two terms.

Equations (1.2) hold on the domain (0.2). Boundary conditions must be imposed on the faces $z = 0, 1$. The most realistic boundary conditions for modeling the experiments with rigid, conducting boundaries are

$$v = 0, \qquad \theta = 0 \quad \text{on} \quad z = 0, 1. \tag{1.3}$$

However, numerical computation is required to solve the equations with these boundary conditions. Therefore, in this case study we consider the simpler conditions

$$\frac{\partial v_1}{\partial z} = \frac{\partial v_2}{\partial z} = v_3 = 0 \quad \text{on} \quad z = 0, 1, \tag{1.4}$$

which describe a sort of stress-free surface on top *and* bottom. The qualitative results are not affected by the change in boundary conditions. See Schluter, Lortz, and Busse [1965] or Chandrasekhar [1961] for a treatment of (1.3). These references also contain a careful derivation of the governing equations (1.2).

Note that $v = 0$, $\theta = 0$ is an equilibrium solution of (1.2, 1.3). Since θ measures the deviation from the pure conduction state, this "trivial solution" is the *pure conduction state* itself. Bifurcation theory is used to find other equilibria. On deleting terms containing a time derivative from (1.2) we obtain the equation characterizing equilibrium solutions. We write this abstractly as

$$\Phi(u, R) = 0 \tag{1.5}$$

where $u = (v, \theta)$ is a shorthand for the state vector of the system. We discuss the uniqueness question for (1.5) by considering its linearization.

Deleting time derivatives in this way and linearizing (1.5) around the zero solution, we obtain the system

$$\begin{aligned} &\text{(a)} \quad \Delta v - \nabla p + \theta \mathbf{g} = 0 \\ &\text{(b)} \quad \operatorname{div} v = 0 \\ &\text{(c)} \quad \Delta \theta + R v_3 = 0. \end{aligned} \tag{1.6}$$

This is a linear system of PDEs with constant coefficients on a domain which is invariant under translations in x and y. Thus we look for solutions of the form

$$e^{i(kx+ly)} f(z). \tag{1.7}$$

Further, (1.7) is invariant under rotations in the (x, y)-plane. Thus there are solutions of the form (1.7) if and only if there are solutions of the form

$$e^{ik'x} f(z) \tag{1.8}$$

where $k' = \sqrt{(k^2 + l^2)}$, In the following analysis we omit the prime in (1.8).

Note that (1.8) does not depend on y. Write out the components of (1.6), omitting all y-derivatives:

$$\begin{aligned} &\text{(a)} \quad \Delta v_1 - \frac{\partial p}{\partial x} = 0 \\ &\text{(b)} \quad \Delta v_2 = 0 \\ &\text{(c)} \quad \Delta v_3 - \frac{\partial p}{\partial z} + \theta = 0 \\ &\text{(d)} \quad \Delta \theta + R v_3 = 0 \\ &\text{(e)} \quad \frac{\partial v_1}{\partial x} + \frac{\partial v_3}{\partial z} = 0. \end{aligned} \tag{1.9}$$

Equation (1.9(b)), together with the boundary condition (1.4), implies that v_2 is constant. By passing to a moving frame, if necessary, we may assume that $v_2 = 0$. To satisfy (1.9(e)) we express the velocity in terms of the stream function ψ:

$$v_1 = \partial\psi/\partial z, \qquad v_3 = -\partial\psi/\partial x.$$

Finally we eliminate the pressure by taking $\partial/\partial z$ of (1.9(a)) minus $\partial/\partial x$ of (1.9(c)). This yields the system

$$\begin{aligned} &\text{(a)} \quad \Delta^2 \psi - \partial\theta/\partial x = 0. \\ &\text{(b)} \quad \Delta \theta - R \partial\psi/\partial x = 0. \end{aligned} \tag{1.10}$$

We now substitute the form (1.8) of the solution into (1.10). More accurately, to avoid working with complex numbers we insert the phase relationship explicitly and look for a solution of (1.10) of the form

$$\psi = \tilde{\psi}(z) \sin kx, \qquad \theta = \tilde{\theta}(z) \cos kx. \tag{1.11}$$

We find that (1.11) satisfies (1.10) provided $\tilde{\psi}$, $\tilde{\theta}$ satisfy the ODE

$$\begin{bmatrix} (D^2 - k^2)^2 & k \\ -Rk & D^2 - k^2 \end{bmatrix} \begin{bmatrix} \tilde{\psi} \\ \tilde{\theta} \end{bmatrix} = 0 \tag{1.12}$$

where $D = \dfrac{d}{dz}$.

To satisfy (1.4) we must require that

$$\psi = d^2\psi/dz^2 = \theta = 0 \quad \text{for } z = 0, 1. \tag{1.13}$$

The beauty of these boundary conditions is that the natural ansatz to substitute into (1.12), namely $e^{i\lambda z}$, automatically leads to functions that satisfy (1.13), provided λ is an integer multiple of π. By comparison, with (1.3) one must take linear combinations of several complex exponentials to satisfy the boundary conditions, and an explicit solution is no longer practical.

We therefore look for solutions of (1.12) depending on z as $\sin n\pi z$, n an integer. We find that there is a nonzero solution of this form if and only if

$$\det \begin{bmatrix} (-n^2\pi^2 - k^2)^2 & k \\ -Rk & -n^2\pi^2 - k^2 \end{bmatrix} = 0$$

which implies that

$$R = \frac{(n^2\pi^2 + k^2)^3}{k^2}. \tag{1.14}$$

Define $R_c(k)$ to be the right-hand side of (1.14) when $n = 1$, and note that $R \geq R_c(k)$. (The subscript c is a mnemonic for *critical*.) The graph of this function is sketched in Figure 1.1.

It is time to interpret our calculations. We are considering the possibility of bifurcation from the zero solution of an equation $\Phi(u, \mathrm{R}) = 0$. We have shown that the linearization of the equation admits bounded exponential solutions if and only if

$$R \geq \min_k R_c(k) = R_*. \tag{1.15}$$

In other words if $R < R_*$ then the linearization $d\Phi$ is invertible, at least on

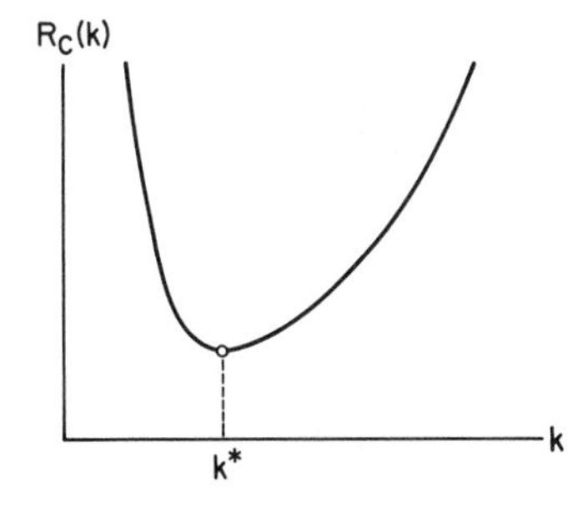

Figure 1.1. Graph of critical Rayleigh numbers as a function of wave number k.

any space amenable to Fourier analysis. This indicates that the zero solution is isolated for $R < R_*$. In contrast, if (1.15) holds, the fact that $\ker d\Phi$ is nonzero indicates that nontrivial solutions bifurcate from the zero solution. However, $\ker d\Phi$ is nonzero for *every* value of R satisfying (1.15), and moreover the kernel is infinite-dimensional since it contains not only solutions with x-dependence e^{ikx} as found previously, but also all the rotations of these as in (1.7). Lacking the technical hypotheses needed to apply Liapunov–Schmidt reduction, we cannot directly conclude that there are bifurcating solutions.

As indicated in §1, we evade these difficulties by restricting attention to periodic solutions. More precisely, consider the lattice spanned by (0.5). Choose the length scale $c = 4\pi/\sqrt{3}k_*$ where k_* is such that $R_c(k_*) = R_*$. The (x, y)-dependence of a function on Ω that is periodic with respect to the lattice may be expanded in a double Fourier series

$$\sum_{j \in \mathbb{Z}^2} C_j(z) e^{i(j_1 k_1 + j_2 k_2)\cdot(x,y)} \tag{1.16}$$

where the fundamental wave numbers for this grid are

$$k_1 = k_*(0, 1), \qquad k_2 = \tfrac{1}{2}k_*(\sqrt{3}, -1).$$

Note that k_1 and k_2 both have length k_* and are inclined at 120°, as in Figure 0.3.

Given the restriction to periodic functions, there are only discrete values of R for which $d\Phi$ fails to be invertible, namely (in an obvious notation)

$$\bigcup_{j \in \mathbb{Z}^2} R_c(j_1 k_1 + j_2 k_2).$$

Moreover, $R_c(k)$ depends only on $|k|^2$ and $\min R_c(k)$ occurs for $|k| = k_*$. Thus $d\Phi$ on the restricted space is invertible for $R < R_* = R_c(k_*)$, is also invertible for $R_* < R < R_* + \varepsilon$ for some $\varepsilon > 0$, while for $R = R_*$, $d\Phi$ has a six-dimensional kernel spanned by the six terms of (1.16) whose wave numbers have length k_*. Thus for R close to R_*, the spatially periodic steady solutions of (1.2, 1.4) correspond by Liapunov–Schmidt reduction to the solutions of a finite system of equations $g(x, R) = 0$, where

$$g\colon \mathbb{R}^6 \times \mathbb{R} \to \mathbb{R}^6.$$

Observe that $\mathbb{R}^6$ is isomorphic to the kernel of the linearized Boussinesq equations and in the preceding calculation may be parametrized by v_3, the vertical component of the velocity field.

(b) Symmetry in the PDE

We define the action of several transformations on solutions of (1.2) as follows. Let $\gamma \in \mathbf{O}(2)$ act on $\mathbb{R}^3$ in the usual way on the horizontal (x, y)-coordinates and trivially on the vertical z-coordinate. Define its action on the PDE by

$$\gamma \cdot (v, \theta)(x) = (\gamma v, \theta)(\gamma^T x). \tag{1.17}$$

If $p \in \mathbb{R}^2$ let

$$p \cdot (v, \theta)(x) = (v, \theta)(x - (p, 0)). \tag{1.18}$$

Finally define reflection μ through the midplane $z = \frac{1}{2}$ by

$$\mu \cdot (v, \theta)(x_1, x_2, x_3) = ((v_1, v_2, v_3), -\theta)(x_1, x_2, 1 - x_3). \tag{1.19}$$

This is the action of the midplane reflection promised in §0(d); see (0.16). In order for this reflection to be a symmetry of the PDE we must also reflect the vertical component of the flow field v_3 and the temperature deviation θ. Applying (1.19) to functions in (0.11) yields the action on $\mathbb{C}^3$ promised in (0.16).

The equations (1.2) are equivariant with respect to these transformations. Thus (1.2, 1.4) has the symmetry group $\mathbf{E}_2 \oplus \mathbf{Z}_2$. As discussed in §0, the restriction to periodic functions cuts this down to the subgroup $\Gamma = (\mathbf{D}_6 \dot{+} \mathbf{T}^2) \oplus \mathbf{Z}_2$. The action of Γ on the six-dimensional kernel of the linearized equations at $R = R_*$ is as in (0.12), and μ acts as minus the identity (0.16). Synthesizing this information, we conclude that periodic solutions of (1.2, 1.4) for R near R_* correspond, via Liapunov–Schmidt reduction, to solutions of a finite system of equations

$$g: \mathbb{R}^6 \times \mathbb{R} \to \mathbb{R}^6,$$

which commutes with the action (0.12, 0.16) of $(\mathbf{D}_6 \dot{+} \mathbf{T}^2) \oplus \mathbf{Z}_2$.

§2. One-Dimensional Fixed-Point Subspaces

In this section we show that the action of $\Gamma = \mathbf{D}_6 \dot{+} \mathbf{T}^2$ has two one-dimensional fixed-point subspaces (corresponding to rolls and hexagons) and the action of $\Gamma \oplus \mathbf{Z}_2$ (which includes the midplane reflection) has four one-dimensional fixed-point subspaces (corresponding to rolls, hexagons, regular triangles, and patchwork quilt). Thus the addition of a single reflectional symmetry alters the types of solution that are expected to occur. We also discuss how the flows corresponding to these solutions may be visualized.

Let $I_R = \mathbf{Z}_2^2 \oplus \mathbf{S}^1$ be the subgroup generated by $(z_1, z_2, z_3) \mapsto (z_1, z_3, z_2)$, $(z_1, z_2, z_3) \mapsto (\bar{z}_1, \bar{z}_2, \bar{z}_3) \in \mathbf{D}_6$ and $(0, t) \in \mathbf{T}^2$, and let $I_H = \mathbf{D}_6$.

Lemma 2.1.
(a) $\mathrm{Fix}(I_R) = \mathbb{R}\{(1, 0, 0)\}$.
(b) $\mathrm{Fix}(I_H) = \mathbb{R}\{(1, 1, 1)\}$.

PROOF. This is straightforward and is left to the reader. A more involved calculation shows that isotropy subgroups of $\mathbf{D}_6 \dot{+} \mathbf{T}^2$ having one-dimensional fixed-point subspaces are conjugate either to $\mathbf{Z}_2^2 \oplus \mathbf{S}^1$ or to $\mathbf{D}_6$. See Buzano and Golubitsky [1983], Theorem 4.4. □

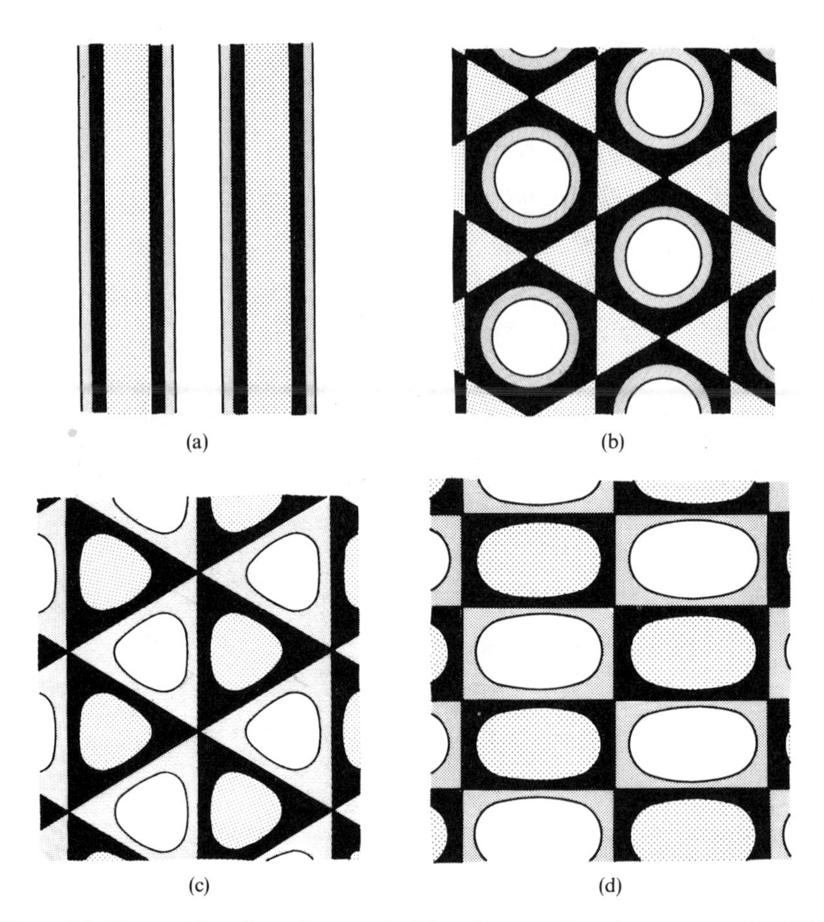

Figure 2.1. Symmetries of steady states in Bénard convection, shown as graphs of the level surfaces of the vertical component of the corresponding eigenfunctions. (a) Rolls; (b) hexagons; (c) regular triangles; (d) patchwork quilt.

The equivariant branching lemma implies that, generically, bifurcation problems with $\mathbf{D}_6 \dot{+} \mathbf{T}^2$ symmetry on $\mathbb{C}^3$ have branches of solutions with isotropy $\mathbf{Z}_2^2 \oplus \mathbf{S}^1$ and $\mathbf{D}_6$. From Figures 0.1 and 0.2 we can see that the symmetries of rolls and hexagons are $\mathbf{Z}_2^2 \oplus \mathbf{S}^1$ and $\mathbf{D}_6$, respectively. In particular, rolls are fixed by translations along the y-axis ($\mathbf{S}^1$), rotation of the (x, y)-plane by 180° $(\bar{z}_1, \bar{z}_2, \bar{z}_3)$, and reflection across the y-axis (z_1, z_3, z_2). Hexagons are clearly fixed by any symmetry in $\mathbf{D}_6$. See Figure 2.1.

Another way to "see" the symmetries of these solutions is to consider the eigenfunctions in the kernel of the linearized Boussinesq equations. Since these eigenfunctions are parametrized by the vertical component v_3 of the velocity field, we may observe the symmetries of solutions by viewing v_3. We do this by restricting v_3 to the midplane ($z = \frac{1}{2}$) and indicating in gray the regions where v_3 is pointing down, and in white the regions where v_3 is pointing up.

See Figure 2.1. The information in this figure is obtained by using the eigenfunctions

$$\begin{aligned} &\text{(a)} \quad e^{ik_1\cdot(x,y)} \\ &\text{(b)} \quad e^{ik_1\cdot(x,y)} + e^{ik_2\cdot(x,y)} + e^{ik_3\cdot(x,y)}. \end{aligned} \tag{2.1}$$

Observe that another planform (or solution type) with $\mathbf{D}_6$ symmetry might also be pictured. The negative of (2.1(b)) corresponds to a hexagonal cell with downwelling at the center and upwelling on the boundary. Busse [1978] calls these two planforms l-hexagons ((2.1(b))) and g-hexagons ($-$(2.1(b))). Here l stands for "liquid" and g for "gas," and the corresponding behavior is typical for these kinds of fluid.

We now turn our attention to Bénard problems having the midplane reflection. Let I_R' be generated by I_R and $(z_1, z_2, z_3) \mapsto (z_1, z_2, z_3)$, the midplane reflection composed with $(\pi, 0) \in \mathbf{T}^2$. Let I_T be the group generated by $\mathbf{D}_3$ (the permutations on (z_1, z_2, z_3)), and $(z_1, z_2, z_3) \mapsto (-\bar{z}_1, -\bar{z}_2, -\bar{z}_3)$ (the midplane reflection composed with conjugation). Let I_P be the eight-element group generated by $(z_1, z_2, z_3) \mapsto (\bar{z}_1, \bar{z}_2, \bar{z}_3)$, (z_2, z_1, z_3), $(z_1, z_2, -z_3)$. (The last generator is just the midplane reflection composed with translation $(\pi, 0)$ through half a period in $\mathbf{T}^2$.)

Lemma 2.2.
(a) $\mathrm{Fix}(I_R') = \mathbb{R}\{(1, 0, 0)\}$.
(b) $\mathrm{Fix}(I_H) = \mathbb{R}\{(1, 1, 1)\}$.
(c) $\mathrm{Fix}(I_T) = \mathbb{R}\{(i, i, i)\}$.
(d) $\mathrm{Fix}(I_P) = \mathbb{R}\{(1, 1, 0)\}$.

PROOF. As with Lemma 2.1, the proof is straightforward and is left to the reader. A more involved calculation shows that isotropy subgroups of $\Gamma \oplus \mathbf{Z}_2$ with one-dimensional fixed-point subspaces are conjugate to precisely one of I_R', I_H, I_T, or I_P. See Golubitsky, Swift, and Knobloch [1984]. □

Thus, when the midplane reflection is an admissible symmetry, the equivariant branching lemma implies that generically there will be four branches of solutions.

To end this section we picture the solutions appearing in the Boussinesq equations in consequence of Lemma 2.2. The pictures of rolls (I_R') and hexagons (I_H) are the same as in Figure 2.1; however, two comments are in order. First, in this case rolls are invariant under midplane reflection coupled with translation by half a period in the x-direction. This symmetry implies that the circulation of rolls in the upper half-space must be identical to the circulation in the lower half-space, except for orientation. Second, the midplane reflection implies that if hexagons with upwelling at the center occur as solutions, then necessarily, for the same parameter values, hexagons with downwelling must also occur.

Regular triangles (I_T) and patchwork quilt (I_P) are also pictured in Figure 2.1.

§3. Bifurcation Diagrams and Asymptotic Stability

We first consider the Bénard problem without the midplane reflection. Observe that the quadratic mapping

$$(z_1, z_2, z_3) \mapsto (\bar{z}_2\bar{z}_3, \bar{z}_1\bar{z}_3, \bar{z}_1\bar{z}_2) \tag{3.1}$$

is $\mathbf{D}_6 \dotplus \mathbf{T}^2$-equivariant. It follows from Theorem XIII, 4.3 (and Remarks XIII, 4.5(a)) that generically solutions found using Lemma 2.1 (namely rolls and hexagons) must be unstable. Therefore, in order to find asymptotically stable solutions using local techniques, one must study degenerate bifurcation problems—those whose second order terms vanish. We shall not launch into such a study here; indeed, the details are quite complicated. We remark that Busse [1962, 1978], using direct calculations based on specific forms of the Boussinesq equations, made substantial progress in this study. Sattinger [1978] put Busse's work into a group-theoretic framework. See also Dancer [1980a]. Finally, Buzano and Golubitsky [1983], using the equivariant singularity theory techniques that will be developed in the next two chapters, classified the least degenerate of these singularities.

When the midplane reflection $(z_1, z_2, z_3) \mapsto (-z_1, -z_2, -z_3)$ is present, the bifurcation equations on $\mathbb{C}^2$ must be odd, and the preceding argument does not preclude the existence of asymptotically stable solutions. Indeed, we shall show that all solutions obtained previously, except patchwork quilt, can in principle be stable. To demonstrate this we must determine explicitly the $\Gamma \oplus \mathbf{Z}_2$-equivariants and use the techniques of XIII, §4, to determine the asymptotic stability of each branch of solutions. Our discussion follows Golubitsky, Swift, and Knobloch [1984].

First we describe the $\mathbf{D}_6 \dotplus \mathbf{T}^2$-invariants and equivariants. Let $u_j = z_j\bar{z}_j$ ($j = 1, 2, 3$) and consider the elementary symmetric polynomials in u_j:

$$\begin{aligned} &\text{(a)} \quad \sigma_1 = u_1 + u_2 + u_3 \\ &\text{(b)} \quad \sigma_2 = u_1u_2 + u_1u_3 + u_2u_3 \\ &\text{(c)} \quad \sigma_3 = u_1u_2u_3. \end{aligned} \tag{3.2}$$

Let $q = z_1z_2z_3 + \bar{z}_1\bar{z}_2\bar{z}_3$.

Theorem 3.1.
(a) *Every* $\mathbf{D}_6 \dotplus \mathbf{T}^2$*-invariant smooth function is a smooth function of* σ_1, σ_2, σ_3, and q.
(b) *The module of* $\mathbf{D}_6 \dotplus \mathbf{T}^2$*-equivariant mappings is freely generated over* $\mathscr{E}(\mathbf{D}_6 \dotplus \mathbf{T}^2)$ *by*

$$\begin{bmatrix} z_1 \\ z_2 \\ z_3 \end{bmatrix} \begin{bmatrix} u_1 z_1 \\ u_2 z_2 \\ u_3 z_3 \end{bmatrix} \begin{bmatrix} u_1^2 z_1 \\ u_2^2 z_2 \\ u_3^2 z_3 \end{bmatrix} \begin{bmatrix} \bar{z}_2 \bar{z}_3 \\ \bar{z}_1 \bar{z}_3 \\ \bar{z}_1 \bar{z}_2 \end{bmatrix} \begin{bmatrix} u_1 \bar{z}_2 \bar{z}_3 \\ u_2 \bar{z}_1 \bar{z}_3 \\ u_3 \bar{z}_1 \bar{z}_2 \end{bmatrix} \begin{bmatrix} u_1^2 \bar{z}_2 \bar{z}_3 \\ u_2^2 \bar{z}_1 \bar{z}_3 \\ u_3^2 \bar{z}_1 \bar{z}_2 \end{bmatrix}. \tag{3.3}$$

We call these generators $X_1, X_2, X_3, Y_1, Y_2, Y_3$, respectively.

Proof. It is straightforward to show that the functions defined here are indeed $\mathbf{D}_6 \dot{+} \mathbf{T}^2$-invariant and -equivariant. A more lengthy, though not particularly more difficult, calculation is needed to show that we have a complete set of generators. See Buzano and Golubitsky [1983], Propositions 2.1 and 3.1. □

The results for $\Gamma \oplus \mathbf{Z}_2$ (where $\Gamma = \mathbf{D}_6 \dot{+} \mathbf{T}^2$) now follow directly. We have the following:

Corollary 3.2.
(a) *Every $\Gamma \oplus \mathbf{Z}_2$-invariant smooth function is a smooth function of $\sigma_1, \sigma_2, \sigma_3$, and $Q = q^2$.*
(b) *Every smooth $\Gamma \oplus \mathbf{Z}_2$-equivariant mapping has the form*

$$g = l_1 X_1 + l_2 X_2 + l_3 X_3 + m_1 q Y_1 + m_2 q Y_2 + m_3 q Y_3 \tag{3.4}$$

where l_j, m_j are $\Gamma \oplus \mathbf{Z}_2$-invariant smooth functions.

Using the notation of (3.4) we list in Table 3.1 the direction of branching and the stability of each of the four primary branches of solutions. The direction of branching is easily determined. First compute (3.4) restricted to each of the fixed-point subspaces listed in Lemma 2.2 (see Table 3.2) and then compute the lowest order terms in these restricted equations to obtain column two of Table 3.1.

Table 3.1. Direction of Branching and Dominant Terms for Eigenvalues of dg. (Note that all functions and derivatives are to be evaluated at the origin.)

Isotropy	Branching Equation	Signs of Eigenvalues of dg
I'_R	$\lambda = -\dfrac{(l_{1,\sigma_1} + l_2)}{l_{1,\lambda}} a^2 + \cdots$	$l_{1,\sigma_1} + l_2, -l_2(4)$
I_T	$\lambda = -\dfrac{(3l_{1,\sigma_1} + l_2)}{l_{1,\lambda}} a^2 + \cdots$	$l_2(2), 3l_{1,\sigma_1} + l_2, m_1$
I_P	$\lambda = -\dfrac{(2l_{1,\sigma_1} + l_2)}{l_{1,\lambda}} a^2 + \cdots$	$-l_2(2), l_2, 2l_{1,\sigma_1} + l_2$
I_H	$\lambda = -\dfrac{(3l_{1,\sigma_1} + l_2)}{l_{1,\lambda}} a^2 + \cdots$	$l_2(2), 3l_{1,\sigma_1} + l_2, -m_1$

Table 3.2. Branching Data

Isotropy	Fix(Σ)	Invariants|Fix(Σ)	$g\|\mathrm{Fix}(\Sigma) \times \mathbb{R}$
I'_R	$(a, 0, 0)$	$\sigma_1 = a^2$; $\sigma_2 = \sigma_3 = q = 0$	$l_1 + a^2 l_2 + a^4 l_3 = 0$
I_T	$i(a, a, a)$	$\sigma_1 = 3a^2$, $\sigma_2 = 3a^4$, $\sigma_3 = a^6$; $q = 0$	$l_1 + a^2 l_2 + a^4 l_3 = 0$
I_P	$(a, a, 0)$	$\sigma_1 = 2a^2$, $\sigma_2 = a^4$; $\sigma_3 = q = 0$	$l_1 + a^2 l_2 + a^4 l_3 = 0$
I_H	(a, a, a)	$\sigma_1 = 3a^2$, $\sigma_2 = 3a^4$, $\sigma_3 = a^6$, $q = 2a^3$	$l_1 + a^2 l_2 + a^4 l_3 + 2a^4(m_1 + a^2 m_2 + a^4 m_3) = 0$

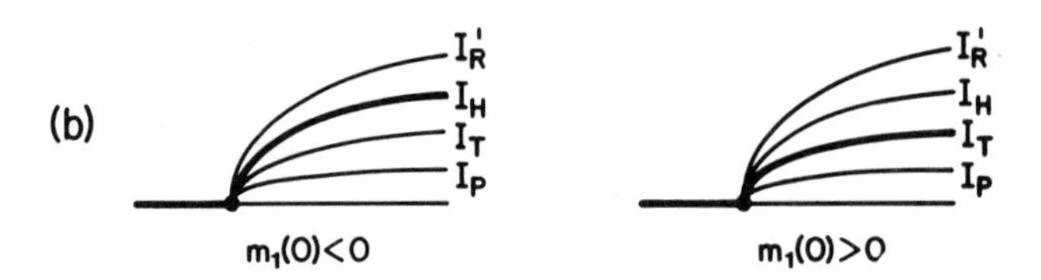

Figure 3.1. Bifurcation diagrams for (3.4) with $l_{1,\lambda}(0) < 0$.

Recall that we write our systems of ODEs as $\dot{z} + g(z, \lambda) = 0$; thus positive eigenvalues indicate stability. We assume that the "trivial" pure conduction solution ($z = 0$) is stable when $\lambda < 0$; that is, $l_{1,\lambda}(0) < 0$. We now indicate which nontrivial solutions can be stable. Note that as long as $l_2(0) \neq 0$, patchwork quilt is unstable.

In order for rolls to be stable, we must have $l_2(0) < 0$ and $l_{1,\sigma_1}(0) + l_2(0) > 0$. It follows that hexagons and regular triangles are unstable and that all branches are supercritical. See Figure 3.1(a).

In order for hexagons or regular triangles to be stable we need both $l_2(0) > 0$ and $3l_{1,\sigma_1}(0) + l_2(0) > 0$. Which of the two solutions is stable depends on the sign of the fifth order term $m_1(0)$. In these cases rolls are unstable and all solution branches are supercritical. See Figure 3.1(b).

We summarize these results. Suppose that the trivial solution is stable when $\lambda < 0$. Then, for any nontrivial branch to be stable, all branches must be supercritical, and then precisely one of rolls, hexagons, or regular triangles is stable. A third order term $l_2(0)$ determines whether rolls are stable. If they are not, then a fifth order term $m_1(0)$ determines which of hexagons or regular triangles is stable. Supercriticality is determined at third order by $l_{1,\sigma_1}(0)$ and $l_2(0)$.

The remainder of this case study is devoted to verifying the asymptotic stability data in Table 3.1, by a complicated calculation that uses the techniques of XIII, §4. For each isotropy subgroup Σ there are four steps:

(a) Determine the restrictions placed on dg by Σ-equivariance,
(b) Find the vectors in $\ker dg$ forced by the action of Γ/Σ,
(c) Compute the eigenvalues of dg directly from the coordinate functions of g,
(d) Compute the lowest order (nonzero) term for each of these eigenvalues.

We derive the results by direct calculation. A more elegant approach to (a) is given in the Exercises.

We begin by setting up the six real coordinates

$$(z_1, z_2, z_3) = (x_1, x_2, x_3, y_1, y_2, y_3) \tag{3.5}$$

where $z_j = x_j + iy_j$. Table 3.3 lists the forms of 6×6 matrices that commute

Table 3.3. Forms of Commuting Matrices

Isotropy	Form of Matrices
I'_R	$\begin{bmatrix} a & 0 & 0 & & & \\ 0 & b & 0 & & 0 & \\ 0 & 0 & b & & & \\ & & & c & 0 & 0 \\ & 0 & & 0 & b & 0 \\ & & & 0 & 0 & b \end{bmatrix}$
I_H, I_T	$\begin{bmatrix} a & b & b & & & \\ b & a & b & & 0 & \\ b & b & a & & & \\ & & & c & d & d \\ & 0 & & d & c & d \\ & & & d & d & c \end{bmatrix}$
I_P	$\begin{bmatrix} a & b & 0 & & & \\ b & a & 0 & & 0 & \\ 0 & 0 & c & & & \\ & & & d & e & 0 \\ & 0 & & e & d & 0 \\ & & & 0 & 0 & f \end{bmatrix}$

with the four isotropy subgroups and answers part (a). We describe the calculations needed for I'_R, the other cases being similar.

The group I'_R is generated by

$$\begin{aligned} &\text{(a)} \quad c(z) = (\bar{z}_1, \bar{z}_2, \bar{z}_3) \\ &\text{(b)} \quad \sigma(z) = (z_1, z_2, -z_3) \\ &\text{(c)} \quad t(z) = (z_1, e^{-it}z_2, e^{it}z_3) \qquad (t \in \mathbb{R}). \end{aligned} \tag{3.6}$$

The matrix form for $c(z)$ in the (x, y)-coordinates of (3.5) is $J = \begin{bmatrix} I & 0 \\ 0 & -I \end{bmatrix}$, and any matrix that commutes with J has the block form $\begin{bmatrix} A & 0 \\ 0 & B \end{bmatrix}$. The matrix form of $\sigma(z)$ is $K = \begin{bmatrix} M & 0 \\ 0 & M \end{bmatrix}$ where

$$M = \begin{bmatrix} 1 & 0 & 0 \\ 0 & 1 & 0 \\ 0 & 0 & -1 \end{bmatrix}.$$

Commuting with K implies that each of A and B has the form

$$\begin{bmatrix} \alpha & \beta & 0 \\ \gamma & \delta & 0 \\ 0 & 0 & \varepsilon \end{bmatrix}.$$

Finally, the matrix form for $t(z)$ is

$$\begin{bmatrix} 1 & 0 & 0 & 0 & 0 & 0 \\ 0 & c & 0 & 0 & s & 0 \\ 0 & 0 & c & 0 & 0 & -s \\ 0 & 0 & 0 & 1 & 0 & 0 \\ 0 & -s & 0 & 0 & c & 0 \\ 0 & 0 & s & 0 & 0 & c \end{bmatrix},$$

where $c = \cos t$ and $s = \sin t$. Commutativity with $t(z)$ yields the form of dg for rolls stated in Table 3.3.

To complete part (b) of the calculation we recall that Γ/Σ forces certain eigenvalues of dg to be zero. More precisely, we obtain vectors in the kernel of dg by differentiating along Γ. In our case the connected component of the identity of Γ is $\mathbf{T}^2$, generated by $(s, 0)$ and $(0, t)$. The possible eigenvectors are then

$$\begin{aligned} &\text{(a)} \quad \frac{d}{ds}(s, 0)\cdot(z_1, z_2, z_3)\bigg|_{s=0} = (iz_1, -iz_2, 0) \\ &\text{(b)} \quad \frac{d}{dt}(0, t)\cdot(z_1, z_2, z_3)\bigg|_{t=0} = (0, -iz_2, iz_3). \end{aligned} \tag{3.7}$$

If we now restrict (3.7) to the various fixed-point subspaces we obtain eigenvectors for dg with eigenvalues 0. These eigenvectors and the restrictions they impose on dg are listed in Table 3.4.

Table 3.4. Null Vectors and Their Implications

Isotropy	Null Vectors	Form of Matrices
I_R'	$\begin{bmatrix} 0 \\ 0 \\ 0 \\ 1 \\ 0 \\ 0 \end{bmatrix}$	$\left[\begin{array}{ccc\|ccc} a & 0 & 0 & & & \\ 0 & b & 0 & & 0 & \\ 0 & 0 & b & & & \\ \hline & & & 0 & 0 & 0 \\ & 0 & & 0 & b & 0 \\ & & & 0 & 0 & b \end{array}\right]$
I_H	$\begin{bmatrix} 0 \\ 0 \\ 0 \\ 1 \\ -1 \\ 0 \end{bmatrix} \begin{bmatrix} 0 \\ 0 \\ 0 \\ 0 \\ 1 \\ -1 \end{bmatrix}$	$\left[\begin{array}{ccc\|ccc} a & b & b & & & \\ b & a & b & & 0 & \\ b & b & a & & & \\ \hline & & & c & c & c \\ & 0 & & c & c & c \\ & & & c & c & c \end{array}\right]$
I_T	$\begin{bmatrix} -1 \\ 1 \\ 0 \\ 0 \\ 0 \\ 0 \end{bmatrix} \begin{bmatrix} 0 \\ 1 \\ -1 \\ 0 \\ 0 \\ 0 \end{bmatrix}$	$\left[\begin{array}{ccc\|ccc} a & a & a & & & \\ a & a & a & & 0 & \\ a & a & a & & & \\ \hline & & & b & c & c \\ & 0 & & c & b & c \\ & & & c & c & b \end{array}\right]$
I_P	$\begin{bmatrix} 0 \\ 0 \\ 0 \\ 1 \\ -1 \\ 0 \end{bmatrix} \begin{bmatrix} 0 \\ 0 \\ 0 \\ 0 \\ 1 \\ 0 \end{bmatrix}$	$\left[\begin{array}{ccc\|ccc} a & b & 0 & & & \\ b & a & 0 & & 0 & \\ 0 & 0 & c & & & \\ \hline & & & 0 & 0 & 0 \\ & 0 & & 0 & 0 & 0 \\ & & & 0 & 0 & d \end{array}\right]$

The eigenvalues of dg can now be read directly from the matrix forms in Table 3.4 and are recorded in Table 3.5. If we write $g = (g_1, g_2, g_3, g_4, g_5, g_6)$ in the coordinates (x, y), then we can determine the eigenvalues directly from certain partial derivatives of the g_j. This information is also included in Table 3.5.

Part (c) of the calculation involves finding the required partial derivatives of g along the four branches of solutions, using the explicit form of g given in (3.4). The last step in computing the eigenvalues of dg is to write down the leading terms of the expressions obtained in part (c). These results are summarized in third column of Table 3.1.

As an example we give the explicit computation of one of these eigenvalues: the coefficient $a = \partial g_1/\partial x_1$ along the branch of rolls. By (3.4),

Table 3.5.

Isotropy	Eigenvalues	Coefficients	
I'_R	$a, b(4), 0$	$a = \partial g_1/\partial x_1$	$b = \partial g_2/\partial x_2$
I_H	$a + 2b, a - b(2), 3c, 0(2)$	$a = \partial g_1/\partial x_1$ $c = \partial g_4/\partial y_1$	$b = \partial g_2/\partial x_1$
I_T	$3a, b + 2c, b - c(2), 0(2)$	$a = \partial g_1/\partial x_1$ $c = \partial g_4/\partial y_2$	$b = \partial g_4/\partial y_1$
I_P	$a + b, a - b, c, d, 0(2)$	$a = \partial g_1/\partial x_1$ $c = \partial g_3/\partial x_3$	$b = \partial g_2/\partial x_1$ $d = \partial g_6/\partial y_3$

$$g_1 = (l_1 + u_1 l_2 + u_1^2 l_3)x_1 + q(m_1 + m_2 u_1 + m_3 u_1^2)(x_2 x_3 - y_2 y_3) \quad (3.8)$$

where $q = z_1 z_2 z_3 + \bar{z}_1 \bar{z}_2 \bar{z}_3$. Along the branch of rolls, $z_2 = z_3 = 0$. Thus $\partial q/\partial x_1 = q = 0$. Moreover along the branch of rolls $l_1 + u_1 l_2 + u_1^2 l_3 = 0$ where, in the notation of Table 3.2, $u_1 = a^2$. From (3.8)

$$\partial g_1/\partial x_1 = x_1 \frac{\partial}{\partial x_1}(l_1 + u_1 l_2 + u_1^2 l_3). \quad (3.9)$$

Now $u_1 \equiv z_1 \bar{z}_1 = x_1^2 + y_1^2$, $\sigma_2 = \sigma_3 = q = \partial\sigma_2/\partial x_1 = \partial\sigma_3/\partial x_1 = \partial q/\partial x_1 = 0$, $\sigma_1 = x_1^2$, and $\partial\sigma_1/\partial x_1 = 2x_1$, along rolls. By (3.9)

$$\begin{aligned}\partial g_1/\partial x_1 &= x_1[l_{1,\sigma_1} 2x_1 + 2x_1 l_2 + x_1^2 l_{2,\sigma_1} 2x_1 + 4x_1^3 l_3 + 2x_1^5 l_{3,\sigma_1}] \\ &= 2x_1^2[l_{1,\sigma_1} + l_2 + x_1^2 l_{2,\sigma_1} + 2x_1^3 l_3 + x_1^4 l_{3,\sigma_1}]. \end{aligned} \quad (3.10)$$

To lowest order $O(x_1^2)$, the expression in (3.10) is $2[l_{1,\sigma_1}(0) + l_2(0)]$, which, up to the positive factor 2, is the first entry in Table 3.1. The remaining calculations are similar.

Exercises

The computation of the eigenvalues in Table 3.5 has just been sketched by using direct "brute force" calculation. In the following exercises we ask the reader to rederive those entries using more abstract notions of representation theory such as isotypic decomposition.

3.1. Verify that the isotypic decompositions for the four maximal isotropy subgroups Σ of $\mathbf{D}_6 \dotplus \mathbf{T}^2$ are as follows: In each case $W_0 = \mathrm{Fix}(\Sigma)$. It may be useful to recall Exercise XII, 1.6.

(a) *Rolls:*

(i) $\mathbb{C}^3 = W_0 \oplus W_1 \oplus W_2$ where

$$W_0 = \mathbb{R}\{(1,0,0)\}$$

$$W_1 = \mathbb{R}\{(i,0,0)\}$$

$$W_2 = \{(0, z_2, z_3)\}.$$

(ii) Show that I'_R acts absolutely irreducibly on W_2.

(b) *Hexagons:*

(i) $\mathbb{C}^2 = W_0 \oplus W_1 \oplus W_2 \oplus W_3$ where

$$W_0 = \mathbb{R}\{(1, 1, 1)\}$$

$$W_1 = \mathbb{R}\{(i, i, i)\}$$

$$W_2 = \mathbb{R}\{(1, -1, 0), (0, 1, -1)\}$$

$$W_3 = \mathbb{R}\{(i, -i, 0), (0, i, -i)\}.$$

(ii) Show that I_H acts absolutely irreducibly on W_2 and W_3.

(c) *Regular Triangles:* After interchanging W_0 and W_1 the isotypic decomposition for regular triangles is identical to that of hexagons in (b).

(d) *Patchwork Quilt:*

(i) $\mathbb{C}^3 = W_0 \oplus W_1 \oplus W_2 \oplus W_3 \oplus W_4 \oplus W_5$ where

$$W_0 = \mathbb{R}\{(1, 1, 0)\}$$

$$W_1 = \mathbb{R}\{(1, -1, 0)\}$$

$$W_2 = \mathbb{R}\{(0, 0, 1)\}$$

$$W_3 = \mathbb{R}\{(i, i, 0)\}$$

$$W_4 = \mathbb{R}\{(i, -i, 0)\}$$

$$W_5 = \mathbb{R}\{(0, 0, i)\}.$$

3.2. Use Theorem XII, 2.5: the isotypic decompositions listed in Exercise 3.1, and the null vectors found in (3.8) to rederive the data of Table 3.5.

CHAPTER XIV

Equivariant Normal Forms

§0. Introduction

From the geometry of equivariant bifurcation problems we move on to their algebra, that is, to singularity theory. Our aim in the next two chapters is to develop Γ-equivariant generalizations of the ideas introduced in Chapters II and III. In particular, in this chapter we develop machinery to solve the recognition problem for Γ-equivariant bifurcation problems. In the next chapter we adapt unfolding theory to the equivariant setting. We also give proofs of the main theorems. When specialized to $\Gamma = \mathbb{1}$ these will provide the promised proof of the Unfolding Theorem III, 2.3.

The idea throughout is to follow the same line of attack as in the non-equivariant case, but to impose suitable symmetry conditions on the various mappings constructed in the proofs. Often these symmetry conditions hold automatically, but occasionally we must average over Γ.

In §1 we introduce the idea of Γ-equivalence and in particular define the Γ-equivariant restricted tangent space $RT(h, \Gamma)$ of a Γ-equivariant bifurcation problem $h \in \vec{\mathscr{E}}(\Gamma)$. We motivate this definition by considering a one-parameter family of Γ-equivalences, just as Chapter II motivated the definition of $RT(g)$. The main result, Theorem 1.3, is a "tangent space constant" result generalizing Theorem II, 2.2. It states that if $RT(h + tp, \Gamma) = RT(h, \Gamma)$ for all $t \in [0, 1]$ then h is Γ-equivalent to $h + tp$, and in particular to $h + p$. We also study the algebraic structure of $RT(h, \Gamma)$, first showing that it is a finitely generated module over $\vec{\mathscr{E}}(\Gamma)$ and then listing generators.

In §2 we prove Theorem 1.3. This is an extended exercise in "putting in the symmetry conditions" and a prototype for several later proofs of deeper results.

In §3 we compute the restricted tangent spaces of general Γ-equivariant bifurcation problems h in five important cases, namely, $\Gamma = \mathbb{1}$, $\mathbf{Z}_2$, $\mathbf{Z}_2 \oplus \mathbf{Z}_2$,

$\mathbf{O}(2)$, and $\mathbf{D}_n$ $(n \geq 3)$. The results are summarized in Table 3.1 for later use. Although detailed computations are given to illustrate the methods, the reader may omit these if so desired.

§4 shows how the machinery developed so far makes it possible to solve certain recognition problems in the equivariant case, that is, to characterize a class of bifurcation problems up to equivalence in terms of conditions on coefficients of the Taylor series. The first examples are the "hilltop" bifurcations on $\mathbb{R}^2$

$$(x^2 - y^2 + \lambda, 2xy)$$

$$(x^2 + \varepsilon\lambda, y^2 + \delta\lambda)$$

where $\Gamma = \mathbb{1}$. The calculations complete the proof of Theorem IX, 2.1. The second example is when $\Gamma = \mathbf{Z}_2 \oplus \mathbf{Z}_2$, and it leads to a normal form for nondegenerate problems with this symmetry, completing the proof of Proposition X, 2.3. The third is $\Gamma = \mathbf{D}_3$, the symmetry group of an equilateral triangle.

In §5 we discuss the preservation—or nonpreservation—of linearized stability by Γ-equivalence. Linearized stability is not, in general, an invariant of Γ-equivalence, but in special cases it is. The main result is as follows: Let Γ be a compact Lie group acting on V, let $g(x, \lambda)$ be a Γ-equivariant bifurcation problem and suppose that Σ is an isotropy subgroup of Γ. Suppose that (x_0, λ_0) is a solution to g in $\mathrm{Fix}(\Sigma)$. Assume that when V is decomposed into irreducible subspaces for Σ,

$$V = V_1 \oplus \cdots \oplus V_k,$$

all V_j are distinct (nonisomorphic) and absolutely irreducible. Then the linearized stability of (x_0, λ_0) is preserved by Γ-equivalence.

§6 is a generalization to the equivariant setting of some results of Chapter II on intrinsic submodules and intrinsic ideals. These ideas are used in §7 to characterize higher order terms in equivariant bifurcation problems (meaning those terms that can be deleted without affecting the problem up to Γ-equivalence). The main theorems are due to Gaffney [1986], based on work of Bruce, du Plessis, and Wall [1987]. We do not give complete proofs since these involve additional algebraic machinery, but we prove a weaker theorem as motivation. The results are used in later chapters to simplify the classification and analysis of equivariant bifurcation problems for various groups, for example $\mathbf{D}_4$.

§1. The Recognition Problem

We recall from Chapter XIII that a steady-state bifurcation problem with symmetry group Γ is a Γ-equivariant germ $g \in \vec{\mathscr{E}}_{x,\lambda}(\Gamma)$ such that $g(0,0) = 0$ and $(dg)_{0,0} = 0$. In this section we define when two such bifurcation problems are Γ-equivalent and give a sufficient condition for Γ-equivalence in Theorem 1.3. This result parallels Theorem II, 2.2, which was basic to the solution of the recognition problem for bifurcation problems in one state variable.

The central idea in this "determinacy" result is the construction of the (formal) tangent space to g under (strong) Γ-equivalence. Roughly speaking, Theorem 1.3 states that if this tangent space remains constant when g is perturbed, then g and its perturbation are Γ-equivalent. The proof of Theorem 1.3 is given in §2.

We divide this section into four subsections:

(a) The definition of Γ-equivalence;
(b) The formulation of the restricted tangent space $RT(g, \Gamma)$ and the statement of the main result, Theorem 1.3;
(c) The algebraic structure of $RT(g, \Gamma)$;
(d) Some remarks on Γ-equivalences.

(a) Γ-Equivalence

Let Γ be a (compact) Lie group acting on a vector space V. Let $\mathscr{L}_\Gamma(V)$ be the vector space of all linear mappings on V that commute with Γ. (Recall from XII, §3, that when V is irreducible, $\mathscr{L}_\Gamma(V)$ is either 1-, 2-, or 4-dimensional over $\mathbb{R}$ and is isomorphic as an algebra to one of $\mathbb{R}$, $\mathbb{C}$, and $\mathbb{H}$.) We define $\mathscr{L}_\Gamma(V)^0$ to be the connected component of $\mathscr{L}_\Gamma(V) \cap \mathbf{GL}(V)$ containing the identity. For example, if Γ acts absolutely irreducibly on V, then $\mathscr{L}_\Gamma(V)$ consists of scalar multiples of the identity I, and

$$\mathscr{L}_\Gamma(V)^0 = \{cI \colon c > 0\}.$$

Definition 1.1. Let $g, h \in \vec{\mathscr{E}}_{x,\lambda}(\Gamma)$ be bifurcation problems with symmetry group Γ. Then g and h are Γ-equivalent if there exists an invertible change of coordinates $(x, \lambda) \mapsto (X(x, \lambda), \Lambda(\lambda))$ and a matrix-valued germ $S(x, \lambda)$ such that

$$g(x, \lambda) = S(x, \lambda) h(X(x, \lambda), \Lambda(\lambda)) \tag{1.1}$$

where

$$X(0, 0) = 0,$$
$$\Lambda(0) = 0, \qquad \Lambda'(0) > 0$$

and for all $\gamma \in \Gamma$

$$\begin{aligned} &\text{(a)} \quad X(\gamma x, \lambda) = \gamma X(x, \lambda), \\ &\text{(b)} \quad S(\gamma x, \lambda)\gamma = \gamma S(x, \lambda), \\ &\text{(c)} \quad S(0, 0), (dX)_{0,0} \in \mathscr{L}_\Gamma(V)^0. \end{aligned} \tag{1.2}$$

We call g and h strongly Γ-equivalent if $\Lambda(\lambda) \equiv \lambda$.

Note that (1.2(a)) can be rephrased as $X \in \vec{\mathscr{E}}_{x,\lambda}(\Gamma)$.

Remarks 1.2.

(a) When $\Gamma = \mathbb{1}$ and $x \in \mathbb{R}$, then Definition 1.1 reduces to the notion of equivalence used in Volume I for bifurcation problems in one state variable.

In particular (1.2(c)) states that $S(0,0) > 0$ and $(\partial X/\partial x)(0,0) > 0$, and (1.2(a, b)) are trivially true.
(b) When $\Gamma = \mathbf{Z}_2$ and $V = \mathbb{R}$ then the definition of equivalence given in Definition 1.1 coincides with that of $\mathbf{Z}_2$-equivalence given in VI, Definition 2.5. Similarly, when $\Gamma = \mathbf{Z}_2 \oplus \mathbf{Z}_2$ and $V = \mathbb{R}^2$, it coincides with that of $\mathbf{Z}_2 \oplus \mathbf{Z}_2$-equivalence given in X, §2. In particular, in this case $\mathscr{L}_{\mathbf{Z}_2 \oplus \mathbf{Z}_2}(\mathbb{R}^2)^0$ consists of 2×2 diagonal matrices whose diagonal entries are both positive. Thus (1.2(c)) is an abstract way of stating (X, 2.7).
(c) When $\Gamma = \mathbb{1}$ and $V = \mathbb{R}^n$, $n > 1$, then the definition of equivalence given in Definition 1.1 differs slightly from that in IX, Definition 1.2. Here $\mathscr{L}_{\mathbb{1}}(\mathbb{R}^n)^0 = \mathbf{GL}(n)^0$, the connected component of the identity in $\mathbf{GL}(n)$. Thus (1.2) may be reformulated as $\det S(0,0) > 0$ and $\det(dX)_{0,0} > 0$. In Chapter IX we required only that these determinants be nonzero. This minor change will cause no problems.

We now comment on the equivariance conditions (1.2). The following is a necessary condition to require of any notion of Γ-equivalence: if h on the right-hand side of (1.1) is Γ-equivariant, then so is g on the left-hand side. Conditions (1.2(a, b)) are sufficient to guarantee this (and moreover, are natural). We ask the reader to verify this in Exercise 1.1.

One of the most important interpretations of the equation $g(x, \lambda) = 0$ is that it determines the steady states of the system of ODEs $dx/dt + g(x, \lambda) = 0$. In this situation it is useful to tell when a given steady state is or is not asymptotically stable, and we wish to do this in terms of a normal form for g under Γ-equivalence. In other words, we want Γ-equivalence to preserve asymptotic stability of solutions. In general, this seems not to be possible; however, when it *is* possible, we need an extra condition, namely (1.2(c)). This is why we include (1.2(c)), instead of the simpler condition $\det S > 0$, $\det dx > 0$, in Definition 1.1. When $\Gamma = \mathbb{1}$ and $\dim V > 0$ then asymptotic stability is *not* preserved even with this extra assumption, which is why it was not included in Definition IX, 1.2. We say more about asymptotic stability in §5.

(b) The Equivariant Restricted Tangent Space

Our next task is to derive the infinitesimal version of (1.1). We define the Γ-*equivariant restricted tangent space* of g to be

$$\begin{aligned} RT(g, \Gamma) = \{S \cdot g + (dg)X \colon & S(x, \lambda) \text{ satisfies } (1.2(b)), \\ & X \text{ satisfies } (1.2a), \text{ and } X(0,0) = 0\}. \end{aligned} \tag{1.3}$$

The motivation for this definition is the same as that for the restricted tangent space in II, §2. Consider a one-parameter family of strong Γ-equivalences of g, namely

$$g(x, \lambda, t) = S(x, \lambda, t) g(X(x, \lambda, t), \lambda) \tag{1.4}$$

where $S(x, \lambda, 0) = I$, $X(0, 0, t) \equiv 0$, and $X(x, \lambda, 0) = x$. Differentiate (1.4) with respect to t and set $t = 0$ to obtain

$$\dot{g}(x, \lambda, 0) = \dot{S}(x, \lambda, 0)g(x, \lambda) + (dg)_{x,\lambda}\dot{X}(x, \lambda, 0). \tag{1.5}$$

Since $X(x, \lambda, t)$ and $S(x, \lambda, t)$ satisfy (1.2(a)) and (1.2(b)), respectively, it follows that $\dot{X}(x, \lambda, 0)$ and $\dot{S}(x, \lambda, 0)$ also satisfy (1.2(a, b)). Moreover, since we demand in Definition 1.1 that X vanish at the origin, it follows that $X(0, 0, t) \equiv 0$. By abuse of notation we set $S(x, \lambda) = \dot{S}(x, \lambda, 0)$ and $X(x, \lambda) = \dot{X}(x, \lambda, 0)$ to arrive at (1.3).

We can now state the basic result of this chapter. Its similarity to Theorem II, 2.2, should be unmistakable.

Theorem 1.3. *Let Γ be a compact Lie group acting on V. Let $h \in \vec{\mathscr{E}}_{x,\lambda}(\Gamma)$ be a Γ-equivariant bifurcation problem and let p be any germ in $\vec{\mathscr{E}}_{x,\lambda}(\Gamma)$. Suppose that*

$$RT(h + tp, \Gamma) = RT(h, \Gamma) \tag{1.6}$$

for all $t \in [0, 1]$. Then $h + tp$ is strongly Γ-equivalent to h for all $t \in [0, 1]$.

Remarks.
(a) We think of $h + tp$ as a family of perturbations of h, indexed by t. In this sense, p is a perturbation of h.
(b) Theorems II, 2.2, and VI, 2.7, are special cases of the preceding result, when $V = \mathbb{R}$ and $\Gamma = \mathbb{1}$, $\mathbf{Z}_2$, respectively.

The proof of Theorem 1.3 is given later in §2.

(c) The Algebraic Structure of $RT(h, \Gamma)$

We begin the discussion by introducing some notation. Define

$$\vec{\mathscr{M}}_{x,\lambda}(\Gamma) = \{g \in \vec{\mathscr{E}}_{x,\lambda}(\Gamma) \colon g(0, 0) = 0\}. \tag{1.7}$$

That is, $\vec{\mathscr{M}}_{x,\lambda}(\Gamma)$ consists of Γ-equivariant mappings that vanish at the origin. (For many group actions $g(0, 0)$ is forced to vanish at zero whenever $g \in \vec{\mathscr{E}}_{x,\lambda}(\Gamma)$, in which case $\vec{\mathscr{M}}_{x,\lambda}(\Gamma) = \vec{\mathscr{E}}_{x,\lambda}(\Gamma)$. By Proposition XIII, 2.2, this occurs precisely when the fixed-point subspace $\mathrm{Fix}(\Gamma) = \{0\}$.) We claim that $\vec{\mathscr{M}}_{x,\lambda}(\Gamma)$ is always finitely generated as a module over $\mathscr{E}_{x,\lambda}(\Gamma)$. This follows since $\vec{\mathscr{E}}_{x,\lambda}(\Gamma)$ is finitely generated; see Exercise 1.2. As standard notation, let

$$X_1, \ldots, X_s \tag{1.8}$$

be a set of generators for the module $\vec{\mathscr{M}}_{x,\lambda}(\Gamma)$ over the ring $\mathscr{E}_{x,\lambda}(\Gamma)$.

Our discussion of the algebraic structure of $RT(h, \Gamma)$ requires a second module, the module of equivariant matrix germs $S(x, \lambda)$. More precisely, following (1.2(b)), define

$$\overleftrightarrow{\mathscr{E}}_{x,\lambda}(\Gamma) = \{n \times n \text{ matrix germs } S(x, \lambda) \colon S(\gamma x, \lambda)\gamma = \gamma S(x, \lambda)\}. \tag{1.9}$$

We claim that $\vec{\mathscr{E}}_{x,\lambda}(\Gamma)$ is a finitely generated module over $\mathscr{E}_{x,\lambda}(\Gamma)$. It is easy to check that the matrix

$$f(x, \lambda)S(x, \lambda)$$

satisfies (1.2(b)) whenever $f \in \mathscr{E}_{x,\lambda}(\Gamma)$ and $S \in \overleftrightarrow{\mathscr{E}}_{x,\lambda}(\Gamma)$. Thus $fS \in \overleftrightarrow{\mathscr{E}}_{x,\lambda}(\Gamma)$, so $\overleftrightarrow{\mathscr{E}}_{x,\lambda}(\Gamma)$ is a module. The proof that $\overleftrightarrow{\mathscr{E}}_{x,\lambda}(\Gamma)$ is finitely generated is outlined in Exercise 1.3. We denote a set of generators for the module $\overleftrightarrow{\mathscr{E}}_{x,\lambda}(\Gamma)$ by

$$S_1, \ldots, S_t. \tag{1.10}$$

We now describe the algebraic structure of $RT(h, \Gamma)$.

Proposition 1.4. *Let Γ be a compact Lie group acting on V and let $h \in \vec{\mathscr{E}}_{x,\lambda}(\Gamma)$. Then $RT(h, \Gamma)$ is a finitely generated submodule of $\vec{\mathscr{E}}_{x,\lambda}(\Gamma)$ over the ring $\mathscr{E}_{x,\lambda}(\Gamma)$. Moreover, $RT(h, \Gamma)$ is generated by*

$$S_1 h, \ldots, S_t h; (dh)(X_1), \ldots, (dh)(X_s) \tag{1.11}$$

where $S_1, \ldots, S_t$ generate $\overleftrightarrow{\mathscr{E}}_{x,\lambda}(\Gamma)$ and $X_1, \ldots, X_s$ generate $\vec{\mathscr{M}}_{x,\lambda}(\Gamma)$.

PROOF. Recall that the Jacobian matrix dh satisfies the symmetry constraint (1.2(b)). More precisely, since $h(\gamma x, \lambda) = \gamma h(x, \lambda)$ the chain rule implies that

$$(dh)_{\gamma x, \lambda}\gamma = \gamma(dh)_{x,\lambda} \tag{1.12}$$

which is (1.2(b)). By (1.3) a typical element of $RT(h, \Gamma)$ has the form

$$Sh + (dh)X \tag{1.13}$$

where $S \in \overleftrightarrow{\mathscr{E}}_{x,\lambda}(\Gamma)$ and $X \in \vec{\mathscr{M}}_{x,\lambda}(\Gamma)$. We noted previously (see also Exercise 1.1) that $Sh \in \vec{\mathscr{E}}_{x,\lambda}(\Gamma)$. By (1.12), $(dh)X$ is also Γ-equivariant. Thus $RT(h, \Gamma) \subset \vec{\mathscr{E}}_{x,\lambda}(\Gamma)$.

To show that $RT(h, \Gamma)$ is a submodule, multiply (1.13) by $f \in \mathscr{E}_{x,\lambda}(\Gamma)$ to obtain

$$(fS)h + (dh)(fX). \tag{1.14}$$

(Since $(dh)_{x,\lambda}$ is linear for each (x, λ), and f takes values in $\mathbb{R}$, we have $f(dh)X = (dh)(fX)$.) Since $\vec{\mathscr{M}}_{x,\lambda}(\Gamma)$ and $\overleftrightarrow{\mathscr{E}}_{x,\lambda}(\Gamma)$ are modules, (1.14) implies that $RT(h, \Gamma)$ is a submodule of $\vec{\mathscr{E}}_{x,\lambda}(\Gamma)$.

Finally, note that S and X in (1.13) may be written as linear combinations over $\mathscr{E}_{x,\lambda}(\Gamma)$ of the S_j and X_j, respectively, so that $RT(h, \Gamma)$ is generated by (1.11). □

(d) Remarks on Γ-Equivalence

In our definition of Γ-equivalence (Definition 1.1) we use a family of *linear* mappings $S(x, \lambda)$. It is fair to ask why we do not consider families of nonlinear mappings. The answer is that nothing is gained from this apparent generalization.

To see why, suppose $Q: V \times \mathbb{R} \times V \to V$ is the germ of a parametrized family of diffeomorphisms on V; that is, for each $(x, \lambda) \in V \times \mathbb{R}$, $Q(x, \lambda, \cdot)$ is viewed as a mapping of V into V. Using Q we can transform the bifurcation problem $g(x, \lambda)$ as follows:

$$h(x, \lambda) = Q(x, \lambda, g(x, \lambda)). \tag{1.15}$$

There are two requirements on h.

(a) The bifurcation diagram of h must be a diffeomorphic image of that of g, and
(b) h must be Γ-equivariant.

To satisfy (a) we demand that

$$Q(x, \lambda, 0) = 0. \tag{1.16a}$$

To satisfy (b) we demand that

$$Q(\gamma x, \lambda, \gamma y) = \gamma Q(x, \lambda, y). \tag{1.16b}$$

Our linear mappings

$$Q(x, \lambda, y) = S(x, \lambda) y$$

meet both demands: (1.16(a)) follows from linearity, and (1.16(b)) is just (1.2(b)).

The following result, due to Mather [1968], shows that considering these more general nonlinear Qs does not, in fact, increase the number of bifurcation problems that are equivalent.

Proposition 1.5. *Let $Q(x, \lambda, y)$ be the germ of a parametrized family of diffeomorphisms on V satisfying* (1.16). *Let $g(x, \lambda)$ be a Γ-equivariant bifurcation problem and let*

$$h(x, \lambda) = Q(x, \lambda, g(x, \lambda)).$$

Then h is strongly Γ-equivalent to g.

Remark. Since we use strong Γ-equivalence, the same proof will work if λ is replaced by a multidimensional parameter, in particular by (λ, α) where $\alpha \in \mathbb{R}^k$ is an unfolding parameter.

PROOF. We show that

$$Q(x, \lambda, g(x, \lambda)) = S(x, \lambda) \cdot g(x, \lambda)$$

where $S(0, 0)$ is an invertible matrix and

$$\gamma^{-1} S(\gamma x, \lambda) \gamma = S(x, \lambda).$$

By (1.16(a) $Q(x, \lambda, 0) = 0$, so we may use Taylor's theorem with parameters (Exercise II, 5.10) to write

$$Q(x, \lambda, y) = A(x, \lambda, y) \cdot y \tag{1.17}$$

where $A(x, \lambda, y)$ is a matrix on V depending smoothly on x, y, and λ. Moreover, $A(0, 0, 0)$ is invertible since $Q(0, 0, \cdot)$ is a diffeomorphism on V.

We claim that it is possible to choose A in (1.17) to satisfy the equivariance property

$$\gamma^{-1} A(\gamma x, \lambda, \gamma y)\gamma = A(x, \lambda, y) \tag{1.18}$$

for all $\gamma \in \Gamma$. Suppose this claim valid, and set

$$S(x, \lambda) = A(x, \lambda, g(x, \lambda)).$$

Then (1.18) implies that S satisfies (1.2(b)); and (1.17) implies that

$$Q(x, \lambda, g(x, \lambda)) = S(x, \lambda) g(x, \lambda)$$

which proves the proposition.

To prove the claim that A satisfying (1.18) can be found we average over Γ. Let

$$B(x, \lambda, y) = \int_\Gamma \gamma^{-1} A(\gamma x, \lambda, \gamma y)\gamma.$$

If $\delta \in \Gamma$ then

$$\begin{aligned} B(\delta x, \lambda, \delta y) &= \int_\Gamma \gamma^{-1} A(\gamma\delta x, \lambda, \gamma\delta y)\gamma \\ &= \int_\Gamma \delta(\gamma\delta)^{-1} A(\gamma\delta x, \lambda, \gamma\delta y)\gamma\delta \cdot \delta^{-1} \\ &= \delta\left(\int_\Gamma (\gamma')^{-1} A(\gamma' x, \lambda, \gamma' y)\gamma'\right)\delta^{-1} \end{aligned}$$

where $\gamma' = \gamma\delta$. Therefore,

$$B(\delta x, \lambda, \delta y) = \delta B(x, \lambda, y)\delta^{-1}$$

for all $\delta \in \Gamma$. Replacing δ by γ we rewrite this as

$$\gamma^{-1} B(\gamma x, \lambda, \gamma y)\gamma = B(x, \lambda, y).$$

Thus B satisfies (1.18).

Finally, the equivariance property (1.16(b)) implies that

$$\int_\Gamma \gamma^{-1} Q(\gamma x, \lambda, \gamma y) = Q(x, \lambda, y).$$

We integrate Equation (1.17) over Γ to obtain

$$Q(x, \lambda, y) = B(x, \lambda, y) y,$$

proving the claim. □

EXERCISES

1.1. Let $h \in \vec{\mathscr{E}}_{x,\lambda}(\Gamma)$ and let g be Γ-equivalent to h. That is,

$$g(x, \lambda) = S(x, \lambda) = S(x, \lambda)h(X(x, \lambda), \Lambda(\lambda))$$

where $S \in \vec{\mathscr{E}}_{x,\lambda}(\Gamma)$ and $x \in \vec{\mathscr{M}}_{x,\lambda}(\Gamma)$. Show that g is also Γ-equivariant.

1.2. Let Γ be a compact Lie group acting on V. Show that $\vec{\mathscr{M}}_{x,\lambda}(\Gamma)$ is a finitely generated module over $\mathscr{E}_{x,\lambda}(\Gamma)$.
Hint: Show that
(a) $\vec{\mathscr{E}}_{x,\lambda}(\Gamma) \supset \vec{\mathscr{M}}_{x,\lambda}(\Gamma) \supset \mathscr{M}_{x,\lambda}(\Gamma)\vec{\mathscr{E}}_{x,\lambda}(\Gamma)$ where $\mathscr{M}_{x,\lambda}(\Gamma) = \{f \in \mathscr{E}_{x,\lambda}(\Gamma): f(0) = 0\}$, and that
(b) There is a finite-dimensional vector subspace W such that

$$\vec{\mathscr{M}}_{x,\lambda}(\Gamma) = W \oplus \mathscr{M}_{x,\lambda}(\Gamma)\vec{\mathscr{E}}_{x,\lambda}(\Gamma).$$

Note that the ideal $\mathscr{M}_{x,\lambda}(\Gamma)$ may be written as

$$\mathscr{M}_{x,\lambda}(\Gamma) = \langle u_1, \ldots, u_r \rangle$$

where $u_1, \ldots, u_r$ are the invariant generators for $\mathscr{E}_{x,\lambda}(\Gamma)$.
(c) Show that the module $\mathscr{M}_{x,\lambda}(\Gamma)\mathscr{E}_{x,\lambda}(\Gamma)$ is generated by

$$\{u_j g_k: 1 \le j \le r, 1 \le k \le s\}$$

where $g_1, \ldots, g_s$ generate $\vec{\mathscr{E}}_{x,\lambda}(\Gamma)$.
(d) Show that $\vec{\mathscr{M}}_{x,\lambda}(\Gamma)$ is generated by the generators for $\mathscr{M}_{x,\lambda}(\Gamma)\vec{\mathscr{E}}_{x,\lambda}(\Gamma)$ augmented by $w_1, \ldots, w_l$ where the w_k form a basis for W.

1.3. Show that the module $\vec{\mathscr{E}}_{x,\lambda}(\Gamma)$ is finitely generated.
(*Hint*: Let $\mathrm{Hom}(V, V)$ be the space of linear mappings $V \to V$. The group Γ acts naturally on $\mathrm{Hom}(V, V)$ by similarities. Specifically, let $\gamma \in \Gamma$, $T \in \mathrm{Hom}(V, V)$, and define

$$\gamma \cdot T = \gamma T \gamma^{-1}. \tag{1.19}$$

Think of $S \in \vec{\mathscr{E}}_{x,\lambda}(\Gamma)$ as a mapping

$$S: V \times \mathbb{R} \to \mathrm{Hom}(V, V). \tag{1.20}$$

Show that the equivariance condition (1.2(b)) implies that S in (1.20) is Γ-equivariant when the action of Γ on $\mathrm{Hom}(V, V)$ is (1.19). Now apply Theorem XII, 6.8.)

§2.* Proof of Theorem 1.3

We now give the proof of Theorem 1.3. It is an essentially routine adaptation of the proof of Theorem 2.2 in Chapter II, §11, keeping track of symmetry properties. We give it in some detail to illustrate this adaptation procedure. We have $h, p \in \vec{\mathscr{E}}_{x,\lambda}(\Gamma)$, and the "tangent space constant" property

$$RT(h + tp, \Gamma) = RT(h, \Gamma) \tag{2.1}$$

for all $t \in [0, 1]$. The objective is to show that $h + tp$ is strongly Γ-equivalent to h for all $t \in [0, 1]$ (and in particular that $h + p$ is, by setting $t = 1$). As in Volume I we begin by localizing the problem.

Proposition 2.1. *Let $h, p \in \vec{\mathscr{E}}_{x,\lambda}(\Gamma)$ be germs such that* (2.1) *holds for t near* 0. *Then $h + tp$ is strongly Γ-equivalent to h for all t sufficiently near* 0.

Remark. Given a proof of Proposition 2.1, Theorem 1.3 follows exactly as in II, §11, by a compactness/connectedness argument.

PROOF OF PROPOSITION 2.1. We proceed as in the proof of Proposition II, 11.1. We seek to construct mappings $X(x, \lambda, t)$ and $S(x, \lambda, t)$ satisfying conditions analogous to those of (II, 11.9). Namely, let

$$H(x, \lambda, t) = h(x, \lambda) + tp(x, \lambda). \tag{2.2}$$

We require:

$$\begin{aligned} &\text{(a)} \quad S(x, \lambda, t)H(X(x, \lambda, t), \lambda, t) = h(x, \lambda), \\ &\text{(b)} \quad X(0, 0, t) \equiv 0, \; X(x, \lambda, 0) \equiv x, \\ &\text{(c)} \quad S(x, \lambda, 0) \equiv I, \end{aligned} \tag{2.3}$$

exactly as in (II, 11.9), together with the appropriate symmetry conditions:

$$\begin{aligned} &\text{(d)} \quad X(\gamma x, \lambda, t) = \gamma X(x, \lambda, t), \\ &\text{(e)} \quad S(\gamma x, \lambda, t)\gamma = \gamma S(x, \lambda, t). \end{aligned} \tag{2.3}$$

If such mappings can be found, then Proposition 2.1 will follow immediately, as before.

Differentiate (2.3(a)) with respect to t, to obtain

$$\begin{aligned} 0 &= h_t(x, \lambda) \\ &= S_t(x, \lambda, t)H(X(x, \lambda, t), \lambda, t) + S(x, \lambda, t)H_t(X(x, \lambda, t), \lambda, t) \\ &= S_t(x, \lambda, t)H(X(x, \lambda, t), \lambda, t) + S(x, \lambda, t)[h(X(x, \lambda, t), \lambda) + tp(X(x, \lambda, t), \lambda)]_t \\ &= S_t(x, \lambda, t)H(X(x, \lambda, t), \lambda, t) + S(x, \lambda, t)(dh)_{(X(x, \lambda, t), \lambda)}X_t(x, \lambda, t) \\ &\quad + S(x, \lambda, t)tp_x(X(x, \lambda, t), \lambda)X_t(x, \lambda, t) + S(x, \lambda, t)p(X(x, \lambda, t), \lambda) \\ &= S_t(x, \lambda, t)H(X(x, \lambda, t), \lambda, t) + S(x, \lambda, t)(dH)_{(X(x, \lambda, t), \lambda, t)}X_t(x, \lambda, t) \\ &\quad + S(x, \lambda, t)p(X(x, \lambda, t), \lambda). \end{aligned} \tag{2.4}$$

Now replace $X(x, \lambda, t)$ by x (and hence x by $X^{-1}(x, \lambda, t)$) and solve (2.4) for $p(x, \lambda)$, obtaining

$$\begin{aligned} p(x, \lambda) = &-S^{-1}(X^{-1}(x, \lambda, t), \lambda, t)S_t(X^{-1}(x, \lambda, t), \lambda, t)H(x, \lambda, t) \\ &- (dH)_{(x, \lambda, t)}X_t(X^{-1}(x, \lambda, t), \lambda, t). \end{aligned} \tag{2.5}$$

Now, as in (II, 11.2), suppose we can write p in the form

$$p(x, \lambda) = -a(x, \lambda, t)H(x, \lambda, t) - (dH)_{(x, \lambda, t)}b(x, \lambda, t) \tag{2.6}$$

where we impose symmetry conditions

$$\begin{aligned} &\text{(a)} \quad a(\gamma x, \lambda, t)\gamma = \gamma a(x, \lambda, t), \\ &\text{(b)} \quad b(\gamma x, \lambda, t) = \gamma b(x, \lambda, t), \end{aligned} \tag{2.7}$$

and also

$$\text{(c)} \quad b(0, 0, t) = 0.$$

Note that (2.6 and 2.7) hold for each t individually using (2.1). The problem is that we do not know that a and b are smooth in t. This issue is addressed in Lemma 2.2.

We now solve the $n \times n$ system of ODEs

$$\begin{aligned} &X_t(x, \lambda, t) = b(X(x, \lambda, t), \lambda, t) \\ &X(x, \lambda, 0) = x \end{aligned} \tag{2.8}$$

noting that $X(0, 0, t) = 0$ is consistent with these equations. To show that X satisfies the symmetry requirement (2.3(d)) we appeal to the uniqueness of solutions to (2.8). Specifically, let

$$Y(x, \lambda, t) = \gamma^{-1} X(\gamma x, \lambda, t).$$

Then

$$\begin{aligned} Y_t(x, \lambda, t) &= \gamma^{-1} X_t(\gamma x, \lambda, t) \\ &= \gamma^{-1} b(X(\gamma x, \lambda, t), \lambda, t) \\ &= b(\gamma^{-1} X(\gamma x, \lambda, t), \lambda, t) \\ &= b(Y(x, \lambda, t), \lambda, t). \end{aligned}$$

Also

$$Y(x, \lambda, 0) = \gamma^{-1} X(\gamma x, \lambda, 0) = \gamma^{-1}\gamma x = x.$$

Hence Y is also a solution to (2.8) with the same initial conditions. Uniqueness implies that $Y(x, \lambda, t) \equiv X(x, \lambda, t)$ as germs, thus establishing (2.3(d)).

Now solve the system of ODEs

$$\begin{aligned} &S_t(x, \lambda, t) = S(x, \lambda, t) a(X(x, \lambda, t), \lambda, t) \\ &S(x, \lambda, 0) = I_n. \end{aligned} \tag{2.9}$$

This is a linear system in n^2 variables, hence has a solution valid for all t. Again we establish the required symmetry property (2.3(e)) by an appeal to uniqueness. Define

$$T(x, \lambda, t) = \gamma^{-1} S(\gamma x, \lambda, t)\gamma.$$

Then

$$\begin{aligned} T_t(x, \lambda, t) &= \gamma^{-1} S_t(\gamma x, \lambda, t)\gamma \\ &= \gamma^{-1} S(\gamma x, \lambda, t) a(X(\gamma x, \lambda, t), \lambda, t)\gamma \end{aligned}$$

$$\begin{aligned}&= \gamma^{-1}S(\gamma x, \lambda, t)a(\gamma X(x, \lambda, t), \lambda, t)\gamma \\ &= \gamma^{-1}S(\gamma x, \lambda, t)\gamma a(X(x, \lambda, t), \lambda, t) \\ &= T(x, \lambda, t)a(X(x, \lambda, t), \lambda, t).\end{aligned}$$

Also

$$T(x, \lambda, 0) = \gamma^{-1}S(\gamma x, \lambda, 0)\gamma = \gamma^{-1}I_n\gamma = I_n.$$

So by uniqueness $T = S$, and (2.3(e)) holds.

With this choice of X and S it follows from (2.8, 2.9, 2.6) that (2.5) holds. Therefore $S(x, \lambda, t)H(X(x, \lambda, t), \lambda, t)$ is independent of t by (2.4). Setting $t = 0$ we find that its value is $h(x, \lambda)$, establishing (2.3(a)).

The result, therefore, follows, provided we show that p can be written in the form 2.6. This is a consequence of the following:

Lemma 2.2. *If* $RT(h + tp, \Gamma) = RT(h, \Gamma)$ *for all* t *near* 0, *then there exist germs* a, b satisfying (2.6) *and* (2.7(a, b, c)).

Proof. Observe that $RT(h, \Gamma)$ has a finite set of generators

$$h = J_1(h), J_2(h), \ldots, J_k(h)$$

which are linear in h. Thus, if

$$RT(h + p, \Gamma) = RT(h, \Gamma) \quad \text{we have}$$

$$\begin{aligned} h + p &= a_{11}h + \sum_{l=2}^{k} a_{1l}J_l(h) \\ J_m(h + p) &= a_{m1}h + \sum_{l=2}^{k} a_{ml}J_l(h) \end{aligned} \tag{2.10}$$

where the $a_{ij} \in \mathscr{E}(\Gamma)$. Therefore, by linearity of J_m, we have

$$\begin{aligned} p &= b_{11}h + \sum b_{1l}J_l(h) \\ J_m(p) &= b_{m1}h + \sum b_{ml}J_l(h) \end{aligned} \tag{2.11}$$

where

$$\begin{aligned} b_{ii} &= a_{ii} - 1 \qquad (i = 1, \ldots, k) \\ b_{ij} &= a_{ij} \qquad (i \neq j). \end{aligned}$$

We let

$$H = h + tp$$

and write (2.11) in matrix form, as

$$\begin{bmatrix} p \\ J_2(p) \\ \vdots \\ J_m(p) \end{bmatrix} = B \begin{bmatrix} h \\ J_2(h) \\ \vdots \\ J_m(h) \end{bmatrix}$$

where $B = (b_{ij})$. Now linearity of the J_m implies that

$$B\begin{bmatrix} h \\ J_2(h) \\ \vdots \\ J_k(h) \end{bmatrix} = B\begin{bmatrix} H - tp \\ J_2(H) - tJ_2(p) \\ \vdots \\ J_k(H) - tJ_k(p) \end{bmatrix},$$

or equivalently that

$$(I + tB)\begin{bmatrix} p \\ J_2(p) \\ \vdots \\ J_k(p) \end{bmatrix} = B\begin{bmatrix} H \\ J_2(H) \\ \vdots \\ J_k(H) \end{bmatrix}.$$

For small t, the matrix $I + tB$ is invertible, so we have

$$\begin{bmatrix} p \\ J_2(p) \\ \vdots \\ J_k(p) \end{bmatrix} = (I + tB)^{-1}B\begin{bmatrix} H \\ J_2(H) \\ \vdots \\ J_k(H) \end{bmatrix}.$$

In particular,

$$p = \sum_{l=1}^{k} c_l J_l(H)$$

where $J_1(H) = H$, for suitable c_l.

Now the generators J_l all have the form

$$J_l(h) = S_l h + (d_x h)(X_l),$$

so $p = -aH - (d_x H)(b)$ as required.

This proves Lemma 2.2, hence Proposition 2.1, and thus completes the proof of Theorem 1.3. □

EXERCISE

2.1. (Equivariant Catastrophe Theory.) Set up the relevant definitions and derive equivariant versions of Exercises II, 11.1, and II, 11.2.

§3. Sample Computations of $RT(h, \Gamma)$

In this section, we compute generators of $RT(h, \Gamma)$ for five specific groups: $\Gamma = \mathbb{1}, \mathbf{Z}_2, \mathbf{Z}_2 \oplus \mathbf{Z}_2, \mathbf{O}(2)$, and $\mathbf{D}_n$ $(n \geq 3)$. The results are summarized in Table 3.1. This lists the group action, a Hilbert basis for the invariant functions $\mathscr{E}(\Gamma)$, and generators for the modules $\vec{\mathscr{E}}(\Gamma)$, $\overleftrightarrow{\mathscr{E}}(\Gamma)$, and—most importantly—$RT(h, \Gamma)$. In each of these examples the modules are free. That is, there is a set of generators $g_1, \ldots, g_r$ such that every $h \in \overleftrightarrow{\mathscr{E}}(\Gamma)$ is *uniquely* of the form

Table 3.1. Data for Γ-Equivariant Restricted Tangent Spaces, $\Gamma = \mathbb{1}, \mathbf{Z}_2, \mathbf{Z}_2 \oplus \mathbf{Z}_2, \mathbf{O}(2), \mathbf{D}_n$

Γ	V	Action	Generators for $\mathcal{E}(\Gamma)$	Generators for $\vec{\mathcal{E}}(\Gamma)$	Generators for $\overleftrightarrow{\mathcal{E}}(\Gamma)$	Generators for $RT(h)$
$\mathbb{1}$	$\mathbb{R}^n$ $x = (x_1, \ldots, x_n)$	—	$x_1, \ldots, x_n$	$e_j = \begin{bmatrix} 0 \\ \vdots \\ 1 \\ \vdots \\ 0 \end{bmatrix} \leftarrow$ jth position $h = (h_1, \ldots, h_n)$	$E_{jk} = n \times n$ matrix with 1 in j, k entry and 0 elsewhere	$h_j e_k \quad 1 \le j, k \le n$ $Y \begin{bmatrix} \partial h_i/\partial x_j \\ \vdots \\ \partial h_n/\partial x_j \end{bmatrix} \quad 1 \le j \le n$ where $Y = x_k$ $(1 \le k \le n)$ or $Y = \lambda$
$\mathbf{Z}_2$ $\begin{bmatrix} 1 & 0 \\ 0 & -1 \end{bmatrix}$	$\mathbb{R}^2$ $x = (y, z)$	$\begin{bmatrix} 1 & 0 \\ 0 & -1 \end{bmatrix}\begin{bmatrix} y \\ z \end{bmatrix}$	$u = y$ $v = z^2$	$g_1 = (1, 0)$ $g_2 = (0, z)$ $h = (p, qz)$ $\equiv [p, q]$	$S_1 = \begin{bmatrix} 1 & 0 \\ 0 & 0 \end{bmatrix}$ $S_2 = \begin{bmatrix} 0 & z \\ 0 & 0 \end{bmatrix}$ $S_3 = \begin{bmatrix} 0 & 0 \\ z & 0 \end{bmatrix}$ $S_4 = \begin{bmatrix} 0 & 0 \\ 0 & 1 \end{bmatrix}$	$[p, 0], [vq, 0]$ $[0, p], [0, q]$ $[up_u, uq_u], [vp_u, vq_u]$ $[\lambda p_u, \lambda q_u], [vp_v, vq_v]$
$\mathbf{Z}_2 \oplus \mathbf{Z}_2$ $\begin{bmatrix} \varepsilon & 0 \\ 0 & \delta \end{bmatrix}$ $\varepsilon = \pm 1$ $\delta = \pm 1$	$\mathbb{R}^2$ $x = (y, z)$	$\begin{bmatrix} \varepsilon & 0 \\ 0 & \delta \end{bmatrix}\begin{bmatrix} y \\ z \end{bmatrix}$	$u = y^2$ $v = z^2$	$g_1 = (y, 0)$ $g_2 = (0, z)$ $h = (py, qz)$ $\equiv [p, q]$	$S_1 = \begin{bmatrix} 1 & 0 \\ 0 & 0 \end{bmatrix}$ $S_2 = \begin{bmatrix} 0 & yz \\ 0 & 0 \end{bmatrix}$ $S_3 = \begin{bmatrix} 0 & 0 \\ yz & 0 \end{bmatrix}$ $S_4 = \begin{bmatrix} 0 & 0 \\ 0 & 1 \end{bmatrix}$	$[p, 0], [0, q]$ $[0, up], [vq, 0]$ $[up_u, uq_u], [vp_v, vq_v]$

Table 3.1. (*Continued*)

Γ	V	Action	Generators for $\mathcal{E}(\Gamma)$	$\vec{\mathcal{E}}(\Gamma)$	$\overleftrightarrow{\mathcal{E}}(\Gamma)$	$RT(h)$
$\mathbf{O}(2)$ $0 \le \theta \le 2\pi$ κ	$\mathbb{C}$	$\theta \cdot z = e^{i\theta} z$ $\kappa \cdot z = \bar{z}$	$u = z\bar{z}$	$g_1 = z$ $h = pz$ $\equiv [p]$	$w \in \mathbb{C}$ $S_1(z)w = w$ $S_2(z)w = z^2\bar{w}$	$[p]$, $[up_u]$
$\mathbf{D}_n$ $(n \ge 3)$ $\theta = 2k\pi/n$ κ	$\mathbb{C}$	$\theta \cdot z = e^{i\theta} z$ $\kappa \cdot z = \bar{z}$	$u = z\bar{z}$ $v = z^n + \bar{z}^n$	$g_1 = z$ $g_2 = \bar{z}^{n-1}$ $h = pz + q\bar{z}^{n-1}$ $\equiv [p, q]$	$w \in \mathbb{C}$ $S_1(z)w = w$ $S_2(z)w = z^2\bar{w}$ $S_3(z)w = \bar{z}^{n-2}\bar{w}$ $S_4(z)w = z^n w$	$[p, q]$, $[2up + vq, 0]$ $[u^{n-2}q, p]$, $[vp + 2u^{n-1}q, 0]$, $[2up_u + nvp_v,$ $nq + (n-1)uq_u + nvq_v]$ $[vp_u + 2nu^{n-1}p_v$ $+ (n-2)u^{n-1}q$ $+ (n-1)u^{n-1}q_u,$ $vq_u + 2nu^{n-1}q_v]$

$$h = f_1 g_1 + \cdots + f_r g_r \tag{3.1}$$

where $f_j \in \mathscr{E}(\Gamma)$. See XII, §5. We use the notation

$$h = [f_1, \ldots, f_r] \tag{3.2}$$

as a shorthand for (3.1). We think of the f_j as "invariant coordinate functions" for h.

The calculations needed to verify Table 3.1 are presented later. It is not necessary for the reader to check every entry, but for completeness we include the required computations.

(a) $\Gamma = \mathbb{1}$, $V = \mathbb{R}^n$

The entries in Table 3.1 for this group action are easily obtained. The ring $\mathscr{E}(\Gamma)$ is just the standard ring of germs $\mathscr{E}_x$. Every function in $\mathscr{E}_x$ is a function of the variables $x_1, \ldots, x_n$. Germs of functions $h: \mathbb{R}^n \times \mathbb{R} \to \mathbb{R}^n$ can be written as

$$h = (h_1, \ldots, h_n) = h_1 e_1 + \cdots + h_n e_n$$

where the h_j are coordinate functions and the e_j form the canonical basis for $\mathbb{R}^n$. The module of matrices $\vec{\mathscr{E}}(\Gamma)$ is clearly generated by the fundamental matrices E_{jk}.

The generators for the module $RT(h, \Gamma)$ are given in (1.11). To apply this, observe that

$$E_{jk} h = h_j e_k$$

which yields the first n^2 generators in Table 3.1. Next note that the module $\vec{\mathscr{M}}_{x,\lambda}(\Gamma)$ consists of mappings $\mathbb{R}^n \times \mathbb{R} \to \mathbb{R}^n$ that vanish at the origin. By Taylor's theorem this module is generated by the $n^2 + n$ mappings

$$\lambda e_j, \; x_k e_j \qquad (1 \leq j, k \leq n).$$

A simple calculation shows that $(dh)(\lambda e_j)$ and $(dh)(x_k e_j)$ yield the remaining generators in Table 3.1.

Remark. When $n = 1$, $RT(h, \Gamma)$ is generated as a module by h, xh_x, λh_x. However, in this case $\vec{\mathscr{E}}_{x,\lambda}(\Gamma) = \mathscr{E}_{x,\lambda}(\Gamma)$ and "modules" are the same as "ideals." Hence $RT(h, \mathbb{1}) = \langle h, xh_x, \lambda h_x \rangle$, which is, of course, the restricted tangent space for bifurcation problems with one state variable (and no symmetry) described in Chapter II.

(b) $\Gamma = \mathbf{Z}_2$, $V = \mathbb{R}^2$

Invariant functions are defined by

$$f(y, z) = f(y, -z)$$

It is clear that $u = y$, $v = z^2$ are generators. Similarly if $h \in \vec{\mathscr{E}}_{x,\lambda}(\mathbf{Z}_2)$ then $h = (h_1, h_2)$ satisfies

$$(h_1(y, -z), h_2(y, -z)) = (h_1(y, z), -h_2(y, z)).$$

Thus h_1 is even and h_2 is odd in z. It follows that

$$h_1(y, z) = p(u, v)$$
$$h_2(y, z) = q(u, v)z.$$

Thus g_1 and g_2 generate $\vec{\mathscr{E}}(\mathbf{Z}_2)$.

Similarly, to find generators for the module $\vec{\mathscr{E}}(\mathbf{Z}_2)$, use the equivariance condition (1.2(b)) to obtain

$$S(y, -z)\begin{bmatrix} 1 & 0 \\ 0 & -1 \end{bmatrix} = \begin{bmatrix} 1 & 0 \\ 0 & -1 \end{bmatrix} S(y, z). \tag{3.3}$$

Let $S = \left[\begin{smallmatrix} s_1 & s_2 \\ s_3 & s_4 \end{smallmatrix}\right]$ and rewrite (3.3) as

$$\begin{bmatrix} s_1 & s_2 \\ s_3 & s_4 \end{bmatrix}(y, -z) = \begin{bmatrix} s_1 & -s_2 \\ -s_3 & s_4 \end{bmatrix}(y, z). \tag{3.4}$$

By (3.4) s_1 and s_4 are $\mathbf{Z}_2$-invariant functions and s_2 and s_3 are odd in z. Hence the generators for $\vec{\mathscr{E}}(\mathbf{Z}_2)$ are

$$S_1 = \begin{bmatrix} 1 & 0 \\ 0 & 0 \end{bmatrix}, \quad S_2 = \begin{bmatrix} 0 & z \\ 0 & 0 \end{bmatrix}, \quad S_3 = \begin{bmatrix} 0 & 0 \\ z & 0 \end{bmatrix}, \quad S_4 = \begin{bmatrix} 0 & 0 \\ 0 & 1 \end{bmatrix}, \tag{3.5}$$

as in Table 3.1.

The generators for $RT(h, \mathbf{Z}_2)$ are obtained from (1.11). The first four generators in Table 3.1 are obtained from $S_j h$, $j = 1, 2, 3, 4$. For example,

$$S_2 h = \begin{bmatrix} 0 & z \\ 0 & 0 \end{bmatrix}\begin{bmatrix} p \\ qz \end{bmatrix} = \begin{bmatrix} qz^2 \\ 0 \end{bmatrix} = \begin{bmatrix} vq \\ 0 \end{bmatrix}.$$

Hence $S_2 h = [vq, 0]$ in invariant coordinates.

The last four generators for $RT(h, \mathbf{Z}_2)$ in Table 3.1 are of the form $(dh)(X_j)$ where the X_j are generators for $\vec{\mathscr{M}}_{x,\lambda}(\mathbf{Z}_2)$. Now a typical mapping in $\vec{\mathscr{M}}_{x,\lambda}(\mathbf{Z}_2)$ has the form

$$X(x, \lambda) = (p(u, v, \lambda), q(u, v, \lambda)z)$$

where $p(0, 0, 0) = 0$. It follows from Taylor's theorem that $\vec{\mathscr{M}}_{x,\lambda}(\mathbf{Z}_2)$ is generated by

$$(u, 0), (v, 0), (\lambda, 0), (0, z). \tag{3.6}$$

Observe that

$$(dh) = \begin{bmatrix} p_u u_y & p_v v_z \\ q_u u_y z & q + q_v v_z z \end{bmatrix} = \begin{bmatrix} p_u & 2p_v z \\ q_u z & q + 2vq_v \end{bmatrix}.$$

Hence (dh) applied to the generators in (3.6) yields

$$(up_u, uq_u z), (vp_u, vq_u z), (\lambda p_u, \lambda q_u z), (2vp_v, (q + 2vq_v)z). \tag{3.7}$$

Write (3.7) in invariant coordinates and use the fact that $[0, q] \in RT(h, \mathbf{Z}_2)$. This yields the last four generators of $RT(h, \mathbf{Z}_2)$ listed in Table 3.1.

(c) $\Gamma = \mathbf{Z}_2 \oplus \mathbf{Z}_2$, $V = \mathbb{R}^2$

This action was discussed in some detail in Chapter X, and generators for $\mathscr{E}(\mathbf{Z}_2 \oplus \mathbf{Z}_2)$, $\vec{\mathscr{E}}(\mathbf{Z}_2 \oplus \mathbf{Z}_2)$, $\overleftrightarrow{\mathscr{M}}(\mathbf{Z}_2 \oplus \mathbf{Z}_2)$, and $\overleftrightarrow{\mathscr{E}}(\mathbf{Z}_2 \oplus \mathbf{Z}_2)$ were given. In particular $\mathscr{E}$ and $\vec{\mathscr{E}}$ were discussed in Lemma X, 1.1, and $\overleftrightarrow{\mathscr{E}}$ in (X, 2.5). The reader may wish to rederive these results; they are straightforward. Since $\mathbf{Z}_2 \oplus \mathbf{Z}_2$-equivariant mappings must vanish at the origin, $\vec{\mathscr{E}} = \vec{\mathscr{M}}$ in this case.

The computation of the generators for $RT(h, \mathbf{Z}_2 \oplus \mathbf{Z}_2)$ listed in Table 3.1 is a simple exercise using (1.11) and can safely be left to the reader.

(d) $\Gamma = \mathbf{O}(2)$, $V = \mathbb{C}$

The generators for $\mathscr{E}(\mathbf{O}(2))$ and $\vec{\mathscr{E}}(\mathbf{O}(2))$ were listed in Example XII, 5.4(b). See (XII, 5.10). We take advantage of the complex notation to compute generators for $\overleftrightarrow{\mathscr{E}}(\mathbf{O}(2))$.

We may write any (real) linear mapping $\mathbb{C} \to \mathbb{C}$ in the form

$$w \mapsto \alpha w + \beta \bar{w} \tag{3.8}$$

where $\alpha, \beta \in \mathbb{C}$. Note that if $w = x_1 + ix_2$, $\alpha = \alpha_1 + i\alpha_2$, and $\beta = \beta_1 + i\beta_2$, then the associated real matrix is

$$\begin{bmatrix} x_1 \\ x_2 \end{bmatrix} \mapsto \begin{bmatrix} \alpha_1 + \beta_1 & \beta_2 - \alpha_2 \\ \alpha_2 + \beta_2 & \alpha_1 - \beta_1 \end{bmatrix} \begin{bmatrix} x_1 \\ x_2 \end{bmatrix}.$$

Conversely, the matrix $\left[\begin{smallmatrix} a & b \\ c & d \end{smallmatrix}\right]$ can be put in the form (3.8) by choosing

$$\begin{aligned} &\text{(a)} \quad \alpha_1 = \tfrac{1}{2}(a + d) \\ &\text{(b)} \quad \alpha_2 = \tfrac{1}{2}(c - b) \\ &\text{(c)} \quad \beta_1 = \tfrac{1}{2}(a - d) \\ &\text{(d)} \quad \beta_2 = \tfrac{1}{2}(c + b). \end{aligned} \tag{3.9}$$

If we now let the coefficients in the linear mapping (3.8) depend on $z \in \mathbb{C}$ then we have written the linear mapping $S(z) \in \overleftrightarrow{\mathscr{E}}(\mathbf{O}(2))$ in the form

$$S(z)w = \alpha(z)w + \beta(z)\bar{w}. \tag{3.10}$$

By Poénaru's theorem XII, 5.3, it is sufficient to find generators for the module of polynomial mappings $S(z)$. That is, we assume, using (3.10), that

$$S(z)w = \sum \alpha_{jk} z^j \bar{z}^k w + \sum \beta_{jk} z^j \bar{z}^k \bar{w}. \tag{3.11}$$

The $\mathbf{O}(2)$-equivariance condition (1.2(b)) takes the form

$$\begin{aligned}&\text{(a)}\quad S(e^{i\theta}z)e^{i\theta}w = e^{i\theta}S(z)w,\\&\text{(b)}\quad \overline{S(\bar{z})\bar{w}} = S(z)w.\end{aligned} \tag{3.12}$$

It follows from (3.12(b)) that the coefficients α_{jk} and β_{jk} in (3.11) must be real. Also, we may rewrite (3.12(a)) as

$$S(z)w = \sum \alpha_{jk}e^{(j-k)i\theta}z^j\bar{z}^k w + \sum \beta_{jk}e^{(j-k-2)i\theta}z^j\bar{z}^k\bar{w}. \tag{3.13}$$

Equating coefficients in (3.13) and (3.11) we find that

$$\begin{aligned}\alpha_{jk} &= 0 \qquad \text{unless } j = k, \quad \text{and}\\ \beta_{jk} &= 0 \qquad \text{unless } j = k + 2.\end{aligned}$$

It follows that $S_1(z)w = w$ and $S_2(z)w = z^2\bar{w}$ generate $\vec{\mathscr{E}}(\mathbf{O}(2))$ over $\mathscr{E}(\mathbf{O}(2))$. This verifies the appropriate entry in Table 3.1.

The computation of the generators of $RT(h, \mathbf{O}(2))$ follows from (1.11) again; however, there is one computational trick that will be useful in future calculations, and we isolate it here. The generators of $RT(h, \mathbf{O}(2))$ of the form Sh are

$$S_1 h = h \quad \text{and} \quad S_2 h = z^2\bar{h}.$$

Since $h = pz$ where $p \in \mathscr{E}(\mathbf{O}(2))$ we have

$$S_1 h = [p] \quad \text{and} \quad S_2 h = [up]. \tag{3.14}$$

Observe that $S_2 h$ is a multiple (by u) of $S_1 h$ and hence is redundant.

In this case $\vec{\mathscr{M}}(\mathbf{O}(2)) = \vec{\mathscr{E}}(\mathbf{O}(2))$ and there is one generator for this module, namely $X(z) = z$. Thus, there is one generator of $RT(h, \mathbf{O}(2))$ of the form $(dh)(X)$ and it is $(dh)(z)$. To compute $(dh)(z)$ explicitly it is useful to calculate the Jacobian matrix in the "real" coordinates z, $\bar{z}$ on $\mathbb{C}$. We exhibit this calculation separately as follows:

Lemma 3.1. *Let $g\colon \mathbb{C} \to \mathbb{C}$ be a real analytic mapping. Then in z, $\bar{z}$ coordinates the 2×2 Jacobian matrix dg of g has the form*

$$(dg)(X) = g_z X + g_{\bar{z}}\bar{X}. \tag{3.15}$$

Assuming (3.15) and recalling that $u = z\bar{z}$ we compute

$$\begin{aligned}(dh)z &= h_z z + h_{\bar{z}}\bar{z} = (p_u\bar{z}z + p)z + p_u z^2\bar{z}\\ &= (p + 2up_u)z = [p + 2up_u].\end{aligned}$$

Hence $RT(h, \mathbf{O}(2))$ is generated by $[p]$ and $[up_u]$, as stated in Table 3.1.

Remark 3.2. There is a substantial similarity between the actions of $\mathbf{Z}_2$ on $\mathbb{R}$ and $\mathbf{O}(2)$ on $\mathbb{R}^2 \equiv \mathbb{C}$. In the first case, $\mathbf{Z}_2$-equivariant mappings are odd functions $h(x, \lambda) = p(x^2, \lambda)x$. In the second case, the $\mathbf{O}(2)$-equivariant mappings are of the form $h(z, \lambda) = p(|z|^2, \lambda)z$. We have shown that both $RT(h, \mathbf{Z}_2)$

and $RT(h, \mathbf{O}(2))$ are generated by $[p]$ and $[up_u]$. It turns out that the singularity structure in these two cases is identical.

To tidy up the remaining loose end for this group, we give the following:

PROOF OF LEMMA 3.1. In z, $\bar{z}$ coordinates we can write g as $(g, \bar{g})$. In these coordinates the Jacobian matrix is

$$dg = \begin{bmatrix} g_z & g_{\bar{z}} \\ \bar{g}_z & \bar{g}_{\bar{z}} \end{bmatrix}.$$

Applying dg to the real 2-vector $(w, \bar{w})$ we get

$$(dg)\begin{bmatrix} w \\ \bar{w} \end{bmatrix} = \begin{bmatrix} g_z w + g_{\bar{z}}\bar{w} \\ \bar{g}_z w + \bar{g}_{\bar{z}}\bar{w} \end{bmatrix}. \tag{3.16}$$

We claim that

$$\begin{aligned} &\text{(a)} \quad \overline{(\partial g/\partial \bar{z})} = \partial \bar{g}/\partial z \\ &\text{(b)} \quad \overline{(\partial g/\partial z)} = \partial \bar{g}/\partial \bar{z}. \end{aligned} \tag{3.17}$$

By (3.17) the second coordinate of the right-hand side of (3.16) is the complex conjugate of the first coordinate, as expected. In particular, the linear map dg is determined by the first coordinate in (3.16), and this establishes (3.15).

Letting $z = x_1 + ix_2$, so $\bar{z} = x_1 - ix_2$, we obtain the formulas

$$\begin{aligned} &\text{(a)} \quad \partial/\partial z = \tfrac{1}{2}(\partial/\partial x_1 - i\partial/\partial x_2) \\ &\text{(b)} \quad \partial/\partial \bar{z} = \tfrac{1}{2}(\partial/\partial x_1 + i\partial/\partial x_2). \end{aligned} \tag{3.18}$$

A direct calculation on $g = g_1 + ig_2$ using (3.18) establishes (3.17). □

(e) $\Gamma = \mathbf{D}_n$ $(n \geq 3)$, $V = \mathbb{C}$

The generators for $\mathscr{E}(\mathbf{D}_n)$ and $\vec{\mathscr{E}}(\mathbf{D}_n) = \vec{\mathscr{M}}(\mathbf{D}_n)$ were given in Examples XII, 4.1, and XII, 5.4. Here we compute generators for the modules $\vec{\mathscr{E}}(\mathbf{D}_n)$ and $RT(h, \mathbf{D}_n)$. These calculations are modeled on those for $\mathbf{O}(2)$ acting on $\mathbb{C}$ given in the previous subsection. They are also the most complicated of the examples considered in this section.

We begin by showing that

$$S_1 \cdot w = w, \qquad S_2 \cdot w = z^2\bar{w}, \qquad S_3 \cdot w = \bar{z}^{n-2}\bar{w}, \qquad S_4 \cdot w = z^n w \tag{3.19}$$

generate $\vec{\mathscr{E}}(\mathbf{D}_n)$, verifying the corresponding entry in Table 3.1. As in the case $\Gamma = \mathbf{O}(2)$ we use complex notation, with S in the form (3.11), For $\mathbf{D}_n$ the invariance conditions (1.2(b)) take the form (3.12) where now $\theta = 2\pi/n$ in (3.12(a)). As with $\mathbf{O}(2)$, (3.12b) implies that the coefficients α_{jk} and β_{jk} are real. Also as previously, (3.12(a)) may be rewritten in the form (3.13); however, θ is no longer arbitrary but equals $2\pi/n$. Equating coefficients in (3.13) and (3.11) now yields

$$\alpha_{jk} = 0 \qquad \text{unless } j \equiv k \pmod{n}, \quad \text{and}$$

$$\beta_{jk} = 0 \qquad \text{unless } j \equiv k + 2 \pmod{n}.$$

Our goal is to determine generators of $\vec{\mathscr{E}}(\mathbf{D}_n)$ as a module over $\mathscr{E}(\mathbf{D}_n)$, which has $u = z\bar{z}$ and $v = z^n + \bar{z}^n$ as generators. Casting out terms in (3.11) where either $\alpha_{jk} = 0$, or $\beta_{jk} = 0$, or $z\bar{z}$ appears as a factor leads to the following result. The generators of $\vec{\mathscr{E}}(\mathbf{D}_n)$ lie in the span of

$$w,\ z^2\bar{w},\ z^{ln}w,\ \bar{z}^{ln}w,\ z^{ln+2}\bar{w},\ \bar{z}^{ln-2}\bar{w} \tag{3.20}$$

where $l = 1, 2, 3, \ldots$.

Note the following identities:

$$\begin{aligned}
&\text{(a)} \quad z^{ln}w = (z^n + \bar{z}^n)z^{(l-1)n}w - (z\bar{z})^n z^{(l-2)n}w \\
&\text{(b)} \quad \bar{z}^{ln}w = (z^{ln} + \bar{z}^{ln})w - z^{ln}w \\
&\text{(c)} \quad z^{ln+2}\bar{w} = (z^n + \bar{z}^n)z^{(l-1)n+2}\bar{w} - (z\bar{z})^n z^{(l-2)n+2}\bar{w} \\
&\text{(d)} \quad \bar{z}^{(l+1)n-2}\bar{w} = (z^{ln} + \bar{z}^{ln})\bar{z}^{n-2}\bar{w} - (z\bar{z})^{n-2}z^{(l-1)n+2}\bar{w} \\
&\text{(e)} \quad z^{n+2}\bar{w} = (z^n + \bar{z}^n)z^2\bar{w} - (z\bar{z})^2\bar{z}^{n-2}\bar{w}.
\end{aligned} \tag{3.21}$$

It follows from (3.21(a)) that the mappings $\bar{z}^{ln}w$ are redundant as generators for $\vec{\mathscr{E}}(\mathbf{D}_n)$, since $z^{ln} + \bar{z}^{ln}$ is $\mathbf{D}_n$-invariant. Similarly (3.21(d)) implies that the mappings $\bar{z}^{ln-2}\bar{w}$ when $l \geq 2$ are redundant as generators. Next, we use (3.21(a, c)) to prove, by induction, that $z^{ln}w$ and $z^{ln+2}\bar{w}$ are redundant as generators when $l \geq 2$. Finally, we use (3.21(e)) to eliminate $z^{n+2}\bar{w}$. Thus (3.19) gives the generators of $\vec{\mathscr{E}}(\mathbf{D}_n)$, as promised.

We can now compute the generators of $RT(h, \mathbf{D}_n)$. As usual, we use (1.11) as a guide and compute the generators in two stages. First we compute $S \cdot h$ and then $(dh)(X)$. Recall that $h = pz + q\bar{z}^{n-1} \equiv [p, q]$. The first stage leads to:

$$\begin{aligned}
&\text{(a)} \quad S_1(z)h = [p, q] \\
&\text{(b)} \quad S_2(z)h = [up + vq, -uq] \\
&\text{(c)} \quad S_3(z)h = [u^{n-2}q, p] \\
&\text{(d)} \quad S_4(z)h = [vp + u^{n-1}q, -up].
\end{aligned} \tag{3.22}$$

The identity

$$z^{n+1} = vz - u\bar{z}^{n-1} \tag{3.23}$$

is useful when checking (3.22). The first four generators of $RT(h, \mathbf{D}_n)$ in Table 3.1 are obtained as follows:

$$S_1 h,\ (S_2 + uS_1)h,\ S_3 h,\ (S_4 + uS_3)h.$$

To determine the remaining generators of $RT(h, \mathbf{D}_n)$ we compute

$$(dh)(z) \quad \text{and} \quad (dh)(\bar{z}^{n-1}).$$

This is done most easily by using Lemma 3.1. See (3.15). Thus, in order to find dh, we compute

$$\begin{aligned} &\text{(a)}\quad h_z = p + up_u + nu^{n-1}q_v + np_v z^n + q_u \bar{z}^n, \\ &\text{(b)}\quad h_{\bar{z}} = p_u z^2 + (nup_v + (n-1)q + nuq_u)\bar{z}^{n-2} + nq_v \bar{z}^{2n-2}. \end{aligned} \tag{3.24}$$

Now use (3.24, 3.15) to compute

$$\begin{aligned} &\text{(a)}\quad (dh)(z) = [p + 2up_u + nvp_v, (n-1)q + (n+1)uq_u + nvq_v], \\ &\text{(b)}\quad (dh)(\bar{z}^{n-1}) = [vp_u + 2nu^{n-1}p_v + (n-1)u^{n-2}q + (n-1)u^{n-1}q_u, \\ &\qquad\qquad p + vq_u + 2nu^{n-1}q_v]. \end{aligned} \tag{3.25}$$

These calculations use the identity (3.23) and the identity

$$\bar{z}^{2n-1} = -u^{n-1}z + v\bar{z}^{n-1}. \tag{3.26}$$

The last two generators of $RT(h, \mathbf{D}_n)$ listed in Table 3.1 are:

$$(dh)(z) - S_1(z)h \quad \text{and} \quad (dh)(\bar{z}^{n-1}) - S_3(z)h.$$

§4. Sample Recognition Problems

In this section we use Theorem 1.3 and the generators of $RT(h, \Gamma)$ listed in Table 3.1 to solve certain simple recognition problems. We consider three examples. In the first two we present theorems whose proofs were promised in Volume I; in the third we prove results needed in Case Study 5.

(a) $\Gamma = \mathbb{1}$, $V = \mathbb{R}^2$

Theorem 4.1. *Let $h(x, y, \lambda)$ be one of the normal forms*

$$\text{(12)}\quad (x^2 - y^2 + \lambda, 2xy),$$

$$\text{(13)}\quad (x^2 + \varepsilon\lambda, y^2 + \delta\lambda),$$

where $\varepsilon, \delta = \pm 1$. Let $g(x, y, \lambda) = h(x, y, \lambda) + p(x, y, \lambda)$ where $p \in (\mathcal{M}^3 + \mathcal{M}\langle\lambda\rangle)\vec{\mathscr{E}}_{x,y,\lambda}$ is a higher order term. Then g is strongly equivalent to h.

Remark. The proof of Theorem 4.1 will complete the "recognition" part of Theorem IX, 2.1 on hilltop bifurcation. The bizarre choice of equation numbers, (12) and (13), is made to conform to the notation of that theorem.

PROOF. We show that

$$RT(h + tp, \mathbb{1}) = RT(h, \mathbb{1})$$

for all $t \in \mathbb{R}$ and then apply Theorem 1.3. In fact we do more. We show that

$$RT(h + tp, \mathbb{1}) = (\mathcal{M}^2 + \langle\lambda\rangle)\vec{\mathcal{E}}. \tag{4.1}$$

Here $\mathcal{M} \subset \mathcal{E}_{x,y,\lambda}$ is the maximal ideal; that is, $\mathcal{M} = \langle x, y, \lambda\rangle$.

Let $g = (g_1, g_2) \in \vec{\mathcal{E}}_{x,y,\lambda}$. From Table 3.1 we see that $RT(g, \mathbb{1})$ is generated (as module over $\mathcal{E}_{x,y,\lambda}$) by the ten mappings

$$\begin{gathered}(g_1, 0), (g_2, 0), (0, g_1), (0, g_2), x(g_{1,x}, g_{2,x}), y(g_{1,x}, g_{2,x}),\\ \lambda(g_{1,x}, g_{2,x}), x(g_{1,y}, g_{2,y}), y(g_{1,y}, g_{2,y}), \lambda(g_{1,y}, g_{2,y}).\end{gathered} \tag{4.2}$$

Observe from (4.2) that when $g = h + tp$, where h and p are chosen as in the statement of Theorem 4.1, then each generator in (4.2) lies in $(\mathcal{M}^2 + \langle\lambda\rangle)\vec{\mathcal{E}}$. Thus to establish (4.1) it suffices to show that

$$(\mathcal{M}^2 + \langle\lambda\rangle)\vec{\mathcal{E}} \subset RT(h + tp, \mathbb{1}). \tag{4.3}$$

We claim that (4.3) holds independently of t or the specific choice of p. To show this we use Nakayama's lemma. The version we need is the one stated in terms of submodules, Fact VI, 2.4(iii). This version of Nakayama's lemma reduces the claim to showing that

$$(\mathcal{M}^2 + \langle\lambda\rangle)\vec{\mathcal{E}} \subset RT(h + tp, \mathbb{1}) + (\mathcal{M}^3 + \mathcal{M}\langle\lambda\rangle)\vec{\mathcal{E}}. \tag{4.4}$$

To do this, observe that $(\vec{\mathcal{M}}^2 + \langle\lambda\rangle)\vec{\mathcal{E}}$ is generated by eight mappings

$$\begin{gathered}(x^2, 0), (xy, 0), (y^2, 0), (\lambda, 0),\\ (0, x^2), (0, xy), (0, y^2), (0, \lambda).\end{gathered} \tag{4.5}$$

Then (4.4) follows provided we show that each mapping in (4.5) can be written as an element of $RT(h + tp, \mathbb{1})$, modulo terms in $(\mathcal{M}^3 + \mathcal{M}\langle\lambda\rangle)\vec{\mathcal{E}}$. Let $g = h + tp$; then (4.2) implies that the term $tp = t(p_1, p_2)$ enters the generators of $RT(h + tp, \mathbb{1})$ only through

$$tp_1, tp_2, ztp_{1,x}, ztp_{1,y}, ztp_{2,x}, ztp_{2,y}, \tag{4.6}$$

where $z = x$, y, or λ. Since $p \in (\mathcal{M}^3 + \mathcal{M}\langle\lambda\rangle)\vec{\mathcal{E}}$ it follows that each term in (4.6) is in $\mathcal{M}^3 + \mathcal{M}\langle\lambda\rangle$. Hence the term tp does not affect the truth or falsity of (4.4), so (4.4) holds, provided

$$(\mathcal{M}^2 + \langle\lambda\rangle)\vec{\mathcal{E}} \subset RT(h, \mathbb{1}) + (\mathcal{M}^3 + \mathcal{M}\langle\lambda\rangle)\vec{\mathcal{E}}. \tag{4.7}$$

This calculation proceeds most naturally by considering which of the ten generators of $RT(h + tp, \mathbb{1})$ listed in (4.2), with $g = h$, may be written as linear combinations of the eight generators of $(\mathcal{M}^2 + \langle\lambda\rangle)\vec{\mathcal{E}}$ listed in 4.5. First we consider the normal form (12), namely

$$h(x, y, \lambda) = (x^2 - y^2 + \lambda, 2xy),$$

and list the relevant information in the 10×8 matrix of Table 4.1. (Zeros are omitted.) Here we have written the generators of $RT(x^2 - y^2 + \lambda, 2xy)$ in terms of (4.5) and have ignored terms in $(\mathcal{M}^3 + \mathcal{M}\langle\lambda\rangle)\vec{\mathcal{E}}$. In particular, this construction leads to two rows in Table 4.1 that are identically zero, namely the λh_x and λh_y rows, indicated by cross-hatching. This leaves an 8×8 matrix. It follows that each generator of $(\mathcal{M}^2 + \langle\lambda\rangle)\vec{\mathcal{E}}$ in (4.5) is a combination of

Table 4.1. Generators of $RT(x^2 - y^2 + \lambda, 2xy)$ Modulo $(\mathcal{M}^3 + \mathcal{M}\langle\lambda\rangle)\vec{\mathscr{E}}$

	$(x^2,0)$	$(xy,0)$	$(y^2,0)$	$(\lambda,0)$	$(0,x^2)$	$(0,xy)$	$(0,y^2)$	$(0,\lambda)$
$(h_1,0)$	1		-1	1				
$(h_2,0)$		2						
$(0,h_1)$					1		-1	1
$(0,h_2)$						2		
xh_x	1					1		
yh_x		1					1	
λh_x	//////	//////	//////	//////	//////	//////	//////	//////
xh_y		-1			1			
yh_y			-1			1		
λh_y	//////	//////	//////	//////	//////	//////	//////	//////

terms on the right-hand side of (4.7), provided the determinant of this matrix is nonzero. Because the matrix contains mostly zeros, this computation is easy and is left to the reader.

The calculations for normal form (13), namely

$$h(x, y, \lambda) = (x^2 + \varepsilon\lambda, y^2 + \delta\lambda),$$

are identical in spirit to those for normal form (12) and are left as an exercise. □

(b) $\Gamma = \mathbf{Z}_2 \oplus \mathbf{Z}_2$, $V = \mathbb{R}^2$

Let $g(x, y, \lambda)$ be a $\mathbf{Z}_2 \oplus \mathbf{Z}_2$-equivariant bifurcation problem. This implies in particular that $(dg)_{0,0,0} = 0$. Thus we can write

$$g(x, y, \lambda) = h(x, y, \lambda) + \varphi(x, y, \lambda)$$

where

$$h(x, y, \lambda) = (Ax^3 + Bxy^2 + \alpha\lambda x, Cx^2y + Dy^3 + \beta\lambda y)$$

and φ consists of higher order terms. Recall from Definition X, 2.2, that g is *nondegenerate* if

$$\begin{aligned} &\text{(a)} \quad A \neq 0, \qquad D \neq 0 \\ &\text{(b)} \quad \alpha \neq 0, \qquad \beta \neq 0 \\ &\text{(c)} \quad A\beta - C\alpha \neq 0, \qquad B\beta - D\alpha \neq 0 \\ &\text{(d)} \quad AD - BC \neq 0. \end{aligned} \tag{4.8}$$

Theorem 4.2. *Suppose that g is a nondegenerate $\mathbf{Z}_2 \oplus \mathbf{Z}_2$-equivariant bifurcation problem. Then g is strongly $\mathbf{Z}_2 \oplus \mathbf{Z}_2$-equivalent to h.*

Remark. The proof of Theorem 4.2 completes that of Proposition X, 2.3, promised in Volume I.

PROOF. The advantages of working in invariant coordinates should become apparent during the proof of this theorem. Once the generators for $RT(h, \Gamma)$ have been computed—as in Table 3.1—then by working in invariant coordinates, the existence of the symmetry group is effectively suppressed.

Our goal is to show that under the assumption of nondegeneracy,

$$RT(h + t\varphi, \mathbf{Z}_2 \oplus \mathbf{Z}_2) = RT(h, \mathbf{Z}_2 \oplus \mathbf{Z}_2) \tag{4.9}$$

for all $t \in \mathbb{R}$. The result then follows by Theorem 1.3.

Invariant coordinates give an isomorphism between $\vec{\mathscr{E}}_{x,y,\lambda}(\mathbf{Z}_2 \oplus \mathbf{Z}_2)$ and $\vec{\mathscr{E}}_{u,v,\lambda}$ where $u = x^2$ and $v = y^2$. That is,

$$g(x, y, \lambda) = (p(x^2, y^2, \lambda)x, q(x^2, y^2, \lambda)y).$$

We therefore write g in the form $[p(u, v, \lambda), q(u, v, \lambda)]$ and henceforth work in $\vec{\mathscr{E}}_{u,v,\lambda}$, which is a module over $\mathscr{E}_{u,v,\lambda}$. In this language, $RT(g, \mathbf{Z}_2 \oplus \mathbf{Z}_2)$ may be viewed as the submodule of $\vec{\mathscr{E}}_{u,v,\lambda}$ whose six generators are

$$[p, 0], [0, q], [0, up], [vq, 0], [up_u, uq_u], [vp_v, vq_v]. \tag{4.10}$$

We begin to verify (4.9) by showing:

$$\mathscr{M}^2_{u,v,\lambda}\vec{\mathscr{E}}_{u,v,\lambda} \subset RT(h + t\varphi, \mathbf{Z}_2 \oplus \mathbf{Z}_2). \tag{4.11}$$

To see this, let $\mathscr{I} \subset RT(g, \mathbf{Z}_2 \oplus \mathbf{Z}_2)$ be the submodule with the fourteen generators

$$z[p, 0], z[0, q], z[0, up], z[vq, 0], z[up_u, uq_u], z[vp_v, vq_v] \tag{4.12}$$

where $z = u, v$, or λ and $g = h + t\varphi$. We claim that

$$\mathscr{M}^2_{u,v,\lambda}\vec{\mathscr{E}}_{u,v,\lambda} = \mathscr{I}. \tag{4.13}$$

If (4.13) holds, then so does (4.11). But more is true. If (4.13) holds then

$$\text{(a)} \quad RT(h + t\varphi, \mathbf{Z}_2 \oplus \mathbf{Z}_2) = \mathscr{M}^2_{u,v,\lambda}\vec{\mathscr{E}}_{u,v,\lambda} + W,$$

where

$$\text{(b)} \quad W = \mathbb{R}\{[p, 0], [0, q], [up_u, uq_u], [vp_v, vq_v]\}. \tag{4.14}$$

It is now easy to compute the basis elements of W modulo terms in $\mathscr{I} = \mathscr{M}^2_{u,v,\lambda}\vec{\mathscr{E}}_{u,v,\lambda}$, i.e., those terms that are quadratic in u, v, λ.

$$\begin{aligned}
&\text{(a)} \quad [p, 0] \equiv [Au + Bv + \alpha\lambda, 0] \quad (\operatorname{mod} \mathscr{I})\\
&\text{(b)} \quad [0, q] \equiv [0, Cu + Dv + \beta\lambda] \quad (\operatorname{mod} \mathscr{I})\\
&\text{(c)} \quad [up_u, uq_u] \equiv [Au, Cu] \quad (\operatorname{mod} \mathscr{I})\\
&\text{(d)} \quad [vp_v, vq_v] \equiv [Bv, Dv] \quad (\operatorname{mod} \mathscr{I}).
\end{aligned} \tag{4.15}$$

By (4.14) and (4.15)

$$RT(h + t\varphi, \mathbf{Z}_2 \oplus \mathbf{Z}_2) = \mathcal{M}^2_{u,v,\lambda}\vec{\mathscr{E}}_{u,v,\lambda} \oplus \mathbb{R}\{[Au + Bv + \alpha\lambda, 0],$$
$$[0, Cu + Dv + \beta\lambda], [Au, Cu], [Bv, Dv]\}. \quad (4.16)$$

Thus (4.16) implies that $RT(h + t\varphi, \mathbf{Z}_2 \oplus \mathbf{Z}_2)$ is independent of $t\varphi$, establishing (4.9).

To finish the proof of (4.9) we must establish (4.13). It is here that the nondegeneracy of h enters. Since the generators of $\mathscr{I}$ in (4.12) all consist of terms of quadratic or higher order in u, v, λ, it follows that

$$\mathcal{M}^2_{u,v,\lambda}\vec{\mathscr{E}}_{u,v,\lambda} \supset \mathscr{I}.$$

Thus (4.13) will follow from

$$\mathcal{M}^2_{u,v,\lambda}\vec{\mathscr{E}}_{u,v,\lambda} \subset \mathscr{I}.$$

By Nakayama's lemma, this is true provided

$$\mathcal{M}^2_{u,v,\lambda}\vec{\mathscr{E}}_{u,v,\lambda} \subset \mathscr{I} + \mathcal{M}^3_{u,v,\lambda}\vec{\mathscr{E}}_{u,v,\lambda}. \quad (4.17)$$

Observe that $t\varphi$ consists of terms of at least quadratic order in u, v, λ. By inspection we see that $t\varphi$ enters the generators of $\mathscr{I}$ in (4.12) only through terms of at least third order in u, v, λ. Hence when verifying (4.17) we may assume that $t\varphi \equiv 0$.

The remainder of the proof proceeds as follows. The module $\mathcal{M}^2_{u,v,\lambda}\vec{\mathscr{E}}_{u,v,\lambda}$ has twelve generators:

$$\begin{aligned} &[u^2, 0], [uv, 0], [v^2, 0], [u\lambda, 0], [v\lambda, 0], [\lambda^2, 0] \\ &[0, u^2], [0, uv], [0, v^2], [0, u\lambda], [0, v\lambda], [0, \lambda^2]. \end{aligned} \quad (4.18)$$

We expand the fourteen generators of $\mathscr{I}$ in (4.12) in terms of the twelve generators of $\mathcal{M}^2_{u,v,\lambda}\vec{\mathscr{E}}_{u,v,\lambda}$ in (4.18). Note that since $t\varphi \equiv 0$ we may write

$$\begin{aligned} &\text{(a)} \quad p = Au + Bv + \alpha\lambda, \\ &\text{(b)} \quad q = Cu + Dv + \beta\lambda. \end{aligned} \quad (4.19)$$

This yields a 14×12 matrix. If we show that the rank of this matrix is 12, then elementary linear algebra implies that each generator in (4.18) can be written in terms of the generators for $\mathscr{I}$ in (4.12), whence (4.17) will follow.

To compute this matrix is straightforward; the result is Table 4.2. We now outline how to use nondegeneracy to prove that this matrix has rank 12. Note that we list generators of the form $[*, 0]$ on the left-hand side of the matrix and $[0, *]$ on the right.

The assumption of nondegeneracy (4.8) implies that $\alpha \neq 0$, $\beta \neq 0$, and $AD - BC \neq 0$. Using these facts we can eliminate four rows ($\lambda[p, 0], \lambda[0, q]$, $uv[p_u, q_u], uv[p_v, q_v]$) and four columns ($[\lambda^2, 0], [0, \lambda^2], [uv, 0], [0, uv]$) to obtain a 10×8 matrix whose rank is 8 precisely when that of the original 14×12 matrix is 12. This matrix is shown in Table 4.3.

Table 4.2.

	$[*, 0]$ where $*$ is						$[0, *]$ where $*$ is					
	u^2	uv	v^2	$u\lambda$	$v\lambda$	λ^2	u^2	uv	v^2	$u\lambda$	$v\lambda$	λ^2
$u[p, 0]$	A	B		α								
$v[p, 0]$		A	B		α							
$\lambda[p, 0]$				A	B	α						
$u[0, q]$							C	D		β		
$v[0, q]$								C	D		β	
$\lambda[0, q]$										C	D	β
$u^2[p_u, q_u]$	A						C					
$uv[p_u, q_u]$		A						C				
$u\lambda[p_u, q_u]$				A						C		
$uv[p_v, q_v]$		B						D				
$v^2[p_v, q_v]$			B						D			
$v\lambda[p_v, q_v]$					B						D	
$[0, up]$							A	B		α		
$[vq, 0]$		C	D		β							

Table 4.3.

	$[*, 0]$ where $*$ is				$[0, *]$ where $*$ is			
	u^2	v^2	$u\lambda$	$v\lambda$	u^2	v^2	$u\lambda$	$v\lambda$
$u[p, 0]$	A		α					
$v[p, 0]$		B		α				
$u[0, q]$					C		β	
$v[0, q]$						D		β
$u^2[p_u, q_u]$	A				C			
$u\lambda[p_u, q_u]$			A				C	
$v^2[p_v, q_v]$		B				D		
$v\lambda[p_v, q_v]$				B				D
$[0, up]$					A		α	
$[vq, 0]$		D		β				

Next we use the nondegeneracy assumptions $B\beta - \alpha D \neq 0$ and $C\alpha - A\beta \neq 0$ to eliminate the four rows $v[p, 0]$, $[vq, 0]$, $u[0, q]$, $[0, up]$ and the four columns $[v^2, 0]$, $[v\lambda, 0]$, $[0, u^2]$, $[0, u\lambda]$. This yields a 6×4 matrix, shown in Table 4.4. Its rank is 4 if and only if that of the original 14×12 matrix is 12.

Finally we use the nondegeneracy conditions $A \neq 0$, $D \neq 0$ (see (4.8)), to show that the matrix in Table 4.4 does have rank 4 (consider the diagonal submatrix formed by the last four rows). The original 14×12 matrix therefore has rank 12, and (4.17) is proved. □

Table 4.4.

	[*, 0]:		[0, *]:	
	u^2	$u\lambda$	v^2	$v\lambda$
$u[p, 0]$	A	α		
$v[0, q]$			D	β
$u^2[p_u, q_u]$	A			
$u\lambda[p_u, q_u]$		A		
$v^2[p_v, q_v]$			D	
$v\lambda[p_v, q_v]$				D

(c) $\Gamma = \mathbf{D}_3$, $V = \mathbb{C}$

From Table 3.1 a $\mathbf{D}_3$-equivariant bifurcation problem g has the form

$$g(z, \lambda) = p(u, v, \lambda)z + q(u, v, \lambda)\bar{z}^2 \tag{4.20}$$

where $p(0, 0, 0) = 0$, $u = z\bar{z}$, and $v = z^3 + \bar{z}^3$. In XIII, §5, we used the equivariant branching lemma (Theorem XIII, 3.2) to show that if $p_\lambda(0, 0, 0) \neq 0$ then there is a nontrivial branch of solutions with isotropy subgroup $\mathbf{Z}_2$. Assuming that $q(0, 0, 0) \neq 0$ we can solve (4.20) explicitly to show that this is the only nontrivial branch of solutions (up to conjugacy). See XIII, §5 (5.8, 5.9). Here we derive a normal form for these bifurcation problems.

Proposition 4.3. *Let g be a $\mathbf{D}_3$-equivariant bifurcation problem as in* (4.20). *Assume*

$$\begin{aligned} &\text{(a)} \quad p(0, 0, 0) = 0 \\ &\text{(b)} \quad p_\lambda(0, 0, 0) \neq 0 \\ &\text{(c)} \quad q(0, 0, 0) \neq 0. \end{aligned} \tag{4.21}$$

Then g is $\mathbf{D}_3$-equivalent to

$$h(z, \lambda) = \varepsilon\lambda z + \delta\bar{z}^2 \tag{4.22}$$

where $\varepsilon = \operatorname{sgn} p_\lambda(0, 0, 0)$ *and* $\delta = \operatorname{sgn} q(0, 0, 0)$.

We give the proof later, but first we state another normal form for a more degenerate $\mathbf{D}_3$-equivariant problem. In Case Study 5 on the traction problem we shall consider a $\mathbf{D}_3$-equivariant bifurcation problem for which $q(0, 0, 0) = 0$. We prove the needed result here.

We begin by specifying the lower order terms in p and q as follows:

$$\begin{aligned} p(u, v, \lambda) &= Au + Bv + \alpha\lambda + \cdots \\ q(u, v, \lambda) &= Cu + Dv + \beta\lambda + \cdots. \end{aligned} \tag{4.23}$$

We call any $\mathbf{D}_3$-equivariant bifurcation problem g satisfying $p(0,0,0) = 0 = q(0,0,0)$ *nondegenerate* if

$$\begin{aligned} &\text{(a)} \quad \alpha \neq 0 \\ &\text{(b)} \quad A \neq 0 \\ &\text{(c)} \quad \alpha C - \beta A \neq 0 \\ &\text{(d)} \quad AD - BC \neq 0. \end{aligned} \tag{4.24}$$

Theorem 4.4. *Let g be a $\mathbf{D}_3$-equivariant bifurcation problem. Assume that $p(0,0,0) = 0 = q(0,0,0)$ and that g is nondegenerate. Then g is $\mathbf{D}_3$-equivalent to the normal form*

$$\text{(a)} \quad N(z, \lambda) = (\varepsilon u + \delta\lambda)z + (\sigma u + mv)\bar{z}^2 \tag{4.25}$$

where $\varepsilon = \operatorname{sgn} A$, $\delta = \operatorname{sgn} \alpha$, $\sigma = \operatorname{sgn}(\alpha C - \beta A) \cdot \operatorname{sgn} \alpha$, *and*

$$\text{(b)} \quad m = \operatorname{sgn}(A) \cdot (AD - BC)\alpha^2/(\alpha C - \beta A)^2.$$

Singularities of $\mathbf{D}_n$-equivariant bifurcation problems have been studied in Buzano, Geymonat, and Poston [1985] in connection with buckling rods of regular polygonal cross section.

Proof of Proposition 4.3. In Table 3.1 we showed that the use of invariant coordinates lets us identify $\vec{\mathscr{E}}_{z,\lambda}(\mathbf{D}_n)$ with $\vec{\mathscr{E}}_{u,v,\lambda}$ by the mapping

$$g \mapsto [p, q].$$

We claim that the hypotheses of Proposition 4.3 imply that

$$RT(g, \mathbf{D}_3) = \mathscr{I} \tag{4.26}$$

where $\mathscr{I}$ is the submodule of $\vec{\mathscr{E}}_{u,v,\lambda}$ defined by

$$\mathscr{I} = [\mathscr{M}_{u,v,\lambda}, \mathscr{E}_{u,v,\lambda}]. \tag{4.27}$$

First we show that (4.27) together with Theorem 1.3 leads to the normal form h. The proof proceeds in two steps. First, write

$$g_t = [p_\lambda(0)\lambda + t(p_u(0)u + p_v(0)v + \varphi(u, v, \lambda)), q(0) + t\psi(u, v, \lambda)]$$

where $g_1 = g$, $\varphi \in \mathscr{M}^2_{u,v,\lambda}$, $\psi \in \mathscr{M}_{u,v,\lambda}$. Theorem 1.3 and the claim (4.26) imply that $g = g_1$ is $\mathbf{D}_3$-equivalent to

$$g_0 = [p_\lambda(0)\lambda, q(0)].$$

This claim also shows that $RT(g_0, \mathbf{D}_3)$ is independent of $p_\lambda(0)$ and $q(0)$, as long as both are nonzero. Thus, by using Theorem 3.1 again, we prove that g_0 is $\mathbf{D}_3$-equivalent to $h = [\varepsilon\lambda, \delta]$ where $\varepsilon = \operatorname{sgn} p_\lambda(0)$ and $\delta = \operatorname{sgn} q(0)$. (Alternatively, we could have scaled g_0 directly to obtain h.)

We now verify (4.26). The six generators of $RT(g, \mathbf{D}_3)$ are given by Table 3.1. It is easy to check that each generator lies in $\mathscr{I}$, since $p(0,0,0) = 0$. Hence

Table 4.5. Matrix for the Generators of $RT(g, \mathbf{D}_3)$

	$[u, 0]$	$[v, 0]$	$[\lambda, 0]$	$[0, 1]$
$[p, q]$	$p_u(0)$	$p_v(0)$	$p_\lambda(0)$	$q(0)$
$[2up + vq, 0]$		$q(0)$		
$[uq, p]$	$q(0)$			
$[vp + 2u^2q, 0]$				
$[2up_u + \cdots, 3q + \cdots]$	$2p_u(0)$	$3p_v(0)$		$3q(0)$
$[vp_u + \cdots, vq_u + \cdots]$		$p_u(0)$		

$RT(g, \mathbf{D}_3) \subset \mathscr{I}$. To show conversely that $\mathscr{I} \subset RT(g, \mathbf{D}_3)$ we first show that

$$\mathscr{I} \subset RT(g, \mathbf{D}_3) + \mathscr{M}_{u,v,\lambda}\mathscr{I} \tag{4.28}$$

and then apply Nakayama's lemma. Note that $\mathscr{I}$ is generated by the four elements

$$[u, 0], [v, 0], [\lambda, 0], [0, 1]. \tag{4.29}$$

In Table 4.5 we write the six generators of $RT(g, \mathbf{D}_3)$, modulo terms in $\mathscr{M}\mathscr{I} = [\mathscr{M}^2_{u,v,\lambda}, \mathscr{M}_{u,v,\lambda}]$, in terms of the four generators (4.29). This produces a 6×4 matrix, which, assuming $q(0) \neq 0$ and $p_\lambda(0) \neq 0$, has rank 4. This proves (4.26). □

Proof of Theorem 4.4. We divide the proof into two parts. In the first part we show that g is $\mathbf{D}_3$-equivalent to

$$h(z, \lambda) = (Au + Bv + \alpha\lambda)z + (Cu + Dv + \beta\lambda)\bar{z}^2. \tag{4.30}$$

We do this by using Theorem 1.3. In the second part we show that h is $\mathbf{D}_3$-equivalent to the normal form N in (4.25). This part is proved by explicitly constructing the $\mathbf{D}_3$-equivalence.

Part I. The most important step in the first part of the proof is to show that

$$\mathscr{J} \subset RT(g, \mathbf{D}_3) \tag{4.31}$$

where

$$\mathscr{J} = \mathscr{M}^2_{u,v,\lambda}\vec{\mathscr{E}}_{u,v,\lambda} = [\mathscr{M}^2_{u,v,\lambda}, \mathscr{M}^2_{u,v,\lambda}].$$

From (4.31) and the explicit form of the generators of $RT(g, \mathbf{D}_3)$ in Table 3.1 it follows that

$$RT(g, \mathbf{D}_3) = \mathscr{J} \oplus \mathbb{R}\{[\varepsilon u + \delta\lambda, \sigma u + mv], [0, \varepsilon u + \delta\lambda], [2\varepsilon u, 5\sigma u + 6mv], [\varepsilon v, \sigma v]\}. \tag{4.32}$$

By (4.32), $RT(g, \mathbf{D}_3)$ is independent of φ. Thus we can use Theorem 1.3 to prove that g is $\mathbf{D}_3$-equivalent to h, where h is defined in (4.30).

To verify (4.31) we use Nakayama's lemma and prove that

Table 4.6.

	u^2	uv	v^2	$u\lambda$	$v\lambda$	λ^2	u^2	uv	v^2	$u\lambda$	$v\lambda$	λ^2
$u[p,q]$	ε			δ			σ	m				
$v[p,q]$		ε			δ			σ	m			
$\lambda[p,q]$				ε		δ				σ	m	
$[2up + vq, 0]$	2ε	σ	m	2δ								
$u[uq,p]$							ε			δ		
$v[uq,p]$								ε			δ	
$\lambda[uq,p]$										ε		δ
$[vp + 2u^2q, 0]$		ε			δ							
$u[2up_u + \cdots]$	2ε						5σ	$6m$				
$v[2up_u + \cdots]$		2ε						5σ	$6m$			
$\lambda[2up_u + \cdots]$				2ε						5σ	$6m$	
$u[vp_u + \cdots]$		ε						σ				
$v[vp_u + \cdots]$			ε						σ			
$\lambda[vp_u + \cdots]$					ε						σ	

Table 4.7.

	uv	v^2	$v\lambda$	uv	v^2	$v\lambda$
$v[p,q]$	ε		δ	σ	m	
$v[uq,p]$				ε		δ
$[vp + 2u^2q, 0]$	ε		δ			
$v[2up_u + \cdots]$	2ε			5σ	$6m$	
$u[vp_u + \cdots]$	ε			σ		
$v[vp_u + \cdots]$		ε			σ	
$\lambda[vp_u + \cdots]$			ε			σ

$$\mathscr{J} \subset RT(g, \mathbf{D}_3) + \mathscr{M}_{u,v,\lambda} \cdot \mathscr{J}. \tag{4.33}$$

The submodule $\mathscr{J}$ has twelve generators over the ring $\mathscr{E}_{u,v,\lambda}$, namely:

$$\begin{gathered} [u^2, 0], [uv, 0], [v^2, 0], [u\lambda, 0], [v\lambda, 0], [\lambda^2, 0], \\ [0, u^2], [0, uv], [0, v^2], [0, u\lambda], [0, v\lambda], [0, \lambda^2]. \end{gathered} \tag{4.34}$$

It is not hard to find fourteen elements of $RT(g, \mathbf{D}_3)$ that are in the submodule $\mathscr{J}$. We record these elements modulo $\mathscr{M}\mathscr{J}$, that is, modulo third order terms, in Table 4.6. In Table 4.6 we use the convention that the columns on the left-hand side headed by "$*$" refer to the generator $[*, 0]$. On the right-hand side those columns headed by "$*$" refer to the generator $[0, *]$. As in previous examples, if the 14×12 matrix has rank 12, then (4.33) will hold.

There is some structure to this 14×12 matrix which enables us to compute its rank reasonably easily. The main observation is that 7 rows have nonzero elements in a total of 6 columns. Isolating these rows leads to the 7×6 matrix in Table 4.7. We first show that this 7×6 submatrix has rank 6. The simplest

Table 4.8.

uv	v^2	$v\lambda$	uv	v^2	$v\lambda$
			σ	m	
			ε		δ
ε		δ			
			σ	$2m$	
ε			σ		
	ε			σ	

Table 4.9.

u^2	$u\lambda$	λ^2	u^2	$u\lambda$	λ^2
ε	δ		σ		
	ε	δ		σ	
2ε	2δ				
			ε	δ	
				ε	δ
2ε			5σ		
	2ε			5σ	

way to complete this step is to cast out the seventh row and show that when $m \neq 0$ then the resulting 6×6 matrix is nonsingular. We record this 6×6 matrix in Table 4.8 after subtracting the third row from the first, subtracting twice the fifth row from the fourth, and dividing the fourth row by 3. It is a simple exercise to show that the determinant of the matrix in Table 4.8 is nonzero if $m \neq 0$.

We can now show that the original 14×12 matrix has rank 12 if, after casting out the seven rows and six columns described previously, the remaining 7×6 matrix has rank 6. This matrix is shown in Table 4.9. To show that the 7×6 matrix in Table 4.9 has rank 6, cast out the last row and compute the determinant.

Part II. In the setting of Theorem 4.4 we assume that g is a nondegenerate $\mathbf{D}_3$-equivariant bifurcation problem. That is, $g = h + \varphi$ where

$$h = (Au + Bv + \alpha\lambda)z + (Cu + Dv + \beta\lambda)\bar{z}^2 \tag{4.35}$$

and φ consists of higher order terms. In this part, we show that h is $\mathbf{D}_3$-equivalent to the normal form N is (4.25), modulo higher order terms. This will complete the proof of Theorem 4.4.

Our approach is to consider how the general $\mathbf{D}_3$-equivalence operates in the intermediate order terms in h. We do this calculation in two steps. The general $\mathbf{D}_3$-equivalence has the form

$$h'(z, \lambda) = S(z, \lambda)h(Z(z, \lambda), \Lambda(\lambda)). \tag{4.36}$$

Write the intermediate terms in h' as

$$h'(z,\lambda) = (A'u + B'v + \alpha'\lambda)z + (C'u + D'v + \beta'\lambda)\bar{z}^2 + \cdots. \tag{4.37}$$

It is possible to write the coefficients of h and the lower order terms of the $\mathbf{D}_3$-equivalence (S, Z, Λ). Specifically, let

$$\begin{aligned} &\text{(a)} \quad Z(z,\lambda) = \zeta z + \eta\bar{z}^2 + \cdots \\ &\text{(b)} \quad \Lambda(\lambda) = \theta\lambda + \cdots \\ &\text{(c)} \quad S = \rho S_1 + \varphi S_2 + \tau S_3 + \psi S_4 + \cdots \end{aligned} \tag{4.38}$$

where the S_j are defined in (3.19). All of ζ, η, θ, ρ, τ, φ, ψ are real constants. The restrictions imposed by being a $\mathbf{D}_3$-equivalence are:

$$\zeta > 0, \qquad \theta > 0, \qquad \rho > 0. \tag{4.39}$$

We claim that

$$\begin{aligned} &\text{(a)} \quad A' = \rho\zeta^3 A \\ &\text{(b)} \quad B' = \rho\zeta^2(2\eta A + \zeta^2 B) \\ &\text{(c)} \quad \alpha' = \rho\zeta\theta\alpha \\ &\text{(d)} \quad C' = (\tau\zeta + \rho\eta)\zeta^2 A + \rho\zeta^4 C \\ &\text{(e)} \quad D' = (\tau\zeta + \rho\eta)(2\zeta\eta A + \zeta^3 B) + \rho\zeta^3(2\eta C + \zeta^2 D) \\ &\text{(f)} \quad \beta' = (\tau\zeta + \rho\eta)\theta\alpha + \rho\zeta^2\theta\beta. \end{aligned} \tag{4.40}$$

Assuming (4.40) let us first show how h may be transformed to the normal form N modulo higher order terms. First choose η and τ so that $B' = \beta' = 0$. Specifically, let

$$\eta = -\zeta^2 B/2A, \qquad \tau\zeta + \rho\eta = -\rho\zeta^2\beta/\alpha. \tag{4.41}$$

Note that we use nondegeneracy when we divide by A and α. See (4.24). Substitute (4.41) into (4.40) to get

$$\begin{aligned} &\text{(a)} \quad A' = \rho\zeta^3 A \\ &\text{(b)} \quad \alpha' = \rho\zeta\theta\alpha \\ &\text{(c)} \quad C' = \rho\zeta^4(\alpha C - \beta A)/\alpha \\ &\text{(d)} \quad D' = \rho\zeta^5(AD - BC)/A \\ &\text{(e)} \quad B' = \beta' = 0. \end{aligned} \tag{4.42}$$

We can now choose ρ so that $|A'| = 1$, ζ so that $|C'| = 1$, and θ so that $|\alpha'| = 1$. Specifically let

$$\begin{aligned} &\text{(a)} \quad \rho = 1/(\zeta^3|A|) \\ &\text{(b)} \quad \theta = \zeta^2|A|/\alpha \\ &\text{(c)} \quad \zeta = |\alpha A/(\alpha C - \beta A)|. \end{aligned} \tag{4.43}$$

Again we use the assumption of nondegeneracy to divide by A, α, and $\alpha C - \beta A$. Substitute (4.43) into (4.42) to obtain

$$\begin{aligned} &\text{(a)} \quad \operatorname{sgn} A' = \operatorname{sgn} A, \qquad |A'| = 1 \\ &\text{(b)} \quad \operatorname{sgn} \alpha' = \operatorname{sgn} \alpha, \qquad |\alpha'| = 1 \\ &\text{(c)} \quad \operatorname{sgn} C' = \operatorname{sgn}(\alpha)\cdot\operatorname{sgn}(\alpha C - \beta A), \qquad |C'| = 1 \\ &\text{(d)} \quad D' = \operatorname{sgn}(A)\cdot\alpha^2(AD - BC)/(\alpha C - \beta A)^2. \end{aligned} \tag{4.44}$$

Set $m = D'$ to identify the intermediate order terms with the normal form $N(z, \lambda)$ of (4.25).

It remains to establish (4.40). First write h in (4.35) in invariant coordinates, so that $h = [a, b]$ where

$$\begin{aligned} a &= Au + Bv + \alpha\lambda \\ b &= Cu + Dv + \beta\lambda. \end{aligned}$$

We want to compute h' in (4.36) in invariant coordinates modulo higher order terms. Let

$$h' = [a', b'].$$

We use the intermediate function

$$h'' = h(Z, \Lambda) = [a'', b'']. \tag{4.45}$$

In Part I of this proof we showed that we can ignore terms in $\mathscr{J} = [\mathscr{M}^2_{u,v,\lambda}, \mathscr{M}^2_{u,v,\lambda}]$; see (4.31). So when we say that we compute modulo higher order terms, we mean that we compute in invariant coordinates modulo terms which are quadratic or higher order in u, v, λ.

For example, the action of the S_j of (4.38) on h'' is given in invariant coordinates by (3.22). Specifically

$$\begin{aligned} &\text{(a)} \quad S_1 h'' = [a'', b''] \\ &\text{(b)} \quad S_2 h'' = [ua'' + vb'', -ub''] \\ &\text{(c)} \quad S_3 h'' = [ub'', a''] \\ &\text{(d)} \quad S_4 h'' = [va'' + u^2 b'', -ua'']. \end{aligned} \tag{4.46}$$

Noting that a'' and b'' vanish at the origin, we compute modulo higher order terms that

$$Sh'' = [\rho a'', \rho b'' + \tau a'']$$

when S is defined as in (4.38). Thus

$$\begin{aligned} &\text{(a)} \quad a' \equiv \rho a'' \\ &\text{(b)} \quad b' \equiv \tau a'' + \rho b''. \end{aligned} \tag{4.47}$$

Next we compute $[a'', b'']$ from $[a, b]$ using the definition of h'' in (4.45).

Specifically, modulo $\mathcal{J}$,

$$h''(z, \lambda) \equiv ((Au(Z) + Bv(Z) + \alpha\Lambda)Z, (Cu(Z) + Dv(Z) + \beta\Lambda)\bar{Z}^2). \quad (4.48)$$

Now we use the form of Z and Λ in (4.38) to obtain

$$\begin{aligned} &\text{(a)} \quad u(Z) = Z\bar{Z} \equiv \zeta^2 u + 2\zeta\eta v \qquad (\operatorname{mod} \mathcal{M}^2_{u,v,\lambda}) \\ &\text{(b)} \quad v(Z) = Z^3 + \bar{Z}^3 \equiv \zeta^3 v \qquad (\operatorname{mod} \mathcal{M}^2_{u,v,\lambda}) \\ &\text{(c)} \quad \Lambda(\lambda) \equiv \theta\lambda \qquad (\operatorname{mod} \mathcal{M}^2_{u,v,\lambda}) \\ &\text{(d)} \quad Z(z, \lambda) \equiv \zeta^2\bar{z}^2 \qquad (\operatorname{mod} \mathcal{M}_{u,v,\lambda}\vec{\mathcal{E}}(\mathbf{D}_3)). \end{aligned} \quad (4.49)$$

The higher order terms in Z as in (4.49(d)) contribute only terms in $\mathcal{J}$ to h''. Thus

$$\begin{aligned} h''(z, \lambda) = {} & (A\zeta^2 u + (2\zeta\eta A + \zeta^3 B)v + \alpha\theta\lambda)(\zeta z + \eta\bar{z}^2) \\ & + (C\zeta^2 u + (2\zeta\eta C + \zeta^3 B)v + \beta\theta\lambda)\zeta^2\bar{z}^2. \end{aligned} \quad (4.50)$$

Finally we combine the results in (4.50) and (4.47) to obtain (4.40). This calculation is left to the reader. □

Exercises

These exercises, based on Melbourne [1988], develop the singularity theory of bifurcation problems commuting with the symmetry group $\Gamma = \mathbb{O} \oplus \mathbf{Z}_2^c$ of the cube in its natural action on $\mathbb{R}^3$; see Exercise XIII, 4.2. Recall that $|\Gamma| = 48$, that the invariants are generated by

$$\begin{aligned} u &= x^2 + y^2 + z^2 \\ v &= x^2y^2 + y^2z^2 + x^2z^2 \\ w &= x^2y^2z^2, \end{aligned}$$

and that the module of equivariants is generated by the mappings

$$X_1 = \begin{bmatrix} x \\ y \\ z \end{bmatrix} \quad X_2 = \begin{bmatrix} x^3 \\ y^3 \\ z^3 \end{bmatrix} \quad X_3 = \begin{bmatrix} y^2z^2x \\ z^2x^2y \\ x^2y^2z \end{bmatrix}.$$

Write an arbitrary Γ-equivariant f in the form

$$f = PX_1 + QX_2 + RX_3 = [P, Q, R],$$

where P, Q, R are functions of u, v, w, λ.

4.1. Define the singularity g_m by

$$g_m(x, y, z, \lambda) = [\delta mu + \varepsilon\lambda + \sigma u^2, \delta, 0]$$

where $\delta = \pm 1$, $\varepsilon = \pm 1$, $\sigma = \pm 1$, and m is a modal parameter. Show that f is Γ-equivalent to g_m if and only if:

$$P(0) = 0,$$

$$P_\lambda(0) \neq 0, \qquad Q(0) \neq 0, \qquad P_u(0)/Q(0) \neq -1, -\tfrac{1}{2}, -\tfrac{1}{3},$$

$$T = \{P_{uu}(0) + (m+1)P_v(0) - 2mQ_u(0) + (m+1)(2m+1)R(0)\} \neq 0$$

with

$$\delta = \operatorname{sgn} Q(0), \qquad \varepsilon = \operatorname{sgn} P_\lambda(0), \qquad \sigma = \operatorname{sgn} T, \qquad m = P_u(0)/Q(0).$$

4.2. Define the singularity h_m by

$$h_m(x, y, z, \lambda) = [\delta m u + \varepsilon\lambda, \delta, 0]$$

where $\delta = \pm 1$, $\varepsilon = \pm 1$, and m is a modal parameter. Show that f is Γ-equivalent to h_m if and only if:

$$P(0) = 0,$$

$$P_\lambda(0) \neq 0, \qquad Q(0) \neq 0, \qquad P_u(0)/Q(0) \neq -1, -\tfrac{1}{2}, -\tfrac{1}{3},$$

$$T = 0 \text{ where } T \text{ is as in Exercise 4.1},$$

with

$$\delta = \operatorname{sgn} Q(0), \qquad \varepsilon = \operatorname{sgn} P_\lambda(0), \qquad m = P_u(0)/Q(0).$$

4.3. Deduce that every Γ-equivariant bifurcation problem satisfying the nondegeneracy conditions $P_\lambda(0) \neq 0$, $Q(0) \neq 0$, $P_u(0)/Q(0) \neq -1, -\tfrac{1}{2}, -\tfrac{1}{3}$ is Γ-equivalent to exactly one of g_m or h_m.

4.4. Let a compact Lie group Γ act absolutely irreducibly on V and let $f\colon V \times \mathbb{R} \to V$ be Γ-equivariant. Suppose that Γ has two isotropy subgroups Σ and Ω with one-dimensional fixed-point subspaces, and corresponding branching equations

$$\lambda = \sigma x^2 + \cdots, \qquad \lambda = \omega y^2 + \cdots,$$

where σ and ω are certain expressions in the Taylor coefficients of f, and x, y parametrize $\mathrm{Fix}(\Sigma)$ and $\mathrm{Fix}(\Omega)$, respectively, with $\|x\| = \|y\|$. Prove that the ratio σ/ω is invariant under Γ-equivalence (and hence acts as a modal parameter).

Interpret the parameter m in Exercises 4.2 and 4.3 in this manner.

(Remark: The parameter T in Exercises 4.2 and 4.3 does not affect the branching directions or stabilities of the bifurcation diagrams. Its significance is singularity-theoretic. For a geometric interpretation see Melbourne [1988]. A more extensive classification of Γ-equivariant bifurcation problems may be found in Melbourne [1987].)

§5. Linearized Stability and Γ-equivalence

We know that the zeros of a Γ-equivariant bifurcation problem $g(x, \lambda)$, where $x \in V$, are preserved (up to change of coordinates) by Γ-equivalence. In this section we discuss when the linearized stability of an equilibrium solution $g(x_0, \lambda_0) = 0$ to the system of ODEs

$$\frac{dx}{dt} + g(x, \lambda) = 0$$

is also preserved by Γ-equivalence. More precisely, suppose that g and h are Γ-equivalent Γ-equivariant bifurcation problems, so that

$$g(x, \lambda) = S(x, \lambda)h(X(x, \lambda), \Lambda(\lambda)). \tag{5.1}$$

Then $(X_0, \Lambda_0) = (X(x_0, \lambda_0), \Lambda(\lambda_0))$ is a zero of h whenever (x_0, λ_0) is a zero of g. The problem we pose is, *when are the signs of the real parts of the eigenvalues of $(dg)_{x_0, \lambda_0}$ the same as those for $(dh)_{X_0, \Lambda_0}$?*

Recall from the definition of Γ-equivalence (Definition 1.1) that X and S in (5.1) satisfy the equivariance conditions

$$\begin{aligned} &\text{(a)} \quad X(\gamma x, \lambda) = \gamma X(x, \lambda) \\ &\text{(b)} \quad S(\gamma x, \lambda)\gamma = \gamma S(x, \lambda). \end{aligned} \tag{5.2}$$

In particular, the chain rule and (5.2(a)) show that the Jacobian matrix dX satisfies (5.2(b)), that is,

$$(dX)_{\gamma x, \lambda}\gamma = \gamma(dX)_{x, \lambda}. \tag{5.3}$$

In addition, Γ-equivalences require that

$$S(0,0) \quad \text{and} \quad (dX)_{0,0} \in \mathscr{L}_\Gamma(V)^0; \tag{5.4}$$

see §1(a); which implies that $\det S > 0$ and $\det(dX)_{0,0} > 0$.

Proposition 5.1. *The linearized stability of the zero (x_0, λ_0) of g is preserved by Γ-equivalence precisely when it is preserved by every Γ-equivalence of the form*

$$S(x, \lambda)g(x, \lambda). \tag{5.5}$$

PROOF. Use the product and chain rules and the fact that $h(X_0, \Lambda_0) = 0$ to differentiate (5.1) and obtain

$$(dg)_{x_0, \lambda_0} = S(x_0, \lambda_0)(dh)_{X_0, \Lambda_0}(dX)_{x_0, \lambda_0}. \tag{5.6}$$

Rewrite this as

$$\tilde{S}(x_0, \lambda_0)(dg)_{x_0, \lambda_0} = (dX)^{-1}_{x_0, \lambda_0}(dh)_{X_0, \Lambda_0}(dX)_{x_0, \lambda_0} \tag{5.7}$$

where

$$\tilde{S}(x_0, \lambda_0) = [S(x_0, \lambda_0)(dX)_{x_0, \lambda_0}]^{-1}. \tag{5.8}$$

Now observe that the eigenvalues of the matrix on the right-hand side of (5.7) are the same as the eigenvalues of $(dh)_{X_0, \Lambda_0}$ (by similarity), and the matrix on the left-hand side of (5.8) represents one of the special kinds of Γ-equivalence indicated in (5.5). To check this last point note that the product of two matrices satisfying (5.2(b)) also satisfies (5.2(b)), with a similar statement for inverses. Hence $\tilde{S}$ satifies (5.2(b)). In addition, $\mathscr{L}_\Gamma(V)^0$ is a group of matrices and is therefore closed under multiplication and inversion, so $\tilde{S}(0,0) \in \mathscr{L}_\Gamma(V)^0$. Thus multiplication by $\tilde{S}$ is a Γ-equivalence.

On the assumption that these special Γ-equivalences preserve linearized stability, it follows that all Γ-equivalences preserve linearized stability. □

Remark 5.2. The sign of the determinant of dg is preserved by Γ-equivalence since (5.6) implies that

$$\det(dg)_{x_0,\lambda_0} = \det S(x_0,\lambda_0)\det(dh)_{X_0,\Lambda_0}\det(dX)_{x_0,\lambda_0},$$

and $\det S > 0$, $\det(dX) > 0$ by (5.4).

We now discuss when the hypothesis of Proposition 5.1 is satisfied. In general, it is *not* satisfied. However, there is one important case when it is. Let Σ be an isotropy subgroup of Γ and suppose that (x_0, λ_0) is a solution of $g = 0$, with isotropy subgroup Σ. Decompose the state space V of xs into irreducible subspaces for the action of Σ,

$$V = V_1 \oplus \cdots \oplus V_k. \tag{5.9}$$

We now state and prove a surprisingly useful criterion for the preservation of linearized stability. It shows that the linearized stability of certain types of solution is preserved by Γ-equivalence, even though the stability of *all* types of solution need not be.

Theorem 5.3. *Assume that the V_j are distinct absolutely irreducible representations of Σ. Then the linearized stability of (x_0, λ_0) is preserved by Γ-equivalence.*

Proof. From Proposition 5.1, it suffices to show that the Jacobian of $h = Sg$ at (x_0, λ_0) and the Jacobian of g at (x_0, λ_0) have eigenvalues with the same sign. Note that

$$(dh)_{x_0,\lambda_0} = S(x_0,\lambda_0)(dg)_{x_0,\lambda_0} \tag{5.10}$$

and that the three matrices in (5.10) all satisfy (5.2(b)).

Suppose that $\sigma \in \Sigma$. Then $\sigma x_0 = x_0$ and (5.2(b)) implies that

$$S(x_0,\lambda_0)\sigma = \sigma S(x_0,\lambda_0),$$

so S commutes with Σ. Because the V_j are distinct, Theorem XII, 3.5, implies that $S(x_0, \lambda_0)$ maps V_j into V_j. The absolute irreduciblity of V_j implies that $S(x_0, \lambda_0)|V_j$ is a multiple of the identity, say $c_j I_{V_j}$. Finally, since $S \in \mathscr{L}_\Gamma(V)^0$, each $c_j > 0$.

Since $(dg)_{x_0,\lambda_0}$ satisfies (5.2(b)) we can argue as previously to conclude that $(dg)_{x_0,\lambda_0}|V_j = d_j I_{V_j}$. Thus all of the eigenvalues of $(dg)_{x_0,\lambda_0}$ are real, since these are the d_j. Further, the eigenvalues of $(dh)_{x_0,\lambda_0}$ are the products $c_j d_j$. Since $c_j > 0$, the signs correspond and the theorem is proved. □

We have already seen a special instance of this theorem when we considered $\mathbf{Z}_2 \oplus \mathbf{Z}_2$-equivariant bifurcation problems in Chapter X. There we proved

(Proposition X, 3.2(a)) that the linearized stability of trivial and pure mode solutions is invariant under $\mathbf{Z}_2 \oplus \mathbf{Z}_2$-equivalence. Theorem 5.3 is the natural generalization of that proposition.

EXERCISE

5.1. In $\mathbf{O}(3)$-equivariant steady-state bifurcation on the space V_l of spherical harmonics, prove that $\mathbf{O}(3)$-equivalence preserves the stability of axisymmetric ($\Sigma \subset \mathbf{SO}(2)$) solutions. (*Hint*: Apply Lemma 3.5 and XIII, §7(c).)

§6. Intrinsic Ideals and Intrinsic Submodules

In Volume I we showed that the higher order terms of a bifurcation problem in one state variable can be characterized as an intrinsic ideal. This result extends to the equivariant case and is presented in §7 later. In this section we discuss general properties of intrinsic ideals and submodules. The main difference here is that we are forced to discuss the notion "intrinsic" with respect to both Γ-equivalence and strong Γ-equivalence.

First, the definitions. Let $g(x, \lambda)$ be a Γ-equivariant bifurcation problem. Recall that a Γ-equivalence has the form

$$g \mapsto S(x, \lambda)g(X(x, \lambda), \Lambda(\lambda)) = \Phi(g),$$

where S and X satisfy the equivariance conditions (1.2). The Γ-equivalence is *strong* if $\Lambda(\lambda) = \lambda$.

Definition 6.1.
(a) An ideal $\mathscr{I} \subset \mathscr{E}(\Gamma)$ is *intrinsic* if $f(X(x, \lambda), \Lambda(\lambda)) \in \mathscr{I}$ for every $f \in \mathscr{I}$ and every Γ-equivariant change of coordinates $(X(x, \lambda), \Lambda(\lambda))$. Similarly $\mathscr{I}$ is *S-intrinsic* if this statement holds under the extra assumption $\Lambda(\lambda) = \lambda$.
(b) A submodule $\mathscr{J} \subset \vec{\mathscr{E}}(\Gamma)$ is *intrinsic* if $\Phi(g) \in \mathscr{J}$ and every Γ-equivalence Φ. Similarly $\mathscr{J}$ is *S-intrinsic* if this statement holds under the extra assumption $\Lambda(\lambda) = \lambda$.

The maximal ideal

$$\mathscr{M} = \{f \in \mathscr{E}(\Gamma): f(0) = 0\} \tag{6.1}$$

and the ideal $\langle \lambda \rangle$ are always intrinsic. Moreover, since sums and products of intrinsic ideals are intrinsic,

$$\mathscr{M}^k + \mathscr{M}^{k_1}\langle \lambda \rangle^{l_1} + \cdots + \mathscr{M}^{k_s}\langle \lambda \rangle^{l_s} \tag{6.2}$$

is intrinsic. We showed in Proposition II, 7.1, that for bifurcation problems in one state variable without symmetry, every intrinsic ideal of finite codimension is of the form (6.2). However, this form is no longer valid when symmetries are present. Indeed, for $\mathbf{Z}_2$-equivariant bifurcation problems in one state variable, every ideal generated by monomials is intrinsic; see Proposition VI, 2.9.

There are certain submodules of $\vec{\mathscr{E}}(\Gamma)$ that are clearly intrinsic. If $\mathscr{I}$ is an intrinsic ideal and $\mathscr{J}$ is an intrinsic submodule, then $\mathscr{I} \cdot \mathscr{J}$ is an intrinsic submodule. Thus $\mathscr{I} \cdot \vec{\mathscr{E}}(\Gamma)$ is intrinsic for every intrinsic ideal $\mathscr{I}$. Again, it is *not* the case that every intrinsic submodule is the sum of intrinsic submodules of this form. The major difficulty in dealing with intrinsic ideals and submodules is that it is hard to determine them when symmetries are present. This computation will be dealt with on a case-by-case basis.

Let $V \subset \vec{\mathscr{E}}(\Gamma)$ be a vector subspace. We define the *intrinsic part* of V to be

$$\operatorname{Itr} V = \sum \{\text{intrinsic submodules of } \vec{\mathscr{E}}(\Gamma) \text{ contained in } V\}. \tag{6.3}$$

Clearly it may also be characterized as the largest intrinsic submodule contained in V. Similarly we define

$$\operatorname{Itr}_s V = \sum \{S\text{-intrinsic submodules of } \vec{\mathscr{E}}(\Gamma) \text{ contained in } V\}.$$

In Theorem II, 8.7, we showed that for bifurcation problems in one state variable we can characterize higher order terms as the intrinsic part of a certain (computable) ideal. In §7 we indicate a generalization to the equivariant case.

Intrinsic submodules $\mathscr{J}$ have a very useful property: if $h \in \mathscr{J}$ then

$$T(h, \Gamma) \subset \mathscr{J}. \tag{6.4}$$

This is proved as follows. Every element in $T(h, \Gamma)$ has the form

$$\frac{d}{dt} \Phi_t(h)|_{t=0} \tag{6.5}$$

where Φ_t is a Γ-equivalence for each t. Since $\mathscr{J}$ is intrinsic, $\Phi_t(h) \in \mathscr{J}$, and since $\mathscr{J}$ is a linear subspace, (6.5) is also in $\mathscr{J}$. Similarly S-intrinsic modules $\mathscr{J}$ satisfy $RT(h, \Gamma) \subset \mathscr{J}$ whenever $h \in \mathscr{J}$.

A partial converse also holds:

Proposition 6.2. *Let $\mathscr{J} \subset \vec{\mathscr{E}}(\Gamma)$ be a submodule of finite codimension. Then $\mathscr{J}$ is intrinsic if and only if every $h \in \mathscr{J}$ satisfies* (6.4).

Proof. The proof that (6.4) implies $\mathscr{J}$ intrinsic uses Lie theory. Since $\mathscr{J}$ has finite codimension there exists k such that $\mathscr{J} \supset \vec{\mathscr{E}}_k(\Gamma)$, where

$$\vec{\mathscr{E}}_k(\Gamma) = \{g \in \vec{\mathscr{E}}_k(\Gamma) \colon g \text{ vanishes to order } k \text{ at the origin}\}.$$

Thus to determine whether $\mathscr{J}$ is intrinsic we need only compute modulo the intrinsic submodule $\vec{\mathscr{E}}_k(\Gamma)$.

The group of Γ-equivalences acts on the finite-dimensional space $\vec{\mathscr{E}}(\Gamma)/\vec{\mathscr{E}}_k(\Gamma)$ as a connected Lie group, which we denote by G_k. Now $\mathscr{J}$ is intrinsic if and only if $\mathscr{J}/\vec{\mathscr{E}}_k(\Gamma)$ is an invariant subspace under G_k. But (6.4) is equivalent to the invariance of $\mathscr{J}$ under the Lie algebra of G_k. It is known that Lie algebra actions have the same invariant subspaces as the actions of the corresponding connected Lie groups. Thus (6.4) implies that $\mathscr{J}$ is intrinsic. □

A similar proposition holds for S-instrinsic modules of finite codimension.

EXERCISE

6.1. Let $\Gamma = \mathbb{1}$.

(a) Show that every S-intrinsic ideal has the form (6.2). (*Hint*: Generalize the proof of Proposition II, 7.1.) Conclude that every S-intrinsic ideal is intrinsic when $\Gamma = \mathbb{1}$.

(b) Show that every S-intrinsic submodule $\mathscr{J}$ has the form $\mathscr{J} = \mathscr{I} \cdot \vec{\mathscr{E}}_n$ with $\mathscr{I}$ and intrinsic ideal. Conclude that S-intrinsic submodules are intrinsic.

§7. Higher Order Terms

In Volume I we showed that the higher order terms of a bifurcation problem in one state variable can be characterized as an intrinsic ideal. In this section we show that a similar result holds for equivariant bifurcation problems; see Proposition 7.5. More importantly, we present a method, due to Gaffney [1986] and Bruce, du Plessis, and Wall [1987] for calculating these higher order terms (Theorems 7.2 and 7.3). Our calculations in §4 show that such abstractions are not needed for many simple examples. However, we shall encounter later examples in which the general results simplify the required calculations.

We begin by stating Gaffney's results. There are two basic theorems: one for strong equivalence and one for equivalence. Recall that the modules of Γ-equivariant mappings $\vec{\mathscr{E}}(\Gamma)$ and matrices $\overleftrightarrow{\mathscr{E}}(\Gamma)$ are finitely generated over the ring $\mathscr{E}(\Gamma)$. The Hilbert–Weyl theorem (Theorem XII, 4.2) lets us choose as generators homogeneous mappings $X_1, \ldots, X_k \in \vec{\mathscr{E}}(\Gamma)$ and homogeneous matrices $S_1, \ldots, S_l \in \overleftrightarrow{\mathscr{E}}(\Gamma)$. That is, we may assume each X_i and each S_j has entries that are homogeneous polynomials of the same degree. We define $\deg X_i$ and $\deg S_j$ to be those common degrees.

Definition 7.1. $\mathscr{K}_s(g, \Gamma)$ is the submodule of $\vec{\mathscr{E}}(\Gamma)$ generated by $\mathscr{M} \cdot RT(g, \Gamma)$, $(dg)(X_i)$ where $\deg X_i \geq 2$, and $S_j g$ where $\deg S_j \geq 1$.

Theorem 7.2. *Let $p \in \operatorname{Itr}_s \mathscr{K}_s(g, \Gamma)$. Then $g + p$ is strongly Γ-equivalent to g.*

Remarks 7.3.

(a) Gaffney proves that when $\operatorname{Fix}(\Gamma) = \{0\}$ Theorem 7.2 is the best possible result for strong Γ-equivalence, in the following sense. Theorem 7.2 characterizes those p such that $h + p$ is strongly Γ-equivalent to h for all h strongly Γ-equivalent to g.

(b) When $\operatorname{Fix}(\Gamma) \neq \{0\}$ the results of Gaffney [1986] and Bruce, du Plessis, and Wall [1987] are actually more general than Theorem 7.2. In particular, they are strong enough to recover Theorem II, 8.7, the main determinacy result for bifurcation problems in one state variable.

(c) In this section we will not prove Gaffney's result in full generality, since the proof relies heavily on machinery from algebraic geometry, beyond the scope of this text. Instead, we prove a simpler result in which p in Theorem 7.2 is assumed to be in $\mathrm{Itr}_s[\mathscr{M}\cdot RT(g,\Gamma)]$.

Gaffney's second result discusses when $g+p$ is Γ-equivalent to g. Here nontrivial changes of the λ-coordinate may be needed. Define

$$\mathscr{K}(g,\Gamma)=\mathscr{K}_s(g,\Gamma)+\mathscr{E}_\lambda\{\lambda^2 g_\lambda\}. \tag{7.1}$$

Note that $\mathscr{K}$ is not in general a submodule.

Theorem 7.4. *Let $p\in\mathrm{Itr}\,\mathscr{K}(g,\Gamma)$. Then $g+p$ is Γ-equivalent to g.*

Remark. Theorem 7.2 is by far the most useful general determinacy result that we know of. In particular, when $\mathrm{Fix}(\Gamma)=\{0\}$, it is best possible.

In the remainder of this section we discuss higher order terms in general and prove the simpler result mentioned in Remark 7.3(c).

Let g be a Γ-equivariant bifurcation problem. The perturbation term $p\in\vec{\mathscr{E}}(\Gamma)$ is *higher order* with respect to g if $h+p$ is Γ-equivalent to g, for every h that is Γ-equivalent to g. By definition, such a perturbation cannot enter into a solution of the recognition problem for g. We denote by $\mathscr{P}(g,\Gamma)$ the set of all higher order terms in this sense; that is,

$$\mathscr{P}(g,\Gamma)=\{p\in\vec{\mathscr{E}}(\Gamma)\colon h\pm p\sim g \text{ for all } h\sim g\} \tag{7.2}$$

where $\sim$ denotes Γ-equivalence. The $\pm$ is needed because of the sign conditions we have imposed on equivalences.

Proposition 7.5. *For each $g\in\vec{\mathscr{E}}(\Gamma)$, the set $\mathscr{P}(g,\Gamma)$ is an intrinsic submodule of $\vec{\mathscr{E}}(\Gamma)$.*

Remark 7.6. This definition of $\mathscr{P}(g,\Gamma)$ is more general than that of $\mathscr{P}(g)$ for problems in one state variable, given in II, §8b. There we include only those terms p for which this equivalence is a strong equivalence obtained by the "restricted tangent space constant" theorem (Theorem II, 2.2). The more general definition (7.2) leads in some respects to a simpler theory. For example, compare the proof of Proposition 7.5 with that of Proposition II, 8.6(b).

PROOF. We first show that $\mathscr{P}(g,\Gamma)$ is a submodule. That is:

$$\begin{aligned}&\text{(a)}\quad p_1,p_2\in\mathscr{P}(g,\Gamma)\Rightarrow p_1\pm p_2\in\mathscr{P}(g,\Gamma)\\&\text{(b)}\quad p\in\mathscr{P}(g,\Gamma),\qquad f\in\mathscr{E}(\Gamma)\Rightarrow fp\in\mathscr{P}(g,\Gamma).\end{aligned} \tag{7.3}$$

To prove (7.3(a)) observe that if $p_1\in\mathscr{P}(g,\Gamma)$ then $h+p_1\sim g$ for all $h\sim g$. By the same token $(h\pm p_1)\pm p_2\sim g$. So $p_1\pm p_2\in\mathscr{P}(g,\Gamma)$.

To prove (7.3(b)) suppose we can prove that

$$p\in\mathscr{P}(g,\Gamma),\qquad f\in\mathscr{E}(\Gamma),\qquad f(0)\neq 0\Rightarrow fp\in\mathscr{P}(g,\Gamma). \tag{7.4}$$

Then (7.3(b)) holds for all f, since if $f(0) = 0$ then $f = (1 + f) - 1$. Now $(1 + f)p \in \mathscr{P}(g, \Gamma)$ by (7.4), and $(1 + f)p - p = fp \in \mathscr{P}(g, \Gamma)$ by (7.3(a)). To verify (7.4) observe that if $h \sim g$ then

$$h \pm fp \sim \frac{h}{|f|} \pm p \sim g$$

since $f(0) \neq 0$, so $(1/|f|)h \sim h \sim g$, and $p \in \mathscr{P}(g, \Gamma)$.

Finally, we show that $\mathscr{P}(g, \Gamma)$ is intrinsic. Suppose $p \in \mathscr{P}(g, \Gamma)$ and Φ is a Γ-equivalence. Since equivalences are composed of diffeomorphisms and invertible matrices, they are always invertible. Thus

$$h \pm \Phi(p) = \Phi(\Phi^{-1}(h) \pm p) \sim \Phi^{-1}(h) \pm p.$$

But if $h \sim g$ then $\Phi^{-1}(h) \sim h \sim g$ and $\Phi(h) \pm p \sim g$ since $p \in \mathscr{P}(g, \Gamma)$. Therefore, $\Phi(p) \in \mathscr{P}(g, \Gamma)$. □

We can now restate Theorems 7.2 and 7.4 as follows:

$$\begin{aligned} &\text{(a)} \quad \operatorname{Itr}_s \mathscr{K}_s(g, \Gamma) \subset \mathscr{P}(g, \Gamma), \\ &\text{(b)} \quad \operatorname{Itr} \mathscr{K}(g, \Gamma) \subset \mathscr{P}(g, \Gamma). \end{aligned} \tag{7.5}$$

Moreover, when $\operatorname{Fix}(\Gamma) = \{0\}$, (7.5(b)) is in fact an equality. Since $\mathscr{M} \cdot RT(g, \Gamma) \subset \mathscr{K}_s(g, \Gamma)$ we have

$$\operatorname{Itr}_s[\mathscr{M} \cdot RT(g, \Gamma)] \subset \mathscr{P}(g, \Gamma). \tag{7.6}$$

See XVII, §6, and XIX, §2, for examples where $\operatorname{Itr} \mathscr{K} \supsetneq \operatorname{Itr} \mathscr{M} \cdot RT$. We prove:

Proposition 7.7. *Let $p \in \operatorname{Itr}_s[\mathscr{M} \cdot RT(g, \Gamma)]$. Then $g + p$ is strongly Γ-equivalent to g.*

We need a preliminary result:

Lemma 7.8. *Let $\mathscr{J}$ be an intrinsic submodule contained in $RT(g, \Gamma)$. Then $\mathscr{J} \subset \mathscr{P}(g, \Gamma)$ if*

$$RT(g, \Gamma) \subset RT(g + p, \Gamma) \tag{7.7}$$

for all $p \in \mathscr{J}$.

Proof. By definition a germ $p \in \mathscr{P}(g, \Gamma)$ if $h \pm p \sim g$ for all $h \sim g$. But to show that an *intrinsic* submodule $\mathscr{J}$ is contained in $\mathscr{P}(g, \Gamma)$, it suffices to show that $g \pm p \sim g$ for every $p \in \mathscr{J}$. To see this, suppose $h = \varphi(g)$ for a Γ-equivalence φ. Then $h \pm p = \varphi(g) \pm p \sim g \pm \varphi^{-1}(p) \sim g$ because $\mathscr{J}$ is intrinsic.

To prove that $g \pm p \sim g$ we may appeal to Theorem 1.3 and show that $RT(g + tp, \Gamma) = RT(g, \Gamma)$ for all $t \in \mathbb{R}$. This condition is in turn satisfied if we show that

$$RT(g + p, \Gamma) = RT(g, \Gamma) \tag{7.8}$$

for all $p \in \mathscr{J}$, since $\mathscr{J}$ is closed under scalar multiplication.

Now (7.7) implies (7.8) provided we show that

$$RT(g + p, \Gamma) \subset RT(g, \Gamma). \tag{7.9}$$

We claim this is true. To see why, recall that $RT(p, \Gamma) \subset \mathscr{I}$ since $\mathscr{I}$ is intrinsic. Since the generators of $RT(g + p, \Gamma)$ are sums of generators from $RT(g, \Gamma)$ and generators of $RT(p, \Gamma) \subset \mathscr{I} \subset RT(g, \Gamma)$, (7.7) follows. □

Proof of Proposition 7.7. Let $p \in \mathrm{Itr}[\mathscr{M} \cdot RT(g, \Gamma)]$. By (6.5) each generator of $RT(p, \Gamma)$ lies in $\mathscr{M} \cdot RT(g, \Gamma)$. We claim that

$$RT(g, \Gamma) \subset RT(g + p, \Gamma) + \mathscr{M} \cdot RT(g, \Gamma).$$

Thus we can apply Nakayama's lemma to obtain (7.7) and then apply Lemma 7.8 to prove the proposition.

To verify the claim note that each generator of $RT(g + p, \Gamma)$ is the sum of a generator of $RT(g, \Gamma)$ and one of $RT(p, \Gamma)$. Since $RT(p, \Gamma) \subset \mathscr{M} \cdot RT(g, \Gamma)$ the claim follows. □

CHAPTER XV

Equivariant Unfolding Theory

§0. Introduction

Unfolding theory is the study of parametrized families of perturbations of a given germ. In the symmetric setting, when a group Γ is acting, we consider only Γ-equivalent perturbations. There is a general theory of Γ-unfoldings, analogous to unfoldings in the nonsymmetric case (Volume I, Chapter III). The heart of the singularity theory approach to bifurcations with symmetry is the equivariant unfolding theorem, which asserts that every Γ-equivariant mapping with finite Γ-codimension has a universal Γ-unfolding and gives a computable test for universality.

We begin in §1 by adapting the basic concepts and definitions of unfolding theory to the equivariant context. In §2 we state the equivariant universal unfolding theorem. This asserts that an unfolding $G(x, \lambda, \alpha)$ of a mapping germ $g(x, \lambda)$ is universal if and only if the partial derivatives $\partial G/\partial \alpha_i$, evaluated at $\alpha = 0$, span a complement to the tangent space $T(g, \Gamma)$ in $\vec{\mathscr{E}}_{x,\lambda}(\Gamma)$.

In §3 we study examples of universal unfoldings. We compute $T(h, \Gamma)$ explicitly for the groups $\Gamma = \mathbb{1}$, $\mathbf{Z}_2$, $\mathbf{Z}_2 \oplus \mathbf{Z}_2$, $\mathbf{O}(2)$, and $\mathbf{D}_n$ in their natural representations. Then we discuss universal unfoldings for specific singularities for the groups $\Gamma = \mathbb{1}, \mathbf{Z}_2 \oplus \mathbf{Z}_2$, and $\mathbf{D}_3$. The first two cases complete the proof of theorems stated in Volume I. The final case is needed later in Case Study 5.

In §4 we continue the study of bifurcations with $\mathbf{D}_3$ symmetry and draw the bifurcation diagrams for the simplest such problems. This example demonstrates a number of useful general principles. The results are used in §5 in connection with the spherical Bénard problem and in Case Study 5.

§5 is devoted to an analysis of the spherical Bénard problem—convection in a spherical annulus. This problem is equivariant under the action of the orthogonal group $\mathbf{O}(3)$, and we interpret many known results within the $\mathbf{O}(3)$-

symmetric context, using results from Chapter XIII. In §6 we study the special case when the bifurcation problem may be posed on the five-dimensional irreducible representation V_2 (the spherical harmonics of order 2) of $\mathbf{O}(3)$. By replacing this space of spherical harmonics by an isomorphic representation, in which $\mathbf{O}(3)$ acts by similarity on the space of 3×3 symmetric matrices over $\mathbb{R}$ of trace zero, we obtain a reduction from $\mathbf{O}(3)$-symmetric problems on V_2 to $\mathbf{D}_3$-symmetric problems on $\mathbb{R}^2$. This remarkable reduction process, which relies on very special properties of the representation, preserves the entire bifurcation structure including orbital asymptotic stability.

The proof of the equivariant universal unfolding theorem is given in §7. In order to make it as accessible as possible we postpone the most abstract result, the equivariant preparation theorem, until §8. In that section we derive the equivariant preparation theorem from the ordinary preparation theorem, proofs of which are readily available.

§1. Basic Definitions

In this section we define the basic notions of unfolding theory in the Γ-equivariant context. These ideas generalize naturally from the corresponding definitions in Volume I, Chapters III and VI, so we shall be brief. Throughout this chapter we assume that Γ is a compact Lie group, acting linearly and orthogonally on the vector space V.

Let $g(x, \lambda) \in \vec{\mathscr{E}}_{x,\lambda}(\Gamma)$ be a Γ-equivariant bifurcation problem. A k-*parameter* Γ-*unfolding* of g is a Γ-equivariant map germ $G(x, \lambda, \alpha) \in \vec{\mathscr{E}}_{x,\lambda,\alpha}(\Gamma)$ where $\alpha \in \mathbb{R}^k$ and

$$G(x, \lambda, 0) = g(x, \lambda).$$

More precisely, $G \in \vec{\mathscr{E}}_{x,\lambda,\alpha}(\Gamma)$ if it is the germ of a mapping

$$G: V \times \mathbb{R} \times \mathbb{R}^k \to V$$

satisfying the equivariance condition

$$G(\gamma x, \lambda, \alpha) = \gamma G(x, \lambda, \alpha) \tag{1.1}$$

for all $\gamma \in \Gamma$. That is, we assume that Γ acts trivially on all parameters λ, α.

Of course, we use Γ-unfoldings to represent families of perturbations of g that preserve the symmetries Γ. In the qualitative theory defined by Γ-equivalences we need to know when the perturbations in a Γ-unfolding H are contained in another unfolding G. We codify this idea using the notion of "factoring." Let $H(x, \lambda, \beta)$ be an l-parameter Γ-unfolding of $g(x, \lambda)$, and let $G(x, \lambda, \alpha)$ be a k-parameter Γ-unfolding of $g(x, \lambda)$. We say that H *factors through* G if

$$H(x, \lambda, \beta) = S(x, \lambda, \beta)G(X(x, \lambda, \beta), \Lambda(\lambda, \beta), A(\beta)) \tag{1.2(a)}$$

where $S(x, \lambda, 0) = I$, $X(x, \lambda, 0) = x$, $\Lambda(\lambda, 0) = \lambda$ and $A(0) = 0$. Equation (1.2(a)) states that for each β, $H(\cdot, \cdot, \beta)$ is equivalent to some member $G(\cdot, \cdot, A(\beta))$ of the family G. We also assume that the equivalence of g with g when $\beta = 0$ is the identity. Moreover, we want these equivalences to be Γ-equivalences. We ensure this by demanding that

$$\begin{aligned} &\text{(i)} \quad S \in \vec{\mathscr{E}}_{x,\lambda,\alpha}(\Gamma), \quad \text{i.e.,} \quad S(\gamma x, \lambda, \alpha)\gamma = \gamma S(x, \lambda, \alpha), \\ &\text{(ii)} \quad X \in \vec{\mathscr{E}}_{x,\lambda,\alpha}(\Gamma), \quad \text{i.e.,} \quad X(\gamma x, \lambda, \alpha) = \gamma X(x, \lambda, \alpha). \end{aligned} \tag{1.2(b)}$$

As expected we define a Γ-unfolding G of g to be *versal* if every Γ-unfolding H of g factors through G. A versal Γ-unfolding G of g is *universal* if G depends on the minimum number of parameters needed for a versal unfolding. This minimum number is called the Γ-*codimension* of g and is denoted by

$$\operatorname{codim}_\Gamma g.$$

In Chapters III, VI, IX, and X we defined these concepts in special cases, namely for $(\Gamma, V) = (\mathbb{1}, \mathbb{R})$, $(\mathbf{Z}_2, \mathbb{R})$, $(\mathbb{1}, \mathbb{R}^n)$, and $(\mathbf{Z}_2 \oplus \mathbf{Z}_2, \mathbb{R}^2)$.

§2. The Equivariant Universal Unfolding Theorem

In this section we derive a necessary condition for a k-parameter Γ-unfolding G of g to be versal. We do this by considering one-parameter unfoldings of the simplest form, just as we did in III, §2, for the nonsymmetric case. This construction also leads naturally to a definition of the Γ-equivariant tangent space $T(g, \Gamma)$.

Let $p(x) \in \vec{\mathscr{E}}_{x,\lambda}(\Gamma)$ and consider the one-parameter unfolding $g + \varepsilon p$ of g. Now suppose that $G(x, \lambda, \alpha)$ is a versal Γ-unfolding of g. Then by definition, $g + \varepsilon p$ factors through G. Explicitly this means that

$$g(x, \lambda) + \varepsilon p(x, \lambda) = S(x, \lambda, \varepsilon) G(X(x, \lambda, \varepsilon), \Lambda(\lambda, \varepsilon), A(\varepsilon)) \tag{2.1}$$

where $S(x, \lambda, 0) = I$, $X(x, \lambda, 0) = x$, $\Lambda(x, 0) = \lambda$, $A(0) = 0$, $S \in \vec{\mathscr{E}}_{x,\lambda,\varepsilon}(\Gamma)$, and $X \in \vec{\mathscr{E}}_{x,\lambda,\varepsilon}(\Gamma)$. Differentiate (2.1) with respect to ε and evaluate at $\varepsilon = 0$ to obtain

$$\begin{aligned} p(x, \lambda) = {} & [\dot{S}(x, \lambda, 0) g(x, \lambda) + (dg)_{x,\lambda} \dot{X}(x, \lambda, 0) + g_\lambda(x, \lambda) \dot{\Lambda}(\lambda, 0)] \\ & + \left[\sum_{j=1}^{k} G_{\alpha_j}(x, \lambda, 0) \dot{A}_j(0) \right]. \end{aligned} \tag{2.2}$$

We make three observations about equation (2.2):

(a) The germ p on the left-hand side of (2.2) is arbitrary.
(b) The germs in the first term on the right-hand side of (2.2) comprise the tangent space $T(g, \Gamma)$; we define this more precisely later;
(c) The germ in the second term on the right-hand side of (2.2) lies in the vector space $\mathbb{R}\{G_{\alpha_1}(x, \lambda, 0), \ldots, G_{\alpha_k}(x, \lambda, 0)\}$.

These three observations together imply that if G is a versal Γ-unfolding of g, then

$$\vec{\mathscr{E}}_{x,\lambda}(\Gamma) = T(g,\Gamma) + \mathbb{R}\{G_{\alpha_1}(x,\lambda,0), \ldots, G_{\alpha_k}(x,\lambda,0)\}. \tag{2.3}$$

As in the special cases of Volume I, condition (2.3) is sufficient as well as necessary.

Theorem 2.1 (Γ-equivariant Universal Unfolding Theorem). *Let Γ be a compact Lie group acting on V. Let $g \in \vec{\mathscr{E}}_{x,\lambda}(\Gamma)$ be a Γ-equivariant bifurcation problem and let $G \in \vec{\mathscr{E}}_{x,\lambda,\alpha}(\Gamma)$ be a k-parameter unfolding of g. Then G is versal if and only if* (2.3) *holds.*

The proof of Theorem 2.1 will be given in §§7–8. For the remainder of this section we describe how to use and interpret the infinitesimal condition (2.3).

As mentioned earlier, the possible entries in the first term on the right-hand side of (2.2) comprise the tangent space $T(g,\Gamma)$. Let us be more explicit. Germs of the form

$$\dot{S}(x,\lambda,0)g(x,\lambda) \quad \text{and} \quad (dg)_{x,\lambda}\dot{X}(x,\lambda,0) \tag{2.4}$$

both lie in $RT(g,\Gamma)$, at least when $\dot{X}(0,0,0) = 0$. Now germs of the form (2.4) lie in

$$\text{(a)} \quad RT(g,\Gamma) + \mathbb{R}\{(dg)_{x,\lambda}(Y_1), \ldots, (dg)_{x,\lambda}(Y_m)\}$$

where

$$\text{(b)} \quad \vec{\mathscr{E}}_{x,\lambda}(\Gamma) = \vec{\mathscr{M}}_{x,\lambda}(\Gamma) \oplus \mathbb{R}\{Y_1, \ldots, Y_m\}. \tag{2.5}$$

That is, the Y_j span the Γ-equivariant germs that do not vanish at the origin, modulo those that do vanish at the origin.

Note. If the fixed-point subspace $\text{Fix}(\Gamma) = \{0\}$, then $\vec{\mathscr{E}}_{x,\lambda}(\Gamma) = \vec{\mathscr{M}}_{x,\lambda}(\Gamma)$. Hence $m = 0$ in this case.

To any finite degree of determinacy, germs of the form $g_\lambda(x,\lambda)\dot{\Lambda}(\lambda,0)$ in (2.2) may be written as germs in

$$\mathbb{R}\{g_\lambda, \lambda g_\lambda, \lambda^2 g_\lambda, \ldots\}. \tag{2.6}$$

Combining (2.5) and (2.6) we get

$$T(g,\Gamma) = RT(g,\Gamma) + \mathbb{R}\{(dg)_{x,\lambda}(Y_1), \ldots, (dg)_{x,\lambda}(Y_m), g_\lambda, \lambda g_\lambda, \lambda^2 g_\lambda, \ldots\}. \tag{2.7}$$

The reader may, in fact, take (2.7) to be the definition of the Γ-equivariant tangent space. Compare (2.7) with the formula for $T(g)$ in III, §2.

Equation (2.7) gives a way to compute the tangent space $T(g,\Gamma)$. When computing universal Γ-unfoldings of g, we need to find complements to $T(g,\Gamma)$ in $\vec{\mathscr{E}}_{x,\lambda}(\Gamma)$. More precisely, we have the following:

Corollary 2.2. *Let $g \in \vec{\mathscr{E}}_{x,\lambda}(\Gamma)$ and let $W \subset \vec{\mathscr{E}}_{x,\lambda}(\Gamma)$ be a vector subspace such that*

$$\vec{\mathscr{E}}_{x,\lambda}(\Gamma) = T(g,\Gamma) \oplus W.$$

Let $p_1(x,\lambda), \ldots, p_k(x,\lambda)$ be a basis for W. Then

$$G(x,\lambda,\alpha) = g(x,\lambda) + \sum_{j=1}^{k} \alpha_j p_j(x,\lambda) \tag{2.8}$$

is a universal Γ-unfolding of g.

Remark. It follows that

$$\operatorname{codim}_\Gamma g = \operatorname{codim} T(g,\Gamma) \tag{2.9}$$

in $\vec{\mathscr{E}}_{x,\lambda}(\Gamma)$.

PROOF. The proof of Corollary 2.2 follows directly from Theorem 2.1; details are left to the reader. □

There is a technical issue concerning (2.7) which should be discussed at this point. The corresponding issue without a symmetry group Γ arose in Chapter III, §2(c). It is often advantageous to know that (2.7) holds with only a *finite* number of elements of the form $\lambda^j g_\lambda$; that is,

$$T(g,\Gamma) = RT(g,\Gamma) + \mathbb{R}\{(dg)_{x,\lambda}(Y_1), \ldots, (dg)_{x,\lambda}(Y_m), g_\lambda, \lambda g_\lambda, \ldots \lambda^s g_\lambda\} \tag{2.10}$$

for some s. Obviously this will be true if

(a) $RT(g,\Gamma)$ has finite codimension in $\vec{\mathscr{E}}_{x,\lambda}(\Gamma)$.

However, the natural condition to expect on aesthetic grounds is the apparently weaker one:

(b) $T(g,\Gamma)$ has finite codimension in $\vec{\mathscr{E}}_{x,\lambda}(\Gamma)$.

As pointed out to us by J. Damon, conditions (a) and (b) are equivalent:

Proposition 2.3. *Let $g \in \vec{\mathscr{E}}_{x,\lambda}(\Gamma)$. Then $RT(g,\Gamma)$ has finite codimension if and only if $T(g,\Gamma)$ has finite codimension.*

We do not prove this proposition for two reasons. First, Damon's argument requires fairly heavy abstract algebraic machinery. Second, as far as specific applications are concerned, there is no difficulty in using condition (a). Indeed our method for computing $T(g,\Gamma)$—a necessary step for verifying (b) in any special case—first requires the computation of $RT(g,\Gamma)$, so that (a) can immediately be checked. Thus the *theoretical* fact stated in Proposition 2.3 is important in order to establish that there will be no loss of generality if we take condition (a) as a hypothesis instead of (b), but its *practical* implications for the actual computation are nil.

A proof of a special case of Proposition 2.3 (when $\Gamma = \mathbb{1}$ and g is a polynomial) was sketched in Lemma III, 2.7.

Exercises

These exercises continue Exercises XIV, 4.1 and 4.2 on $\mathbb{O} \oplus \mathbf{Z}_2^c$ bifurcation and use the same notation.

2.1. Prove that
   (a) The codimension of g_m is 1 and its topological codimension is 0.
   (b) A universal Γ-unfolding is $G(x, y, z, n, \lambda) = [\delta nu + \varepsilon\lambda + \sigma u^2, \delta, 0]$ where n is close to m.

2.2. Prove that
   (a) The codimension of h_m is 2 and its topological codimension is 1.
   (b) A universal Γ-unfolding is given by $H(x, y, z, n, t, \lambda) = [\delta nu + \varepsilon\lambda + tu^2, \delta, 0]$ where n is close to m and t close to 0.

§3. Sample Universal Γ-unfoldings

In this section we compute $T(h, \Gamma)$ for five examples: $\Gamma = \mathbb{1}, \mathbf{Z}_2, \mathbf{Z}_2 \oplus \mathbf{Z}_2, \mathbf{O}(2)$ and $\mathbf{D}_n (n \geq 3)$. This information is summarized in Table 3.1 and uses the notation and setting of Chapter XIV, Table 3.1.

We then compute universal unfoldings for several examples: $\Gamma = \mathbb{1}$, $\mathbf{Z}_2 \oplus \mathbf{Z}_2$, $\mathbf{D}_3$. The first two results verify theorems stated in Volume I; the last example will be needed in Case Study 5.

All of the entries in Table 3.1 are computed by using (2.7). For the last three cases, $\mathbf{Z}_2 \oplus \mathbf{Z}_2, \mathbf{O}(2)$, and $\mathbf{D}_n$, we know that $\text{Fix}(\Gamma) = \{0\}$. Hence (2.7) reduces to

$$T(h, \Gamma) = RT(h, \Gamma) + \mathscr{E}_\lambda\{h_\lambda\}.$$

We have computed h_λ in invariant coordinates for these examples. When $\Gamma = \mathbb{1}$, $V = \mathbb{R}^n$, the complementary subspace W to $\vec{\mathscr{M}}_{x,\lambda}(\Gamma)$ in $\vec{\mathscr{E}}_{x,\lambda}(\Gamma)$ is $\mathbb{R}\{e_1, \ldots, e_n\}$. When $\Gamma = \mathbf{Z}_2$, $V = \mathbb{R}^2$ and $W = \mathbb{R}\{(1,0)\}$. This observation coupled with (2.7) completes Table 3.1.

Table 3.1. Sample Equivariant Tangent Spaces

Γ	$T(h, \Gamma)$
$\mathbb{1}$	$\mathscr{E}_{x,\lambda}\{h_j e_k, \partial h/\partial x_k\} + \mathscr{E}_\lambda\{h_\lambda\}$
$\mathbf{Z}_2$	$RT(h, \Gamma) + \mathbb{R}\{[p_u, q_u]\} + \mathscr{E}_\lambda\{[p_\lambda, q_\lambda]\}$
$\mathbf{Z}_2 \oplus \mathbf{Z}_2$	$RT(h, \Gamma) + \mathscr{E}_\lambda\{[p_\lambda, q_\lambda]\}$
$\mathbf{O}(2)$	$RT(h, \Gamma) + \mathscr{E}_\lambda\{p_\lambda\}$
$\mathbf{D}_n$	$RT(h, \Gamma) + \mathscr{E}_\lambda\{[p_\lambda, q_\lambda]\}$

(a) $\Gamma = \mathbb{1}$, $V = \mathbb{R}^n$ $(n \geq 2)$

In Chapter IX we discussed two results about the codimension of bifurcation problems with many state variables. We needed both of these to complete the classification of bifurcation problems up to codimension 3, in Theorem IX, 2.1. The first of these two results, Proposition IX, 1.3, states that a bifurcation problem with n state variables has codimension at least $n^2 - 1$. The second states that hilltop bifurcations have codimension 3. In the next paragraph we derive the estimate $n^2 - 1$; then we prove the result for hilltop bifurcation in Theorem 3.1 later.

Recall that a bifurcation problem with n state variables is a mapping

$$h: \mathbb{R}^n \times \mathbb{R} \to \mathbb{R}^n$$

satisfying

$$h(0,0) = 0, \qquad (dh)_{0,0} = 0. \tag{3.1}$$

Condition (3.1) implies that the only possible nonzero linear terms in h are multiples of λ. Using Table 3.1 we see that

$$T(h, \mathbb{1}) \subset \{\mathscr{M}^2 + \lambda\}\vec{\mathscr{E}}_{x,\lambda} + \mathbb{R}\{\partial h/\partial x_1, \ldots, \partial h/\partial x_n, \partial h/\partial_\lambda\}. \tag{3.2}$$

Now the codimension of $\{\mathscr{M}^2 + \lambda\}\vec{\mathscr{E}}_{x,\lambda}$ in $\vec{\mathscr{E}}_{x,\lambda}$ is $n^2 + n$, since there are n^2 linear terms involving the x-coordinates and n constant terms. Hence the codimension of the right-hand side of (3.2) is at least $n^2 + n - (n + 1) = n^2 - 1$. This verifies the stated bound.

Theorem 3.1. *Hilltop bifurcations have codimension* 3. *In particular*:

(a) $(x^2 - y^2 + \lambda + 2\alpha x - 2\beta y, 2xy + \gamma)$ *is a universal unfolding of* $(x^2 - y^2 + \lambda, 2xy)$.

(b) $(x^2 + \varepsilon(\lambda + \gamma) + 2\alpha y, y^2 - \lambda + 2\beta x + \gamma)$ *is a universal unfolding of* $(x^2 + \varepsilon\lambda, y^2 - \lambda)$.

Remark. The proof of this theorem verifies (IX, 3.2). Bifurcation diagrams for hilltop bifurcation were given in Chapter IX, Figures 3.1–3.5.

PROOF. In the proof of Theorem XIV, 4.1, we showed that

$$RT(h, \mathbb{1}) = (\mathscr{M}^2 + \langle\lambda\rangle)\vec{\mathscr{E}}.$$

See also Exercise XIV, 2. We prove part (a) of Theorem 3.1, leaving part (b) as an exercise.

The computation of $T(h, \mathbb{1})$ proceeds by recalling that in this case $W = \mathbb{R}\{(1,0), (0,1)\}$. In particular,

$$(dh)\begin{bmatrix}1\\0\end{bmatrix} = h_x = (2x, 2y)$$

$$(dh)\begin{bmatrix}0\\1\end{bmatrix} = h_y = (-2y, 2x).$$

Also $h_\lambda = (1,0)$ and $\lambda h_\lambda \in RT(h, \mathbb{1})$. Hence

$$T(h, \mathbb{1}) = (\mathcal{M}^2 + \langle\lambda\rangle)\vec{\mathscr{E}}_{x,\lambda} \oplus \mathbb{R}\{(x, y), (-y, x), (1,0)\}.$$

It is now easy to find a complementary subspace to $T(h, \mathbb{1})$, spanned by

$$(0, 1), (x, 0), (y, 0)$$

yielding the universal unfolding in part (a). □

(b) $\Gamma = \mathbf{Z}_2 \oplus \mathbf{Z}_2$, $V = \mathbb{R}^2$

Suppose that h is a nondegenerate $\mathbf{Z}_2 \oplus \mathbf{Z}_2$-equivariant bifurcation problem in normal form as in (X, 2.10):

$$h(x, y, \lambda) = [\varepsilon_1 u + mv + \varepsilon_2\lambda, nu + \varepsilon_3 v + \varepsilon_4\lambda] \tag{3.3}$$

where $m \neq \varepsilon_2\varepsilon_3\varepsilon_4$, $n \neq \varepsilon_1\varepsilon_3\varepsilon_4$, and $mn \neq \varepsilon_1\varepsilon_3$.

Theorem 3.2.

$$H(x, y, \lambda, \alpha, \tilde{n}, \tilde{m}) = [\varepsilon_1 u + \tilde{m}v + \varepsilon_2\lambda, \tilde{n}u + \varepsilon_3 v + \varepsilon_4(\lambda - \alpha)]$$

is a universal $\mathbf{Z}_2 \oplus \mathbf{Z}_2$-unfolding of h.

Remark. Theorem 3.2 restates Theorem X, 2.4, whose proof was deferred to Volume II.

PROOF. We showed in (XIV, 4.14) that

$$RT(h, \mathbf{Z}_2 \oplus \mathbf{Z}_2) = \mathcal{M}^2_{u,v,\lambda}\vec{\mathscr{E}}_{u,v,\lambda} + \mathbb{R}\{[\varepsilon_1 u + mv + \varepsilon_2\lambda, 0],$$
$$[0, nu + \varepsilon_3 v + \varepsilon_4\lambda], [\varepsilon_1 u, nu], [mv, \varepsilon_3 v]\}, \tag{3.4}$$

where we use the particular form of h in (3.3). From Table 3.1 we see that in this case

$$T(h, \mathbf{Z}_2 \oplus \mathbf{Z}_2) = RT(h, \mathbf{Z}_2 \oplus \mathbf{Z}_2) + \mathbb{R}\{[\varepsilon_2, \varepsilon_4], [\varepsilon_2\lambda, \varepsilon_4\lambda]\}. \tag{3.5}$$

To prove Theorem 3.2 we show that $[v, 0]$, $[0, u]$, and $[0, -\varepsilon_4]$ span a complementary subspace to $T(h, \mathbf{Z}_2 \oplus \mathbf{Z}_2)$ in $\vec{\mathscr{E}}_{x,y,\lambda}(\mathbf{Z}_2 \oplus \mathbf{Z}_2)$. This calculation is best made modulo $\mathcal{M}^2_{u,v,\lambda}\vec{\mathscr{E}}_{u,v,\lambda}$, which is contained in $RT(h, \mathbf{Z}_2 \oplus \mathbf{Z}_2)$. The vectors that contribute to this calculation are listed in Table 3.2.

Simple linear algebra shows that if the 9×8 matrix in that table has rank 8, then we have found the correct vectors to span a complementary subspace. Note that the sum of rows 1 and 2 equals the sum of rows 3, 4, and 6. Hence the codimension of $T(h, \mathbf{Z}_2 \oplus \mathbf{Z}_2)$ is 3 and three vectors are needed to span a complement to $T(h, \mathbf{Z}_2 \oplus \mathbf{Z}_2)$. Cast out row 6, and eliminate rows 7, 8, and 9 by row reduction. We are left with the 5×5 matrix shown in Table 3.3. Since this clearly has rank 5, the original matrix has rank 8 as required. □

Table 3.2. Calculation of $T(h, \mathbf{Z}_2 \oplus \mathbf{Z}_2)$ for a Nondegenerate $\mathbf{Z}_2 \oplus \mathbf{Z}_2$-Bifurcation Problem h in Normal Form

	1	u	v	λ	1	u	v	λ
Vectors in		ε_1	m	ε_2				
$RT(h, \mathbf{Z}_2 \oplus \mathbf{Z}_2)$						n	ε_3	ε_4
See (3.3)		ε_1				n		
			m				ε_3	
Vectors in	ε_2				ε_4			
$T(h, \mathbf{Z}_2 \oplus \mathbf{Z}_2)/$				ε_2				ε_4
$RT(h, \mathbf{Z}_2 \oplus \mathbf{Z}_2)$								
See (3.4)								
Vectors to			1					
span complement						1		
to $T(h, \mathbf{Z}_2 \oplus \mathbf{Z}_2)$					ε_4			

Table 3.3. 5×5 Reduced Matrix Obtained from Table 3.2.

1	u	λ	v	λ
	ε_1	ε_2		
			ε_3	ε_4
	ε_1			
			ε_4	
ε_2				

(c) $\Gamma = \mathbf{D}_3$, $V = \mathbb{C}$

In this section we find universal $\mathbf{D}_3$-unfoldings for the $\mathbf{D}_3$ normal forms discussed in XIV, §§3, 4. Recall that a $\mathbf{D}_3$-bifurcation problem has the form

$$h(z, \lambda) = p(u, v, \lambda)z + q(u, v, \lambda)\bar{z}^2 = [p, q] \tag{3.6}$$

where

$$\text{(a)} \quad u = z\bar{z}$$
$$\text{(b)} \quad v = z^3 + \bar{z}^3. \tag{3.7}$$

We prove the following:

Theorem 3.3.

(a) $h(z, \lambda) = \varepsilon\bar{z}^2 + \delta\lambda z$, *where* $\varepsilon, \delta = \pm 1$ *has* $\mathbf{D}_3$*-codimension* 0.
(b) $h(z, \lambda) = (\varepsilon u + \delta\lambda)z + (\sigma u + mv)\bar{z}^2$, *where* $\varepsilon, \delta, \sigma = \pm 1$ and $m \neq 0$, *has* $\mathbf{D}_3$*-codimension* 2 *and modality* 1. *A universal unfolding of* h *is*

$$H(z,\lambda,\tilde{m},\alpha)=(\varepsilon u+\delta\lambda)z+(\sigma u+\tilde{m}v+\alpha)\bar{z}^2 \tag{3.8}$$

where $(\tilde{m},\alpha)$ *varies near* $(m,0)$.

PROOF.

(a) In the proof of Proposition XIV, 4.3, we showed that

$$RT(\varepsilon\bar{z}^2+\delta\lambda z,\mathbf{D}_3)=[\mathscr{M}_{u,v,\lambda},\mathscr{E}_{u,v,\lambda}].$$

See (XIV, 4.26) and (XIV, 4.27). From Table 3.1 we conclude that

$$T(\varepsilon\bar{z}^2+\delta\lambda z,\mathbf{D}_3)=[\mathscr{M}_{u,v,\lambda},\mathscr{E}_{u,v,\lambda}]+\mathscr{E}_\lambda[\delta,0]=\vec{\mathscr{E}}_{u,v,\lambda}. \tag{3.9}$$

It follows that $\operatorname{codim}_{\mathbf{D}_3}(\varepsilon\bar{z}^2+\delta\lambda z)=0$ as claimed.

(b) In the proof of Theorem XIV, 4.4, we showed that for this particular h,

$$\begin{aligned}RT(h,\mathbf{D}_3)=[\mathscr{M}^2_{u,v,\lambda},\mathscr{M}^2_{u,v,\lambda}]\oplus\mathbb{R}\{&[\varepsilon u+\delta\lambda,\sigma u+mv],\\ &[0,\varepsilon u+\delta\lambda],[2\varepsilon u,5\sigma u+6mv],[\varepsilon v,\sigma v]\}.\end{aligned} \tag{3.10}$$

From Table 3.1 and (3.10) we see that $T(h,\mathbf{D}_3)$ consists of all quadratic terms in u, v, λ plus the vector space spanned by

$$\begin{aligned}&\text{(a)}\quad[\varepsilon u+\delta\lambda,\sigma u+mv],\\ &\text{(b)}\quad[0,\varepsilon u+\delta\lambda],\\ &\text{(c)}\quad[2\varepsilon u,5\sigma u+6mv]\\ &\text{(d)}\quad[\varepsilon v,\sigma v],\\ &\text{(e)}\quad[\delta,0],\text{ and}\\ &\text{(f)}\quad[\delta\lambda,0].\end{aligned} \tag{3.11}$$

We claim that the vector space spanned by

$$\begin{aligned}&\text{(a)}\quad[0,1]\text{ and}\\ &\text{(b)}\quad[0,v]\end{aligned} \tag{3.12}$$

yields a complementary subspace to $T(h,\mathbf{D}_3)$ in $\vec{\mathscr{E}}_{x,\lambda}(\mathbf{D}_3)$. To verify this, we must check that the 8×8 matrix in Table 3.4 has rank 8. This is left to the reader as an exercise.

Table 3.4.

	$[1,0]$	$[u,0]$	$[v,0]$	$[\lambda,0]$	$[0,1]$	$[0,u]$	$[0,v]$	$[0,\lambda]$
(3.11(a))		ε		δ		σ	m	
(3.11(b))						ε		δ
(3.11(c))		2ε				5σ	$6m$	
(3.11(d))			ε				σ	
(3.11(e))	δ							
(3.11(f))				δ				
(3.12(a))					1			
(3.12(b))							1	

§4. Bifurcation with $\mathbf{D}_3$ Symmetry

In XIII, §5, and XIV, §4, we considered several aspects of bifurcation with triangular ($\mathbf{D}_3$) symmetry. In this section we complete the unfolding analysis and draw the associated bifurcation diagrams for the simplest $\mathbf{D}_3$ singularities. These results are used in the next section in a discussion of the spherical Bénard problem, in Case Study 5, and in the discussion of buckling of a triangular beam by Buzano, Geymonat, and Poston [1985]. We begin with a summary of previous results.

Assume that the action of $\mathbf{D}_3$ on $\mathbb{R}^2 \equiv \mathbb{C}$ is generated by $z \mapsto \bar{z}$ and $z \mapsto e^{2\pi i/3} z$. The form of the general $\mathbf{D}_3$-equivariant bifurcation problem is

$$g(z, \lambda) = p(u, v, \lambda)z + q(u, v, \lambda)\bar{z}^2 \equiv [p, q] \tag{4.1}$$

where $u = z\bar{z}$, $v = z^3 + \bar{z}^3$, and $p(0, 0, 0) = 0$. See Chapter XIV, Table 3.1. We showed in XIII, §5, that the equivariant branching lemma implies that the simplest such bifurcation problems ($p_\lambda(0) \neq 0$) have branches of solutions. However, generically ($q(0) \neq 0$) the solutions obtained in this way are *unstable*. Thus, when analyzing stable solutions for $\mathbf{D}_3$-bifurcation problems using local techniques, one must study degenerate singularities and unfold. This will be our approach.

Normal forms for the simplest $\mathbf{D}_3$-singularities were discussed in detail in XIV, §4. They are:

$$k(z, \lambda) = \varepsilon\lambda z + \delta\bar{z}^2 \tag{4.2}$$

(see Proposition XIV, 4.3) and

$$h(z, \lambda) = (\varepsilon u + \delta\lambda)z + (\sigma u + mv)\bar{z}^2 \tag{4.3}$$

(see Theorem XIV, 4.4) where $\varepsilon, \delta, \sigma = \pm 1$ and $m \neq 0$. The normal form (4.2) was obtained under the assumptions

$$p(0) = 0,\ p_\lambda(0) \neq 0, \qquad q(0) \neq 0$$

and represents the simplest $\mathbf{D}_3$ singularity. Normal form (4.3) is the simplest singularity satisfying the defining condition

$$q(0) = 0.$$

Of course, $q(0)$ is the coefficient of the only $\mathbf{D}_3$-equivariant quadratic in (4.1), and it is the coefficient which guarantees that solutions to (4.2) will be unstable. The list of nondegeneracy conditions needed to obtain the normal form (4.3) may be found in the statement of Theorem XIV, 4.4.

By Theorem 3.3 the universal unfoldings of (4.2) and (4.3) are as follows:

(a) $k(z, \lambda)$ in (4.2) has $\mathbf{D}_3$-codimension 0.

(b) $h(z, \lambda)$ in (4.3) has $\mathbf{D}_3$-codimension 2, topological codimension 1, and universal unfolding

$$H(z, \lambda, \mu, \alpha) = (\varepsilon u + \delta\lambda)z + (\sigma u + \mu v + \alpha)\bar{z}^2$$

where $\mu \sim m$ and $\alpha \sim 0$. (4.4)

We now derive the bifurcation diagrams for these $\mathbf{D}_3$-equivariant normal forms. Recall from XIII, §5, that orbits of solutions to $g = 0$ in (4.1) are

$$\begin{aligned} &\text{(a)}\quad z = 0 \\ &\text{(b)}\quad p(x^2, 2x^3, \lambda) + q(x^2, 2x^3, \lambda)x = 0, \qquad x \neq 0 \\ &\text{(c)}\quad p = q = 0. \end{aligned} \tag{4.5}$$

See Chapter XIII, Table 5.1. We claim that the eigenvalues of dg along these solutions are:

$$\begin{aligned} &\text{(a)}\quad p(0,0,\lambda) \quad [\text{twice}] \\ &\text{(b)}\quad qx + 2p_u x^2 + (6p_v + 2q_u)x^3 + 6q_v x^4 - 3xq \\ &\text{(c)}\quad \det dg = 3(z^3 - \bar{z}^3)^2[p_u q_v - p_v q_u] \\ &\qquad\ \operatorname{trace}(dg) = 2[up_u + (3p_v + 2q_u)v + 3u^2 q_v]. \end{aligned} \tag{4.6}$$

Note that in (4.6(c)) the eigenvalues are given only implicitly. To derive (4.6) we use the calculations in XIII, §5. In particular, the eigenvalues along the $\mathbf{Z}_2$ branch were shown to be $g_z \pm g_{\bar{z}}$ (XIII, 5.14), and these derivatives were computed in (XIII, 5.13). A short calculation yields (4.6(b)). To compute (4.6(c)), which determines the eigenvalues of dg when $p = q = 0$, recall by (XIII, 5.12) that in complex notation

$$(dg)w = g_z w + g_{\bar{z}} \bar{w} \tag{4.7}$$

and that in this form

$$\begin{aligned} &\text{(a)}\quad \det(dg) = |g_z|^2 - |g_{\bar{z}}|^2 \\ &\text{(b)}\quad \operatorname{trace}(dg) = 2 \operatorname{Re} g_z. \end{aligned} \tag{4.8}$$

Now compute (4.6(c)), using (4.8), (XIII, 5.13), and $p = q = 0$. (The calculation of $\det(dg)$ is somewhat long, but straightforward.)

We can now draw the bifurcation diagrams for the normal forms k and h. To simplify the discussion we assume that the trivial solution is asymptotically stable when $\lambda < 0$. In addition, by transforming $g(z, \lambda)$ to $-g(-z, \lambda)$ we may fix another choice of signs. This change of coordinates preserves the asymptotic stability of solutions. Thus we may assume the normal forms are:

$$\begin{aligned} &\text{(a)}\quad k(z, \lambda) = \bar{z}^2 - \lambda z \\ &\text{(b)}\quad h(z, \lambda) = (\varepsilon u - \lambda)z + (u + mv)\bar{z}^2. \end{aligned} \tag{4.9}$$

Using (4.5 and 4.6) we can now draw the unperturbed bifurcation diagrams, the result being Figure 4.1. We show only the case $\varepsilon = +1$ in (4.9(b)); this choice corresponds to supercritical bifurcation.

Note that in the simplest bifurcation problems with $\mathbf{D}_3$ symmetry the only asymptotically stable solution is the subcritical trivial solution, whereas for the degenerate singularity one of the two supercritical $\mathbf{Z}_2$ solutions is asymptotically stable.

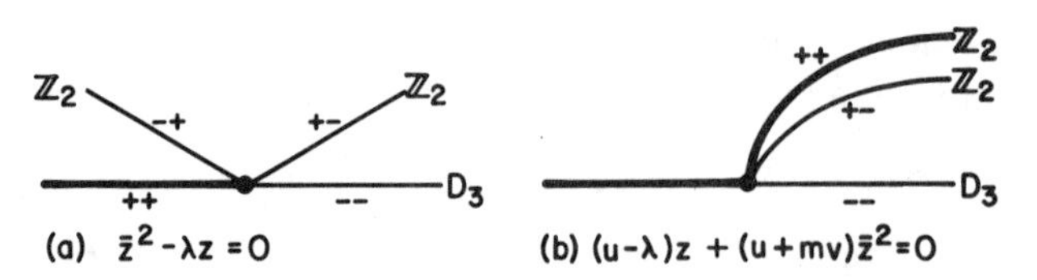

Figure 4.1. Unperturbed $\mathbf{D}_3$-symmetric bifurcation diagrams for the two normal forms (4.9(a, b)).

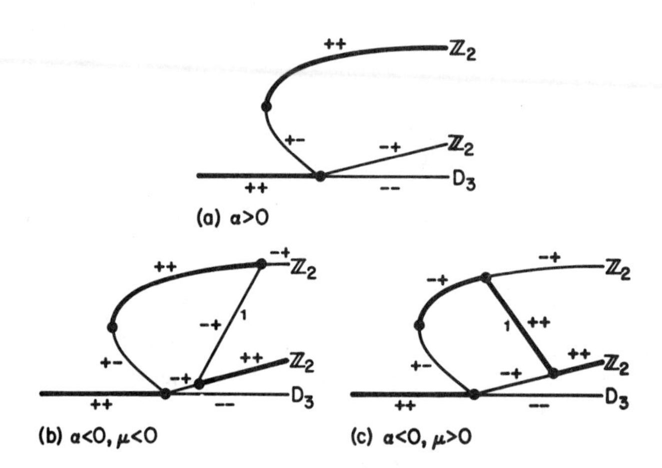

Figure 4.2. Bifurcation diagrams for (4.11): $(u - \lambda)z + (u + \mu v - \alpha)\bar{z}^2 = 0$.

The $\mathbf{Z}_2$ solutions in Figure 4.1(b) are determined by the equation

$$x(x^2 - \lambda + x^3 + 2mx^4) = 0; \tag{4.10}$$

see (4.5(b)). Thus (4.10) determines an asymmetric pitchfork bifurcation, yielding *two* supercritical branches. Since no group element in $\mathbf{D}_3$ takes $(x, 0) \mapsto (-x, 0)$ there is no reason why the two nontrivial branches of the pitchfork should be identified, and they are not. It might seem that the difference between the two branches is minimal since the asymmetry only results from the x^3 term in (4.10) or the $u\bar{z}^2$ term in h. However, the fact that solutions on one branch are stable and on the other are unstable indicates that the difference is genuine.

Next we discuss the perturbed bifurcation diagrams contained in the universal unfolding

$$H(z, \lambda, \mu, \alpha) = (u - \lambda)z + (u + \mu v + \alpha)\bar{z}^2 \tag{4.11}$$

of (4.19(b)). We assume $\varepsilon = +1$ so that the $\mathbf{Z}_2$ branches are supercritical. The resulting bifurcation diagrams are given in Figure 4.2. Not surprisingly, the bifurcation diagram for the universal unfolding $F(\cdot, \cdot, \mu, \alpha) = 0$ has a transcritical bifurcation (on the $\mathbf{Z}_2$ branch) when $\alpha \neq 0$, since locally the bifurcation from the trivial solution is equivalent to Figure 4.1(a). In fact, the equation for the $\mathbf{Z}_2$ branch may be obtained from (4.5(b)) and (4.11):

$$\lambda = \alpha x + x^2 + x^3 + 2\mu x^4. \tag{4.12}$$

Thus a limit point occurs when $\partial\lambda/\partial x = \alpha + 2x + 3x^2 + 8\mu x^3 = 0$ along the $\mathbf{Z}_2$ branch. Near the origin this occurs at

$$x \approx -\alpha/2. \tag{4.13}$$

This limit point is indicated in Figure 4.2.

Although it is less obvious, taking $\alpha \neq 0$ can lead to secondary bifurcation to a solution with trivial isotropy. To see this, write equations (4.5(c)) for (4.11) as

$$\begin{aligned} &\text{(a)} \quad \lambda = x^2 + y^2 \\ &\text{(b)} \quad -\alpha = x^2(1 + 2\mu x) + y^2(1 - 6\mu x) \end{aligned} \tag{4.14}$$

where $z = x + iy$, $u = x^2 + y^2$, and $v = 2(x^3 - 3xy^2)$. Provided $\alpha \neq 0$, (4.14(b)) has real solutions in a neighborhood of the origin.

We now show how these solutions with trivial isotropy fit into Figure 4.2(b, c). Observe that for small $\alpha < 0$ the solutions of (4.14(b)) (near the origin) form a circlelike curve in the (x, y)-plane of radius approximately $\sqrt{|\alpha|}$. This is exactly true for $\mu = 0$, and approximately true for small $\mu \neq 0$. It follows that in (x, y, λ)-space this curve intersects the $y = 0$ plane at two points (x_-, λ_-) and (x_+, λ_+) where $x_- < 0 < x_+$. These correspond to intersection points of the branch with trivial isotropy and solutions with isotropy $\mathbf{Z}_2$. Moreover, the group element $z \mapsto \bar{z}$ (or $(x, y) \mapsto (x, -y)$) identifies the two semicircles $y > 0$ and $y < 0$. Thus, since we are plotting only orbits of solutions with trivial isotropy, this circlelike curve shows that there is a unique orbit of such solutions for each λ between λ_- and λ_+.

Recall from (4.13) that when $\alpha < 0$ the limit point on the $\mathbf{Z}_2$ branch occurs on the branch where $x > 0$. It follows that the unique branch of solutions with trivial isotropy connects the two branches of $\mathbf{Z}_2$ solutions, and we can tell whether this branch slants from left to right, or vice versa, by computing the sign of $\lambda_- - \lambda_+$. We claim that

$$\operatorname{sgn}(\lambda_- - \lambda_+) = \operatorname{sgn}(\mu). \tag{4.15}$$

Assuming (4.15) we see that $\lambda_- < \lambda_+$ when $\mu < 0$. Thus the intersection of the secondary branch with the $\mathbf{Z}_2$ branch that has no limit point, occurs at a value of λ (namely λ_-) less than the value of λ (namely λ_+) at which it intersects the $\mathbf{Z}_2$-branch that has a limit point. See Figure 4.2(b). A similar discussion holds for Figure 4.2(c).

To establish (4.15) we show that

$$\lambda_- - \lambda_+ = -\alpha\mu(x_+ - x_-)/[(1 + \mu x_-)(1 + \mu x_+)]. \tag{4.16}$$

Observe that $(-\alpha)(x_+ - x_-)$ and $(1 + \mu x_-)(1 + \mu x_+)$ are positive (when $\alpha \sim 0$) so that (4.15) follows from (4.16). Now (4.14(a)) implies that $\lambda_+ = x_+^2$ and $\lambda_- =$

x_-^2. Substitute into (4.12) to obtain

$$\begin{aligned} &\text{(a)} \quad \alpha + x_+^2 + \mu x_+^3 = 0 \\ &\text{(b)} \quad \alpha + x_-^2 + \mu x_-^3 = 0. \end{aligned} \tag{4.17}$$

Rewrite (4.17) using $\lambda_\pm = x_\pm^2$ to find

$$\begin{aligned} &\text{(a)} \quad \lambda_+ = -\alpha/(1 + \mu x_+) \\ &\text{(b)} \quad \lambda_- = -\alpha/(1 + \mu x_-). \end{aligned} \tag{4.18}$$

Now compute $\lambda_- - \lambda_+$ from (4.18) to establish (4.16).

We end the description of Figure 4.2 by discussing the asymptotic stability of solutions. Using (4.11) and (4.6) we may determine the eigenvalues along the three types of branch. In this case we have

$$\begin{aligned} &\text{(a)} \quad -\lambda \quad \text{[twice]} \\ &\text{(b)} \quad \alpha x + 2x^2 + 3x^3 + 8\mu x^4 - 3(\alpha x + x^3 + 2\mu x^4) \\ &\text{(c)} \quad \operatorname{sgn} \det dg = \operatorname{sgn} \mu \\ &\qquad \operatorname{trace} dg = (2 + 4x + 3\mu x^2)x^2. \end{aligned} \tag{4.19}$$

The stability of the trivial solution and the solutions with trivial isotropy follow directly from (4.19(a, c)). Moreover, the eigenvalues of the $\mathbf{Z}_2$ solutions in the unperturbed diagram ($\alpha = 0$, Figure 4.1(b)) and near the origin in the perturbed cases ($\alpha \neq 0$) follow directly from (4.19(b)). To complete the discussion we must determine when the eigenvalues in (4.19(b)) change sign. To do this, use (4.12) to rewrite the eigenvalues in (4.19(b)) as

$$\begin{aligned} &\text{(a)} \quad x\, d\lambda/dx \\ &\text{(b)} \quad 3(x^2 - \lambda). \end{aligned} \tag{4.20}$$

Thus the first eigenvalue (4.20(a)) changes sign at limit points on the $\mathbf{Z}_2$ branch, and the second eigenvalue (4.20(b)) changes sign of the point of bifurcation with the branch having trivial isotropy.

We end this section with a discussion of the bifurcation diagrams associated with universal unfoldings of singularities that are $\mathbf{D}_3$-equivalent to (4.11). Since $\mathbf{D}_3$-equivalence preserves zero sets (up to changes in coordinates) the only issue is the asymptotic stability of solutions. Of the three kinds of solution listed in (4.5), we may analyze the first two by using Theorem XIV, 5.3. This states that Γ-equivalence preserves stability of a solution whenever the isotropy subgroup of that solution decomposes the space into a direct sum of distinct absolutely irreducible representations. We claim that the isotropy subgroups $\mathbf{D}_3$ and $\mathbf{Z}_2$ satisfy this hypothesis. Indeed $\mathbf{D}_3$ acts absolutely irreducibly on $\mathbb{C}$. Also $\mathbf{Z}_2$, generated by $z \mapsto \bar{z}$, decomposes $\mathbb{C}$ into $\mathbb{R}\{1\} \oplus \mathbb{R}\{i\}$, and the action on these spaces is, respectively, by the identity and by minus the identity.

Thus it suffices to show that $\mathbf{D}_3$-equivalence preserves the asymptotic stability of the solutions (4.5(c)) with trivial isotropy. We do this by an ad hoc argument. From Remark XIV, 5.2, we know that the sign of $\det(dH)$ is an invariant of $\mathbf{D}_3$-equivalence. Hence, when $\operatorname{sgn} \det(dH) < 0$ at a solution with trivial isotropy, the eigenvalues of dH have opposite signs, and this fact is preserved by $\mathbf{D}_3$-equivalence.

Finally we assume that $\operatorname{sgn} \det(dH) > 0$ at a solution with trivial isotropy. Then the signs of the real parts of the eigenvalues of dH, and hence the asymptotic stability of the solution, are determined by the sign of trace dH. In (4.6(c)) we computed this trace, obtaining

$$2[up_u + (3p_v + q_u)v + 3u^2 q_v]. \tag{4.21}$$

Now this is dominated, modulo higher order terms, by $2up_u(0)$. Thus, if the sign of $p_u(0)$ is an invariant of $\mathbf{D}_3$-equivalence, then the asymptotic stability of solutions with trivial isotropy is preserved by $\mathbf{D}_3$-equivalence. Now recall Theorem XIV, 4.4, in which the recognition problem for the $\mathbf{D}_3$ singularity (4.3) was solved. We showed (XIV, 4.25, 4.24) that $\varepsilon = \operatorname{sgn} p_u(0)$ where $\varepsilon = \pm 1$ is the sign in the normal form (4.3). Therefore, $\operatorname{sgn} p_u(0)$ is invariant under $\mathbf{D}_3$-equivalence.

§5.† The Spherical Bénard Problem

The Bénard problem concerns convection in a viscous fluid when it is heated from below; see Case Study 4. In this section we discuss the Bénard problem in spherical geometry; that is, we assume that the fluid is confined in a spherical shell. We ask what types of convection pattern appear by bifurcation from the "trivial" pure conduction state. The discussion will be descriptive, based on the $\mathbf{O}(3)$ symmetry inherent in the problem.

We consider the Bénard problem in the Boussinesq approximation. In this model there is a trivial solution representing pure heat conduction directed radially outward. As the nondimensionalized temperature on the inner sphere (the *Rayleigh number* R) is increased, the trivial solution loses stability, say at $R = R_0$. The Bénard problem is the study of the resulting bifurcation. This problem has been widely studied, and we survey here results of Busse [1975], Busse and Riahi [1982], Chossat [1979, 1982], Golubitsky and Schaeffer [1982], and Young [1974].

The Boussinesq equations are posed on the three-dimensional annulus

$$\Omega = \{x \in \mathbb{R}^3 \mid r_i < |x| < r_o\}.$$

We call $\eta = r_0/r_i$ the *radius ratio*. After subtraction of the pure conduction solution (so that the bifurcation is from zero) the Boussinesq equations become

$$\begin{aligned} &\text{(a)} \quad v_t = -\nabla p + \Delta v + R\theta g(r) - (v \cdot \nabla) r \\ &\text{(b)} \quad \theta_t = \frac{1}{P}\{\Delta\theta + R\nabla T_0 \cdot r\} - (v \cdot \nabla)\theta \\ &\text{(c)} \quad \nabla \cdot v = 0 \end{aligned} \tag{5.1}$$

where $v \in \mathbb{R}^3$ is the velocity, p is the pressure, θ is temperature, R the Rayleigh number, and P the Prandtl number. We assume that the gravity vector $g(r)$ and the equilibrium temperature gradient ∇T_0 have the form

$$\begin{aligned} &\text{(a)} \quad g(r) = (\gamma_1/r^3 + \gamma_2)\mathbf{r} \\ &\text{(b)} \quad \nabla T_0 = (\beta_1/r^3 + \beta_2)\mathbf{r} \end{aligned} \tag{5.2}$$

where $\mathbf{r}$ is the vector of length r in the outward radial direction. Of course, (5.1) must be supplemented by appropriate boundary conditions on $\partial\Omega$: a typical choice is Dirichlet or *rigid* boundary conditions.

We consider only small-amplitude steady-state solutions to (5.1), which we regard as a bifurcation problem with the Rayleigh number R as bifurcation parameter. The first step in finding small-amplitude solutions is to consider the Boussinesq equations linearized about the trivial solution. Values of R where these linearized equations have nonzero solutions are called *eigenvalues*, and the critical Rayleigh number R_0 is just the smallest eigenvalue.

We let V denote the kernel of the linearized Boussinesq equations at R_0. The second step is to determine V. Now the Boussinesq equations are invariant under the natural action of the orthogonal group $\mathbf{O}(3)$ on the annulus Ω. This action induces a representation of $\mathbf{O}(3)$ on V. An abstract description of V can be obtained by writing V as a direct sum of irreducible subspaces.

Now that we have done this, there are two basic ways to proceed. Either we can seek "generic" solution branches, using the equivariant branching lemma, or we can perform a Liapunov–Schmidt reduction onto V and solve the reduced equations, perhaps by singularity theory methods, to determine precisely the small amplitude solutions. Of course, the second approach is more difficult, but when it can be applied it yields more detailed information.

Results concerning the first two steps (the critical Rayleigh number R_0 and the structure of the kernel V) are summarized and proved in Chossat [1979]. Generically (in the radius ratio η) $\mathbf{O}(3)$ acts irreducibly on V, and $V \cong V_l$, the spherical harmonics of degree l with the natural $(-1)^l$ action of $-I \in \mathbf{O}(3)$. The exact value of l depends on η. In particular, the value of l is monotonic in η and approaches ∞ as $\eta \to 1$ (the shell becomes infinitesimally thin). Since l is an integer, there is an infinite sequence of values η at which l jumps in value by 1. These values of η correspond to "multiple eigenvalues" or "mode interactions" where the $\mathbf{O}(3)$-action on V is reducible. In our discussion we (unfortunately have to) avoid these special values of η. For details see Chossat [1979], Theorem 1, p. 631.

The reader should note that the kind of local bifurcation analysis we are about to give is meaningful only when η is far from 1, since as $\eta \to 1$ many eigenvalues of the linearized Boussinesq equations pile up at R_0. It follows that the range of validity (in R) of a local analysis is very small when $\eta \approx 1$. Thus our main interest is in cases where l is small, say $l \leq 10$.

In their analysis of the spherical Bénard problem Busse [1975] and Busse and Riahi [1982] have used (implicitly) the first approach, based on the equivariant branching lemma. They study the cases $l = 1, 2, 3, 4$, and 6, finding

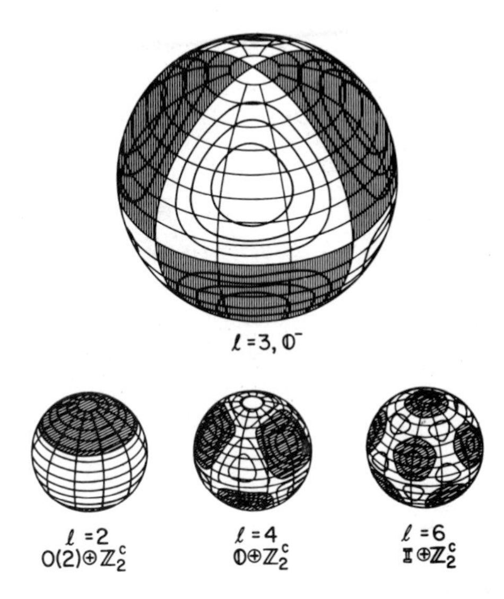

Figure 5.1. Patterns of convection in spherical shells. The sign of the radial velocity is opposite in shaded and unshaded areas. From Busse and Riahi [1982].

the solutions predicted by the general $\mathbf{O}(3)$ theory. See Theorem XIII, 9.8. As well as the axisymmetric solutions (with isotropy $\mathbf{O}(2)$) that exist for all l, they find:

octahedral solutions with isotropy subgroup $\mathbb{O}^-$ when $l = 3$ and $\mathbb{O} \oplus \mathbf{Z}_2^c$ when $l = 4, 6$,

sector solutions with isotropy subgroup $\mathbf{D}_6^d$ when $l = 3$,

icosahedral solutions with isotropy subgroup $\mathbb{I} \oplus \mathbf{Z}_2^c$ when $l = 6$.

Some of their solutions are pictured in Figure 5.1. Shaded regions represent areas where the fluid is flowing up, and unshaded regions show where the fluid is flowing down (or vice versa).

The general $\mathbf{O}(3)$ theory summarized in Theorem XIII, 8.8, allows us to find a number of solutions to the Boussinesq equations when $l \leq 10$. Moreover, the existence of these solutions does not depend on the particular nonlinearities in the Boussinesq equations, just on the $\mathbf{O}(3)$ symmetry and details of the linearized equations.

Remark. When l is even there is a nonzero equivariant quadratic mapping on V_l (which is the gradient of an $\mathbf{O}(3)$-invariant cubic). Hence Theorem XIII, 4.4., applies, showing that when l is even all of the solutions that we, Busse, and Riahi have found are unstable to perturbations in V_l. Chossat [1979] considers specifically the case $l = 2$, where the axisymmetric solutions found previously must be unstable. Note that there are two physically distinct types of axisymmetric solution when $l = 2$: those in which the fluid is upwelling at the poles and downwelling at the equator, and those in which the upwelling is at the equator and the downwelling is at the poles. Each flow has isotropy $\mathbf{O}(2) \oplus \mathbf{Z}_2^c$, so the isotropy subgroup does not distinguish all physical characteristics of the flow.

Chossat's results are best understood in terms of a mathematical fact which is proved in the next section. The analysis of $\mathbf{O}(3)$-symmetric bifurcation problems defined on V_2 is *identical* to that of $\mathbf{D}_3$-symmetric problems defined on $\mathbb{C}$, considered in §4. The only difference is that the $\mathbf{Z}_2$-symmetric branches in the $\mathbf{D}_3$ analysis correspond to the axisymmetric solutions in the $\mathbf{O}(3)$ analysis.

Assuming this identification of two different contexts, we describe Chossat's results. The simplest and generic bifurcation occurs to a transcritical axisymmetric branch of solutions; see Figure 4.1(a). The super- and subcritical branches correspond to up- and downwelling at the poles, in some order. To determine whether this singularity occurs in the Boussinesq equations, Chossat explicitly computes, by Liapunov–Schmidt reduction, the second order term $q(0)$ in (4.1). He then shows that this quadratic term vanishes for certain choices of the gravity vector and equilibrium temperature gradient in (5.2). These choices are defined by the relation

$$\gamma_1 \beta_2 - \gamma_2 \beta_1 = 0 \tag{5.3}$$

where the γ_j, β_j are as in (5.2). When (5.3) holds we say we are in the *self-adjoint* case. This terminology reflects the fact that the linearized Boussinesq equations are self-adjoint when (5.3) holds; moreover, this self-adjointness is what is used to prove that $q(0) = 0$.

Chossat shows that the cubic term $p_u(0)$ (or ε) is positive, so that the two branches of axisymmetric solutions are supercritical as in Figure 4.1(b). He then assumes that certain expressions involving fourth and fifth order terms are nonzero—these correspond exactly to the nondegeneracy conditions in Theorem XIV, 4.4—to conclude that one of the branches of axisymmetric solutions is asymptotically stable, at least to perturbations in V_l. In fact, only the fourth order coefficient is needed to determine whether upwelling or downwelling at the poles is stable. The analytic calculation of these coefficients is not possible by hand. However, Chossat [1982] uses numerical calculations to compute the fourth order coefficient when the radius ratio $\eta = 0.3$, a value for which the kernel at the critical Rayleigh number is isomorphic to V_2. He finds that the stable solutions are those with upwelling at the poles.

In conclusion, we consider the slightly non-self-adjoint case. It is shown in Golubitsky and Schaeffer [1982] that perturbing the β_j and γ_j leads to a universal unfolding of Chossat's degenerate bifurcation problem. Hence, according to Figure 4.2(b) or (c), there is a perturbation yielding nonaxisymmetric solutions, and also a change in the direction of welling at the poles, as the Rayleigh number is increased. Whether this nonaxisymmetric solution can be stable depends on the fifth order coefficient, whose computation has not yet been attempted. Young [1974] has found nonaxisymmetric flows by direct numerical integration of the Boussinesq equations. However, these flows are not near the self-adjoint case, where our analysis is valid.

§6.† Spherical Harmonics of Order 2

In this section we concentrate on bifurcation problems commuting with the action of $\mathbf{O}(3)$ on the spherical harmonics V_2 of order 2. We show that these bifurcation problems are identical in structure with problems in the plane commuting with $\mathbf{D}_3$, already studied in §4. More precisely, we show that there is a (natural) one-to-one correspondence between bifurcation problems $G: V_2 \times \mathbb{R} \to V_2$ commuting with $\mathbf{O}(3)$ and bifurcation problems $g: \mathbb{C} \times \mathbb{R} \to \mathbb{C}$ commuting with $\mathbf{D}_3$.

When l is even, $-I \in \mathbf{O}(3)$ acts as the identity on V_l. Thus, when considering V_2, we may as well consider only the action of $\mathbf{SO}(3)$. Recall that V_2 is five-dimensional and that all five-dimensional irreducible representations of $\mathbf{SO}(3)$ are isomorphic. We use this fact to give a simpler presentation of this five-dimensional representation of $\mathbf{SO}(3)$. Let

$$V = \{3 \times 3 \text{ symmetric matrices } A: \operatorname{tr}(A) = 0\}. \tag{6.1}$$

Observe that V is five-dimensional and that $\mathbf{SO}(3)$ acts on V in a natural way, by similarity of matrices. That is,

$$\gamma \cdot A = \gamma A \gamma^t \tag{6.2}$$

where $\gamma \in \mathbf{SO}(3)$ and $A \in V$. (Note that $\gamma \cdot A$ denotes the action of γ on A, whereas $\gamma A \gamma^t$ indicates matrix multiplication.)

Let $W \subset V$ be the two-dimensional subspace of diagonal matrices of trace zero. It is easy to see that W is the fixed-point subspace of the group Σ generated by

$$\sigma_1 = \begin{bmatrix} -1 & 0 & 0 \\ 0 & -1 & 0 \\ 0 & 0 & 1 \end{bmatrix} \quad \text{and} \quad \sigma_2 = \begin{bmatrix} 1 & 0 & 0 \\ 0 & -1 & 0 \\ 0 & 0 & -1 \end{bmatrix}. \tag{6.3}$$

Note that Σ has four elements and is isomorphic to $\mathbf{D}_2 = \mathbf{Z}_2 \oplus \mathbf{Z}_2 \subset \mathbf{SO}(3)$.

Let $\mathbf{S}_3$ denote the six-element group of 3×3 permutation matrices. Observe that each element of $\mathbf{S}_3$ leaves the space of diagonal matrices W invariant. That is, if $\tau \in \mathbf{S}_3$ and $D \in W$, then $\tau D \tau^t$ is also diagonal.

Our main result is as follows:

Theorem 6.1. *Let $G: V \to V$ be $\mathbf{SO}(3)$-equivariant. Then $g = G|W$ is an $\mathbf{S}_3$-equivariant mapping $W \to W$. Moreover, every (smooth) $\mathbf{S}_3$-equivariant mapping on W extends uniquely to an $\mathbf{SO}(3)$-equivariant mapping on V.*

Remark. Since $W = \text{Fix}(\Sigma)$ is a fixed-point subspace, G maps W to itself. Since $\mathbf{S}_3 \subset \mathbf{O}(3)$ leaves W invariant, g is $\mathbf{S}_3$-equivariant. The content of this theorem is that every such g extends to an $\mathbf{SO}(3)$-equivariant, and that this extension is unique.

Besides Theorem 6.1 there are two results which are needed to complete the identification of $\mathbf{SO}(3)$ and $\mathbf{D}_3$ bifurcation problems. The first shows that $\mathbf{S}_3$-equivalence on W is the same as $\mathbf{SO}(3)$-equivalence on V. The second shows that asymptotic stability on W corresponds to orbital asymptotic stability on V. We prove these in Lemma 6.7 and Theorem 6.9.

In order to prove Theorem 6.1 we need four preliminary lemmas.

Lemma 6.2. *Let $N(\Sigma)$ be the normalizer of Σ in $\mathbf{SO}(3)$. Then $N(\Sigma) = \mathbb{O}$, the octahedral group, and $N(\Sigma)/\Sigma \cong \mathbf{S}_3$.*

Proof. Both Σ and $\mathbf{S}_3$ consist of group elements which are symmetries of the cube. So Σ and $\mathbf{S}_3$ are contained in the twenty-four-element group $\mathbb{O}$. Let T be the subgroup of $\mathbf{SO}(3)$ generated by the four-element group Σ and the six-element group $\mathbf{S}_3$. Now $T \subset \mathbb{O}$, and T has at least twenty-four elements since $\mathbf{S}_3 \cap \Sigma = \mathbb{1}$. Therefore, $T = \mathbb{O}$.

By direct calculation $\mathbf{S}_3 \subset N(\Sigma)$; in other words, $\tau\Sigma\tau^{-1} \subset \Sigma$ for every permutation matrix τ. Therefore, $N(\Sigma) \supset \mathbb{O}$, since $N(\Sigma)$ contains both Σ and $\mathbf{S}_3$. Now $\mathbb{O}$ is a maximal closed subgroup of $\mathbf{SO}(3)$. Hence either $N(\Sigma) = \mathbb{O}$ or $N(\Sigma) = \mathbf{SO}(3)$. But the latter is easily shown to be false, whence $N(\Sigma) = \mathbb{O}$ as claimed.

Finally, since $\mathbf{S}_3 \cap \Sigma = \mathbb{1}$, the canonical projection $N(\Sigma) \to N(\Sigma)/\Sigma$ yields an isomorphism of $\mathbf{S}_3$ with $N(\Sigma)/\Sigma$. □

Lemma 6.3. *The groups $\mathbf{S}_3$ and $\mathbf{D}_3$ are isomorphic. The action of $\mathbf{S}_3$ on W is isomorphic to the standard action of $\mathbf{D}_3$ on $\mathbb{C}$.*

Proof. Identify $\mathbb{C}$ with W as follows: let $z = x + iy \in \mathbb{C}$ and let

$$D(z) = \begin{bmatrix} x & 0 & 0 \\ 0 & \frac{1}{2}(-x + \sqrt{3}y) & 0 \\ 0 & 0 & \frac{1}{2}(-x - \sqrt{3}y) \end{bmatrix}.$$

Observe that

$$\tau = \begin{bmatrix} 1 & 0 & 0 \\ 0 & 0 & 1 \\ 0 & 1 & 0 \end{bmatrix}$$

acts on $D(z)$ by

$$\tau D(z)\tau^t = D(\bar{z})$$

and that

$$\rho = \begin{bmatrix} 0 & 1 & 0 \\ 0 & 0 & 1 \\ 1 & 0 & 0 \end{bmatrix}$$

acts on $D(z)$ by

$$\rho D(z) \rho^t = D(e^{i\theta} z)$$

where $\theta = 2\pi/3$. Since τ and ρ generate $\mathbf{S}_3$, and $z \mapsto \bar{z}$ and $z \mapsto e^{i\theta} z$ generate $\mathbf{D}_3$, these calculations show both that $\mathbf{S}_3 \cong \mathbf{D}_3$ and that the actions on W and $\mathbb{C}$ are isomorphic.

Lemma 6.4. *Let $D \in W$ be nonzero, and let Σ_D be its isotropy subgroup. If two of the diagonal entries in D are equal, then $\Sigma_D \cong \mathbf{O}(2)$. If all the diagonal entries in D are distinct, then $\Sigma_D = \Sigma$.*

Proof. Suppose that $\gamma \cdot D = D$ for some $\gamma \in \mathbf{SO}(3)$. We may rewrite this identity as

$$\gamma D = D\gamma. \tag{6.4}$$

It is a straightforward exercise to check that (6.4) implies the stated result. To do this, write

$$\gamma = \begin{bmatrix} A & B & C \\ E & F & G \\ H & J & K \end{bmatrix} \quad \text{and} \quad D = \begin{bmatrix} x & 0 & 0 \\ 0 & y & 0 \\ 0 & 0 & z \end{bmatrix}.$$

Then (6.4) implies that

$$Bx = By, \quad Cx = Cz, \quad Ex = Ey, \quad Gy = Gz, \quad Hx = Hz, \quad Jy = Jz. \tag{6.5}$$

If x, y, z are distinct then γ is diagonal. Since $\gamma \in \mathbf{SO}(3)$ it follows that $\gamma \in \Sigma$.

Since $D \neq 0$ and $\operatorname{tr} D = 0$, at most two entries of D are equal. Suppose $x = y$. Then (6.5) implies that

$$\gamma = \begin{bmatrix} A & B & 0 \\ E & F & 0 \\ 0 & 0 & K \end{bmatrix}.$$

Orthogonality implies that $\left[\begin{smallmatrix} A & B \\ E & F \end{smallmatrix}\right] \in \mathbf{O}(2)$, and $K = \det\left[\begin{smallmatrix} A & B \\ E & F \end{smallmatrix}\right]$. □

Remark 6.5. The points in W with isotropy subgroup $\mathbf{O}(2)$ in $\mathbf{SO}(3)$ are precisely those which have isotropy subgroup $\mathbf{Z}_2$ in $\mathbf{D}_3$. Those with isotropy subgroup Σ in $\mathbf{SO}(3)$ are precisely those with isotropy subgroup $\mathbb{1}$ in $\mathbf{D}_3$.

Lemma 6.6. *Let $D \in W$ be a diagonal matrix with isotropy subgroup Σ in $\mathbf{SO}(3)$. Suppose that $\gamma \cdot D \in W$ for some $\gamma \in \mathbf{SO}(3)$. Then $\gamma \in N(\Sigma) = \mathbb{O}$.*

Remark. This is (a special case of) a simple converse to Lemma XIII, 10.1(a), and is proved the same way. For completeness, we give a proof here.

PROOF. Let $\sigma \in \Sigma$. Then $\gamma^{-1}\sigma\gamma D = \gamma^{-1}\gamma D$ since $\gamma D \in W$ so σ fixes γD. Thus $\gamma^{-1}\sigma\gamma \in \Sigma_D = \Sigma$. Therefore, $\gamma\Sigma\gamma^{-1} = \Sigma$ so $\gamma \in N(\Sigma)$. □

We have now assembled the lemmas needed for the following:

PROOF OF THEOREM 6.1. By the Remark after the statement of the theorem, it suffices to prove that every $\mathbf{S}_3$-equivariant mapping $g: W \to W$ extends uniquely to a (smooth) $\mathbf{SO}(3)$-equivariant mapping $G: V \to V$.

First we show that if an extension exists, it is unique. Let A be a symmetric 3×3 matrix. Since every symmetric matrix can be diagonalized by an orthogonal change of coordinates, there exists $\gamma \in \mathbf{SO}(3)$ such that $\gamma \cdot A$ is diagonal. Therefore, if G exists and commutes with $\mathbf{SO}(3)$, we have

$$g(\gamma \cdot A) = G(\gamma \cdot A) = \gamma G(A),$$

so

$$G(A) = \gamma^t g(\gamma \cdot A). \tag{6.6}$$

Therefore, G is unique.

Now we prove existence, using the module structure of $\mathbf{S}_3$-equivariant mappings over the ring of $\mathbf{S}_3$-invariants. We claim that every equivariant mapping g on W has the form

$$g(D) = pD + q[D^2 - \tfrac{1}{3}\operatorname{tr}(D^2)I] \tag{6.7}$$

where the $\mathbf{S}_3$-invariants p and q are functions of $\operatorname{tr}(D^2)$ and $\det D$. If (6.7) holds then g extends to G by using the same formula:

$$G(A) = p(\operatorname{tr}(A^2), \det A)A + q(\operatorname{tr}(A^2), \det A)[A^2 - \tfrac{1}{3}\operatorname{tr}(A^2)I]. \tag{6.8}$$

(It is easy to check that such a mapping G is $\mathbf{SO}(3)$-equivariant.)

To establish (6.7) we recall from (XII, 5.12) that every $\mathbf{D}_3$-equivariant mapping on $\mathbb{C}$ is of the form

$$p(z\bar{z}, \operatorname{Re} z^3)z + q(z\bar{z}, \operatorname{Re} z^3)\bar{z}^2. \tag{6.9}$$

The identification of the preceding $\mathbf{S}_3$ and $\mathbf{D}_3$ actions can be used to derive (6.7) explicitly. Alternatively, a simple counting argument suffices. The module of $\mathbf{D}_3$-equivariant mappings is generated by unique linear and quadratic equivariants; the same is asserted in (6.7) for $\mathbf{S}_3$. Similarly the quadratic and cubic generators for the invariants of $\mathbf{D}_3$ must match those for $\mathbf{S}_3$. □

Next we show that the correspondence between $\mathbf{SO}(3)$ and $\mathbf{S}_3$ preserves equivalence.

Theorem 6.7. *Let $g, h: W \to W$ be $\mathbf{S}_3$-equivalent, and let $G, H: V \to V$ be their $\mathbf{SO}(3)$-equivariant extensions. Then G and H are $\mathbf{SO}(3)$-equivalent.*

Proof. Let $S(A) \in \vec{\mathscr{E}}(\mathbf{SO}(3))$. Let $D_1, D_2 \in W$. Then $S(D_1)D_2 \in W$. To see why, let $\sigma \in \Sigma$. The equivariance condition on S states that

$$S(\sigma A)\sigma B = \sigma S(A)B.$$

Hence, since $\sigma D_1 = D_1$ and $\sigma D_2 = D_2$, we have

$$S(D_1)D_2 = \sigma S(D_1)D_2.$$

It follows that $S(D_1)D_2 \in \mathrm{Fix}(\Sigma) = W$.

Now suppose that G and H are $\mathbf{SO}(3)$-equivalent, so that

$$G(A) = S(A)H(X(A)), \tag{6.10}$$

where $X\colon V \to V$ is $\mathbf{SO}(3)$-equivariant. Restrict (6.10) to W, to obtain

$$g(D) = S(D)h(X(D)),$$

since $X\colon W \to W$ and $S(D)\colon W \to W$. Thus g and h are $\mathbf{S}_3$-equivalent.

Conversely suppose that g and h are $\mathbf{S}_3$-equivalent, so that

$$g(D) = s(D)h(x(D)) \tag{6.11}$$

where $x\colon W \to W$ and $s(D)\colon W \to W$ satisfy the appropriate $\mathbf{S}_3$-equivariance conditions. We claim that $s\colon W \times W \to W$ extends to an $\mathbf{SO}(3)$-equivariant mapping $S\colon V \times V \to V$ that is linear in the second factor. If this claim is valid, then the uniqueness of extensions shows that

$$G(A) = S(A)H(X(A)) \tag{6.12}$$

where X is the unique $\mathbf{SO}(3)$-equivariant extension of x. (Each side of (6.12) is an extension of the corresponding side of (6.11), but these are equal.) Note that we have *not* assumed that s extends uniquely to S. In fact, Exercise 6.1 shows that this is not the case.

To prove the claim, recall that $\vec{\mathscr{E}}(\mathbf{D}_3)$ is generated as a module over $\mathscr{E}(\mathbf{D}_3)$ by

$$\begin{aligned} t_0(z,w) &= w,\\ t_1(z,w) &= \bar{z}\bar{w},\\ t_2(z,w) &= z^2\bar{w},\\ t_3(z,w) &= z^3 w. \end{aligned} \tag{6.13}$$

We claim that $\vec{\mathscr{E}}(\mathbf{S}_3)$ is generated over $\mathscr{E}(\mathbf{S}_3)$ by

$$\begin{aligned} s_0(D,E) &= E,\\ s_1(D,E) &= DE - \tfrac{1}{3}\mathrm{tr}(DE)I,\\ s_2(D,E) &= D^2E - \tfrac{1}{3}\mathrm{tr}(D^2E)I,\\ s_3(D,E) &= \mathrm{tr}(D^2E)D. \end{aligned} \tag{6.14}$$

This is easy to see. By direct computation each $s_j \in \vec{\mathscr{E}}(\mathbf{S}_3)$. Since the $\mathbf{D}_3$- and

$\mathbf{S}_3$-actions are isomorphic, and the degrees of homogeneity match up in (6.13) and (6.14), the result follows. (We must check directly that

$$s_2, \operatorname{tr}(D^2)s_0$$

and

$$s_3, \operatorname{tr}(D^2)s_1, \det(D)s_2$$

are linearly independent.)

Finally, observe that each of s_0, s_1, s_2, s_3 extends to a matrix in $\vec{\mathscr{E}}(\mathbf{SO}(3))$ by using the same formula, whence it follows by linearity over $\mathscr{E}(\mathbf{SO}(3))$ that every matrix in $\vec{\mathscr{E}}(\mathbf{S}_3)$ extends (smoothly) to one in $\vec{\mathscr{E}}(\mathbf{SO}(3))$. □

Remark 6.8. The proof of Theorem 6.7 is valid if g and h (and G and H) depend smoothly on parameters. Therefore, the restricted tangent spaces, and the tangent spaces, are the same in the $\mathbf{S}_3$ and $\mathbf{SO}(3)$ contexts. Hence the recognition problems and unfolding theories are identical for these two group actions.

Finally in this section we discuss the relation between stabilities of solution branches in the two cases.

Theorem 6.9. *Let $G: V \to V$ be an $\mathbf{SO}(3)$-equivariant mapping and let $g = G|W$ be its $\mathbf{S}_3$-equivariant restriction. Let $D \in V$ satisfy $G(D) = 0$. Then D is orbitally asymptotically stable if and only if D is asymptotically stable for the system $\dot{x} + g(x) = 0$.*

PROOF. In Lemma 6.4 we showed that there are three types of isotropy subgroup for the action of $\mathbf{SO}(3)$ on V, which correspond to isotropy subgroups for the action of $\mathbf{S}_3$ on W as shown in Figure 6.1. Since $\mathbf{SO}(3)$ acts absolutely irreducibly on V it follows that $(dG)_0$ is a multiple of the identity, say cI_V. Thus $(dg)_0 = cI_W$ so the result holds for the trivial solution.

Since $\mathbf{SO}(3)$ is three-dimensional, $(dG)_D$ has three zero eigenvalues at points with trivial ($\mathbf{D}_2$) isotropy. Thus the orbital asymptotic stability of D is determined by two possibly nonzero eigenvalues in the fixed-point subspace of $\mathbf{D}_2$, which is just W. These eigenvalues are those of $(dg)_D$, and the theorem is valid for solutions with trivial isotropy.

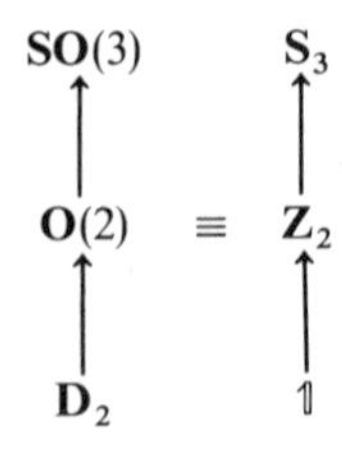

Figure 6.1. Lattice of isotropy subgroups of the five-dimensional irreducible representation of $\mathbf{SO}(3)$, from Lemma 6.4.

Finally we assume that the isotropy subgroup of D is $\mathbf{O}(2)$. Now $\mathbf{O}(2)$ is one-dimensional, so two eigenvalues of $(dG)_D$ are forced by the group action to be zero. Two of the remaining eigenvalues are given by $(dg)_D$. Our problem is to compute the fifth eigenvalue of $(dG)_D$.

First, note that the fixed-point subspaces $\text{Fix}_W(\mathbf{Z}_2)$ and $\text{Fix}_V(\mathbf{O}(2))$ are one-dimensional and equal for $D \in W$. Therefore, one eigenvalue of $(dg)_D$ is real and has an eigenvector in $\text{Fix}_W(\mathbf{Z}_2)$. Let μ be the other (real) eigenvalue of $(dg)_D$ in W. We claim that μ is a double eigenvalue for $(dg)_D$, thus accounting for the fifth eigenvalue and proving the theorem.

This claim may be verified directly; see Golubitsky and Schaeffer [1982], p. 101. Alternatively, we can use the theory developed in XIII, §6. Recall that $\mathbf{SO}(2)$ is a maximal torus in $\mathbf{SO}(3)$ and its Cartan decomposition splits V into irreducible subspaces

$$V = V_0 \oplus V_1 \oplus V_2$$

where $V_0 = \text{Fix}_V(\mathbf{SO}(2))$ is one-dimensional, and $\dim V_1 = \dim V_2 = 2$. Moreover, $\theta \in \mathbf{SO}(2)$ acts as rotation by θ on V_1, and rotation by 2θ on V_2. Now $\mathbf{O}(2) \supset \mathbf{SO}(2)$ acts absolutely irreducibly on each V_j; see Proposition XIII, 6.2. By Theorem XII, 3.5, $(dg)_D$ leaves each V_j invariant and is a multiple of the identity on each. The eigenvalue of $(dG)_D$ on V_0 is just the eigenvalue on $\text{Fix}_W(\mathbf{Z}_2)$. The group action shows that $(dG)_D \equiv 0$ on V_1 or V_2 (V_2, as it happens). The eigenvalues of $(dG)_D$ on V_1 are equal and one of them is μ. This proves the claim. □

EXERCISE

6.1. Show that the equivariant matrices in $\vec{\mathscr{E}}(\mathbf{SO}(3))$,

$$S_1(A, B) = \text{tr}(AB)A$$

$$S_2(A, B) = -\tfrac{1}{2}\text{tr}(A^2)B + 3[A^2B - \tfrac{1}{3}\text{tr}(A^2B)I]$$

are distinct but restrict to $W \times W$ to give identical equivariant matrices in $\vec{\mathscr{E}}(\mathbf{S}_3)$. Hence $\mathbf{S}_3$-equivalences on W do not extend uniquely to $\mathbf{SO}(3)$-equivalences on V.

§7.* Proof of the Equivariant Universal Unfolding Theorem

In this section we prove the equivariant universal unfolding theorem (Theorem 2.1). The proof relies on the equivariant preparation theorem, which is stated and proved in §8. This theorem was first proved in Poénaru [1976]. Setting $\Gamma = \mathbb{1}$ and $V = \mathbb{R}$ we obtain the universal unfolding theorem for bifurcation problems in one state variable (Theorem III, 2.3) whose proof was promised in Chapter II, §2.

The proof is based on the exposition by Martinet [1982] for ordinary singularities, as extended by Golubitsky and Schaeffer [1979a, b] to take account of a distinguished parameter λ. For the equivariant case a similar approach may be used. The only change, aside from checking the appropriate equivariance conditions, is based on an observation which follows from Schwarz's theorem (Theorem XII, 4.3).

Recall the statement of the equivariant universal unfolding theorem. Suppose that $g \in \vec{\mathscr{E}}_{x,\lambda}(\Gamma)$ has a k-parameter unfolding $G \in \vec{\mathscr{E}}_{x,\lambda,\alpha}(\Gamma)$, so that $G(x, \lambda, 0) = g(x, \lambda)$ and $\alpha \in \mathbb{R}^k$. We wish to show that G is versal if and only if the algebraic condition

$$\vec{\mathscr{E}}_{x,\lambda}(\Gamma) = T(g, \Gamma) + \mathbb{R}\{G_{\alpha_1}(x, \lambda, 0), \ldots, G_{\alpha_k}(x, \lambda, 0)\} \tag{7.1}$$

holds.

We proved in §2 that the condition (7.1) is necessary for versality, by considering one-parameter Γ-unfoldings of g. The hard part is to show that (7.1) implies versality.

To do this, let $H(x, \lambda, \beta)$ be any l-parameter unfolding of $g(x, \lambda)$. We must show that H factors through G. We begin by forming the sum unfolding of g,

$$K(x, \lambda, \alpha, \beta) = G(x, \lambda, \alpha) + H(x, \lambda, \beta) - g(x, \lambda).$$

Then H factors through K in a trivial way: just set $\alpha = 0$. We seek to eliminate the β-parameters one by one, appealing to (7.1), in such a way that H still factors through the resulting germ. The required condition on K will be proved from the equivariant preparation theorem. The result needed is as follows:

Lemma 7.1. *Let N be a finitely generated module over $\mathscr{E}_{x,\lambda,\delta}(\Gamma)$, where $\delta \in \mathbb{R}^m$. Then the following are equivalent:*

$$\begin{aligned} &\text{(a)} \quad N \text{ is generated over } \mathscr{E}_\delta \text{ by } n_1, \ldots, n_t; \text{ i.e., } N = \mathscr{E}_\delta\{n_1, \ldots, n_t\}, \\ &\text{(b)} \quad N = \langle \delta_1, \ldots, \delta_m \rangle N + \mathbb{R}\{n_1, \ldots, n_t\}. \end{aligned} \tag{7.2}$$

Remark. In (b) we may clearly replace $n_1, \ldots, n_t$ by any $n'_1, \ldots, n'_t$ such that $n_i - n'_i \in \langle \delta_1, \ldots, \delta_m \rangle N$ for $i = 1, \ldots, t$. Condition (b) is equivalent to

$$N_0 = \mathbb{R}\{\tilde{n}_1, \ldots, \tilde{n}_t\}$$

where $N_0 = N/\mathscr{M}_\delta N$ and $\tilde{n}_j$ is the projection of n_j into N_0.

This lemma follows directly from the equivariant preparation theorem (Theorem 8.1) as we show in §8. In order to apply Lemma 7.1 we need some notation. Let $g \in \vec{\mathscr{E}}_{x,\lambda}(\Gamma)$ and suppose that $K \in \vec{\mathscr{E}}_{x,\lambda,\delta}(\Gamma)$ is an m-parameter unfolding of g. A technical problem which complicates the analysis is the fact that the tangent space $T(g, \Gamma)$ is not an $\mathscr{E}_{x,\lambda}$-module. However, it contains the restricted tangent space $RT(g, \Gamma)$, which is an $\mathscr{E}_{x,\lambda}$-module. Moreover, when g has finite codimension these are "almost" the same: $T(g, \Gamma)$ differs from $RT(g, \Gamma)$ by a finite-dimensional subspace. This is formula (2.9), which holds since Proposition 2.3 is true. It is, however, more convenient to use an

$\mathscr{E}_{x,\lambda}(\Gamma)$-module that is larger than $RT(g, \Gamma)$ but still lies inside $T(g, \Gamma)$. Namely, define

$$\tilde{T}(g, \Gamma) = \{(dg)(X) + Sg \colon X \in \vec{\mathscr{E}}_{x,\lambda}(\Gamma) \text{ and } S \in \vec{\mathscr{E}}_{x,\lambda}(\Gamma)\}. \tag{7.3}$$

In the definition of $RT(g, \Gamma)$ we require that X vanish at the origin; we do not make this requirement in $\tilde{T}(g, \Gamma)$. In particular, $\tilde{T}(g, \Gamma)$ is just the module appearing in (2.5(a)), and

$$RT(g, \Gamma) \subset \tilde{T}(g, \Gamma) \subset T(g, \Gamma).$$

By Proposition 2.3, if any of these three spaces has finite codimension in $\vec{\mathscr{E}}_{x,\lambda}(\Gamma)$, then so do the other two.

Analogously we define for the unfolding K:

$$\tilde{T}^u(K, \Gamma) = \{S(x, \lambda, \delta)K + (d_x K)(X(x, \lambda, \delta))\}, \tag{7.4}$$

$$T^u(K, \Gamma) = \tilde{T}^u(K, \Gamma) + \mathscr{E}_{\lambda,\delta}\{\partial K/\partial \lambda\}, \tag{7.5}$$

with the appropriate equivariance conditions on S and X holding as usual. That is,

$$X(x, \lambda, \delta) \in \vec{\mathscr{E}}_{x,\lambda,\delta}(\Gamma), \qquad S(x, \lambda, \delta) \in \vec{\mathscr{E}}_{x,\lambda,\delta}(\Gamma).$$

Again, $T^u(K, \Gamma)$ is not an $\mathscr{E}_{x,\lambda}$-module, but $\tilde{T}^u(K, \Gamma)$ is. We now prove the following:

Corollary 7.2. *Let $g \colon V \times \mathbb{R} \to V$ have finite Γ-codimension. Let $K \colon V \times \mathbb{R} \times \mathbb{R}^r \to V$ be an r-parameter unfolding of g with parameters $\delta = (\delta_1, \ldots, \delta_r)$. Let $q_1, \ldots, q_r \in \mathscr{E}_{x,\lambda,\delta}(\Gamma)$. Then the following are equivalent:*

(a) $T(g, \Gamma) + \mathbb{R}\{q_1(x, \lambda, 0), \ldots, q_r(x, \lambda, 0)\} = \vec{\mathscr{E}}_{x,\lambda}(\Gamma)$.

(b) $T^u(K, \Gamma) + \mathscr{E}_\delta\{q_1, \ldots, q_r\} = \vec{\mathscr{E}}_{x,\lambda,\delta}(\Gamma)$.

Proof. That (b) implies (a) is obvious: set $\delta = 0$. For the converse, define

$$N = \vec{\mathscr{E}}_{x,\lambda,\delta}(\Gamma)/\tilde{T}^u(K, \Gamma), \qquad N_0 = N/\mathscr{M}_\delta N.$$

Take $n_1 = q_1, \ldots, n_r = q_r$; $n_{r+1} = K_\lambda$, $n_{r+2} = \lambda K_\lambda, \ldots, n_{r+s} = \lambda^{s-1} K_\lambda$, where s is as in (2.10), and set $t = r + s$. Let $n_{i,0} = n_i|(\delta = 0)$. Now $n_{i,0}$ can be shown to be the projection of n_i in N_0 by the following argument: Choose $p_i \in \vec{\mathscr{E}}_{x,\lambda,\delta}(\Gamma)$ such that n_i is the projection of p_i in N. By Taylor's theorem,

$$p_i(x, \lambda, \delta) = p_i(x, \lambda, 0) + \sum_1^m \delta_j h_j(x, \delta, \lambda) \tag{7.6}$$

for some $h_j \in \vec{\mathscr{E}}_{x,\lambda,\delta}(\Gamma)$. On projecting both sides of (7.6) into N, we obtain

$$n_i(x, \lambda, \delta) = n_i(x, \lambda, 0) + \sum \delta_j l_j(x, \lambda, \delta)$$

where l_j is the projection of h_j in N. Hence $n_{i,0}$ is the projection n_i' of n_i, and (a) can be rewritten as

$$\vec{\mathscr{E}}_{x,\lambda}(\Gamma) = \tilde{T}(g,\Gamma) + \mathbb{R}\{n'_1,\ldots,n'_t\} \tag{7.7}$$

since $\lambda^j K_\lambda(x,\lambda,0) = \lambda^j g_\lambda(x,\lambda)$. Therefore,

$$N_0 = \mathbb{R}\{n'_1,\ldots,n'_t\}.$$

Further, $\vec{\mathscr{E}}_{x,\lambda,\delta}(\Gamma)$, and hence N, is a finitely generated module over $\mathscr{E}_{x,\lambda,\delta}(\Gamma)$ by Poénaru's theorem, Theorem XII, 5.4. So by Lemma 7.1

$$N = \mathscr{E}_\delta\{n_1,\ldots,n_t\}.$$

In other words,

$$\begin{aligned}\vec{\mathscr{E}}_{x,\lambda,\delta}(\Gamma) &= \tilde{T}^u(K,\Gamma) + \mathscr{E}_\delta\{n_1,\ldots,n_t\}\\ &= \tilde{T}^u(K,\Gamma) + \mathscr{E}_\delta\{q_1,\ldots,q_r; K_\lambda, \lambda K_\lambda,\ldots,\lambda^{s-1}K_\lambda\}\\ &\subset \tilde{T}^u(K,\Gamma) + \mathscr{E}_{\lambda,\delta}\{K_\lambda\} + \mathscr{E}_\delta\{q_1,\ldots,q_r\}\\ &= T^u(K,\Gamma) + \mathscr{E}_\delta\{q_1,\ldots,q_r\}.\end{aligned}$$ □

Some observations about unfoldings are now required. Suppose that $G(x,\lambda,\alpha)$ is a k-parameter Γ-unfolding of $g(x,\lambda)$, and that $H(x,\lambda,\beta)$ is an l-parameter Γ-unfolding. Recall from §2 that H factors through G if

$$H(x,\lambda,\beta) = S(x,\lambda,\beta)G(X(x,\lambda,\beta),\Lambda(\lambda,\beta),A(\beta)) \tag{7.8}$$

for suitable S, X, Λ, A (subject to appropriate equivariance conditions). We say that G and H are Γ-*isomorphic* if (7.8) holds with A a diffeomorphism. Then

$$H \text{ factors through } G \quad\text{and}\quad G \text{ factors through } H \tag{7.9}$$

and further $k = l$.

Note that (7.9) alone is *not* enough to show that G and H are Γ-isomorphic. For example, consider $\mathbf{Z}_2$-unfoldings

$$G(x,\lambda,\alpha) = x^3 - \lambda x + \alpha$$

$$H(x,\lambda,\beta,\gamma) = x^3 - \lambda x + \beta + \gamma x.$$

Then

$$H(x,\lambda,\beta,\gamma) = G(x,\lambda-\gamma,\beta)$$

and

$$G(x,\lambda,\alpha) = H(x,\lambda,\alpha,0)$$

so each factors through the other. But the unfolding dimensions are 1 for G and 2 for H, so they cannot be Γ-isomorphic. We return to this phenomenon at the end of the section.

Next, suppose that $G\colon V \times \mathbb{R} \times \mathbb{R}^k \to V$ is a k-parameter Γ-unfolding of g, and let $A\colon \mathbb{R}^l \to \mathbb{R}^k$ be a smooth map such that $A(0) = 0$. Then we can define the *pullback unfolding*

$$A^*G\colon V \times \mathbb{R} \times \mathbb{R}^l \to V$$

by

$$(A^*G)(x, \lambda, \beta) = G(x, \lambda, A(\beta)). \tag{7.10}$$

Clearly this is an unfolding, since

$$(A^*G)(x, \lambda, 0) = G(x, \lambda, A(0)) = G(x, \lambda, 0) = g(x, \lambda).$$

But A^*G has l unfolding parameters, rather than the original k of G.

EXAMPLES.

(a) Let $V = \mathbb{R}$, $\Gamma = \mathbb{1}$, and

$$G(x, \lambda, \alpha) = x^3 - \lambda x + \alpha.$$

Let $A: \mathbb{R}^2 \to \mathbb{R}$, $A(\alpha, \beta) = \alpha$. Then

$$(A^*G)(x, \lambda, \alpha, \beta) = G(x, \lambda, A(\alpha, \beta)) = G(x, \lambda, \alpha) = x^3 - \lambda x + \alpha.$$

Here A^*G has the same *formula* as G, but with a new "dummy" unfolding parameter β. Even though β does not explicitly enter into the formula, A^*G is still a two-parameter unfolding: it is just that its value is independent of β.

(b) Take G as before, but let $A(\alpha, \beta) = \alpha + \beta^2$. Then

$$(A^*G)(x, \lambda, \alpha, \beta) = x^3 - \lambda x + \alpha + \beta^2. \tag{7.11}$$

Notice that A^*G always factors through G, for (7.10) is the same as (7.8) with $S = Id$, $X = x$, $\Lambda = \lambda$.

The notion of pullback is introduced precisely because of the possible presence of "unnecessary" unfolding parameters. In particular, suppose that $A: \mathbb{R}^l \to \mathbb{R}^k$ is the germ of a *submersion*, that is, $\operatorname{rank}(dA)_0 = k$. (Note that necessarily $l \geq k$ in this case.) Then, as we show in Proposition 7.3 later, there exist coordinate systems $(\beta_1, \ldots, \beta_l)$ on $\mathbb{R}^l$ and $(\alpha_1, \ldots, \alpha_k)$ on $\mathbb{R}^k$ such that

$$A(\beta_1, \ldots, \beta_l) = (\beta_1, \ldots, \beta_k).$$

In this case A^*G is just G with extra dummy parameters $\beta_{k+1}, \ldots, \beta_l$;

$$(A^*G)(x, \lambda, \beta_1, \ldots, \beta_k, \beta_{k+1}, \ldots, \beta_l) = G(x, \lambda, \beta_1, \ldots, \beta_k).$$

So "form the pullback via a submersion" is just a coordinate-free way of saying "add dummy unfolding parameters." The advantage of the coordinate-free approach is that (as in (7.11) earlier) the *formula* for a submersion A, in the original coordinates, may involve all l parameters. In a more complicated case than (7.11) it would not be obvious that some of the l parameters are dummy parameters.

To complete the preceding discussion we establish the well-known normal form for submersions (see, e.g., Abraham, Marsden, and Ratiu [1983]).

Proposition 7.3. *Let $A: \mathbb{R}^l \to \mathbb{R}^k$ be a germ of a submersion, and let $\pi: \mathbb{R}^l \to \mathbb{R}^k$ be projection onto the first k coordinates. Then there is a diffeomorphism germ ρ on $\mathbb{R}^l$ such that $A(x) = \pi(\rho(x))$.*

PROOF. The matrix $(dA)_0$ has rank k so some $k \times k$ submatrix is nonsingular. By permuting coordinates (a linear diffeomorphism) we may assume it is the leading $k \times k$ submatrix $(\partial A_i/\partial x_j)_{i,j \leq k}$. Define $\rho: \mathbb{R}^l \to \mathbb{R}^l$ by

$$\rho(x_1, \ldots, x_\lambda) = (A_1(x), \ldots, A_k(x), x_{k+1}, \ldots, x_\lambda).$$

Then $(d\rho)_0$ has matrix

$$\left[\begin{array}{c|c} \partial A_i/\partial x_j & * \\ \hline 0 & I_{l-k} \end{array}\right] \begin{array}{l} \}k \\ \}l-k \end{array}$$

which is nonsingular. Therefore, ρ is a diffeomorphism germ, by the inverse function theorem. But clearly $A(x) = \pi(\rho(x))$. □

We have now amassed the machinery needed to prove the equivariant universal unfolding theorem.

PROOF OF THEOREM 2.1. Let g, G be as stated in the hypotheses of Theorem 2.1, and suppose that (7.1) holds. We must prove that G is versal. To this end let $H: \mathbb{R}^n \times \mathbb{R} \times \mathbb{R}^l \to \mathbb{R}^n$ be any l-parameter Γ-unfolding with parameters $\beta = (\beta_1, \ldots, \beta_l)$. We must show that H factors (equivariantly) through G.

Consider the sum unfolding

$$K(x, \lambda, \alpha, \beta) = G(x, \lambda, \alpha) + H(x, \lambda, \beta) - g(x, \lambda).$$

Clearly $K \in \vec{\mathscr{E}}_{x,\lambda,\alpha,\beta}(\Gamma)$ and is a $(k + l)$-parameter Γ-unfolding of g. We claim:

$$\text{There exists a submersion } A: \mathbb{R}^k \times \mathbb{R}^l \to \mathbb{R}^k \text{ such that } K \text{ is } \Gamma\text{-isomorphic to } A^*G. \tag{7.12}$$

First let us show that (7.12) establishes the theorem. If (7.12) holds then

$$K(x, \lambda, \alpha, \beta) = S(x, \lambda, \alpha, \beta)G(X(x, \lambda, \alpha, \beta), \Lambda(\lambda, \alpha, \beta), A(\alpha, \beta))$$

where S and X satisfy the usual equivariance conditions. Set $\alpha = 0$ to get

$$\begin{aligned} H(x, \lambda, \beta) &= K(x, \lambda, 0, \beta) \\ &= S(x, \lambda, 0, \beta)G(X(x, \lambda, 0, \beta), \Lambda(\lambda, 0, \beta), A(0, \beta)). \end{aligned}$$

So H factors through G.

It remains to prove (7.12). We show that A exists by induction on l. The case $l = 0$ is obvious: let $A = Id$. Assume the result holds for $l - 1$, where $\mathbb{R}^{l-1}$ is embedded in $\mathbb{R}^l$ by $(\beta_1, \ldots, \beta_{l-1}) \mapsto (\beta_1, \ldots, \beta_{l-1}, 0)$. Let

$$L(x, \lambda, \alpha, \beta_1, \ldots, \beta_{l-1}) = K(x, \lambda, \alpha, \beta_1, \ldots, \beta_{l-1}, 0).$$

Then by induction there is a submersion

$$B: \mathbb{R}^k \times \mathbb{R}^{l-1} \to \mathbb{R}^k$$

such that B^*G is Γ-isomorphic to L. It suffices to find a submersion

$$C: \mathbb{R}^k \times \mathbb{R}^l \to \mathbb{R}^k \times \mathbb{R}^{l-1}$$

such that C^*L is Γ-isomorphic to K. For then we may set $A = B \circ C$.

To ease notational problems, set $\delta = (\delta_1, \ldots, \delta_{k+l})$ where $\delta_1 = \alpha_1, \ldots, \delta_k = \alpha_k$; $\delta_{k+1} = \beta_1, \ldots, \delta_{k+l} = \beta_l$. In other words, $\delta = (\alpha, \beta)$. Let $m = k + l$. From (7.1) we have

$$\vec{\mathscr{E}}_{x,\lambda}(\Gamma) = T(g, \Gamma) + \mathbb{R}\{\partial K/\partial\alpha_1, \ldots, \partial K/\partial\alpha_k\}(\alpha, \beta) = 0.$$

Therefore, by Corollary 7.2 we have

$$\vec{\mathscr{E}}_{x,\lambda,\delta}(\Gamma) = T^u(K, \Gamma) + \mathscr{E}_\delta\{\partial K/\partial\alpha_1, \ldots, \partial K/\partial\alpha_k\}.$$

Now by (7.4, 7.5) we have

$$T^u(K, \Gamma) = \{S(x, \lambda, \delta)K + (d_x K)(X(x, \lambda, \delta))\} + \mathscr{E}_{\lambda,\delta}\{\partial K/\partial\lambda\}.$$

Further, $-\partial K/\partial\delta_m = -\partial K/\partial\beta_l \in \vec{\mathscr{E}}_{x,\lambda,\delta}(\Gamma)$, so we have

$$\begin{aligned} -\partial K/\partial\delta_m = {} & S(x, \lambda, \delta)K(x, \lambda, \delta) + \sum_{j=1}^{n} X_j(x, \lambda, \delta)\partial K/\partial x_j \\ & + \Lambda(\lambda, \delta)\partial K/\partial\lambda + \sum_{i=1}^{m-1} \xi_i(\delta)\partial K/\partial\delta_i \end{aligned} \tag{7.13}$$

(where, in fact, $\xi_i(\delta) = 0$ for $i > k$). In (7.13) each of S and $X = (X_1, \ldots, X_n)$ satisfies appropriate equivariance conditions.

Consider the system of ODEs

$$\begin{aligned} d\delta_m/dt &= 1 \\ dx_j/dt &= X_j(x, \lambda, \delta) \qquad (j = 1, \ldots, n) \\ d\lambda/dt &= \Lambda(\lambda, \delta) \\ d\delta_i/dt &= \xi_i(\delta) \qquad (i = 1, \ldots, m-1) \end{aligned} \tag{7.14}$$

and suppose that $(\delta_m(t), x_j(t), \lambda(t), \delta_i(t))$ is a solution. Then (7.13) may be rewritten in the form

$$\frac{d}{dt}K(x(t), \lambda(t), \delta(t)) = -S(x(t), \lambda(t), \delta(t))K(x(t), \lambda(t), \delta(t)). \tag{7.15}$$

Take initial conditions

$$\delta_m = 0, \qquad x_j = x_{j0}, \qquad \lambda = \lambda_0, \qquad \delta_i = \delta_{i0} \qquad (i = 1, \ldots, m-1)$$

when $t = 0$. Then the solution curve $P(t)$ of (7.14) through

$$P_0 = (0, x_{10}, \ldots, x_{n0}, \lambda_0, \delta_{10}, \ldots, \delta_{m-1,0})$$

is transverse to the hyperplane $\delta_m = 0$ because $d\delta_m/dt = 1$. See Figure 7.1. Define a map

$$\varphi: V \times \mathbb{R} \times \mathbb{R}^m \to V \times \mathbb{R} \times \mathbb{R}^{m-1}$$

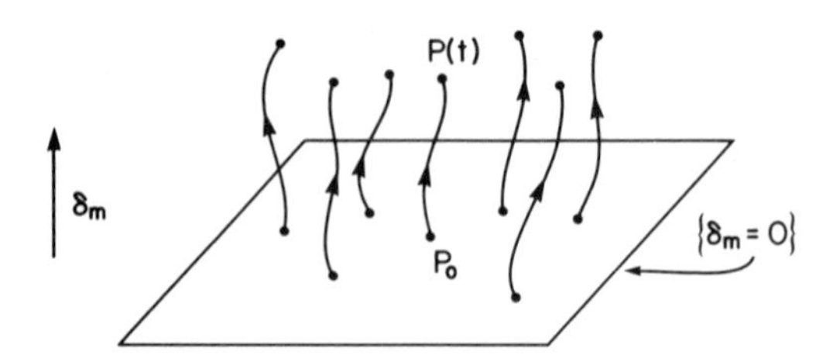

Figure 7.1. The flow defined by (7.14) near the hyperplane $\delta_m = 0$.

by projecting (x, λ, δ) along these integral curves until $\delta_m = 0$. By standard properties of ODEs (Abraham, Marsden and Ratiu [1983]) φ is smooth; further, the restriction of φ to the hyperplane $\delta_m = 0$ is the identity. Therefore, the Jacobian matrix of φ at 0 is of the form

$$\left[\begin{array}{c|c} I_{n+m} & * \end{array}\right]$$

and rank $(d\varphi)_0 = n + m$. Therefore, φ is a submersion.

We can clearly write φ in the form

$$\varphi(x, \lambda, \delta) = (\rho_{\lambda\delta}(x), \Lambda_\delta(\lambda), C(\delta)) \tag{7.16}$$

because (7.14) implies that the subspaces $0 \times 0 \times \mathbb{R}^l$ and $0 \times \mathbb{R} \times \mathbb{R}^l$ are invariant under the flow. Since $C\colon \mathbb{R}^m \to \mathbb{R}^{m-1}$, the rank of $(dC)_0$ is at most $m - 1$. Now $(d\varphi)_0$ can be written as

$$\begin{array}{c} \\ x \\ \lambda \\ \delta_1 \\ \vdots \\ \delta_{m-1} \end{array} \begin{array}{c} \begin{array}{ccc} x & \lambda & \delta_1, \ldots, \delta_m \end{array} \\ \left[\begin{array}{c|c|c} (d_x\rho_{\lambda\delta})_0 & * & * \\ \hline 0 & (d_\lambda\Lambda_\delta)_0 & * \\ \hline 0 & 0 & (d_\delta C)_0 \end{array}\right] \end{array}.$$

Since φ is a submersion with rank $(d\varphi)_0 = n + m$, it follows that

(a) rank $(d_\delta C)_0 = m - 1$ so C is a submersion;
(b) $(d_\lambda \Lambda_\delta)_0$ is nonsingular, so each Λ_δ is a diffeomorphism;
(c) $(d_x \rho_{\lambda\delta})_0$ is nonsingular, so each $\rho_{\lambda\delta}$ is a diffeomorphism.

Further, by the usual argument appealing to uniqueness of solutions of ODEs, the equivariance condition

$$\rho_{\lambda\delta}(\gamma x) = \gamma\rho_{\lambda\delta}(x) \qquad (\gamma \in \Gamma)$$

holds. Note also that (7.14) implies $\delta_m = t$, so we may identify t with δ_m.

For a given point $P = (x, \lambda, \delta)$ select an integral curve $(x(t), \lambda(t), \delta(t))$ of (7.14) passing through P. Now consider the nonautonomous ODE

$$dy/dt = -S(x(t), \lambda(t), \delta(t))y \tag{7.17}$$

where $y \in \mathbb{R}^n$. Note that (7.17) really depends only on $(x(0), \lambda(0), \delta(0))$ together with the initial condition $y = y_0$, since $(x(t), \lambda(t), \delta(t))$ is obtained by integrating (7.14). The solution of (7.17) is therefore of the form

$$y(t) = Y(y_0, t; x(0), \lambda(0), \delta(0)).$$

We claim that for fixed t, $x(0)$, $\lambda(0)$, $\delta(0)$ near zero, the map $y_0 \mapsto y(t)$ is a diffeomorphism. To see this, set $x(0) = \lambda(0) = \delta(0) = 0$. Then (7.14) leads to

$$x(t) = 0, \qquad \lambda(t) = 0, \qquad \delta_i(t) = 0 \qquad (i = 1, \ldots, m-1), \qquad \delta_m(t) = t.$$

Therefore, $(d_y Y)_0$ is nonsingular for $(x(0), \lambda(0), \delta(0)) = 0$, hence by continuity for $(x(0), \lambda(0), \delta(0))$ near 0. Clearly Y depends smoothly on its arguments.

Now (7.15) says that $K(x(t), \lambda(t), \delta(t))$ satisfies (7.17). So

$$K(x(t), \lambda(t), \delta(t)) = Y(K(x(0), \lambda(0), \delta(0)), t, x(0), \lambda(0), \delta(0)). \tag{7.18}$$

Further, if $\tilde{\delta} = (\delta_1, \ldots, \delta_{m-1})$ then

$$\varphi(x(t), \lambda(t), \delta(t)) = (x(0), \lambda(0), \tilde{\delta}(0))$$

by definition of φ. Since any point (x, λ, δ) near 0 has an integral curve of (7.14) passing through it at time $t = \delta_m$, we can restate (7.18) as

$$K(x, \lambda, \delta) = E_{x\lambda\delta}(K(\varphi(x, \lambda, \delta), 0))$$

where

$$E_{x\lambda\delta}(y) = Y(y, \delta_m, \varphi(x, \lambda, \delta), 0))$$

is a family of diffeomorphisms on V.

By (7.16) this yields

$$\begin{aligned} K(x, \lambda, \delta) &= E_{x\lambda\delta} K(\rho_{\lambda\delta}(x), \Lambda_\delta(\lambda), C(\delta), 0)) \\ &= E_{x\lambda\delta} C^* L(\rho_{\lambda\delta}(x), \Lambda_\delta(\lambda), \delta) \end{aligned}$$

since $C(\delta) \in \mathbb{R}^{m-1}$. By Proposition XIV, 1.5, we can replace the nonlinear diffeomorphism $E_{x\lambda\delta}$ by matrices $S_{x\lambda\delta}$. Therefore, C^*L is Γ-isomorphic to K, as required. □

Theorem 2.1 provides a necessary and sufficient condition for versality. We may also inquire to what extent a universal Γ-unfolding is unique.

Theorem 7.4. *Two versal Γ-unfoldings $G(x, \lambda, \alpha)$ and $H(x, \lambda, \beta)$ are Γ-isomorphic if and only if they have the same number of unfolding parameters.*

Remark. A Γ-unfolding $G(x, \lambda, \alpha)$ of g is said to be *universal* if it is versal and the number of unfolding parameters $(\alpha_1, \ldots, \alpha_k)$ is as small as possible. Clearly we have the following:

(a) Any two universal Γ-unfoldings of $g \in \vec{\mathscr{E}}_{x,\lambda}(\Gamma)$ are Γ-isomorphic.

(b) A versal Γ-unfolding is just a universal Γ-unfolding "with dummy parameters." More precisely, let H be an l-parameter versal Γ-unfolding of g, and let G be a (k-parameter) universal Γ-unfolding. Let $A: \mathbb{R}^l \to \mathbb{R}^k$ be projection. Then H is Γ-isomorphic to A^*G.

PROOF. Necessity is obvious; we prove sufficiency. We may assume both G and H have the *same* unfolding parameters $\alpha \in \mathbb{R}^k$. First, assume that $k = l$ where $l = \operatorname{codim}_\Gamma g$. Since G is versal, H factors through it, so

$$H(x, \lambda, \alpha) = S(x, \lambda, \alpha)G(X(x, \lambda, \alpha), \Lambda(\lambda, \alpha), A(\alpha))$$

where X and S satisfy

$$\gamma^{-1}S(\gamma x, \lambda, \alpha)\gamma = S(x, \lambda, \alpha)$$
$$X(\gamma x, \lambda, \alpha) = \gamma X(x, \lambda, \alpha).$$

There is no equivariance condition on A, so to prove Γ-isomorphism we must show that A is a diffeomorphism germ. To do this we show that $(dA)_0$ is invertible. Now $H(x, \lambda, \alpha)$ is Γ-isomorphic to the unfolding

$$K(x, \lambda, \alpha) = G(x, \lambda, A(\alpha))$$

and K is therefore versal. Further,

$$\partial K/\partial \alpha_i|_{\alpha=0} = \sum_{j=1}^{l} \partial A_j/\partial \alpha_i|_{\alpha=0} \cdot \partial G/\partial \alpha_j|_{\alpha=0}. \tag{7.19}$$

Therefore,

$$\{\partial K/\partial \alpha_i|_{\alpha=0}\} \text{ and } \{\partial G/\partial \alpha_i|_{\alpha=0}\}$$

both span l-dimensional spaces, because $l = \operatorname{codim}_\Gamma g$ and both K, G are versal, so Theorem 2.1 applies. Therefore,

$$(dA)_0 = [\partial A_j/\partial \alpha_i|_{\alpha=0}]_{ij}$$

is invertible, and A is a diffeomorphism germ.

For the general case, suppose that G is a k-parameter versal unfolding and let L be a fixed l-parameter versal unfolding where $l = \operatorname{codim}_\Gamma g$. Then G factors through L, so

$$G(x, \lambda, \alpha) = S(x, \lambda, \alpha)L(X(x, \lambda, \alpha), \Lambda(\lambda, \alpha), A(\alpha)).$$

We have rank $(dA)_0 = l$, so $A: \mathbb{R}^k \to \mathbb{R}^l$ is a submersion and G is Γ-isomorphic to the k-parameter unfolding $L(x, \lambda, A(\alpha))$. Similarly H is Γ-isomorphic to $L(x, \lambda, B(\alpha))$ for a submersion $B: \mathbb{R}^k \to \mathbb{R}^l$. By the implicit function theorem there is a diffeomorphism $\sigma: \mathbb{R}^k \to \mathbb{R}^k$ such that $B(\sigma(\alpha)) = A(\alpha)$. Hence $L(x, \lambda, A(\alpha))$ and $L(x, \lambda, B(\alpha))$ are Γ-isomorphic. (There is no equivariance condition on unfolding parameters, so we do not require any equivariance for σ.) Therefore, G and H are Γ-isomorphic. □

§8.* The Equivariant Preparation Theorem

In this section we must consider a situation in which a compact Lie group Γ acts linearly on two spaces V and W. To distinguish between Γ-invariant germs on V and W we shall use x to denote coordinates in V and y to denote coordinates in W. Thus $\mathscr{E}_x(\Gamma)$ is the ring of Γ-invariant germs on V, and $\mathscr{E}_y(\Gamma)$ the ring of invariant germs on W.

Given a Γ-equivariant mapping $\varphi: V \to W$, it is possible to interpret any $\mathscr{E}_x(\Gamma)$-module N as an $\mathscr{E}_y(\Gamma)$-module, by a method described later. The equivariant preparation theorem provides a test for deciding when N is finitely generated as an $\mathscr{E}_y(\Gamma)$-module. In this section we first state the equivariant preparation theorem and then show how this rather abstract algebraic result can be used to prove Lemma 7.1. Then we prove the theorem itself, by quoting the ordinary preparation theorem.

(a) Proof of Lemma 7.1

Let $\varphi: V \to W$ be (germ of a) a smooth Γ-equivariant mapping satisfying $\varphi(0) = 0$. Then φ induces a map

$$\begin{aligned} \varphi^*: \mathscr{E}_y(\Gamma) &\to \mathscr{E}_x(\Gamma) \\ f &\mapsto f(\varphi) \end{aligned} \tag{8.1}$$

called the *pullback* of φ. The map φ^* is a ring homomorphism.

Now suppose that N is an $\mathscr{E}_x(\Gamma)$-module. In particular, we can multiply $n \in N$ by any $h \in \mathscr{E}_x(\Gamma)$ to obtain another element of N, denoted $h \cdot n$. We can use φ^* to view N as an $\mathscr{E}_y(\Gamma)$-module as follows. Let $f \in \mathscr{E}_y(\Gamma)$, $n \in N$, and define

$$f \cdot n = \varphi^*(f) \cdot n. \tag{8.2}$$

This is a module action because φ^* is a ring homomorphism.

We claim that a simple condition must hold if N is to be a finitely generated $\mathscr{E}_y(\Gamma)$-module. To state it we need some notation. Recall that $\mathscr{M}_y(\Gamma)$ is the maximal ideal in $\mathscr{E}_y(\Gamma)$, consisting of those equivariant mappings on W that vanish at the origin. By

$$\mathscr{M}_y(\Gamma) \cdot N \tag{8.3}$$

we indicate the submodule of N generated by elements $f \cdot n$ where $f(0) = 0$.

Now suppose that N is finitely generated over $\mathscr{E}_y(\Gamma)$ and let $n_1, \ldots, n_t$ be a set of generators. For every $n \in N$ there exist invariant functions $f_1, \ldots, f_t \in \mathscr{E}_y(\Gamma)$ such that

$$n = f_1 n_1 + \cdots + f_t n_t. \tag{8.4}$$

By Taylor's theorem we can write

$$f_j(y) = c_j + \tilde{f}_j(y) \tag{8.5}$$

where $c_j = f_j(0) \in \mathbb{R}$ and $\tilde{f} \in \mathscr{M}_y(\Gamma)$. Combining (8.4 and 8.5) we get

$$n = c_1 n_1 + \cdots + c_t n_t + \tilde{n} \tag{8.6}$$

where $\tilde{n} \in \mathscr{M}_y(\Gamma) \cdot N$. Therefore, if N is a finitely generated $\mathscr{E}_y(\Gamma)$-module then

$$N = \mathbb{R}\{n_1, \ldots, n_t\} + \mathscr{M}_y(\Gamma) \cdot N.$$

In the language of quotients, this means that

$$\dim N/\mathscr{M}_y(\Gamma) \cdot N \leq t < \infty. \tag{8.7}$$

The remarkable conclusion of the equivariant preparation theorem is that this necessary condition (8.7) is also sufficient.

Theorem 8.1 (Equivariant Preparation Theorem). *Let N be a finitely generated $\mathscr{E}_x(\Gamma)$-module. Let $\varphi: V \to W$ be a smooth Γ-equivariant mapping satisfying $\varphi(0) = 0$. Then, via φ^*, N is a finitely generated $\mathscr{E}_y(\Gamma)$-module if and only if*

$$\dim(N/\mathscr{M}_y(\Gamma) \cdot N) < \infty.$$

The proof of this theorem will be given in subsection (b). Theorem 8.1 can be used to determine explicit generators for the module N over the ring $\mathscr{E}_y(\Gamma)$.

Corollary 8.2. *Let N be a finitely generated module over $\mathscr{E}_x(\Gamma)$. Let $\varphi: V \to W$ be Γ-equivariant and satisfy $\varphi(0) = 0$. Let $n_1, \ldots, n_t \in N$. Then $n_1, \ldots, n_t$ generate N as a module over $\mathscr{E}_y(\Gamma)$ if and only if*

$$N = \mathbb{R}\{n_1, \ldots, n_t\} + \mathscr{M}_y(\Gamma) \cdot N. \tag{8.8}$$

Remark. Clearly we may replace n_j by any n_j' such that $n_j' - n_j \in \mathscr{M}_y(\Gamma) \cdot N$.

Proof. The necessity of (8.8) was proved earlier in checking (8.7). To prove sufficiency let $\tilde{N}$ be the submodule of N generated by $n_1, \ldots, n_t$. We must show that $N = \tilde{N}$. However, we know that $N \supset \tilde{N}$, so it remains to prove that $N \subset \tilde{N}$. Since, by Theorem 8.1, N is a finitely generated module over $\mathscr{E}_y(\Gamma)$ we can prove this by showing that

$$N \subset \tilde{N} + \mathscr{M}_y(\Gamma) \cdot N. \tag{8.9}$$

But (8.9) follows directly from (8.8). □

We end this subsection by showing that Lemma 7.1 follows from Corollary 8.2.

Proof of Lemma 7.1. Let N be a finitely generated module over $\mathscr{E}_{x,\lambda,\alpha}(\Gamma)$. Define $\varphi: V \times \mathbb{R} \times \mathbb{R}^m$ by $\varphi(x, \lambda, \alpha) = \alpha$. Since Γ acts trivially on $\mathbb{R}^m$, it follows that φ is equivariant. By Corollary 8.2,

$$N = \mathscr{E}_\alpha\{n_1, \ldots, n_t\}$$

if and only if

$$N/\mathcal{M}_y(\Gamma)\cdot N = \mathbb{R}\{m_1, \ldots, m_t\} \tag{8.10}$$

where m_j is the image of n_j under the natural projection. In this case $y = \alpha$ and $\mathcal{M}_\alpha(\Gamma) = \langle \alpha_1, \ldots, \alpha_m \rangle$. Thus (8.10) is just (7.2(b)).

(b) Proof of Theorem 8.1

We prove the equivariant preparation theorem by coupling the standard Malgrange preparation theorem (the case $\Gamma = \mathbb{1}$) with Schwarz's theorem (XII, 4.3). We will not prove the Malgrange preparation theorem since it is standard in the literature; see, for example, Golubitsky and Guillemin [1973], Theorem IV, 3.6.

Theorem 8.3 (Malgrange Preparation Theorem). *Let N be a finitely generated $\mathcal{E}_v$-module and let ψ: $\mathbb{R}^s \to \mathbb{R}^t$ be a smooth map satisfying $\psi(0) = 0$ where v and w denote coordinates on $\mathbb{R}^s$ and $\mathbb{R}^t$, respectively. Then, via ψ^*, N is a finitely generated $\mathcal{E}_w$-module if and only if*

$$\dim_{\mathbb{R}} N/\mathcal{M}_w N < \infty.$$

Our proof of the equivariant preparation theorem follows Damon [1984], in which, in fact, a much more general version of the preparation theorem is presented.

The proof uses two consequences of Schwarz's theorem. Let $v_1, \ldots, v_s$ be a Hilbert basis for $\mathcal{E}_x(\Gamma)$. Define the mapping ρ_V: $V \to \mathbb{R}^s$ by

$$\rho_V(x) = (v_1(x), \ldots, v_s(x)).$$

Schwarz's theorem states that

$$\rho_V^*: \mathcal{E}_v \to \mathcal{E}_x(\Gamma)$$

is a surjective ring homomorphism. Similarly we can define ρ_W: $W \to \mathbb{R}^t$ using the Hilbert basis $w_1, \ldots, w_t$ in $\mathcal{E}_y(\Gamma)$.

The first consequence is that any module N over $\mathcal{E}_x(\Gamma)$ is automatically a module over $\mathcal{E}_v$ since a function η on $\mathbb{R}^s$ can act on N via multiplication by $\rho_V^*(\eta)$. Moreover, since ρ_V^* is surjective, N is a finitely generated $\mathcal{E}_x(\Gamma)$-module if and only if it is a finitely generated $\mathcal{E}_v$-module.

The second consequence is that smooth Γ-equivariant mappings φ: $V \to W$ can be identified in a natural way with smooth mappings ψ: $\mathbb{R}^s \to \mathbb{R}^t$ as follows. There exists a smooth mapping ψ satisfying

$$\psi\rho_V = \rho_W\varphi. \tag{8.11}$$

Moreover, if $\varphi(0) = 0$ then $\psi(0) = 0$. We define ψ by observing that $w_j(\varphi)$ is a Γ-invariant function on V. Hence there exists a smooth function ψ_j: $\mathbb{R}^s \to \mathbb{R}$ such that $w_j(\varphi) = \psi_j(\rho_V)$. Now define $\psi = (\psi_1, \ldots, \psi_t)$.

To prove Theorem 8.1 we let N be a finitely generated $\mathscr{E}_x(\Gamma)$-module and suppose that

$$\dim_{\mathbb{R}} N/\mathscr{M}_y(\Gamma)N < \infty. \tag{8.12}$$

We want to show that N is a finitely generated $\mathscr{E}_y(\Gamma)$-module. Observe that via ρ_V, N is a finitely generated $\mathscr{E}_v$-module, and that via ρ_W, N is an $\mathscr{E}_w$-module. (This follows from the first consequence of Schwarz's theorem.) To prove Theorem 7.1 it is sufficient to show that N is a finitely generated $\mathscr{E}_w$-module.

Next, (8.11) shows that the module structure of N over the ring $\mathscr{E}_w$ is obtained from the module structure of N over the ring $\mathscr{E}_v$ via ψ^*. Since N is finitely generated as a module over $\mathscr{E}_v$ we can use the Malgrange preparation theorem to show that N is a finitely generated $\mathscr{E}_w$-module, provided we can show that

$$\dim_{\mathbb{R}} N/\psi^*(\mathscr{M}_w)N < \infty. \tag{8.13}$$

This will follow from (8.12). However, the module action on N for functions $\mathbb{R}^s \to \mathbb{R}$ is defined using ρ_V^*. Thus $\psi^*(\mathscr{M}_w)N$ has finite codimension in N precisely when $\rho_V^*(\psi^*(\mathscr{M}_w))N$ has finite codimension in N. By (8.11) $\rho_V^*(\psi^*(\mathscr{M}_w))N = \varphi^*(\rho_W^*(\mathscr{M}_w))N$. Again by Schwarz's theorem $\rho_W^*(\mathscr{M}_w) = \mathscr{M}_y(\Gamma)$. Since the action of $\mathscr{M}_y(\Gamma)$ in (8.12) is given via φ^*, the dimensions in (8.12) and (8.13) are identical. This completes the proof of Theorem 7.1. □

CASE STUDY 5

The Traction Problem for Mooney–Rivlin Material

§0. Introduction

In this case study we analyze the Rivlin cube, a bifurcation problem introduced in Chapter XI. Our purpose there was to illustrate the phenomenon of spontaneous symmetry-breaking and to describe the kinds of results that can be obtained by a singularity-theoretic analysis. This case study has a different aim: to present complete calculations supporting the singularity theory analysis of a specific bifurcation problem. The Rivlin cube is an ideal example since the calculations are tractable. This case study has been written so as to be independent of Chapter XI.

The Rivlin cube problem (Rivlin [1948, 1974]) is to describe the equilibria of a homogeneous isotropic elastic material subjected to equal tensions perpendicular to each face; see Figure 0.1. We assume that the undeformed equilibrium configuration is a cube. We consider only homogeneous deformations

$$(x_1, x_2, x_3) \mapsto (l_1 x_1, l_2 x_2, l_3 x_3) \tag{0.1}$$

where $l_j > 0$. These map the cube to a rectangular parallelepiped. We call the l_j the *principal stretches*.

Ball and Schaeffer [1983] show that the same analysis applies to noncubical bodies and consider more general homogeneous deformations, including rotations. (A general discussion of a different class of bifurcation problems with the symmetry of the cube, permitting deformations into shapes other than rectangular parallelepipeds and including applications to the deformation of crystal lattices under temperature changes, is given in Melbourne [1986, 1987, 1988].)

We assume that the tension on each face is a *dead load*, meaning that whatever the deformation, the same total force λ acts, uniformly distributed over the current area of the face. In (0.1) suppose the body to be *incompressible*,

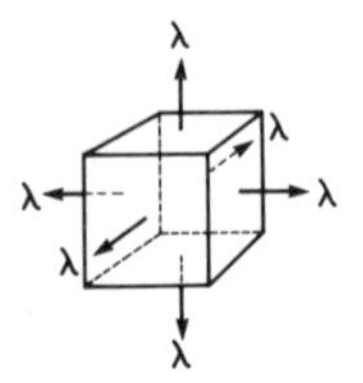

Figure 0.1. The Rivlin cube problem with tension λ.

that is, that

$$l_1 l_2 l_3 = 1.$$

Let $\Phi(l_1, l_2, l_3)$ be the stored energy function of the material. Because of incompressibility, Φ need be defined only on the surface $\{l_1 l_2 l_3 = 1\}$, but we shall assume it is defined for all positive l_j. Our main concern is with *Mooney–Rivlin material*, for which

$$\Phi = \mu \sum_{j=1}^{3} l_j^2 + \nu \sum_{j=1}^{3} l_j^{-2}, \tag{0.2}$$

where μ and ν are constants. The special case $\nu = 0$ is noteworthy and has its own name: *neo-Hookean material.*

The analysis of the Rivlin cube leads to the following problem:

$$\text{Minimize} \quad \Phi(l_1, l_2, l_3) - \lambda(l_1 + l_2 + l_3) \quad \text{on} \quad \{l_1 l_2 l_3 = 1\}. \tag{0.3}$$

The second term in (0.3) represents the work done by the external forces; it assumes this simple form because λ is a dead load. Differentiating (0.3) we obtain a necessary condition for a minimum:

$$\frac{\partial \Phi}{\partial l_j} - \lambda = p \frac{\partial}{\partial l_j}(l_1 l_2 l_3) \qquad (j = 1, 2, 3) \tag{0.4}$$

where p is a Lagrange multiplier associated with the constraint (physically, the pressure). We combine the preceding relations to obtain a bifurcation problem with state variables l_1, l_2, l_3, p and bifurcation parameter λ:

$$\begin{aligned} &\text{(a)} \quad \frac{\partial \Phi}{\partial l_j} - \lambda = \frac{p}{l_j} \qquad (j = 1, 2, 3) \\ &\text{(b)} \quad l_1 l_2 l_3 = 1. \end{aligned} \tag{0.5}$$

As described in Chapter XI, this bifurcation problem is equivariant with respect to the group $\mathbf{S}_3$, acting by permutations of l_1, l_2, l_3. In the next section we reduce (0.5) to a $\mathbf{D}_3$-equivariant problem in two state variables, where $\mathbf{D}_3$ acts by its standard representation on $\mathbb{R}^2 \equiv \mathbb{C}$.

Rivlin [1948] analyzed (0.5) for neo-Hookean material. He found that the trivial solution $l_1 = l_2 = l_3 = 0$ loses stability when $\lambda = 2\mu$ and that new solutions bifurcate, both sub- and supercritically, from this point. The bifurcating solutions have two principal stretches equal, say $l_1 = l_2$. We call these solutions *platelike* if $l_1 = l_2 > 1 > l_3$, *rodlike* if $l_1 = l_2 < 1 < l_3$. All the bifur-

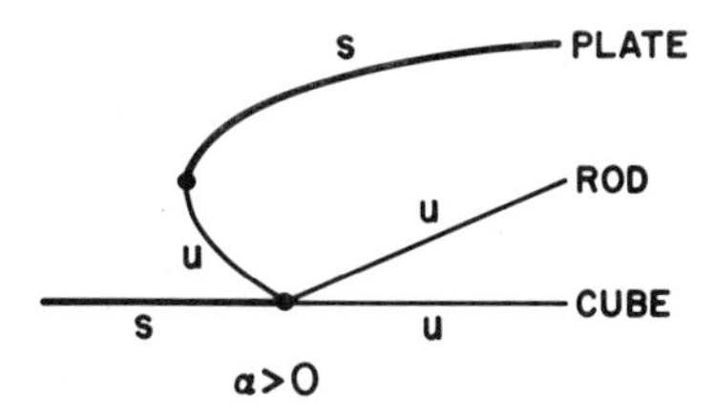

Figure 0.2. The global bifurcation diagram: neo-Hookean material.

cating solutions are unstable near the bifurcation point. (With 20-20 hindsight, we see that this bifurcation is the generic $\mathbf{D}_3$-symmetric one described in Proposition XIV, 4.3.) Rivlin's analysis, which was global, found that the subcritical branch of platelike solutions turned around in parameter space to regain stability as in Figure 0.2.

In neo-Hookean material μ is an inessential parameter: it merely sets the scale of the problem. By contrast, in Mooney–Rivlin material the parameter ν/μ is essential, that is, independent of scaling. By Theorem XIV, 4.4, greater degeneracy can be expected in a one-parameter family of $\mathbf{D}_3$-symmetric bifurcation problems than occurs for a single such problem.

With this motivation, Ball and Schaeffer [1983] analyzed the bifurcations of (0.5) for a Mooney–Rivlin stored energy function Φ. They found that the degenerate bifurcation problem of Theorem XIV, 4.4, with normal form

$$(\varepsilon u + \delta\lambda)z + (\sigma u + mv)\bar{z}^2, \tag{0.6}$$

occurs when $\nu = \mu/3$. They also showed that ν/μ universally unfolds this degeneracy; see Figure 3.1 later. These results provide a striking illustration of the way local methods can elucidate global behavior: the global turnaround of the platelike branch found by Rivlin can be analyzed by local methods.

Ball and Schaeffer [1983] also found that the neo-Hookean case is somewhat singular. Specifically, as soon as ν is positive, the platelike solutions that Rivlin found lose stability in the limit $\lambda \to \infty$, and the system evolves to rodlike solutions, passing through a secondary branch of solutions with all three l_j different.

Singularity theory made three contributions to this problem. First, it motivated considering a more general material than Rivlin did. Second, the normal form (0.6) made the analysis of the degeneracy $\nu = \mu/3$ much easier. Finally, the persistence of the universal unfolding could be invoked to show that the results remained true if the equations were perturbed. For example, introducing a slight degree of compressibility into the system leads to equations that cannot be solved explicitly, but by persistence, the qualitative behavior must be unchanged.

This case study is divided into three sections. In §1 we transform the equilibrium equations (0.5) into the standard form for a $\mathbf{D}_3$-invariant bifurcation problem, namely a mapping $g: \mathbb{C} \times \mathbb{R} \to \mathbb{C}$ commuting with the standard action of $\mathbf{D}_3$. In §2 we relate the invariant coefficient functions p and q, which occur in the general representation (XII, 5.12) for such mappings, to the stored energy function. (Incidentally, we consider a slightly more general energy function than (0.2).) Finally in §3 we compute those derivatives of p and q that determine the normal form (0.6), thereby verifying Figure 3.1.

We shall consider only symmetric loads, equal on all faces. A result of Sawyers [1976] suggests that a small symmetry-breaking perturbation might lead to some surprising effects, but we shall not pursue the matter here.

§1. Reduction to $\mathbf{D}_3$ Symmetry in the Plane

In this section we show how to reduce the solution of the equilibrium equations for the Rivlin cube,

$$\begin{aligned} &\text{(a)} \quad \frac{\partial \Phi}{\partial l_j} - \lambda = \frac{p}{l_j} \qquad (j = 1, 2, 3), \\ &\text{(b)} \quad l_1 l_2 l_3 = 1, \qquad l_j > 0, \end{aligned} \tag{1.1}$$

to finding the zeros of a bifurcation problem

$$g: \mathbb{C} \times \mathbb{R} \to \mathbb{C} \tag{1.2}$$

with $\mathbf{D}_3$ symmetry. In making the reduction we assume as always that the stored energy function $\Phi(l_1, l_2, l_3)$ is invariant under the permutation group $\mathbf{S}_3$.

The reduction is performed in three stages. First, we change coordinates so that (1.1(a)) is posed on a linear subspace W rather than the nonlinear surface defined by (1.1(b)). Then we eliminate the Lagrange multiplier p, replacing (1.2(b)) by $\tilde{g} = 0$ where $\tilde{g}$: $W \times R \to W$ commutes with $\mathbf{S}_3$. Finally, we identify W with $\mathbb{C}$ so that the action of $\mathbf{S}_3$ on W is identified with the standard action of $\mathbf{D}_3$ on $\mathbb{C}$. Writing $\tilde{g}$ in these coordinates on $\mathbb{C}$ yields the mapping g of (1.2).

Throughout we use the notation

$$\varphi_j(l) = \frac{\partial \Phi}{\partial l_j}(l) \qquad (j = 1, 2, 3) \tag{1.3}$$

for the derivatives of Φ.

The first step is to convert the incompressibility constraint (1.1(b)) to a linear condition by defining $w = (w_1, w_2, w_3)$ where

$$w_j = \log l_j. \tag{1.4}$$

Then the set of w satisfying (1.1(b)) is the subspace

$$W = \{w: w_1 + w_2 + w_3 = 0\}. \tag{1.5}$$

We rewrite (2.1(a)) as

$$h(w_1, w_2, w_3, \lambda) = p \begin{bmatrix} 1 \\ 1 \\ 1 \end{bmatrix} \tag{1.6}$$

where

$$h = \begin{bmatrix} e^{w_1}(\varphi_1(e^{w_1}, e^{w_2}, e^{w_3}) - \lambda) \\ e^{w_2}(\varphi_2(e^{w_1}, e^{w_2}, e^{w_3}) - \lambda) \\ e^{w_3}(\varphi_3(e^{w_1}, e^{w_2}, e^{w_3}) - \lambda) \end{bmatrix}. \tag{1.7}$$

Since Φ is invariant under $\mathbf{S}_3$, the mapping h: $W \times \mathbb{R} \to W$ commutes with $\mathbf{S}_3$.

The second step is to show that solving (1.1) is equivalent to finding the zeros of a mapping $\tilde{g}$: $W \times \mathbb{R} \to W$ where $\tilde{g}$ commutes with the action of $\mathbf{S}_3$ on W. Let ρ: $\mathbb{R}^3 \to W$ be orthogonal projection:

$$\rho(u) = u - \tfrac{1}{3}(u_1 + u_2 + u_3)(1, 1, 1), \tag{1.8}$$

where $u = (u_1, u_2, u_3)$. Note that

$$\ker \rho = \mathbb{R}\{(1, 1, 1)\}.$$

If we define

$$\tilde{g}(w, \lambda) = \rho h(w, \lambda) \tag{1.9}$$

then $\tilde{g}(w, \lambda) = 0$ if and only if $h(w, \lambda)$ is a scalar multiple of $(1, 1, 1)$, that is, if there exists p satisfying (1.6).

The final step is to identify W with $\mathbb{C}$ so that the action of $\mathbf{S}_3$ becomes the standard action of $\mathbf{D}_3$. To do this, define T: $W \to \mathbb{C}$ by

$$T(w) = \frac{1}{2}\left[w_1 + \frac{i}{\sqrt{3}}(w_2 - w_3)\right] \tag{1.10}$$

and let

$$g(z, \lambda) = T\tilde{g}(T^{-1}z, \lambda). \tag{1.11}$$

The reader should check that

$$\begin{aligned} &\text{(a)} \quad T^{-1}(x + iy) = (2x, -x + \sqrt{3}y, -x - \sqrt{3}y) \\ &\text{(b)} \quad T(w_1, w_3, w_2) = \overline{T(w_1, w_2, w_3)} \\ &\text{(c)} \quad T(w_3, w_1, w_2) = e^{2\pi i/3}T(w_1, w_2, w_3). \end{aligned} \tag{1.12}$$

Thus T defines a linear isomorphism between W and $\mathbb{C}$ whereby the action of $\mathbf{S}_3$ becomes the standard action of $\mathbf{D}_3$.

§2. Taylor Coefficients in the Bifurcation Equation

Proposition XIV, 4.3, and Theorem XIV, 4.4, solve the recognition problem for two of the simplest bifurcation problems with $\mathbf{D}_3$ symmetry. The first singularity is stable to small perturbations, and all nontrivial branches of solutions have $\mathbf{Z}_2$ symmetry. These correspond to the rod- and platelike states of the Rivlin cube. See Chapter XV, Figure 4.1(a). However, for group-theoretic reasons alone (see XIII, §4c, and XV, §4), these nontrivial solutions must be unstable. The second singularity admits the possibility of stable rod- or platelike states; moreover, its universal unfolding admits the possibility of stable solutions with trivial isotropy, depending on the signs of certain coefficients.

In this section we show how to compute the data needed to recognize whether these singularities occur in (1.2) and to determine which branch of solutions of the more degenerate singularity is stable. We perform these calculations for a stored energy function Φ of the form

$$\Phi(l_1, l_2, l_3) = \varphi(l_1) + \varphi(l_2) + \varphi(l_3) \tag{2.1}$$

where $\varphi(l)$ is a smooth function defined near $l = 1$. The form (2.1) is special, but it includes the stored energy functions (0.2) for neo-Hookean and Mooney–Rivlin materials. There is experimental evidence (Jones and Treloar [1975]) that (2.1) is a reasonable assumption for rubber.

First we review our analysis of the simplest $\mathbf{D}_3$ bifurcation problems. By (XIV, 4.20) any $\mathbf{D}_3$-symmetric bifurcation problem must have the form

$$g(z, \lambda) = p(u, v, \lambda)z + q(u, v, \lambda)\bar{z}^2 \tag{2.2}$$

where $u = z\bar{z}$ and $v = z^3 + \bar{z}^3$. We assume that (2.2) has a singularity at $(0, 0, \lambda_0)$. Then the normal forms for the simplest $\mathbf{D}_3$ singularities are

$$\begin{aligned} &\text{(a)} \quad \varepsilon(\lambda - \lambda_0)z + \delta\bar{z}^2 \\ &\text{(b)} \quad (\varepsilon u + \delta(\lambda - \lambda_0))z + (\sigma u + mv)\bar{z}^2 \end{aligned} \tag{2.3}$$

where $\varepsilon, \delta, \sigma = \pm 1$ and $m \neq 0$ is a modal parameter.

Assuming that $p(0, 0, \lambda_0) = 0$, the recognition problems for (2.3(a, b)) are solved as follows. The $\mathbf{D}_3$ bifurcation problem g is $\mathbf{D}_3$-equivalent to (2.3(a)) if and only if

$$\varepsilon \equiv \operatorname{sgn}(p_\lambda) \neq 0, \qquad \delta \equiv \operatorname{sgn}(q) \neq 0 \tag{2.4}$$

at $(0, 0, \lambda_0)$. When the degeneracy $q(0, 0, \lambda_0) = 0$ occurs, g is $\mathbf{D}_3$-equivalent to (2.3b) if and only if

$$\begin{aligned} &\text{(a)} \quad \varepsilon \equiv \operatorname{sgn}(p_u) \neq 0, \qquad \delta \equiv \operatorname{sgn}(p_\lambda) \neq 0, \\ &\text{(b)} \quad \sigma = \delta \operatorname{sgn}(p_\lambda q_u - q_\lambda p_u) \neq 0, \\ &\text{(c)} \quad m \equiv \varepsilon \frac{p_u q_v - p_v q_u}{(p_\lambda q_u - q_\lambda p_u)^2} p_\lambda^2 \neq 0, \end{aligned} \tag{2.5}$$

where all derivatives are evaluated at $(0, 0, \lambda_0)$.

In order to use (2.4, 2.5) to see whether these $\mathbf{D}_3$-singularities occur in (1.2), we must compute the eight coefficients

$$p, p_u, p_v, p_\lambda, q, q_u, q_v, q_\lambda \tag{2.6}$$

at $(u, v, \lambda) = (0, 0, \lambda_0)$, in terms of Φ. In §3 we will choose λ_0 so that $p(0, 0, \lambda_0) = 0$.

We describe the results of these calculations in terms of the Taylor expansion of $\varphi(l)$ at $l = 1$ and that of

$$\mathscr{X}(t) = e^t \varphi'(e^t) \tag{2.7}$$

at $t = 0$. The reason for introducing $\mathscr{X}$ will become apparent. Suppose that

$$\begin{aligned} &\text{(a)}\quad \varphi(l) = \sum_n c_n(l-1)^n/n! \\ &\text{(b)}\quad \mathscr{X}(t) = \sum_n \mathscr{X}_n t^n/n!. \end{aligned} \tag{2.8}$$

By repeated differentiation of (2.7) we find that

$$\begin{aligned} &\text{(a)}\quad \mathscr{X}_0 = c_1 \\ &\text{(b)}\quad \mathscr{X}_1 = c_1 + c_2 \\ &\text{(c)}\quad \mathscr{X}_2 = c_1 + 3c_2 + c_3 \\ &\text{(d)}\quad \mathscr{X}_3 = c_1 + 7c_2 + 6c_3 + c_4 \\ &\text{(e)}\quad \mathscr{X}_4 = c_1 + 15c_2 + 25c_3 + 10c_4 + c_5 \\ &\text{(f)}\quad \mathscr{X}_5 = c_1 + 31c_2 + 90c_3 + 65c_4 + 15c_5 + c_6. \end{aligned} \tag{2.9}$$

We will show that the eight coefficients (2.6) take on very simple forms in terms of the $\mathscr{X}_n$, namely:

$$\begin{aligned} &\text{(a)}\quad p = \mathscr{X}_1 - \lambda_0 = c_1 + c_2 - \lambda_0 \\ &\text{(b)}\quad q = (\mathscr{X}_2 - \lambda_0)/2 \\ &\text{(c)}\quad p_u = (\mathscr{X}_3 - \lambda_0)/2 \\ &\text{(d)}\quad p_v = (\mathscr{X}_4 - \lambda_0)/24 \\ &\text{(e)}\quad p_\lambda = -1 \\ &\text{(f)}\quad q_u = (\mathscr{X}_4 - \lambda_0)/8 \\ &\text{(g)}\quad q_v = (\mathscr{X}_5 - \lambda_0)/120 \\ &\text{(h)}\quad q_\lambda = -\tfrac{1}{2} \end{aligned} \tag{2.10}$$

evaluated at $(u, v, \lambda) = (0, 0, \lambda_0)$. We establish these formulas in the rest of this section; their derivation may be omitted on a first reading.

We begin by expanding (2.2) to degree 5 in z, obtaining

$$\begin{aligned} g \equiv\ & (p + p_u u + p_v v + p_{uu}u^2/2 + p_\lambda(\lambda - \lambda_0))z \\ & + (q + q_u u + q_v v + q_\lambda(\lambda - \lambda_0))\bar{z}^2 \\ & + O(|z|^3, |z|^2|\lambda - \lambda_0|, |z||\lambda - \lambda_0|^2). \end{aligned} \tag{2.11}$$

To compute the coefficients (2.10) we evaluate $g(x, 0)$ and $g(iy, \lambda)$ for real variables x, y. Recall that $u = z\bar{z}$ and $v = z^3 + \bar{z}^3$, and compute

$$\begin{aligned} &\text{(a)}\quad g(x, 0) = px + qx^2 + p_u x^3 + (2p_v + q_u)x^4 + (\tfrac{1}{2}p_{uu} + 2q_v)x^5 + \cdots \\ &\text{(b)}\quad g(iy, \lambda) = i(py + p_u y^3 + \tfrac{1}{2}p_{uu}y^5 + p_\lambda(\lambda - \lambda_0)y) \\ &\qquad - (qy^2 + q_u y^4 + q_\lambda(\lambda - \lambda_0)y^2) + \cdots. \end{aligned} \tag{2.12}$$

By (2.12) the computation of (2.10) depends only on finding $g(iy, \lambda)$ and the fourth and fifth order terms in $g(x, 0)$.

We find g explicitly from (1.9 and 1.11):

$$g(z, \lambda) = T\rho h(T^{-1}z, \lambda). \tag{2.13}$$

Now h and ρ are defined in (1.7 and 1.8), and $l_j = e^{w_j}$ by (1.4), so

$$\rho h(l_1, l_2, l_3, \lambda) = \begin{bmatrix} l_1\varphi'(l_1) - \lambda v_1 - K \\ l_2\varphi'(l_2) - \lambda v_2 - K \\ l_3\varphi'(l_3) - \lambda v_3 - K \end{bmatrix}$$

where

$$K = \tfrac{1}{3}[l_1\varphi'(l_1) + l_2\varphi'(l_2) + l_3\varphi'(l_3) - \lambda(l_1 + l_2 + l_3)].$$

By the definition (1.10) of T we have

$$T\rho h = X + iY$$

where

$$\begin{aligned} &\text{(a)} \quad X = \frac{1}{6}[2l_1\varphi'(l_1) - l_2\varphi'(l_2) - l_3\varphi'(l_3) - \lambda(2l_1 - l_2 - l_3)] \\ &\text{(b)} \quad Y = \frac{1}{2\sqrt{3}}[l_2\varphi'(l_2) - l_3\varphi'(l_3) - \lambda(l_2 - l_3)]. \end{aligned} \tag{2.14}$$

By (1.12(a))

$$T^{-1}(x) = (2x, -x, -x), \qquad T^{-1}(iy) = (0, \sqrt{3}y, -\sqrt{3}y).$$

Thus, to compute $g(x, 0)$ and $g(iy, 0)$ we must evaluate (2.14) at

$$\begin{aligned} &\text{(a)} \quad (l_1, l_2, l_3) = (e^{2x}, e^{-x}, e^{-x}) \\ &\text{(b)} \quad (l_1, l_2, l_3) = (1, e^{\sqrt{3}y}, e^{-\sqrt{3}y}), \end{aligned} \tag{2.15}$$

respectively.

Therefore,

$$\begin{aligned} &\text{(a)} \quad g(iy, \lambda) = \frac{1}{6}[2\varphi'(1) - (\mathscr{X}(\sqrt{3}y) + \mathscr{X}(-\sqrt{3}y) - ((\lambda - \lambda_0) + \lambda_0) \\ &\qquad \times (2 - (e^{\sqrt{3}y} + e^{-\sqrt{3}y})) + \frac{i}{2\sqrt{3}}[\mathscr{X}(\sqrt{3}y) - \mathscr{X}(-\sqrt{3}y) \\ &\qquad - ((\lambda - \lambda_0) + \lambda_0)(e^{\sqrt{3}y} - e^{-\sqrt{3}y})] \\ &\text{(b)} \quad g(x, \lambda) = \tfrac{1}{3}[\mathscr{X}(2x) - \mathscr{X}(-x) - ((\lambda - \lambda_0) + \lambda_0)(e^{2x} - e^{-x})] \end{aligned} \tag{2.16}$$

where $\mathscr{X}(t) = e^t\varphi(e^t)$ as in (2.7).

Next we evaluate (2.16(a)) in terms of the Taylor coefficients $\mathscr{X}_j$:

$$\begin{aligned} g(iy, \lambda) = &-[(\mathscr{X}_2 - \lambda_0)y^2/2 + \mathscr{X}_4 - \lambda_0)y^4/8 - (\lambda - \lambda_0)y^2/2] \\ &+ i[(\mathscr{X}_1 - \lambda_0)y + (\mathscr{X}_3 - \lambda_0)y^3/2 + 9(\mathscr{X}_5 - \lambda_0)y^5/120 \\ &- (\lambda - \lambda_0)y] + \cdots. \end{aligned} \tag{2.17}$$

Equating coefficients in (2.12(b) and 2.17) we get the formulas for p, p_λ, p_u, q, q_λ, q_u in (2.10). Further,

$$p_{uu} = \tfrac{3}{20}[\mathscr{X}_5 - \lambda_0]. \tag{2.18}$$

Finally, we evaluate the fourth and fifth order coefficients in (2.16(b)) and compare with (2.12(a)) to obtain

$$\begin{aligned} &\text{(a)}\quad 2p_v + q_u = \tfrac{5}{24}[\mathscr{X}_4 - \lambda_0] \\ &\text{(b)}\quad \tfrac{1}{2}p_{uu} + 2q_v = \tfrac{11}{120}[\mathscr{X}_5 - \lambda_0]. \end{aligned} \tag{2.19}$$

From (2.18 and 2.10(f)) we obtain (2.10(d, g)). This completes the calculation.

§3. Bifurcations of the Rivlin Cube

Now we apply the preceding results to obtain bifurcations occuring in the Rivlin cube problem for Mooney–Rivlin material. In bifurcation problems with $\mathbf{D}_3$ symmetry a necessary condition for bifurcation to occur at $\lambda = \lambda_0$ is

$$p(0, 0, \lambda_0) = 0.$$

By (2.10(a))

$$p(0, 0, \lambda_0) = c_1 + c_2 - \lambda_0.$$

Hence a singularity along the trivial solution occurs when

$$\lambda_0 = c_1 + c_2. \tag{3.1}$$

The simplest $\mathbf{D}_3$ singularity occurs when q and p_λ are both nonzero at λ_0. By (2.10(e)) p_λ is always nonzero, so the nondegeneracy condition is $q \neq 0$. By (2.9(c), 2.10(b) and 3.1) this condition is $2c_2 + c_3 \neq 0$.

The more complicated $\mathbf{D}_3$ singularity occurs when

$$2c_2 + c_3 = 0, \tag{3.2}$$

in which case we must compute the six remaining coefficients in (2.10). We do this, assuming (3.1 and 3.2) hold, obtaining:

$$\begin{aligned} &\text{(a)}\quad p_u = (3c_3 + c_4)/2 \\ &\text{(b)}\quad p_v = (18c_3 + 10c_4 + c_5)/24 \\ &\text{(c)}\quad p_\lambda = -1 \\ &\text{(d)}\quad q_u = (18c_3 + 10c_4 + c_5)/8 \\ &\text{(e)}\quad q_v = (75c_3 + 65c_4 + 15c_5 + c_6)/120 \\ &\text{(f)}\quad q_\lambda = -\tfrac{1}{2}. \end{aligned} \tag{3.3}$$

To determine the normal form we compute the data in (2.5). We have

$$\begin{aligned} &\text{(a)} \quad \varepsilon = \operatorname{sgn}(3c_3 + c_4) \\ &\text{(b)} \quad \delta = -1 \\ &\text{(c)} \quad \sigma = -\operatorname{sgn}(18c_3 + 10c_4 + c_5) \\ &\text{(d)} \quad \operatorname{sgn}(m) = \varepsilon \operatorname{sgn}(p_u q_v - p_v q_u) \end{aligned} \tag{3.4}$$

where (d) may be evaluated by using (3.3).

We can now calculate the bifurcations for Mooney–Rivlin material. The stored energy function is given by

$$\varphi(l) = \tfrac{1}{2}(\mu l^2 + \nu l^{-2}) \tag{3.5}$$

where μ, $\nu > 0$. When $\nu = 0$ we have the stored energy function for neo-Hookean material. We determine the c_j from (2.8(a)) by using the Taylor expansion of $\varphi(l)$ at $l = 1$. The result is:

$$\begin{aligned} &\text{(a)} \quad c_1 = \mu - \nu \\ &\text{(b)} \quad c_2 = \mu + 3\nu \\ &\text{(c)} \quad c_3 = -12\nu \\ &\text{(d)} \quad c_4 = 60\nu \\ &\text{(e)} \quad c_5 = -360\nu \\ &\text{(f)} \quad c_6 = 2520\nu. \end{aligned} \tag{3.6}$$

From (3.1) bifurcation from the undeformed cubic state occurs when

$$\lambda_0 = 2(\mu + \nu). \tag{3.7}$$

The associated singularity is the simplest normal form when

$$0 \neq 2c_2 + c_3 = 2\mu - 6\nu. \tag{3.8}$$

For neo-Hookean material $2\mu - 6\nu = 2\mu > 0$ so only this simplest singularity occurs (locally). Hence, the local bifurcation analysis given here does not lead to the existence of nontrivial stable equilibria. For Mooney–Rivlin material, however, the more degenerate singularity occurs when

$$\mu = 3\nu. \tag{3.9}$$

By (3.7), at such a point,

$$\lambda_0 = 4\nu. \tag{3.10}$$

The data (3.4) are

$$\begin{aligned} &\text{(a)} \quad \varepsilon = \operatorname{sgn}(\nu) = 1 \\ &\text{(b)} \quad \delta = -1 \\ &\text{(c)} \quad \sigma = -\operatorname{sgn}(24\nu) = -1 \\ &\text{(d)} \quad \operatorname{sgn}(m) = 1. \end{aligned} \tag{3.11}$$

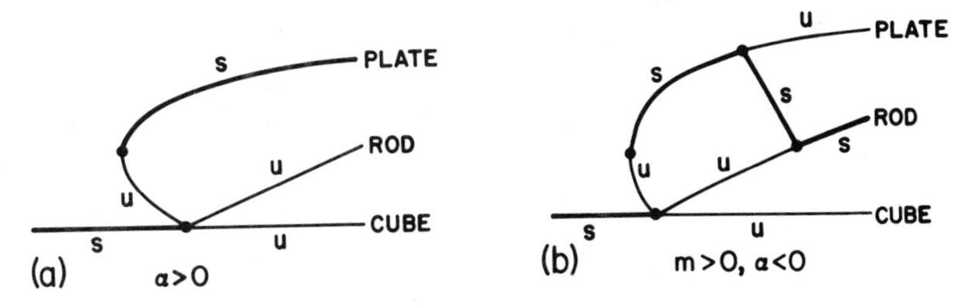

Figure 3.1. Bifurcation diagrams in the unfolding (3.13) when $m > 0$.

By (2.3 and 3.11) the normal form for the degenerate $\mathbf{D}_3$-equivariant singularity occurring for Mooney–Rivlin material is

$$(u - (\lambda - \lambda_0))z + (-u + mv)\bar{z}^2 \tag{3.12}$$

where $m > 0$.

The universal unfolding for (3.12),

$$H(z, \lambda, \alpha, \tilde{m}) = (u - (\lambda - \lambda_0))z + (-u + \tilde{m}v + \alpha)\bar{z}^2, \tag{3.13}$$

was discussed in XV, §4, and the bifurcation diagrams were computed. We recall the results in Figure 3.1.

The only difference between Figure 3.1 and Figure 4.2 of Chapter XV is the explicit labeling of the $\mathbf{Z}_2$ solutions as rod- and platelike. These branches are identified as follows. In XV, §4, we specified $\mathbf{Z}_2$ to be the group generated by $z \to \bar{z}$. Thus $\text{Fix}(\mathbf{Z}_2) = \{z : z = x \in \mathbb{R}\}$. Using T^{-1} as defined in (1.12(a)) we identify x with

$$T^{-1}(x) = (2x, -x - x) \tag{3.14}$$

in W. Finally, points in W are identified with deformations of the cube, by exponentiation. Thus (3.14) corresponds to a deformed cube with sides (e^{2x}, e^{-x}, e^{-x}). Thus solutions with $x > 0$ are rodlike, and those with $x < 0$ are platelike.

Recall that the universal unfolding H in (3.13), when restricted to $\text{Fix}(\mathbf{Z}_2)$, has the form

$$H(x, \lambda) = (x^2 - x^3 + 2mx^4 - (\lambda - \lambda_0) + \alpha)x \tag{3.15}$$

and the eigenvalues of dH along solutions $H(x, \lambda) = 0$ are

$$\begin{aligned} &\text{(a)} \quad 2x^2 - 3x^3 + 4mx^4 + \alpha x \\ &\text{(b)} \quad 3(x^3 - mx^4 - \alpha x). \end{aligned} \tag{3.16}$$

See (XV, 4.22(b)). When $\alpha = 0$, (3.16(a)) is always positive, whereas (3.16(b)) is positive only when $x > 0$. The two branches in Figure 3.1(b) with $\mathbf{Z}_2$ symmetry correspond to solutions of $H(x, \lambda) = 0$ with $x > 0$, $x < 0$, respectively. Thus the branch with $x > 0$ (rodlike solutions) is stable. When $\alpha \neq 0$ the types of solutions are identified in a similar way.

CHAPTER XVI

Symmetry-Breaking in Hopf Bifurcation

§0. Introduction

Until now the theory developed has applied largely to steady-state bifurcation, the exception being the study of degenerate Hopf bifurcation without symmetry in Chapter VIII. There a dynamic phenomenon—the occurrence of periodic trajectories—was *reduced* to a problem in singularity theory by applying the Liapunov–Schmidt procedure. In the remainder of this volume we will show that this is a far-reaching idea and that dynamic phenomena in many different contexts can be studied by similar methods.

In this chapter we generalize the Liapunov–Schmidt treatment of *nondegenerate* Hopf bifurcation to the case where the vector field possesses symmetry. Linear and nonlinear degeneracies will be studied in later chapters, but only in special cases (trivial or $\mathbf{O}(2)$ symmetry) and by special methods. At the time of writing a comprehensive theory of degenerate Hopf bifurcation with symmetry remains to be developed, but some preliminary work is under way.

The chapter divides into four main parts, each with its own theme. These are:

(a) Existence of periodic solutions (§§1–4),
(b) Stability of solutions in Birkhoff normal form (§§5–6),
(c) General group-theoretic results (§§7–9),
(d) Removal of the assumption of Birkhoff normal form from stability calculations, and related theoretical fine points (§§10–11).

We now sketch the flow of ideas in each of these parts.

(a) Existence of Periodic Solutions (§§1–4)

We will say that an ODE

$$\dot{v} + f(v, \lambda) = 0, \qquad f(0, 0) \equiv 0$$

undergoes a *Hopf bifurcation* at $\lambda = 0$ if $(df)_{0,0}$ has a purely imaginary eigenvalue. Under additional hypotheses of nondegeneracy, this condition implies the occurrence of a branch of periodic solutions. The fundamental nondegeneracy hypothesis in the standard Hopf theorem is that the eigenvalue be simple. However, as in static bifurcation, symmetry can force multiple eigenvalues, so the standard Hopf theorem does not apply directly.

In §1 we investigate the occurrence of imaginary eigenvalues in equations with symmetry group Γ and show that there are nontrivial restrictions on the corresponding imaginary eigenspace. Specifically, it must contain a Γ-simple invariant subspace, where a subspace X is Γ-simple if it is either nonabsolutely irreducible or isomorphic to the direct sum of two copies of the same absolutely irreducible representation. Moreover, generically the imaginary eigenspace itself is Γ-simple. Thus for purely imaginary eigenvalues, "Γ-simple" is the equivariant analogue of "simple eigenvalue."

As motivation for our main result we show in §2 that if Σ is a subgroup of Γ with $\dim \mathrm{Fix}(\Sigma) = 2$ then, subject to conditions analogous to those of the standard Hopf theorem, there exists a branch of periodic solutions with isotropy subgroup containing Σ. The proof is easy: restrict the vector field f to $\mathrm{Fix}(\Sigma)$ and apply the standard Hopf theorem. However, this is a rather weak result: for example, in the oscillating hosepipe (XI, §1) it captures the standing wave but not the rotating wave. This suggests that we must take *temporal phase-shift symmetries* into account as well as spatial (Γ) symmetries. In §3 we describe the introduction of phase-shift symmetries in terms of actions of the circle group $\mathbf{S}^1$, much as in Chapter VIII. The main existence result, the equivariant Hopf theorem, is proved in §4 and generalizes §2 to subgroups Σ of $\Gamma \times \mathbf{S}^1$ such that $\dim \mathrm{Fix}(\Sigma) = 2$. The underlying philosophy for the remainder of this chapter is that Hopf bifurcation for Γ-equivariant systems reduces to static bifurcation for $\Gamma \times \mathbf{S}^1$-equivariant systems. However, this should be considered as a guiding analogy rather than a rigid rule.

(b) Stability of Solutions in Birkhoff Normal Form (§§5–6)

After existence, the next basic problem is stability. Our main technical tool for calculating stabilities is Birkhoff normal form, and we discuss this in §5. The idea is that by making suitable coordinate changes, then—up to any given order k—the vector field f can be made to commute not only with Γ but with a group $\mathbf{S}$ determined by the linearization $(df)_{0,0}$. In Hopf bifurcation $\mathbf{S}$ is the circle group $\mathbf{S}^1$. This effectively introduces extra symmetry, which can be exploited. The approach here is due to Elphick et al. [1986].

In §6 we describe the application of Floquet theory to the solutions whose existence follows from the equivariant Hopf theorem. In particular we show that for such solutions the Floquet operator M has at least $d_\Sigma = \dim \Gamma + 1 - \dim \Sigma$ eigenvalues forced to zero by symmetry. We also show that if f is in Birkhoff normal form to all orders—that is, if we ignore the order k "tail" in the Birkhoff normal form procedure—then the Floquet equation can be solved explicitly, and the Floquet exponents are determined by the eigenvalues of df.

(c) General Group-Theoretic Results (§§7–9)

In order to apply the results to specific symmetry groups, we need methods for finding isotropy subgroups Σ of $\Gamma \times \mathbf{S}^1$ with two-dimensional fixed-point subspaces. In §7 we show that every such Σ can be written as a "twisted" subgroup $H^\theta = \{(h, \theta(h)\}$ where $\theta \colon H \to \mathbf{S}^1$ is a group homomorphism and H is a subgroup of Γ. This lets us classify potential isotropy subgroups according to the subgroups of the spatial part Γ alone.

In §8 we prove two formulas for the dimensions of $\mathrm{Fix}(\Sigma)$. One is a trace formula; the other is more overtly representation-theoretic. The results of §§7 and 8 will be used in later sections.

In §9 we briefly consider the relationship between the invariant theory of Γ and that of $\Gamma \times \mathbf{S}^1$, and we explain a general procedure that will be used to find invariants and equivariants for $\Gamma \times \mathbf{S}^1$, needed in later chapters.

(d) Removal of the Assumption of Birkhoff Normal Form (§§10–11)

The main aim of the final part of the chapter is to show that the stability conditions obtained by the methods of §6 remain true when f, rather than being in Birkhoff normal form to all orders, is only in Birkhoff normal form up to an explicitly determined order k. By §5 every Γ-equivariant vector field, at a Hopf bifurcation point, can be put into Birkhoff normal form up to any assigned order k, by a change of coordinates, so in principle the results apply to arbitrary Γ-equivariant vector fields.

We begin in §10 by making explicit the relation between Birkhoff normal form and Liapunov–Schmidt reduction. In §11 we apply a "perturbation" approach to establish the aforementioned result, that if the Birkhoff normal form is truncated at an appropriate order k, then the stability results remain valid.

§1. Conditions for Imaginary Eigenvalues

Suppose we have a system of ODEs

$$\dot{v} + f(v, \lambda) = 0 \tag{1.1}$$

where $v \in \mathbb{R}^n$, $\lambda \in \mathbb{R}$ is a bifurcation parameter, and $f: \mathbb{R}^n \times \mathbb{R} \to \mathbb{R}^n$ is a smooth (C^∞) mapping commuting with the action of a compact Lie group Γ on $\mathbb{R}^n$. That is,

$$f(\gamma v, \lambda) = \gamma f(v, \lambda) \qquad \text{for all } \gamma \in \Gamma.$$

Further assume that

$$f(0, \lambda) \equiv 0, \tag{1.2}$$

so there is a trivial Γ-invariant equilibrium solution $v = 0$.

Let $(df)_{v,\lambda}$ be the $n \times n$ Jacobian matrix of derivatives of f with respect to the variables v_j, evaluated at (v, λ). The most important hypothesis of the standard Hopf theorem (cf. Theorem VIII, 3.1) is that $(df)_{0,0}$ should have a pair of *simple* purely imaginary eigenvalues (cf. VIII, 1.2). In the presence of a symmetry group Γ, it is not always possible to arrange for eigenvalues of df to be purely imaginary. For example, if Γ acts absolutely irreducibly on $\mathbb{R}^n$, then $(df)_{0,\lambda}$ is a multiple of the identity and hence has all eigenvalues real. In certain circumstances, however, df can have purely imaginary eigenvalues. The group Γ often forces these eigenvalues to be multiple, so the standard Hopf theorem cannot be applied directly. Although the symmetries complicate the analysis by forcing multiple eigenvalues, they also potentially simplify it by placing restrictions on the form of the mapping f. We have already seen this in static bifurcation theory, and we exploit such a phenomenon here for Hopf-type bifurcation to periodic solutions.

In this section we do three things:

1. We find conditions on the action of Γ that allow $(df)_{0,0}$ to have purely imaginary eigenvalues; namely, there must be a Γ-invariant subspace of $\mathbb{R}^n$ that is either:

   (a) of the form $V \oplus V$ where V is absolutely irreducible,

   or (1.3)

   (b) irreducible but not absolutely irreducible.

   (Recall from XII, §3, that a representation of Γ is *absolutely irreducible* if the only linear maps commuting with Γ are real multiples of the identity. Henceforth we use the phrase *non–absolutely irreducible* to mean (b).)
2. We show that generically the eigenspace corresponding to the purely imaginary eigenvalues is of one of the two forms in (1.3). The latter result is the analogue for Hopf bifurcation of Proposition XIII, 3.4, on static

bifurcation, where the zero eigenspace is generically absolutely irreducible. The proof is similar.

3. We obtain a simple "normal form" for $(df)_{0,0}$ in the generic case, showing that we may assume that it has the block form

$$(df)_{0,0} = \begin{bmatrix} 0 & -I \\ I & 0 \end{bmatrix}.$$

We begin with some examples that motivate the dichotomy expressed in (1.3).

EXAMPLES 1.1.

(a) $\Gamma = \mathbf{O}(2)$. The natural action of $\mathbf{O}(2)$ on $\mathbb{R}^2 \equiv \mathbb{C}$ is absolutely irreducible. Hence the commuting matrices are of the form

$$\begin{bmatrix} a & 0 \\ 0 & a \end{bmatrix}$$

($a \in \mathbb{R}$) with *real* eigenvalues a (twice). Hence for a 2×2 ODE $\dot{x} + f(x, \lambda) = 0$, which commutes with the standard action of $\mathbf{O}(2)$, Hopf bifurcation cannot occur.

Suppose, however, that $\mathbf{O}(2)$ acts on $\mathbb{R}^4 = \mathbb{C}^2$ by the diagonal action

$$\gamma(z_1, z_2) = (\gamma z_1, \gamma z_2).$$

Then the commuting linear mappings are of the form

$$\begin{bmatrix} aI_2 & bI_2 \\ cI_2 & dI_2 \end{bmatrix}$$

where $a, b, c, d \in \mathbb{R}$ and I_2 is the 2×2 identity matrix. The eigenvalues of this matrix are those of the 2×2 matrix $\left[\begin{smallmatrix} a & b \\ c & d \end{smallmatrix}\right]$, repeated twice. If we take $a = d = 0$, $b = -1$, $c = 1$, we obtain eigenvalues $\pm i$ (twice). Hence for an ODE posed in this "doubled" representation, $\mathbf{O}(2)$ Hopf bifurcation is possible. Note that we are forced to have *multiple* eigenvalues $\pm i$ in this case.

If instead we form the direct sum of two nonisomorphic irreducible representations of $\mathbf{O}(2)$ (each being absolutely irreducible), then Theorem XII, 3.5, implies that the commuting linear mappings are of the form

$$\begin{bmatrix} aI_2 & 0 \\ 0 & bI_2 \end{bmatrix}$$

with *real* eigenvalues a, b (twice). Again no Hopf bifurcation can occur.

(b) $\Gamma = \mathbf{SO}(2)$. Consider the natural representation on $\mathbb{R}^2 \equiv \mathbb{C}$, given by

$$\theta \cdot z = e^{i\theta} z.$$

This is non-absolutely irreducible. The action of the rotation $\pi/2 \in \mathbf{SO}(2)$ induces the mapping

$$\begin{bmatrix} 0 & -1 \\ 1 & 0 \end{bmatrix}$$

with eigenvalues $\pm i$, and this mapping commutes with $\mathbf{SO}(2)$ since $\mathbf{SO}(2)$ is abelian. Thus Hopf bifurcation may occur for a 2×2 ODE with $\mathbf{SO}(2)$ symmetry.

The preceding examples are simple archetypes for the two basic cases (1.3(a, b)) and should be borne in mind throughout this chapter. We now embark upon the analysis of the general case by considering an arbitrary linear mapping L commuting with the group Γ. Note that any such L can be realized in the form $(df)_{0,0}$ for a Γ-equivariant f—for example, $f(x, \lambda) = Lx$. Thus Hopf bifurcation can occur only when some commuting L has purely imaginary eigenvalues. By Theorem XII, 2.2, we may decompose $\mathbb{R}^n$ into a direct sum of irreducible Γ-invariant subspaces

$$\mathbb{R}^n = V_1 \oplus \cdots \oplus V_k. \tag{1.4}$$

Lemma 1.2. *Let L: $\mathbb{R}^n \to \mathbb{R}^n$ be a linear map having a nonreal eigenvalue and commuting with Γ. Then either*

(a) *Some absolutely irreducible representation of Γ occurs at least twice (up to Γ-isomorphism) in the decomposition* (1.4), *or* (1.5)

(b) *The action of Γ on some V_j is not absolutely irreducible.*

Proof. If not, all the V_j are absolutely irreducible and nonisomorphic. Theorem XII, 3.5, implies that $L(V_j) \subset V_j$ for all j, and by absolute irreducibility $L|V_j = \mu_j I$ where $\mu_j \in \mathbb{R}$. Hence the eigenvalues of L are just the μ_j, and these are real, contrary to assumption. □

The preceding result motivates the following definition:

Definition 1.3. A representation W of Γ is Γ-*simple* i' either

(a) $W \cong V \oplus V$ where V is absolutely irreducible for Γ, or
(b) W is non–absolutely irreducible for Γ.

We also require some notation for eigenspaces. Suppose that L: $\mathbb{R}^n \to \mathbb{R}^n$ is a linear map and $\mu \in \mathbb{C}$. We define the (real) *eigenspace* E_μ and *generalized eigenspace* G_μ of L as follows:

$$E_\mu = \begin{cases} \{x \in \mathbb{R}^n \colon (L - \mu I)x = 0\} & (\mu \in \mathbb{R}) \\ \{x \in \mathbb{R}^n \colon (L - \mu I)(L - \bar{\mu} I)x = 0\} & (\mu \notin \mathbb{R}) \end{cases}$$

$$G_\mu = \begin{cases} \{x \in \mathbb{R}^n \colon (L - \mu I)^n x = 0\} & (\mu \in \mathbb{R}) \\ \{x \in \mathbb{R}^n \colon (L - \mu I)^n (L - \bar{\mu} I)^n x = 0\} & (\mu \notin \mathbb{R}). \end{cases}$$

We also define the *imaginary eigenspace* of L to be the sum of all E_μ for which μ is purely imaginary.

In studying the bifurcation problem (1.1) we also wish to consider how the eigenvalues of $(df)_{0,\lambda}$ cross the imaginary axis at $\lambda = 0$ and to describe the structure of the associated eigenspace. (We shall see in §3 that the Liapunov–Schmidt technique reduces the problem to one posed on this eigenspace.) We may restate Lemma 1.2 as follows: if L has a purely imaginary eigenvalue, then $\mathbb{R}^n$ must contain a Γ-simple invariant subspace. Further, the proof shows that this subspace must lie in the imaginary eigenspace of L. The main observation is the following analogue of Proposition XIII, 3.4, which shows that in the generic situation the imaginary eigenspace is itself Γ-simple:

Proposition 1.4. *Let $f: \mathbb{R}^n \times \mathbb{R} \to \mathbb{R}^n$ be a Γ-equivariant bifurcation problem. Suppose that $(df)_{0,0}$ has purely imaginary eigenvalues $\pm i\omega$. Let $G_{i\omega}$ be the corresponding real generalized eigenspace of $(df)_{0,0}$. Then generically $G_{i\omega}$ is Γ-simple. Moreover, $G_{i\omega} = E_{i\omega}$.*

PROOF. We follow the first part of the proof of Proposition XIII, 3.4, taking the requirement on the eigenvalues of $(df)_{0,0}$ into account. Decompose $\mathbb{R}^n$ as the sum of $G_{i\omega}$ and all remaining generalized eigenspaces:

$$\mathbb{R}^n = G_{i\omega} \oplus G_{\mu_1} \oplus \cdots \oplus G_{\mu_r}.$$

By Lemma 1.2, $G_{i\omega}$ contains a Γ-invariant subspace U of the form (a) or (b). If $G_{i\omega}$ is not of the desired form, then there is a nonzero Γ-invariant complement W to U, so that $G_{i\omega} = U \oplus W$. Define $M: \mathbb{R}^n \to \mathbb{R}^n$ to be the unique linear mapping such that

$$\begin{aligned} M|U &= 0 \\ M|G_{\mu_i} &= 0 \qquad (i = 1, \ldots, r) \\ M|W &= I_W. \end{aligned}$$

Let $\varepsilon \in \mathbb{R}$ and consider the perturbation $f_\varepsilon(x, \lambda) = f(x, \lambda) + \varepsilon Mx$. The eigenvalues of $(df_\varepsilon)_{0,0}$ are $\pm i\omega$ on U, $-\varepsilon \pm i\omega$ on W, and μ_i on G_{μ_i}. Hence the $i\omega$ generalized eigenspace of $(df_\varepsilon)_{0,0}$ is U. Therefore, generically we have $G_{i\omega}$ of the form (a) or (b). By considering possible Γ-invariant subspaces it is easy to see that $G_{i\omega}$, considered as a complex vector space, is spanned by (complex) eigenvectors. Therefore, $G_{i\omega}$ is the real part of the $i\omega$ eigenspace. □

The import of the final statement of the theorem is that generically $(df)_{0,0}$ on $G_{i\omega}$ is semisimple; that is, its real Jordan normal form is block-diagonal with 2×2 blocks

$$\begin{bmatrix} 0 & -\omega \\ \omega & 0 \end{bmatrix}.$$

Remark. In the abstract discussion of symmetric Hopf bifurcation we may, without any real loss of generality, assume that *all* the eigenvalues of $(df)_{0,0}$ are on the imaginary axis. This assumption may be justified either by the center manifold theorem or by Liapunov–Schmidt reduction, as discussed at the end of §3. Moreover, since we are focusing here on Hopf bifurcations, we assume that 0 is not an eigenvalue of $(df)_{0,0}$. The proof of Proposition 1.4 can easily be adapted to show that generically $(df)_{0,0}$ has only one pair of complex conjugate eigenvalues on the imaginary axis, perhaps of high multiplicity, so this is a generic assumption. More general situations are dealt with in Chapters XIX and XX on mode interactions, but only for very simple symmetries.

We end this section by showing that we may assume $(df)_{0,0} = J$, where

$$J = \begin{bmatrix} 0 & -I_m \\ I_m & 0 \end{bmatrix}, \tag{1.6}$$

where $m = n/2$. More precisely, we prove the following:

Lemma 1.5. *Assume that $\mathbb{R}^n$ is Γ-simple, f is Γ-equivariant, and $(df)_{0,0}$ has i as an eigenvalue. Then*
(a) *The eigenvalues of $(df)_{0,\lambda}$ consist of a complex conjugate pair $\sigma(\lambda) \pm i\rho(\lambda)$, each of multiplicity m. Moreover, σ and ρ are smooth functions of λ.*
(b) *There is an invertible linear map $S\colon \mathbb{R}^n \to \mathbb{R}^n$, commuting with Γ, such that*

$$(df)_{0,0} = SJS^{-1}.$$

Proof. We first prove this lemma under the assumption that $\mathbb{R}^n = V \oplus V$ with V absolutely irreducible, as in Definition 1.3(a). Then we consider the case of Definition 1.3(b), where $\mathbb{R}^n$ is non-absolutely irreducible.

Let L be a linear map $V \oplus V \to V \oplus V$, commuting with Γ. Write L in block form as

$$L = \begin{bmatrix} A & B \\ C & D \end{bmatrix}$$

for $m \times m$ matrices A, B, C, D. Since L commutes with the diagonal action of Γ, each of A, B, C, D commutes with the action of Γ on V. By absolute irreducibility we have

$$L = \begin{bmatrix} aI & bI \\ cI & dI \end{bmatrix}. \tag{1.7}$$

If A, B, C, D commute (as here) then

$$\det\begin{bmatrix} A & B \\ C & D \end{bmatrix} = \det(AD - BC);$$

see Halmos [1974], p. 102, ex. 9. Therefore, the characteristic polynomial of L is

$$\det(L - \mu I) = [(a - \mu)(d - \mu) - bc]^m. \tag{1.8}$$

Thus each eigenvalue of L occurs with multiplicity at least m. Now $(df)_{0,0}$ commutes with Γ and has a pair of (nonzero) complex conjugate purely imaginary eigenvalues. Therefore, each occurs with multiplicity m. The smoothness of σ and ρ also follows from (1.8).

$$\text{Let } (df)_{0,0} = \begin{bmatrix} aI & bI \\ cI & dI \end{bmatrix}.$$

For i to be an eigenvalue, (1.8) implies that $a + d = 0$, $ad - bc = 1$. We now conjugate $(df)_{0,0}$ so that $a = d = 0$. Assuming $a \neq 0$, define

$$R_\theta = \begin{bmatrix} \cos\theta\, I & -\sin\theta\, I \\ \sin\theta\, I & \cos\theta\, I \end{bmatrix},$$

which commutes with Γ. Choose θ so that $\cot(2\theta) = (b + c)/2a$. Then

$$R_\theta (df)_{0,0} R_\theta^{-1} = \begin{bmatrix} 0 & hI \\ -h^{-1}I & 0 \end{bmatrix}$$

for $h \in \mathbb{R}$. Finally note that

$$J = \begin{bmatrix} I & 0 \\ 0 & -hI \end{bmatrix} \begin{bmatrix} 0 & hI \\ -h^{-1}I & 0 \end{bmatrix} \begin{bmatrix} I & 0 \\ 0 & -h^{-1}I \end{bmatrix}.$$

Thus $S = \left[\begin{smallmatrix} I & 0 \\ 0 & -hI \end{smallmatrix}\right]$ provides the required similarity.

The second possibility is that $\mathbb{R}^n$ is non–absolutely irreducible. By XII, §3, the set of commuting matrices $\mathscr{D}$ is isomorphic either to $\mathbb{C}$ or to $\mathbb{H}$. Moreover, the action of $\mathscr{D}$ turns $\mathbb{R}^n$ into a vector space over $\mathscr{D}$, so we may assume $\mathbb{R}^n \cong \mathscr{D}^k$ (where $k = n/2$ if $\mathscr{D} \cong \mathbb{C}$ and $k = n/4$ if $\mathscr{D} \cong \mathbb{H}$). Further, $d \in \mathscr{D}$ acts on $\mathscr{D}^k$ by coordinate multiplication; that is, $d(d_1, \ldots, d_k) = (dd_1, \ldots, dd_k)$. The eigenvalues of d acting on $\mathscr{D}^k$ are those of d acting on $\mathscr{D}$, repeated k times. The action of Γ is $\mathscr{D}$-linear since $\mathscr{D}$ commutes with Γ.

Since $(df)_{0,\lambda}$ commutes with Γ we may identify it with some $d(\lambda) \in \mathscr{D}$. Moreover, $d(\lambda)$ varies smoothly with λ (since $(df)_{0,\lambda}$ does).

First suppose $\mathscr{D} \cong \mathbb{C}$. Then the eigenvalues of d acting on $\mathbb{C}$ are d and $\bar{d}$, proving part (a) of the lemma in this case. It also follows that $d(0) = \pm i$, and it is easy to choose a basis on $\mathbb{C}^n$ for which the matrix of i is J, proving part (b).

Similarly, suppose $\mathscr{D} = \mathbb{H}$. Then the eigenvalues of $d = \alpha + \beta i + \gamma j + \delta k$ acting on $\mathbb{H}$ are $\alpha \pm i\sqrt{\beta^2 + \gamma^2 + \delta^2}$ repeated twice, proving part (a). It also follows that $d(0) = \beta i + \gamma j + \delta k$ where $\beta^2 + \gamma^2 + \delta^2 = 1$. There exists a unit quaternion q such that $qd(0)q^{-1} = i$. Now multiplication by i on $\mathbb{H}$ has the matrix form

$$\left[\begin{array}{cc|cc} 0 & -1 & \multicolumn{2}{c}{\multirow{2}{*}{0}} \\ 1 & 0 & & \\ \hline \multicolumn{2}{c|}{\multirow{2}{*}{0}} & 0 & -1 \\ & & 1 & 0 \end{array}\right].$$

which by a coordinate change may be transformed to J. □

Henceforth we will continue to write $(df)_{0,0} = L$ when W is arbitrary, but when W is assumed to be Γ-simple we will write $(df)_{0,0} = J$ for emphasis. The canonical form (1.6) for J simplifies the arguments occasionally but is not used extensively.

EXERCISES

1.1. For a steady-state $\mathbf{D}_n$-equivariant bifurcation problem on $\mathbb{R}^2$, show that along branches with $\mathbf{Z}_2$ isotropy, Hopf bifurcation cannot occur.

1.2. More generally, consider a Γ-equivariant bifurcation problem on V. Show that if an isotropy subgroup Σ decomposes V into a direct sum of distinct absolutely irreducible representations, then Hopf bifurcation from solutions corresponding to Σ is not possible. (*Note*: This is the same condition on Σ as that for which we can prove that Γ-equivalence preserves stability.)

§2. A Simple Hopf Theorem with Symmetry

The difficulty in applying the Hopf theorem to the system of ODEs described in §1 is that generically the purely imaginary eigenvalues of $(df)_{0,0}$ may have high multiplicity. Our method for finding periodic solutions to such a system rests on prescribing in advance the symmetry of the solutions we seek. This can often be used to select a subspace on which the eigenvalues are simple and is analogous to the steady-state situation described in the equivariant branching lemma. The crux of the argument is to describe precisely what we mean by a symmetry of a periodic solution and to find the correct context in which to select the subspace. There is a simplified context which we discuss now as motivation. Later we describe a more general context which captures more of the periodic behavior.

Definition 2.1. The periodic solution $x(t)$ has a *spatial symmetry* $\gamma \in \Gamma$ if for every t,

$$\gamma x(t) = x(t).$$

Theorem 2.2. *Let Σ be an isotropy subgroup of Γ such that the fixed-point subspace* $\mathrm{Fix}(\Sigma)$ *is two-dimensional. Assume that* $(df)_{0,0}$ *is as in* (1.7) *earlier and* *that the eigenvalue-crossing condition*

$$\sigma'(0) \neq 0 \tag{2.1}$$

holds, where σ is defined in Lemma 1.5(a). *Then there is a unique branch of small-amplitude periodic solutions to* (1.1), *of period near 2π, whose spatial symmetries are Σ.*

PROOF. Since f commutes with Γ, it maps $\mathrm{Fix}(\Sigma) \times \mathbb{R}$ to $\mathrm{Fix}(\Sigma)$ by Lemma XIII, 2.1. Restricting the $n \times n$ system (1.1) to the two-dimensional space $\mathrm{Fix}(\Sigma)$ yields a 2×2 system that satisfies the hypotheses of the standard Hopf theorem. Solutions in this plane are precisely those with the required spatial symmetries, by the definition of $\mathrm{Fix}(\Sigma)$. □

Remarks 2.3.

(a) Indeed $\mathrm{Fix}(\Sigma)$ is an invariant subspace for the *dynamics* of (1.1), for any $\Sigma \subset \Gamma$.

(b) In the case (1.6(a)) when $\mathbb{R}^n \cong V \oplus V$, it is easy to show that

$$\mathrm{Fix}_{V \oplus V}(\Sigma) = \mathrm{Fix}_V(\Sigma) \oplus \mathrm{Fix}_V(\Sigma);$$

see Exercise 2.1. Thus each one-dimensional fixed-point subspace for the action of Γ on V yields a branch of periodic solutions in Hopf bifurcation on $V \oplus V$. These fixed-point subspaces are precisely those calculated for use in the static equivariant branching lemma.

(c) If the trivial solution is stable subcritically then subcritical branches obtained in this way are unstable. This follows from exchange of stability in the standard Hopf theorem (Theorem VIII, 4.1). However, supercritical branches may or may not be stable, since directions not in $\mathrm{Fix}(\Sigma)$ may be involved.

Let us briefly apply this theorem to the oscillating hosepipe mentioned in XI, §1(c). Here the group is $\mathbf{O}(2)$, and we consider only those modes corresponding to $\mathbf{O}(2)$ acting on the plane $\mathbb{R}^2 \equiv \mathbb{C}$ in its standard representation. The isotropy lattice is

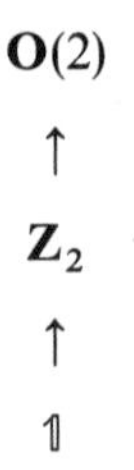

where $\mathbf{Z}_2 = \{1, \kappa\}$. The group $\mathbf{Z}_2$ has a one-dimensional fixed-point subspace, the real axis $\mathbb{R} \subset \mathbb{C}$. Therefore, we obtain a mode of oscillation with $\mathbf{Z}_2$ symmetry. This obviously corresponds to oscillation in a vertical plane. Thus

Theorem 2.2 detects *one* of the two oscillatory modes observed in this model, but it fails to detect the second rotating wave mode.

Indeed this is the main defect of Theorem 1.2: it fails to detect those periodic solutions whose symmetries combine both space and time. Our main objective in this chapter is to develop a context, involving both spatial and temporal symmetries, in which an analogous but more powerful theorem can be proved. The simple Hopf theorem acts as motivation for the approach that we shall take, with emphasis being placed on two-dimensional fixed-point subspaces.

§3. The Circle Group Action

In §1 we discussed a setting for Hopf bifurcation with symmetry. The problem is to find periodic solutions to a system of ODEs

$$\dot{v} + f(v, \lambda) = 0 \tag{3.1}$$

where $f\colon \mathbb{R}^n \times \mathbb{R} \to \mathbb{R}^n$ is smooth and commutes with Γ and $(df)_{0,0}$ has purely imaginary eigenvalues. In §2 we proved a simple version of the Hopf theorem for spatial symmetries and noted the need to incorporate temporal symmetries as well.

In this section we discuss the following ideas:

1. The required temporal symmetries are phase shifts and may be thought of as elements of a circle group $\mathbf{S}^1$, acting on the infinite-dimensional space of 2π-periodic functions.
2. The proof of the Hopf bifurcation theorem by Liapunov–Schmidt reduction generalizes to the Γ-equivariant context, incorporating these temporal phase-shift symmetries in a natural way.
3. The Liapunov–Schmidt reduction induces a related but different action of $\mathbf{S}^1$ on a finite-dimensional space, which can be identified with the exponential of the linearization $L = (df)_{0,0}$, acting on the imaginary eigenspace E_i of L. *Both* actions of $\mathbf{S}^1$ are needed to state the equivariant version of the Hopf bifurcation theorem; see Theorem 4.1.
4. Γ-simple subspaces of E_i are precisely the subspaces that are irreducible under the action of group $\Gamma \times \mathbf{S}^1$ generated by Γ and $\mathbf{S}^1$. This leads to analogies between steady-state bifurcation and Hopf bifurcation, in which Γ is replaced by $\Gamma \times \mathbf{S}^1$.

First we describe what we mean by a symmetry of a periodic solution $v(t)$. For simplicity we suppose that $v(t)$ is 2π-periodic in t. (If not we can rescale time to make the period 2π). Identify the circle $\mathbf{S}^1$ with $\mathbb{R}/2\pi\mathbb{Z}$. Then a *symmetry* of the periodic function $v(t)$ is an element $(\gamma, \theta) \in \Gamma \times \mathbf{S}^1$ such that

$$\gamma v(t) = v(t - \theta); \tag{3.2}$$

that is, the spatial action of γ on V may be exactly compensated by a phase shift. In this sense the symmetry (γ, θ) is a mixture of spatial and temporal symmetries. (The term *spatial* refers to the vector field symmetry of the original equations rather than actual physical space.) Note that $\mathbf{S}^1$ acts on the space of 2π-periodic mappings $v(t)$, not on $\mathbb{R}^n$. We call this action of $\mathbf{S}^1$ the *phase-shift* action.

The collection of all symmetries for $v(t)$ forms a subgroup

$$\Sigma_{v(t)} = \{(\gamma, \theta) \in \Gamma \times \mathbf{S}^1 : \gamma v(t) = v(t - \theta)\} \subset \Gamma \times \mathbf{S}^1. \tag{3.3}$$

There is a natural action of $\Gamma \times \mathbf{S}^1$ on the space $\mathscr{C}_{2\pi}$ of 2π-periodic mappings $\mathbb{R} \to \mathbb{R}^n$, defined by

$$(\gamma, \theta) \cdot v(t) = \gamma v(t + \theta). \tag{3.4}$$

That is, the Γ-action is induced from its spatial action on $\mathbb{R}^n$, and the $\mathbf{S}^1$-action is by phase shift. We can rewrite (3.2) as $(\gamma, \theta) \cdot v(t) = v(t)$. This shows that $\Sigma_{v(t)}$ is just the isotropy subgroup of $v(t)$ with respect to this action.

We now consider generalizing the proof of the Hopf theorem by Liapunov–Schmidt reduction (cf. VIII, §2) to the Γ-equivariant context; cf. Sattinger [1983]. The details are given in §4. For simplicity we choose the time scale so that $L = (df)_{0,0}$ has eigenvalues $\pm i$. We look for periodic solutions to (3.1) with period approximately 2π by rescaling time as

$$s = (1 + \tau)t$$

for a new *period-scaling parameter* τ near 0. This yields the system

$$(1 + \tau)\frac{du}{ds} + f(u, \lambda) = 0 \tag{3.5}$$

where $u(s) = v((1 + \tau)t)$. Then 2π-periodic solutions to (3.5) correspond to $2\pi/(1 + \tau)$-periodic solutions to (3.1). To find 2π-periodic solutions to (3.5) we define the operator

$$\Phi: \mathscr{C}^1_{2\pi} \times \mathbb{R} \times \mathbb{R} \to \mathscr{C}_{2\pi} \tag{3.6}$$

by

$$\Phi(u, \lambda, \tau) = (1 + \tau)\frac{du}{ds} + f(u, \lambda) \tag{3.7}$$

as in (VIII, 2.5). Here $\mathscr{C}_{2\pi}$ and $\mathscr{C}^1_{2\pi}$ are Banach spaces of continuous, respectively, once-differentiable 2π-periodic maps $\mathbb{R} \to \mathbb{R}^n$. A solution (u, λ, τ) to $\Phi = 0$ corresponds to a $2\pi/(1 + \tau)$-periodic solution to (3.5).

Now $\mathscr{L} = (d\Phi)_{0,0,0}$ is the linear operator

$$\mathscr{L}u = \frac{du}{ds} + Lu. \tag{3.8}$$

Thus Liapunov–Schmidt reduction lets us find solutions to $\Phi = 0$ by solving a reduced set of equations $\varphi = 0$ where

$$\varphi: \ker \mathscr{L} \times \mathbb{R} \times \mathbb{R} \to \operatorname{coker} \mathscr{L}. \tag{3.9}$$

The crucial point here is that if Φ commutes with a group action, then, provided the reduction is done in a reasonable manner (i.e., by choosing Γ-invariant complements), φ commutes with the action of this group on the kernel and cokernel of $(d\Phi)_{0,0,0}$. (In our applications these spaces will be equal.) See VII, §3.

Our immediate objective is to describe how $\mathbf{S}^1$ acts on ker $\mathscr{L}$. First we compute ker $\mathscr{L}$ explicitly:

Lemma 3.1. *Assume that L has no eigenvalues ki where $k \in \mathbb{Z}$, $k \neq \pm 1$. Then the kernel of $\mathscr{L}$ may naturally be identified with the eigenspace of the $n \times n$ matrix L for the eigenvalue i.*

PROOF. A function $v(s)$ lies in ker $\mathscr{L}$ if and only if it is 2π-periodic and satisfies the linear ODE

$$\frac{dv}{ds} + Lv = 0.$$

The general solution is $v(s) = e^{-sL}v_0$ for any $v_0 \in \mathbb{R}^n$. However, such a solution is 2π-periodic only when $v_0 \in E_{ki}$, the ki eigenspace of L acting on $\mathbb{C}^n$, for some $k \in \mathbb{Z}$. By hypothesis $E_{ki} = \{0\}$ unless $k = \pm 1$. Therefore, identifying $v(s)$ with its initial point v_0, we may identify ker $\mathscr{L}$ with E_i. □

Next we show that the action (3.4) of $\Gamma \times \mathbf{S}^1$ on $\mathscr{C}_{2\pi}$ induces an action on ker $\mathscr{L}$ and describe this action.

Lemma 3.2.
(a) *Φ commutes with the action of $\Gamma \times \mathbf{S}^1$ on $\mathscr{C}_{2\pi}$ defined in* (3.4).
(b) *The action of Γ on* ker $\mathscr{L} \equiv E_i$ *is the restriction of the Γ-action on $\mathbb{R}^n$. An element $s \in \mathbf{S}^1$ acts as e^{-sJ} where $J = L|E_i$.*

PROOF.
(a) We know from standard Hopf theory that Φ commutes with the phase-shift symmetries in $\mathbf{S}^1$ (see VIII, 2.8). By Γ-equivariance of f and linearity of the Γ-action,

$$\Phi(\gamma u, \lambda, \tau) = \gamma\Phi(u, \lambda, \tau).$$

(b) The map Φ commutes with $\Gamma \times \mathbf{S}^1$, so ker $\mathscr{L}$ is invariant under $\Gamma \times \mathbf{S}^1$. In Lemma 3.1 we identified $v \in E_i$ with the function $v(s) = e^{-sL}v$. We may replace this by $e^{-sJ}v$ since $J = L|E_i$. Then $\theta \in \mathbf{S}^1$ acts as follows:

$$\theta \cdot v \equiv \theta \cdot v(s) = v(s + \theta) = e^{-(s+\theta)J}v = e^{-sJ}(e^{-\theta J}v) \equiv e^{-\theta J}v.$$

Further, the Γ-action is straightforward:

$$\gamma \cdot v \equiv \gamma \cdot v(s) = \gamma \cdot e^{-sJ}v = e^{-sJ}\gamma v,$$

which is the periodic solution identified with γv. □

Remarks 3.3.

(a) By Proposition 1.4, generically we may assume E_i is Γ-simple, that is, either $E_i \cong V \oplus V$ where V is absolutely Γ-irreducible, or $E_i \cong W$ where W is non–absolutely Γ-irreducible. We see later that the Liapunov–Schmidt procedure yields a reduced bifurcation problem φ posed on $\ker \mathscr{L} \times \mathbb{R} \times \mathbb{R}$, so generically the problem of Γ-equivariant Hopf bifurcation reduces to finding zeros of $\Gamma \times \mathbf{S}^1$-equivariant mappings defined on a space of the type $(V \oplus V) \times \mathbb{R} \times \mathbb{R}$ or $W \times \mathbb{R} \times \mathbb{R}$.

(b) The action of $\Gamma \times \mathbf{S}^1$ in the $V \oplus V$ case may be described more explicitly. By Lemma 1.5, J may be put in the form

$$\begin{bmatrix} 0 & -I_m \\ I_m & 0 \end{bmatrix}$$

where $m = n/2$. Then direct computation of e^{-sJ} leads to the following description: Let $(x, y) \in V \oplus V$, $(\gamma, \theta) \in \Gamma \times \mathbf{S}^1$. Then

$$(\gamma, \theta) \cdot (x, y) = \gamma[x|y]R_\theta$$

where $[x|y]$ is the $m \times 2$ matrix whose columns are the m-vectors x and y, and

$$R_\theta = \begin{bmatrix} \cos\theta & -\sin\theta \\ \sin\theta & \cos\theta \end{bmatrix}$$

is the usual rotation matrix.

(c) The action in (b) can also be described more abstractly as the action of $\Gamma \times \mathbf{S}^1$ on $V \otimes \mathbb{C}$ defined by

$$(\gamma, \theta)(v \otimes z) = (\gamma v) \otimes (e^{-i\theta} z)$$

where $v \in V$, $z \in \mathbb{C}$, $\gamma \in \Gamma$, $\theta \in \mathbf{S}^1$. We shall not use this form of the action here, but it accounts for some mathematical features of the analysis that might otherwise appear "accidental."

(d) Another way to describe (a) is to take a basis for V as a real vector space and consider $V \oplus V$ to be the vector space over $\mathbb{C}$ with this basis. Elements of Γ act on $V \oplus V$ by the same matrices as for V (but now thought of as matrices over $\mathbb{C}$ that just happen to have real entries), and $\theta \in \mathbf{S}^1$ acts as scalar multiplication by $e^{-i\theta}$. In other words, J is identified with i, and $V \oplus V$ is the "complexification" of V, so might perhaps be better thought of as $V \oplus iV$.

(e) If $E_i = W$ is non–absolutely irreducible and W is of "complex" type, $\mathscr{D} \cong \mathbb{C}$, there are, abstractly speaking, two distinct possible actions of $\mathbf{S}^1$. One is identified with multiplication by $e^{i\theta}$, the other with $e^{-i\theta}$. So one is the time reversal of the other. Which action occurs depends on how the commuting matrices are identified with $\mathbb{C}$, there always being two possibilities, one the complex conjugate of the other.

(f) We have not described the $\mathbf{S}^1$-action in the case $\mathscr{D} \cong \mathbb{H}$. It will be by a circle subgroup of the unit quaternions, without loss of generality the one passing through i, and consisting of the quaternions $\{\cos\theta + i\sin\theta\}$. Since $+i$ and $-i$ are conjugate quaternions, the "clockwise" and "counterclockwise" actions are equivalent. Since we do not require this case in any later applications, we pursue the matter no further.

The following lemma will prove useful when analyzing the details of the Liapunov–Schmidt reduction in §4. One consequence is that the action of $\Gamma \times \mathbf{S}^1$ on $\ker \mathscr{L}$ is non–absolutely irreducible (of "complex" type).

Lemma 3.4.
(a) *In the generic case, when* $\ker \mathscr{L}$ *is* Γ*-simple, the matrices* I *and* J *form a basis for the vector space of all linear mappings on* $\mathbb{R}^n$ *that commute with the action of* $\Gamma \times \mathbf{S}^1$.
(b) *In the generic case,* $\Gamma \times \mathbf{S}^1$ *acts non–absolutely irreducibly on* $\ker \mathscr{L}$, *and this action is of "complex" type.*

PROOF.
(a) When $\ker \mathscr{L}$ is nonabsolutely irreducible and $\mathscr{D} \cong \mathbb{C}$, this is obvious since the commuting mappings identify with $\mathbb{C}$, I identifies with 1, and J identifies with i.

In the case when $\ker \mathscr{L} \cong V \oplus V$, where V is absolutely irreducible, let $A: V \oplus V \to V \oplus V$ be a linear map commuting with Γ. From (1.8) we see that there exist scalars a, b, c, d such that

$$A(v, w) = (av + bw, cv + dw)$$

for $v, w \in V$. Since A commutes with $\mathbf{S}^1$, the 2×2 matrix $\left[\begin{smallmatrix} a & b \\ c & d \end{smallmatrix}\right]$ must commute with all rotation matrices, whence $d = a$ and $c = -b$. Then

$$A(v, w) = a(v, w) + b(w, -v)$$

and $A = aI - bJ$ as claimed.

When $\ker \mathscr{L}$ is non–absolutely irreducible and $\mathscr{D} \cong \mathbb{H}$, the $\mathbf{S}^1$-action identifies with a circle subgroup of $\mathbb{H}$, without loss of generality $\{\cos\theta + i\sin\theta\}$. Although there are additional linear Γ-equivariants j and k, these do not commute with this $\mathbf{S}^1$-action since they do not commute with i. So the commuting mappings are spanned by $\{1, i\}$ which identify with I, J.
(b) The only case in which something must be proved is when $\ker \mathscr{L} = V \oplus V$, for absolutely irreducible V. A nontrivial Γ-invariant subspace of $V \oplus V$ is of the form $V' = \{(v, Av): v \in V\}$ where A commutes with Γ. By (a) we have $A = aI + bJ$. This allows us to show that V' is not $\Gamma \times \mathbf{S}^1$-irreducible, whence $\Gamma \times \mathbf{S}^1$ acts irreducibly on $V \oplus V$. For $J(v, Av) = (v, AJv)$, and in general $Jv \neq v$. Thus $J(v, Av) \notin V'$. By (a) $V \oplus V$ must be of "complex" type. $\square$

The preceding results show that when considering Hopf bifurcation with symmetry it is the $\Gamma \times \mathbf{S}^1$-action on $\ker \mathscr{L}$, rather than the Γ-action, that is crucial. The next result provides some unification; however, it plays no other role and can be omitted if desired.

Proposition 3.5. *Let X be irreducible under an action of $\Gamma \times \mathbf{S}^1$, and suppose that $\mathbf{S}^1$ acts nontrivially. Then X is Γ-simple.*

PROOF. First consider the action of $\mathbf{S}^1$. Decompose X into isotypic parts for the $\mathbf{S}^1$-action,

$$X = \bigoplus X_k$$

where on X_k, $\mathbf{S}^1$ acts by copies of the k-fold action $\theta \cdot z = e^{ki\theta}z$. We claim that only one k can occur. This is clear since Γ commutes with $\mathbf{S}^1$, hence leaves each X_k invariant. By factoring out the kernel of the $\mathbf{S}^1$-action we can without loss of generality assume that $\mathbf{S}^1$ is acting on X by a direct sum of copies of its standard action. Therefore, if the matrix J is defined by the action of $\pi/2$, then a general $\theta \in \mathbf{S}^1$ acts as $\cos\theta + \sin\theta\, J$. Also $J^2 = -I$.

Suppose first that X is Γ-irreducible. Now $\mathbf{S}^1$ cannot act nontrivially by real multiples of the identity, hence $\pi/2$ induces a commuting mapping not of the form rI, $r \in \mathbb{R}$. Thus X is non–absolutely irreducible.

If X is Γ-reducible, let Y be a Γ-irreducible subspace. The action of $\mathbf{S}^1$ is by $\cos\theta + \sin\theta\, J$, so the subspace $Y + JY$ is $\mathbf{S}^1$-invariant. It is also Γ-invariant since J commutes with Γ. Therefore, by $\Gamma \times \mathbf{S}^1$-irreducibility $X = Y + JY$. The sum is direct since $Y \cap JY = 0$. Clearly $Y \cong JY$ under the Γ-action, so $X = Y \oplus JY \cong Y \oplus Y$.

It remains to prove that Y is absolutely irreducible. By the proof of Lemma 3.4:

$$\text{the set of linear mappings on } X \text{ that commute with } \Gamma \times \mathbf{S}^1 \text{ is two-dimensional.} \tag{3.10}$$

Since $X = Y \oplus JY$ we can write elements of X uniquely in the form $y + Jz$, where $y, z \in Y$. Let $\alpha\colon Y \to Y$ commute with Γ, and extend it to a map $\beta\colon X \to X$ by

$$\beta(y + Jz) = \alpha y + J\alpha z.$$

Then β commutes with J, since

$$J\beta(y + Jz) = J(\alpha y + J\alpha z) = J\alpha y - \alpha z,$$

$$\beta(J(y + Jz)) = \beta(Jy - z) = J\alpha y - \alpha z.$$

Here we use the fact that $J^2 = -I$. Clearly β commutes with Γ, since J and α do. Suppose that α is not a real multiple of the identity. Then the three maps I, J, β on X commute with $\Gamma \times \mathbf{S}^1$ and are linearly independent. This contradicts (3.10). Therefore, Y is absolutely irreducible for Γ. □

Remark. The proof shows that all spaces $V \oplus V$ (V absolutely Γ-irreducible) and W (W non–absolutely Γ-irreducible) can arise as irreducible representations of $\Gamma \times \mathbf{S}^1$.

There is thus a full analogy with the results for steady state bifurcations:

For steady-state bifurcation generically E_0 is (absolutely) irreducible for Γ;
For Hopf bifurcation generically E_i is (non-absolutely) irreducible for $\Gamma \times \mathbf{S}^1$.

(Further, in the static case the commuting mappings for Γ must be $\mathbb{R}$; in the Hopf case the commuting mappings for $\Gamma \times \mathbf{S}^1$ must be $\mathbb{C}$.) We were not able to characterize the generic situation earlier in these terms, since the relevance of the $\mathbf{S}^1$-action on E_i was not then sufficiently clear.

§4. The Hopf Theorem with Symmetry

The main result of this section is a generalization of the simple Hopf theorem given in §2. The essential point of this generalization is the possibility of periodic solutions with mixed spatiotemporal symmetries.

As usual we consider a system of ODEs

$$\frac{dv}{dt} + f(v, \lambda) = 0 \tag{4.1}$$

where $f: \mathbb{R}^n \times \mathbb{R} \to \mathbb{R}^n$ is smooth and commutes with a compact Lie group Γ. We make the generic hypothesis that $\mathbb{R}^n$ is Γ-simple and choose coordinates so that

$$(df)_{0,0} = J \equiv \begin{bmatrix} 0 & -I_m \\ I_m & 0 \end{bmatrix}, \tag{4.2}$$

where $m = n/2$. (These assumptions are not strictly necessary, but they simplify the proof; see Remark 4.2(c).) By Lemma 1.5, the eigenvalues of $(df)_{0,\lambda}$ are $\sigma(\lambda) \pm i\rho(\lambda)$, each of multiplicity m. Assumption (4.2) implies that $\sigma(0) = 0$, $\rho(0) = 1$. As in the standard Hopf theorem (VIII, 3.1) we assume that the eigenvalues of df cross the imaginary axis with nonzero speed; that is,

$$\sigma'(0) \neq 0. \tag{4.3}$$

Temporal symmetries enter the equivariant Hopf theorem through an isotropy subgroup $\Sigma \subset \Gamma \times \mathbf{S}^1$ acting on $\mathbb{R}^n$ as in Lemma 3.2(b).

Theorem 4.1 (Equivariant Hopf Theorem). *Let the system of ODEs* (4.1) *satisfy* (4.2) *and* (4.3). *Suppose that*

$$\dim \operatorname{Fix}(\Sigma) = 2.$$

Then there exists a unique branch of small-amplitude periodic solutions to (4.1) *with period near* 2π, *having* Σ *as their group of symmetries.*

Remarks 4.2.
(a) As usual, when we say the branch is unique we are not distinguishing between solutions with the same trajectory. Although periodic trajectories with isotropy subgroup Σ are unique, Γ-equivariance implies that there are $(\Gamma \times \mathbf{S}^1)/\Sigma$ different periodic solutions. These have isotropy subgroups conjugate to Σ in $\Gamma \times \mathbf{S}^1$.
(b) Several examples applying Theorem 4.1 are given in the next two chapters. The remainder of this section is devoted to a proof of Theorem 4.1.
(c) There is a more general version of Theorem 4.1, which can be proved by the same method. Instead of assuming that $\mathbb{R}^n$ is the entire imaginary eigenspace and is Γ-simple, we assume only that E_i is Γ-simple and that the system is *nonresonant* in the sense that no eigenvalues ki occur for $k \in \mathbb{Z}$, $k \neq \pm 1$. If in Theorem 3.1 we assume that $\dim \mathrm{Fix}_{E_i}(\Sigma) = 2$, then the conclusion still applies. This is the case even if there are other imaginary eigenvalues, provided they are not integer multiples of i.

Before proving the theorem we set up the generalities that will establish the existence of the Liapunov–Schmidt reduced mapping φ in (3.9). Then we use φ to demonstrate the existence of periodic solutions with a suitable mixture of spatial and temporal symmetries.

As in VIII, §2a, $\mathscr{C}_{2\pi}$ is a Banach space with norm

$$\|u\| = \max_s |u(s)|$$

and $\mathscr{C}_{2\pi}^1$ is a Banach space with norm

$$\|u\|_1 = \|u\| + \|du/ds\|.$$

We are interested in the operator

$$\Phi\colon \mathscr{C}_{2\pi}^1 \times \mathbb{R} \times \mathbb{R} \to \mathscr{C}_{2\pi} \tag{4.4}$$

defined by

$$\Phi(u, \lambda, \tau) = (1 + \tau)\frac{du}{ds} + f(u, \lambda).$$

Recall that (using the notational convention $(df)_{0,0} = J$ when $\mathbb{R}^n$ is Γ-simple)

$$\mathscr{L} = (d\Phi)_{0,0,0} = \frac{d}{ds} + J.$$

We claim that there is a $\Gamma \times \mathbf{S}^1$-invariant splitting

$$\mathscr{C}_{2\pi} = \text{range}\, \mathscr{L} \oplus \ker \mathscr{L} \tag{4.5}$$

inducing the splitting

$$\mathscr{C}_{2\pi}^1 = \ker \mathscr{L} \oplus M \tag{4.6}$$

where $M = (\text{range}\, \mathscr{L}) \cap \mathscr{C}_{2\pi}^1$. If this is true, we have

$$\text{coker}\,\mathcal{L} \cong \ker \mathcal{L}$$

and the Liapunov–Schmidt reduced mapping has the form

$$\varphi\colon \ker \mathcal{L} \times \mathbb{R} \times \mathbb{R} \to \ker \mathcal{L} \tag{4.7}$$

where φ commutes with the action of $\Gamma \times \mathbf{S}^1$ on $\ker \mathcal{L}$ described in Lemma 3.2.

We prove (4.6 and 4.7). The formal adjoint of $\mathcal{L}$ with respect to the inner product

$$\langle u, v \rangle = \frac{1}{2\pi} \int_0^{2\pi} v(s)^t u(s) ds$$

is

$$\mathcal{L}^* = -\frac{d}{ds} + J^t = -\frac{d}{ds} - J = -\mathcal{L}; \tag{4.8}$$

see Appendix 4, p. 332, Vol. I. By the Fredholm alternative

$$\mathscr{C}_{2\pi} = \ker \mathcal{L}^* \oplus \text{range}\,\mathcal{L} = \ker \mathcal{L} \oplus \text{range}\,\mathcal{L}$$

as claimed in (4.6).

The formalities of the Liapunov–Schmidt reduction follow the standard pattern. Let

$$E\colon \mathscr{C}_{2\pi} \to \text{range}\,\mathcal{L} \tag{4.9}$$

by projection with kernel $\ker \mathcal{L}$. Split the equation $\Phi = 0$ into the pair of equations

$$\begin{aligned} &\text{(a)} \quad E\Phi(u, \lambda, \tau) = 0 \\ &\text{(b)} \quad (I - E)\Phi(u, \lambda, \tau) = 0. \end{aligned} \tag{4.10}$$

Write $u = v + w$ where $v \in \ker \mathcal{L}$ and $w \in M$, and solve (4.10a) by the implicit function theorem for $w = W(v, \lambda, \tau)$. Substitute into (4.10(b)) to obtain

$$\varphi(v, \lambda, \tau) = (I - E)\Phi(v + W(v, \lambda, \tau), \lambda, \tau). \tag{4.11}$$

The generalities of Liapunov–Schmidt reduction imply that

$$(d\varphi)_{0,0,0} = 0. \tag{4.12}$$

Because of $\Gamma \times \mathbf{S}^1$-irreducibility, the action of $\Gamma \times \mathbf{S}^1$ on $\ker \mathcal{L}$ satisfies $\text{Fix}_{\ker \mathcal{L}}(\Gamma \times \mathbf{S}^1) = 0$. Thus there is a "trivial solution" $\varphi(0, \lambda, \tau) \equiv 0$.

Proof of Theorem 4.1. The assumptions of Theorem 4.1 let us apply the Liapunov–Schmidt procedure described earlier. In particular, small-amplitude periodic solutions of (4.1), of period near 2π, correspond to zeros of the reduced equation $\varphi = 0$ in (4.11). Finding periodic solutions of (4.1) with symmetries Σ is equivalent to finding zeros of φ with isotropy subgroup

Σ. Such zeros are found by solving

$$\varphi|\text{Fix}(\Sigma) = 0.$$

Since φ commutes with $\Gamma \times \mathbf{S}^1$, and not just Γ, we know that φ maps $\text{Fix}(\Sigma)$ into itself. The main point of this proof is that φ has zeros on the two-dimensional subspace $\text{Fix}(\Sigma)$ in precisely the same way that the Liapunov–Schmidt reduced equation for standard Hopf bifurcation has zeros. See Theorem VIII, 3.1.

We might say that the two-dimensional subspace has split off a pair of simple eigenvalues, to which the standard Hopf theorem applies, but this is slightly misleading since the original vector field f is equivariant only under Γ, not $\Gamma \times \mathbf{S}^1$, so $\text{Fix}(\Sigma)$ need not be invariant subspace for the *dynamics*. At the level of the reduced equation, which captures periodic behavior but not all the dynamics, there is full $\Gamma \times \mathbf{S}^1$ symmetry and the idea works.

We use the $\mathbf{S}^1$ symmetry to show that $\varphi|\text{Fix}(\Sigma) \times \mathbb{R}^2$ has the form

$$\varphi(v, \lambda, \tau) = p(|v|^2, \lambda, \tau)v + q(|v|^2, \lambda, \tau)Jv. \tag{4.13}$$

To do this we first observe that $\varphi|\text{Fix}(\Sigma) \times \mathbb{R} \times \mathbb{R}$ commutes with the standard action of $\mathbf{S}^1$. This is a special case of Lemma XIII, 10.1, but can be seen directly as follows. If $\sigma \in \Sigma$, $\theta \in \mathbf{S}^1$, and $w \in \text{Fix}(\Sigma)$, then

$$\theta \cdot w = \theta \cdot \sigma \cdot w = \sigma \cdot \theta w.$$

Hence $\theta w \in \text{Fix}(\Sigma)$, and $\varphi|\text{Fix}(\Sigma) \times \mathbb{R} \times \mathbb{R}$ must commute with θ since $\theta \in \mathbf{S}^1 \subset \Gamma \times \mathbf{S}^1$.

By Remark 3.3(b), $\mathbb{R}^n$ is a direct sum of m two-dimensional irreducible representations of $\mathbf{S}^1$, all isomorphic to the standard representation of $\mathbf{S}^1$ on $\mathbb{R}^2$. By Theorem XII, 2.5, the action of $\mathbf{S}^1$ on any two-dimensional $\mathbf{S}^1$-invariant subspace of $\mathbb{R}^n$, in particular $\text{Fix}(\Sigma)$, is itself standard.

Let (α, β) be coordinates on $\text{Fix}(\Sigma)$, so that $\mathbf{S}^1$ acts in these coordinates as multiplication by

$$R_\theta = \begin{bmatrix} \cos\theta & -\sin\theta \\ \sin\theta & \cos\theta \end{bmatrix}.$$

Let $\tilde{\varphi}$ be $\varphi|\text{Fix}(\Sigma) \times \mathbb{R}^2$, written in the (α, β) coordinates. Since $\varphi|\text{Fix}(\Sigma)$ commutes with $\mathbf{S}^1$, Lemma VIII, 2.5, implies that $\tilde{\varphi}$ has the form

$$\tilde{\varphi}(\alpha, \beta, \lambda, \tau) = P(\alpha^2 + \beta^2, \lambda, \tau)\begin{bmatrix} \alpha \\ \beta \end{bmatrix} + Q(\alpha^2 + \beta^2, \lambda, \tau)\begin{bmatrix} -\beta \\ \alpha \end{bmatrix}.$$

In the (α, β) coordinates $|v|^2 = \alpha^2 + \beta^2$, the identity map is $\left[\begin{smallmatrix} \alpha \\ \beta \end{smallmatrix}\right]$, and $J|\text{Fix}(\Sigma)$ is $\left[\begin{smallmatrix} -\beta \\ \alpha \end{smallmatrix}\right]$. Thus (4.13) follows from the form of $\tilde{\varphi}$.

Moreover, since the linear terms vanish in the reduced equation (see (4.12)), we know that

$$p(0,0,0) = q(0,0,0) = 0.$$

We claim the following:

Lemma 4.3.

$$\begin{aligned} &\text{(a)} \quad p_\tau(0,0,0) = 0, \qquad q_\tau(0,0,0) = -1, \\ &\text{(b)} \quad p_\lambda(0,0,0) = \sigma'(0) \neq 0. \end{aligned} \tag{4.14}$$

The remainder of the proof of Theorem 4.1 then proceeds exactly as in Theorem VIII, 2.1 (Volume I, p. 344). □

Before proving Lemma 4.3 we state and prove the following:

Lemma 4.4. *Let A be an $n \times n$ matrix and let $v \in \ker \mathscr{L}$. Then*
(a) *If A commutes with J, then $Av \in \ker \mathscr{L}$.*
(b) *$(I - E)$: $\mathscr{C}_{2\pi} \to \ker \mathscr{L}$ is orthogonal projection.*
(c) *$(I - E)Av = \bar{A}v$*
where

$$\bar{A} = \frac{1}{2\pi} \int_0^{2\pi} e^{tJ} A e^{-tJ}\, dt. \tag{4.15}$$

PROOF.
(a) Recall that $\mathscr{L} = (d/ds) + J$. Hence $\mathscr{L}Av = ((d/ds) + J)Av = A((d/ds) + J)v = A\mathscr{L}v = 0$.
(b) $(I - E)$: $\mathscr{C}_{2\pi} \to \ker \mathscr{L}$ is a projection since E: $\mathscr{C}_{2\pi} \to$ range $\mathscr{L}$ is projection with kernel $\ker \mathscr{L}$. To show that $I - E$ is orthogonal projection we must show that $\ker \mathscr{L}$ and range $\mathscr{L}$ are orthogonal with respect to the inner product

$$\langle u, v \rangle = \frac{1}{2\pi} \int_0^{2\pi} v(s)^T u(s)\, ds,$$

where we use T for transpose to avoid confusion with time. This follows since the formal adjoint $\mathscr{L}^* = -\mathscr{L}$ by (4.8), so if $v \in \ker \mathscr{L}$

$$0 = -\langle \mathscr{L}v, u \rangle = -\langle v, \mathscr{L}^* u \rangle = \langle v, \mathscr{L}u \rangle$$

whence v is orthogonal to range $\mathscr{L}$.
(c) Clearly $\bar{A}$ commutes with the $\mathbf{S}^1$-action, hence with J, so $\bar{A}v \in \ker \mathscr{L}$. Since $(I - E)$ is orthogonal projection, in order to show that $\bar{A}v = (I - E)Av$ it suffices to show that, for all $x \in \ker \mathscr{L}$,

$$\langle x, \bar{A}v \rangle = \langle x, Av \rangle.$$

We compute:

$$\begin{aligned} \langle x, \bar{A}v \rangle &= \frac{1}{2\pi} \int_{s=0}^{2\pi} x(s)^T \bar{A}v(s)\, ds \\ &= \frac{1}{2\pi} \int_{s=0}^{2\pi} x(s)^T \left(\frac{1}{2\pi} \int_{t=0}^{2\pi} e^{tJ} Av(s) e^{-tJ}\, dt \right) v(s)\, ds \\ &= \frac{1}{4\pi^2} \int_{s=0}^{2\pi} \int_{t=0}^{2\pi} (e^{-tJ} x(s))^T A (e^{-tJ} v(s))\, ds\, dt. \end{aligned} \tag{4.16}$$

If $z(s) \in \ker \mathscr{L}$ then $z(s) = e^{-sJ}z(0)$ and $z(s+t) = e^{-tJ}z(s)$. Therefore, (4.16) becomes

$$\frac{1}{4\pi^2}\int_{s=0}^{2\pi}\int_{t=0}^{2\pi} x(s+t)^T Av(s+t)\,ds\,dt. \tag{4.17}$$

Let $s' = s + t$, $t' = t$. Write (4.17) as

$$\begin{aligned}\langle x, \bar{A}v\rangle &= \frac{1}{2\pi}\int_{t'=0}^{2\pi}\left(\frac{1}{2\pi}\int_{s'=0}^{2\pi} x(s')^T Av(s')\,ds'\right)dt' \\ &= \frac{1}{2\pi}\int_{t'=0}^{2\pi}\langle x, Av\rangle\,dt' \\ &= \langle x, Av\rangle\end{aligned}$$

as required. □

Proof of Lemma 4.3. It remains to verify (4.14). The computation of p_τ, q_τ, and p_λ at $(0,0,0)$ involves only linear terms in φ. By $\Gamma \times \mathbf{S}^1$-equivariance we have

$$\varphi(v,\lambda,\tau) = a(\lambda,\tau)v + b(\lambda,\tau)Jv + O(\|v\|^2), \tag{4.18}$$

where $v \in \ker \mathscr{L}$, and $a(\lambda,\tau), b(\lambda,\tau) \in \mathbb{R}$. Here

$$a(\lambda,\tau) = p(0,\lambda,\tau), \qquad b(\lambda,\tau) = q(0,\lambda,\tau).$$

We will find explicit formulas for a, b.

Define

$$A(\lambda) = (df)_{0,\lambda}.$$

This has eigenvalues $\sigma(\lambda) \pm i\rho(\lambda)$ of multiplicity $n/2$, and $A(0) = J$. We have $\sigma(0) = 0$, $\rho(0) = 1$, $\sigma'(0) \neq 0$. By Taylor expansion with respect to $u \in \mathbb{R}^n$,

$$f(u,\lambda) = A(\lambda)u + O(\|u\|^2).$$

Thus

$$\Phi(u,\lambda,\tau) = \left((1+\tau)\frac{d}{ds} + A(\lambda)\right)u + O(\|u\|^2). \tag{4.19}$$

By generalities of the Liapunov–Schmidt procedure, the implicitly defined function $W(v,\lambda,\tau)$ vanishes through first order in v. Hence

$$\Phi(v + W(v,\lambda,\tau),\lambda,\tau) = \Phi(v,\lambda,\tau) + O(\|v\|^2).$$

By Taylor expansion

$$\begin{aligned}\Phi(v,\lambda,\tau) &= \Phi(0,\lambda,\tau) + (d_v\Phi)_{(0,\lambda,\tau)}v + O(\|v\|^2) \\ &= (d_v\Phi)_{(0,\lambda,\tau)}v + O(\|v\|^2)\end{aligned}$$

since $\Phi(0,\lambda,\tau) = 0$. Now

$$(d_v\Phi)_{(0,\lambda,\tau)} = \left[(1+\tau)\frac{d}{ds} + A(\lambda)\right]\Bigg|_{(0,\lambda,\tau)}. \tag{4.20}$$

Since $v \in \ker \mathscr{L}$ we have $(dv/ds) + Jv = 0$, so (4.20) becomes

$$(d_v\Phi)_{(0,\lambda,\tau)} = [-(1+\tau)J + A(\lambda)]v,$$

and

$$\Phi(v, \lambda, \tau) = [-(1+\tau)J + A(\lambda)]v + O(\|v\|^2). \tag{4.21}$$

By (4.11)

$$\varphi = (I - E)\Phi, \tag{4.22}$$

where $(I - E)\colon \mathscr{C}_{2\pi} \to \ker \mathscr{L}$ is orthogonal projection. By Lemma 4.4 and (4.22),

$$\varphi(v, \lambda, \tau) = [-(1+\tau)J + \bar{A}(\lambda)]v + O(\|v\|^2)$$

where

$$\bar{A}(\lambda) = \frac{1}{2\pi}\int_{t=0}^{2\pi} e^{tJ}A(\lambda)e^{-tJ}\,dt. \tag{4.23}$$

Now $\bar{A}(\lambda)$ commutes with $\Gamma \times \mathbf{S}^1$ so is of the form

$$\bar{A}(\lambda) = \bar{a}(\lambda)I + \bar{b}(\lambda)J$$

for functions $\bar{a}, \bar{b}\colon \mathbb{R} \to \mathbb{R}$; see Lemma 3.4. By (4.18)

$$\begin{aligned} a(\lambda, \tau) &= \bar{a}(\lambda) \\ b(\lambda, \tau) &= -(1+\tau) + \bar{b}(\lambda). \end{aligned}$$

Thus we have the identities

$$\begin{aligned} a_\tau &= 0 \\ b_\tau &= -1 \end{aligned}$$

and in particular these equations hold at $(0, 0)$.

Finally we claim that at $(0, 0)$

$$a_\lambda = \sigma'(0).$$

To see why, observe that $\operatorname{tr} J = 0$ so $\bar{a}(\lambda) = \frac{1}{n}\operatorname{tr}\bar{A}(\lambda)$. By (4.23),

$$\operatorname{tr}\bar{A}(\lambda) = \operatorname{tr} A(\lambda).$$

But the eigenvalues of $A(\lambda)$ are $\sigma(\lambda) \pm i\rho(\lambda)$ of multiplicity $n/2$, so

$$\operatorname{tr} A(\lambda) = n\sigma(\lambda)$$

Thus

$$\bar{a}(\lambda) = \sigma(\lambda),$$

so

$$a_\lambda(\lambda, \tau) = \bar{a}_\lambda(\lambda) = \sigma'(\lambda)$$

and

$$a_\lambda(0, 0) = \sigma'(0).$$

But $a(\lambda, \tau) = p(0, \lambda, \tau)$ and $b(\lambda, \tau) = q(0, \lambda, \tau)$. Since no v-derivatives are involved, Lemma 4.3 follows. □

The hypothesis of Theorem 4.1, that $\dim \mathrm{Fix}(\Sigma) = 2$, is analogous to that of the equivariant branching lemma, where $\dim \mathrm{Fix}(\Sigma) = 1$. However, Fiedler [1987] has shown, using equivariant index theory, that the condition "$\dim \mathrm{Fix}(\Sigma) = 2$" can be weakened to "$\Sigma$ is a maximal isotropy subgroup of $\Gamma \times \mathbf{S}^1$." The proof is beyond the scope of this book, but we state the result for reference:

Theorem 4.5 (Fiedler [1987]). *Let the system of ODEs* (4.1) *satisfy* (4.2) *and* (4.3). *Suppose that Σ is a maximal isotropy subgroup of $\Gamma \times \mathbf{S}^1$. Then there exist small-amplitude periodic solutions to* (4.1) *with period near 2π, having Σ as their group of symmetries.*

Exercises

4.1. Let $\Gamma = \mathbf{O}(2)$ with its standard action on $\mathbb{R}^2$. Let $\mathbf{O}(2) \times \mathbf{S}^1$ act on $\mathbb{R}^2 \oplus \mathbb{R}^2$ as in Remark 3.3(b); that is,

$$(\gamma, \theta) \cdot [x|y] = \gamma [x|y] R_\theta$$

where $[x|y]$ is the 2×2 matrix with columns x, y. Define

$$\widetilde{\mathbf{SO}}(2) = \{(\theta, -\theta) \in \mathbf{O}(2) \times \mathbf{S}^1\}.$$

Show that $\mathrm{Fix}(\widetilde{\mathbf{SO}}(2)) = \{\begin{bmatrix} a & -b \\ b & a \end{bmatrix}\}$. Thus $\dim \mathrm{Fix}(\widetilde{\mathbf{SO}}(2)) = 2$. Apply Theorem 4.1 to conclude that there is a branch of periodic solutions u whose symmetries include

$$R_\theta u(s - \theta) = u(s);$$

that is,

$$u(\theta) = R_\theta u(0). \tag{4.24}$$

Abstract solutions satisfying (4.24) are usually called *rotating waves*. Indeed, in the hosepipe example of XI, §1(c), this solution corresponds to the rotating wave mode.

4.2. Let the groups $\mathbf{SU}(2)$ of unit quaternions act on $\mathbb{R}^4 \equiv \mathbb{H}$ by left multiplication and let $g: \mathbb{R}^4 \times \mathbb{R} \to \mathbb{R}^4$ be $\mathbf{SU}(2)$-equivariant. Consider the ODE $\dot{x} + g(x, \lambda) = 0$.

(a) Show that Hopf bifurcation can occur and that the corresponding action of $\theta \in \mathbf{S}^1$ can be assumed to be multiplication on the right by $\cos\theta + i \sin\theta$, $i \in \mathbb{H}$.

(b) Show that the isotropy lattice is then

$$\begin{array}{c} \mathbf{SU}(2) \times \mathbf{S}^1 \\ \uparrow \\ \widetilde{\mathbf{SO}}(2) \\ \uparrow \\ \mathbb{1} \end{array}$$

where $\widetilde{\mathbf{SO}}(2) = \{(\cos\theta - i\sin\theta, \cos\theta + i\sin\theta): 0 \leq \theta < 2\pi\}$. Find the fixed-point subspace of $\widetilde{\mathbf{SO}}(2)$.

(c) Apply the equivariant Hopf theorem to conclude that generically, in such an $\mathbf{SU}(2)$ Hopf bifurcation, there exists a branch of periodic solutions with $\widetilde{\mathbf{SO}}(2)$ isotropy.

(d) Show that the orbit of such a solution under $\mathbf{SU}(2)$ fibers a 3-sphere according to the Hopf fibration. [Compare Exercise XIII, 10.8(f).]

4.3. (See Cicogna and Gaeta [1987].) Consider the ODE $\dot{x} + f(x, \lambda) = 0$ where $x \in \mathbb{R}^4$. Suppose that $(df)_{0,0}$ has the form

$$\left[\begin{array}{cc|cc} 0 & 1 & 0 & 0 \\ -1 & 0 & 0 & 0 \\ \hline 0 & 0 & 0 & 1 \\ 0 & 0 & -1 & 0 \end{array}\right]$$

and that f commutes with the action of $\mathbf{Z}_2$ on $\mathbb{R}^4$ generated by a linear transformation with matrix

$$\begin{bmatrix} 0 & S^{-1} \\ S & 0 \end{bmatrix}$$

where S is any nonsingular 2×2 matrix. Show that $\mathbf{Z}_2$ has a two-dimensional fixed-point subspace and deduce the occurrence of a periodic solution branch with *spatial* isotropy $\mathbf{Z}_2$. [Note that the $\mathbf{Z}_2$ action here is *reducible*.]

4.4. Let $H: \mathbb{R}^n \to \mathbb{R}$ be a smooth function. Consider the Hamiltonian system $\dot{x} + J\,dH(x) = 0$ where

$$J = \begin{bmatrix} 0 & -I \\ I & 0 \end{bmatrix}.$$

Suppose this ODE has an equilibrium at $x = 0$. The Liapunov center theorem states that if the Hessian $d^2H|_0$ has a pair of simple purely imaginary eigenvalues $\pm i\omega$ then there is a smooth two-dimensional submanifold of $\mathbb{R}^n$, passing through the equilibrium point, foliated by periodic solutions of period near $2\pi/\omega$. It is well known that the Liapunov center theorem can be derived from the Hopf bifurcation theorem by the following trick. Consider the system $\dot{x} + (J + \lambda)dH(x) = 0$, and show that:

(a) The Hopf bifurcation theorem implies that this has a branch of periodic solutions bifurcating from $\lambda = 0$,

(b) If $\lambda \neq 0$ then H either increases strictly or decreases strictly along a solution, hence there are no periodic solutions unless $\lambda = 0$.

Use the same method, applied to the equivariant Hopf theorem, to prove an equivariant analogue of the Liapunov center theorem. For an improved result—an equivariant Weinstein–Moser theorem—see Montaldi, Roberts, and Stewart [1988].

§5. Birkhoff Normal Form and Symmetry

The idea of Birkhoff normal form is to simplify a system of ODEs

$$\dot{x} + f(x) = 0, \qquad f(0) = 0, \qquad x \in \mathbb{R}^n \tag{5.1}$$

by using successive polynomial changes of coordinates to set to zero many terms in the Taylor expansion of f at degree k. More precisely, let

$$f(x) = \tilde{f}_k(x) + h_k(x) + \cdots \tag{5.2}$$

where $\tilde{f}_k$ is a polynomial mapping of degree $< k$, h_k is a homogeneous polynomial mapping of degree k, and $\cdots$ indicates terms of degree $k + 1$ or higher. The main question is, How can we simplify h_k by changing coordinates in (5.1) while leaving $\tilde{f}_k$ unchanged? Since the terms of degree $> k$ impose no constraint, the process of putting a system of ODEs into Birkhoff normal form is a recursive one. Rather than launching directly into the statement of the theorem, we present a calculation that both motivates and proves it.

The idea is straightforward. Consider coordinate changes of the form

$$x = y + P_k(y) \tag{5.3}$$

where P_k is a homogeneous polynomial of degree k. Then $\dot{x} = \dot{y} + (dP_k)_y\dot{y}$. In the new coordinates (5.1) becomes

$$\dot{y} = (I + (dP_k)_y)^{-1}f(y + P_k(y)). \tag{5.4}$$

Modulo higher order terms (5.4) is

$$\dot{y} = (I - (dP_k)_y)f(y + P_k(y)) + \cdots. \tag{5.5}$$

Our object is to put the right-hand side of (5.5) in the form (5.2) by calculating (5.5) modulo higher order terms. This is easier if we set

$$\tilde{f}_k(x) = Lx + f_k(x) \tag{5.6}$$

where L is the linear part of f, and f_k consists of terms of degree between 2 and $k - 1$. Since P_k contains only terms of degree k, it follows that

$$\begin{aligned} &\text{(a)} \quad f_k(y + P_k(y)) = f_k(y) + \cdots \\ &\text{(b)} \quad h_k(y + P_k(y)) = h_k(y) + \cdots. \end{aligned} \tag{5.7}$$

Substituting (5.6 and 5.7) into (5.5) yields

$$\begin{aligned} \dot{y} &= (I - (dP_k)_y)[L(y + P_k(y)) + f_k(y) + h_k(y)] + \cdots \\ &= Ly + f_k(y) + h_k(y) + LP_k(y) - (dP_k)_yLy + \cdots. \end{aligned} \tag{5.8}$$

Thus any homogeneous term of degree k in f of the form

$$LP_k(y) - (dP_k)_y Ly \tag{5.9}$$

may be eliminated from (5.1) by a change of coordinates of the type (5.3). Moreover, this change of coordinates does not disturb terms of degree $< k$.

This conclusion may be abstracted as follows. Let $\mathscr{P}_k$ be the space of homogeneous polynomial mappings of degree k on $\mathbb{R}^m$. Then ad_L, defined by

$$\mathrm{ad}_L(P_k)(y) = LP_k(y) - (dP_k)_y Ly$$

as in (5.9), is a linear map $\mathscr{P}_k \to \mathscr{P}_k$. Thus the terms in the Taylor expansion of f that can be eliminated by this process are precisely those in the subspace $\mathrm{ad}_L(\mathscr{P}_k) \subset \mathscr{P}_k$. For each k we choose a complementary subspace $\mathscr{G}_k \subset \mathscr{P}_k$, so that

$$\mathscr{P}_k = \mathscr{G}_k \oplus \mathrm{Im}\,\mathrm{ad}_L. \tag{5.10}$$

Then we have proved the following:

Theorem 5.1 (Poincaré–Birkhoff Normal Form Theorem). *Let $L = (df)_0$ and choose a value of k. Then there exists a polynomial change of coordinates of degree k such that in the new coordinates the system* (5.1) *has the form*

$$\dot{y} = Ly + g_2(y) + \cdots + g_k(y) + \cdots$$

where $g_j \in \mathscr{G}_j$, and $\cdots$ indicates terms of degree at least $k + 1$.

We call the system

$$\dot{x} = Lx + g_2(x) + \cdots + g_k(x) \tag{5.11}$$

the (kth order) *truncated Birkhoff normal form* of (5.1).

The dynamics of the truncated Birkhoff normal form are related to, but not identical with, the local dynamics of the system (5.1) around the equilibrium point $x = 0$. The question of determining precisely which qualitative features of the dynamics are preserved in the kth order truncated Birkhoff normal form for some k, is still open. One technical problem is that the process of putting a vector field f into Birkhoff normal form is a formal one. Suppose that f is analytic. Although there exists a change of coordinates that puts f in normal form for each k, it is not possible to find a single change of coordinates that puts f into normal form to all orders. The problem is one of "small divisors." The power series defined by successive coordinate changes may have zero radius of convergence; see, for example, Siegel and Moser [1956]. If f is C^∞ (and the successive coordinate changes are invertible on a common domain) then there exists a C^∞ change of coordinates putting f in Birkhoff normal form to all orders. Then, however, one must deal explicitly with the flat "tail." Despite this truncation problem, the Birkhoff normal form is an important tool for discussing Hopf bifurcation, as we show below.

There are many possible choices for the complement $\mathscr{G}_k$ in (5.10). However, it has been proved by Cushman and Sanders [1986] and Elphick et al. [1986] that there exists a canonical choice in which the elements of $\mathscr{G}_k$ commute with a one-parameter group **S** of mappings defined in terms of the linear part L of f. The main objectives of this section are to prove this result and to show that it has a simple generalization to the equivariant setting. The group **S** effectively introduces extra symmetry into the mathematical analysis, with important consequences. Recall that E_i is the real eigenspace associated with the eigenvalues $\pm i$. In §7 we show that for Hopf bifurcation the action of **S**, restricted to E_i, may be interpreted as the symmetries induced by phase shifts already discussed in the context of Liapunov–Schmidt reduction. However, it is (an approximation to) the vector field in the original space $\mathbb{R}^n$, rather than the reduced bifurcation equation on E_i, that inherits the new symmetries. This will turn out to be useful in stability calculations.

We now define the group **S** and present the proof of the theorem by Elphick et al. [1986]. The linear part $L = (df)_0$ acts on $\mathbb{R}^n$, and we have a one-parameter group of transformations

$$\mathbf{R} = \{\exp(sL^t)\colon s \in \mathbb{R}\}.$$

This is a group since

$$\exp(s_1 L^t)\exp(s_2 L^t) = \exp((s_1 + s_2)L^t),$$

but it may not be a Lie subgroup since it may not be closed; see Remark (b). We let **S** be the closure of this group in $\mathbf{GL}(n)$,

$$\mathbf{S} = \bar{\mathbf{R}} = \overline{\{\exp(sL^t)\}}.$$

Then **S**, being closed, is a Lie subgroup of $\mathbf{GL}\,(n)$.

Remark 5.2.

(a) **R** is an abelian group since the exponentials commute. Hence its closure **S** is also an abelian group.

(b) A skew line on a torus provides an example where **R** is not closed. For instance, let

$$L = \begin{bmatrix} 0 & -1 & 0 & 0 \\ 1 & 0 & 0 & 0 \\ 0 & 0 & 0 & -\sqrt{2} \\ 0 & 0 & \sqrt{2} & 0 \end{bmatrix}.$$

Then

$$\exp(sL^t) = \begin{bmatrix} R_s & 0 \\ 0 & R_{\sqrt{2}s} \end{bmatrix}$$

which is dense on the torus $\mathbf{T}^2 = \mathbf{S}^1 \times \mathbf{S}^1 \subset \mathbb{R}^2 \times \mathbb{R}^2 = \mathbb{R}^4$, where $\mathbf{S}^1$ is the unit circle in $\mathbb{R}^2$. Thus **R** is not closed. In this case we have $\mathbf{S} = \mathbf{T}^2$.

(c) **S** need not in general be compact. For example, suppose that

$$L = \begin{bmatrix} 0 & 1 \\ 0 & 0 \end{bmatrix}.$$

Then

$$\exp(sL^t) = \begin{bmatrix} 1 & 0 \\ s & 1 \end{bmatrix}$$

which is unbounded for $s \in \mathbb{R}$.
(d) The general form of **S** is either $\mathbf{T}^k$ or $\mathbf{T}^k \times \mathbb{R}$ as we show in Proposition 5.7(a).

Having made these preliminary remarks, we can now state the first main result:

Theorem 5.3 (Elphick et al. [1986]). *Let* **S** *be as earlier. Define*

$$\mathscr{P}_k(\mathbf{S}) = \{p \in \mathscr{P}_k\colon p \text{ commutes with } \mathbf{S}\}.$$

Then for $k \geq 2$,

$$\mathscr{P}_k = \mathscr{P}_k(\mathbf{S}) \oplus \operatorname{Im} \operatorname{ad}_L. \tag{5.12}$$

The proof of Theorem 5.3 depends on two lemmas, and we state these first.

Lemma 5.4.

(a)
$$\operatorname{ad}_L p(x) = \frac{d}{ds} e^{-sL} p(e^{sL}x)|_{s=0}. \tag{5.13}$$

(b) $\operatorname{ad}_L p = 0$ *if and only if* p *commutes with* **S**. *That is,* $\mathscr{P}_k(\mathbf{S}) = \ker \operatorname{ad}_L$.

Proof.
(a) Compute

$$\frac{d}{ds} e^{-sL} p(e^{sL}x) = e^{-sL}[Lp(e^{sL}x) + (dp)_{e^{sL}x} Le^{sL}x]$$

$$= e^{-sL}[\operatorname{ad}_L p](e^{sL}x). \tag{5.14}$$

Evaluate (5.14) at $s = 0$ to get (5.13(a)).
(b) By (5.13(a)), if p commutes with **S**, then $\operatorname{ad}_L p = 0$. Conversely, (5.14) implies that if $\operatorname{ad}_L p \equiv 0$, then $e^{-sL}p(e^{sL}x)$ is a constant function of s, hence equals its value at $s = 0$, which is $p(x)$. Thus p commutes with **S**. □

The second and most interesting point is as follows:

Lemma 5.5. *There exists an inner product* $\langle\langle\ ,\ \rangle\rangle$ *on* $\mathscr{P}_k$ *such that*

$$\langle\langle \operatorname{ad}_L P, Q \rangle\rangle = \langle\langle P, \operatorname{ad}_{L^t} Q \rangle\rangle \tag{5.15}$$

for all $P, Q \in \mathscr{P}_k$.

We postpone the proof of Lemma 5.5 and first show how the main result follows:

PROOF OF THEOREM 5.3. By the Fredholm alternative

$$\operatorname{Im} A = (\ker A^t)^\perp$$

for any linear mapping of a vector space V into itself. Therefore,

$$\begin{aligned}\operatorname{Im} \operatorname{ad}_L &= [\ker(\operatorname{ad}_L)^t]^\perp \\ &= [\ker \operatorname{ad}_{L^t}]^\perp,\end{aligned}$$

the second equality following from Proposition 5.5. Hence

$$\operatorname{Im} \operatorname{ad}_L \oplus \ker \operatorname{ad}_{L^t} = \mathscr{P}_k.$$

Now apply Lemma 5.4(b) to conclude that $\ker \operatorname{ad}_{L^t} = \mathscr{P}_k(\mathrm{S})$. □

In order to prove Lemma 5.5, we first define the scalar product to be used. Let α, β be multi-indices and let x^α, x^β be the associated monomials. Define

$$\langle x^\alpha, x^\beta \rangle = \delta_{\alpha\beta}\alpha!. \tag{5.16}$$

Use (5.16) and linearity to define an inner product on the space of real-valued polynomials. Finally, let $P = (p_1, \ldots, p_n)$ and $Q = (q_1, \ldots, q_n)$ be in $\mathscr{P}_k$, and define

$$\langle\!\langle P, Q \rangle\!\rangle = \sum_{j=i}^{n} \langle p_j, q_j \rangle. \tag{5.17}$$

There is a slightly different way to define the inner product $\langle\ ,\ \rangle$, and hence $\langle\!\langle\ ,\ \rangle\!\rangle$. Rewrite (5.16) as

$$\langle x^\alpha, x^\beta \rangle = \frac{\partial^{|\alpha|}}{\partial x^\alpha} x^\beta |_{x=0}.$$

Linearity now implies that

$$\langle x^\alpha, q(x) \rangle = \left[\frac{\partial^{|\alpha|}}{\partial x^\alpha} q\right](0)$$

and that

$$\langle p(x), q(x) \rangle = p(\partial/\partial x_1, \ldots, \partial/\partial x_n) q(0). \tag{5.18}$$

PROOF OF LEMMA 5.5. We let $\langle\!\langle\ ,\ \rangle\!\rangle$ be defined as in (5.17). Let $A\colon \mathbb{R}^n \to \mathbb{R}^n$ be linear. We claim that

$$\begin{aligned}&\text{(a)}\quad \langle\!\langle AP(x), Q(x) \rangle\!\rangle = \langle\!\langle P(x), A^t Q(x) \rangle\!\rangle \\ &\text{(b)}\quad \langle\!\langle P(Ax), Q(x) \rangle\!\rangle = \langle\!\langle P(x), Q(A^t x) \rangle\!\rangle\end{aligned} \tag{5.19}$$

for all $P, Q \in \mathscr{P}_k$. It follows that

$$\text{(c)}\quad \langle\langle e^{-sL}P(e^{sL}x), Q(x)\rangle\rangle = \langle\langle P(x), e^{-sL^t}Q(e^{sL^t}x)\rangle\rangle.$$

Differentiating (5.20) with respect to s, evaluating at $s = 0$, and applying Lemma 5.4(b) yield (5.15).

To verify (5.19(a)) we may assume, using linearity, that $P(x) = x^\alpha e_j$ and $Q(x) = x^\beta e_m$ where $|\alpha| = |\beta| = k$ and e_i is the unit vector in the ith direction. Let $A = (a_{ij})$ and compute using (5.17)

$$\begin{aligned}\langle\langle A(x^\alpha e_j), x^\beta e_m\rangle\rangle &= \left\langle\!\!\left\langle x^\alpha \sum_l a_{lj}e_l, x^\beta e_m\right\rangle\!\!\right\rangle \\ &= \sum_m a_{lj}\langle x^\alpha, x^\beta\rangle \delta_{lm} \\ &= a_{mj}\langle x^\alpha, x^\beta\rangle.\end{aligned}$$

Similarly

$$\begin{aligned}\langle\langle x^\alpha e_j, A^t(x^\beta e_m)\rangle\rangle &= a^t_{jm}\langle x^\beta, x^\alpha\rangle \\ &= a_{mj}\langle x^\alpha, x^\beta\rangle.\end{aligned}$$

Thus (5.19(a)) holds.

By linearity and (5.17), (5.19(b)) is verified if we can show that

$$\langle p(Ax), q(x)\rangle = \langle p(x), q(A^t x)\rangle \tag{5.20}$$

where p and q are homogeneous polynomials of degree k. Consider now the change of coordinates $x = A^t y$. By the chain rule $\partial/\partial x = A^{-1}(\partial/\partial y)$. Using (5.18), (5.20) becomes

$$p\left(\frac{\partial}{\partial y}\right)q(A^t y)|_{y=0} = p\left(\frac{\partial}{\partial x}\right)q(A^t x)|_{x=0},$$

which is trivially true. □

Examples 5.6.

(a) Suppose $f(x) = Lx + \cdots$ where

$$L = \begin{bmatrix} 0 & 1 \\ -1 & 0 \end{bmatrix}.$$

Then $\exp(sL^t) = R_s$, a rotation in the plane; and $\mathbf{S} = \mathbf{S}^1$. In Lemma VIII, 2.5, we showed that $\mathscr{P}_k(\mathbf{S}^1) = 0$ if k is even and is generated by

$$(x_1^2 + x_2^2)^l \begin{bmatrix} x_1 \\ x_2 \end{bmatrix} \quad \text{and} \quad (x_1^2 + x_2^2)^l \begin{bmatrix} -x_2 \\ x_1 \end{bmatrix}$$

if $k = 2l + 1$.

(b) Suppose that $f(x) = Lx + \cdots$ where

$$L = \begin{bmatrix} 0 & 1 \\ 0 & 0 \end{bmatrix}.$$

As in Remark 5.2(c), **S** consists of all mappings

$$\begin{bmatrix} 1 & 0 \\ s & 1 \end{bmatrix}$$

so $\mathscr{P}_k(\mathbf{S})$ consists of polynomial mappings $P = (p_1, p_2)$ satisfying

$$\begin{aligned} &\text{(a)} \quad p_1(x_1, x_2) = p_1(x_1, x_2 + sx_1) \\ &\text{(b)} \quad p_2(x_1, x_2) = p_2(x_1, x_2 + sx_1) - sp_1(x_1, x_2 + sx_1). \end{aligned} \tag{5.21}$$

Since s is arbitrary, (5.21(a)) holds only when p_1 is independent of x_2. Since p_1 is homogeneous of degree k it must equal cx_1^k. Identity (5.21(b)) now takes the form

$$p_2(x_1, x_2) = p_2(x_1, x_2 + sx_1) - scx_1^k. \tag{5.22}$$

Differentiating (5.22) with respect to s we see that $p_{2,x_2} = cx_1^{k-1}$. Therefore,

$$p_2 = cx_1^{k-1}x_2 + dx_1^k.$$

Thus $\mathscr{P}_k(\mathbf{S})$ is two-dimensional, being generated by

$$\begin{bmatrix} 0 \\ x_1^k \end{bmatrix} \quad \text{and} \quad \begin{bmatrix} x_1^k \\ x_1^{k-1}x_2 \end{bmatrix}.$$

The general form of **S** may be deduced from an observation of Van der Meer [1985]. By linear algebra (cf. Hoffman and Kunze [1971]) we may decompose

$$L = D + N$$

where D is semisimple, N is nilpotent, and $DN = ND$. Since D is semisimple and all its eigenvalues are imaginary we may choose coordinates so that D has block diagonal form with each block D_j being either ω or $\left[\begin{smallmatrix} 0 & \omega \\ -\omega & 0 \end{smallmatrix}\right]$ for some $\omega \in \mathbb{R}$. In these coordinates $\overline{\exp(sD)}$ is a k-torus where k is the number of algebraically independent eigenvalues (or frequencies) in D.

Proposition 5.7.

(a)
$$\mathbf{S} = \begin{cases} \mathbf{T}^k & \text{if } N = 0 \\ \mathbf{T}^k \times \mathbb{R} & \text{if } N \neq 0 \end{cases}$$

where $\mathbf{T}^k = \overline{\exp(sD)}$ *and* $\mathbb{R} = \exp(sN^t)$.

(b) *The truncated Birkhoff normal form commutes with* $\mathbf{T}^k$.

PROOF. Statement (b) follows from (a) since the linear part L of the Birkhoff normal form commutes with $\mathbf{T}^k$ (but not with N^t if $N \neq 0$).

To prove (a), we claim that

$$\ker L = \ker D \cap \ker N. \tag{5.23}$$

Since $L = D + N$, clearly $\ker L \supset \ker D \cap \ker N$. If we show that $\ker L \subset \ker D$, then $\ker L \subset \ker N$ as well since $N = L - D$, and (5.23) is proved. To show that $\ker L \subset \ker D$, let $v \in \ker L$. Then $D^l v = (L - N)^l v = (-N)^l v$ since L and N commute and $Lv = 0$. Since N is nilpotent, $N^l = 0$ for some l, so $D^l v = 0$. But D is semisimple, so $Dv = 0$, as required.

Now we claim that

$$\mathrm{ad}_{L^t} = \mathrm{ad}_{D^t} + \mathrm{ad}_{N^t} \tag{5.24}$$

is the semisimple/nilpotent decomposition of ad_{L^t} when viewed as a linear map on $\mathscr{P}_k$. Direct calculation shows that

$$\text{(a)} \quad \mathrm{ad}_A \, \mathrm{ad}_B = \mathrm{ad}_B \, \mathrm{ad}_A$$

whenever A and B commute, and (5.25)

$$\text{(b)} \quad \mathrm{ad}_{A^l} = (\mathrm{ad}_A)^l.$$

By (5.25(a)) ad_{D^t} and ad_{N^t} commute, and by (5.25(b)) ad_{N^t} is nilpotent and ad_{D^t} is semisimple. This proves (5.24).

Apply (5.23) to (5.24) to obtain

$$\ker \mathrm{ad}_{L^t} = \ker \mathrm{ad}_{D^t} \cap \ker \mathrm{ad}_{N^t}.$$

But $D^t = -D$, so $\ker \mathrm{ad}_{D^t} = \ker \mathrm{ad}_D$. Lemma 5.4(b) now proves Proposition 5.7(a). □

The next main result in this section is the equivariant version of Theorem 5.3. Suppose that $f(x)$ is Γ-equivariant. Write

$$f(x) = Lx + f_2(x) + \cdots + f_k(x) + \cdots \tag{5.26}$$

and observe that each f_j is also Γ-equivariant. The Birkhoff normal form for (5.26) is obtained using near-identity changes of coordinates (5.3), which themselves are Γ-equivariant; that is, we may assume $P_k(y)$ in (5.3) is Γ-equivariant. As in the proof of Theorem 5.1, we can eliminate any term in (5.26) in the image of ad_L. Since $\mathrm{ad}_L(P_k)(y) = LP_k(y) - (dP_k)_y Ly$ it is easy to see that $\mathrm{ad}_L(P_k)$ is also Γ-equivariant. Thus we have proved the following:

Theorem 5.8. *Let f be Γ-equivariant, let $L = (df)_0$, and choose a value of k. Then there exists a Γ-equivariant change of coordinates of degree k such that in the new coordinates the system* (5.1) *has the form*

$$Y = Ly + g_2(y) + \cdots + g_k(y) + h$$

where $g_j \in \mathscr{G}_j$, h is of order $k + 1$, and

$$\mathscr{P}_k(\Gamma) = \mathscr{G}_j \oplus \mathrm{ad}_L(\mathscr{P}_k(\Gamma)). \tag{5.27}$$

The equivariant version of Theorem 5.3 is as follows:

Theorem 5.9. *Let* **S** *be as defined before Remark 5.2. Then for $k \geq 2$,*

$$\mathscr{P}_k(\Gamma) = \mathscr{P}_k(\Gamma \times \mathbf{S}) \oplus \mathrm{ad}_L(\mathscr{P}_k(\Gamma)). \tag{5.28}$$

Remark. The actions of Γ and $\mathbf{S}$ on $\mathbb{R}^n$ commute, thus yielding an action of $\Gamma \times \mathbf{S}$. To see this recall that Γ is assumed to act orthogonally and commutes with L. Now $\gamma L = L\gamma$ implies $L^t\gamma^t = \gamma^t L^t$ for all $\gamma \in \Gamma$. But $\gamma^t = \gamma^{-1}$ and γ is arbitrary. Hence $L^t\gamma = \gamma L^t$ for all $\gamma \in \Gamma$. Therefore, Γ commutes with $\exp(sL^t)$, hence with $\mathbf{S}$.

Before proving Theorem 5.9 we present a preliminary result. Let A, Y: $\mathbb{R}^n \to \mathbb{R}^n$ and assume that A is linear. Define

$$A_* Y(x) = A^{-1} Y(Ax). \tag{5.29}$$

For a compact group Γ we can define

$$\rho(Y) = \int_\Gamma \gamma_*(Y) d\gamma \tag{5.30}$$

using the normalized invariant Haar measure.

Lemma 5.10. *ρ: $\mathscr{P}_k \to \mathscr{P}_k(\Gamma)$ is a linear projection.*

Proof. The mapping ρ is clearly linear. Note that if Y is a homogeneous polynomial of degree k, then so is $\gamma_* Y$. Since ρ just averages these polynomials, we see that $\rho(Y)$ is also in $\mathscr{P}_k$. Moreover, for $\tau \in \Gamma$ we have

$$\begin{aligned}\tau_*\rho(Y) &= \int_\Gamma \tau_*\gamma_*(Y)d\gamma \\ &= \int_\Gamma (\tau\gamma)_*(Y)d\gamma \\ &= \int_\Gamma (\tau\gamma)_*(Y)d(\tau\gamma) \\ &= \rho(Y),\end{aligned}$$

showing that $\rho(Y) \in \mathscr{P}_k(\Gamma)$. If $Y \in \mathscr{P}_k(\Gamma)$ then $\gamma_* Y = Y$ so that $\rho(Y) = Y$. Therefore, $\rho(\rho(Y)) = \rho(Y)$, and ρ is a projection. □

Proof of Theorem 5.8. Since ρ commutes with the action of $\mathbf{S}$, the proof of Lemma 5.10 adapts to prove that

$$\rho: \mathscr{P}_k(\mathbf{S}) \to \mathscr{P}_k(\Gamma \times \mathbf{S})$$

is a projection. By Theorem 5.2

$$\mathscr{P}_k = \mathscr{G}_k \oplus \operatorname{Im} \operatorname{ad}_L.$$

Applying ρ to this we get

$$\begin{aligned}\mathscr{P}_k(\Gamma) &= \rho[\mathscr{P}_k(\mathbf{S})] \oplus \rho(\operatorname{Im} \operatorname{ad}_L) \\ &= \mathscr{P}_k(\Gamma \times \mathbf{S}) \oplus \rho(\operatorname{Im} \operatorname{ad}_L).\end{aligned}$$

But $\gamma_* \operatorname{ad}_L(Y) = \operatorname{ad}_L(\gamma_* Y)$ since

$$\begin{aligned}\gamma_* \operatorname{ad}_L(Y)(x) &= \gamma_*(LY(x) - (dY)_x Lx) \\ &= \gamma^{-1}[LY(\gamma x) - (dY)_{\gamma x} L\gamma x] \\ &= L\gamma_* Y(x) - \gamma^{-1}(dY)_{\gamma x}\gamma Lx \\ &= L\gamma_* Y(x) - (d(\gamma_* Y))_x Lx \\ &= \operatorname{ad}_L(\gamma_* Y)(x).\end{aligned}$$

Therefore $\rho(\operatorname{ad}_L Y) = \operatorname{ad}_L(\rho Y)$, so $\rho(\operatorname{Im}\operatorname{ad}_L) = \operatorname{ad}_L(\mathscr{P}_k(\Gamma))$, as desired. □

EXERCISES

5.1. Show that the proof of the "easy" equivariant Hopf theorem 2.2 yields the "hard" equivariant Hopf theorem 4.1 if the vector field f is in Birkhoff normal form. (*Note*: The problem is then to show that the periodic solutions so obtained survive breaking the normal form.)

5.2. Let $\mathbf{O}(2)$ act standardly on $\mathbb{C} \oplus \mathbb{C}$, and let $(df)_0$ be

$$\left[\begin{array}{c|c} 0 & I_2 \\ \hline 0 & 0 \end{array}\right].$$

Find the equivariant normal form. See Dangelmayr and Knobloch [1987a].

§6. Floquet Theory and Asymptotic Stability

In this section we take the first steps towards computing the asymptotic stability of the periodic solutions $u(s)$ to

$$(1 + \tau)\frac{dw}{ds} + f(w, \lambda) = 0 \tag{6.1}$$

obtained in Theorem 4.1. (Recall that τ is the period-scaling parameter, and $s = (1/1 + \tau)t$ is rescaled time.) We do this by using an equivariant version of Floquet theory. A heuristic description of Floquet theory was given in VIII, §4(b).

In this section we address three issues:

(a) The number of Floquet multipliers that are forced by symmetry to equal 1;
(b) The explicit solution of the Floquet equation when f in (6.1) is in Birkhoff normal form, and the implications for stability;
(c) The effect of symmetry on this explicit solution.

A more delicate analysis, developed in §§10 and 11, exploits these symmetry restrictions on the Floquet multipliers to obtain information on the stability of the periodic solution $u(s)$, even when f is not in Birkhoff normal form.

(a) Floquet Theory

Let $u(s)$ be a 2π-periodic solution to (6.1). The *Floquet equation* for $u(s)$ is the linearization of (6.1) about $u(s)$; that is,

$$\frac{dz}{ds} + \frac{1}{1+\tau}(df)_{u(s),\lambda} z = 0. \tag{6.2}$$

The *Floquet operator* (or *monodromy operator*) $M_u \colon \mathbb{R}^n \to \mathbb{R}^n$ is defined as follows. Let $z_0 \in \mathbb{R}^n$ and let $z(s)$ be a solution to (6.2) such that $z(0) = z_0$. Define

$$M_u z_0 = z(2\pi). \tag{6.3}$$

Since (6.2) is linear, M_u is linear. The eigenvalues of M_u are the *Floquet multipliers* of $u(s)$.

Because the periodic solution $u(s)$ is "neutrally stable" to phase shift, at least one Floquet multiplier is always forced to be 1. The standard Floquet theorem, Proposition VIII, 4.4, states that if the remaining eigenvalues of M_u lie strictly inside the unit circle, then $u(s)$ is an asymptotically stable periodic solution to (5.1).

As in the case of steady-state bifurcation (XIII, 4.4) symmetry may force many eigenvalues of M_u to 1. More precisely, let $\Sigma \subset \Gamma \times \mathbf{S}^1$ be the isotropy subgroup of $u(s)$, and define

$$d_\Sigma = \dim \Gamma + 1 - \dim \Sigma. \tag{6.4}$$

Then we have the following:

Proposition 6.1. *The Floquet operator M_u has d_Σ eigenvalues equal to* 1.

Before proving this, we state—without proof—the basic result of equivariant Floquet theory.

Theorem 6.2. *A small-amplitude 2π-periodic solution $u(s)$ to* (6.1) *is orbitally asymptotically stable if the $n - d_\Sigma$ eigenvalues of M_u, not forced to* 1 *by symmetry have modulus less than* 1. *The solution $u(s)$ is unstable if an eigenvalue of M_u has modulus greater than* 1. □

Proof of Proposition 6.1. Recall from standard Floquet theory that there is one eigenvalue of M_u, with eigenvector $\dot{u}(0)$, that is forced to 1 by phase shifts. To see this evaluate (6.1) at $u(s+\theta)$ obtaining

$$u(s+\theta) + \frac{1}{1+\tau} f(u(s+\theta), \lambda) = 0. \tag{6.5}$$

Differentiate (6.5) with respect to θ and let $\theta = 0$, to obtain

$$\ddot{u}(s) + \frac{1}{1+\tau}(df)_{u(s),\lambda}\dot{u}(s) = 0. \tag{6.6}$$

Since u is 2π-periodic, so is $\dot{u}$. By (6.6) $M_u\dot{u}(0) = \dot{u}(0)$ as claimed.

This argument can be adapted for any smooth arc γ_t in Γ with $\gamma_0 = 1$. The Γ-equivariance of f implies that

$$\gamma_t\dot{u}(s) + \frac{1}{1+\tau}f(\gamma_t u(s), \lambda) = 0. \tag{6.7}$$

Differentiate (6.7) with respect to t and set $t = 0$:

$$\frac{d}{ds}[\dot{\gamma}_0 u(s)] + \frac{1}{1+\tau}(df)_{u(s),\lambda}[\dot{\gamma}_0 u(s)] = 0. \tag{6.8}$$

Since $\dot{\gamma}_0 u(s)$ is 2π-periodic, (6.8) implies that

$$M_u[\dot{\gamma}_0 u(0)] = \dot{\gamma}_0 u(0).$$

If $\dot{\gamma}_0 u(0) \neq 0$ we have another eigenvector of M_u with eigenvalue equal to 1.

We claim that these constructions yield d_Σ independent eigenvectors of M_u. First, the two preceding constructions may be incorporated into one: let (γ_t, θ_t) be a smooth arc in $\Gamma \times \mathbf{S}^1$ with $(\gamma_0, \theta_0) = (1, 0)$. Then

$$\frac{d}{dt}[(\gamma_t, \theta_t)\cdot u(s)]|_{t=0} \tag{6.9}$$

is an eigenvector of M_u with eigenvalue 1.

Define $\alpha: \Gamma \times \mathbf{S}^1 \to \mathbb{R}^n$ by $\alpha(\gamma, \theta) = \gamma u(\theta)$. The number of independent eigenvectors in (6.9) is

$$\dim \operatorname{Im}(d\alpha)_{(1,0)} = \dim \Gamma \times \mathbf{S}^1 - \dim \ker(d\alpha)_{(1,0)}. \tag{6.10}$$

Since $\alpha^{-1}(u(0)) = \Sigma$, the isotropy subgroup of u, we have

$$\dim \ker(d\alpha)_{(1,0)} = \dim \Sigma.$$

Thus the right-hand side of (6.10) is d_Σ. □

Remark. Proposition 6.2 may also be proved by using the methods of Montaldi, Roberts, and Stewart [1988]. They show that the Floquet operator commutes with an action of Σ that is isomorphic to the action obtained by restriction from $\Gamma \times \mathbf{S}^1$.

(b) Floquet Theory in Birkhoff Normal Form

For the rest of this section we assume that $\mathbb{R}^n$ is Γ-simple and f in (6.1) is in Birkhoff normal form. Subsection (b) establishes two main points. When f is in Birkhoff normal form all periodic solutions $u(s)$ to (6.1) are $\mathbf{S}^1$-orbits, hence geometric circles. The Floquet equation can then be solved *exactly* by using the $\mathbf{S}^1$ symmetry of Birkhoff normal form, and the Floquet operator can be found.

The first point is established in the following theorem; see Renardy [1982]:

Theorem 6.3. *Suppose that $\mathbb{R}^n$ is Γ-simple, and that f in* (6.1) *commutes with $\Gamma \times \mathbf{S}^1$, where $s \in \mathbf{S}^1$ acts by $s \cdot x = e^{-sJ}x$, $J = (df)_{0,0}$. Then every solution is of the form*

$$u(s) = e^{-sJ}u(0), \qquad u(0) \in \mathbb{R}^n. \tag{6.11}$$

Moreover, $u(0)$ must satisfy the steady-state equation

$$\frac{1}{1+\tau} f(u(0), \lambda) - Ju(0) = 0. \tag{6.12}$$

Proof. In the Liapunov–Schmidt reduction the $\mathbf{S}^1$-action on a solution $u(s)$ is identified with phase shift. Hence $u(s) = e^{-sJ}u(0)$ and (6.11) holds. To prove (6.12) substitute (6.11) in (6.1). Then

$$\begin{aligned} 0 &= du/ds + \frac{1}{1+\tau} f(u, \lambda) \\ &= -Je^{-sJ}u(0) + \frac{1}{1+\tau} f(e^{-sJ}u(0), \lambda) \\ &= e^{-sJ}\left(-Ju(0) + \frac{1}{1+\tau} f(u(0), \lambda)\right). \end{aligned}$$

Multiplying by e^{sJ} we obtain (6.12). □

Note that $u(0)$, and hence $u(s)$, depends on λ and τ. The theorem implies that when f is in Birkhoff normal form, the periodic orbits $u(s)$ occurring via Hopf bifurcation are geometric circles. The second point is that we can write down the Floquet operator M_u explicitly:

Proposition 6.4. *For a periodic solution $(u(s), \lambda_0, \tau_0)$ to* (6.1) *of the form* (6.11), *the Floquet operator is*

$$M_u = \exp\left[-2\pi\left(\frac{1}{1+\tau}(df)_{u_0,\lambda_0} - J\right)\right]. \tag{6.13}$$

Proof. Recall from (6.2) that the Floquet equation is

$$dz/ds + \frac{1}{1+\tau}(df)_{u(s),\lambda} z = 0. \tag{6.14}$$

We use the group action to rewrite (6.14) as a constant coefficient equation. In particular, let

$$z(s) = e^{-sJ}w(s). \tag{6.15}$$

Observe that $e^{2\pi J} = I_n$. This follows by direct calculation, or by using the fact that the $\mathbf{S}^1$ phase shift action by e^{sJ} is 2π-periodic. It follows that

$$z(0) = w(0) \quad \text{and} \quad z(2\pi) = w(2\pi). \tag{6.16}$$

Now compute (6.14) as follows:

$$\begin{aligned} 0 &= dz/ds + \frac{1}{1+\tau_0}(df)_{u(s),\lambda_0} z(s) \\ &= -Je^{-sJ}w + e^{-sJ}dw/ds + \frac{1}{1+\tau_0}(df)_{e^{-sJ}u(0),\lambda_0} e^{-sJ}w \\ &= e^{-sJ}\left[-Jw + dw/ds + \frac{1}{1+\tau_0}(df)_{u(0),\lambda_0} w\right]. \end{aligned}$$

Thus $z(s)$ satisfies the Floquet equation precisely when

$$dw/ds + \left[\frac{1}{1+\tau_0}(df)_{u(0),\lambda_0} - J\right]w = 0.$$

This linear equation has the explicit solution

$$w(s) = \exp\left[-s\left(\frac{1}{1+\tau_0}(df)_{u(0),\lambda_0} - J\right)\right]w(0). \tag{6.17}$$

By (6.16, 6.17) and the definition (6.3) of M_u we obtain (6.13). □

We can use Proposition 6.4 to give a criterion for stability in Birkhoff normal form:

Theorem 6.5. *Assume that* (6.1) *is in Birkhoff normal form. Then a small-amplitude 2π-periodic solution $u(s)$ to* (6.1) *is orbitally asymptotically stable if the $n - d_\Sigma$ eigenvalues of*

$$(df)_{u_0,\lambda_0,\tau_0} - (1+\tau_0)J$$

that are not forced to zero by the group action have positive real parts. The solution is unstable if one of these eigenvalues has negative real part.

Proof. By (6.13), if the $n - d_\Sigma$ eigenvalues of $df - (1+\tau)J$ that are not constrained by the group to be zero have positive real parts, then the corresponding Floquet multipliers lie inside the unit circle and Theorem 6.1 implies that $u(s)$ is orbitally asymptotically stable. □

(c) Isotropy Subgroups and Eigenvalues of df

This subsection explains how the $\Gamma \times \mathbf{S}^1$ symmetry simplifies finding the eigenvalues of the Floquet operator in Birkhoff normal form. However, the title could have been "Reread XIII, §4(b)." We have reduced finding periodic solutions $u(s)$ of (6.1) to finding static equilibria u_0 of a map $\varphi\colon \mathbb{R}^n \times \mathbb{R} \to \mathbb{R}^n$ given by (6.12), namely

$$\varphi(u, \lambda) \equiv \frac{1}{1+\tau} f(u, \lambda) - Ju,$$

that commutes with $\Gamma \times \mathbf{S}^1$. We have also shown that when (6.1) is in Birkhoff normal form the (orbital) asymptotic stability of $u(s)$ is given by the (orbital) asymptotic stability of u_0, in relation to the ODE

$$\frac{du}{ds} + \varphi(u, \lambda) = 0.$$

Moreover, when properly interpreted, the isotropy subgroups of $u(s)$ and u_0 are identical. In XIII, §4(b), we described the restrictions placed on $d\varphi$ at (u_0, λ_0, τ_0) by the isotropy subgroup Σ of u_0. For emphasis, we repeat those restrictions here.

Form the isotypic decomposition $\mathbb{R}^n = W_1 \oplus \cdots \oplus W_l$. In particular the fixed-point subspace $\mathrm{Fix}(\Sigma)$ consists of all subspaces on which Σ acts trivially, so $\mathrm{Fix}(\Sigma) = W_j$ for some j. For definiteness set

$$W_1 = \mathrm{Fix}(\Sigma).$$

Using this notation, Theorem XII, 3.5, implies

$$(d\varphi)_{u_0, \lambda_0, \tau_0}(W_j) \subset W_j, \qquad j = 1, \ldots, l. \tag{6.18}$$

The block diagonal form induced on $d\varphi$ by (6.18) is often helpful when computing the eigenvalues of $d\varphi$.

When $\dim \mathrm{Fix}(\Sigma) = 2$, the assumption of Birkhoff normal form implies that we can apply the standard Hopf theorem to (6.1) restricted to $\mathrm{Fix}(\Sigma) \times \mathbb{R}$. In this case exchange of stability holds, so that if the steady-state solution is stable subcritically, then a subcritical branch of periodic solutions with isotropy subgroup Σ is unstable. As stated earlier, supercritical branches may be either stable or unstable depending on the signs of the real parts of the eigenvalues on $\mathrm{Fix}(\Sigma)^{\perp}$.

Exercise

6.1. Let $\Gamma = \mathbf{O}(2)$ in its standard action on $\mathbb{R}^2$, with $\mathbf{O}(2) \times \mathbf{S}^1$ acting on $\mathbb{R}^2 \oplus \mathbb{R}^2$ as in Exercise 4.1. There are two subgroups $\mathbf{Z}_2 = \langle \kappa \rangle$, where κ is the flip, and $\widetilde{\mathbf{SO}}(2) = \{(\theta, -\theta)\}$, that have two-dimensional fixed-point subspaces. In fact

$$\mathrm{Fix}(\mathbf{Z}_2) = \left\{\begin{bmatrix} a & b \\ 0 & 0 \end{bmatrix}\right\} \qquad \text{[standing waves]}$$

$$\mathrm{Fix}(\widetilde{\mathbf{SO}}(2)) = \left\{\begin{bmatrix} a & -b \\ b & a \end{bmatrix}\right\} \qquad \text{[rotating waves].}$$

Using the methods of subsections (b, c) find the eigenvalues of $(df)_{u_0, \lambda_0, \tau_0}$ at such periodic solutions of (6.1).

(*Hints*: In more detail: show the following:

(a) In the isotypic decomposition of $\mathbb{R}^4$ we have

Σ	W_2
$\mathbf{Z}_2$	$\begin{bmatrix} 0 & 0 \\ c & d \end{bmatrix}$
$\widetilde{\mathbf{SO}}(2)$	$\begin{bmatrix} a & b \\ b & -a \end{bmatrix}$

(b) Two eigenvalues of df are zero for standing waves, and one eigenvalue is zero for rotating waves.

(c) When $\dim \mathrm{Fix}(\Sigma) = 2$, precisely one eigenvalue of $(df)|\mathrm{Fix}(\Sigma)$ is zero. Conclude that for standing waves the two (possibly) nonzero eigenvalues of df are $\mathrm{tr}\, df|\mathrm{Fix}(\Sigma)$ and $\mathrm{tr}\, df|W_2$.

(d) The action of $(\theta, -\theta)$ on W_2 is rotation through 2θ. For rotating waves, $df|W_2$ is a rotation matrix, hence the sign of the real part of its complex conjugate pair of eigenvalues is also $\mathrm{tr}\, df|W_2$.)

Remark. By using the invariant theory of $\Gamma \times \mathbf{S}^1$, the asymptotic stability of the standing and rotating waves can be determined from the third order terms in f; see, e.g., Golubitsky and Stewart [1985]. We will present this information in Chapter XVII in more convenient coordinates.

§7. Isotropy Subgroups of $\Gamma \times \mathbf{S}^1$

In order to apply the equivariant Hopf theorem we require information on what form isotropy subgroups Σ of $\Gamma \times \mathbf{S}^1$ can take, and which of them have two-dimensional fixed-point subspaces. In this section we discuss the form of the isotropy subgroups, and in the next we discuss general methods for determining the dimensions of their fixed-point subspaces. Both of these sections require more familiarity with group-theoretic techniques than has been demanded before.

We obtain two simple but useful results. We characterize isotropy subgroups of $\Gamma \times \mathbf{S}^1$ as "twisted" subgroups, and we present a method for determining the conjugacy classes of (closed) twisted subgroups of $\Gamma \times \mathbf{S}^1$ using only group-theoretic information about Γ.

Definition 7.1. Let $H \subset \Gamma$ be a subgroup and let $\theta\colon H \to \mathbf{S}^1$ be a group homomorphism. We call

$$H^\theta = \{(h, \theta(h)) \in \Gamma \times \mathbf{S}^1 : h \in H\}$$

a *twisted* subgroup of $\Gamma \times \mathbf{S}^1$.

Our first proposition states that all proper isotropy subgroups of $\Gamma \times \mathbf{S}^1$ are twisted subgroups. Let $\pi\colon \Gamma \times \mathbf{S}^1 \to \Gamma$ be projection.

Proposition 7.2. *Let Σ be an isotropy subgroup of $\Gamma \times \mathbf{S}^1$ acting on a Γ-simple space $\mathbb{R}^n$, with $\Sigma \neq \Gamma \times \mathbf{S}^1$. Let $H = \pi(\Sigma)$. Then*
(a) *$\pi: \Sigma \to H$ is an isomorphism,*
(b) *There is a homomorphism $\theta: H \to \mathbf{S}^1$ such that $\Sigma = H^\theta$.*

PROOF.
(a) By Remark 3.3, or direct calculation, when $\mathbb{R}^n$ is Γ-simple the only vector in $\mathbb{R}^n$ fixed by some $\theta \neq 0$ in $\mathbf{S}^1$ is the zero vector. (Technically, we say that the action of $\mathbf{S}^1$ on $\mathbb{R}^n$ is *fixed-point-free*.) Thus $\Sigma \cap \mathbf{S}^1 = \mathbb{1}$. Since $\mathbf{S}^1 = \ker \pi$, we have $\Sigma \cap \ker \pi = \mathbb{1}$ and thus $\pi: \Sigma \to H$ is an isomorphism.
(b) Part (a) implies that every $\sigma \in \Sigma$ may be written uniquely as $(h, \theta(h))$ for a map $\theta: H \to \Sigma$, where $h \in \pi(\Sigma) = H$. We must show that θ is a group homomorphism. Since Σ is a subgroup of $\Gamma \times \mathbf{S}^1$, we must have

$$(h, \theta(h))(k, \theta(k)) = (hk, \theta(h)\theta(k)).$$

Therefore,

$$\theta(h)\theta(k) = \theta(hk), \quad \text{and} \quad \theta \text{ is a homomorphism.}$$ □

Remark. In representation theory a homomorphism $\theta: H \to \mathbf{S}^1$ is sometimes called a *character* of H. To each character there corresponds an orthogonal representation of H on $\mathbb{R}^2 \equiv \mathbb{C}$ defined by $h \mapsto e^{i\theta(h)}$, and conversely. However, this terminology more properly refers to *abelian* (or commutative) groups H. In the next section we shall use the term *character* in its usual sense in the representation theory of nonabelian groups, which is more general, so we prefer the term *twist* here.

Intuitively we think of elements of Γ as *spatial* symmetries (acting on $\mathbb{R}^n$), and elements of $\mathbf{S}^1$ as *temporal* symmetries, acting on periodic solutions by phase shift. In this sense, an element $\sigma = (h, \theta(h)) \in \Gamma \times \mathbf{S}^1$ is a spatial symmetry of $\theta(h) = 0$ and a combined *spatiotemporal* symmetry if $\theta(h) \neq 0$. For a given isotropy subgroup $H^\theta \subset \Gamma \times \mathbf{S}^1$, the spatial symmetries form a normal subgroup

$$K = \ker \theta. \tag{7.1}$$

To apply Theorem 4.1 it is necessary to enumerate the conjugacy classes of isotropy subgroups and then compute the dimensions of the corresponding fixed-point subspaces. Since the isotropy subgroups of $\Gamma \times \mathbf{S}^1$ are twisted subgroups, it is possible to reduce the problem of finding all twisted subgroups of $\Gamma \times \mathbf{S}^1$ to a problem about subgroups of Γ. A key step is to answer the question, When are two twisted subgroups conjugate in $\Gamma \times \mathbf{S}^1$? This is a slightly delicate question. We have two sufficient conditions, one stronger than the other:

Lemma 7.3.
(a) *Let H^θ and L^ψ be conjugate twisted subgroups in $\Gamma \times \mathbf{S}^1$. Then H and L are conjugate subgroups of Γ.*

(b) *Let H^θ and H^ψ be twisted subgroups of $\Gamma \times \mathbf{S}^1$ with spatial subgroups $K_\theta = \ker\theta$ and $K_\psi = \ker\psi$. If H^θ and H^ψ are conjugate in $\Gamma \times \mathbf{S}^1$ then there exists $\gamma \in N_\Gamma(H)$ such that $K_\psi = \gamma K_\theta \gamma^{-1}$.*

PROOF.
(a) Suppose H^θ and H^ψ are conjugate in $\Gamma \times \mathbf{S}^1$. Then there exists $(\gamma, \varphi) \in \Gamma \times \mathbf{S}^1$ such that

$$(\gamma, \varphi)^{-1} H^\theta (\gamma, \varphi) = H^\psi.$$

Let $(h, \theta(h)) \in H^\theta$. Then

$$(\gamma, \varphi)^{-1}(h, \theta(h))(\gamma, \varphi) = (\gamma^{-1} h \gamma, \theta(h)) \tag{7.2}$$

since $\mathbf{S}^1$ commutes with all of $\Gamma \times \mathbf{S}^1$. So $(\gamma^{-1} h\gamma, \theta(h)) \in L^\psi$, whence $\gamma^{-1} h \gamma \in L$. Reversing the roles of H and L we obtain $\gamma^{-1} H \gamma = L$, and H and L are conjugate in Γ.
(b) Suppose that H^θ and H^ψ are conjugate twisted subgroups of $\Gamma \times \mathbf{S}^1$. Then there exists $(\gamma, \varphi) \in \Gamma \times \mathbf{S}^1$ that conjugates H^θ to H^ψ. By (7.2), $\gamma^{-1} H \gamma = H$, whence $\gamma \in N_\Gamma(H)$. Moreover, (7.2) implies that

$$(\gamma^{-1} h \gamma, \theta(h)) = (h', \psi(h')).$$

Thus $\theta(h) = \psi(\gamma^{-1} h \gamma)$ and $\ker\theta = \gamma(\ker\psi)\gamma^{-1}$; that is, $K_\theta = \gamma K_\psi \gamma^{-1}$, as claimed. □

However, the converse to (b) is *false*. Before giving an example to show this, we introduce some terminology. Let H^θ be a twisted subgroup of $\Gamma \times \mathbf{S}^1$, with $K = \ker\theta$ as earlier. Since K is the kernel of a homomorphism, it is a normal subgroup of H, and H/K is isomorphic to a closed subgroup of $\mathbf{S}^1$, namely $\operatorname{Im}\theta$. Now the only closed subgroups of $\mathbf{S}^1$ are

$$\mathbb{1}, \mathbf{Z}_n (n = 2, 3, 4, \ldots) \quad \text{and} \quad \mathbf{S}^1. \tag{7.3}$$

We say that a twist θ is *trivial*, $\mathbf{Z}_n$, or $\mathbf{S}^1$, according as $\operatorname{Im}\theta = \mathbb{1}$, $\mathbf{Z}_n$, or $\mathbf{S}^1$.

Now we prove the falsity of the converse to part (b) of Lemma 7.3. Suppose that $\Gamma = \mathbf{Z}_3 = \langle\omega\rangle$. Let $H = \mathbf{Z}_3$ and define θ, ψ: $H \to \mathbf{S}^1$ by $\theta(\omega) = 2\pi/3$, $\psi(\omega) = -2\pi/3$. Since $\mathbf{Z}_3 \times \mathbf{S}_1$ is abelian, H^θ is conjugate to H^ψ only if $H^\theta = H^\psi$, i.e., if $\theta = \psi$, which is false. But $\ker\theta = \ker\psi = \mathbb{1}$, so the condition in (b) is trivially true.

The source of the difficulty is that effectively the condition in (b) determines the conjugacy class of the pair (H, K), $K = \ker\theta$, in $\Gamma \times \mathbf{S}^1$. But (H, K) does not determine H^θ uniquely, because it does not determine θ uniquely. However, the source of nonuniqueness can be controlled. A brief description follows: the reader may wish to expand it. Suppose θ is some homomorphism $H \to \mathbf{S}^1$ with kernel K. Then all others are of the form $\alpha \circ \theta$ where α is an automorphism of $\operatorname{Im}\theta$. Nonidentity α may or may not define a twisted group that is not conjugate to H^θ. It depends on whether α can be "induced" by conjugation in Γ. For example, if we apply the preceding construction with $\Gamma = \mathbf{O}(2)$

rather than $\mathbf{Z}_3$, but still take $H = \mathbf{Z}_3$, then H^θ and H^ψ are now conjugate, because conjugation by $\kappa \in \mathbf{O}(2)$ induces the automorphism $\alpha(y) = -y$ of $\operatorname{Im}\theta = \mathbf{Z}_3$. More precisely, $\kappa^{-1}\omega\kappa = -\omega$, so

$$\theta(\kappa^{-1}\omega\kappa) = \theta(-\omega) = -2\pi/3 = \psi(\omega).$$

The conjugacy classes of (closed) twisted subgroups of $\Gamma \times \mathbf{S}^1$ can be enumerated as follows: First, find the conjugacy classes of closed subgroups of Γ. For each conjugacy class choose a representative H. Second, find all closed normal subgroups $K \subset H$ such that H/K is isomorphic to $\mathbb{1}$, $\mathbf{Z}_n$, or $\mathbf{S}^1$. (Any such K must contain the *commutator subgroup*

$$H' = \langle h^{-1}k^{-1}hk \colon h, k \in H\rangle$$

and hence can be found from the structure of H/H'. Here $\langle\ \rangle$ indicates "group generated by.") Choose one representative of each conjugacy class of Ks under the action of $N_\Gamma(H)/H$. This gives a list of all pairs (H, K). Find the possible θ for each pair by listing the automorphisms of H/K, modulo those that are induced by conjugation by elements $\gamma \in N_\Gamma(H)$. (Since $\mathbf{S}^1$ is central its elements act trivially by conjugation and hence can be ignored.) This procedure gives a complete list. Two simplifications are often useful:

For twist types $\mathbb{1}$ and $\mathbf{Z}_2$ there are no such automorphisms.

If, as in the case of $\mathbf{O}(2)$, there exists an element κ which acts by conjugation to invert each element of H, then for twists of type $\mathbf{Z}_3$ and $\mathbf{S}^1$ the only nontrivial automorphism of H/K is inversion, but this is induced by conjugation by κ and hence can be eliminated.

EXERCISE

7.1. Let $u\colon \mathbf{S}^1 \to \mathbb{R}^n$ be a periodic solution with isotropy subgroup $\Sigma \subset \Gamma \times \mathbf{S}^1$. Let $\pi\colon \Gamma \times \mathbf{S}^1 \to \Gamma$ be projection. Let $T_u = \{u(t) \colon t \in \mathbf{S}^1\}$ be the trajectory of u. Show that

$$\pi(\Sigma) = \{\gamma \in \Gamma \colon \gamma T_u = T_u\}.$$

§8.* Dimensions of Fixed-Point Subspaces

In this section we derive criteria for determining the dimension of the fixed-point subspace of a twisted subgroup H_θ of $\Gamma \times \mathbf{S}^1$ acting on a Γ-simple space $\mathbb{R}^n$. We do this in two different ways. For twist types $\mathbb{1}$, $\mathbf{Z}_2$, $\mathbf{Z}_3$, $\mathbf{Z}_4$, and $\mathbf{Z}_6$ (which appear frequently in specific examples) we can find this dimension in terms of the dimensions of fixed-point subspaces of subgroups of Γ acting on V. See Proposition 8.4. The method is peculiar to these types of twist. The second method expresses the dimension in terms of the number of times a certain representation of H, related to the twist, appears in the given representation of H on $V \oplus V$. See Proposition 8.5.

We begin by making a distinction between "spatial" and "temporal" isotropy subgroups. An isotropy subgroup $H^\theta \subset \Gamma \times \mathbf{S}^1$ is *spatial* if

$$\mathrm{Fix}(H^\theta) = \mathrm{Fix}(K)$$

where $K = \ker\theta \subset H$. Otherwise it is *temporal.*

This classification into spatial and temporal depends on the representation of $\Gamma \times \mathbf{S}^1$, and not just on H^θ. At first sight the sensible definition of a spatial isotropy subgroup would be one containing only spatial symmetries (that is, contained in Γ). However, this leads to certain technical difficulties in the implementation of Theorem 4.1. In particular it may happen that a subgroup K of Γ has $\dim \mathrm{Fix}(K) = 2$, but K is not an isotropy subgroup because a larger twisted group H^θ also fixes $\mathrm{Fix}(\Sigma)$. An example is $K = \mathbf{Z}_2$ in §7c, where $H^\theta = \mathbf{Z}_2 \oplus \mathbf{Z}_2^c$. It is, therefore, convenient to consider H^θ as being spatial, even though the periodic solutions associated with it do have some additional spatiotemporal symmetry.

We prove two main criteria for $\dim \mathrm{Fix}(H^\theta)$ to equal 2:

Theorem 8.1. Let H^θ be a twisted subgroup of $\Gamma \times \mathbf{S}^1$ acting on $V \oplus V$ and let $K = \ker\theta$. Then $\dim \mathrm{Fix}(H^\theta) = 2$ if any of the following holds:
(a) $\theta(H) = \mathbb{1}$ and $\dim \mathrm{Fix}(H) = 1$.
(b) $\theta(H) = \mathbf{Z}_2$ and $\dim \mathrm{Fix}(K) - \dim \mathrm{Fix}(H) = 1$.
(c) $\theta(H) = \mathbf{Z}_3$ and $\dim \mathrm{Fix}(K) - \dim \mathrm{Fix}(H) = 2$.
(d) $\theta(H) = \mathbf{Z}_4$ and $\dim \mathrm{Fix}(K) - \dim \mathrm{Fix}(L) = 2$ where L is the unique subgroup such that $K \subset L \subset H$ and $|H/L| = 2$.
(e) $\theta(H) = \mathbf{Z}_6$ and $\dim \mathrm{Fix}(H) + \dim \mathrm{Fix}(K) - \dim \mathrm{Fix}(L) - \dim \mathrm{Fix}(M) = 2$ where L, M are the unique subgroups between K and H such that $|H/L| = 3$, $|H/M| = 2$.

We postpone the proof in order to state the second criterion. For each twist $\theta\colon H \to \mathbf{S}^1$ we define an irreducible representation ρ_θ as follows. If $\theta(H) = \mathbb{1}$ or $\mathbf{Z}_2$ then H acts on $\mathbb{R}$ by

$$\rho_\theta(h)\cdot x = (\cos\theta(h))x, \qquad x \in \mathbb{R}. \tag{8.1}$$

If $\theta(H) \neq \mathbb{1}, \mathbf{Z}_2$, then H acts on $\mathbb{C}$ by

$$\rho_\theta(h)\cdot z = e^{i\theta(h)}z, \qquad z \in \mathbb{C}. \tag{8.2}$$

Theorem 8.2. $\dim \mathrm{Fix}(H^\theta) = 2$ *precisely when* ρ_θ *occurs exactly once in the action of* H *on* V.

Before proving Theorem 8.2 we make the statement more precise by defining the multiplicity of an irreducible representation. Let G be a Lie group acting on a vector space W over $\mathbb{R}$. Decompose W into irreducible subspaces for G:

$$W = W_1 \oplus \cdots \oplus W_k.$$

Let ρ be an irreducible representation of G. Then the *multiplicity* of ρ in the action of G on W, written $\mu(\rho)$, is the number of W_j on which the representation of G is equivalent to ρ.

Theorem 8.1 is an immediate consequence of the following formulas, which use the same notation:

Proposition 8.3.
(a) *If* $\theta(H) = \mathbb{1}$ *then* $\dim \operatorname{Fix}(H^\theta) = 2 \dim \operatorname{Fix}(H)$.
(b) *If* $\theta(H) = \mathbf{Z}_2$ *then* $\dim \operatorname{Fix}(H^\theta) = 2(\dim \operatorname{Fix}(K) - \dim \operatorname{Fix}(H))$.
(c) *If* $\theta(H) = \mathbf{Z}_3$ *then* $\dim \operatorname{Fix}(H^\theta) = \dim \operatorname{Fix}(K) - \dim \operatorname{Fix}(H)$.
(d) *If* $\theta(H) = \mathbf{Z}_4$ *then* $\dim \operatorname{Fix}(H^\theta) = \dim \operatorname{Fix}(K) - \dim \operatorname{Fix}(L)$.
(e) *If* $\theta(H) = \mathbf{Z}_6$ *then* $\dim \operatorname{Fix}(H^\theta) = \dim \operatorname{Fix}(H) + \dim \operatorname{Fix}(K) - \dim \operatorname{Fix}(L) - \dim \operatorname{Fix}(M)$.

Theorem 8.2 is an equally direct consequence of the following:

Proposition 8.4. $\dim \operatorname{Fix}(H^\theta)$ *is equal to twice the multiplicity of* ρ_θ *in the action of* H *on* V.

Both these propositions follow from the trace formula, which we recall from XIII, §2(b). See Theorem XIII, 2.3. Let H be a closed subgroup of a group G acting on the vector space W. The trace formula is

$$\dim \operatorname{Fix}(H) = \int_H \operatorname{Tr}(h) \tag{8.3}$$

where the integral is with respect to normalized Haar measure on H.

We apply the trace formula as follows: Instead of considering the action of H^θ on $\mathbb{R}^n$ we consider the corresponding "twisted" action of $H \subset \Gamma$ on $\mathbb{R}^n$ in which

$$h \cdot (v, w) = (h \cdot v, \theta(h) \cdot w). \tag{8.4}$$

Then $\operatorname{Fix}(H^\theta)$ for the original action is the same as $\operatorname{Fix}(H)$ for the twisted action, so they have the same dimensions. We show that the trace formula yields

$$\dim \operatorname{Fix}(H^\theta) = 2 \int_H \operatorname{Tr}(h) \cos \theta(h). \tag{8.5}$$

To check this we use (8.4). We can write the action of h alone in block diagonal form

$$\begin{bmatrix} h & 0 \\ 0 & h \end{bmatrix}$$

since Γ acts diagonally on $\mathbb{R}^n$. The action of θ on $\mathbb{R}^n$ can be written in block form as

$$\begin{bmatrix} \cos\theta\, I & -\sin\theta\, I \\ \sin\theta\, I & \cos\theta\, I \end{bmatrix}.$$

Since $(h, \theta(h)) = (h, 0)\cdot(1, \theta(h))$ we can find the action of $(h, \theta(h))$ by matrix multiplication, obtaining

$$\begin{bmatrix} \cos\theta(h) h & -\sin\theta(h) h \\ \sin\theta(h) h & \cos\theta(h) h \end{bmatrix}.$$

The trace of this matrix is clearly $2\cos\theta(h)\,\mathrm{Tr}(h)$, and (8.5) follows from (8.3).

PROOF OF PROPOSITION 8.3. Let $\int$ denote the normalized Haar integral on H. If $P \subset H$ and the index $|H/P| = p$ then $\int$ is $1/p$ times the normalized Haar integral on P.

(a) This is trivial since $H^\theta = H \subset \Gamma$.

(b) By (8.5)

$$\dim \mathrm{Fix}(H^\theta) = 2\left[\int_K \mathrm{Tr}(h) - \int_{H \sim K} \mathrm{Tr}(h)\right]$$

since $\cos\theta(h) = 1$ when $h \in K$ and $\cos\theta(h) = -1$ when $h \in H \sim K$.

On the other hand,

$$\dim \mathrm{Fix}(H) = \int_K \mathrm{Tr}(h) + \int_{H \sim K} \mathrm{Tr}(h),$$

and

$$\dim \mathrm{Fix}(K) = 2\int_K \mathrm{Tr}(h)$$

since $[H:K] = 2$. Putting all these together we obtain the required formula

$$\dim \mathrm{Fix}(H^\theta) = 2\dim \mathrm{Fix}(K) - 2\dim \mathrm{Fix}(H).$$

The other cases are similar but more elaborate.

(c) When $\theta(H) = \mathbf{Z}_3$ we have $\cos\theta(h) = 1$ on K, but $-\frac{1}{2}$ on $H \sim K$. Now

$$\dim \mathrm{Fix}(H^\theta) = 2\left[\int_K \mathrm{Tr}(h) - \tfrac{1}{2}\int_{H \sim K} \mathrm{Tr}(h)\right],$$

$$\dim \mathrm{Fix}(H) = \int_K \mathrm{Tr}(h) + \int_{H \sim K} \mathrm{Tr}(h),$$

$$\dim \mathrm{Fix}(K) = 3\int_K \mathrm{Tr}(h),$$

since $[H:K] = 3$. Eliminate the integrals to obtain the stated result.

(d) Denote the elements of $\mathbf{Z}_4$ by ζ^k where $k = 0, 1, 2, 3$. The cosets of K in H correspond to these four elements: let $H_k = \theta^{-1}(\zeta^k)$. Then

$$H = H_0 \cup H_1 \cup H_2 \cup H_3$$
$$K = H_0$$

and the unique subgroup L between H and K is

$$L = H_0 \cup H_2.$$

On H_0 we have $\cos\theta(h) = 1$; on H_1 and H_3 it is 0; and on H_2 it is -1. Let $I_k = \int_{H_k} \mathrm{Tr}(h)$. Then

$$\dim \mathrm{Fix}(H^\theta) = 2[I_0 - I_2],$$
$$\dim \mathrm{Fix}(H) = I_0 + I_1 + I_2 + I_3,$$
$$\dim \mathrm{Fix}(L) = 2[I_0 + I_2],$$
$$\dim \mathrm{Fix}(K) = 4I_0.$$

Eliminate the I_k.

(e) Denote the elements of $\mathbf{Z}_6$ by ζ^k where $k = 0, 1, 2, 3, 4, 5$. The cosets of K in H correspond to these six elements: let $H_k = \theta^{-1}(\zeta^k)$. Then

$$H = H_0 \cup H_1 \cup H_2 \cup H_3 \cup H_4 \cup H_5,$$
$$K = H_0.$$

The unique subgroup L between H and K with $|H/L| = 3$ is

$$L = H_0 \cup H_3.$$

The unique subgroup M such that $|H/M| = 2$ is

$$M = H_0 \cup H_2 \cup H_4.$$

On H_0 we have $\cos\theta(h) = 1$; on H_1 and H_5 it is $\frac{1}{2}$; on H_2 and H_4 it is $-\frac{1}{2}$, and on H_3 it is -1. Let $I_k = \int_{H_k} \mathrm{Tr}(h)$. Then

$$\dim \mathrm{Fix}(H^\theta) = 2[I_0 + \tfrac{1}{2}(I_1 + I_5) - \tfrac{1}{2}(I_2 + I_4) - I_3],$$
$$\dim \mathrm{Fix}(H) = [I_0 + I_1 + I_2 + I_3 + I_4 + I_5],$$
$$\dim \mathrm{Fix}(L) = 3[I_0 + I_3],$$
$$\dim \mathrm{Fix}(M) = 2[I_0 + I_2 + I_4],$$
$$\dim \mathrm{Fix}(K) = 6[I_0].$$

Eliminate the I_k. □

Remark. The method of proof used earlier works for $\mathbf{Z}_k$ only when $k = 1, 2, 3, 4, 6$; see Exercise 8.1. Presumably no similar formulas exist for the other values of k.

The proof of Proposition 8.4 uses orthogonality of characters. We summarize the required results first. Let a compact Lie group G act on W, so that $\gamma \in G$ acts on W by a linear mapping $\rho(\gamma)$. In representation theory the function

$\chi: G \to \mathbb{R}$

$$\chi(\gamma) = \operatorname{Tr} \rho(\gamma)$$

is called the *character* of ρ, and it determines ρ uniquely up to equivalence. Let χ_1 and χ_2 be the characters of irreducible representations ρ_1, ρ_2 of G. Then the *orthogonality relations* (see Adams [1969]) for real representations state that:

$$\int_G \chi_1(\gamma)\chi_2(\gamma) = \begin{cases} 0 & (\rho_1, \rho_2 \text{ inequivalent}) \\ 1 & (\rho_1 \sim \rho_2, \text{ absolutely irreducible}) \\ 2 & (\rho_1 \sim \rho_2, \text{ non-absolutely irreducible}). \end{cases} \tag{8.6}$$

Here $\sim$ denotes equivalence of representations, that is, isomorphism of the corresponding actions.

Proof of Proposition 8.4. Let χ_θ be the character of ρ_θ, and χ the character of $\rho|H$. Let $\mu(\theta)$ be the multiplicity of ρ_θ in $\rho|H$. We must show that $\dim \operatorname{Fix}(H^\theta) = 2\mu(\theta)$.

Now

$$\chi_\theta = \operatorname{Tr} \theta = \begin{cases} \cos\theta & \text{if } |\theta(H)| \le 2, \\ 2\cos\theta & \text{otherwise.} \end{cases}$$

By (8.4),

$$\dim \operatorname{Fix}(H^\theta) = 2 \int_H \chi(h) \cos\theta(h). \tag{8.7}$$

If $|\theta(H)| \le 2$ then (8.7) equals

$$2 \int_H \chi(h)\chi_\theta(h).$$

By the orthogonality relations (8.6) this is $2\mu(\theta)$ since ρ_θ is absolutely irreducible. If on the other hand, $|\theta(H)| > 2$, then (8.7) is equal to

$$\int_H \chi(h)\chi_\theta(h).$$

But ρ_θ is now non-absolutely irreducible, so this is also $2\mu(\theta)$. □

EXERCISE

8.1. Show that the method used to prove Proposition 8.3 will not work for $\mathbf{Z}_k$ twists unless $k = 1, 2, 3, 4$, or 6. (*Hint*: Identify H/K with $\mathbf{Z}_k$ and for each subgroup L with $K \subset L \subset H$ define $\varphi_L: \mathbf{Z}_k \to \{0, 1\}$ by $\varphi_L(r) = 1$ if $r \in L$, $\varphi_L(r) = 0$ if $r \notin L$. Let $\chi: \mathbf{Z}_k \to \mathbb{R}$ be defined by $\chi(r) = \cos(2\pi r/k)$. Define integrals analogous to the I_k in part (e), and show that a formula of the type derived in Proposition 8.3 holds if and only if χ is a linear combination of the φ_L. Show that this is impossible if there are two distinct numbers prime to k lying between k and $k/4$ (k even) or k and $k/2$ (k odd), and that this rules out all k except 1, 2, 3, 4, 6.)

§9. Invariant Theory for $\Gamma \times \mathbf{S}^1$

There is a general approach which sometimes simplifies the calculation of $\Gamma \times \mathbf{S}^1$-invariants and -equivariants. The idea is to compute the $\mathbf{S}^1$-invariants first (easy with Lemma 9.2) and then use these to compute the $\Gamma \times \mathbf{S}^1$-invariants. The theoretical basis for the second step is given in Lemma 9.1.

Lemma 9.1. *Let $\Gamma \times S$ be a Lie group acting on V. Let $f: V \to \mathbb{R}$ be an S-invariant function and let $\gamma \in \Gamma$. Then*

$$(\gamma \cdot f)(x) \equiv f(\gamma^{-1} x)$$

is also S-invariant. Similarly if $g: V \to V$ is S-equivariant, then so is

$$(\gamma \cdot g)(x) \equiv g(\gamma^{-1} x).$$

PROOF. Use routine computations, and note that S commutes with Γ. □

We apply this in particular to Hopf bifurcation, taking $S = \mathbf{S}^1$. We have $\Gamma \times \mathbf{S}^1$ acting on a Γ-simple space $V = \mathbb{R}^n$. By Lemma 1.5 we can assume that $\theta \in \mathbf{S}^1$ acts as $e^{-\theta J}$ where

$$J = \begin{bmatrix} 0 & -I_m \\ I_m & 0 \end{bmatrix},$$

$m = n/2$. Let the corresponding basis be $\{x_1, \ldots, x_m; y_1, \ldots, y_m\}$ and set $z_j = x_j + iy_j$. This identifies V with $\mathbb{C}^m$ in such a way that $\theta \in \mathbf{S}^1$ acts on $z = (z_1, \ldots, z_m) \in \mathbb{C}^m$ by

$$\theta \cdot z(e^{i\theta} z_1, \ldots, e^{i\theta} z_m). \tag{9.1}$$

(Alternatively this follows abstractly since the $\mathbf{S}^1$-action is fixed-point-free.)

For $z = (z_1, \ldots, z_m)$ define $\bar{z} = (\bar{z}_1, \ldots, \bar{z}_m)$. The next lemma reduces the problem of finding $\mathbb{R}$-valued invariants to that of finding $\mathbb{C}$-valued ones:

Lemma 9.2. *Let Γ act on $\mathbb{C}^m$. Suppose that $N_1, \ldots, N_s$ generate (over $\mathbb{C}$) the $\mathbb{C}$-valued invariants in z, $\bar{z}$. Then* $\mathrm{Re}(N_1), \ldots, \mathrm{Re}(N_s), \mathrm{Im}(N_1), \ldots, \mathrm{Im}(N_s)$ *generate (over $\mathbb{R}$) the $\mathbb{R}$-valued invariants.*

PROOF. The $\mathbb{R}$-valued invariants are those $\mathbb{C}$-valued invariants whose values happen to lie in $\mathbb{R}$. Hence they are generated by the real and imaginary parts of monomials $N_1^{\alpha_1}, \ldots, N_s^{\alpha_s}$. But if p and q are polynomials in $z, \bar{z}$ over $\mathbb{C}$ then

$$\mathrm{Re}(pq) = \mathrm{Re}(p)\mathrm{Re}(q) - \mathrm{Im}(p)\mathrm{Im}(q)$$

$$\operatorname{Im}(pq) = \operatorname{Re}(p)\operatorname{Im}(q) + \operatorname{Im}(p)\operatorname{Re}(q),$$

and an induction completes the proof. □

We now compute the invariants and equivariants for the action (9.1) of $\mathbf{S}^1$:

Lemma 9.3.
(a) *A Hilbert basis for the* $\mathbf{S}^1$*-invariant functions is given by the* m^2 *quadratics*

$$u_j = z_j\bar{z}_j \qquad (1 \le j \le m)$$
$$\operatorname{Re} v_{ij}, \operatorname{Im} v_{ij} \qquad (1 \le i < j \le m) \qquad \text{where } v_{ij} = z_i\bar{z}_j.$$

Relations are given by

$$v_{ij}\bar{v}_{ij} = u_i u_j.$$

(b) *Let* $g = (g_1, \ldots, g_m)\colon \mathbb{C}^m \to \mathbb{C}^m$ *be* $\mathbf{S}^1$*-equivariant. Then each* g_j *satisfies*

$$g_j(\theta \cdot z) = e^{i\theta} g_j(z). \tag{9.2}$$

The module of such $g_j\colon \mathbb{C}^m \to \mathbb{C}$ *is generated over the invariants by the* $2m$ *mappings*

$$X_j(z) = z_j$$
$$Y_j(z) = iz_j$$

for $1 \le j \le m$. *Thus the module of* $\mathbf{S}^1$*-equivariants* $\mathbb{C}^m \to \mathbb{C}^m$ *has* $2m^2$ *generators of the form* $(0, \ldots, 0, X_j, 0, \ldots, 0)$ *and* $(0, \ldots, 0, Y_j, 0, \ldots, 0)$.

Proof. As usual, by the theorems of Schwarz and Poénaru we may restrict attention to polynomials. The rest is routine, but we give details for completeness.
(a) Let f be invariant. Using multi-indices write

$$f(z) = \sum a_{\alpha\beta} z^\alpha \bar{z}^\beta \qquad (a_{\alpha\beta} \in \mathbb{C}).$$

Since $f(\theta \cdot z) = f(z)$ we have $a_{\alpha\beta} = 0$ unless $|\alpha| = |\beta|$. By pairing the z_is with $\bar{z}_j$s we can always write $z_\alpha \bar{z}_\beta$ as a monomial in the u_j and v_{ij}. Now use Lemma 9.2 and note that $\operatorname{Im} u_j = 0$ is superfluous.
(b) Let $g_j\colon \mathbb{C}^m \to \mathbb{C}$ satisfy (9.2). Then

$$g_j(z) = \sum b_{\alpha\beta} z_\alpha \bar{z}_\beta, \qquad b_{\alpha\beta} \in \mathbb{C}.$$

By (9.2) $b_{\alpha\beta} = 0$ unless

$$|\alpha| = |\beta| + 1. \tag{9.3}$$

In particular $|\alpha| > 0$. We can thus divide out a z_j from each monomial $z_\alpha \bar{z}_\beta$, and the quotient is a $\mathbb{C}$-valued $\mathbf{S}^1$-invariant. □

We now describe an approach to finding the $\Gamma \times \mathbf{S}^1$-invariants $f\colon \mathbb{C}^m \to \mathbb{R}$. By Lemma 9.3

$$f(z) = k(u_j, v_{ij}) \tag{9.4}$$

for some polynomial $k\colon \mathbb{R}^m \times \mathbb{C}^{(1/2)m(m-1)} \to \mathbb{R}$.

By Lemma 9.1, $\gamma \cdot u_j$, $\gamma \cdot \mathrm{Re}(v_{ij})$, and $\gamma \cdot \mathrm{Im}(v_{ij})$ are also quadratic, $\mathbf{S}^1$-invariant polynomials, hence can be expressed as real linear combinations of the generators u_j, $\mathrm{Re}(v_{ij})$, $\mathrm{Im}(v_{ij})$. This calculation yields an action of Γ on $\mathbb{R}^{m^2}$. A Hilbert basis for the Γ-invariants of this action, when written in the original z-coordinates and after eliminating redundancies, yields a Hilbert basis for the $\Gamma \times \mathbf{S}^1$-invariants on $\mathbb{C}^m$.

Similarly we can use Lemmas 9.1 and 9.3(b) to compute the $\Gamma \times \mathbf{S}^1$-equivariants. We use these ideas in later chapters.

§10. Relationship Between Liapunov–Schmidt Reduction and Birkhoff Normal Form

Consider as usual the system of ODEs

$$\dot{v} + f(v, \lambda) = 0 \tag{10.1}$$

where $v \in \mathbb{R}^n$, $f\colon \mathbb{R}^n \times \mathbb{R} \to \mathbb{R}^n$ commutes with the action of a compact Lie group Γ on $\mathbb{R}^n$, and $(df)_{0,0} = J$ as in (1.7).

Both the Liapunov–Schmidt and the Birkhoff normal form procedure introduce $\mathbf{S}^1$ symmetry into the analysis, but in different ways. The Liapunov–Schmidt method introduces the $\mathbf{S}^1$ symmetries in a rigorous way as phase shifts on periodic solutions. This allows us to describe symmetries of periodic solutions using subgroups of $\Gamma \times \mathbf{S}^1$ and to detect periodic solutions by using their symmetries. The Birkhoff normal form method lets us gain control not just of periodic solutions, but of the entire vector field f, by providing a change of coordinates to a vector field $\tilde{f}$ that commutes with $\Gamma \times \mathbf{S}^1$. This would obviously be superior, were it not for one technical problem: the Birkhoff normal form procedure is valid only to any finite order.

The main advantage of Birkhoff normal form is that (to any finite order) we control, in theory, all of the dynamics of (10.1), not just periodic trajectories. The main disadvantage is the formal nature of the process. The occurrence of certain dynamics in the normal form equation does not immediately imply that the same dynamics persist when the $\mathbf{S}^1$ symmetry is broken. With the Liapunov–Schmidt procedure we gain rigor when discussing periodic solutions but lose control over the rest of the dynamics. This rigor lets us discuss degenerate Hopf bifurcation in Chapter VIII. However, in this situation the dominant feature of the dynamics is the periodic orbit, and the technique matches the problem.

In later chapters we shall see a number of instances in which a Birkhoff normal form analysis yields information about invariant tori, and not just periodic trajectories. This happens, for example, in degenerate $\mathbf{O}(2)$ Hopf bifurcation and in mode interactions. On the whole, our use of Birkhoff normal form will be formal: we discuss the dynamics of the system, ignoring terms of high order. Questions concerning the effect of these high order $\mathbf{S}^1$-symmetry-breaking terms in f lie outside the scope of this volume.

In this section we show that there is an intimate connection between the Birkhoff normal form of f and the Liapunov–Schmidt reduced equation of f, both of which commute with $\Gamma \times \mathbf{S}^1$. In the discussion we shall assume that f is already in Birkhoff normal form and derive the reduced equation. That is, we assume

$$f \text{ commutes with } \Gamma \times \mathbf{S}^1. \tag{10.2}$$

Theorem 10.1. *Suppose that the vector field f in* (10.1) *is in Birkhoff normal form. Then it is possible to perform a Liapunov–Schmidt reduction on* (10.1) *such that the reduced equation φ has the form*

$$\varphi(v, \lambda, \tau) = f(v, \lambda) - (1 + \tau)Jv \tag{10.3}$$

where τ is the period-scaling parameter.

Remark. Theorem 10.1 shows that in general (aside from the way τ enters) there are no additional restrictions on the form of the reduced mapping φ obtained by a Liapunov–Schmidt reduction, other than those required by $\Gamma \times \mathbf{S}^1$ symmetry and the occurrence of purely imaginary eigenvalues in $(df)_{0,0}$.

Corollary 10.2. *Suppose that the vector field f in* (10.1) *is in Birkhoff normal form and that $\varphi(v, \lambda, \tau)$ is the mapping obtained by using the Liapunov–Schmidt procedure. Let (v_0, λ_0, τ_0) be a solution to $\varphi = 0$, and let $v(s)$ be the corresponding periodic solution of* (10.1). *Then $v(s)$ is orbitally asymptotically stable if the $n - d_\Sigma$ eigenvalues of $(d\varphi)_{v_0, \lambda_0, \tau_0}$ which are not forced to zero by the group action have positive real parts.*

Proof. Apply Theorems 6.5 and 10.1. □

The basic observation needed in the proof of Theorem 10.1 is that when f is in Birkhoff normal form all small-amplitude periodic solutions to (10.1) are also solutions to the linearized system $\dot{w} + Jw = 0$, as long as a rescaling of time is made so that the solutions to (10.1) become 2π-periodic. This is the content of Theorem 6.1.

Proof of Theorem 10.1. We begin by reviewing the Liapunov–Schmidt procedure as described in §2. First, we rescale time as $s = (1 + \tau)t$ to obtain

$$(1 + \tau)\frac{du}{ds} + f(u, \lambda) = 0. \tag{10.4}$$

This lets us assume the periodic solutions have period 2π. Then we form the operator $\Phi: \mathscr{C}_{2\pi}^1 \times \mathbb{R} \times \mathbb{R} \to \mathscr{C}_{2\pi}$ defined by

$$\Phi(u, \lambda, \tau) = (1 + \tau)\frac{du}{ds} + f(u, \lambda) \tag{10.5}$$

and seek zeros of Φ. The Liapunov–Schmidt procedure finds zeros of Φ by looking at the reduced equation

$$\varphi: \ker \mathscr{L} \times \mathbb{R} \times \mathbb{R} \to \ker \mathscr{L}$$

where $\mathscr{L} = (d/ds) + J$ is the linearization of (10.1). To define φ we use the splitting $\mathscr{C}_{2\pi} = \text{range}\, \mathscr{L} \oplus \ker \mathscr{L}$ in §4 and decouple the equation $\Phi = 0$ into the pair of equations (4.10)

$$\begin{aligned} &\text{(a)} \quad E\Phi(u, \lambda, \tau) = 0 \\ &\text{(b)} \quad (I - E)\Phi(u, \lambda, \tau) = 0, \end{aligned} \tag{10.6}$$

where $E: \mathscr{C}_{2\pi} \to \text{range}\, \mathscr{L}$ is projection with kernel $\ker \mathscr{L}$. Write $u = v + w$ where $v \in \ker \mathscr{L}$, $w \in \text{range}\, \mathscr{L}$, and solve (10.6(a)) by the implicit function theorem to express w as $W(v, \lambda, \tau)$. Then

$$\varphi = (I - E)\Phi(v + W(v, \lambda, \tau), \lambda, \tau). \tag{10.7}$$

We claim that:

> when f is in Birkhoff normal form, the implicitly defined function W is identically zero. (10.8)

Since $I - E$ is the identity on $\ker \mathscr{L}$ it follows from (10.5) and (10.7) that

$$\varphi = (1 + \tau)\frac{dv}{ds} + f(v, \lambda). \tag{10.9}$$

Note also that $v(s) \in \ker \mathscr{L}$ satisfies, by definition, the linearized equation

$$\frac{dv}{ds} + Jv = 0. \tag{10.10}$$

Now (10.9 and 10.10) show that φ has the form claimed in (10.3).

It remains to show that $W \equiv 0$. By Theorem 6.3 $v \in \ker \mathscr{L}$ if and only if

$$v(s) = e^{-sJ} v_0 \tag{10.11}$$

for some fixed vector v_0. Now (10.11) implies that

$$\begin{aligned} &\text{(a)} \quad dv/ds = e^{-sJ} J v_0 \\ &\text{(b)} \quad f(v, \lambda) = e^{-sJ} f(v_0, \lambda) \end{aligned} \tag{10.12}$$

are both in $\ker \mathscr{L}$.

To see that $W \equiv 0$ is a solution, calculate (10.6(a)) when $v(s) \in \ker \mathscr{L}$:

$$E\Phi(v, \lambda, \tau) = E\left[(1 + \tau)\frac{dv}{ds} + f(v, \lambda)\right] = 0,$$

since $E = 0$ on $\ker \mathscr{L}$. The uniqueness of implicitly defined functions shows that $W \equiv 0$ is the only possible solution. □

Corollary 10.3. *If, with the preceding notation, $f = \tilde{f} + \hat{f}$ where $\tilde{f}$ is in Birkhoff normal form and $\hat{f}(v, \lambda) = O(\|v\|^k)$, then the implicitly defined function W in the Liapunov–Schmidt reduction satisfies*

$$W(v, \lambda, \tau) = O(\|v\|^k).$$

Proof. To order $k - 1$, the function W depends only on the terms of f of order $k - 1$. But these are the corresponding terms of $\tilde{f}$, for which $W \equiv 0$ by (10.8). □

We use Corollaries 10.2 and 10.3 in the next section to determine the asymptotic stability of periodic solutions to (10.1), even when f is not in Birkhoff normal form.

§11.* Stability in Truncated Birkhoff Normal Form

In §6 we showed how to compute stabilities of bifurcating branches in symmetric Hopf bifurcations for ODEs

$$\dot{x} + f(x, \lambda) = 0 \tag{11.1}$$

when the Γ-equivariant vector field f is in Birkhoff normal form to all orders, that is, commutes with $\mathbf{S}^1$ as well as Γ. In this section we show that the results remain true if $f(x, \lambda)$ is of the form $\tilde{f}(x, \lambda) + o(\|x\|^k)$, where $\tilde{f}$ commutes with $\Gamma \times \mathbf{S}^1$ but the perturbation $o(\|x\|^k)$ commutes only with Γ, provided k is large enough. Here, as usual, $g(x) = o(\|x\|^k)$ means that $g(x)/\|x\|^k \to 0$ as $\|x\| \to 0$.

The precise theorem requires a technical hypothesis, referred to later as *finitely determined stability*, that holds in all the examples so far studied. We suspect that this hypothesis, or some similar condition under which the following proof still works, holds for all compact Γ.

By Theorem 5.9 there always exists a polynomial coordinate change putting f into such a form, so in principle this result completes the stability analysis of Hopf bifurcation for general Γ-equivariant vector fields f in (11.1).

The basic strategy, motivated and outlined in subsection (a), is to compare (11.1) with its order k truncated Birkhoff normal form

$$\dot{x} + \tilde{f}(x, \lambda) = 0. \tag{11.2}$$

We show that the relevant mathematical objects for (11.1) are *small perturbations* of the cooresponding objects for (11.2). Thus in subsection (b) we investigate the way the Floquet operator M_u behaves under such perturbations. In (c) we describe the way its eigenvalues change under perturbation.

In (d) we compare the ODEs (11.1) and (11.2) and show that we have enough control of the movement of eigenvalues to compute the stability when f is not in Birkhoff normal form.

(a) Outline of the Ideas

We begin with a motivating example. The group $\mathbf{O}(2)$ is perhaps a little too well-behaved for comfort, so we consider $\mathbf{D}_n$ Hopf bifurcation on $\mathbb{R}^2 \oplus \mathbb{R}^2$. This will be studied in XVIII, §§1–3, but here we require only an anticipatory sketch of the results. For simplicity we assume n is odd and ≥ 3.

There are three branches of solutions corresponding to isotropy subgroups $\Sigma \subset \mathbf{D}_n \times \mathbf{S}^1$ with $\dim \mathrm{Fix}(\Sigma) = 2$. To quadratic order the branching equations are

$$\lambda + Ka^2 \approx 0 \tag{11.3}$$

where K depends on Σ. (Here $\approx$ means "is approximately equal to.") Further, K can be written in terms of Taylor coefficients of f at 0 and is generically nonzero. By the implicit function theorem, on each solution branch λ is a smooth function of a, and $a \sim \sqrt{\lambda}$, in the sense that $a/\sqrt{\lambda}$ tends to a nonzero constant as $\lambda \to 0$.

The Floquet exponents μ_j along a given branch either are forced to 0 by symmetry or can be expressed in the form

$$\mu_j = \alpha_j a^{p_j} + o(a^{p_j}).$$

Here p_j is the *leading order* of μ_j in a, and the complex-valued coefficient α_j depends on a finite number of coefficients of the Taylor expansion of f. Therefore, the sign of $\mathrm{Re}(\alpha_j)$ determines the sign of $\mathrm{Re}(\mu_j)$ when λ, hence a, is near 0. We must, of course, assume that $\mathrm{Re}(\alpha_j) \neq 0$: this, by definition, is a nondegeneracy condition.

For example, consider the isotropy subgroup $\Sigma = \mathbf{Z}_2(\kappa)$ in $\mathbf{D}_n$ Hopf bifurcation. One eigenvalue is forced to zero by symmetry. The other three are

$$\begin{aligned}
\mu_1 &= \alpha_1 a^2 + o(a^2) \\
\mu_2 &= \alpha_2 a^2 + o(a^2) \\
\mu_3 &= \alpha_3 a^{2n-2} + o(a^{2n-2})
\end{aligned}$$

for certain coefficients α_j.

Our hypotheses for the general theorem that follows are based on such examples. It is no accident that the branching equations take the form (11.3): this follows from our observation that when $\dim \mathrm{Fix}(\Sigma) = 2$ we effectively have standard nondegenerate Hopf bifurcation on $\mathrm{Fix}(\Sigma)$, for which the branching is parabolic. In subsection (d) we show that when $\dim \mathrm{Fix}(\Sigma) = 2$, the periodic solutions to (11.2) given by the equivariant Hopf theorem are circles of radius $a \sim \sqrt{\lambda}$. These considerations motivate the following:

Definitions 11.1.
(a) Suppose $\dim \mathrm{Fix}(\Sigma) = 2$. Then Σ has *p-determined stability* if all eigenvalues of $d\tilde{f} - (1 + \tau_0)J$, other than those forced to zero by Σ, have the form

$$\mu_j = \alpha_j a^{m_j} + o(a^{m_j})$$

on a periodic solution $u(s)$ of (11.2) such that $\|u(s)\| = a$, where α_j is a $\mathbb{C}$-valued function of the Taylor coefficients of terms of degree $\leq p$ in $\tilde{f}$.
(b) If the exact value of p is unimportant we say that Σ has has *finitely determined stability*.
(c) We say that $\tilde{f}$ is *nondegenerate for* Σ if all α_j have nonzero real parts.

Since the α_j are expressions in Taylor coefficients of $\tilde{f}$ at the origin, we expect their real parts to be generically nonzero (though we do not claim a general proof of this). The conditions for (c) are the natural *nondegeneracy conditions* obtained when computing stabilities along branches.

For example, the preceding sketch shows that in standard $\mathbf{D}_n$ Hopf bifurcation (n odd), the isotropy subgroup $\mathbf{Z}_2(\kappa)$ has $(2n - 1)$-determined stability. The other two maximal isotropy subgroups for $\mathbf{D}_n$, and those for $\mathbf{O}(2)$, also have finitely determined stability. Note also that the number of eigenvalues forced to zero is $d_\Sigma = \dim \Gamma - \dim \Sigma + 1$, by Proposition 6.4, and this number is derived without assuming f to be in Birkhoff normal form.

The main result is as follows:

Theorem 11.2. *Suppose that the hypotheses of Theorem* 4.1 *hold, and the isotropy subgroup* $\Sigma \subset \Gamma \times \mathbf{S}^1$ *has p-determined stability. Let* $k \geq p$ *and assume that* $\tilde{f}$ *is nondegenerate for* Σ. *Then for* λ *sufficiently near* 0, *the stabilities of a periodic solution of* (11.1) *with isotropy subgroup* Σ *are given by the same expressions in the coefficients of f as those that define the stability of a solution of the truncated Birkhoff normal form* (11.2) *with isotropy subgroup* Σ.

We sketch the proof of Theorem 11.2; details follow in subsections (b, c, d). For given λ near 0 let $u_\lambda(s)$, $\tilde{u}_\lambda(s)$ be 2π-periodic solutions to (11.1, 11.2), respectively, having the same isotropy subgroup Σ. Suppose that the period-scaling parameters are τ_λ and $\tilde{\tau}_\lambda$, respectively. The corresponding Floquet equations are

$$\frac{dv}{ds} + (1 + \tau_\lambda)(df)_{u_\lambda(s)}v = 0, \tag{11.4}$$

$$\frac{dv}{ds} + (1 + \tilde{\tau}_\lambda)(d\tilde{f})_{\tilde{u}_\lambda(s)}v = 0. \tag{11.5}$$

Let M_u, $\tilde{M}_{\tilde{u}}$ be the corresponding Floquet operators.

Suppose we can show that there exist choices of phase in $u_\lambda(s)$, $\tilde{u}_\lambda(s)$ such that for λ near 0, and all $s \in [0, 2\pi)$,

$$\|(1 + \tau_\lambda)(df)_{u_\lambda(s)} - (1 + \tilde{\tau}_\lambda)(d\tilde{f})_{\tilde{u}_\lambda(s)}\| = o(\tilde{a}^l), \tag{11.6}$$

where $\tilde{a} = \|\tilde{u}_\lambda(s)\| = \|\tilde{u}_\lambda(0)\|$ and $l \to \infty$ as $k \to \infty$. Then (11.4) is "close" to (11.5). Hence M_u is "close" to $\tilde{M}_{\tilde{u}}$. Therefore, the eigenvalues of M_u are "close" to those of $\tilde{M}_{\tilde{u}}$. But the latter can be computed by Proposition 6.4. Now Σ has p-determined stability. Suppose that "close to" means $o(a^q)$ where $q \to \infty$ as $l \to \infty$. Then for large enough k (and in any case making sure that $k \geq p$) the eigenvalues of M_u are determined, to leading order in a, by the expressions α_j in coefficients of $\tilde{f}$. But f *and* $\tilde{f}$ agree to order k, so the stability of $u_\lambda(s)$ is determined by the same expression α_j in the Taylor coefficients of f. Since only terms of degree $\leq p$ occur in the α_j, and the process of putting f into Birkhoff normal form to degree $j + 1$ does not alter the coefficients up to degree j, it in fact suffices to take $k = p$.

Thus it remains to prove, with a suitably precise interpretation of "close," that:

(a) If (11.6) holds then M_u is "close" to $\tilde{M}_{\tilde{u}}$,
(b) If M_u is "close" to $\tilde{M}_{\tilde{u}}$ then the eigenvalues of M_u are "close to" those of $\tilde{M}_{\tilde{u}}$,
(c) (11.6) holds.

These three points are dealt with in turn in subsections (b, c, d), culminating in a proof of Theorem 11.2 in (d).

(b) Perturbing the Floquet Operator

In this section we study how the Floquet operator M_u varies as we make small changes to the vector field f in (11.1) and to the periodic solution $u(s)$ $(= u_\lambda(s))$ being studied. In fact, we look at a general 2π-periodic linear system

$$\frac{dx}{ds} + A(s)x = 0 \tag{11.7}$$

and consider the effect of perturbing $A(s)$ slightly. (The general T-periodic case can be obtained by scaling time suitably.) We are most interested in the case when $A(s) = (df)_{u(s)}$, which lies in the space

$$\mathscr{C}_n = \{2\pi\text{-periodic } C^\infty \text{ maps } \mathbb{R} \to \mathbf{GL}(n)\}.$$

However, our proofs will require $A(s)$ to lie in a Banach space, and there are well-known difficulties in giving spaces of C^∞ mappings a Banach space structure. Instead we work with C^r mappings for some r and define

$$\mathscr{C}_n^r = \{2\pi\text{-periodic } C^r \text{ maps } \mathbb{R} \to \mathbf{GL}(n)\}.$$

A suitable norm on $\mathscr{C}_n^r$ is introduced as follows: For any matrix $C = (c_{ij}) \in \mathbf{GL}(n)$ define its norm to be

$$\|C\| = \max_{ij}|c_{ij}|,$$

with a similar definition for multilinear maps. Then for $A = A(s) \in \mathscr{C}_n^r$ define

$$\|A\| = \max_k \sup_{s \in \mathbb{R}} \|d_s^r A(s)\|.$$

The sup exists since all derivatives $d_s^r A(s)$ are 2π-periodic and continuous, hence bounded.

Associated with each $A \in \mathscr{C}_n^r$ is its *Floquet operator* M_A defined by

$$M_A x(0) = x(2\pi)$$

where $x(s)$ satisfies (11.7).

We wish to appeal to a general result about vector fields on a Banach space, which makes precise the idea of "smooth dependence on initial conditions and parameters." Suppose that $\mathscr{B}$ is a Banach space, and F is a vector field on $\mathscr{B}$ defining a flow x satisfying the ODE

$$\frac{dx}{ds} + F(x) = 0. \tag{11.8}$$

Let $\Phi_s(x_0)$ denote the integral curve of (11.8) passing through x_0 at $s = 0$.

Lemma 11.3. *Suppose that the vector field F in (11.8) is of class C^r ($1 \leq r \leq \infty$). Then the map $\Phi_s(x)$ is class C^r in x and C^{r+1} in s.*

PROOF. See Abraham, Marsden, and Ratiu [1983], Lemma 4.1.9, p. 190. □

Proposition 11.4. *As A runs through $\mathscr{C}_n^r$, the Floquet operator M_A depends in C^r fashion on A.*

PROOF. We "suspend" the ODE (11.7) in order to obtain a differential equation to which Lemma 11.3 may be applied. Consider the Banach space

$$\mathscr{B} = \mathbb{R}^n \oplus \mathbb{R} \oplus \mathscr{C}_n^r$$

with variables (x, u, B), subject to the ODE

$$\begin{aligned} &\text{(a)} \quad dx/ds = -B(u)x \\ &\text{(b)} \quad du/ds = 1 \\ &\text{(c)} \quad dB/ds = 0. \end{aligned} \tag{11.9}$$

We claim that (11.9) defines a C^r vector field on $\mathscr{B}$. This follows by considering the variables separately. The right-hand side of (a) is continuous and linear, hence C^∞, in B when u and s are held constant. It is C^r in u, when B and s are held constant, since by assumption B is C^r. It is continuous and linear, hence C^∞, in x if B and u are held constant. By Dieudonné [1960], p. 167, Theorem (8.9.1) the vector field $(-B(u)x, 1, 0)$ is C^r.

Consider a solution $x(s)$ to (11.9) with the following initial conditions at $s = 0$:

$$x(0) = x_0,$$

$$u(0) = 0,$$

$$B(0) = A.$$

By (11.9(b)) $u(s) = s$ for all s, hence by (11.9(c)) $B(s) = A$ for all s. Thus (11.9(a)) becomes

$$dx/dt + A(s)x = 0$$

with $x(0) = x_0$. This is the same as equation (11.7).

This step in the argument may appear confusing, because (11.9(c)) appears to say that B is constant. In fact B is "constant" *as an element of* $\mathscr{C}_n^r$. Indeed B is the single element A of $\mathscr{C}_n^r$. But since elements of $\mathscr{C}_n^r$ are matrices of functions, we can still *evaluate* B at times s, which is what (11.9(a)) does. In other words, in the first stage of the argument s is *not* the variable in $\mathbb{R}$ that defines B as a function in $\mathscr{C}_n^r$, and $B(s)$ is just some path of Bs parametrized by s. However, at the next stage (11.9(b)) identifies the dummy variable u with s, and then (11.9(a)) produces the desired equation.

By Lemma 11.3 the flow map $\Phi_s(x, u, B)$ for (11.9) is C^r in (x, u, B). Hence the flow map $x(s)$ for (11.7) is C^r in the parameter A, locally in x and s. Therefore, the time 2π flow M_A is C^r in A, for x near 0, but since M_A is linear in x, M_A depends in C^r fashion on A. □

Let $\mathscr{C}_{2\pi}^r$ be the Banach space of periodic mappings $w: \mathbb{R} \to \mathbb{R}^n$ with the C^r topology

$$\|w\|_r = \max_r\{\|d^r w\|\}.$$

We may consider u as a member of $\mathscr{C}_{2\pi}^r$ since, in fact, u is C^∞. Then $(df)_u$ belongs to $\mathscr{C}_n^r$, and we have the following:

Corollary 11.5. *Let $r \geq 0$ and let $M: \mathscr{C}_{2\pi}^r \to \mathscr{C}_n^r$ be the map sending u to the Floquet operator M_u for the equation*

$$\frac{dx}{ds} + (df)_{u(s)}x = 0.$$

Then M is C^r.

Note that the Banach spaces involved in Corollary 11.5 depend on r, so we are not asserting that the dependence is C^∞. Later we need only the case $r = 1$ of Corollary 11.5: M_u depends continuously–differentiably on f and u.

(c) Estimates on the Movement of Eigenvalues

In this section we state estimates on the extent to which the eigenvalues of a matrix A move when A is subjected to a small perturbation. This estimate is valid even when A has multiple eigenvalues and is given in Lemma 11.6. For a geometric interpretation see Figure 11.1.

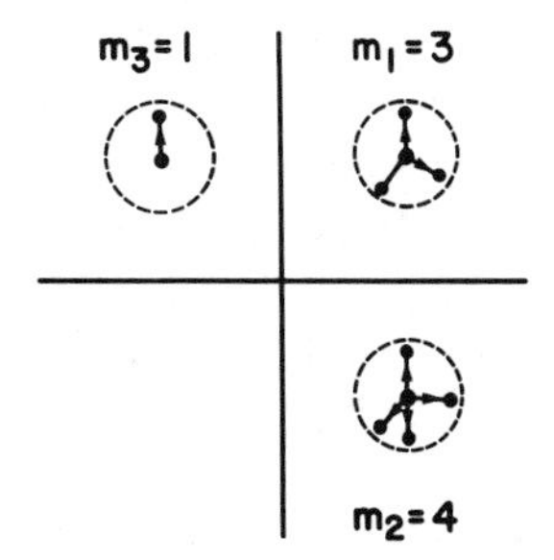

Figure 11.1. Behavior of (possibly multiple) eigenvalues of a matrix under small perturbations.

Lemma 11.6. *Let A be an $n \times n$ matrix having distinct eigenvalues $\alpha_1, \ldots, \alpha_\sigma \in \mathbb{C}$ of multiplicities $m_1, \ldots, m_\sigma$. Then there exist constants $K, \varepsilon > 0$ such that, for all $n \times n$ matrices B with $\|A - B\| < \varepsilon$, the eigenvalues of B can be divided into subsets*

$$\{\beta_{jk}: j = 1, \ldots, \sigma, k = 1, \ldots, m_j\}$$

such that for all j, k in these ranges

$$|\beta_{jk} - \alpha_j| < K\|A - B\|^{1/n}.$$

Proof. See Chatelin [1983] or Wilkinson [1965]. □

(d) Perturbing Periodic Orbits

In this section we study how $(1 + \tilde{\tau}_\lambda)(d\tilde{f})_{\tilde{u}_\lambda(s)}$ perturbs into $(1 + \tau)(df)_{u_\lambda(s)}$ and complete the proof of Theorem 11.2.

First consider the result of applying Liapunov–Schmidt reduction to an arbitrary Γ-equivariant vector field at a point of nondegenerate Hopf bifurcation. This yields a reduced bifurcation equation

$$\psi(u, \lambda, \tau) = 0$$

where

$$\psi: \mathbb{R}^n \times \mathbb{R} \times \mathbb{R} \to \mathbb{R}^n$$

is $\Gamma \times \mathbf{S}^1$-equivariant. Suppose that $\Sigma \subset \Gamma \times \mathbf{S}^1$ is an isotropy subgroup with $\dim \mathrm{Fix}(\Sigma) = 2$. Now $\mathbf{S}^1$ normalizes Σ, hence by Lemma XIII, 10.2, it leaves $\mathrm{Fix}(\Sigma)$ invariant. But the action of $\mathbf{S}^1$ is fixed-point-free. Hence we can identify $\mathrm{Fix}(\Sigma)$ with $\mathbb{C}$ in such a way that the $\mathbf{S}^1$-action on $z \in \mathrm{Fix}(\Sigma)$ is

$$\theta \cdot z = e^{i\theta} z.$$

If an inner product is chosen on $\mathbb{R}^n$ to make $\Gamma \times \mathbf{S}^1$ act orthogonally, then the norm induced on $\text{Fix}(\Sigma)$ is identified with the usual norm in $\mathbb{C}$. Now ψ is $\Gamma \times \mathbf{S}^1$-equivariant, so $\psi|\text{Fix}(\Sigma)$ is $\mathbf{S}^1$-equivariant. Therefore, its zeros are geometric circles $|u| = a$, where a depends on λ.

The standard Hopf bifurcation obtained by restriction on $\text{Fix}(\Sigma)$ is nondegenerate, hence has parabolic branching of the form

$$\lambda = Ka^2 + o(a^2) \tag{11.10}$$

for nonzero K. Thus in particular $a \sim \sqrt{\lambda}$.

Having established these facts, let us return to the ODEs (11.1 and 11.2). We find periodic solutions by considering the operators

$$\Phi(v, \lambda, \tau) = (1 + \tau)\frac{dv}{ds} + f(u, \lambda)$$

$$\tilde{\Phi}(\tilde{v}, \lambda, \tau) = (1 + \tilde{\tau})\frac{d\tilde{v}}{ds} + \tilde{f}(\tilde{v}, \lambda),$$

and these have the same linearization $\mathscr{L}$. By Liapunov–Schmidt reduction, with the same choice of splitting (4.6), we obtain two reduced bifurcation equations

$$\varphi(u, \lambda, \tau) = 0 \tag{11.11}$$

$$\tilde{\varphi}(u, \lambda, \tau) = 0. \tag{11.12}$$

Now f and $\tilde{f}$ agree up to and including order k. By inspection of the Liapunov–Schmidt reduction process, we see that terms of degree l in φ are determined by those of degree $\leq l$ in f, and similarly for $\tilde{\varphi}$. Therefore,

$$\|\gamma(u, \lambda, \tau) - \tilde{\varphi}(u, \lambda, \tau)\| = o(\|u\|^k) \tag{11.13}$$

for all λ, τ near 0.

By §4 the solutions to (11.4 and 11.5) are given, respectively, by $v(t) = u(t) + w(t)$ and $\tilde{v}(t) = \tilde{u}(t) + \tilde{w}(t)$, where u satisfies (11.11), and $\tilde{u}$ satisfies (11.12). See Figure 11.2. By (10.8) $\tilde{w}(t)$ is identically zero because $\tilde{f}$ is in Birkhoff normal form. By Corollary 10.3, $w(t) = o(\|u\|^k)$.

Because zeros of φ and $\tilde{\varphi}$ are circles, we have

$$u(t) = e^{it}u(0),$$

$$\tilde{u}(t) = e^{it}\tilde{u}(0).$$

To make $u(t)$ and $\tilde{u}(t)$ agree in phase we choose $u(0), \tilde{u}(0) \in \mathbb{R}^+ \subset \mathbb{C} \equiv \text{Fix}(\Sigma)$. This ensures that if $u(0)$ and $\tilde{u}(0)$ are "close," then so are $u(t)$ and $\tilde{u}(t)$ for all t, since they are both circles in the plane $\text{Fix}(\Sigma)$, described at uniform speed, and have the same period 2π. Let $\tilde{a} = \|\tilde{u}(0)\| \sim \|u(0)\| = a$. To establish (11.6) with $l = k - 1$ it is enough to show that

$$\tau_\lambda = \tilde{\tau}_\lambda + o(\tilde{a}^k), \tag{11.14}$$

$$u(0) = \tilde{u}(0) + o(\tilde{a}^k). \tag{11.15}$$

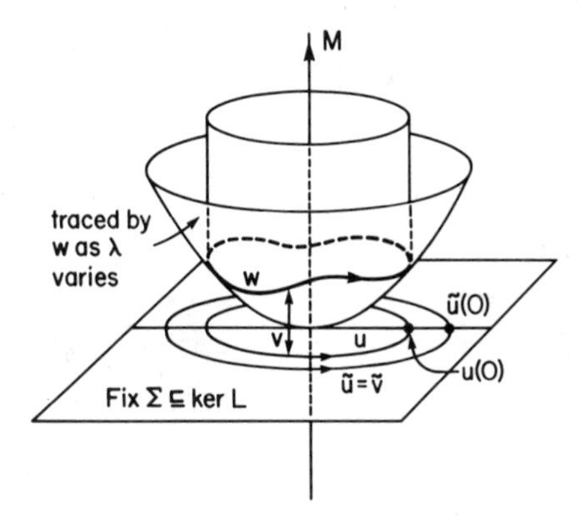

Figure 11.2. Relation of the periodic mappings u, $\tilde{u}$, v, and $\tilde{v}$.

(Note that df and $d\tilde{f}$ agree only to order $k-1$ if f and $\tilde{f}$ agree to order k.) Equation (11.14) follows from the implicit function theorem and the way the equations (11.11 and 11.12) are solved for τ in §4. To prove (11.15) we note (ignoring τ as we may now do) that (11.11 and 11.12) lead to branching equations of the form

$$\lambda = \omega(a) \tag{11.16}$$

$$\lambda = \tilde{\omega}(\tilde{a}) \tag{11.17}$$

where $a = \|u(0)\|$, $\tilde{a} = \|\tilde{u}(0)\|$. The functions ω and $\tilde{\omega}$ agree to order k in a, are even (the result of $\mathbf{Z}_2$ symmetry remaining from the $\mathbf{S}^1$-equivariance; compare standard Hopf bifurcation), and have the same nonzero quadratic terms. Hence

$$\sqrt{\lambda} \sim a \sim \tilde{a}$$

and

$$a/\tilde{a} \to 1 \quad \text{as} \quad \lambda \to 0.$$

By nondegeneracy for Σ we can solve (11.16 and 11.17) in the form

$$a^2 = \Lambda(\lambda)$$

$$\tilde{a}^2 = \tilde{\Lambda}(\lambda)$$

where, by the implicit function theorem, Λ and $\tilde{\Lambda}$ agree to order $k/2$ in λ (that is, to order k in $\tilde{a}$). Hence a and $\tilde{a}$ agree to order k in $\tilde{a}$, as required.

We can now complete the following:

Proof of Theorem 11.2. The preceding shows that (12.6) holds with $l = k - 1$. That is, for any s,

$$\|(1 + \tau_\lambda)(df)_{u_\lambda(s)} - (1 + \tilde{\tau}_\lambda)(d\tilde{f})_{\tilde{u}_\lambda(s)}\| = o(\tilde{a}^{k-1}). \tag{11.18}$$

But effectively (since the us are 2π-periodic) s runs through the circle $\mathbb{R}/2\pi\mathbb{Z}$, which is compact, so (11.18) implies that, as matrices of periodic functions,

$$\|(1+\tau_\lambda)(df)_{u_\lambda(\cdot)} - (1+\tilde{\tau}_\lambda)(d\tilde{f})_{\tilde{u}_\lambda(\cdot)}\| = o(\tilde{a}^{k-1}). \tag{11.19}$$

By Proposition 11.4 with $r = 1$ the Floquet operator M_A is a C^1 function of A. The appropriate norm here is the C^1 norm, $\| \ \|_1$, so we find that there are constants K_1, ε_1 such that for $|\lambda| < \varepsilon_1$

$$\|M_u - \tilde{M}_{\tilde{u}}\|_1 < K_1 \|(1+\tau_\lambda)(df)_{u_\lambda(s)} - (1+\tilde{\tau}_\lambda)(d\tilde{f})_{\tilde{u}_\lambda(s)}\|_1.$$

We claim that the C^0 norm $\|M_u - \tilde{M}_{\tilde{u}}\|$ is $o(\tilde{a}^{k-2})$. Now $\|M_u - \tilde{M}_{\tilde{u}}\| \leq \|M_u - \tilde{M}_{\tilde{u}}\|_1$. Also, since $du/ds + f(u,\lambda) = 0 = d\tilde{u}/ds + \tilde{f}(\tilde{u},\lambda)$, (11.19) easily implies that

$$\|(1+\tau_\lambda)(df)_{u_\lambda(\cdot)} - (1+\tilde{\tau}_\lambda)(d\tilde{f})_{\tilde{u}_\lambda(\cdot)}\|_1 = o(\tilde{a}^{k-2}).$$

This proves the claim.

Let the eigenvalues of M_u be $\mu_1, \ldots, \mu_n$, and those of $\tilde{M}_{\tilde{u}}$ be $\tilde{\mu}_1, \ldots, \tilde{\mu}_n$. Choose the ordering of the indices j so that $\tilde{\mu}_j$ is deformed continuously into μ_j when $\tilde{M}_{\tilde{u}}$ is deformed continuously into M_u (compare Corollary 11.7). By Corollary 11.7

$$|\tilde{\mu}_j - \mu_j| = o(\tilde{a}^{(k-1)/n}) = o(\lambda^{(k-1)/2n})$$

since $\tilde{a} \sim \sqrt{\lambda}$ along the Σ branch.

Now Σ has p-determined stability, so

$$\tilde{\mu}_j = \alpha_j a^{m_j} + o(\tilde{a}^{m_j}) = \alpha'_j \lambda^{m_j/2} + o(\lambda^{m_j/2})$$

where $\alpha'_j = \alpha_j K^{-m_j/2}$, with K as in (11.10). Here α_j, hence α'_j, depends only on Taylor coefficients in $\tilde{f}$ of degree $\leq p$. If we choose k large enough so that

$$(k-2)/2n > m_j/2,$$

that is, $k > nm_j + 2$, then

$$\mu_j = \alpha'_j \lambda^{m_j/2} + o(\lambda^{m_j/2}),$$

and hence the sign of μ_j is determined by α'_j, that is, by α_j. If also $k \geq p$ then the coefficients occurring in α_j are the same as the corresponding coefficients of f. Thus $k = \max(n^2 + 3, p)$ suffices.

Since the process of putting f into Birkhoff normal form is an inductive one, and the coefficients of degree $\leq j$ are unchanged when the terms of degree $j+1$ are put into Birkhoff normal form, the required coefficients may, in fact, be found by taking $k = p$. □

Remarks.

(a) Instead of using expressions for the eigenvalues of $d\tilde{f}$ themselves, we can use equivalent expressions that determine their signs (such as the traces and determinants that will be used in both $\mathbf{O}(2)$ and $\mathbf{D}_n$), provided these expressions are *finitely determined*, that is, have nonzero leading terms in $\tilde{a}$. In practice the expressions in this form are often simpler.

(b) The final remark in §6(b) states that if the trivial branch is stable subcritically and f is in Birkhoff normal form, then subcritical branches of periodic solutions obtained by the equivariant Hopf theorem are asymptotically unstable. By Theorem 11.2 the same result holds even when f is not in Birkhoff normal form.

CHAPTER XVII

Hopf Bifurcation with **O**(2) Symmetry

§0. Introduction

The object of this chapter is to study Hopf bifurcation with **O**(2) symmetry in some depth, including a formal analysis—that is, assuming Birkhoff normal form—of nonlinear degeneracies. The most important case, to which most others reduce, is the standard action of **O**(2) on $\mathbb{R}^2$. Since this representation is absolutely irreducible the corresponding Hopf bifurcation occurs on $\mathbb{R}^2 \oplus \mathbb{R}^2$. We repeat the calculations of XVI, §7(c), in a more convenient coordinate system and in greater detail. In §1 we find that there are two maximal isotropy subgroups, corresponding to standing and rotating waves, as in the example of a circular hosepipe. We also give a brief discussion of nonstandard actions of **O**(2), for which the standing and rotating waves acquire extra spatial symmetry. In §2 we derive the generators for the invariants and equivariants of $\mathbf{O}(2) \times \mathbf{S}^1$ acting on $\mathbb{R}^2 \oplus \mathbb{R}^2$. In §3 we apply these results to analyze the branching directions of these solutions in terms of the Taylor expansion of the vector field.

In §4 we reformulate Hopf bifurcation with **O**(2) symmetry in terms of phase/amplitude equations. To do this we assume that the vector field is in Birkhoff normal form to all orders and introduce suitable polar coordinates. Eliminating the phases leads to amplitude equations on $\mathbb{R}^2$ which turn out to be the most general equations that commute with the standard action of the dihedral group $\mathbf{D}_4$. We apply the results to obtain the stabilities of the rotating and standing wave solutions, showing that one of these two branches is stable only if both are supercritical, in which case (generically) precisely one branch is stable. Which branch is stable is determined by cubic terms in the vector field.

In §5 we show that these results generalize without difficulty to Hopf bifurcation for $\mathbf{O}(n)$ in its standard action on $\mathbb{R}^n$.

We return to $\mathbf{O}(2)$ Hopf bifurcation in §6 but now allow the bifurcation to be degenerate. The amplitude equation method reduces this problem to degenerate static bifurcation with $\mathbf{D}_4$ symmetry, and we use the methods of Chapters XIV and XV to classify such bifurcations in low codimension. An interesting feature is that modal parameters enter at topological codimension 0. The corresponding bifurcation diagrams are described in §7. In codimension 1 we find 2-tori with linear flow, and in codimension 2 we find 3-tori.

Finally in §8 we consider the simpler cases of Hopf bifurcation with $\mathbf{SO}(2)$ or $\mathbf{Z}_n$ symmetry, acting on $\mathbb{R}^2$. These illustrate the way the analysis works for non–absolutely irreducible representations. Here the *existence* of solutions follows from the usual Hopf theorem, but the analysis of their *symmetries* is more naturally carried out within the framework set up in Chapter XVI. In particular we show that for nonstandard actions the generic branches are rotating waves with additional spatial symmetry.

§1. The Action of $\mathbf{O}(2) \times \mathbf{S}^1$

We begin by investigating the group-theoretic and invariant-theoretic generalities of $\mathbf{O}(2)$ Hopf bifurcation. In subsection (a) we summarize the results of this section and §§2 and 3. In (b) we write the group action in convenient coordinates. In (c) we find the isotropy subgroups and related data, and in (d) we consider how to modify the results for nonstandard actions of $\mathbf{O}(2)$. The results include the verification, in the new coordinate system, of the results described without proof in XVI, §7c.

(a) The Main Results

The first example we consider is the *standard* action of $\Gamma = \mathbf{O}(2)$ on $\mathbb{R}^2 \equiv \mathbb{C}$. The other actions (in which $\mathbf{SO}(2)$ acts by k-fold rotations, $k > 1$) reduce to this case, but some care is needed when interpreting the results; see subsection (d). In the philosophy of Chapter XVI, the first step in studying nondegenerate Hopf bifurcation is to find the conjugacy classes of isotropy subgroups of $\mathbf{O}(2) \times \mathbf{S}^1$ acting on $\mathbb{C} \oplus \mathbb{C}$. We will show that the isotropy lattice consists of four conjugacy classes as in Figure 1.1. Since the fixed-point subspaces of $\widetilde{\mathbf{SO}}(2)$ and $\mathbf{Z}_2 \oplus \mathbf{Z}_2^c$ are each two-dimensional, Theorem 4.1 implies that there exist two distinct branches of periodic solutions corresponding to these isotropy subgroups, provided the usual transversality condition, that eigenvalues cross the imaginary axis with nonzero speed, holds.

This group-theoretic structure has several implications for Hopf bifurcation

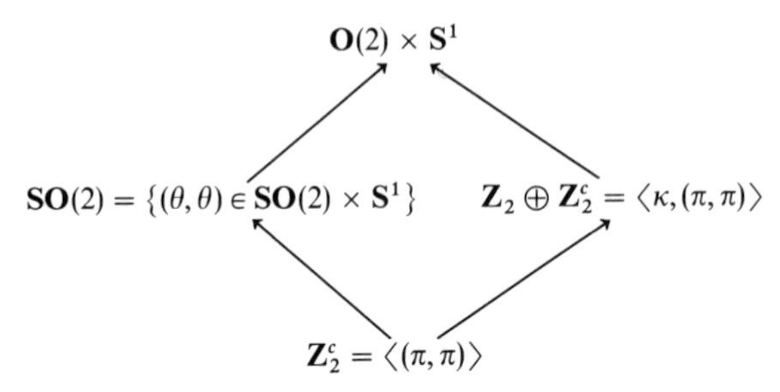

Figure 1.1. The isotropy lattice of $\mathbf{O}(2) \times \mathbf{S}^1$.

to periodic solutions in a system of ODEs with $\mathbf{O}(2)$ symmetry, which we now list.

Since the element $(\pi, \pi) \in \mathbf{O}(2) \times \mathbf{S}^1$ acts trivially on $\mathbb{C} \oplus \mathbb{C}$, every periodic solution $x(t)$ obtained in this analysis has the following property:

> Spatial rotation of $x(t)$ through the angle π has the same effect as shifting the phase of $x(t)$ by half a period.

That is,

$$R_\pi x(t) = x(t + \pi). \tag{1.1}$$

Periodic solutions $x(t)$ with $\widetilde{\mathbf{SO}}(2)$ symmetry have a more general property:

> Spatial rotation of $x(t)$ through any angle θ has the same effect as shifting the phase of $x(t)$ by θ.

That is,

$$R_\theta x(t) = x(t + \theta).$$

In particular, such solutions satisfy

$$x(\theta) = R_\theta x(0). \tag{1.2}$$

Periodic solutions satisfying (1.2) are called *rotating waves*. In rotating waves there is a coupling between spatial and temporal symmetries. There have been numerous studies of bifurcation to (and from) rotating waves. We mention Renardy [1982] and, in the context of reaction–diffusion equations, Erneux and Herschkowitz-Kaufman [1977] and Auchmuty [1979].

Solutions with $\mathbf{Z}_2 \oplus \mathbf{Z}_2^c$ isotropy possess a purely spatial symmetry in addition to (1.1). Let $\kappa \in \mathbf{O}(2)$ be the flip, acting by complex conjugation on $\mathbb{C}$. Then periodic solutions $x(t)$ with isotropy subgroup $\mathbf{Z}_2 \oplus \mathbf{Z}_2^c$ satisfy $\kappa x(t) = x(t)$. We call this solution a *standing wave*. As we have already seen in XVI, §7, these solutions correspond to the two modes of an oscillating hosepipe mentioned in XI, §1.

The submaximal isotropy subgroup $\mathbf{Z}_2^c$ acts trivially and thus has a four-

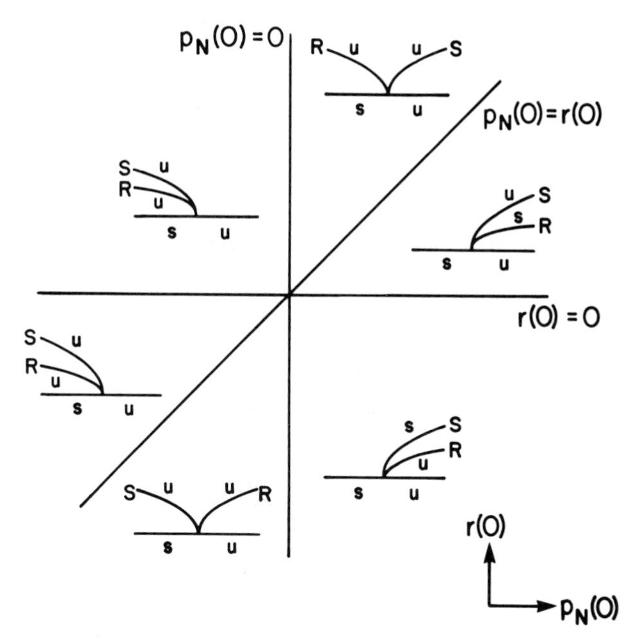

Figure 1.2. Branching and stability in nondegenerate Hopf bifurcation with $\mathbf{O}(2)$ symmetry. S = standing wave; R = rotating wave.

dimensional fixed-point subspace. Solutions of this type occur only in *degenerate* $\mathbf{O}(2)$ Hopf bifurcation and correspond (for vector fields in Birkhoff normal form) to invariant tori with linear flow. We discuss such degeneracies in §6.

Further, by using the invariant theory of $\mathbf{O}(2) \times \mathbf{S}^1$, the stabilities of the two primary branches can be determined. The stabilities depend on the coefficients of two third order terms, which we call $p_N(0)$ and $r(0)$, in the Birkhoff normal form. The possibilities are shown in Figure 1.2, on the assumption that the trivial state is stable subcritically and unstable supercritically. Observe that for either branch to consist of (orbitally) asymptotically stable periodic solutions, both branches must be supercritical. If so, then one branch is stable and the other unstable. This relationship contrasts dramatically with the usual exchange of stability in standard Hopf bifurcation, see XIII, §4.

(b) The Group Action

We start by rewriting the action of $\mathbf{O}(2) \times \mathbf{S}^1$ in a more convenient form. We claim that there exist coordinates (z_1, z_2) on $\mathbb{C}^2$ for which the action is given by

$$\begin{aligned} &\text{(a)}\quad \theta(z_1,z_2)=(e^{i\theta}z_1,e^{i\theta}z_2) \qquad (\theta\in\mathbf{S}^1)\\ &\text{(b)}\quad \varphi(z_1,z_2)=(e^{-i\varphi}z_1,e^{i\varphi}z_2) \qquad (\varphi\in\mathbf{SO}(2))\\ &\text{(c)}\quad \kappa(z_1,z_2)=(z_2,z_1) \qquad (\kappa=\text{flip in }\mathbf{O}(2)). \end{aligned} \tag{1.3}$$

These coordinates were used by van Gils [1984] and pointed out to us by A. Vanderbauwhede. We derive them from those given by the general theory. There $\Gamma = \mathbf{O}(2)$ acts diagonally on $\mathbb{C} \oplus \mathbb{C}$; that is, we have

$$\begin{aligned} &\text{(a)}\quad \varphi(w_1,w_2)=(e^{i\varphi}w_1,e^{i\varphi}w_2) \qquad (\varphi\in\mathbf{SO}(2))\\ &\text{(b)}\quad \kappa(w_1,w_2)=(\bar{w}_1,\bar{w}_2) \end{aligned} \tag{1.4}$$

in the usual coordinates (w_1, w_2). We construct new coordinates (z_1, z_2) by finding a two-dimensional subspace $Z_1 \subset \mathbb{C}^2$ such that

$$\begin{aligned} &\text{(a)}\quad \mathbf{SO}(2)\times\mathbf{S}^1 \quad \text{leaves } Z_1 \text{ invariant},\\ &\text{(b)}\quad \mathbb{C}^2 = Z_1\oplus Z_2 \quad \text{where } Z_2=\kappa Z_1. \end{aligned} \tag{1.5}$$

Concretely, we have $Z_1 = \mathbb{C}\{(1,i)\}$. However, it is instructive to deduce the existence of Z_1 abstractly: we "diagonalize" the action of $\mathbf{SO}(2) \times \mathbf{S}^1$. This is a torus group and it acts nontrivially. Theorem XII, 7.1, states that nontrivial irreducible representations of tori are two-dimensional. Let Z_1 be a two-dimensional irreducible subspace; then Z_1 satisfies (1.5(a)). We claim that $Z_2 = \kappa Z_1$ is also invariant under $\mathbf{SO}(2) \times \mathbf{S}^1$. Suppose that $\theta \in \mathbf{S}^1$. Then $\theta\kappa = \kappa(\theta)$, whence $\theta Z_2 \subset Z_2$. Similarly for $\varphi \in \mathbf{SO}(2)$ we have $\varphi\kappa = \kappa(-\varphi)$, which implies that $\varphi Z_2 \subset Z_2$. Now $Z_1 \cap Z_2 = \{0\}$ since it is invariant under $\mathbf{O}(2) \times \mathbf{S}^1$ and this acts irreducibly on $\mathbb{C}^2$; recall Lemma XVI, 3.4(b). Thus (1.5(b)) holds.

If we choose a complex coordinate z_1 on Z_1 and let $z_2 = \kappa z_1$ then (1.3(c)) holds. Since $\mathbf{S}^1$ acts on $\mathbb{C}^2$ without fixed points and commutes with κ, we can choose z_1 so that (1.3(a)) holds. Since $\varphi \in \mathbf{SO}(2)$ acts standardly on $\mathbb{C}$ and diagonally on $\mathbb{C}^2$, it acts on Z_1 by $e^{\pm i\varphi}$. Since $\kappa\varphi\kappa = -\varphi$ the action is by $e^{\mp i\varphi}$ on Z_2. Interchanging Z_1 and Z_2 if necessary, we have (1.3(b)). Alternatively, a concrete calculation yields the same result. Namely, in the coordinates $z_2(1,i) + \bar{z}_1(1,-i)$,

$$\begin{aligned} &\varphi(z_1,z_2)=(e^{-i\varphi}z_1,e^{i\varphi}z_2) \qquad (\varphi\in\mathbf{SO}(2))\\ &\kappa(z_1,z_2)=(z_2,z_1). \end{aligned} \tag{1.6}$$

(c) The Isotropy Lattice

We can now compute the (conjugacy classes of) isotropy subgroups for the preceding action of $\mathbf{O}(2) \times \mathbf{S}^1$.

Proposition 1.1. *There are four conjugacy classes of isotropy subgroups for the standard action of* $\mathbf{O}(2) \times \mathbf{S}^1$ *on* $\mathbb{C}^2$. *They are listed, together with their orbit representatives and fixed-point subspaces, in Table* 1.1.

Table 1.1. Group-Theoretic Data for the Standard Action of $\mathbf{O}(2) \times \mathbf{S}^1$ on $\mathbb{C}^2$

Orbit Representative	Isotropy Subgroup	Fixed-Point Subspace	Dimension
$(0,0)$	$\mathbf{O}(2) \times \mathbf{S}^1$	$\{0\}$	0
$(a,0)$, $a > 0$	$\widetilde{\mathbf{SO}}(2) = \{(\theta, \theta)\}$	$\{(z_1, 0)\}$	2
(a,a), $a > 0$	$\mathbf{Z}_2 \oplus \mathbf{Z}_2^c = \{(0,0), \kappa, (\pi, \pi), \kappa(\pi, \pi)\}$	$\{(z_1, z_1)\}$	2
(a,b), $a > b > 0$	$\mathbf{Z}_2^c = \{(0,0), (\pi, \pi)\}$	$\{(z_1, z_2)\}$	4

Remarks 1.2.

(a) The manifold $(\mathbf{O}(2) \times \mathbf{S}^1)/\widetilde{\mathbf{SO}}(2)$ has two connected components, each a circle (see Remark XIV, 1.3). The orbit representatives for these two components are $(a, 0)$ and $(0, a)$. The isotropy subgroup for $(0, a)$ is the conjugate $\{(\theta, -\theta)\}$ of $\widetilde{\mathbf{SO}}(2)$. Physically, the solutions corresponding to these points are both rotating waves; however, they correspond to the two possible senses of rotation—"clockwise" and "counterclockwise."

(b) The manifold $(\mathbf{O}(2) \times \mathbf{S}^1)/(\mathbf{Z}_2 \oplus \mathbf{Z}_2^c)$ is a connected 2-torus, foliated by circular trajectories. Each circle corresponds to a particular choice of axis for the reflectional "flip" symmetry, which can be any *conjugate* $\kappa\theta$ of κ. That is, in the fixed-point subspace for each isotropy subgroup conjugate to $\mathbf{Z}_2 \oplus \mathbf{Z}_2^c$ we find a periodic solution. All such solutions lie in the same group orbit. They glue together to form the 2-torus.

(c) We consider (a) and (b) for the "hosepipe" example of XI, §1. As already remarked, the rotating wave oscillations correspond to the isotropy subgroup $\widetilde{\mathbf{SO}}(2)$, as in Remark (a). There are two distinct senses of rotation, and there is a unique solution (up to choice of phase) in each sense. The standing wave solutions have isotropy subgroup $\mathbf{Z}_2 \oplus \mathbf{Z}_2^c$. The first $\mathbf{Z}_2$, generated by the flip, imposes mirror symmetry in some axis (and confines the oscillation to a fixed vertical plane). The $\mathbf{Z}_2^c$ symmetry means that, just as for a pendulum, the oscillations to left and right are identical except for a phase shift of π. The axis of reflection for a flip symmetry can be any radial line throught the origin: different radial lines correspond to different (conjugate) choices of flip. There is a circle's worth of radial lines corresponding to the circle's worth of (circular) trajectories foliating the 2-torus.

Proof of Proposition 1.1. Elements in the same orbit have conjugate isotropy subgroups. Hence by (1.2) we may assume that $(z_1, z_2) = (a, b)$ where $a, b \geq 0$ are real. By applying κ we may assume that $a \geq b \geq 0$. The computation of isotropy subgroups and fixed-point subspaces is now routine, but as an illustration we give details.

Clearly $(0, 0)$ is fixed by the whole of $\mathbf{O}(2) \times \mathbf{S}^1$ and is the only point so fixed.

If $a > b > 0$ then (a, b) cannot be fixed by any group element not in $\mathbf{SO}(2) \times \mathbf{S}^1$ (that is, involving κ) since $|z_1|$ and $|z_2|$ must be preserved. Now

$$(\varphi, \theta)(a, b) = (e^{i(\theta-\varphi)}a, e^{i(\theta+\varphi)}b)$$

so (φ, θ) fixes (a, b) if and only if $\theta \pm \varphi \equiv 0 \pmod{2\pi}$. Hence $(\varphi, \theta) = (0, 0)$ or (π, π).

Similarly if $a > 0$, the point $(a, 0)$ can be fixed only by elements (φ, θ) of $\mathbf{SO}(2) \times \mathbf{S}^1$, and now we need $\theta - \varphi \equiv 0 \pmod{2\pi}$, that is, $\varphi = \theta$.

Finally (a, a) is fixed by κ. Hence its isotropy subgroup is generated by κ, together with a subgroup of $\mathbf{SO}(2) \times \mathbf{S}^1$. It can be fixed by $(\varphi, \theta) \in \mathbf{SO}(2) \times \mathbf{S}^1$ only if $\theta \pm \varphi \equiv 0 \pmod{2\pi}$, which as earlier leads to $\{(0, 0), (\pi, \pi)\}$. Together with κ these generate $\mathbf{Z}_2 \oplus \mathbf{Z}_2^c$ as claimed. □

(d) Nonstandard Actions of **O**(2)

We now indicate how to modify the preceding calculations for a (nontrivial) nonstandard action of **O**(2). Consider the representation ρ_l for which

$$\varphi \cdot z = e^{li\varphi}z \qquad (\varphi \in \mathbf{SO}(2))$$

$$\kappa \cdot z = \bar{z},$$

where $l > 1$ is an integer. We can obtain this representation by composing the standard representation ρ_0 with the group homomorphism

$$\alpha\colon \mathbf{O}(2) \to \mathbf{O}(2)$$

$$\varphi \mapsto l\varphi$$

$$\kappa \mapsto \kappa$$

whose kernel is

$$\mathbf{Z}_l = \{2m\pi/l\colon m = 0, \ldots, l-1\} \subset \mathbf{O}(2).$$

Then the (nonstandard) action $\rho_l(\gamma)$ of $\gamma \in \mathbf{O}(2)$ on $\mathbb{C}$ is the same as the standard action of $\alpha(\gamma)$ on $\mathbb{C}$. Thus the representation ρ_l behaves just as the standard representation ρ_0 does except that $\mathbf{Z}_1$ acts trivially.

The same is true of the corresponding action of $\mathbf{O}(2) \times \mathbf{S}^1$ on $\mathbb{C}^2$, in which $\mathbf{O}(2)$ acts diagonally by $(w_1, w_2) \mapsto (e^{li\varphi}w_1, e^{li\varphi}w_2)$. As in subsection (a) we can choose coordinates (z_1, z_2) so that this action takes the form

$$\theta(z_1, z_2) = (e^{i\theta}z_1, e^{i\theta}z_2)$$

$$\varphi(z_1, z_2) = (e^{-li\varphi}z_1, e^{li\varphi}z_2)$$

$$\kappa(z_1, z_2) = (z_2, z_1).$$

Again this is just like the standard action, except that $\mathbf{Z}_l$ acts trivially. The orbit data are the same as in Table 1.2, except that $\mathbf{Z}_l$ must be added to every isotropy subgroup. Thus the rotating wave solutions have an additional cyclic symmetry $\mathbf{Z}_l$ of order l, and the standing waves have dihedral group symmetry $\mathbf{D}_l = \langle \kappa, \mathbf{Z}_l \rangle$ (plus the original kernel $\mathbf{Z}_2^c$).

The invariants and equivariants for nonstandard actions of $\mathbf{O}(2)$ are the same as those for the standard action (since the kernel $\mathbf{Z}_l$ acts trivially and hence does not change the invariance or equivariance conditions). The preceding considerations apply generally to nonstandard representations of $\mathbf{O}(2)$ or $\mathbf{SO}(2)$, and similar ideas apply to nonstandard representations of $\mathbf{D}_n$ and $\mathbf{Z}_n$.

§2. Invariant Theory for $\mathbf{O}(2) \times \mathbf{S}^1$

In order to determine the direction of branching (super- or subcritical) and the stability of the branches of periodic solutions, we must compute the $\mathbf{O}(2) \times \mathbf{S}^1$ invariants and equivariants for the preceding action. (As just noted, these are identical for all nontrivial actions of $\mathbf{O}(2)$.) The results are as follows:

Proposition 2.1.
(a) *Every $\mathbf{O}(2) \times \mathbf{S}^1$-invariant germ f has the form*

$$f(z_1, z_2) = P(N, \Delta)$$

where $N = |z_1|^2 + |z_2|^2$, $\Delta = \delta^2$, and $\delta = |z_2|^2 - |z_1|^2$.
(b) *Every $\mathbf{O}(2) \times \mathbf{S}^1$-equivariant germ g has the form*

$$g(z_1, z_2) = (p + iq)\begin{bmatrix} z_1 \\ z_2 \end{bmatrix} + (r + is)\delta\begin{bmatrix} z_1 \\ -z_2 \end{bmatrix}, \tag{2.1}$$

where p, q, r, and s are $\mathbf{O}(2) \times \mathbf{S}^1$-invariant germs.

Proof. As usual Schwarz's theorem lets us assume that f and g are polynomial. First we consider the invariance of f under $\mathbf{SO}(2) \times \mathbf{S}^1$, which implies that

$$f(e^{i(\theta-\varphi)}z_1, e^{i(\theta+\varphi)}z_2) = f(z_1, z_2).$$

Define $\psi_1 = (\psi/2, \psi/2)$, $\psi_2 = (-\psi/2, \psi/2) \in \mathbf{SO}(2) \times \mathbf{S}^1$. Then

$$f(z_1, e^{i\psi_2}z_2) = f(z_1, z_2) = f(e^{i\psi_1}z_1, z_2). \tag{2.2}$$

Thus $f = h(u, v)$ where $u = |z_1|^2$, $v = |z_2|^2$. By κ-invariance $h(u, v) = h(v, u)$, whence $h(u, v) = k(u + v, uv)$. Then $N = u + v$ and $\Delta = (v - u)^2 = N^2 - 4uv$ provide an alternative Hilbert basis. Note that $\delta = v - u$.

Now suppose that $g(z_1, z_2) = (g_1(z_1, z_2), g_2(z_1, z_2))$ is equivariant under $\mathbf{O}(2) \times \mathbf{S}^1$. Then (again using ψ_1 and ψ_2)

$$\begin{aligned} &\text{(a)} \quad g_1(z_1, z_2) = e^{-i\psi_1}g_1(e^{i\psi_1}z_1, z_2) \\ &\text{(b)} \quad g_1(z_1, z_2) = g_1(z_1, e^{i\psi_2}z_2) \\ &\text{(c)} \quad g_2(z_1, z_2) = g_1(z_2, z_1). \end{aligned} \tag{2.3}$$

Identity (2.3(b)) implies that $g_1 = h(z_1, |z_2|^2)$, and (2.3(a)) implies that $h(z_1, |z_2|^2) = k(|z_1|^2, |z_2|^2)z_1$.

Recall that $u = |z_1|^2$, $v = |z_2|^2$. Using the coordinate change $(u, v) \mapsto (u + v, u - v)$ we can write

$$k(u, v) = l(u + v, v - u)$$

and decompose l into an even and an odd function in the second coordinate. Thus

$$k(u, v) = l_0(u + v, (v - u)^2) + l_1(u + v, (v - u)^2)(v - u).$$

Summarizing the preceding results we have

$$g_1(z_1, z_2) = P(N, \Delta)z_1 + R(N, \Delta)\delta z_1.$$

Then (2.3(c)) just specifies g_2 in terms of g_1:

$$g_2(z_1, z_2) = P(N, \Delta)z_2 - R(N, \Delta)\delta z_2.$$

Finally, note that P and R are complex-valued invariant functions. Thus we complete the proof of (2.1) by setting $P = p + iq$ and $R = r + is$. □

§3. The Branching Equations

Suppose that we have a system of ODEs on $\mathbb{R}^4$,

$$\dot{x} + X(x, \lambda) = 0 \tag{3.1}$$

where X is smooth and $\mathbf{O}(2)$-invariant. Suppose also that

$$(dX)_{0,0} = \begin{bmatrix} 0 & I \\ -I & 0 \end{bmatrix}$$

and that the eigenvalues of $(dX)_{0,\lambda}$ cross the imaginary axis with nonzero speed. Then XVI, §4 implies that there is a Liapunov–Schmidt reduction to a mapping $g(z_1, z_2, \lambda, \tau)$ of $\mathbb{C}^2 \times \mathbb{R} \times \mathbb{R} \to \mathbb{C}^2$ commuting with $\mathbf{O}(2) \times \mathbf{S}^1$, whose zeros are in one-to-one correspondence with periodic solutions of (3.1) of period near 2π. Here τ is the period-perturbing parameter.

By Proposition 2.1, g has the form

$$g = (p + iq)\begin{bmatrix} z_1 \\ z_2 \end{bmatrix} + (r + is)\delta\begin{bmatrix} z_1 \\ -z_2 \end{bmatrix} \tag{3.2}$$

where p, q, r, s are functions of N, Δ, λ, and τ. The Liapunov–Schmidt reduction shows that

$$\begin{aligned} &\text{(a)} \quad p(0) = 0, \qquad p_\tau(0) = 0, \\ &\text{(b)} \quad q(0) = 0, \qquad q_\tau(0) = -1, \\ &\text{(c)} \quad p_\lambda(0) \neq 0 \qquad \text{(eigenvalue crossing condition).} \end{aligned} \tag{3.3}$$

To find the zeros of g, it suffices to look at representative points on $\mathbf{O}(2) \times \mathbf{S}^1$-

Table 3.1. Branching Equations for $\mathbf{O}(2) \times \mathbf{S}^1$ Hopf Bifurcation

Orbit Representative		Isotropy Subgroup	Branching Equations
(a) $(0,0)$		$\mathbf{O}(2) \times \mathbf{S}^1$	—
(b) $(a,0)$,	$a > 0$	$\widetilde{\mathbf{SO}}(2)$	$p - a^2 r = 0$ $q - a^2 s = 0$
(c) (a,a),	$a > 0$	$\mathbf{Z}_2 \oplus \mathbf{Z}_2^c$	$p = 0$ $q = 0$
(d) (a,b),	$a > b > 0$	$\mathbf{Z}_2^c$	$p = q = r = s = 0$

Table 3.2. Branches of Periodic Solutions for $\mathbf{O}(2)$ Hopf Bifurcation

Name	Isotropy Subgroup	Branching Equation
Trivial solution	$\mathbf{O}(2) \times \mathbf{S}^1$	$z = 0$
Rotating wave	$\widetilde{\mathbf{SO}}(2)$	$\lambda = \dfrac{(-p_N(0) + r(0))}{p_\lambda(0)} a^2 + \cdots$
Standing wave	$\mathbf{Z}_2 \oplus \mathbf{Z}_2^c$	$\lambda = \dfrac{-2p_N(0)}{p_\lambda(0)} a^2 + \cdots$

orbits. That is, when solving $g = 0$ we may assume that $z_1 = a$ and $z_2 = b$ are real. It is now easy to check that the equation $g = 0$ reduces to the entries of Table 3.1.

By (3.3(b)) we can always use the equation involving q to solve for τ. This is a specific instance of general arguments in the proof by Liapunov–Schmidt reduction. Generically $r(0)$ and $s(0)$ are nonzero; thus generically there are no solutions with isotropy subgroup $\mathbf{Z}_2^c$.

It is now easy to solve for the leading terms of the branching equations for rotating waves ($\widetilde{\mathbf{SO}}(2)$) and standing waves ($\mathbf{Z}_2 \oplus \mathbf{Z}_2^c$). These are given in Table 3.2. Assuming that $p_\lambda(0) < 0$, or equivalently that the trivial solution is stable subcritically, it is now possible to establish the directions of branching for rotating and standing waves shown in Figure 1.2 earlier.

§4. Amplitude Equations, $\mathbf{D}_4$ Symmetry, and Stability

Return now to the system of ODEs (3.1) and assume that it is in Birkhoff normal form. As we saw in XVI, §6, this means that X commutes with $\mathbf{O}(2) \times \mathbf{S}^1$ rather than just $\mathbf{O}(2)$. By Proposition 2.1,

$$X = (p + iq)\begin{bmatrix} z_1 \\ z_2 \end{bmatrix} + (r + is)\delta \begin{bmatrix} z_1 \\ -z_2 \end{bmatrix} \tag{4.1}$$

where p, q, r, s are functions of N, Δ, λ, and $p(0) = 0$, $q(0) = 1$. By Theorem XVI, 10.1, the Liapunov–Schmidt reduced function g of (3.2) for this X has the explicit form

$$g(z_1, z_2, \lambda, \tau) = X(z_1, z_2, \lambda) - (1 + \tau)i\begin{bmatrix} z_1 \\ z_2 \end{bmatrix}. \tag{4.2}$$

Thus the branching equations for X are precisely those given in Tables 3.1 and 3.2, since the τ-dependence enters in a simple way in those equations.

(a) Amplitude Equations

One of the remarkable facts about the Birkhoff normal form (4.1) is that it lets us separate the 4×4 system of ODEs into amplitude and phase equations, which in turn permit a simple analysis of the stability of the rotating and standing waves. Write

$$\begin{aligned} z_1 &= xe^{i\psi_1} \\ z_2 &= ye^{i\psi_2}. \end{aligned} \tag{4.3}$$

Then (4.1) implies that the system (3.1) becomes

$$\begin{aligned} \dot{z}_1 + (p + iq + (r + is)\delta)z_1 &= 0 \\ \dot{z}_2 + (p + iq - (r + is)\delta)z_2 &= 0. \end{aligned} \tag{4.4}$$

In the amplitude/phase variables (4.3) these become

$$\begin{aligned} \text{(a)} \quad & \dot{x} + (p + r\delta)x = 0 \\ & \dot{y} + (p - r\delta)y = 0 \\ \text{(b)} \quad & \dot{\psi}_1 + (q + s\delta) = 0 \\ & \dot{\psi}_2 + (q - s\delta) = 0 \end{aligned} \tag{4.5}$$

where p, q, r, s are functions of $N = x^2 + y^2$, $\Delta = (y^2 - x^2)^2$, and λ; and $\delta = y^2 - x^2$. (This calculation may be performed by differentiating the identity $x^2 = z_1\bar{z}_1$ to obtain

$$x\dot{x} = \operatorname{Re}(z_1 \cdot \bar{z}_1) = -(p + r\delta)z_1\bar{z}_1.$$

Similarly one uses the identity $y^2 = z_2\bar{z}_2$.)

Nontrivial zeros of the amplitude equations correspond to invariant circles and invariant tori. In particular when $y = 0$ (the rotating waves branch) zeros correspond to circles; when $x = y$ (the standing waves branch) zeros correspond to invariant 2-tori in the original four-dimensional system. On such a torus we see from (4.3(b)) that $\dot{\psi}_1 = \dot{\psi}_2$ since $\delta = 0$. Therefore, trajectories on this 2-torus are all circles. Compare this observation with the group-theoretic Remark 1.2(b). In particular, this qualitative feature of the flow persists even when the vector field is not in Birkhoff normal form. Finally, a zero $x > y > 0$ of the amplitude equations corresponds to an invariant 2-torus with linear flow ($\dot{\psi}_1$ and $\dot{\psi}_2$ constant). This linear flow is in general quasiperiodic. As remarked earlier, however, generically $r(0) \neq 0$, so there are (locally) no zeros of the amplitude equations with $x > y > 0$. So such tori do not occur in nondegenerate $\mathbf{O}(2)$-symmetric Hopf bifurcation. It is not surprising, however, that when considering degenerate cases such as $r(0) = 0$, such tori do occur. This observation of Erneux and Matkowsky [1984] will be discussed in more detail later.

Lemma 4.1. *A zero of the amplitude equations is asymptotically stable if and only if the corresponding steady-state, periodic trajectory, or invariant 2-torus is (orbitally) asymptotically stable in the four-dimensional system.*

PROOF. A zero (x_0, y_0) of the amplitude equations (4.5(a)) is asymptotically stable if every trajectory $(x(t), y(t))$ with initial point sufficiently close to (x_0, y_0) stays near (x_0, y_0) for all $t > 0$, and $\lim_{t\to\infty}(x(t), y(t)) = (x_0, y_0)$. Let M be the connected component of the orbit of $\mathbf{O}(2) \times \mathbf{S}^1$ that contains (x_0, y_0): this consists of points (z_1, z_2) with absolute values x_0, y_0. By definition this means that any trajectory $(z_1(t), z_2(t))$ of the four-dimensional system, with $(z_1(0), z_2(0))$ sufficiently close to M, converges in norm to M. On M, $\dot{\psi}_1$ and $\dot{\psi}_2$ are constant, so the trajectory $(z_1(t), z_2(t))$ converges to a single trajectory on M. This is what is meant by orbital asymptotic stability. □

(b) $\mathbf{D}_4$-Symmetry

Let the dihedral group $\mathbf{D}_4$ act on $\mathbb{C}$ as symmetries of the square. That is, the action is generated by

$$\begin{aligned} &\text{(a)} \quad z \mapsto \bar{z} \\ &\text{(b)} \quad z \mapsto iz. \end{aligned} \tag{4.6}$$

By Examples XII, 4.1(c), the general $\mathbf{D}_4$-equivariant mapping has the form $pz + q\bar{z}^3$ where p and q are functions of $u = z\bar{z}$ and $v = \mathrm{Re}(z^4)$. If we let $z = x + iy$, then we have $u = x^2 + y^2 = N$ and $v = x^4 - 6x^2y^2 + y^4 = -(x^2 + y^2)^2 + 2(y^2 - x^2)^2 = 2\Delta - N^2$. Moreover, as mappings $\mathbb{R}^2 \to \mathbb{R}^2$, the mappings

$$z \mapsto z \text{ and } z \mapsto \bar{z}^3$$

correspond to

$$\begin{bmatrix} x \\ y \end{bmatrix} \mapsto \begin{bmatrix} x \\ y \end{bmatrix} \quad \text{and} \quad \begin{bmatrix} x \\ y \end{bmatrix} \mapsto \begin{bmatrix} x^3 - 3xy^2 \\ y^3 - 3x^2y \end{bmatrix} = -N\begin{bmatrix} x \\ y \end{bmatrix} - 2\delta\begin{bmatrix} x \\ -y \end{bmatrix}.$$

It follows that *the general form of the amplitude equations is exactly the same as the general form of the* $\mathbf{D}_4$*-equivariant mappings from* $\mathbb{R}^2$ *into* $\mathbb{R}^2$. The $\mathbf{D}_4$ symmetry may be thought of as what remains of the original $\mathbf{O}(2) \times \mathbf{S}^1$ symmetry after reduction to the amplitude equations.

This has two consequences: one mildly useful and one extremely important. Suppose we let

$$h(x, y, \lambda) = p(N, \Delta, \lambda)\begin{bmatrix} x \\ y \end{bmatrix} + r(N, \Delta, \lambda)\delta\begin{bmatrix} x \\ -y \end{bmatrix}. \tag{4.7}$$

Then the amplitude equations (4.5) have the form

$$\begin{bmatrix} \dot{x} \\ \dot{y} \end{bmatrix} + h(x, y, \lambda) = 0. \tag{4.8}$$

The mildly useful observation is that the $\mathbf{D}_4$ symmetries restrict the form of dh at solutions, so that the asymptotic stability of steady states can easily be computed. This we do later; however, these results can also be obtained by direct calculation without knowledge of the underlying $\mathbf{D}_4$ symmetry.

The important observation is that it is now possible to classify the *degenerate* $\mathbf{O}(2)$-equivariant Hopf bifurcations, by classifying the $\mathbf{D}_4$-equivariant germs h up to $\mathbf{D}_4$-equivalence. We do this in §6. Moreover, the solutions so obtained include not only the periodic solutions of rotating and standing waves, but also the invariant 2-tori with linear flow.

Remark. This situation has already occurred in our study of degenerate Hopf bifurcation in Chapter VIII. There we had a simple conjugate pair of purely imaginary eigenvalues, with no spatial symmetry. The Birkhoff normal form commutes only with $\mathbf{S}^1$ acting on $\mathbb{C}$ and has the form

$$\dot{z} + p(z\bar{z}, \lambda)z + q(z\bar{z}, \lambda)iz = 0.$$

This equation splits into amplitude and phase equations on setting $z = xe^{i\psi}$, giving

$$\dot{x} + p(x^2, \lambda)x = 0$$
$$\dot{\psi} + q(x^2, \lambda) = 0.$$

The form of the amplitude equation is just that of the general $\mathbf{Z}_2$-equivariant mapping. Indeed we studied degenerate Hopf bifurcation in Chapter VIII by using the classification of $\mathbf{Z}_2$-equivariant mappings under $\mathbf{Z}_2$-equivalence given in Chapter VI.

This kind of reduction (Γ-equivariant Hopf bifurcation $\to \Gamma \times \mathbf{S}^1$ Birkhoff normal form $\to \Sigma$-equivariant amplitude equations) seldom happens in such a nice way, but when it does, we have a method for studying the degenerate Γ-equivariant Hopf bifurcations. For an invariant-theoretic interpretation, see Exercise 4.1.

We end this section with a discussion of the asymptotic stability of rotating and standing waves. These results are summarized in Table 4.1.

Table 4.1. Stabilities of Rotating and Standing Waves in $\mathbf{O}(2)$ Hopf Bifurcation

Name	Orbit Representative	Isotropy Subgroup	Signs of Eigenvalues
Trivial solution	0	$\mathbf{D}_4$	$p(0, 0, \lambda)$ [twice]
Rotating wave	$y = 0$	$(x, y) \mapsto (x, -y)$	r
			$p_N - r + x^2(2p_\Delta - rN) - 2x^4 r_\Delta$
Standing wave	$x = y$	$(x, y) \mapsto (y, x)$	$-r$
			p_N

A rotating wave ($y = 0$) has isotropy subgroup in $\mathbf{D}_4$ generated by the reflection $(x, y) \mapsto (x, -y)$. At a rotating wave solution the Jacobian dh must commute with the matrix $\begin{bmatrix} 1 & 0 \\ 0 & -1 \end{bmatrix}$. Thus dh is diagonal and the eigenvalues of dh are just $(p + \delta r)_x$ and $p - \delta r$. Since $\delta = -x^2$ and $p + \delta r = 0$ on this branch, we obtain the information in Table 4.1.

A standing wave ($x = y$) has isotropy subgroup generated by $(x, y) \mapsto (y, x)$. Thus dh commutes with the matrix $\begin{bmatrix} 0 & 1 \\ 1 & 0 \end{bmatrix}$ and so has the form $\begin{bmatrix} a & b \\ b & a \end{bmatrix}$, which has eigenvalues $a \pm b$. Now

$$a = [(p + \delta r)x]_x$$

$$b = (p + \delta r)_y x$$

and $p = \delta = 0$ along the standing waves branch. This information lets us compute the signs of eigenvalues of dh at the standing wave solution, also listed in Table 4.1.

Table 4.1 leads to the stability assignments given in Figure 1.2, since in nondegenerate $\mathbf{O}(2)$-equivariant Hopf bifurcation we assume that

$$p_\lambda(0) \neq 0, \qquad p_N(0) \neq 0, \qquad r(0) \neq 0, \qquad p_N(0) - r(0) \neq 0. \tag{4.9}$$

As we shall see, these are precisely the nondegeneracy conditions needed to classify the least degenerate $\mathbf{D}_4$-equivariant bifurcation problems.

Exercises

4.1. This exercise provides an alternative viewpoint on the introduction of amplitude/phase variables. Consider the system (3.1), $\dot{x} + X(x, \lambda) = 0$, where X is as in (4.1). Obtain expressions for dN/dt and $d\Delta/dt$ as functions of N and Δ, where N and Δ are the invariant generators of Proposition 2.1(a). Note that orbits of $\mathbf{O}(2) \times \mathbf{S}^1$ are parametrized by the values of invariants, so these expressions may be intepreted as the dynamics of *orbits*. Show that the resulting equations are equivalent to the amplitude equations (4.3(a)). Interpret the $\mathbf{D}_4$-equivariance of the amplitude equations in terms of the geometry of the image of the mapping $(z_1, z_2) \mapsto (N, \Delta)$.

4.2. More generally, suppose that $\dot{x} + X(x, \lambda)$ is a Γ-equivariant ODE, and let $\{I_1, \ldots, I_r\}$ be a system of invariant generators for Γ. Show that dI_j/dt is Γ-

invariant, and deduce that the dynamics of the orbits of Γ can be described by a system of equations

$$dI_j/dt = \psi_j(I_1, \ldots, I_r)$$

defined on the image of the mapping $x \mapsto (I_1, \ldots, I_r)$, that is, the discriminant variety of Γ. Note that this image usually has singularities.

§5.[†] Hopf Bifurcation with O(n)-Symmetry

We now prove that the analysis of Hopf bifurcation for $\mathbf{O}(2)$ acting on $\mathbb{R}^2$ generalizes to $\mathbf{O}(n)$ acting on $\mathbb{R}^n$ by its standard representation, with essentially identical results. The argument is group-theoretic and uses only some results from classical invariant theory.

For $n \geq 2$ let $\mathbf{O}(n)$ act on $\mathbb{R}^n$ by matrix multiplication, and on $\mathbb{R}^n \oplus \mathbb{R}^n$ by the diagonal action. Let $\langle \; , \; \rangle$ be the standard inner product on $\mathbb{R}^n$. Define coordinates (x, y) on $\mathbb{R}^n \oplus \mathbb{R}^n$.

Proposition 5.1.

(a) *Let* $f: \mathbb{R}^n \oplus \mathbb{R}^n \to \mathbb{R}$ *be* $\mathbf{O}(n)$*-invariant. Then*

$$f(x, y) = p(|x|^2, |y|^2, \langle x, y \rangle).$$

(b) *Let* $g: \mathbb{R}^n \oplus \mathbb{R}^n \to \mathbb{R}^n \oplus \mathbb{R}^n$ *be* $\mathbf{O}(n)$*-equivariant. Then*

$$g(x, y) = p_1 \begin{bmatrix} x \\ 0 \end{bmatrix} + p_2 \begin{bmatrix} y \\ 0 \end{bmatrix} + q_1 \begin{bmatrix} 0 \\ x \end{bmatrix} + q_2 \begin{bmatrix} 0 \\ y \end{bmatrix} \tag{5.1}$$

where p_1, p_2, q_1, *and* q_2 *are* $\mathbf{O}(n)$*-invariant.*

PROOF. The main theoretical point in this proof is that the result for general n follows directly from that for $n = 2$. Moreover, this is "the reason" why the $\mathbf{O}(n)$ Hopf theory mimics the $\mathbf{O}(2)$ theory.

Let

$$\begin{aligned} &\text{(a)} \quad W = \{x \in \mathbb{R}^n: x_3 = \cdots = x_n = 0\}, \\ &\text{(b)} \quad V = W \oplus W. \end{aligned} \tag{5.2}$$

We make three claims.

(a) Every group orbit of $\mathbf{O}(n)$ on $\mathbb{R}^n \oplus \mathbb{R}^n$ intersects V.
(b) V is the fixed-point subspace of a subgroup Σ of $\mathbf{O}(n)$.
(c) Let T be the subgroup of $\mathbf{O}(n)$ which leaves V invariant. Then (5.3)

$$T/\Sigma \cong \mathbf{O}(2).$$

Assume that these claims are valid and that $g: \mathbb{R}^n \oplus \mathbb{R}^n \to \mathbb{R}^n \oplus \mathbb{R}^n$ is $\mathbf{O}(n)$-equivariant. By (5.3(b)) g maps V into V, and by (5.3(c)) $g|V$ commutes with $\mathbf{O}(2)$. Assuming the proposition holds for $n = 2$ we see that $g|V$ has the form

(5.1). Further, (5.3(a)) implies that g is determined uniquely by $g|V$. However, the form (5.1) for $n = 2$ has an obvious $\mathbf{O}(n)$-equivariant extension to $\mathbb{R}^n \oplus \mathbb{R}^n$, namely (5.1) itself. Hence by uniqueness g has the form (5.1), proving the proposition for general n.

To conclude the proof we must establish the claims (5.3) and then prove the proposition when $n = 2$.

We verify (5.3(a)). Let $(x, y) \in \mathbb{R}^n \oplus \mathbb{R}^n$, and let $W' \subset \mathbb{R}^n$ be the subspace spanned by $\{x, y\}$. For simplicity we assume that W' is two-dimensional, augmenting it by an independent vector if necessary. There exists $\gamma \in \mathbf{O}(n)$ mapping W to W': just choose an orthonormal basis $\{w_1', \ldots, w_n'\}$ for $\mathbb{R}^n$ with $w_1', w_2' \in W'$ and let γ be the matrix with columns $w_1', \ldots, w_n'$. Thus $\gamma^{-1}(x, y) \in W \oplus W = V$, proving (5.3(a)).

To verify (5.3(b)) note that the group Σ which fixes $W \subset \mathbb{R}^n$ pointwise consists of matrices of the form

$$\left[\begin{array}{c|c} I_2 & 0 \\ \hline 0 & \sigma \end{array}\right]$$

where $\sigma \in \mathbf{O}(n-2)$. Moreover, the only vectors in $\mathbb{R}^n$ fixed by Σ are those in W. Therefore, $\mathrm{Fix}_{\mathbb{R}^n \oplus \mathbb{R}^n}(\Sigma) = V$.

To verify (5.3(c)) note that the only matrices in $\mathbf{O}(n)$ that, in the diagonal action of $\mathbf{O}(n)$ on $\mathbb{R}^n \oplus \mathbb{R}^n$, map V into itself are those that map W into W in the standard action on $\mathbb{R}^n$. These matrices have the form

$$\left[\begin{array}{c|c} \tau & 0 \\ \hline 0 & \sigma \end{array}\right]$$

where $\tau \in \mathbf{O}(2)$ and $\sigma \in \mathbf{O}(n-2)$. Therefore, $T/\Sigma \cong \mathbf{O}(2)$ as claimed.

Finally we let $n = 2$ and show that (5.1) holds. Identify $\mathbb{R}^2$ with $\mathbb{C}$. Let $f: \mathbb{C}^2 \to \mathbb{R}$ be an $\mathbf{O}(2)$-invariant polynomial, so that

$$\begin{aligned} &\text{(a)} \quad f(e^{i\theta}z_1, e^{i\theta}z_2) = f(z_1, z_2) \\ &\text{(b)} \quad f(\bar{z}_1, \bar{z}_2) = f(z_1, z_2). \end{aligned} \tag{5.4}$$

We can write any polynomial $f: \mathbb{C}^2 \to \mathbb{R}$ in the form

$$f(z_1, z_2) = \sum a_{\alpha\beta\gamma\delta} z_1^\alpha \bar{z}_1^\beta z_2^\gamma \bar{z}_2^\delta$$

where $\bar{a}_{\alpha\beta\gamma\delta} = a_{\beta\alpha\delta\gamma}$ since f is real-valued. The identities (5.4) imply

$$\begin{aligned} &\text{(a)} \quad \alpha - \beta = \delta - \gamma, \\ &\text{(b)} \quad a_{\alpha\beta\gamma\delta} = a_{\beta\alpha\delta\gamma}. \end{aligned} \tag{5.5}$$

Eliminating factors $z_1\bar{z}_1$ and $z_2\bar{z}_2$ yields possible generators for the invariants:

$$p_k = z_1^k \bar{z}_2^k + \bar{z}_1^k z_2^k.$$

A simple induction argument shows that only $k = 1$ is needed in the Hilbert basis, since

$$p_{k+1} = p_1 p_k - z_1\bar{z}_1 z_2\bar{z}_2 p_{k-1}.$$

Thus a Hilbert basis for **O**(2)-invariant functions is

$$|z_1|^2, |z_2|^2, \operatorname{Re}(z_1\bar{z}_2).$$

In real coordinates, these are the desired generators.

Let $g: \mathbb{C}^2 \to \mathbb{C}^2$ be an **O**(2)-equivariant polynomial with $g = (g_1, g_2)$ in coordinates. Each g_j satisfies

$$\begin{aligned} &\text{(a)}\quad g_j(e^{i\theta}z_1, e^{i\theta}z_2) = e^{i\theta}g_j(z_1, z_2)\\ &\text{(b)}\quad g_j(\bar{z}_1, \bar{z}_2) = \bar{g}_j(z_1, z_2). \end{aligned} \tag{5.6}$$

Since g is a polynomial we may write

$$g_1(z_1, z_2) = \sum b_{\alpha\beta\gamma\delta} z_1^\alpha \bar{z}_1^\beta z_2^\gamma \bar{z}_2^\delta.$$

Identities (5.6) imply

$$\begin{aligned} &\text{(a)}\quad \alpha - \beta + \gamma - \delta = 1,\\ &\text{(b)}\quad b_{\alpha\beta\gamma\delta} = \bar{b}_{\alpha\beta\gamma\delta}, \end{aligned} \tag{5.7}$$

so the coefficients b are real. Eliminating factors $z_1\bar{z}_1$ and $z_2\bar{z}_2$ yields module generators

$$r_k = z_1^{k+1}\bar{z}_2^k, \qquad s_k = \bar{z}_1^k z_2^{k+1}.$$

The identities

$$\begin{aligned} &\text{(a)}\quad r_k = \operatorname{Re}(z_1\bar{z}_2)^{k-1}z_1 - |z_1|^2 s_{k-1}\\ &\text{(b)}\quad s_k = \operatorname{Re}(\bar{z}_1 z_2)^k z_2 - |z_2|^2 r_{k-1} \end{aligned} \tag{5.8}$$

show by induction that z_1 and z_2 are the only module generators required for g_1. Hence $(z_1, 0)$, $(z_2, 0)$, $(0, z_1)$ and $(0, z_2)$ are the module generators for the equivariants, as claimed. □

Corollary 5.2. *The* **O**$(n) \times$ **S**^{1}*-equivariant mappings on* $\mathbb{R}^n \oplus \mathbb{R}^n$ *are the unique extensions of the* **O**$(2) \times$ **S**^{1}*-equivariant mappings on* V.

Proof. The **S**^{1}-action leaves $V = W \oplus W$ invariant. The **O**$(n) \times$ **S**^{1}-equivariants comprise that subset of the **O**$(n) \times$ **S**^{1}-equivariants that are also **S**^{1}-equivariant; similarly for **O**$(2) \times$ **S**^{1}. The rest is easy. □

Corollary 5.2 implies that the branches of periodic solutions in the **O**(n) Hopf theory are identical to those in the **O**(2) Hopf theory. The n-dimensional analogues of rotating and standing waves have the isotropy subgroups indicated in Figure 5.1.

Finally we discuss the asymptotic stability of these periodic solutions in the **O**(n) Hopf theory. Each periodic solution for an **O**(n)-equivariant vector field can be conjugated to a periodic orbit in V by an element of **O**(n). Clearly the

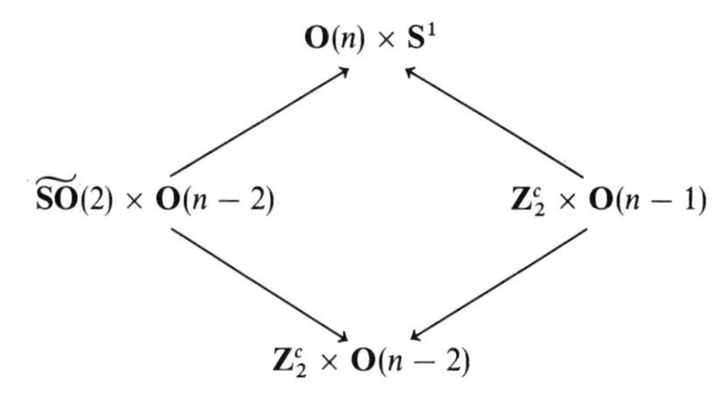

Figure 5.1. Lattice of isotropy subgroups of $\mathbf{O}(n) \times \mathbf{S}^1$.

stability to perturbation within V, of a periodic solution in V, is accounted for by the $\mathbf{O}(2)$ theory. The only question concerns perturbations in the $2n - 4$ directions transverse to V. However, these are all accounted for by the group action, since

$$\dim \mathbf{O}(n)/T = \dim \mathbf{O}(n) - \dim \mathbf{O}(n-2) - \dim \mathbf{O}(2) = 2n - 4,$$

because $\dim \mathbf{O}(n) = n(n-1)/2$. Here $T = \mathbf{O}(n-2) \times \mathbf{O}(2)$ is the subgroup of $\mathbf{O}(n)$ that leaves V invariant.

Thus a periodic solution is orbitally asymptotically stable in the $\mathbf{O}(n)$ theory if and only if it is stable in the $\mathbf{O}(2)$ theory, and the two theories correspond exactly.

§6.† Bifurcation with $\mathbf{D}_4$-Symmetry

In §4 we showed that the classification of degenerate Hopf bifurcations with $\mathbf{O}(2)$ symmetry can be reduced to that of degenerate steady-state bifurcations with $\mathbf{D}_4$ symmetry. This reduction is made by writing the Birkhoff normal form equations in amplitude-phase variables and studying the zeros of the amplitude equations.

Recall from §4(b) that the general $\mathbf{D}_4$-equivariant bifurcation problem may be written as

$$h(x, y, \lambda) = p(N, \Delta, \lambda)\begin{bmatrix} x \\ y \end{bmatrix} + r(N, \Delta, \lambda)\delta\begin{bmatrix} x \\ -y \end{bmatrix} \tag{6.1}$$

where $N = x^2 + y^2$, $\delta = y^2 - x^2$, and $\Delta = \delta^2$. We assume $\lambda = 0$ is a bifurcation point, so $p(0, 0, 0) = 0$. For simplicity of notation we identify h in (6.1) with the pair of $\mathbf{D}_4$-invariant functions $[p, r]$.

Table 6.1 lists the degenerate $\mathbf{D}_4$-singularities up to topological $\mathbf{D}_4$-codimension two, where by topological $\mathbf{D}_4$-codimension we mean the standard $\mathbf{D}_4$-codimension minus the number of modal parameters. The simplest of these singularities has $\mathbf{D}_4$-codimension one and one modal parameter. In

Table 6.1. Classification of $\mathbf{D}_4$ Bifurcation Problems up to Topological Codimension Two

	Defining Conditions	Nondegeneracy Conditions	Normal Form	Coefficients in Normal Form	Universal Unfolding	codim
I	—	$p_N \neq 0,\quad r \neq 0$ $p_N \neq r,\quad p_\lambda \neq 0$	$(\varepsilon_0\lambda + mN, \varepsilon_1)$ $m \neq 0, \varepsilon_1$	$\varepsilon_0 = \operatorname{sgn} p_\lambda,\quad \varepsilon_1 = \operatorname{sgn} r$ $m = p_N/\lvert r\rvert$	—	0
II	$p_N = 0$	$r \neq 0,\quad p_\lambda \neq 0$ $p_{NN} \neq 0$	$(\varepsilon_0\lambda + \varepsilon_1 N^2, \varepsilon_2)$	$\varepsilon_0 = \operatorname{sgn} p_\lambda,\quad \varepsilon_1 = \operatorname{sgn} p_{NN}$ $\varepsilon_2 = \operatorname{sgn} r$	$\alpha(N, 0)$	1
III	$p_N = r$	$p_N \neq 0,\quad p_\lambda \neq 0$ $p_{NN} + 2p_\Delta - 2r_N \neq 0$	$(\varepsilon_0\lambda + \varepsilon_1 N + \varepsilon_2\Delta, \varepsilon_1)$	$\varepsilon_0 = \operatorname{sgn} p_\lambda,\quad \varepsilon_1 = \operatorname{sgn} p_N$ $\varepsilon_2 = \operatorname{sgn}(p_{NN} + 2p_\Delta - 2r_N)$	$\alpha(N, 0)$	1
IV	$r = 0$	$p_N \neq 0,\quad p_\lambda \neq 0$ $p_N r_\Delta - p_\Delta r_N \neq 0$ $p_\lambda r_N - p_N r_\lambda \neq 0$	$(\varepsilon_0\lambda + \varepsilon_1 N + m\Delta, \varepsilon_2 N)$ $m \neq 0$	$\varepsilon_0 = \operatorname{sgn} p_\lambda,\quad \varepsilon_1 = \operatorname{sgn} p_N$ $\varepsilon_2 = \varepsilon_0 \operatorname{sgn}(p_\lambda r_N - p_N r_\lambda)$ $m = \varepsilon_2 p_\lambda^2 \dfrac{p_N r_\Delta - p_\Delta r_N}{(p_\lambda r_N - p_N r_\lambda)^2}$	$\alpha(0, 1)$	1
V	$p_\lambda = 0$	$p_N \neq 0,\quad p_N \neq r$ $r \neq 0,\quad p_{\lambda\lambda} \neq 0$ $p_{\lambda N} r - p_N r_\lambda \neq 0$	$(\varepsilon_0\lambda^2 + mN + \varepsilon_1\lambda N, \varepsilon_2)$ $m \neq 0, \varepsilon_2$	$\varepsilon_0 = \operatorname{sgn} p_{\lambda\lambda},\quad \varepsilon_2 = \operatorname{sgn} r$ $\varepsilon_1 = \varepsilon_2 \operatorname{sgn}(p_\lambda r_N - p_N r_\lambda)$ $m = p_N/r$	$\alpha(1, 0)$	1
VI	$p_N = 0$ $p_{NN} = 0$	$r \neq 0,\quad p_\lambda \neq 0$ $p_{NNN} \neq 0$	$(\varepsilon_0\lambda + \varepsilon_1 N^3, \varepsilon_2)$	$\varepsilon_0 = \operatorname{sgn} p_\lambda,\quad \varepsilon_1 = \operatorname{sgn} p_{NNN}$ $\varepsilon_2 = \operatorname{sgn} r$	$\alpha(N, 0)$ $+ \beta(N^2, 0)$	2
VII	$p_N = 0$ $p_{NN} + 2p_\Delta - 2r_N = 0$	$p_N \neq 0,\quad p_\lambda \neq 0$ $p_{NNN} + 6p_{N\Delta} - 6r_\Delta - 3r_{NN} \neq 0$	$(\varepsilon_0\lambda + \varepsilon_1 N + \varepsilon_2 N\Delta, \varepsilon_1)$	$\varepsilon_0 = \operatorname{sgn} p_\lambda,\quad \varepsilon_1 = \operatorname{sgn} p_N$ $\varepsilon_2 = \operatorname{sgn}(p_{NNN} + 6p_{N\Delta} - 6r_\Delta$ $- 3r_{NN})$	$\alpha(N, 0)$ $+ \beta(\Delta, 0)$	2
VIII	$r = 0$ $p_N r_\Delta - p_\Delta r_N = 0$	$p_N \neq 0,\quad p_\lambda \neq 0$ $p_\lambda r_N - p_N r_\lambda \neq 0$ $\xi_1 \neq 0$*	$(\varepsilon_0\lambda + \varepsilon_1 N + m\Delta^2, \varepsilon_2 N)$ $m \neq 0$	$\varepsilon_0 = \operatorname{sgn} p_\lambda,\quad \varepsilon_1 = \operatorname{sgn} p_N$ $\varepsilon_2 = \varepsilon_0 \operatorname{sgn}(p_\lambda r_N - p_N r_\lambda)$ $m = \varepsilon_2 \xi_2$*	$\alpha(0, 1)$ $+ \beta(\Delta, 0)$	2
IX	$r = 0$ $p_\lambda r_N - p_N r_\lambda = 0$	$p_N \neq 0,\quad p_\lambda \neq 0$ $p_N r_\Delta - p_\Delta r_N \neq 0$ $\xi_2 \neq 0$*	$(\varepsilon_0\lambda + \varepsilon_1 N, \varepsilon_2\Delta + m\lambda^2)$ $m \neq 0$	$\varepsilon_0 = \operatorname{sgn} p_\lambda,\quad \varepsilon_1 = \operatorname{sgn} p_N$ $\varepsilon_2 = \operatorname{sgn}(p_N r_\Delta - p_\Delta r_N)$	$\alpha(N, 0)$ $+ \beta(0, N)$	2

X	$p_N = 0$ $r = 0$	$p_\lambda \neq 0$, $p_{NN} \neq 0$, $p_\Delta \neq 0$ $r_N \neq 0$, $p_{NN} + 2p_\Delta - 2r_N \neq 0$ $\xi_3 \neq 0$, $p_{NN} + 2p_\Delta - r_N \neq 0$	$(\varepsilon_0\lambda + mN^2 + n\Delta, \varepsilon_1 N + \varepsilon_2\Delta)$ $m \neq 0$, $n \neq 0$ $m + n \neq \varepsilon_1, \varepsilon_1/2$	$\varepsilon_0 = \operatorname{sgn} p_\lambda$, $\varepsilon_1 = \operatorname{sgn} r_N$ $\varepsilon_2 = \varepsilon_0 \operatorname{sgn} \xi_3$* $m = p_{NN}/2\lvert r_N\rvert$, $n = p_\Delta/\lvert r_N\rvert$	$\alpha(N,0)$ $+ \beta(0,1)$	2
XI	$p_N = 0$ $p_\lambda = 0$	$r \neq 0$, $p_{NN} \neq 0$, $p_{\lambda\lambda} \neq 0$ $p_{\lambda N}^2 \neq p_{\lambda\lambda}p_{NN}$	$(\varepsilon_0\lambda^2 + \varepsilon_1 N^2 + m\lambda N, \varepsilon_2)$ $m^2 \neq 4\varepsilon_0\varepsilon_1$	$\varepsilon_0 = \operatorname{sgn} p_{\lambda\lambda}$, $\varepsilon_2 = \operatorname{sgn} r$ $\varepsilon_1 = \operatorname{sgn} p_{NN}$ $m = 2p_{\lambda N}/\sqrt{\lvert p_{\lambda\lambda}p_{NN}\rvert}$	$\alpha(1,0)$ $+ \beta(N,0)$	2
XII	$p_N = r$ $p_\lambda = 0$	$p_N \neq 0$, $p_{\lambda\lambda} \neq 0$ $p_{NN} + 2p_\Delta - 2r_N \neq 0$ $(p_{\lambda N} - r_\lambda)^2 \neq$ $p_{\lambda\lambda}(p_{NN} + 2p_\Delta - 2r_N)$	$(\varepsilon_0\lambda^2 + \varepsilon_1 N + \varepsilon_2\Delta + m\lambda N, \varepsilon_1)$ $m^2 \neq 4\varepsilon_0\varepsilon_2$	$\varepsilon_0 = \operatorname{sgn} p_{\lambda\lambda}$, $\varepsilon_1 = \operatorname{sgn} p_N$ $\varepsilon_2 = \operatorname{sgn}(p_{NN} + 2p_\Delta - 2r_N)$ $m = 2\varepsilon_1(p_{\lambda N} - r_\lambda)/$ $\sqrt{\lvert p_{\lambda\lambda}\rvert\lvert p_{NN} + 2p_\Delta - 2r_N\rvert}$	$\alpha(1,0)$ $+ \beta(N,0)$	2
XIII	$r = 0$ $p_\lambda = 0$	$p_N \neq 0$, $p_{\lambda\lambda} \neq 0$ $r_\lambda \neq 0$, $p_N r_\Delta - p_\Delta r_N \neq 0$	$(\varepsilon_0\lambda^2 + \varepsilon_1 N, \varepsilon_2\lambda + m\Delta)$ $m \neq 0$	$\varepsilon_0 = \operatorname{sgn} p_{\lambda\lambda}$, $\varepsilon_1 = \operatorname{sgn} p_N$ $\varepsilon_2 = \operatorname{sgn} r_\lambda$ $m = \varepsilon_1 p_{\lambda\lambda}^2(p_N r_\Delta - p_\Delta r_N)/r_\lambda^2$	$\alpha(1,0)$ $+ \beta(0,1)$	2
XIV	$p_\lambda = 0$ $p_{\lambda\lambda} = 0$	$p_N \neq 0$, $p_N \neq r$, $r \neq 0$ $p_{\lambda\lambda\lambda} \neq 0$, $p_{\lambda N}r - p_N r_\lambda \neq 0$	$(\varepsilon_0\lambda^3 + mN + \varepsilon_1\lambda N, \varepsilon_2)$ $m \neq 0, \varepsilon_2$	$\varepsilon_0 = \operatorname{sgn} p_{\lambda\lambda\lambda}$, $\varepsilon_2 = \operatorname{sgn} r$ $\varepsilon_1 = \varepsilon_2 \operatorname{sgn}(p_{\lambda N}r - p_N r_\lambda)$ $m = p_N/\lvert r\rvert$	$\alpha(1,0)$ $+ \beta(\lambda,0)$	2
XV	$p_\lambda = 0$ $p_{\lambda N}r - p_N r_\lambda = 0$	$p_N \neq 0$, $p_N \neq r$ $r \neq 0$, $p_{\lambda\lambda} \neq 0$	$(\varepsilon_0\lambda^2 + mN, \varepsilon_1)$ $m \neq 0, \varepsilon_1$	$\varepsilon_0 = \operatorname{sgn} p_{\lambda\lambda}$, $\varepsilon_1 = \operatorname{sgn} r$ $m = p_N/r$	$\alpha(1,0)$ $+ \beta(\lambda N,0)$	2

* Here

$$\xi_1 = p_N p_\lambda^3\{p_\Delta^2 r_\Delta p_\lambda + p_\lambda p_\Delta(r_\Delta p_{NN} - p_\Delta r_{NN}) + 2p_N(p_\Delta r_\lambda r_{N\Delta} - r_\Delta p_\lambda p_{N\Delta}) + p_\lambda p_N(r_N p_{\Delta\Delta} - p_N r_{\Delta\Delta})\}/2(p_\lambda r_N - p_N r_\lambda)^4$$

$$\xi_2 = \{p_N(p_N r_{\lambda\lambda} - r_N p_{\lambda\lambda}) + p_\lambda(p_\lambda r_{NN} - r_\lambda p_{NN}) - 2p_\lambda(p_N r_{\lambda N} - r_N p_{\lambda N})\}/2p_N p_\lambda^2(r_\Delta p_N - p_\Delta r_N)$$

$$\xi_3 = (p_\lambda r_\Delta - p_\Delta r_\lambda) - r_N\{2p_\lambda p_{NNN}(p_\Delta - r_N) + 6p_{NN}(r_\lambda r_N - p_\lambda p_{N\Delta}) + 3p_{NN}(p_\lambda r_{NN} - r_\lambda p_{NN})\}/6p_{NN}(2p_\Delta + p_{NN} - r_N).$$

the remainder of this section we sketch how this table is derived. More details are in Golubitsky and Roberts [1987], from which the results are taken. Buzano, Geymonat, and Poston [1985] study singularities I–IV in the context of buckling of a beam with square cross section. In addition to Table 6.1 we list the defining and nondegeneracy conditions and the universal unfolding for the various normal forms. The algebraic data needed for a singularity theory analysis are listed in Table 6.2. The generators for $\vec{\mathscr{E}}(\mathbf{D}_4)$ were given using different coordinates in Chapter XIV, Table 3.1.

The higher order terms for each of the fifteen singularities listed in Table 6.1 are given in Table 6.3. We have indicated explicitly those cases (XI–XIV) where equivalences other than strong equivalences are needed to compute these higher order terms. Knowing $\operatorname{Itr}\mathscr{K}(\cdot,\mathbf{D}_4)$ makes the calculation of $\mathscr{K}(\cdot,\mathbf{D}_4)$, $RT(\cdot,\mathbf{D}_4)$, and $T(\cdot,\mathbf{D}_4)$ rather easy. Thus it is a relatively straightforward calculation, using Table 6.3, to verify the universal unfoldings in Table 6.1. To compute the entries in Table 6.3, one needs to know the form of the intrinsic submodules for the action of $\mathbf{D}_4$. This information is given in Proposition 6.1.

The logic needed to classify $\mathbf{D}_4$ singularities is represented as a flowchart in Table 6.4. To derive the normal forms and nondegeneracy conditions of Table 6.1 we must make explicit changes of coordinates, modulo terms in $\mathscr{P}(\cdot,\mathbf{D}_4)$, of a type we now describe.

Consider the $\mathbf{D}_4$-equivalence defined by

$$
\begin{aligned}
&\text{(a)}\quad Z = a\begin{bmatrix} x \\ y \end{bmatrix} + b\delta\begin{bmatrix} x \\ -y \end{bmatrix},\\
&\text{(b)}\quad S = AS_1 + BS_2 + CS_3 + DS_4,\\
&\text{(c)}\quad \Lambda = \Lambda(\lambda),
\end{aligned}
\tag{6.2}
$$

where $a, b, A, B, C, D \in \mathscr{E}_{u,\lambda}(\mathbf{D}_4)$ and $a(0) > 0$; $S(0) = s_0 I$; $s_0 > 0$.

Composing N, δ, and Δ with Z gives

$$
\begin{aligned}
&\text{(a)}\quad \tilde{N} = N \circ Z = a^2 N - 2ab\Delta + b^2 N\Delta\\
&\text{(b)}\quad \tilde{\delta} = \delta \circ Z = (a^2 - 2abN + b^2\Delta)\delta\\
&\text{(c)}\quad \tilde{\Delta} = \Delta \circ Z = (a^2 - 2abN + b^2\Delta)^2\Delta.
\end{aligned}
\tag{6.3}
$$

The result of applying the coordinate changes Z and Λ to $f = [p, r]$ is

$$
(a\tilde{p} + b(a^2 - 2abN + b^2\Delta)\Delta\tilde{r}, b\tilde{p} + a(a^2 - 2abN + b^2\Delta)\tilde{r}) \tag{6.4}
$$

where $\tilde{p} = p(\tilde{N}, \tilde{\Delta}, \Lambda)$, $\tilde{r} = r(\tilde{N}, \tilde{D}, \Lambda)$.

Finally, applying S to (6.4), using (6.2), yields $(\hat{p}, \hat{r})$ where

$$
\begin{aligned}
\hat{p} &= \{Aa + BaN - (Bb + Da)\Delta + DaN^2\}\tilde{p}\\
&\quad + \{[Ab - Ba + BbN + Db(N^2 - \Delta)][a^2 - 2abN + b^2\Delta]\}\Delta\tilde{r}\\
\hat{r} &= \{Ab + Ca - CbN - Db(N^2 - \Delta)\}\tilde{p}\\
&\quad + \{[Aa - CaN + (Da + Cb)\Delta - DaN^2][a^2 - 2abN + B^2\Delta]\}\tilde{r}.
\end{aligned}
\tag{6.5}
$$

Table 6.2. Algebraic Data for $\mathbf{D}_4$ Singularities

Object	Generators
$\mathscr{E}(\mathbf{D}_4)$	$N = x^2 + y^2, \quad \Delta = \delta^2 \quad$ where $\delta = y^2 - x^2$
$\vec{\mathscr{E}}(\mathbf{D}_4)$	$\begin{bmatrix} x \\ y \end{bmatrix}, \delta \begin{bmatrix} x \\ -y \end{bmatrix}$
$\overset{\rightrightarrows}{\mathscr{E}}(\mathbf{D}_4)$	$S_1 = \begin{bmatrix} 1 & 0 \\ 0 & 1 \end{bmatrix}, \quad S_2 = \begin{bmatrix} x^2 & xy \\ xy & y^2 \end{bmatrix},$ $S_3 = \begin{bmatrix} -x^2 & xy \\ xy & -y^2 \end{bmatrix}, \quad S_4 = \begin{bmatrix} 0 & x^3y \\ xy^3 & 0 \end{bmatrix}$
$RT([p,r],\mathbf{D}_4)$	$a_1 = [p,r], \quad a_2 = [Np - \Delta r, 0],$ $a_3 = [0, p - Nr], \quad a_4 = (N^2 - \Delta)[p, -r]$ $b_1 = [p + 2Np_N + 4\Delta p_\Delta, 3r + 2Nr_N + 4\Delta r_\Delta]$ $b_2 = [-2\Delta p_N - 4N\Delta p_\Delta + \Delta r_\Delta, p - 2Nr - 2\Delta r_N - 4N\Delta r_\Delta]$
$\mathscr{K}([p,r],\mathbf{D}_4)$	$Na_1, \Delta a_1, \lambda a_1, a_2, a_3, a_4, Nb_1, \Delta b_1, \lambda b_1, b_2$
$\mathscr{P}([p,r],\mathbf{D}_4)$	$\operatorname{Itr}(\mathscr{K}([p,r],\mathbf{D}_4) + \mathscr{E}_\lambda\{\lambda^2[p_\lambda, r_\lambda]\})$

Table 6.3. Intrinsic Part of $\mathscr{K}$, and $\mathscr{P}$, for $\mathbf{D}_4$ Singularities

No.	$\operatorname{Itr}\mathscr{K}(\cdot,\mathbf{D}_4)$	$\mathscr{P}(\cdot,\mathbf{D}_4)$ if $\neq \operatorname{Itr}\mathscr{K}(\cdot,\mathbf{D}_4)$
I	$(\mathscr{M}^2 + \langle\Delta\rangle, \mathscr{M})$	—
II	$(\mathscr{M}^3 + \mathscr{M}\langle\lambda,\Delta\rangle, \mathscr{M})$	—
III	$(\mathscr{M}^3 + \mathscr{M}\langle\lambda,\Delta\rangle, \mathscr{M}^2 + \langle\lambda,\Delta\rangle)$	—
IV	$(\mathscr{M}^2, \mathscr{M}^2)$	—
V	$(\mathscr{M}^3 + \langle N,\Delta\rangle^2 + \langle\Delta\rangle, \mathscr{M}^2 + \langle N,\Delta\rangle)$	—
VI	$(\mathscr{M}^4 + \mathscr{M}\langle\lambda\rangle + \langle\Delta\rangle, \mathscr{M})$	—
VII	$(\mathscr{M}^4 + \mathscr{M}^2\langle\Delta\rangle + \langle\Delta\rangle^2 + \mathscr{M}\langle\lambda\rangle,$ $\mathscr{M}^3 + \mathscr{M}\langle\Delta\rangle + \langle\lambda\rangle)$	—
VIII	$(\mathscr{M}^3 + \mathscr{M}\langle\lambda\rangle, \mathscr{M}^3 + \mathscr{M}\langle\lambda\rangle)$	—
IX	$(\mathscr{M}^3 + \mathscr{M}\langle\Delta\rangle, \mathscr{M}^3 + \mathscr{M}\langle\Delta\rangle)$	—
X	$(\mathscr{M}^4 + \mathscr{M}^2\langle\Delta\rangle + \langle\Delta\rangle^2 + \mathscr{M}\langle\lambda\rangle,$ $\mathscr{M}^3 + \mathscr{M}\langle\lambda,\Delta\rangle)$	—
XI	$(\mathscr{M}^4 + \langle\Delta\rangle, \mathscr{M}^2 + \langle N,\Delta\rangle)$	$(\mathscr{M}^3 + \langle\Delta\rangle, \mathscr{M})$
XII	$(\mathscr{M}^4 + \mathscr{M}^2\langle\Delta\rangle + \langle\Delta\rangle^2, \mathscr{M}^3 + \mathscr{M}\langle\Delta\rangle)$	$(\mathscr{M}^3 + \mathscr{M}\langle\Delta\rangle, \mathscr{M}^2 + \langle\Delta\rangle$
XIII	$(\mathscr{M}^3 + \langle N,\Delta\rangle^2, \mathscr{M}^3 + \mathscr{M}\langle N,\Delta\rangle)$	$(\mathscr{M}^3 + \mathscr{M}\langle N,\Delta), \mathscr{M}^2)$
XIV	$(\mathscr{M}^4 + \langle N,\Delta\rangle^2 + \langle\Delta\rangle, \mathscr{M}^3 + \langle N,\Delta\rangle)$	$(\mathscr{M}^4 + \mathscr{M}^2\langle N,\Delta\rangle + \langle N,\Delta\rangle^2, \mathscr{M}^2 + \langle N,\Delta\rangle)$
XV	$(\mathscr{M}^3 + \langle N,\Delta\rangle^2 + \langle\Delta\rangle, \mathscr{M}^2 + \langle N,\Delta\rangle)$	—

Table 6.4. Flowchart for Recognition and Classification of $\mathbf{D}_4$ Bifurcation Problems

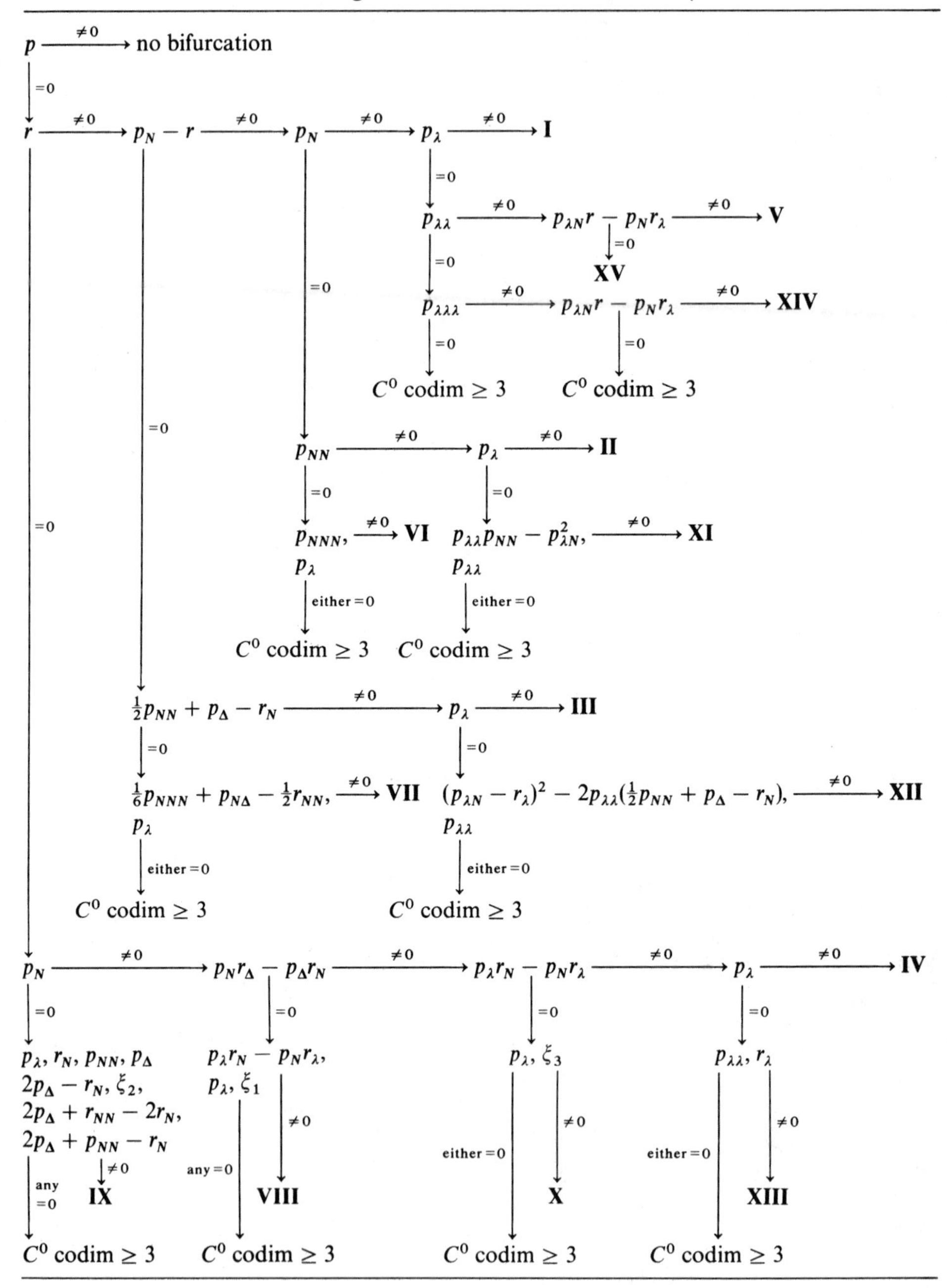

Table 6.5. Low Order Terms of Bifurcation Problems $\mathbf{D}_4$-Equivalent to $[p, r]$.

Terms in $[p, r]$	Corresponding Terms in $[\hat{p}, \hat{r}]$
$[\lambda, 0]$	$Aa\Lambda_\lambda p_\lambda$
$[N, 0]$	$Aa^3 p_N$
$[\Delta, 0]$	$-2Aa^2 b p_N + Aa^5 p_\Delta + (Aa^2 b - Ba^3)r$
$[\lambda^2, 0]$	$(Aa)_\lambda \Lambda_\lambda p_\lambda + \frac{1}{2}Aa(\Lambda_\lambda)^2 p_{\lambda\lambda}$
$[\lambda N, 0]$	$((Aa)_N + Ba)\Lambda_\lambda p_\lambda + (Aa)_\lambda a^2 p_N + Aa^3 \Lambda_\lambda p_{\lambda N}$
$[N^2, 0]$	$(Ba^3 + (Aa)_N a^2)p_N + \frac{1}{2}Aa^5 p_{NN}$
$[\lambda\Delta, 0]$	$((Aa)_\Delta - (Bb + Da))\Lambda_\lambda p_\lambda - 2(Aa)_\lambda ab p_N + (Aa)_\lambda a^4 p_\Delta - 2Aa^2 b\Lambda_\lambda p_{\lambda N}$ $+ Aa^5 \Lambda_\lambda p_{\lambda\Delta} + (Aa^2 b - Ba^3)_\lambda r + (Aa^2 b - Ba^3)\Lambda_\lambda r_\lambda$
$[N\Delta, 0]$	$(Aab^2 - 3Ba^2 b - Da^3 - 2(Aa)_N ab + (Aa)_\Delta a^2)p_N$ $+ (-4Aa^4 b + Ba^5 + (Aa)_N a^4)p_\Delta + Aa^7 p_{N\Delta}$ $+ (-2Aab^2 + 3Ba^2 b + (Aa^2)_N)r + (Aa^4 b - Ba^5)r_N$
$[0, 1]$	$Aa^3 r$
$[0, \lambda]$	$(Ab + Ca)\Lambda_\lambda p_\lambda + (Aa^3)_\lambda r + Aa^3 \Lambda_\lambda r_\lambda$
$[0, N]$	$(Aa^2 b + Ca^3)p_N - (2Aa^2 b + Ca^3)r + Aa^5 r_N$
$[0, \Delta]$	$-2(Aab^2 + Ca^2 b)p_N + (Aa^4 b + Ca^5)p_\Delta$ $+ (Aab^2 + Ca^2 b + Da^3)r - 2Aa^4 b r_N + Aa^7 r_\Delta$

That is, any germ $\mathbf{D}_4$-equivalent to $[p, r]$ can be written as $[\hat{p}, \hat{r}]$ for some a, b, A, B, C, D, Λ.

By taking the Taylor expansions of p and r we can extract from (6.5) the coefficients of low order terms of all bifurcation problems $\mathbf{D}_4$-equivalent to $[p, r]$. Those we need for the examples in Table 6.1 are given in Table 6.5. The expressions p_λ, $r_{N\Delta}$, $(Aa)_N$, etc., are partial derivatives with respect to the subscripts, evaluated at 0. Given (6.1) we assume throughout that $p(0) = 0$.

We can now solve the recognition problem up to topological $\mathbf{D}_4$-codimension two, as indicated in Table 6.1. As stated previously. the information presented here constitutes only a sketch and details are in Golubitsky and Roberts [1987]. However, the calculations needed to reach codimension one (singularities I–IV) are relatively easy; see Exercise 6.1.

We end this section with the promised discussion of intrinsic ideals and submodules for this action of $\mathbf{D}_4$.

Proposition 6.1.
(a) *Sums and products of the intrinsic ideals* $\langle\lambda\rangle$, $\langle\Delta\rangle$, *and* $\langle N, \Delta\rangle$ *are* $\mathbf{D}_4$-*intrinsic.*
(b) *A finite-codimension submodule* $[\mathscr{I}, \mathscr{J}] \subset \vec{\mathscr{E}}_{u,\lambda}(\mathbf{D}_4)$ *is intrinsic if and only if*
 (i) $\mathscr{I}$ *and* $\mathscr{J}$ *are intrinsic ideals,*
 (ii) $\mathscr{I} \subset \mathscr{J}$,
 (iii) $\langle\Delta\rangle\mathscr{J} \subset \mathscr{I}$.

Remark. It follows that $[\mathscr{I}, \mathscr{I}]$ and $[\langle\Delta\rangle\mathscr{I}, \mathscr{I}]$ are intrinsic submodules whenever $\mathscr{I}$ is an intrinsic ideal. Moreover, every intrinsic submodule is a sum of intrinsic submodules of this form since

$$[\mathscr{I}, \mathscr{J}] = [\mathscr{I}, \mathscr{I}] + [\langle\Delta\rangle\mathscr{J}, \mathscr{J}].$$

PROOF. Using (6.3(a, c)) it is easy to check that the ideals $\langle\lambda\rangle$, $\langle\Delta\rangle$, and $\langle N, \Delta\rangle$ are intrinsic, and it is easy to check that submodules $[\mathscr{I}, \mathscr{J}]$ satisfying (i–iii) are intrinsic.

Conversely, suppose that $\mathscr{N} \subset \vec{\mathscr{E}}_{u,\lambda}(\mathbf{D}_4)$ is intrinsic. Then we may write $\mathscr{N} = [\mathscr{I}, \mathscr{J}]$ where $\mathscr{I}$ and $\mathscr{J}$ are ideals in $\mathscr{E}_{u,\lambda}(\mathbf{D}_4)$. We also know that if $f \in \mathscr{N}$ then $RT(f, \mathbf{D}_4) \subset \mathscr{N}$ (see Proposition XIV, 6.2). Thus if $f = [p, r]$ then we may use a_3 of Table 6.2 to conclude that $[0, p] \in \mathscr{N}$ and hence that $\mathscr{I} \subset \mathscr{J}$. Similarly if $f = [0, r] \in \mathscr{N}$ then we use a_2 of Table 6.2 to conclude that $[\Delta r, 0] \in \mathscr{N}$ and hence that $\langle\Delta\rangle\mathscr{J} \subset \mathscr{I}$. Finally we use (6.4) to show that $\mathscr{I}$ and $\mathscr{J}$ are intrinsic ideals. In the notation of (6.4) we must show that if $[p, 0] \in \mathscr{N}$ then so is $[\tilde{p}, 0]$. Now (6.4) with $r = 0$ shows that $[a\tilde{p}, b\tilde{p}] \in \mathscr{N}$. But we know from (ii) that $b\tilde{p} \in \mathscr{J}$, hence $a\tilde{p} \in \mathscr{I}$. But $a(0) \neq 0$ (by the definition of equivalence) and $\tilde{p} \in \mathscr{I}$, as desired. Similarly, if $[0, r] \in \mathscr{N}$ then (6.4) shows that

$$[b(a^2 - 2abN + b^2\Delta)\Delta\tilde{r}, a(a^2 - 2abN + b^2\Delta)\tilde{r}] \in \mathscr{N},$$

where $\tilde{r} \in \mathscr{J}$. □

EXERCISES

6.1. Using the tables in this section as a guide, give a complete proof of the classification of $\mathbf{D}_4$-equivariant bifurcation problems up to topological codimension one.

6.2. Show that the ideal $\langle N^2 - \Delta\rangle$ is intrinsic. (*Hint*: Use (6.2) to show that $N^2 - \Delta$ transforms to $(N^2 - \Delta)(a^2 - b^2\Delta)^2$.)

§7. The Bifurcation Diagrams

We now exhibit the bifurcation diagrams, including stabilities, for the universal unfoldings of the degenerate $\mathbf{D}_4$ singularities of Table 6.1. As noted in §4 these results have a direct interpretation for degenerate Hopf bifurcations with $\mathbf{O}(2)$ symmetry. In that interpretation we find 2-tori with linear flow in codimension one, 3-tori in codimension two, and various possibilities for multiplicity of solutions and exchanges of stability between branches. The existence of the 2-torus was observed by Erneux and Matkowsky [1984]; the proof that the flow on the 2-torus is linear even when Birkhoff normal form is not assumed is due to Chossat [1986]; and the existence of the 3-torus was observed by Knobloch [1986].

Table 7.1. Branching and Stability for $\mathbf{D}_4$-Equivariants

Solution	Branching Equations	Signs of Eigenvalues
Trivial	$x = y = 0$	p [twice]
R	$y = 0$	$p_N - r + x^2(2p_\Delta - r_N) - 2x^4 r_\Delta$
	$p - x^2 r = 0$	r
S	$x = y$	p_N
	$p = 0$	$-r$
T	$p = 0$	$\operatorname{sgn} \operatorname{tr} dh = Np_N + 2\Delta p_\Delta - \Delta r_N - 2N\Delta r_\Delta$
	$r = 0$	$\operatorname{sgn} \det dh = p_\Delta r_N - p_N r_\Delta$

The computations needed to obtain the bifurcation diagrams of this section are extensive. Our aim here is to describe how the diagrams are derived; the details may be found in Golubitsky and Roberts [1987], and we omit most of them here.

By §4 nontrivial zeros of the $\mathbf{D}_4$-equivariant amplitude equations

$$h(x, y, \lambda) \equiv p(N, \Delta, \lambda)\begin{bmatrix} x \\ y \end{bmatrix} + r(N, \Delta, \lambda)\delta\begin{bmatrix} x \\ -y \end{bmatrix} = 0 \tag{7.1}$$

(where we may assume $x \geq y \geq 0$) correspond to rotating waves $(y = 0)$, standing waves $(x = y)$, and 2-tori $(x \neq y, x \neq 0)$. We denote these equilibria of (7.1) by R, S, T, respectively. The equations for such solutions and the computations needed for the signs of the eigenvalues are given in Table 7.1. The information for R and S was already given in Table 4.1. The branching equations for T follow from (7.1) since when $x > y > 0$ the vectors $[\begin{smallmatrix} x \\ y \end{smallmatrix}]$ and $[\begin{smallmatrix} x \\ -y \end{smallmatrix}]$ are independent and δ is nonzero. The calculation of $\det dh$ and $\operatorname{tr} dh$ is left to the reader.

According to Lemma 4.11 the asymptotic stability of equilibria of (7.1) corresponds to the asymptotic stability of the periodic solutions R and S and the 2-tori T. We wish to use the normal forms in Table 6.1 to analyze stabilities, so we must discuss the effect of $\mathbf{D}_4$-equivalence on the asymptotic stability of equilibria. The signs of the eigenvalues of dh at solutions R and S are invariants of $\mathbf{D}_4$-equivalence. This follows from XIII, §4(b), since the isotropy subgroup for R, $(x, y) \mapsto (x, -y)$, and that for S, $(x, y) \mapsto (y, x)$, force dh to be diagonalizable at such solutions. Further, the sign of $\det dh$ is invariant under $\mathbf{D}_4$-equivalence. Thus only when $\det dh > 0$ is there a possible ambiguity, and we discuss this point later. A similar situation with $\mathbf{Z}_2 \oplus \mathbf{Z}_2$ symmetry is examined in Lemma X, 3.1.

The number of normal forms in Table 6.1 is quite large. There are two reasons: each case has several choices of signs, and most cases have at least one modal parameter. To reduce the number of cases we will consider as equivalent normal forms related by time reversal $h \mapsto -h$ and reversals of the bifurcation parameter $\lambda \mapsto -\lambda$. Usually this lets us specify two sign choices in

Table 7.2. Universal Unfoldings and Topological Codimension for $\mathbf{D}_4$ Bifurcation Problems. Note the Sign Choices Specified

	Universal Unfolding $[p, r]$		Topological Codimension
I	$[-\lambda + mN, 1]$	$m \neq 0, 1$	0
II	$[-\lambda + N^2 + \alpha N, \varepsilon_2]$		1
III	$[-\lambda + (1 + \alpha)N + \varepsilon_2 \Delta, 1]$		1
IV	$[-\lambda + \varepsilon_1 N + m\Delta, N + \alpha]$	$m \neq 0$	1
V	$[\lambda^2 + mN + \lambda N + \alpha, \varepsilon_2]$	$m \neq 0, \varepsilon_2$	1
VI	$[-\lambda + N^3 + \beta N^2 + \alpha N, \varepsilon_2]$		2
VII	$[-\lambda + N + \varepsilon_2 N\Delta + \alpha N + \beta\Delta, 1]$		2
VIII	$[-\lambda + N + m\Delta^2 + \beta\Delta, \varepsilon_2 N + \alpha]$	$m \neq 0$	2
IX	$[-\lambda + N, \varepsilon_2 \Delta + m\lambda^2 + \beta N + \alpha]$	$m \neq 0$	2
X	$[-\lambda + mN^2 + n\Delta + \alpha N, N + \varepsilon_2 \Delta + \beta]$	$mn \neq 0, \quad m + n \neq \frac{1}{2}$	2
XI	$[\lambda^2 + \varepsilon_1 N^2 + m\lambda N + \alpha + \beta N, \varepsilon_2]$	$m \geq 0, \quad m^2 \neq 4\varepsilon_1$	2
XII	$[\lambda^2 + \varepsilon_1 N + \varepsilon_2 \Delta + m\lambda N, \varepsilon_1]$	$m \geq 0, \quad m^2 \neq 4\varepsilon_2$	2
XIII	$[\lambda^2 + \varepsilon_1 N + \alpha, \lambda + m\Delta + \beta]$	$m \neq 0$	2
XIV	$[-\lambda^3 + mN + \varepsilon_1 \lambda N + \alpha + \beta\lambda, 1]$	$m \neq 0, 1$	2
XV	$[\lambda^2 + mN + \alpha + \beta\lambda N, \varepsilon_1]$	$m \neq 0, \varepsilon_1$	2

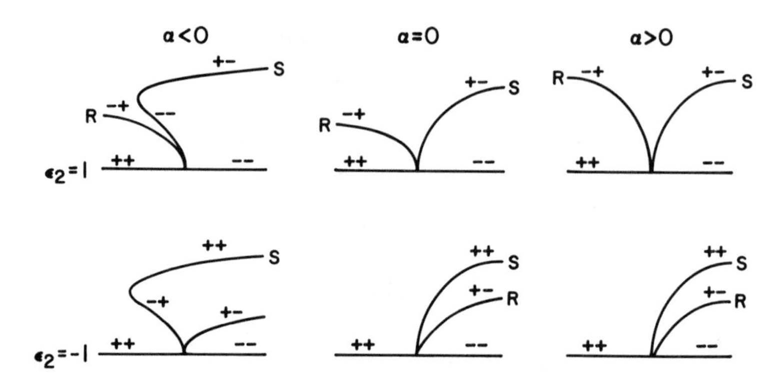

Figure 7.1. Transition variety and bifurcation diagrams for normal form II: $[-\lambda + N^2 + \alpha N, \varepsilon_2]$.

the normal forms. We have made these choices arbitrarily, except that we have required the trivial steady state $x = y = 0$ to be stable subcritically ($\lambda < 0$). Table 7.2 lists the sign choices.

Normal form I corresponds to the nondegenerate case; the possible bifurcation diagrams have already been given in Figure 1.2. The simplest degenerate situations are those with topological $\mathbf{D}_4$-codimension one, cases II–V. The corresponding bifurcation diagrams are displayed in Figures 7.1–7.3. The bifurcation diagram for case III is simple to derive and is left as an exercise.

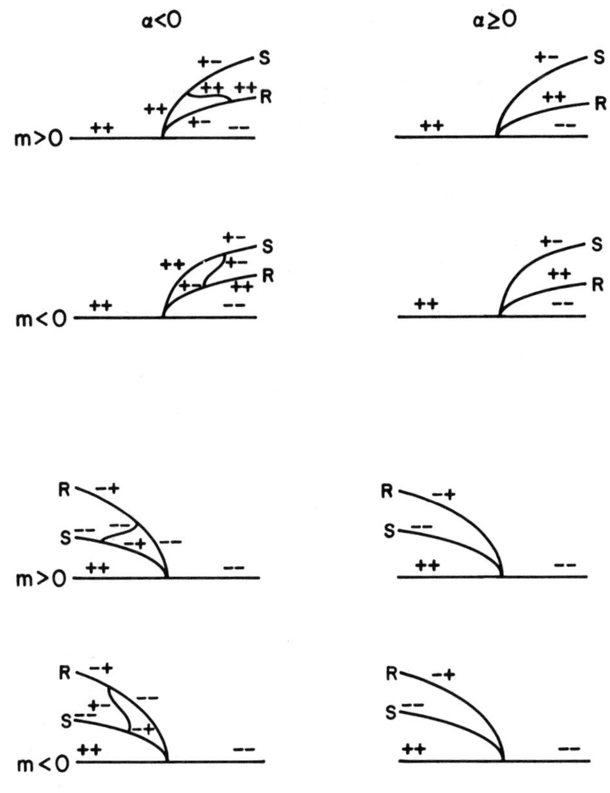

Figure 7.2. Transition variety and bifurcation diagrams for normal form IV: $[-\lambda + \varepsilon_1 N + m\Delta, N + \alpha]$.

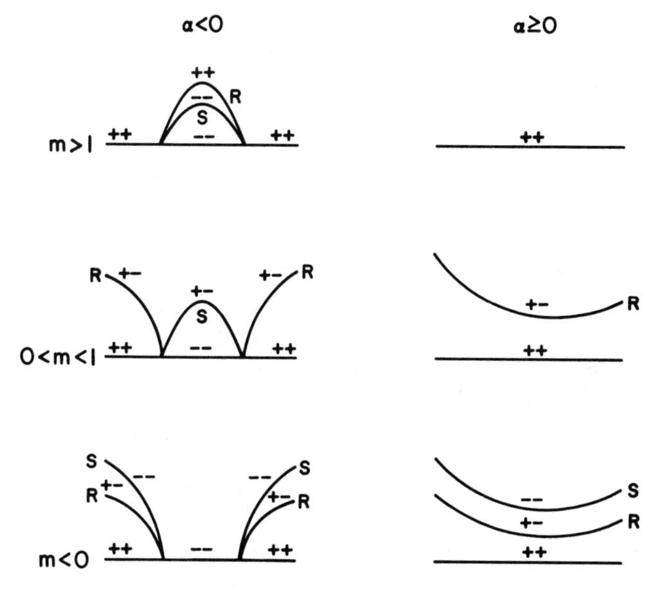

Figure 7.3. Transition variety and bifurcation diagrams for normal form V: $[\lambda^2 + \lambda N + mN + \alpha, 1]$.

Table 7.3. Solutions and Stability for Normal Form IV

Solution Type	Branching Equations	Stability	
R	$y = 0$	$\varepsilon_1 + \cdots$	[1]
	$\lambda = (\varepsilon_1 - \alpha)x^2 + (m - 1)x^4$	$\alpha + x^2$	[1]
S	$y = x$	ε_1	[1]
	$\lambda = 2\varepsilon_1 x^2$	$-(\alpha + 2x^2)$	[1]
T	$N = -\alpha$	$\operatorname{sgn} \operatorname{tr} dh = \varepsilon_1 N + \cdots$	
	$\lambda = \varepsilon_1 N + m\Delta$	$\operatorname{sgn} \det dh = m\Delta$	

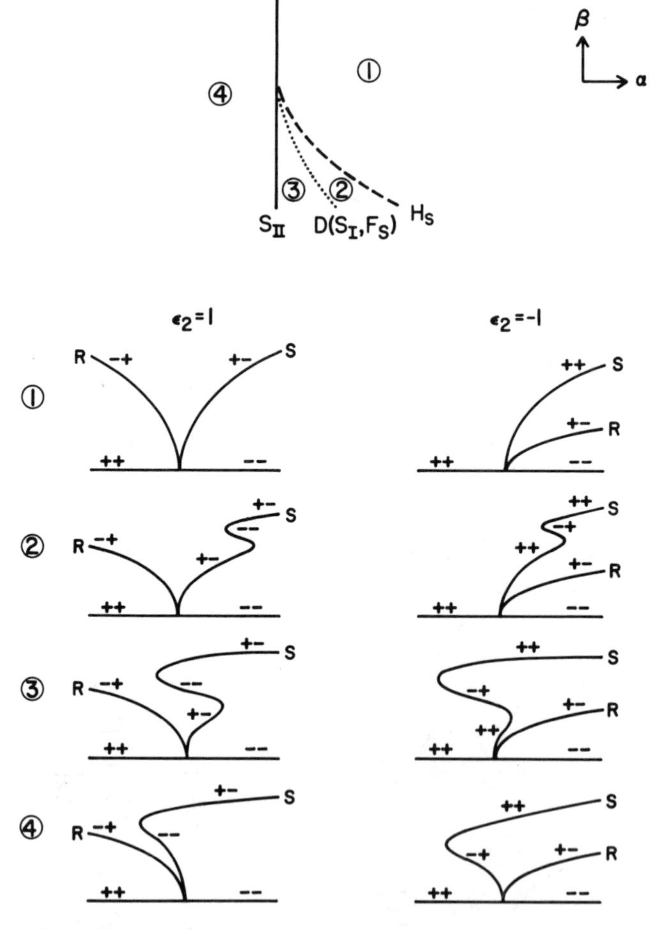

Figure 7.4. Transition variety and bifurcation diagrams for normal form VI: $[-\lambda + N^3 + \beta N^2 + \alpha N, \varepsilon_2]$.

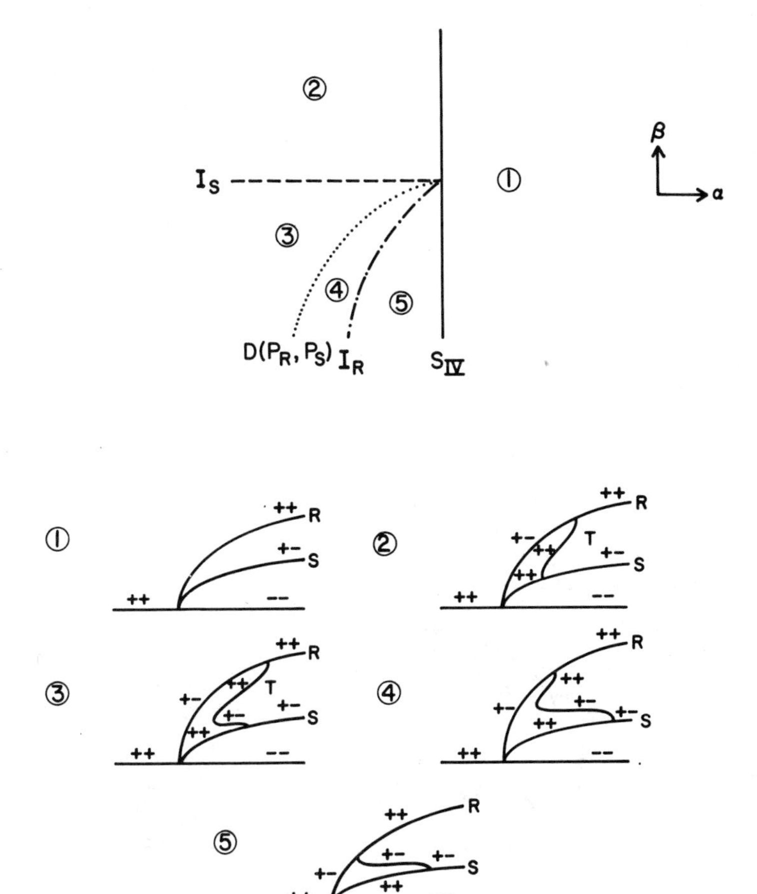

Figure 7.5. Transition variety and bifurcation diagrams for normal form VIII: $[-\lambda + N + m\Delta^2 + \beta\Delta, N + \alpha]$.

The most interesting example is IV, where solutions corresponding to 2-tori can occur and be asymptotically stable.

The branching and stability information for this case is given in Table 7.3. From this table we see that T solutions exist when $\alpha < 0$ and are asymptotically stable when $m > 0$ ($\det dh > 0$) and $\varepsilon_1 = 1$ ($\operatorname{tr} dh > 0$). This information is recorded in Figure 7.2.

Finding the bifurcation diagrams for topological $\mathbf{D}_4$-codimension two is harder. The main complication is in determining the sources of nonpersistence in $\mathbf{D}_4$ universal unfoldings. This is done (in an ad hoc way) in Golubitsky and Roberts [1987] and will not be reproduced here. Some of the transition varieties (a) and bifurcation diagrams (b) for normal forms VI–XIV are presented in Figures 7.4–7.10. Part (a) of these figures gives the transition varieties and part (b) the persistent bifurcation diagrams. The results are not

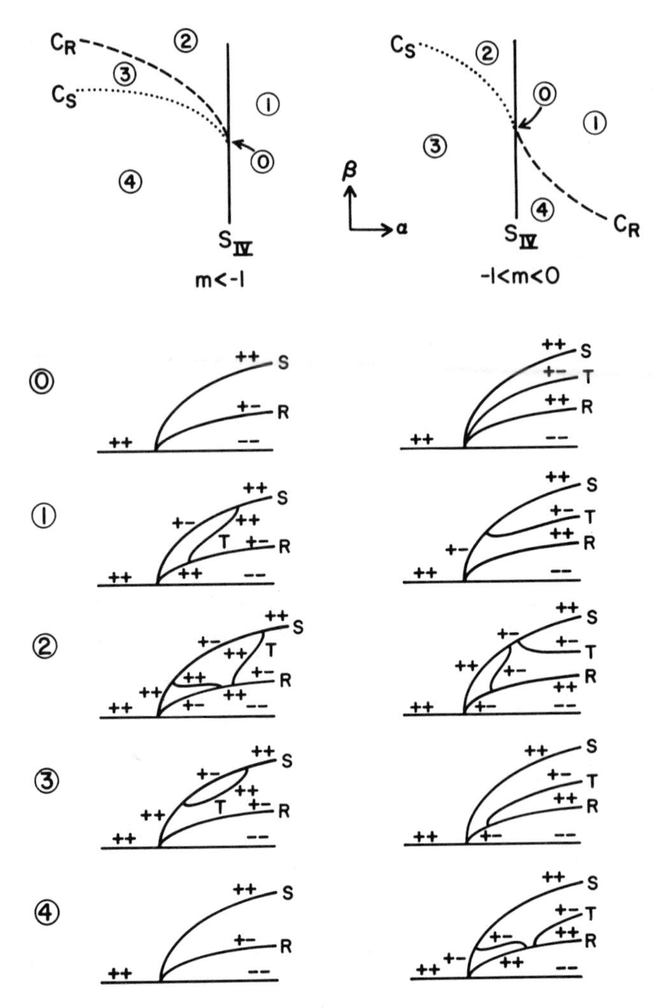

Figure 7.6. Transition variety and bifurcation diagrams for normal form IX: $[-\lambda + N, \Delta + m\lambda^2 + \beta N + \alpha]$.

complete, especially for case X, which has two modal parameters. See Table 7.4 for the branching and stability data in case X.

In X, Figure 7.7, we have drawn a subcase in which an exchange of stability is forced to occur along the T branch. Here we get a Hopf bifurcation in the amplitude equations, representing a bifurcation to an invariant 3-torus in the original **O**(2)-equivariant system.

Suppose now that the R branch bifurcates supercritically whereas the S branch bifurcates subcritically. Thus

$$0 > \alpha > \beta. \tag{7.2}$$

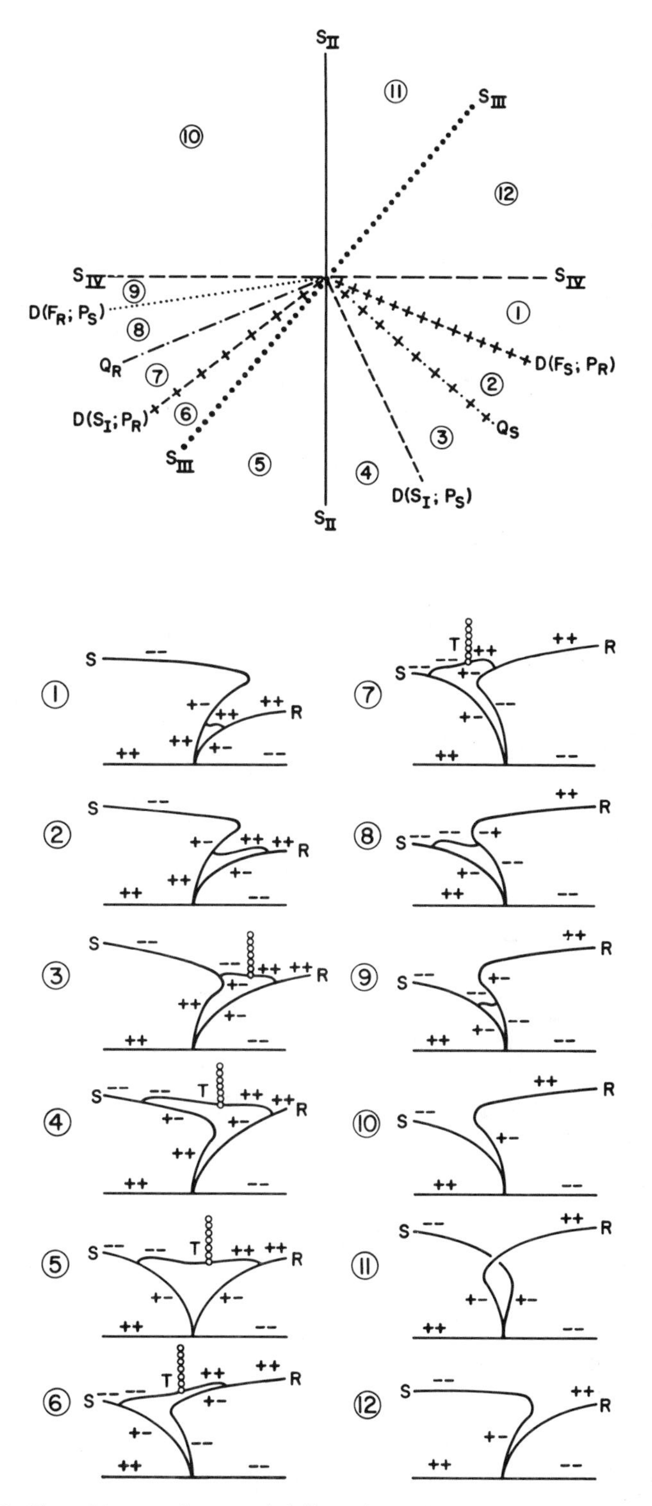

Figure 7.7. Transition variety and bifurcation diagrams for normal form X: $[-\lambda + mN^2 + n\Delta + \alpha N, N + \Delta + \beta]$, $m + n - 1 > 0$, $m < 0$. Note the Hopf bifurcation to a three-frequency torus.

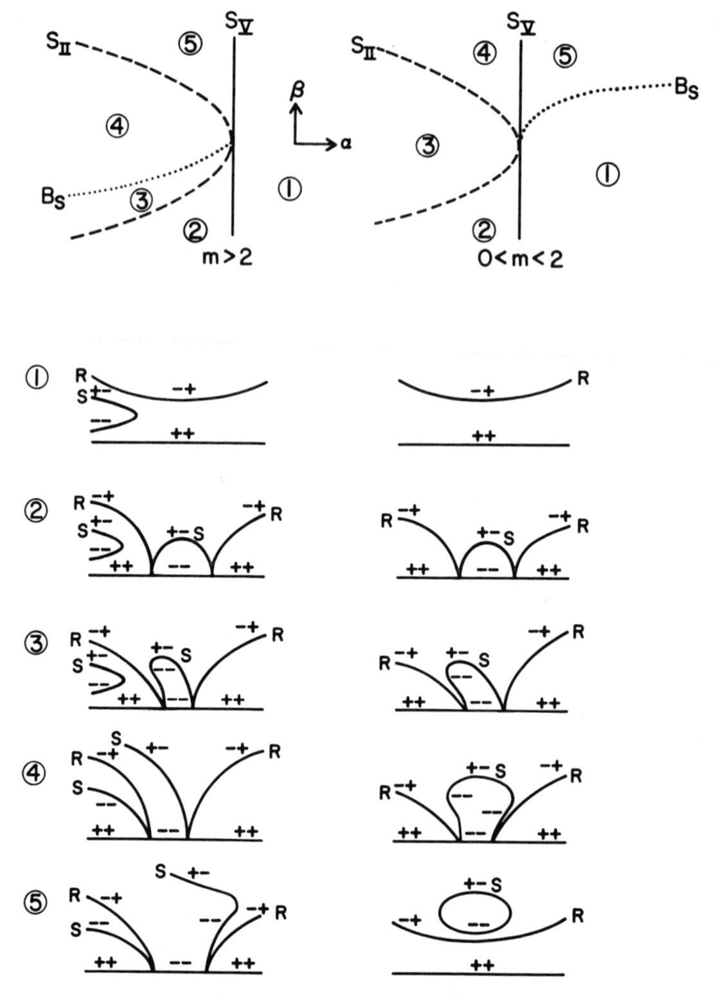

Figure 7.8. Transition variety and bifurcation diagrams for normal form XI: $[\lambda^2 + N^2 + m\lambda N + \alpha + \beta N, 1]$, $m > 0$.

If (7.2) holds, then a T branch exists and the Jacobian dh has positive determinant when

$$n > 0, \tag{7.3}$$

which we now assume. At the intersection of the S and T branches ($x = y$ and hence $\Delta = 0$)

$$\operatorname{sgn} \operatorname{tr} dh = \operatorname{sgn}(\alpha + 2mN).$$

Along the T branch, however, $N = -\beta$. Thus $\operatorname{sgn} \operatorname{tr} dh < 0$ when the S and T branches intersect, if

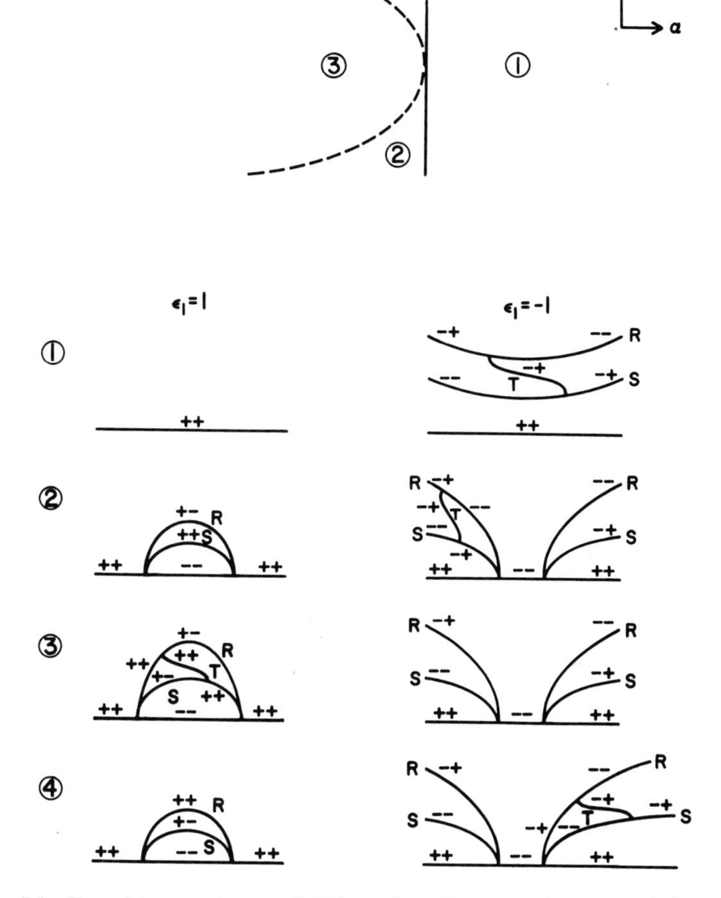

Figure 7.9. Transition variety and bifurcation diagrams for normal form XIII: $[\lambda^2 + \varepsilon_1 N + \alpha, \lambda + m\Delta + \beta]$, $m < 0$.

$$\alpha - 2m\beta < 0, \tag{7.4}$$

which we assume. Finally, at the intersection of the R and T branches

$$\operatorname{sgn} \operatorname{tr} dh = \operatorname{sgn}(\alpha + (2m + 2n - 1)x^2 - 2\varepsilon_2 x^4).$$

Since at such an intersection $y = 0$ and $\beta + x^2 + \varepsilon_2 x^4 = 0$, we can show that for $|\beta|$ small

$$\operatorname{sgn} \operatorname{tr} dh = \operatorname{sgn}(\alpha - 4\beta(m + n - 1)).$$

Thus, if we assume

$$\alpha - 4\beta(m + n - 1) > 0 \tag{7.5}$$

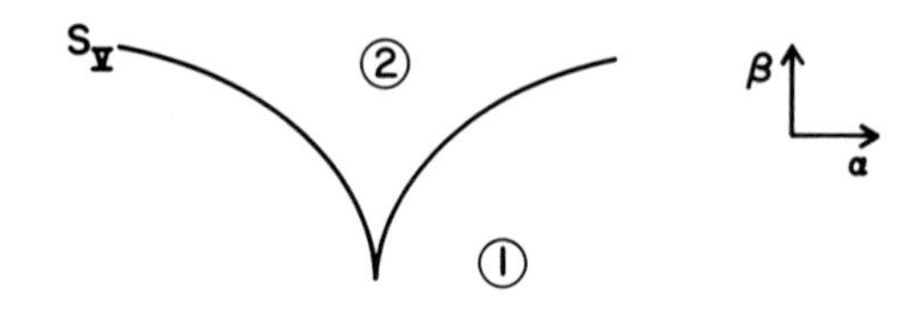

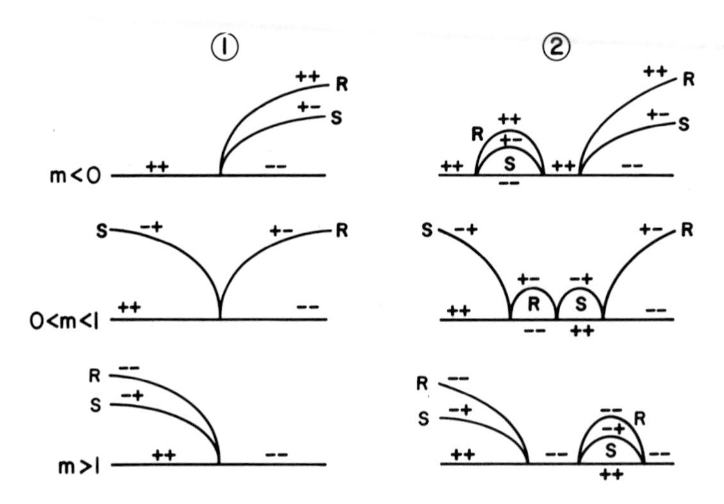

Figure 7.10. Transition variety and bifurcation diagrams for normal form XIV: $[-\lambda^3 + mN + \lambda N + \alpha + \beta\lambda, 1]$.

Table 7.4. Normal Form X

Type	Equations	Signs of Eigenvalues	
R	$y = 0$	$(\alpha - \beta) + 2(m + n - 1) - 3\varepsilon_2 x^4$	[1]
	$\lambda = (\alpha - \beta)x^2 + (m + n - 1)x^4$ $- \varepsilon_2 x^6$	$\beta + x^2 + \varepsilon_2 x^4$	[1]
S	$x = y$	$\alpha + 4mx^2$	[1]
	$\lambda = 2\alpha x^2 + 4mx^4$	$-(\beta + 2x^2)$	[1]
T	$\lambda = \alpha N + n\Delta + mN^2 = 0$	sgn tr dh:	
	$\beta + N + \varepsilon_2 \Delta = 0$	$\alpha N + (2n - 1)\Delta + 2mN^2 - 2\varepsilon_2 N\Delta$	
		sgn det dh:	
		$n - \alpha\varepsilon_2 - 2m\varepsilon_2 N$	

then the T branch has two positive eigenvalues where it meets the R branch. The eigenvalues along the T branch must change sign and cannot go through 0, so there is a Hopf bifurcation in the amplitude equations. This corresponds to the creation of an invariant 3-torus in the original ODEs.

We leave it to the reader to verify that when

$$n > 0 \quad \text{and} \quad m + n > 1 \tag{7.6}$$

there exists a nonempty open set of α, β simultaneously satisfying (7.2, 7.4, and 7.5).

Any periodic solution of the amplitude equation created by a Hopf bifurcation from the T branch can exist only for the bounded range of values for which the T branch itself exists. Of course, there may be more than one Hopf bifurcation from the T branch, but since there is a net change of stability during its existence there must also be a net production of periodic orbits. Thus there must be some other means whereby such a periodic orbit is destroyed. The only possibility for a planar system is some form of infinite period bifurcation involving the collision of the periodic orbit with one or more separatrices of the amplitude equations. The existence of this infinite period bifurcation is preserved by $\mathbf{D}_4$-equivalence. A further study of the normal form X should reveal more details of its dynamics, but it seems probable that most of the dynamic behavior is not preserved by $\mathbf{D}_4$-equivalence.

§8.† Rotating Waves and **SO**(2) or $\mathbf{Z}_n$ Symmetry

In Proposition XVI, 1.4, we showed that for Γ-symmetric Hopf bifurcation there are two generic cases. In one, the group Γ acts absolutely irreducibly on a space V and diagonally on $V \oplus V$; the underlying ODE is posed on $V \oplus V$. In the other, Γ acts irreducibly but not absolutely irreducibly on W, and the ODE is posed on W. The action of **O**(2) discussed earlier exemplifies the first case.

In this section we consider the second case, specifically, irreducible actions of the groups **SO**(2) and $\mathbf{Z}_n$. Because the irreducible representations of these groups occur on $\mathbb{R}^2$ (or exceptionally on $\mathbb{R}$) the only isotropy group with a two-dimensional fixed-point subspace is the kernel of the $\Gamma \times \mathbf{S}^1$-action. The application of the general theory is then trivial, but its implications are not. As we see later, the theory predicts Hopf bifurcation to a unique branch of *rotating waves* (having additional spatial symmetry when the representation is not standard), whose stability is determined by the usual exchange of stability rule in Hopf bifurcation.

Of course, the standard Hopf theorem applies directly to this situation and predicts the occurrence of a unique periodic branch. It is easy to obtain the additional spatial symmetry, given by the kernel of the Γ-action. But the rotating wave nature of the solution cannot be obtained so trivially.

(a) SO(2) Symmetry

Suppose that Γ is a circle group, which we denote by $\mathbf{SO}(2)$ to prevent confusion with the phase-shift group $\mathbf{S}^1$. For each integer $m > 0$ there is an irreducible $\mathbf{SO}(2)$-action on $\mathbb{C}$ given by

$$\zeta \cdot z = e^{im\zeta} z,$$

and all nontrivial irreducible representations are of this form. The algebra $\mathscr{D}$ of commuting linear mappings consists of all mappings

$$z \mapsto wz$$

for $w \in \mathbb{C}$. Hence these representations are not absolutely irreducible: they are of "complex" type.

Since the phase-shift group $\mathbf{S}^1$ commutes with Γ, the $\mathbf{S}^1$-action is by mappings in $\mathscr{D}$, hence by

$$\theta \cdot z = e^{\pm i\theta} z. \tag{8.1}$$

Reversing the direction of time if necessary (or equivalently performing a reflection in $\mathbb{C}$) we may assume the sign is $+$. The full $\mathbf{SO}(2) \times \mathbf{S}^1$-action is thus given by

$$(\zeta, \theta) \cdot z = e^{i(m\zeta + \theta)} z. \tag{8.2}$$

Modulo the kernel we have a standard circle group action, so the invariant functions are generated by $N = |z|^2$ and the equivariants are generated as a module over the invariants by $z \mapsto z$ and $z \mapsto iz$. A general equivariant is thus of the form

$$p(N, \lambda, \tau)z + q(N, \lambda, \tau)iz. \tag{8.3}$$

This is exactly the form of the reduced equation for ordinary Hopf bifurcation, so the direction of branching and the stability calculations are identical to standard Hopf bifurcation.

The space $V = \mathbb{C}$ is two-dimensional, so the only isotropy group Σ with $\dim \mathrm{Fix}(\Sigma) = 2$ is the kernel of the action, which by definition fixes all of $\mathbb{C}$. From (8.2) this is

$$\Sigma = \{(\zeta, -m\zeta): \zeta \in \mathbf{SO}(2)\}.$$

Despite the triviality of these calculations, the results have important consequences. The isotropy group Σ represents a *rotating wave* having a spatial $\mathbf{Z}_m$ symmetry. (In the standard action $m = 1$ and the spatial symmetry is trivial.) Such a wave takes the same form if its phase is shifted by θ and it is spatially rotated by $-\theta/m$. For a mental picture, imagine an m-armed spiral, rotating uniformly. The spatial $\mathbf{Z}_m$ symmetry occurs because the subgroup

$$\mathbf{SO}(2) \cap \Sigma = \{(2k\pi/m, 0)\}$$

is isomorphic to $\mathbf{Z}_m$.

Assume in (8.3) that $p_\lambda(0) \leq 0$, so that the trivial solution is stable subcritically. Then we have proved the following:

Theorem 8.1. *For generic* $\mathbf{SO}(2)$*-Hopf bifurcation corresponding to the action* $z \mapsto e^{mi\theta}z$, *there is a unique branch of periodic solutions consisting of rotating waves with* $\mathbf{Z}_m$ *spatial symmetry. In terms of the reduced equation* (8.3) *this branch is super- or subcritical according to whether* $p_N(0) < 0$ *or* > 0. *It is orbitally asymptotically stable if and only if it is supercritical.*

Remark. The direction of rotation may be clockwise or counterclockwise, depending on the sign of the $\mathbf{S}^1$-action in (8.1). In the preceding discussion this sign was set to 1, the price being the possibility of a time reversal, or a reflection of $\mathbb{C}$, either of which reverses the direction of rotation.

However, unlike for the $\mathbf{O}(2)$ rotating waves, only one direction of rotation will occur at any generic $\mathbf{SO}(2)$ Hopf bifurcation.

(b) $\mathbf{Z}_n$ Symmetry

Next we consider the case $\Gamma = \mathbf{Z}_n$. Let ζ be a generator of $\mathbf{Z}_n$ (and write the group multiplicatively, so that its elements are ζ^t), and let $\omega = e^{2\pi i/n}$. The irreducible representations of $\mathbf{Z}_n$ are either one- or two-dimensional. The one-dimensional irreducible representations are the trivial representation ρ_0 on $\mathbb{R}$, and when n is even the nontrivial representation $\rho_{n/2}$ given by

$$\zeta \cdot x = -x.$$

The two-dimensional irreducible representations ρ_m on $\mathbb{C}$ are of the form

$$\zeta \cdot z = \omega^m z \tag{8.4}$$

where

$$m = \begin{cases} 1, 2, \ldots, (n-1)/2 & \text{if } n \text{ is odd,} \\ 1, 2, \ldots, n/2 - 1 & \text{if } n \text{ is even.} \end{cases}$$

For Hopf bifurcation we must consider the diagonal action on $\mathbb{R} \oplus \mathbb{R} \equiv \mathbb{C}$ for ρ_0 and $\rho_{n/2}$, and the preceding action on $\mathbb{C}$ for all other ρ_m. Note that in all cases the relevant action of $\mathbf{Z}_n$ on $\mathbb{C}$ is now given by (8.4). The corresponding action of $\mathbf{Z}_n \times \mathbf{S}^1$ on $\mathbb{C}$ is then

$$(\zeta, \theta) = \omega^m e^{i\theta} z$$

in all cases, even when $m = 0$ or $n/2$. (As earlier we may have to reverse time to obtain the positive sign on θ.)

The equivariants are again given by (8.3), and the direction of criticality and the stability results are the same as for standard Hopf bifurcation.

The only way for Σ to have a two-dimensional fixed-point subspace is if Σ is the kernel of the $\mathbf{Z}_n \times \mathbf{S}^1$-action. This kernel consists of the elements (ζ^t, θ)

such that $\omega^{mt}e^{i\theta} = 1$; that is, $2mt\pi/n + \theta = 0$. Hence

$$\Sigma = \{(\zeta^t, -2mt\pi/n)\}.$$

This group is isomorphic to $\mathbf{Z}_n$. Its spatial part $\Sigma \cap \mathbf{Z}_n$ consists of those elements for which mt is divisible by n, so that t is a multiple of n/d where $d = gcd(m, n)$. That is, $\Sigma \cap \mathbf{Z}_n \cong \mathbf{Z}_d$. Therefore, (4) represents a rotating wave with spatial symmetry $\mathbf{Z}_d$, and we have proved the following:

Theorem 8.2. *For generic $\mathbf{Z}_n$-Hopf bifurcation corresponding to the action $z \mapsto \omega^m z$, $(n \geq 2, n > m \geq 1)$, there is a unique branch of periodic solutions consisting of rotating waves with $\mathbf{Z}_d$ spatial symmetry, $d = gcd(m, n)$. In terms of the reduced equation (8.3) this branch is super- or subcritical according to whether $p_N(0) < 0$ or > 0. Assuming $p_\lambda(0) < 0$ it is orbitally asymptotically stable if and only if it is supercritical.*

Again either sense of rotation is possible, but not both.

CHAPTER XVIII

Further Examples of Hopf Bifurcation with Symmetry

§0. Introduction

In this chapter we give three illustrations of the general theory of symmetric Hopf bifurcation developed in Chapter XVI. We study systems with dihedral group symmetry $\mathbf{D}_n$, systems with $\mathbf{O}(3)$ symmetry (corresponding to any irreducible representation), and systems with the symmetry $\mathbf{T}^2 \dotplus \mathbf{D}_6$ of the hexagonal lattice. For $\mathbf{D}_n$ and $\mathbf{T}^2 \dotplus \mathbf{D}_6$ we consider the stability of bifurcating branches. These examples illustrate several features of specific applications that have not yet appeared in our discussions, and they show the different levels at which the methods can be used.

The case of $\mathbf{D}_n$ is discussed in greater detail and is applied to equations modeling a ring of coupled cells. We study how the coupling of cells affects the type of oscillation that may be observed in the system. The analysis was inspired by work of Alexander and Auchmuty [1986] on coupled oscillators, in particular, three identical cells with symmetric coupling, shown schematically in Figure 0.1.

We sketch the main ideas and results here. Suppose that the state of oscillator p ($p = 0, 1, 2$) may be described by two state variables (x_p, y_p). Let λ be a bifurcation parameter. Suppose that there is a matrix

$$K(\lambda) = \begin{bmatrix} k_{11} & k_{12} \\ k_{21} & k_{22} \end{bmatrix}$$

of "coupling strengths" which depends only on λ (or is constant). Let

$$\delta_{ij} = x_i - x_j, \qquad \varepsilon_{ij} = y_i - y_j.$$

Then this system may be modeled by the following equations:

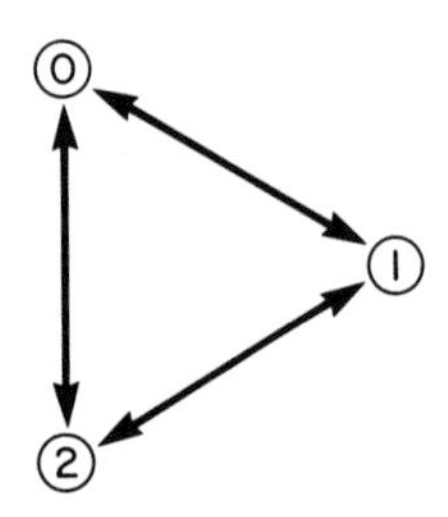

Figure 0.1. Schematic picture of three symmetrically coupled identical cells.

$$\begin{aligned}
\frac{d}{dt}(x_0, y_0) &= F(x_0, y_0, \lambda) + K(\lambda)\cdot(\delta_{01} + \delta_{02}, \varepsilon_{01} + \varepsilon_{02}) \\
\frac{d}{dt}(x_1, y_1) &= F(x_1, y_1, \lambda) + K(\lambda)\cdot(\delta_{12} + \delta_{10}, \varepsilon_{12} + \varepsilon_{10}) \\
\frac{d}{dt}(x_2, y_2) &= F(x_2, y_2, \lambda) + K(\lambda)\cdot(\delta_{20} + \delta_{21}, \varepsilon_{20} + \varepsilon_{21}).
\end{aligned} \tag{0.1}$$

The form of the coupling is chosen so that it vanishes if the oscillators are behaving identically and is linear in the x_p and y_p.

It is easy to check that these equations are equivariant with respect to the action of the group $\mathbf{D}_3$. The subgroup $\mathbf{Z}_3$ permutes the variables cyclically:

$$(x_0, y_0) \mapsto (x_1, y_1) \mapsto (x_2, y_2) \mapsto (x_0, y_0),$$

whereas the flip interchanges

$$(x_1, y_1) \leftrightarrow (x_2, y_2)$$

and leaves (x_0, y_0) fixed. This behavior is to be expected on geometric grounds because of the "triangular" symmetry of the physical system, evident in Figure 0.1.

More generally we consider nonlinear coupling, assuming only that the symmetry is maintained, and we allow the state of each oscillator to be described by k variables rather than two. The most general $\mathbf{D}_3$-equivariant system of this kind is

$$\begin{aligned}
dx_0/dt &= g(x_2, x_0, x_1; \lambda) \\
dx_1/dt &= g(x_0, x_1, x_2; \lambda) \\
dx_2/dt &= g(x_1, x_2, x_0; \lambda)
\end{aligned} \tag{0.2}$$

where $x_p \in \mathbb{R}^k$; $p = 0, 1, 2$; λ is a bifurcation parameter; and

$$g(u, v, w; \lambda) = g(w, v, u; \lambda).$$

A similar system can be set up representing n oscillators, by taking $p = 0, 1, \ldots, n - 1$. It is then $\mathbf{D}_n$-equivariant.

The first step in analyzing these systems is to consider generic Hopf bifurcation for a vector field commuting with the *standard* action of $\mathbf{D}_n$ on $\mathbb{R}^2 \equiv \mathbb{C}$. Later we shall see that nonstandard actions also enter the analysis, but (as for nonstandard actions of $\mathbf{O}(2)$ in XVII, §1(d) the behavior in those cases may be derived from that for the standard action by a simple group-theoretic trick.

The general theory of XVI shows that generically in the case of $\mathbf{D}_n$-symmetry, the purely imaginary eigenvalues associated with a Hopf bifurcation are either simple or double. When these eigenvalues are simple, the standard Hopf theorem implies the existence of a unique branch of periodic solutions; here this corresponds to all cells oscillating with the same waveform. The analysis in §§1–4 deals exclusively with the double eigenvalue case. We review these results here.

In §1 we show that, for the standard action of $\mathbf{D}_n$ on $\mathbb{C}$, generically there are (at least) three branches of periodic solutions, bifurcating from the trivial solution at $\lambda = 0$. Similar results have been obtained independently by van Gils and Valkering [1986]. These solutions are found by using the equivariant Hopf theorem, Theorem XVI, 4.1. They thus correspond to three (conjugacy classes of) isotropy subgroups of $\mathbf{D}_n \times \mathbf{S}^1$ acting on $\mathbb{C} \oplus \mathbb{C}$, each having a two-dimensional fixed-point subspace. One subgroup is cyclic of order n, and the others are isomorphic to $\mathbf{Z}_2$ (or $\mathbf{Z}_2 \oplus \mathbf{Z}_2$ when n is even), but the detailed list is slightly different in the cases n odd, $n \equiv 2 \pmod 4$, and $n \equiv 0 \pmod 4$. Interpreted for coupled oscillators, these isotropy subgroups lead to the oscillation patterns shown in Figure 0.2.

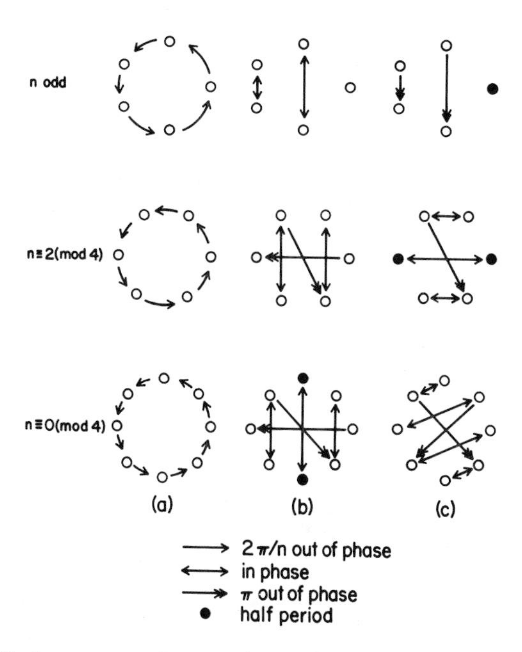

Figure 0.2. Oscillation patterns for generic Hopf bifurcation in a system of n identical coupled cells.

These patterns are deduced from the form of the isotropy subgroup, and their interpretation depends on the fact that $\mathbf{D}_n$ acts by permutations of the variables. In other $\mathbf{D}_n$-equivariant systems the same isotropy subgroups will occur, but the interpretation may be different. Here the isotropy subgroups describe various types of phase relationship between oscillators and in some cases show that a particular oscillator has half the period of the others.

In §2 we describe the most general possible form of a $\mathbf{D}_n \times \mathbf{S}^1$-equivariant mapping f. The ring of invariants has four generators, and the module of equivariants has eight. In §3(a) we apply the results of Chapter XVI, §7, and show how this restricted form of f lets us find some solutions of the bifurcation equations, by the standard method of prescribing in advance the required symmetries of solutions. In §3(b) we state conditions on appropriate coefficients in the general form of f which determine the direction of criticality and the asymptotic stability of these solutions. Coefficients of high order terms, not just those of degree 3, are involved here; certain terms of intermediate degree are irrelevant to the stability assignments.

The results for $n = 4$ differ markedly from those for other n. Specifically, if $n \geq 3$ and $n \neq 4$ then (subject to certain nondegeneracy conditions stated in §3(b)) in order for any of the three branches to be stable, all three must be supercritical; further, exactly one branch is then stable. When $n = 4$, however, some branches can be stable when others are subcritical; further, two distinct branches may be stable for the same values of the coefficients in f. The detailed situation is summarized in §3(b). Swift [1986] has investigated the dynamics in the case $n = 4$ to obtain a more detailed understanding of what occurs.

Also in §3(b) we relate our results for $\mathbf{D}_n$ symmetry, as n becomes large, to the standard results for $\mathbf{O}(2)$ symmetry—the "limiting case" as $n \to \infty$. In particular, we explain how the three distinct types of oscillation occurring for $\mathbf{D}_n$ merge to give only two distinct types for $\mathbf{O}(2)$.

We apply these results in §4 to a system of n nonlinear cells coupled together in a ring, with symmetric nearest-neighbor coupling, as described previously. This section is divided into five parts, as follows.

In §4(a) we introduce the general equations corresponding to an n-cell system and observe their equivariance under $\mathbf{D}_n$. In §4(b) we specialize temporarily to the case of three oscillators, to avoid combinatorial complications and illustrate the basic ideas. We give conditions on the Jacobian df that guarantee the existence of the branches of symmetry-breaking oscillations predicted by the general theory and describe the corresponding oscillation patterns.

When $n = 3$ there are two types of generic behavior, depending on the eigenspace corresponding to the purely imaginary eigenvalues that produce the Hopf bifurcation. One is that there is a unique branch of periodic (orbits of) solutions, on which all three oscillators have identical waveforms, in phase. The other is that there are three branches of symmetry-breaking oscillations. On one branch the oscillations have the same waveform for each cell but are phase-shifted by $2\pi/3$. On the second, two cells undergo oscillations that are identical and in phase, the third behaving differently. On the third branch, two cells have identical waveforms but are out of phase by π, and the third has double the frequency. We also support these conclusions with numerical simulations, in §4(c).

In §4(d) we discuss the case of general n, and in §4(e) we consider particular examples when $n = 2, 4, 5$. Again the case $n = 4$ has several peculiarities of its own.

The oscillator problem raises one important issue that we have largely ignored hitherto. Suppose that a vector field on $\mathbb{R}^n$, with symmetry group Γ, undergoes a Hopf bifurcation. Then the corresponding purely imaginary eigenspace E is invariant under Γ and generically is Γ-simple. If, as is often the case, E is a proper subspace of $\mathbb{R}^n$, then there may be several different possibilities for this generic Γ-action, depending on the precise eigenspace and the accompanying representation of $\Gamma \times \mathbf{S}^1$. These represent different "modes" of behavior. (This situation is responsible for the two types of generic Hopf bifurcation with $\mathbf{D}_3$ symmetry, mentioned previously.) The generic behavior in each case depends on an analysis of the *nonlinear* system (reduced to the appropriate eigenspace), but which eigenspace occurs in a given application depends on the *linear* analysis of the Jacobian df.

We might add that in practice, if there are several modes, there will probably be interactions between them, leading to complicated dynamics. We do not pursue this problem here, but emphasize that our theorems assert the existence of certain types of periodic solution, not the nonexistence of any other dynamics.

The remaining sections of the chapter are less detailed. In §5 we sketch the case of $\mathbf{O}(3)$ symmetry. The main aim is to classify all possible isotropy subgroups of $\mathbf{O}(3) \times \mathbf{S}^1$ having two-dimensional fixed-point subspaces, and we do this for all possible irreducible representations of $\mathbf{O}(3)$ (on spherical harmonics V_l). The results of XIII, §§7 and 9, are combined with the methods developed in XVI, §§8, 9, to compute the dimensions of fixed-point subspaces of twisted subgroups. The stability of the solutions found by this method is *not* discussed.

Finally in §6 we reproduce some results of Roberts, Swift, and Wagner [1986] on periodic solutions of doubly diffusive convection on the hexagonal lattice. This applies our methods for studying symmetric Hopf bifurcation, in conjunction with results on the static Bénard problem, Case Study 4. Eleven classes of time-periodic solution for problems posed on the hexagonal lattice are obtained. The major result here is the existence of time-periodic solutions whose spatial patterns change periodically in time.

§1. The Action of $\mathbf{D}_n \times \mathbf{S}^1$

In this section we do two things. We choose coordinates that make the action of $\mathbf{D}_n \times \mathbf{S}^1$ on $\mathbb{R}^4$ easy to compute with, and we classify the isotropy subgroups that occur, together with their fixed-point subspaces.

(a) Definition of the Group Action

We begin by assuming that $\mathbf{D}_n$ $(n \geq 3)$ acts on $\mathbb{C}$ in the standard way as symmetries of the regular n-gon, and on $\mathbb{C}^2$ by the diagonal action

$$\gamma(z_1, z_2) = (\gamma z_1, \gamma z_2).$$

Although $\mathbf{D}_n$ will in general have many distinct irreducible representations (there are $(n+3)/2$ when n is odd, $(n+6)/2$ when n is even, all of dimension 1 or 2) there is no real loss of generality in making this assumption. Essentially it is possible to arrange for a standard action by relabeling the group elements and dividing by the kernel of the action. See §4 for further discussion.

We recall notation for the elements of $\mathbf{D}_n$. Its cyclic subgroup $\mathbf{Z}_n$ consists of rotations of the plane through $0, \zeta, 2\zeta, \ldots, (n-1)\zeta$ where $\zeta = 2\pi/n$. The *flip* κ is reflection in the x-axis. In complex notation $\mathbf{D}_n$ acts on $\mathbb{C}$ as follows:

$$(m\zeta) \cdot z = e^{im\zeta} z,$$

$$\kappa \cdot z = \bar{z}.$$

We denote a typical element of $\mathbf{S}^1$ by θ.

To make it easier to analyze $\mathbf{D}_n$-equivariant Hopf bifurcation we choose a simple form for the action of $\mathbf{D}_n \times \mathbf{S}^1$ on $\mathbb{C}^2$. In fact, we use the action of $\mathbf{O}(2) \times \mathbf{S}^1$ on $\mathbb{R}^4 \equiv \mathbb{C}^2$ obtained in XVII, §1, and restrict it to $\mathbf{D}_n \times \mathbf{S}^1$. This is permissible since the standard action of $\mathbf{D}_n$ on $\mathbb{R}^2$ is the restriction of that of $\mathbf{O}(2)$, so the action of $\mathbf{D}_n \times \mathbf{S}^1$ on $\mathbb{R}^2 \oplus \mathbb{R}^2$ is the restriction of that of $\mathbf{O}(2) \times \mathbf{S}^1$. Then the elements of $\mathbf{D}_n \times \mathbf{S}^1$ act on $(z_1, z_2) \in \mathbb{C}^2$ as follows:

$$\begin{aligned}
&\text{(a)} \quad \gamma(z_1, z_2) = (e^{i\gamma} z_1, e^{-i\gamma} z_2) \qquad (\gamma \in \mathbf{Z}_n)\\
&\text{(b)} \quad \kappa(z_1, z_2) = (z_2, z_1) \qquad\qquad\qquad\qquad (1.1)\\
&\text{(c)} \quad \theta(z_1, z_2) = (e^{i\theta} z_1, e^{i\theta} z_2) \qquad (\theta \in \mathbf{S}^1).
\end{aligned}$$

(b) Isotropy Subgroups of $\mathbf{D}_n \times \mathbf{S}^1$

We compute, up to conjugacy, the isotropy subgroups of $\mathbf{D}_n \times \mathbf{S}^1$. There are three distinct cases, depending on whether n is odd, $n \equiv 2 \pmod 4$, or $n \equiv 0 \pmod 4$. The results are given in Tables 1.1–1.3.

Table 1.1. Isotropy Subgroups of $\mathbf{D}_n \times \mathbf{S}^1$ Acting on $\mathbb{C}^2$, When n Is Odd

Orbit Type	Isotropy Subgroup	Fixed-Point Space	Dimension
$(0,0)$	$\mathbf{D}_n \times \mathbf{S}^1$	$\{(0,0)\}$	0
$(a,0)$	$\tilde{\mathbf{Z}}_n = \{(\gamma, -\gamma) \mid \gamma \in \mathbf{Z}_n\}$	$\{(z_1,0)\}$	2
(a,a)	$\mathbf{Z}_2(\kappa)$	$\{(z_1,z_1)\}$	2
$(a,-a)$	$\mathbf{Z}_2(\kappa,\pi)$	$\{(z_1,-z_1)\}$	2
(a,z_2) $z_2 \neq \pm a, 0$	$\mathbb{1}$	$\mathbb{C}^2$	4

Table 1.2. Isotropy Subgroups of $\mathbf{D}_n \times \mathbf{S}^1$ Acting on $\mathbb{C}^2$, When $n \equiv 2 \pmod 4$. Note That $\mathbf{Z}_2^c = \{(0,0),(\pi,\pi)\}$

Orbit Type	Isotropy Subgroup	Fixed-Point Space	Dimension
$(0,0)$	$\mathbf{D}_n \times \mathbf{S}^1$	$\{(0,0)\}$	0
$(a,0)$	$\tilde{\mathbf{Z}}_n = \{(\gamma,-\gamma) \mid \gamma \in \mathbf{Z}_n\}$	$\{(z_1,0)\}$	2
(a,a)	$\mathbf{Z}_2(\kappa) \oplus \mathbf{Z}_2^c$	$\{(z_1,z_1)\}$	2
$(a,-a)$	$\mathbf{Z}_2(\kappa,\pi) \oplus \mathbf{Z}_2^c$	$\{(z_1,-z_1)\}$	2
(a,z_2) $z_2 \neq \pm a, 0$	$\mathbf{Z}_2^c$	$\mathbb{C}^2$	4

Table 1.3. Isotropy Subgroups of $\mathbf{D}_n \times \mathbf{S}^1$ Acting on $\mathbb{C}^2$, When $n \equiv 0 \pmod 4$. Note That $\mathbf{Z}_2^c = \{(0,0),(\pi,\pi)\}$

Orbit Type	Isotropy Subgroup	Fixed-Point Space	Dimension
$(0,0)$	$\mathbf{D}_n \times \mathbf{S}^1$	$\{(0,0)\}$	0
$(a,0)$	$\tilde{\mathbf{Z}}_n = \{(\gamma,-\gamma) \mid \gamma \in \mathbf{Z}_n\}$	$\{(z_1,0)\}$	2
(a,a)	$\mathbf{Z}_2(\kappa) \oplus \mathbf{Z}_2^c$	$\{(z_1,z_1)\}$	2
$(a,e^{2\pi i/n}a)$	$\mathbf{Z}_2(\kappa\zeta) \oplus \mathbf{Z}_2^c$	$\{(z_1,e^{2\pi i/n}z_1)\}$	2
(a,z_2) $z_2 \neq \pm a, 0$	$\mathbf{Z}_2^c$	$\mathbb{C}^2$	4

We have also listed the fixed-point subspaces for the isotropy subgroups. Since three of these fixed-point subspaces are two-dimensional, it follows from the equivariant Hopf theorem, Theorem XVI, 4.1, that there are (at least) three branches of periodic solutions occurring generically in Hopf bifurcation with $\mathbf{D}_n$ symmetry.

We now verify the entries in Table 1.1. If $(z_1,z_2)=(0,0)$ then trivially the isotropy subgroup is $\mathbf{D}_n \times \mathbf{S}^1$, so we may assume $(z_1,z_2) \neq (0,0)$. By use of θ and κ, if necessary, we may assume that $z_1 = a > 0$, and that $(z_1,z_2)=(a,re^{i\psi})$. We claim that we may assume $0 \leq \psi \leq \zeta/2 = \pi/n$ [n odd]; $0 \leq \psi < \zeta = 2\pi/n$ [n even]. This is trivial if $r=0$, so assume $r>0$. The group elements in $\mathbf{D}_n \times \mathbf{S}^1$ have the form $(m\zeta,\theta)$ and $(\kappa(m\zeta),\theta)$ where $m = 0, 1, \ldots, n-1$. These group elements transform $(a,re^{i\psi})$ to:

$$\begin{aligned} &\text{(a)} \quad (ae^{i[m\zeta+\theta]}, re^{i[\psi-m\zeta+\theta]}) \\ &\text{(b)} \quad (re^{i[\psi-m\zeta+\theta]}, ae^{i[m\zeta+\theta]}), \end{aligned} \tag{1.2}$$

respectively. For these group elements to preserve the form (a,z_2) we must assume in (1.2(a)) that

$$m\zeta + \theta = 2\pi k$$

and in (1.2(b)) that

$$\psi - m\zeta + \theta = 2\pi k$$

for some integer k. In addition, in (1.2(b)) it is convenient to interchange the labels on r and a. In this way we obtain

$$\begin{aligned} &\text{(a)} \quad (a, re^{i[\psi - 2m\zeta]}) \\ &\text{(b)} \quad (a, re^{i[2m\zeta - \psi]}). \end{aligned} \tag{1.3}$$

Using (1.3) we can translate ψ by 2ζ and flip ψ to $-\psi$. Now when n is odd, rotations by 2ζ generate the whole group $\mathbf{Z}_n$. Hence we may actually translate ψ by ζ, not just 2ζ. It is now easy to show that every ψ may be assumed in the interval $0 \le \psi \le \zeta/2$ as claimed. On the other hand, when n is even we can only assume ψ is in the interval $0 \le \psi \le \zeta$.

We now consider the three cases: n odd, $n \equiv 2 \pmod 4$, $n \equiv 0 \pmod 4$.

(i) *n odd.* We first show that if $r \ne a$, 0 and $0 < \psi < \zeta/2$, then the isotropy subgroup of $(a, re^{i\psi})$ is $\mathbb{1}$. If $r \ne a$, then (1.2(b)) shows that no element of the form $(\kappa(m\zeta), \theta)$ can be in the isotropy subgroup. If $r \ne 0$, then our previous calculations showed that the form $(a, re^{i\psi})$ is fixed only by $\theta \equiv -m\zeta \pmod{2\pi}$. Thus $(a, re^{i\psi})$ is fixed only when $\theta = m = 0$, or $\theta = -\pi$, $m = n/2$ when n is even. Since n is odd here, the isotropy subgroup is $\mathbb{1}$.

It also follows from these calculations that $(a, 0)$ is fixed precisely by $(m\zeta, -m\zeta) \in \mathbf{Z}_n \times \mathbf{S}^1$. We have now reduced to the case $r = a$ and $\psi = 0$, $\zeta/2$. When $\psi = 0$ we have a point (a, a) and its isotropy subgroup is $\mathbf{Z}_2(\kappa) = \{(0,0), \kappa\}$. Similarly, since n is odd, $(a, ae^{i\zeta/2})$ is in the same orbit as $(a, -a)$ and the isotropy subgroup is $\mathbf{Z}_2(\kappa, \pi) = \{(0,0), (\kappa, \pi)\} \subset \mathbf{D}_n \times \mathbf{S}^1$.

Finally, the fixed-point subspaces are easily computed once the isotropy subgroups are known.

(ii) *$n \equiv 2 \pmod 4$.* This is similar. In fact $\mathbf{Z}_2^c = \{(0,0), (\pi, \pi)\}$ acts trivially on $\mathbb{C}^2$ and hence is contained in every isotropy subgroup. Now $(\mathbf{D}_n \times \mathbf{S}^1)/\mathbf{Z}_2^c \cong \mathbf{D}_{n/2} \times \mathbf{S}^1$ since $\mathbf{D}_{n/2} \times \mathbf{S}^1 \subset \mathbf{D}_n \times \mathbf{S}^1$ and $(\mathbf{D}_{n/2} \times \mathbf{S}^1) \cap \mathbf{Z}_2^c = \mathbb{1}$. There is one subtle point: the induced action of $\mathbf{S}^1$ (when $\mathbf{Z}_2^c$ is thus factored out) is by $e^{2i\theta}$, not $e^{i\theta}$. The same arguments work (with θ replaced by $\theta/2$), but all isotropy subgroups are augmented by $\mathbf{Z}_2^c$.

(iii) *$n \equiv 0 \pmod 4$.* Again (π, π) fixes $\mathbb{C}^2$, so every isotropy subgroup contains $\mathbf{Z}_2^c$. The previous analysis shows that we may assume

$$(z_1, z_2) = (a, re^{i\psi}) \qquad (0 \le \psi \le \zeta). \tag{1.4}$$

If $r = 0$ then we have $(a, 0)$ and the isotropy subgroup is $\tilde{\mathbf{Z}}_n$ as before. Otherwise $r \ne 0$. We claim that the only elements (1.4) with $r \ne 0$ that have isotropy subgroup larger than $\mathbf{Z}_2^c$ are (a, a) and $(a, ae^{2\pi i/n})$.

It follows from (1.3) that for such elements any isotropy subgroup larger than $\mathbf{Z}_2^c$ must contain an element of the form $(\kappa(m\zeta), \theta)$ and hence interchange the coordinates. Therefore, $r = a$. From (1.3) we must have

$$2m\zeta - \psi = \psi + 2k\pi$$

so that

$$\psi = m\zeta - k\pi.$$

But $\pi \in \mathbf{D}_n$ when n is even, so $\psi = m\zeta$. Therefore, from (1.4) $\psi = 0$ or $\psi = \zeta$. This leads to the two cases (a, a) and $(a, e^{2\pi i/n}a)$. It is easy to check that the isotropy subgroups are as stated in Table 1.3, and that since $n \equiv 0 \pmod 4$ the two isotropy subgroups $\mathbf{Z}_2(\kappa) \oplus \mathbf{Z}_2^c$ and $\mathbf{Z}_2(\kappa\zeta) \oplus \mathbf{Z}_2^c$ are not conjugate.

Remark. When $n \equiv 2 \pmod 4$ the element $\kappa\zeta$ is conjugate to $\kappa\pi$. To see this, let $q = (n-2)/4$ and compute

$$(-q\zeta)(\kappa\zeta)(q\zeta) = \kappa(q\zeta + \zeta + q\zeta) = \kappa(\tfrac{1}{2}n\zeta) = \kappa\pi.$$

It follows that $\mathbf{Z}_2(\kappa\zeta) \oplus \mathbf{Z}_2^c$ is conjugate to $\mathbf{Z}_2(\kappa, \pi) \oplus \mathbf{Z}_2^c$. Therefore, the entries in Table 1.3 also apply when $n \equiv 2 \pmod 4$ and provide an alternative description of the orbit structure in that case.

Exercises

These exercises sketch a method, due to Montaldi (unpublished), for finding two-dimensional fixed-point subspaces for $\Gamma \times \mathbf{S}^1$ acting on a space of the form $V \oplus V$, *without* classifying the orbits of $\Gamma \times \mathbf{S}^1$.

1.1. Suppose that Γ acts absolutely irreducibly on a space V, and let $\Gamma \times \mathbf{S}^1$ act on $V \oplus V$ as in Remark XVI, 3.3(d).
   (a) Show that the circle group action induces on $V \oplus V$ the structure of a complex vector space, in which $re^{i\theta} \in \mathbb{C}$ acts as scalar multiplication by r composed with the action of $\theta \in \mathbf{S}^1$. (See Remark XVI, 3.3(c), for a more abstract statement of this fact.)
   (b) Show that every subspace of the form $\mathrm{Fix}(\Sigma)$ is a *complex* vector subspace.
   (c) If $\dim_{\mathbb{R}} \mathrm{Fix}(\Sigma) = 2$ then $\dim_{\mathbb{C}} \mathrm{Fix}(\Sigma) = 1$. Hence show that $\mathrm{Fix}(\Sigma) = \mathbb{C}\{z_0\}$ where z_0 is a simultaneous eigenvector for all $\sigma \in \Sigma$, with eigenvalue 1.

1.2. Suppose that Σ is a twisted subgroup of the form $H^\theta = \{(\gamma, \theta(\gamma)) \colon \gamma \in H \subset \Gamma\}$, and z_0 is as in Exercise 1.1(c). Show that $\gamma \cdot z_0 = e^{-i\theta(\gamma)} z_0$, that is, z_0 is a simultaneous eigenvector for all $\gamma \in H$, with eigenvalues $e^{-i\theta(\gamma)}$.

1.3. Deduce that the isotropy subgroups Σ of $\Gamma \times \mathbf{S}^1$ with two-dimensional fixed-point subspaces (over $\mathbb{R}$) can be found as follows:
   (a) Select a representative γ_j from each conjugacy class of Γ.
   (b) Compute the complex eigenvectors z_{jk} for γ_j.
   (c) For each z_{jk} compute the subgroup H_{jk} of Γ consisting of all γ for which $\gamma \cdot z_{jk} = e^{i\theta(\gamma)} z_{jk}$.
   (d) Show that the twisted subgroup $\Sigma_{jk} = H_{jk}^{-\theta}$ is the isotropy subgroup of z_{jk} in $\Gamma \times \mathbf{S}^1$.
   (e) Calculate $\mathrm{Fix}(\Sigma_{jk})$ and discard any Σ_{kj} for which this has (real) dimension greater than 2.
   (f) Eliminate any repetitions due to conjugacy.

1.4. Apply the preceding procedure to $\Gamma = \mathbf{O}(2)$ acting on $\mathbb{R}^2$, as follows:
   (a) Representatives of conjugacy classes are κ and any $\theta \in \mathbf{SO}(2)$.

(b) The eigenvectors of κ acting on $\mathbb{C}^2$ are $(1, 0)$ with eigenvalue 1 and $(0, 1)$ with eigenvalue -1.

(c) The eigenvectors of θ acting on $\mathbb{C}^2$ are $(1, i)$ with eigenvalue $e^{i\theta}$ and $(1, -i)$ with eigenvalue $e^{-i\theta}$.

(d) Compute the subgroups H_{jk} as in Exercise 1.3(c) for $z_{jk} = (1, 0), (0, 1), (1, i)$ and $(1, -i)$, and conclude that every isotropy subgroup with a two-dimensional fixed-point subspace is conjugate to $\mathbf{Z}_2(\kappa) \oplus \mathbf{Z}_2^c$ or $\widetilde{\mathbf{SO}}(2)$.

1.5. Apply the procedure to $\Gamma = \mathbf{D}_n$ acting on $\mathbb{R}^2$, and recover the classification of isotropy subgroups with two-dimensional fixed-point subspaces obtained from Tables 1.1–1.3.

1.6. Can a similar method be used for equivariant Hopf bifurcation on a non–absolutely irreducible space W? If so, what changes are necessary?

§2. Invariant Theory for $\mathbf{D}_n \times \mathbf{S}^1$

In this section we find a Hilbert basis for the invariant functions $\mathbb{C}^2 \to \mathbb{R}$ and a module basis for the equivariant mappings $\mathbb{C}^2 \to \mathbb{C}^2$. The results here depend only on the parity of n. We therefore define

$$m = \begin{cases} n & [n \text{ odd}] \\ n/2 & [n \text{ even}]. \end{cases} \tag{2.1}$$

Proposition 2.1. *Let $n \geq 3$ and let m be as in (2.1). Then*

(a) *Every smooth $\mathbf{D}_n \times \mathbf{S}^1$-invariant germ $f: \mathbb{C}^2 \to \mathbb{R}$ has the form*

$$f(z_1, z_2) = h(N, P, S, T)$$

where $N = |z_1|^2 + |z_2|^2$, $P = |z_1|^2|z_2|^2$, $S = (z_1\bar{z}_2)^m + (\bar{z}_1 z_2)^m$, and

$$T = i(|z_1|^2 - |z_2|^2)((z_1\bar{z}_2)^m - (\bar{z}_1 z_2)^m).$$

(b) *Every smooth $\mathbf{D}_n \times \mathbf{S}^1$-equivariant map germ $g: \mathbb{C}^2 \to \mathbb{C}^2$ has the form*

$$g(z_1, z_2) = A\begin{bmatrix} z_1 \\ z_2 \end{bmatrix} + B\begin{bmatrix} z_1^2\bar{z}_1 \\ z_2^2\bar{z}_2 \end{bmatrix} + C\begin{bmatrix} \bar{z}_1^{m-1} z_2^m \\ z_1^m \bar{z}_2^{m-1} \end{bmatrix} + D\begin{bmatrix} z_1^{m+1}\bar{z}_1^m \\ \bar{z}_1^m z_2^{m+1} \end{bmatrix}$$

where A, B, C, D are complex-valued $\mathbf{D}_n \times \mathbf{S}^1$-invariant functions.

Remark. The $\mathbf{D}_n \times \mathbf{S}^1$-invariants do not form a polynomial ring. There is a relation

$$T^2 = (4P - N^2)(S^2 - 4P^m). \tag{2.2}$$

Proof. We find the invariants using Lemma XVI, 9.2, which lets us consider the simpler situation of complex-valued invariants.

There is a chain of subgroups

$$\mathbf{S}^1 \subset \mathbf{Z}_n \times \mathbf{S}^1 \subset \mathbf{D}_n \times \mathbf{S}^1.$$

We find the $\mathbf{D}_n \times \mathbf{S}^1$-invariants by climbing the chain: first finding the $\mathbf{S}^1$-invariants, then the $\mathbf{Z}_n \times \mathbf{S}^1$-invariants, and finally the $\mathbf{D}_n \times \mathbf{S}^1$-invariants.

Since $\mathbf{S}^1$ acts on $\mathbb{C}^2$ by $\theta \cdot (z_1, z_2) = (e^{i\theta} z_1, e^{i\theta} z_2)$ it is clear that the $\mathbf{S}^1$-invariants are generated by

$$u_1 = z_1 \bar{z}_1, \qquad u_2 = z_2 \bar{z}_2, \qquad v = z_1 \bar{z}_2, \quad \text{and} \quad \bar{v}, \tag{2.3}$$

with the relation

$$u_1 u_2 = v\bar{v}. \tag{2.4}$$

Next we compute the action of $\mathbf{D}_n$ on the three-dimensional space (u_1, u_2, v). Recall that $\gamma = 2\pi/n \in \mathbf{Z}_n$ acts on (z_1, z_2) by $(e^{i\gamma} z_1, e^{-i\gamma} z_2)$. Thus

$$\gamma \cdot (u_1, u_2, v) = (u_1, u_2, e^{2i\gamma} v).$$

When n is odd, $e^{2i\gamma}$ generates $\mathbf{Z}_n$, whereas when n is even it generates $\mathbf{Z}_{n/2}$. Thus, with m defined as in (2.1), $\mathbf{Z}_n \subset \mathbf{D}_n$ acts on $v \in \mathbb{C}$ as $\mathbf{Z}_m$. Therefore, the $\mathbf{Z}_n \times \mathbf{S}^1$-invariants are generated by u_1, u_2, and

$$w = v^m, \quad \bar{w}, \quad \text{and} \quad x = v\bar{v}. \tag{2.5}$$

Recall, however, that $x = v\bar{v} = u_1 u_2$, so that x is redundant.

The next step is to compute the action of κ on the $\mathbf{Z}_n \times \mathbf{S}^1$-invariants, obtaining

$$\kappa \cdot (u_1, u_2, w) = (u_2, u_1, \bar{w}). \tag{2.6}$$

It is now a straightforward exercise to show that the κ-invariants on (u, w)-space are generated by

$$\begin{aligned} &\text{(a)} \quad u_1 + u_2, \quad u_1 u_2 \\ &\text{(b)} \quad w + \bar{w}, \quad w\bar{w} \\ &\text{(c)} \quad (u_1 - u_2)(w - \bar{w}). \end{aligned} \tag{2.7}$$

Having found generators for the $\mathbf{D}_n \times \mathbf{S}^1$-invariants in (u, w)-space, we now translate these generators into (z_1, z_2)-coordinates. Note that $w\bar{w} = (v\bar{v})^m = (u_1 u_2)^m$ is redundant. We are left with the four generators

$$\begin{aligned} &\text{(a)} \quad u_1 + u_2 = |z_1|^2 + |z_2|^2 = N \\ &\text{(b)} \quad u_1 u_2 = |z_1|^2 |z_2|^2 = P \\ &\text{(c)} \quad w + \bar{w} = (z_1 \bar{z}_2)^m + (\bar{z}_1 z_2)^m = S \\ &\text{(d)} \quad (u_1 - u_2)(w - \bar{w}) = [z_1^{m+1} \bar{z}_1 \bar{z}_2^m + \bar{z}_1^m z_2^{m+1} \bar{z}_2 - z_1 \bar{z}_1^{m+1} z_2^m - z_1^m z_2 \bar{z}_2^{m+1}] \\ &\qquad = -iT. \end{aligned} \tag{2.8}$$

Part (a) of the proposition follows from Lemma XVI, 10.2, noting that N, P, S are real-valued, whereas $-iT$ is purely imaginary. Also (2.8) implies that

$$T^2 = -(u_1 - u_2)^2(w - \bar{w})^2 = (4P - N^2)(S^2 - 4P^m),$$

yielding the relation (2.2).

We now turn to part (b), the derivation of the $\mathbf{D}_n \times \mathbf{S}^1$-equivariants. Suppose that $g(z_1, z_2) = (\Phi_1(z_1, z_2), \Phi_2(z_1, z_2))$ commutes with $\mathbf{D}_n \times \mathbf{S}^1$. Commutativity with κ implies that $\Phi_2(z_1, z_2) = \Phi_1(z_2, z_1)$. Thus we must determine the mappings $\Phi: \mathbb{C}^2 \to \mathbb{C}$ that commute with $\mathbf{Z}_n \times \mathbf{S}^1$. The $\mathbf{S}^1$-equivariants have the form

$$\Phi(z_1, z_2) = p(u, v)z_1 + q(u, v)z_2 \tag{2.9}$$

where $u = (u_1, u_2) \in \mathbb{R}^2$ and $v \in \mathbb{C}$ are defined as in (2.3).

The action of $\mathbf{Z}_n$ on (2.9) produces

$$\Phi(e^{i\gamma}z_1, e^{i\gamma}z_2) = p(u, e^{2i\gamma}v)e^{i\gamma}z_1 + q(u, e^{2i\gamma}v)e^{-i\gamma}z_2. \tag{2.10}$$

Commutativity with $\mathbf{Z}_n$ implies that

$$\Phi(e^{i\gamma}z_1, e^{i\gamma}z_2) = e^{i\gamma}\Phi(z_1, z_2). \tag{2.11}$$

From (2.10, 2.11) we obtain

$$\begin{aligned} &\text{(a)} \quad p(u, e^{2i\gamma}v) = p(u, v) \\ &\text{(b)} \quad q(u, e^{2i\gamma}v) = e^{2i\gamma}q(u, v). \end{aligned} \tag{2.12}$$

Identity (2.12(a)) states that p is $\mathbf{Z}_m \times \mathbf{S}^1$-invariant in v, with u as a parameter, hence has the form

$$p(u, v) = A(u, w) \tag{2.13}$$

with w as in (2.5). Similarly q commutes with $\mathbf{Z}_m \times \mathbf{S}^1$ in v, with u as a parameter, so has the form

$$q(u, v) = \alpha(u, v\bar{v}, v^m)v + \beta(u, v\bar{v}, v^m)\bar{v}^{m-1}. \tag{2.14}$$

Since $v\bar{v} = u_1u_2$ is redundant, and $v^m = w$ we may rewrite (2.14) as

$$q(u, v) = B(u_1, u_2, w)v + C(u_1, u_2, w)\bar{v}^{m-1}. \tag{2.15}$$

Use (2.9, 2.13, and 2.15) to write

$$\Phi(z_1, z_2) = Az_1 + B|z_2|^2 z_1 + C\bar{z}_1^{m-1}z_2^m \tag{2.16}$$

where A, B, C are functions of $u_1 = |z_1|^2$, $u_2 = |z_2|^2$, and $w = (z_1\bar{z}_2)^m$.

The next step is to rewrite A, B, C in terms of the $\mathbf{D}_n \times \mathbf{S}^1$-invariants. Every polynomial in u_1, u_2, w has the form

$$\alpha(u_1, u_2, u_1u_2, w) + \beta(u_1, u_2, u_1u_2, w)u_1.$$

Since $u_1|z_2|^2 z_1 = Pz_1$ and $u_1\bar{z}_1^{m-1}z_2^m = Sz_1 - z_1^{m+1}\bar{z}_2^m$, we may use (2.8(a,b)) to rewrite (2.16) as

$$\Phi = (A_1 + A_2|z_1|^2)z_1 + B_1|z_2|^2 z_1 + C_1\bar{z}_1^{m-1}z_2^m + C_2 z_1^{m+1}\bar{z}_2^m. \tag{2.17}$$

Next, rewrite $A_2|z_1|^2 + B_1|z_2|^2 = \alpha(|z_1|^2 + |z_2|^2) + \beta|z_1|^2$. Then (2.17) has the form

$$\Phi = Az_1 + B|z_1|^2 z_1 + C\bar{z}_1^{m-1} z_2^m + D z_1^{m+1} \bar{z}_2^m \tag{2.18}$$

where A, B, C, and D are functions of N, P, S, and w.

Finally, every function of N, P, S, and w may be written as

$$\alpha(N, P, S, w + \bar{w}, (w - \bar{w})^2) + \beta(N, P, S, w + \bar{w}, (w - \bar{w})^2)(w - \bar{w}).$$

Now $w + \bar{w} = S$ and $(w - \bar{w})^2 = S^2 - 4P^m$, so that α and β are $\mathbf{D}_n \times \mathbf{S}^1$-invariants. Thus the $\mathbf{Z}_n \times \mathbf{S}^1$-equivariants are given by the eight generators

$$\begin{aligned} &\text{(a)} \quad z_1, |z_1|^2 z_1, \bar{z}_1^{m-1} z_2^m, z_1^{m+1} \bar{z}_2^m, \\ &\text{(b)} \quad (w - \bar{w}) z_1, (w - \bar{w})|z_1|^2 z_1, (w - \bar{w})\bar{z}_1^{m-1} z_2^m, (w - \bar{w}) z_1^{m+1} \bar{z}_2^m. \end{aligned} \tag{2.19}$$

Since $w - \bar{w} = v^m - \bar{v}^m = (z_1 \bar{z}_2)^m - (\bar{z}_1 z_2)^m$ the generators in (2.19(b)) are redundant. In particular, the identities

$$\begin{aligned} &\text{(a)} \quad (w - \bar{w}) z_1 = 2 z_1^{m+1} \bar{z}_2^m - S z_1 \\ &\text{(b)} \quad (w - \bar{w})|z_1|^2 z_1 = N z_1^{m+1} \bar{z}_2^m + \tfrac{1}{2}(-iT) z_1 - \tfrac{1}{2} N S z_1 \\ &\text{(c)} \quad (w - \bar{w})\bar{z}_1^{m-1} z_2^m = -N P^{m-1} z_1 + P^{m-1}|z_1|^2 z_1 - S \bar{z}_1^{m-1} z_2^m \\ &\text{(d)} \quad (w - \bar{w}) z_1^{m+1} \bar{z}_2^m = S z_1^{m+1} \bar{z}_2^m - 2P^m z_1 \end{aligned} \tag{2.20}$$

hold.

We have proved that every $\mathbf{D}_n \times \mathbf{S}^1$-equivariant has the form

$$A\begin{bmatrix} z_1 \\ z_2 \end{bmatrix} + B\begin{bmatrix} z_1^2 \bar{z}_1 \\ z_2^2 \bar{z}_2 \end{bmatrix} + C\begin{bmatrix} \bar{z}_1^{m-1} z_2^m \\ z_1^m \bar{z}_2^{m-1} \end{bmatrix} + D\begin{bmatrix} z_1^{m+1} \bar{z}_2^m \\ \bar{z}_1^m z_2^{m+1} \end{bmatrix}$$

where A, B, C, and D depend on N, P, S, and in addition, A depends on T. This statement is slightly stronger than part (b) of Proposition 2.1. □

When $n = 4$ there are three independent cubic equivariants

$$(|z_1|^2 + |z_2|^2)\begin{bmatrix} z_1 \\ z_2 \end{bmatrix}, \begin{bmatrix} z_1^2 \bar{z}_1 \\ z_2^2 \bar{z}_2 \end{bmatrix}, \begin{bmatrix} \bar{z}_1 z_2^2 \\ z_1^2 \bar{z}_2 \end{bmatrix}.$$

However, when $n \geq 3$, $n \neq 4$, there are only two equivariant cubics (the first two). This has an effect on the analysis of branching and stability in $\mathbf{D}_4$-Hopf bifurcation in §3. This effect also appears in $\mathbf{D}_{4k}$, Hopf bifurcation for representations having kernel $\mathbf{D}_k$.

To tie up one loose end, note that when $n = 2$ the nontrivial irreducible representations of $\mathbf{D}_2 \cong \mathbf{Z}_2 \oplus \mathbf{Z}_2$ are one-dimensional and have kernels K such that $\mathbf{D}_2/K \cong \mathbf{Z}_2$, so effectively the problem reduces to $\mathbf{Z}_2 \times \mathbf{S}^1$ acting on $\mathbb{R} \oplus \mathbb{R} \equiv \mathbb{C}$ where $\mathbf{Z}_2$ acts as minus the identity. The invariants and equivariants for this action are the same as those of $\mathbf{S}^1$ on $\mathbb{C}$, namely one invariant generator $x^2 + y^2$ and two equivariants (x, y) and $(-y, x)$. This happens because the $\mathbf{Z}_2$-action is the same as that of the rotation $\pi \in \mathbf{S}^1$.

§3. Branching and Stability for $\mathbf{D}_n$

We now apply the results of the previous two sections to obtain the generic behavior of $\mathbf{D}_n$-equivariant Hopf bifurcation: in particular, the branching equations for the three maximal isotropy subgroups found in §1, and the stabilities along those branches. We perform the calculations in subsection (a) and summarize the generic behavior in subsection (b) by drawing the possible bifurcation diagrams.

(a) Branching and Eigenvalues

The results of §2, together with Chapter XVI, imply that when $n \geq 3$ the branching equations for $\mathbf{D}_n$-equivariant Hopf bifurcation may be written

$$g = A\begin{bmatrix} z_1 \\ z_2 \end{bmatrix} + B\begin{bmatrix} z_1^2\bar{z}_1 \\ z_2^2\bar{z}_2 \end{bmatrix} + C\begin{bmatrix} \bar{z}_1^{m-1} z_2^m \\ z_1^m \bar{z}_2^{m-1} \end{bmatrix} + D\begin{bmatrix} z_1^{m+1}\bar{z}_2^m \\ \bar{z}_1^m z_2^{m+1} \end{bmatrix} = 0.$$

(If, further, the original vector field is in Birkhoff normal form, that is, commutes with $\mathbf{D}_n \times \mathbf{S}^1$, then $A = A' - (1 + \tau)i$ where $A'(0) = i$ and τ is the period-scaling parameter.) As usual we solve the equations $g = 0$ by restricting g to each fixed-point subspace $\mathrm{Fix}(\Sigma)$. In Tables 3.1 and 3.2 we list the equations for each of the three maximal isotropy subgroups Σ when $n \neq 4$; in Table 3.3 we list them for $n = 4$. Note that each of the branching equations consists of a real and an imaginary part. The imaginary parts of these equations may be solved for τ. The real parts contain the branching information.

By Theorem XVI, 7.3, the asymptotic stability of the solutions is determined by the eigenvalues of dg. The $\mathbf{S}^1$ symmetry forces one eigenvalue of dg to be zero; the signs of the real parts of the remaining three eigenvalues determine

Table 3.1. Branching Equations for $\mathbf{D}_n$ Hopf Bifurcation, $n \geq 3$, n Odd or $n \equiv 2 \pmod 4$. Here $m = n$ (n odd), $n/2$ (n even)

Orbit Type	Branching Equations	Signs of Eigenvalues
$(0, 0)$	—	$\mathrm{Re}\, A(0, \lambda)$
$(a, 0)$	$A + Ba^2 = 0$	$\mathrm{Re}(A_N + B) + O(a)$ $-\mathrm{Re}(B)$ [twice]
(a, a)	$A + Ba^2 + Ca^{2m-2} + Da^{2m} = 0$	$\mathrm{Re}(2A_N + B) + O(a)$ $\begin{cases}\text{trace} = \mathrm{Re}(B) + O(a) \\ \det = -\mathrm{Re}(B\bar{C}) + O(a)\end{cases}$
$(a, -a)$	$A + Ba^2 - Ca^{2m-2} - Da^{2m} = 0$	$\mathrm{Re}(2A_N + B) + O(a)$ $\begin{cases}\text{trace} = \mathrm{Re}(B) + O(a) \\ \det = \mathrm{Re}(B\bar{C}) + O(a)\end{cases}$

Table 3.2. Branching Equations for $\mathbf{D}_n$ Hopf Bifurcation, $n \equiv 0 \pmod 4$, $n \neq 4$. Here $m = n/2$

Orbit Type	Branching Equations	Signs of Eigenvalues
$(0,0)$	—	$\operatorname{Re} A(0,\lambda)$
$(a,0)$	$A + Ba^2 = 0$	$\operatorname{Re}(A_N + B) + O(a)$ $-\operatorname{Re}(B)$ [twice]
(a,a)	$A + Ba^2 + Ca^{2m-2} + Da^{2m} = 0$	$\operatorname{Re}(2A_N + B) + O(a)$ $\begin{cases}\text{trace} = \operatorname{Re}(B) + O(a)\\ \det = -\operatorname{Re}(B\bar{C}) + O(a)\end{cases}$
$(a, e^{2\pi i/n}a)$	$A + Ba^2 - Ca^{2m-2} - Da^{2m} = 0$	$\operatorname{Re}(2A_N + B) + O(a)$ $\begin{cases}\text{trace} = \operatorname{Re}(B) + O(a)\\ \det = \operatorname{Re}(B\bar{C}) + O(a)\end{cases}$

Table 3.3. Branching Equations for $\mathbf{D}_4$ Hopf Bifurcation

Orbit Type	Branching Equations	Signs of Eigenvalues
$(0,0)$	—	$\operatorname{Re} A(0,\lambda)$
$(a,0)$	$A + Ba^2 = 0$	$\operatorname{Re}(A_N + B) + O(a)$ $\begin{cases}\text{trace} = -\operatorname{Re}(B) + O(a)\\ \det = \|B\|^2 - \|C\|^2 + O(a)\end{cases}$
(a,a)	$A + (B + C)a^2 + Da^4 = 0$	$\operatorname{Re}(2A_N + B + C) + O(a)$ $\text{trace} = \operatorname{Re}(B - 3C) + O(a)$ $\det = \|C\|^2 - \operatorname{Re}(B\bar{C}) + O(a)$
(a, ia)	$A + (B - C)a^2 - Da^4 = 0$	$\operatorname{Re}(2A_N + B - C) + O(a)$ $\text{trace} = \operatorname{Re}(B + 3C) + O(a)$ $\det = \|C\|^2 + \operatorname{Re}(B\bar{C}) + O(a)$

the asymptotic stability. We list here, and derive later, the signs of these eigenvalues.

The isotropy subgroup of a solution restricts the form of dg at that solution since for $z = (z_1, z_2)$ and γ in the isotropy subgroup we have

$$(dg)_z\gamma = \gamma(dg)_z. \tag{3.1}$$

For each of the three maximal isotropy subgroups Σ, the action of Σ implies that dg has two two-dimensional invariant subspaces, namely the fixed-point subspace V_0 on which Σ acts trivially and an invariant complement V_1. See Tables 3.4 and 3.5.

Of course, the zero eigenvalue has an eigenvector in V_0; hence the other eigenvalue of $dg|V_0$ is given by trace $dg|V_0$. The remaining two eigenvalues of dg are those of $dg|V_1$.

Table 3.4. Decomposition of $\mathbb{R}^4$ into Invariant Subspaces for Σ, When $n \equiv 1, 2, 3 \pmod 4$

Isotropy	Orbit Representative	Invariant Subspaces
$\tilde{\mathbf{Z}}_n = \{(-\gamma, \gamma)\}$	$(a, 0)$	$V_0 = \{(w, 0)\}$ $V_1 = \{(0, w)\}$; $(-\gamma, \gamma)$ acts by $e^{-2i\gamma}$
$\mathbf{Z}_2(\kappa)$ or $\mathbf{Z}_2(\kappa) \oplus \mathbf{Z}_2^c$	(a, a)	$V_0 = \{(w, w)\}$ $V_1 = \{(w, -w)\}$; κ acts as $-Id$
$\mathbf{Z}_2(\kappa, \pi)$ or $\mathbf{Z}_2(\kappa, \pi) \oplus \mathbf{Z}_2^c$	$(a, -a)$	$V_0 = \{(w, -w)\}$ $V_1 = \{(w, w)\}$; κ acts as $-Id$

Table 3.5. Decomposition of $\mathbb{R}^4$ into Invariant Subspaces for Σ, When $n \equiv 0 \pmod 4$

Isotropy	Orbit Representative	Invariant Subspaces
$\tilde{\mathbf{Z}}_n = \{(-\gamma, \gamma)\}$	$(a, 0)$	$V_0 = \{(w, 0)\}$ $V_1 = \{(0, w)\}$; $(-\gamma, \gamma)$ acts by $e^{-2i\gamma}$
$\mathbf{Z}_2(\kappa) \oplus \mathbf{Z}_2^c$	(a, a)	$V_0 = \{(w, w)\}$ $V_1 = \{(w, -w)\}$; κ acts as $-Id$
$\mathbf{Z}_2(\kappa\zeta) \oplus \mathbf{Z}_2^c$	$(a, e^{2\pi i/n} a)$	$V_0 = \{(w, e^{2\pi i/n} w)\}$ $V_1 = \{(w, e^{-2\pi i/n} w)\}$; κ acts as $-Id$

To compute these it is convenient to use the complex coordinates (z_1, z_2). Recall that an $\mathbb{R}$-linear mapping on $\mathbb{C} \equiv \mathbb{R}^2$ has the form

$$w \to \alpha w + \beta \bar{w}$$

where $\alpha, \beta \in \mathbb{C}$. A simple calculation shows that

$$\text{trace} = 2\,\text{Re}(\alpha), \qquad \det = |\alpha|^2 - |\beta|^2. \tag{3.2}$$

We are now in a position to compute the eigenvalues of dg for each maximal isotropy subgroup Σ. For this purpose we write g in coordinates:

$$\begin{aligned} &\text{(a)} \quad Z_1 = Az_1 + Bz_1^2\bar{z}_1 + C\bar{z}_1^{m-1} z_2^m + Dz_1^{m+1}\bar{z}_2^m, \\ &\text{(b)} \quad Z_2 = Az_2 + Bz_2^2\bar{z}_2 + Cz_1^m\bar{z}_2^{m-1} + D\bar{z}_1^m z_2^{m+1}. \end{aligned} \tag{3.3}$$

In these coordinates dg takes the form

$$dg\begin{bmatrix} w_1 \\ w_2 \end{bmatrix} = \begin{bmatrix} Z_{1,z_1} w_1 + Z_{1,\bar{z}_1}\bar{w}_1 + Z_{1,z_2} w_2 + Z_{1,\bar{z}_2}\bar{w}_2 \\ Z_{2,z_1} w_1 + Z_{2,\bar{z}_1}\bar{w}_1 + Z_{2,z_2} w_2 + Z_{2,\bar{z}_2}\bar{w}_2 \end{bmatrix}. \tag{3.4}$$

Tables 3.4 and 3.5 show that for all n the computations will be very similar for the isotropy subgroups $\tilde{\mathbf{Z}}_n$, and for $\mathbf{Z}_2(\kappa)[\oplus \mathbf{Z}_2^c]$, but the third case, namely $\mathbf{Z}_2(\kappa, \pi)[\oplus \mathbf{Z}_2^c]$ and $\mathbf{Z}_2(\kappa\zeta) \oplus \mathbf{Z}_2^c$, will differ according to the value of $n\,(\text{mod } 4)$. Further, when $n = 4$ additional low order terms occur, so the results will be different in that case.

Case $\tilde{\mathbf{Z}}_n$. The first eigenvalue is the trace of $dg|V_0$, which is the mapping

$$w \mapsto (dg)(w, 0) = Z_{1,z_1} w + Z_{1,\bar{z}_1} \bar{w}.$$

From (3.3) the trace is $2\,\mathrm{Re}\, Z_{1,z_1}$. A computation yields

$$Z_{1,z_1}(a, 0) = A + A_N a^2 + 2Ba^2 + B_N a^4.$$

Since $A + Ba^2 = 0$ along $\tilde{\mathbf{Z}}_n$ solutions we obtain the first entry in Tables 3.1, 3.2, and 3.3.

To obtain the remaining eigenvalues in this case, note that when $n \neq 4$, $dg|V_1$ must be a (scalar multiple of a) rotation matrix since it commutes with $e^{-2i\gamma}$. Thus the eigenvalues of $dg|V_1$ are either complex conjugates or real and equal. In either case the required sign is equal to that of the trace. Now $dg|V_1$ is

$$w \mapsto (dg)(0, w) = Z_{2,z_2} w + Z_{2,\bar{z}_2} \bar{w}.$$

Using (3.3) we see that the trace of this map is $2\,\mathrm{Re}\, Z_{2,z_2}$. As earlier we compute $Z_{2,z_2}(a, 0) = A = -Ba^2$.

However, if $n = 4$ then $e^{-2i\gamma} = -1$, so we cannot assume that $dg|V_1$ is a rotation matrix. Thus we must compute both the trace and determinant of $dg|V_1$. We find:

$$\mathrm{trace} = 2\,\mathrm{Re}(-B)a^2,$$
$$\det = (|B|^2 - |C|^2)a^4.$$

Case $Z_2(\kappa)$ or $Z_2(\kappa) \oplus Z_2^c$. The first eigenvalue in this case is the trace of the mapping $dg|V_0$, which is

$$w \mapsto (dg)(w, w).$$

In coordinates this is

$$w \to (Z_{1,z_1} + Z_{1,z_2})w + (Z_{1,\bar{z}_1} + Z_{1,\bar{z}_2})\bar{w}.$$

Its trace is $2\,\mathrm{Re}(Z_{1,z_1} + Z_{1,z_2})$. To evaluate this we compute

$$\begin{aligned}
\text{(a)}\quad Z_{1,z_1}(a, a) &= A + A_{z_1} a + 2Ba^2 + B_{z_1} a^3 + C_{z_1} a^{2m-1} + D_{z_1} a^{2m+1} \\
&\quad + (m+1)Da^{2m} \\
&= A_{z_1} a + Ba^2 + B_{z_1} a^3 - Ca^{2m-2} + C_{z_1} a^{2m-1} + mDa^{2m} \qquad (3.5)\\
&\quad + D_{z_1} a^{2m+1}.
\end{aligned}$$

$$\text{(b)}\quad Z_{1,z_2}(a, a) = A_{z_2} a + B_{z_2} a^3 + mCa^{2m-2} + C_{z_2} a^{2m-1} + D_{z_2} a^{2m+1}.$$

Observe that $N_{z_1}(a, a) = a = N_{z_2}(a, a)$; $P_{z_1}(a, a) = a^3 = P_{z_2}(a, a)$; $S_{z_1}(a, a) =$

$ma^{2m-1} = S_{z_2}(a, a)$; and $T_{z_1}(a, a) = 0 = T_{z_2}(a, a)$. Thus $f_{z_1} = f_{z_2}$ for any invariant function f, evaluated at (a, a). It follows from (3.6) that trace $dg|V_0 =$

$$2A_{z_1}a + Ba^2 + 2B_{z_1}a^3 + (m-1)Ca^{2m-2} + 2C_{z_1}a^{2m-1} + mDa^{2m} + 2D_{z_1}a^{2m+1}$$

$$= \begin{cases} (2A_N + B)a^2 + O(a^3) & \text{if } n \neq 4, \\ (2A_N + B + C)a^2 + O(a^3) & \text{if } n = 4. \end{cases}$$

This gives the corresponding entry in Tables 3.1, 3.2, and 3.3.

To compute the remaining two eigenvalues in this case, we must find det and trace of $dg|V_1$. In coordinates, this mapping is

$$w \mapsto (dg)(w, -w) = (Z_{1,z_1} - Z_{1,z_2})w + (Z_{1,\bar{z}_1} - Z_{1,\bar{z}_2})\bar{w}.$$

Its trace is $2\,\mathrm{Re}(Z_{1,z_1} - Z_{1,z_2})$, which we can compute directly from (3.5), obtaining

$$Z_{1,z_1}(a, a) - Z_{1,z_2}(a, a) = Ba^2 - (m+1)Ca^{2m-2} + mDa^{2m}$$

$$= \begin{cases} Ba^2 + O(a^3) & \text{if } n \neq 4, \\ (B - 3C)a^2 + O(a^3) & \text{if } n = 4. \end{cases} \tag{3.6}$$

To evaluate $\det dg|V_1$ we must first compute

$$Z_{1,\bar{z}_1}(a, a) - Z_{1,\bar{z}_2}(a, a) = Ba^2 + (m-1)Ca^{2m-2} - mDa^{2m}. \tag{3.7}$$

From (3.3, 3.6, and 3.7) we have

$\det dg|V_1$

$$= |Ba^2 - (m+1)Ca^{2m-2} + mDa^{2m}|^2 - |Ba^2 + (m-1)Ca^{2m-2} - mDa^{2m}|^2$$

$$= \begin{cases} -4m\,\mathrm{Re}(B\bar{C})a^{2m} + O(a^{2m+1}) & \text{if } n \neq 4, \\ 8(|C|^2 - \mathrm{Re}(B\bar{C}))a^4 + O(a^5) & \text{if } n = 4. \end{cases}$$

Case $Z_2(\kappa, \pi)$ or $Z_2(\kappa, \pi) \oplus Z_2^c$. The calculations in this case, which holds when n is odd or $n \equiv 2 \pmod 4$, are almost identical with those in the previous case, hence we shall be brief.

$$dg|V_0 = (Z_{1,z_1} - Z_{1,z_2})w + (Z_{1,\bar{z}_1} - Z_{1,\bar{z}_2})\bar{w},$$

$$dg|V_1 = (Z_{1,z_1} + Z_{1,z_2})w + (Z_{1,\bar{z}_1} + Z_{1,\bar{z}_2})\bar{w}.$$

The first eigenvalue is trace $dg|V_0 = 2\,\mathrm{Re}(Z_{1,z_1} - Z_{1,z_2})$ evaluated at $(a, -a)$. Calculate

$$\begin{aligned} Z_{1,z_1}(a, -a) &= A + A_{z_1}a + 2Ba^2 + B_{z_1}a^3 \\ &\quad - [C_{z_1}a^{2m-1} + D_{z_1}a^{2m+1} + (m+1)Da^{2m}] \\ &= A_{z_1}a + Ba^2 + B_{z_1}a^3 + Ca^{2m-2} \\ &\quad - [C_{z_1}a^{2m-1} + D_{z_1}a^{2m+1} + mDa^{2m}], \\ Z_{1,z_2}(a, -a) &= A_{z_1}a + B_{z_2}a^3 + mCa^{2m-2} - C_{z_2}a^{2m-1} - D_{z_2}a^{2m+1}. \end{aligned}$$

At $(a, -a)$ we have $N_{z_1} = a = -N_{z_2}$; $P_{z_1} = a^3 = -P_{z_2}$; $S_{z_1} = (-1)^m a^{2m-1} = -S_{z_2}$; and $T_{z_1} = 0 = T_{z_2}$. Therefore, $f_{z_1}(a, -a) = -f_{z_2}(a, -a)$ for any invariant function f. It follows that

$$\operatorname{trace} dg | V_0 = 2(A_N + B)a^2 + O(a^3).$$

Also

$$Z_{1,z_1}(a, -a) + Z_{1,z_2}(a, -a) = Ba^2 + (m+1)Ca^{2m-2} - mDa^{2m}.$$

Hence

$$\operatorname{trace} dg | V_1 = 2\operatorname{Re}(B)a^2 + O(a^3).$$

To compute $\det dg | V_1$ evaluate

$$Z_{1,\bar{z}_1}(a, -a) + Z_{1,\bar{z}_2}(a, -a) = Ba^2 - (m-1)Ca^{2m-2} + mDa^{2m}.$$

Thus $\det dg | V_1 =$

$$\begin{aligned} &|Ba^2 + (m+1)Ca^{2m-2} - mDa^{2m}|^2 - |Ba^2 - (m-1)Ca^{2m-2} - mDa^{2m}|^2 \\ &\quad = 8m\operatorname{Re}(B\bar{C})a^{2m} + O(a^{2m+1}). \end{aligned}$$

Case $Z_2(\kappa\zeta) \oplus Z_2^c$. This is the final case; it applies for $n \equiv 0 \pmod 4$. Define $\omega = e^{2\pi i/n}$, so that $\omega^m = -1$. Then $V_0 = \{(z, \omega z)\}$ and $V_1 = \{(z, -\omega z)\}$, where the action of $\kappa\zeta$ on V_1 is by $-Id$. For this calculation

$$\begin{aligned} dg | V_0 &= (Z_{1,z_1} + \omega Z_{1,z_2})w + (Z_{1,\bar{z}_1} + \omega^{-1} Z_{1,\bar{z}_2})\bar{w}, \\ dg | V_1 &= (Z_{1,z_1} + \omega^{-1} Z_{1,z_2})w + (Z_{1,\bar{z}_1} + \omega Z_{1,\bar{z}_2})\bar{w}. \end{aligned}$$

On the orbit $(a, \omega a)$ we have $N_{z_1}(a, \omega a) = a = \omega N_{z_2}(a, \omega a)$; $P_{z_1}(a, \omega a) = a^3 = \omega P_{z_2}(a, \omega a)$; $S_{z_1}(a, \omega a) = -ma^{2m-1} = \omega S_{z_2}(a, \omega a)$; and $T_{z_1}(a, \omega a) = 0 = \omega T_{z_2}(a, \omega a)$. Thus $f_{z_1} = \omega f_{z_2}$ for any invariant function f, evaluated at $(a, \omega a)$.

When $n > 4$ the calculations are much the same as in the previous case, except for higher order terms. We omit the details.

When $n = 4$ we have $\omega = i$. On V_0 we need only the trace of dg, which is

$$2\operatorname{Re}(2A_N + B - C)a^2 + O(a^3).$$

On V_1 we need both trace and determinant. The trace is

$$2\operatorname{Re}(Z_{1,z_1} - iZ_{1,z_2}) = (B + 3C)a^2 + O(a^3).$$

The determinant is

$$\begin{aligned} &|Z_{1,z_1} - iZ_{1,z_2}|^2 - |Z_{1,\bar{z}_1} + iZ_{1,\bar{z}_2}|^2 \\ &\quad = (|B + 3C|^2 - |B - C|^2)a^4 + O(a^5) \\ &\quad = 8(|C|^2 + 2\operatorname{Re}(B\bar{C}))a^4 + O(a^5). \end{aligned}$$

This completes the calculations for Tables 3.1, 3.2, and 3.3.

(b) Bifurcation Diagrams

In this subsection we use the information contained in Tables 3.1(a, b) to derive bifurcation diagrams describing the generic $\mathbf{D}_n$-equivariant Hopf bifurcations. When $n \neq 4$ we assume the nondegeneracy conditions

$$\begin{aligned} &\text{(a)} \quad \operatorname{Re}(A_N + B) \neq 0, \\ &\text{(b)} \quad \operatorname{Re}(B) \neq 0, \\ &\text{(c)} \quad \operatorname{Re}(2A_N + B) \neq 0, \\ &\text{(d)} \quad \operatorname{Re}(B\bar{C}) \neq 0, \\ &\text{(e)} \quad \operatorname{Re}(A_\lambda) \neq 0, \end{aligned} \tag{3.8}$$

where each term is evaluated at the origin. When $n = 4$ we assume the nondegeneracy conditions

$$\begin{aligned} &\text{(a)} \quad \operatorname{Re}(A_N + B) \neq 0, \\ &\text{(b)} \quad \operatorname{Re}(B) \neq 0, \\ &\text{(c)} \quad \operatorname{Re}(2A_N + B + C) \neq 0, \\ &\text{(d)} \quad \operatorname{Re}(B - 3C) \neq 0, \\ &\text{(e)} \quad \operatorname{Re}(B + 3C) \neq 0, \\ &\text{(f)} \quad |B|^2 - |C|^2 \neq 0, \\ &\text{(g)} \quad |C|^2 - \operatorname{Re}(B\bar{C}) \neq 0, \\ &\text{(h)} \quad |C|^2 + \operatorname{Re}(B\bar{C}) \neq 0, \\ &\text{(i)} \quad \operatorname{Re}(A_\lambda) \neq 0. \end{aligned} \tag{3.9}$$

The main result is as follows:

Theorem 3.1. *Assuming the nondegeneracy conditions* (3.8) *or* (3.9), *there exists precisely one branch of small amplitude, near-2π-periodic solutions, for each of the isotropy subgroups* $\tilde{\mathbf{Z}}_n$; $\mathbf{Z}_2(\kappa)$ [n odd] *or* $\mathbf{Z}_2(\kappa) \oplus \mathbf{Z}_2^c$ [*n even*]; *and* $\mathbf{Z}_2(\kappa, \pi)$ [*n odd*], $\mathbf{Z}_2(\kappa, \pi) \oplus \mathbf{Z}_2^c$ [$n \equiv 2 \pmod 4$], *or* $\mathbf{Z}_2(\kappa\zeta) \oplus \mathbf{Z}_2^c$ [$n \equiv 0 \pmod 4$].

Assume that the trivial branch is stable subcritically and loses stability as λ passes through 0. *If $n \geq 3$, $n \neq 4$, then*:

(a) *The $\tilde{\mathbf{Z}}_n$ branch is super- or subcritical according to whether* $\operatorname{Re}(A_N(0) + B(0))$ *is positive or negative. It is stable if* $\operatorname{Re}(A_N(0) + B(0)) > 0$, $\operatorname{Re} B(0) < 0$.

(b) *The* $\mathbf{Z}_2(\kappa)[\oplus \mathbf{Z}_2^c]$ *branch is super- or subcritical according to whether* $\operatorname{Re}(2A_N(0) + B(0))$ *is positive or negative. It is stable if* $\operatorname{Re}(2A_N(0) + B(0)) > 0$, $\operatorname{Re} B(0) > 0$, *and* $\operatorname{Re}(B(0)\bar{C}(0)) < 0$.

(c) *The* $\mathbf{Z}_2(\kappa, \pi)[\oplus \mathbf{Z}_2^c]$ *or* $\mathbf{Z}_2(\kappa\zeta) \oplus \mathbf{Z}_2^c$ *branch is super- or subcritical according to whether* $\operatorname{Re}(2A_N(0) + B(0))$ *is positive or negative. It is stable if* $\operatorname{Re}(2A_N(0) + B(0)) > 0$, $\operatorname{Re} B(0) > 0$, and $\operatorname{Re}(B(0)\bar{C}(0)) > 0$.

If $n = 4$ *then:*
(d) *The* $\tilde{\mathbf{Z}}_4$ *branch is super- or subcritical according to whether* $\operatorname{Re}(A_N(0) + B(0))$ *is positive or negative. It is stable if* $\operatorname{Re}(A_N(0) + B(0)) > 0$, $\operatorname{Re} B(0) < 0$ and $|B(0)|^2 > |C(0)|^2$.
(e) *The* $\mathbf{Z}_2(\kappa) \oplus \mathbf{Z}_2^c$ *branch is super- or subcritical according to whether* $\operatorname{Re}(2A_N(0) + B(0) + C(0))$ *is positive or negative. It is stable if* $\operatorname{Re}(2A_N(0) + B(0) + C(0)) > 0$, $\operatorname{Re}(B(0) - 3C(0)) > 0$, *and* $|C(0)|^2 - \operatorname{Re}(B(0)\bar{C}(0)) > 0$.
(f) *The* $\mathbf{Z}_2(\kappa\zeta) \oplus \mathbf{Z}_2^c$ *branch is super- or subcritical according to whether* $\operatorname{Re}(2A_N(0) + B(0) - C(0))$ *is positive or negative. It is stable if* $\operatorname{Re}(2A_N(0) + B(0) - C(0)) > 0$, $\operatorname{Re}(B(0) + 3C(0)) > 0$, *and* $|C(0)|^2 + \operatorname{Re}(B(0)\bar{C}(0)) > 0$.

PROOF. The existence of the stated branches follows from the equivariant Hopf theorem, Theorem XVI, 4.1. We consider their directions of criticality and their stabilities, using the results of §3. First suppose $n \geq 3$, $n \neq 4$. To lowest order the three nontrivial branches are given by:

$$\begin{aligned}
&\text{(a)} \quad \lambda = -a^2[\operatorname{Re}(A_N(0) + B(0))]/\operatorname{Re} A_\lambda(0) + \cdots \qquad \tilde{\mathbf{Z}}_n \\
&\text{(b)} \quad \lambda = -a^2[\operatorname{Re}(2A_N(0) + B(0))]/\operatorname{Re} A_\lambda(0) + \cdots \qquad \mathbf{Z}_2(\kappa)[\oplus \mathbf{Z}_2^c] \qquad (3.10) \\
&\text{(c)} \quad \lambda = -a^2[\operatorname{Re}(2A_N(0) + B(0))]/\operatorname{Re} A_\lambda(0) + \cdots \qquad \mathbf{Z}_2(\kappa, \pi)[\oplus \mathbf{Z}_2^c].
\end{aligned}$$

Eigenvalues of dg with positive real part indicate stability. We have assumed that the steady state $z = 0$ is stable for $\lambda < 0$ and loses stability when $\lambda > 0$. Hence $\operatorname{Re} A_\lambda(0) < 0$. With these assumptions we can draw the bifurcation diagrams in Figure 3.1. The two $\mathbf{Z}_2$ branches are either both supercritical or both subcritical.

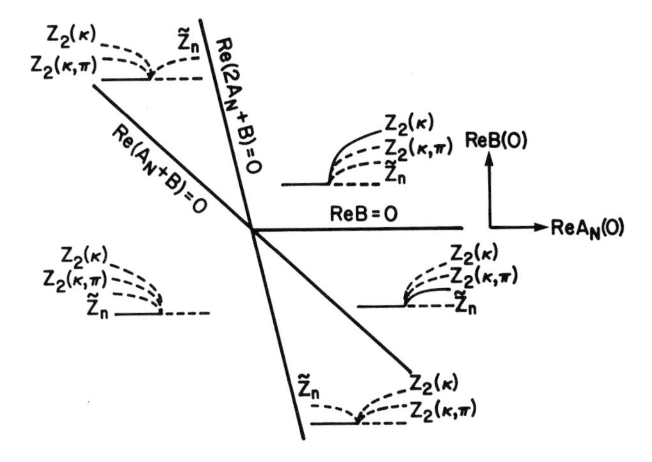

Figure 3.1. Generic branching for $\mathbf{D}_n$ Hopf bifurcation, $n \geq 3$, $n \neq 4$. For any branch to be stable, all three must be supercritical. Exactly one branch is then stable. Note that higher order terms can interchange the $\mathbf{Z}_2$ branches. When $n \equiv 0 \pmod 4$ the label $\mathbf{Z}_2(\kappa\pi)$ should be changed to $\mathbf{Z}_2(\kappa\zeta)$ and when n is even the $\mathbf{Z}_2^c$ label is suppressed.

(a)

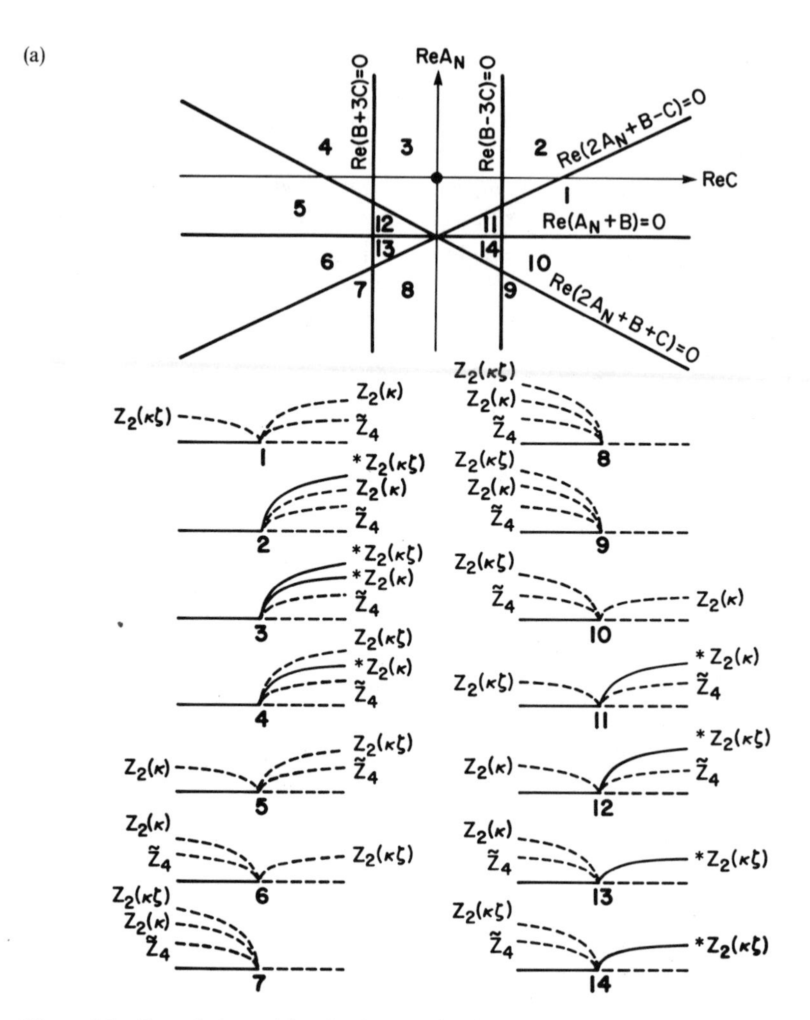

Figure 3.2. Generic branching for $\mathbf{D}_4$ Hopf bifurcation. An asterisk denotes a branch whose stability depends on other third order terms in Table 3.3, namely $\mathrm{Im}(B)$ and $\mathrm{Im}(C)$. (a) $\mathrm{Re}(B) < 0$; (b) $\mathrm{Re}(B) > 0$.

For any periodic solution to be asymptotically stable, all three branches must be supercritical. Then either the $\tilde{\mathbf{Z}}_n$ branch is stable (if $\mathrm{Re}\, B(0) < 0$); and precisely one of the $\mathbf{Z}_2$ branches is stable (if $\mathrm{Re}\, B(0) > 0$). Which of these branches is stable depends on the sign of $\mathrm{Re}(B(0)\bar{C}(0))$. Note that $A_N(0)$ and $B(0)$ are cubic coefficients, but $\bar{C}(0)$ is the coefficient of a term of degree $2m - 1$. Moreover, that term is needed to determine which of the $\mathbf{Z}_2$ branches is stable, *independent of the many lower order terms that may exist.*

When $n = 4$, however, there are *three* linearly independent cubic terms in g. In this case the branching equations take the form

(b)

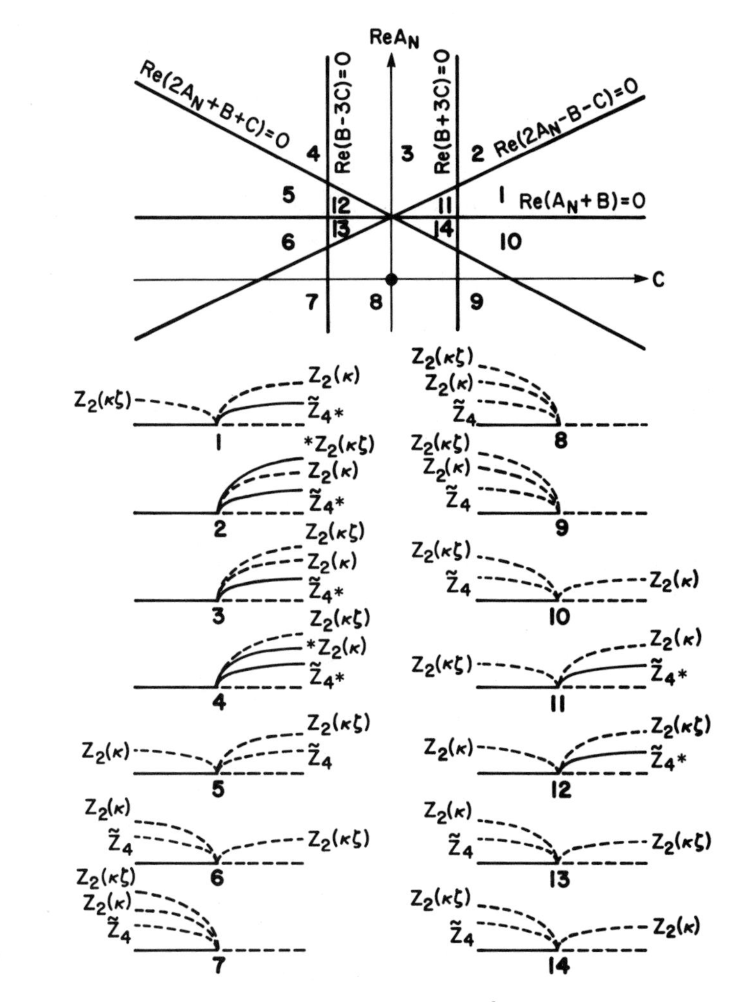

Figure 3.2 *Continued*

$$\begin{array}{lll}
\text{(a)} & \lambda = -a^2[\operatorname{Re}(A_N(0) + B(0))]/\operatorname{Re} A_\lambda(0) + \cdots & \tilde{\mathbf{Z}}_4 \\
\text{(b)} & \lambda = -a^2[\operatorname{Re}(2A_N(0) + B(0) + C(0))]/\operatorname{Re} A_\lambda(0) + \cdots & \mathbf{Z}_2(\kappa) \oplus \mathbf{Z}_2^c] \\
\text{(c)} & \lambda = -a^2[\operatorname{Re}(2A_N(0) + B(0) - C(0))]/\operatorname{Re} A_\lambda(0) + \cdots & \mathbf{Z}_2(\kappa\zeta) \oplus \mathbf{Z}_2^c.
\end{array} \tag{3.11}$$

The possible configurations of branches that involve *stable* solutions are shown in Figure 3.2. (Differences in stability assignments of unstable branches are not indicated, for simplicity, but may be derived from Table 3.1(b).) □

We note two interesting features where the case $n = 4$ differs from all others (assuming the standard action of $\mathbf{D}_n$):

(a) The two $\mathbf{Z}_2$ branches need not have the same direction of criticality.
(b) More than one branch may be stable for the same values of the coefficients of g. That is, the system may have nonunique (orbits of) stable states. Indeed for any choice of two out of the three isotropy subgroups we are considering, there are parameter values that make both of these two branches stable, but the third unstable. It is *not* possible to have all three branches stable simultaneously.

See Swift [1986] for a more complete discussion of the dynamics contained in Hopf bifurcation with $\mathbf{D}_4$ symmetry.

Remarks.
(a) In Figure 3.2 those branches marked with an asterisk are stable or unstable depending on other coefficients noted in Table 3.3. Both possibilities can occur with suitable choices of coefficients.

In one case, namely case 3 of Figure 3.2(b), there are two branches whose stabilities depend upon higher order coefficients. Either one of these, or both, may be stable, but it is easy to see that they cannot both be unstable, since this would require $|C|^2$ to be negative.
(b) The preceding results are obtained on the assumption that a center manifold reduction has already been performed and the vector field is in Birkhoff normal form. By XVI, §11, they remain true when it is in Birkhoff normal form up to order $2m - 1$. We have not considered the explicit computation of such a Birkhoff normal form for a general $\mathbf{D}_n$-equivariant vector field: it is presumably exceedingly complicated when n is large. Also, coefficients of terms of order less than $2m - 1$ will presumably appear in the expression for $\bar{C}(0)$.

Relation with $\mathbf{O}(2)$*-Symmetric Systems.* Dihedral group symmetry $\mathbf{D}_n$ often arises when a continuous system, having circular $\mathbf{O}(2)$ symmetry, is approximated by a discrete one. It may seem curious that there are *three* classes of isotropy subgroups with two-dimensional fixed-point spaces for $\mathbf{D}_n$, however large n is, but only *two* classes for the "limit" $\mathbf{O}(2)$. An analysis of this sheds some light on the approximation of a continuous system by a discrete one.

In algebraic terms what happens is that the solutions with the two $\mathbf{Z}_2$ isotropy subgroups "merge" as $n \to \infty$. Solutions of these types differ by a vanishingly small amount for large n. For example, consider for definiteness a tower

$$\mathbf{D}_4 \subset \mathbf{D}_8 \subset \mathbf{D}_{16} \subset \cdots .$$

The $\mathbf{Z}_2$ solutions are those with isotropy subgroups $\mathbf{Z}_2(\kappa) \oplus \mathbf{Z}_2^c$ and $\mathbf{Z}_2(\kappa\zeta) \oplus \mathbf{Z}_2^c$. Now κ and $\kappa\zeta$ are both reflections: the first in the real axis, the second in a line making an angle π/n with the real axis. These lines approach each other for large n.

In addition, the coefficient $C(0)$ that determines which of the two branches is stable is attached to a term of increasingly high order as $n \to \infty$. Thus the

Figure 3.3. A torus on which n stable periodic trajectories are separated by n unstable ones. Each family of n trajectories corresponds to a $\mathbf{D}_n$-orbit under one of the two $\mathbf{Z}_2$ solutions.

distinction between the stabilities becomes more delicate. Since this coefficient is the one attached to the lowest order $\mathbf{D}_n \times \mathbf{S}^1$-equivariant term that is not $\mathbf{O}(2) \times \mathbf{S}^1$-equivariant, it is not surprising that it governs the distinction between $\mathbf{D}_n$ and $\mathbf{O}(2)$ Hopf bifurcation.

Geometrically, we can picture the relevant periodic solutions in the following way. For $\mathbf{D}_n$ think of a torus, on which n stable periodic trajectories are separated by n unstable ones, as in Figure 3.3. These correspond to the pair of $\mathbf{Z}_2$ branches. As $n \to \infty$ we get increasingly many closed trajectories, and the degree of instability weakens. We approach a torus foliated by a continuous family of periodic orbits, neutrally stable to displacements around the torus. Nagata [1986] has studied this picture of the dynamics in detail.

§4. Oscillations of Identical Cells Coupled in a Ring

We now apply the results on generic $\mathbf{D}_n$ Hopf bifurcation to a system of n identical cells coupled in a ring, as in Figure 4.1. (The literature often refers to "coupled oscillators," but we shall see later that the entire system may oscillate under conditions in which the individual cells, if uncoupled, would not; hence we prefer a more neutral term.)

If we assume that the coupling between neighboring cells is symmetric, that is, invariant under interchanging the cells, then the entire system has $\mathbf{D}_n$ symmetry. The cyclic subgroup $\mathbf{Z}_n$ permutes the cells cyclically, sending cell p to $p + 1 \pmod n$; the flip sends cell p to $-p \pmod n$.

Such problems have been studied by a number of authors, including Alexander and Auchmuty [1986], and van Gils and Valkering [1986]. Alexander [1986] considers the behavior of a *plexus*, or network of identical cells, possibly without symmetry. Applications include, in particular, chemical oscillators and biological oscillators (cells coupled via membrane transport of ions). The situation is considered in the seminal paper by Turing [1952] on morphogenesis. Tsotsis [1981] discusses the problem in the context of chemical reactors. Surveys in the literature include Winfree [1980]; De Kleine, Kennedy, and MacDonald [1982]; and Kopell [1983]. The case of two cells

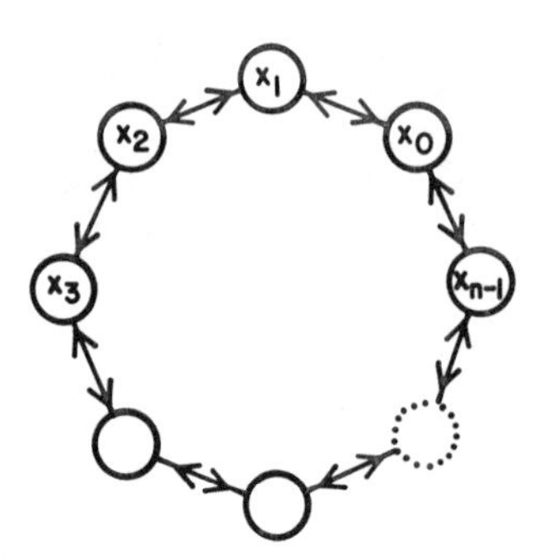

Figure 4.1. Schematic picture of a ring of n coupled identical cells.

has been discussed by Othmer and Scriven [1971], Smale [1974], and Howard [1979]. We shall describe the generic behavior, paying particular attention to the cases $n \leq 5$. The results should prove applicable in a number of physical contexts.

The analysis is divided into five parts. In subsection (a) we set up the equations and discuss their symmetries. In (b) we analyze the three-cell case, finding conditions under which purely imaginary eigenvalues can occur, deriving the corresponding representations of $\mathbf{D}_3$, and listing the patterns of oscillation that generically can occur. In (c) we present some numerical simulations of the three-cell case. In (d) we generalize the methods to the general case of n cells. Finally in (e) we discuss as examples the cases $n = 2, 4, 5$ and compare our results with those of Alexander and Auchmuty [1986].

For simplicity we assume that the coupling is "nearest-neighbor" and that it is symmetrical in the sense that the interaction between any neighboring pair of cells takes the same form. The "nearest-neighbor" assumption may be relaxed, with minor changes in our conclusions, provided the symmetry is retained, and the method may be applied to systems having other kinds of symmetry—for example, four cells with any two interacting identically, a system with tetrahedral symmetry. If the coupling is asymmetric but preserves the cyclic symmetry, our results from Chapter XVII, §8, on $\mathbf{S}^1$ Hopf bifurcation can be applied instead. For various extensions of the results, see the Exercises.

(a) The Equations

For purposes of *illustration* we shall use a fairly concrete system of equations with linear interaction, but we shall *prove* our results for any system of equations with the appropriate symmetry. Cell p is described by two state variables (x_p, y_p) and p is taken mod n. Then we have a system of n equations

$$\frac{d}{dt}(x_p, y_p) = F(x_p, y_p, \lambda) + K(\lambda)\cdot(2x_p - x_{p-1} - x_{p+1}, 2y_p - y_{p-1} - y_{p+1}). \tag{4.1}$$

Here $F: \mathbb{R}^2 \to \mathbb{R}^2$ is an arbitrary smooth function and λ is a bifurcation parameter. For each λ, the matrix K is constant:

$$K = \begin{bmatrix} k_{11} & k_{12} \\ k_{21} & k_{22} \end{bmatrix}.$$

The k_{ij} (which depend only on λ and may be constant) represent *coupling strengths*. We have (e.g.)

$$2x_p - x_{p-1} - x_{p+1} = (x_p - x_{p-1}) + (x_p - x_{p+1}).$$

Thus the form of the equations is chosen so that the coupling is linear, nearest-neighbor, and symmetric, and the coupling term in K vanishes if all cells behave identically. The bifurcation parameter λ may be present in F, or K, or both, depending on interpretation.

For theoretical discussion and proofs we abstract from this system its symmetries and consider the more general system

$$dx_p/dt = g(x_{p-1}, x_p, x_{p+1}; \lambda). \tag{4.2}$$

Here p is taken mod n and runs from 0 to $n-1$, and $x_p \in \mathbb{R}^k$ for some k, so the system (4.2) is on $\mathbb{R}^{nk}$. (We shall see later that $k \geq 2$ is required for Hopf bifurcation, and henceforth we assume this.) We require $g: \mathbb{R}^k \times \mathbb{R}^k \times \mathbb{R}^k \times \mathbb{R} \to \mathbb{R}^k$ to be a smooth function with the symmetry property

$$g(u, v, w; \lambda) = g(w, v, u; \lambda). \tag{4.3}$$

It is easy to check that (4.1) has this symmetry when written in the form (4.2). Note that in (4.3) we do *not* assume linearity of the coupling.

(b) Three Cells

In this section we find conditions under which Equation (4.2) can possess purely imaginary eigenvalues and use group theory to analyze the resulting generic patterns of oscillation. To prevent complicating the argument with group-theoretic generalities we first consider the case of three cells, which illustrates the main principles. The general case is similar and will be easier to describe once the method has been illustrated for $n = 3$.

Equation (4.2) becomes

$$\begin{aligned} dx_0/dt &= g(x_2, x_0, x_1; \lambda) \\ dx_1/dt &= g(x_0, x_1, x_2; \lambda) \\ dx_2/dt &= g(x_1, x_2, x_0; \lambda) \end{aligned} \tag{4.4}$$

where $x_p \in \mathbb{R}^k$ and λ is a bifurcation parameter. Suppress the λ-dependence in the notation and write (4.4) as

$$dx/dt = G(x)$$

where $x = (x_0, x_1, x_2) \in \mathbb{R}^{3k}$. Then we summarize the results of this section as follows:

Theorem 4.1. *Suppose that $L = (dG)_{(0,\lambda)}$ has a pair of purely imaginary eigenvalues $\pm i$ (without loss of generality at $\lambda = 0$) which cross the imaginary axis with nonzero speed as λ passes through 0. Let A and B be the matrices of partial derivatives $d_{x_1}G$ and $d_{x_0}G$, restricted to the spaces of x_1, x_0-variables, respectively. Then generically one of the following occurs:*

(a) *The purely imaginary eigenvalues are those of $A + 2B$ and are simple. There is a Hopf bifurcation at $\lambda = 0$, and in the resulting periodic solution all three cells have the same waveform and the same phase.*

(b) *The purely imaginary eigenvalues are those of $A - B$ and have multiplicity 2. There are three branches of symmetry-breaking oscillations, with the following patterns:*

Isotropy subgroup $\tilde{\mathbf{Z}}_3$: The cells have the same waveforms but with phase shifts of $2\pi/3$ from one to the next.

Isotropy subgroup $\mathbf{Z}_2(\kappa)$: Two cells have the same waveform and same phase; the third oscillates with the same period but a different waveform.

Isotropy subgroup $\mathbf{Z}_2(\kappa, \pi)$: Two cells have the same waveform but are π out of phase; the third oscillates with half the period.

Proof. The first step is to find conditions under which the linearization $L = (dG)_{(0,\lambda)}$ can possess purely imaginary eigenvalues. Now $g(x_0, x_1, x_2) = g(x_2, x_1, x_0)$ and we have

$$d_{x_1}G = Ax_1$$
$$d_{x_0}G = Bx_0$$
$$d_{x_2}G = Bx_2$$

for certain $k \times k$ matrices A and B. Hence in $k \times k$ block form,

$$L = \begin{bmatrix} A & B & B \\ B & A & B \\ B & B & A \end{bmatrix}.$$

There is a k-dimensional space V_0 of vectors $[v, v, v]$ ($v \in \mathbb{R}^k$) that is invariant under L. Indeed

$$L[v, v, v] = L\begin{bmatrix} v \\ v \\ v \end{bmatrix} = \begin{bmatrix} A & B & B \\ B & A & B \\ B & B & A \end{bmatrix} = \begin{bmatrix} (A + 2B)v \\ (A + 2B)v \\ (A + 2B)v \end{bmatrix}. \tag{4.5}$$

Hence the eigenvalues of $L|V_0$ are those of $A + 2B$.

Let $\omega = e^{2\pi i/3}$ be a cube root of unity in $\mathbb{C}$. Then, complexifying from $\mathbb{R}^{nk}$ to $\mathbb{C}^{nk}$, we can find two other subspaces invariant under L:

$$V_1 = \{[v, \omega v, \omega^2 v] | v \in \mathbb{R}^k\},$$
$$V_2 = \{[v, \omega^2 v, \omega v] | v \in \mathbb{R}^k\}.$$

A similar calculation to (4.5) shows that the eigenvalues of $L|V_1$ are those of $A + \omega B + \omega^2 B$, which is $A - B$ since $1 + \omega + \omega^2 = 0$. Similarly $L|V_2$ has the eigenvalues of $A + \omega^2 B + \omega B = A - B$. We conclude that the eigenvalues of L are:

$$\begin{aligned} &\text{The eigenvalues of } A + 2B, \\ &\text{The eigenvalues of } A - B, \text{ repeated twice.} \end{aligned} \tag{4.6}$$

Hence L has purely imaginary eigenvalues, giving the possibility of Hopf bifurcation in (4.2), if and only if either $A + 2B$ or $A - B$ has purely imaginary eigenvalues. For this to occur, we must have $k \geq 2$. We shall see that the two cases correspond to different representations of $\mathbf{D}_3$ and hence lead to different patterns of oscillation.

In our standard notation, we have $\mathbf{D}_3 \supset \mathbf{Z}_3 = \{1, \zeta, \zeta^2\}$, and the flip is $\kappa \in \mathbf{D}_3$. The standard action of ζ on the plane $\mathbb{C}$ is rotation through $2\pi/3$, that is, multiplication by $\omega = e^{2\pi i/3}$. The actions on $\mathbb{R}^{3k}$ are as follows:

$$\begin{aligned} \zeta(x_0, x_1, x_2) &= (x_1, x_2, x_0) \\ \kappa(x_0, x_1, x_2) &= (x_0, x_2, x_1). \end{aligned} \tag{4.7}$$

We seek to decompose $\mathbb{R}^{3k}$ into irreducible subspaces for the action of $\mathbf{D}_3$, noting that every eigenspace for L is $\mathbf{D}_3$-invariant. Clearly both ζ and κ act trivially on vectors $[v, v, v] \in V_0$, so $\mathbf{D}_3$ acts on V_0 by k copies of the trivial action on $\mathbb{R}$. We assert that $\mathbb{R}^{3k} = V_0 \oplus W_0$ where W_0 is the sum of k copies of the nontrivial action of $\mathbf{D}_3$ on $\mathbb{R}^2 \equiv \mathbb{C}$. This can easily be seen directly, because $\mathbf{D}_3$ has only two distinct irreducible representations, and the trivial one occurs only on V_0. That is, if $[u, v, w]$ is fixed by $\mathbf{D}_3$ then $u = v = w$, as is obvious from (4.5). (For $\mathbf{D}_n$ with $n > 3$ a little more care must be taken at this stage of the analysis since there are several nontrivial representations; see subsection (c).)

To summarize: $\mathbb{R}^{3k}$ breaks up as $\mathbb{R}^k \oplus \mathbb{R}^{2k}$ where $\mathbf{D}_3$ acts on $\mathbb{R}^k$ by k copies of the trivial representation, and on $\mathbb{R}^{2k}$ by k copies of the nontrivial representation. Explicitly, $V_0 = \mathbb{R}^k$ is spanned by all $[v, v, v]$ and $W_0 = \mathbb{R}^{2k}$ by all $[v, w, -v - w]$ (since this is obviously an invariant complement to V_0).

Thus generically there are two cases. In the first case the purely imaginary eigenvalue of L (which, of course, must be part of a complex conjugate pair) comes from $A + 2B$. It is then (generically) simple, and there is a standard Hopf bifurcation. Since $\mathbf{D}_3$ acts trivially on V_0, all three cells behave identically (same waveform, same phase). To put it another way, $\mathbf{D}_3$ lies in the isotropy subgroup of such a solution.

In the second case the purely imaginary eigenvalue of L comes from $A - B$. By (4.4) this will have multiplicity at least 2, and generically (in the world of $\mathbf{D}_3$ symmetry) exactly 2 by Proposition XVI, 1.4. We are then in the situation of Theorem XVI, 4.1, with $\Gamma = \mathbf{D}_3$ in its standard representation. We conclude that there are (at least) three branches of oscillations, with isotropy subgroups $\tilde{\mathbf{Z}}_3$, $\mathbf{Z}_2(\kappa)$, and $\mathbf{Z}_2(\kappa, \pi)$. We describe the interpretation of each in turn.

The first, $\tilde{\mathbf{Z}}_3$, is the discrete analogue of a rotating wave. The waveform is fixed under cyclic permutation of the cells, provided phase is shifted by $2\pi/3$. The oscillations thus have the identical waveform in all three cells, but a phase lag of $2\pi/3$ from each cell to the next. (The phase lag may also be $-2\pi/3$, which is a physically distinct solution only when the numbering of the cells has been chosen: it is in the same orbit under $\mathbf{D}_3 \times \mathbf{S}^1$.) By *waveform* we mean the trajectory of x_p in phase space $\mathbb{R}^k$.

On the $\mathbf{Z}_2(\kappa)$ branch, the waveform is identical when x_1 and x_2 are interchanged. In other words, cells 1 and 2 behave identically, and cell 0 has a different waveform (not prescribed by the symmetry).

On the $\mathbf{Z}_2(\kappa, \pi)$ branch the waveform is identical if x_1 and x_2 are interchanged *and* the phase is shifted by π. In other words, cells 1 and 2 have the same waveform but are exactly π out of phase, whereas cell 0 is "π out of phase with itself." That is, a phase shift of π produces the same waveform in cell 0, which is therefore oscillating with *double* the frequency of cells 1 and 2. The exact shape of the waveform is not prescribed by the symmetry approach, but will be nearly sinusoidal close to the bifurcation point. □

At least with our current techniques, it is a more complicated matter to determine whether the bifurcation is super- or subcritical, and which branch (if any) is stable. Before the results of §3 can be used, G must be Liapunov–Schmidt or center manifold reduced to the appropriate eigenspace. Certain cubic terms will distinguish stability of $\tilde{\mathbf{Z}}_n$ or some $\mathbf{Z}_2$ branch; suitable fifth order terms will determine whether $\mathbf{Z}_2(\kappa)$ or $\mathbf{Z}_2(\kappa, \pi)$ is stable. In addition, the signs of the real parts of the other eigenvalues of $(dG)_{0,0}$ must be computed in order to determine asymptotic stability. We shall not pursue this problem here.

Remark. In the notation of (4.1) we have

$$A = (dF)_{(0,\lambda)} + 2K$$

$$B = -K$$

so the relevant eigenvalues are those of

$$(dF)_{(0,\lambda)} \quad \text{(once)}, \qquad (dF)_{(0,\lambda)} + 3K \quad \text{(twice)}.$$

In the second case, where the bifurcation breaks symmetry to $\tilde{\mathbf{Z}}_3$, $\mathbf{Z}_2(\kappa)$, or $\mathbf{Z}_2(\kappa, \pi)$, it is possible for $(dF)_{(0,\lambda)}$ to have eigenvalues with negative real part. That is, if the coupling is removed ($K = 0$), then the individual cells need not be capable of oscillating on their own. This effect was noticed for two coupled cells by Smale [1974]. Loosely speaking, the coupling, rather than instabilities of individual components, can be the source of the oscillation.

(c) Numerical Simulations

We have performed some numerical experiments on a system of the form (4.1), taking

$$F(x, y, \lambda) = \begin{bmatrix} -4 & 1 \\ -1 & -4 \end{bmatrix} \begin{bmatrix} x \\ y \end{bmatrix} + p(x^2 + y^2) \begin{bmatrix} x \\ y \end{bmatrix}$$

$$+ q(x^2 + y^2) \begin{bmatrix} -y \\ x \end{bmatrix} - 2K(\lambda) \begin{bmatrix} x \\ y \end{bmatrix} \tag{4.8}$$

where

$$K(\lambda) = -\lambda \begin{bmatrix} -4 & 2 \\ -2 & -4 \end{bmatrix}.$$

When $\lambda = 1.05$, $p = -5$, $q = 30$ we observe $\tilde{\mathbf{Z}}_3$ symmetry. The three waveforms (plotted for the x-variable) are shown in Figure 4.2, and the fact that they are identical but phase-shifted by $2\pi/3$ is clear. Plotting trajectories in the three phase planes (x_p, y_p) we observe identical limit cycles, with the three phase points traversing them at three roughly equally spaced points.

When $\lambda = 1.2$, $p = -5$, $q = -50$ a stable periodic solution with $\mathbf{Z}_2(\kappa, \pi)$ symmetry is observed. (This observation does not imply that this solution is stable at the point of Hopf bifurcation. In fact, a calculation by M. Silber using the specific form of the equation (4.8) shows that it is not.) Now two cells traverse the same limit cycle π out of phase, and the third remains *steady* at the origin. By varying additional fifth order terms the $\mathbf{Z}_2(\kappa)$ solution should be obtainable, although we have not attempted this. (We remark that other kinds of behavior appear to be observed in the numerical solutions, notably quasiperiodic oscillation. This might be expected as a secondary bifurcation linking distinct primary branches.)

The fact that one cell is steady in this simulation deserves comment. Suppose that F is odd, so that $F(-x, -y) = -F(x, y)$. Then $\mathbf{D}_3 \times \mathbf{Z}_2 \subset \mathbf{D}_3 \times \mathbf{S}^1$ commutes with the vector field, where $\mathbf{Z}_2 = \{0, \pi\}$. It follows that the fixed-point suspace for $\Sigma = \mathbf{Z}_2(\kappa, \pi) \subset \mathbf{D}_3 \times \mathbf{Z}_2$ is invariant under the flow. But this subspace is the set of vectors $[0, v, -v]$. In other words, cell 0 stays at the origin, hence is steady.

Even if F is not odd, we can put the vector field into Birkhoff normal form

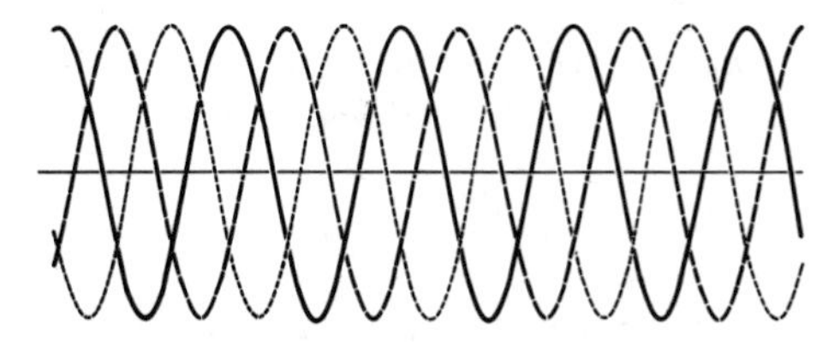

Figure 4.2. Numerical solution of (4.8) showing three identical waveforms, $2\pi/3$ out of phase. Here $\lambda = 1.05$, $p = -5$, $q = -50$.

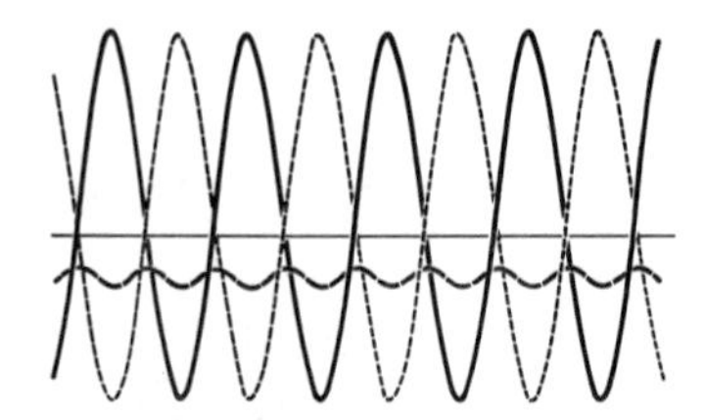

Figure 4.3. Numerical solution of (4.8) showing two identical waveforms, π out of phase, and a third with half the period. Here $\lambda = 1.1, p = -5, q = -50, r = 10, s = 0$.

up to arbitrarily high order so that in particular the normal form commutes with $\mathbf{Z}_2$. It is tempting to conclude that for the $\mathbf{Z}_2(\kappa, \pi)$ branch one cell is *always* steady. That this is *not* the case is shown by Figure 4.3, obtained from (4.8) by adding terms

$$r\begin{bmatrix}(x^2+y^2)^2\\0\end{bmatrix} + s\begin{bmatrix}0\\(x^2+y^2)y\end{bmatrix}$$

to F, again at $\lambda = 1.1$, $p = -5$, $q = -50$, $r = 10$, $s = 0$. Instead of the third cell's being steady, it exhibits *a small-amplitude oscillation at double the frequency* of the other two, and these latter are out of phase with each other. This is precisely what is predicted by the symmetry analysis. Thus in this example the dynamic behavior can definitely be changed by a symmetry-breaking "tail" of arbitrarily high order occurring in the reduction to Birkhoff normal form.

(d) The General Case

We now turn to the general case of n cells, described by (4.1). We will prove the following:

Theorem 4.2. *Suppose that* $L = (dG)_{(0,\lambda)}$ *has a pair of purely imaginary eigenvalues* $\pm i$ *(without loss of generality at* $\lambda = 0$*) which cross the imaginary axis with nonzero speed as* λ *passes through* 0. *Let A and B be the matrices of partial derivatives* $d_{x_1}G$ *and* $d_{x_0}G$, *restricted to the spaces of* x_1*- and* x_0*-variables, respectively. Then generically the purely imaginary eigenvalues of L are those of* $A + l_jB$ *where* $l_j = 2\cos 2\pi j/n$, $0 \le j \le n-1$. *They are simple if* $j = 0$, *or if* $j = n/2$ [*n even*], *double otherwise. The resulting oscillation patterns may be described by using the results of* §3, *applied to a particular representation* ρ_j *of* $\mathbf{D}_n$.

Remarks.

(a) To avoid group-theoretic generalities we describe the use of ρ_j and give examples in subsection (e).

(b) The double eigenvalue does actually occur, for all n; see Exercise 4.6.

PROOF. Let $\zeta = e^{2\pi i/n}$ be a primitive nth root of unity in $\mathbb{C}$. (We also use ζ to denote a generator of $\mathbf{D}_n$, but since this acts on $\mathbb{C}$ as multiplication by $e^{2\pi i/n}$ no confusion should arise.) Recall that

$$\zeta^{-1} = \bar{\zeta} = \zeta^{n-1}, \qquad \zeta^{-j} = (\overline{\zeta^j}) = \zeta^{n-j}.$$

To find the eigenvalues of L we complexify. Define the space of vectors

$$V_j = \{[v, \zeta^j v, \zeta^{2j} v, \ldots, \zeta^{(n-1)j} v] \colon v \in \mathbb{R}^k\}.$$

Then

$$\mathbb{C}^{nk} = V_0 \oplus V_1 \oplus \cdots \oplus V_{n-1}.$$

The linearization $L = (dG)_{(0,\lambda)}$ is

$$n \left\{ \begin{bmatrix} A & B & & & & & & B \\ B & A & B & & & & & \\ & B & A & B & & & & \\ & & B & A & B & & & \\ & & & & \cdots & & & \\ & & & & & \cdots & & \\ & & & & & B & A & B \\ B & & & & & & B & A \end{bmatrix} \right. \tag{4.9}$$

where A and B are $k \times k$ matrices. This banded ("circulant") structure of course comes from the nearest-neighbor coupling. Now

$$L[v, \zeta^j v, \ldots, \zeta^{(n-1)j} v] = (A + (\zeta^j + \zeta^{(n-1)j})B)[v, \zeta^j v, \ldots, \zeta^{(n-1)j} v],$$

whence the eigenvalues of $L|V_j$ are those of $A + (\zeta^j + \zeta^{(n-1)j})B$, that is, of $A + (\zeta^j + \bar{\zeta}^j)B = A + l_j B$ where $l_j = 2\cos 2\pi j/n$.

For example, when $n = 3$ we have

$$\begin{aligned} j = 0&: A + 2\cos 0 \cdot B &&= A + 2B \\ j = 1&: A + 2\cos(2\pi/3)B &&= A - B \\ j = 2&: A + 2\cos(4\pi/3)B &&= A - B \end{aligned}$$

as before.

Since $\cos(-\theta) = \cos\theta$ these occur in pairs, $l_j = l_{n-j}$, except for $j = 0$ [n odd] and $j = 0, n/2$ [n even].

Provided $k \geq 2$ there can be purely imaginary eigenvalues of $A + l_j B$ for any j. Generically these will be simple on V_j. Because of the way the l_j pair up, purely imaginary eigenvalues of L will generically be simple for $j = 0$ [n odd], and $j = 0, n/2$ [n even]; and of multiplicity 2 for all other $j = 1, 2, \ldots, [n/2]$ [n odd], $1, 2, \ldots, \frac{1}{2}n - 1$ [n even]. The simple eigenvalue case corresponds to ordinary Hopf bifurcation when $j = 0$ and to a type of $\mathbf{Z}_2$ Hopf bifurcation when $j = n/2$ (see following discussion); the double eigenvalue case is symmetric Hopf bifurcation of the type studied in Chapter XVI. □

The relevant action of $\mathbf{D}_n$ depends on the value of j, and we describe the results in the following. They may be proved by a complexification argument, by direct calculation, or by abstract representation theory. They are sufficiently natural that a proof would be superfluous. Let ρ_j be the representation of $\mathbf{D}_n$ on $\mathbb{C}$ given by

$$z \mapsto \zeta^j z \qquad \text{(rotation due to } \zeta \in \mathbf{Z}_n)$$

$$z \mapsto \bar{z} \qquad \text{(flip } \kappa)$$

when $j \neq 0, n/2$. Let ρ_0 be the trivial representation on $\mathbb{R}$. If n is even let $\rho_{n/2}$ be the representation on $\mathbb{R}$ given by

$$z \mapsto -z \qquad \text{(rotation)}$$

$$z \mapsto z \qquad \text{(flip).}$$

Suppose that a Hopf-type bifurcation occurs when eigenvalues of $A + l_j B$ cross the imaginary axis with nonzero speed, with generic assumptions on multiplicity (simple for $j = 0, n/2$; double for all other j). Then the action of $\mathbf{D}_n$ on the corresponding imaginary eigenspace is (isomorphic to) ρ_j. From this we can predict the possible patterns of oscillation. To do this, observe that ρ_j can be obtained by composing the standard representation of a suitable dihedral group $\mathbf{D}_n$ with a homomorphism

$$\varphi_j: \mathbf{D}_n \to \mathbf{D}_n$$

sending the rotation generator $\zeta \in \mathbf{D}_n$ to ζ^j and sending κ to itself. The group $\mathbf{D}$ is the image of $\mathbf{D}_n$ under φ_j and is a dihedral group $\mathbf{D}_q$ where

$$q = n/h, \qquad h = gcd(n,j).$$

Thus we may apply the general theory for $\mathbf{D}_q$ and reinterpret the results for the original system of n cells. Essentially the effect is to bunch them into sets of size h (corresponding to cosets of $\mathbf{Z}_h$ in $\mathbf{Z}_n$), all cells in one such set behaving identically; in addition the waveforms on the q sets of cells so obtained behave like the waveforms for a system of q cells with standard $\mathbf{D}_q$-action. Rather than prove these assertions in general (they are combinatorially somewhat complicated exercises in elementary group theory) we describe in the next section examples, for low n, which exhibit the typical features of the analysis.

(e) Examples

We have already studied the case $n = 3$; here we look at $n = 2$, 4, and 5.

$n = 2$. There are exceptional features to $\mathbf{D}_2$—for example, its irreducible representations are all one-dimensional—but the preceding general theory still applies provided we take care with the group actions. We have $l_0 = 2$, $l_1 = -2$, so the eigenvalues of L are those of $A \pm 2B$. The corresponding representations of $\mathbf{D}_2$ on $\mathbb{R}^2 \equiv \mathbb{C}$ are trivial ($\zeta \cdot z = z$) or "standard" ($\zeta \cdot z = -z$, $\kappa \cdot z = \bar{z}$). In the first case the isotropy group is $\mathbf{Z}_2$ (both cells have identical waveforms and are in phase). In the second case it is $\mathbf{Z}_2^c = \{(0,0),(\pi,\pi)\}$. So a spatial rotation by π (interchange cells) is compensated for by a phase shift of π. In other words, they oscillate with identical waveforms but π out of phase. We thus recover a result of Othmer and Scriven [1971].

$n = 4$. Now

$$l_0 = 2\cos 0 = 2 \qquad l_1 = 2\cos \pi/2 = 0$$
$$l_2 = 2\cos \pi = -2 \qquad l_3 = 2\cos 3\pi/2 = 0.$$

There are three cases: purely imaginary eigenvalues of $A + 2B$ (simple, yielding the representation ρ_0 of $\mathbf{D}_4$); A (double, ρ_1); and $A - 2B$ (simple, ρ_2). The typical oscillation patterns in these cases are given in Golubitsky and Stewart [1986b], Fig. 8.1.

The case ρ_2 deserves comment as it is a prototype for ρ_l on $\mathbf{D}_n$ when n and l have a common factor. The invariant subspace on which ρ_2 occurs consists of all vectors $[v, -v, v, -v]$ for $v \in \mathbb{R}^k$. On this, ζ acts as $-Id$ and κ acts trivially. So we get k copies of the one-dimensional representation

$$\zeta \cdot x = -x, \qquad \kappa \cdot x = x$$

on $\mathbb{R}$. There is a unique isotropy subgroup generated by ζ^2, κ, and (ζ, π).

$n = 5$. Here

$$l_0 = 2 \qquad l_1 = l_4 = 2\cos 2\pi/5 = \tau - 1$$
$$l_2 = l_3 = 2\cos 4\pi/5 = -\tau$$

where τ is the golden number $\frac{1}{2}(\sqrt{5} + 1)$. Again there are three cases: purely imaginary eigenvalues of $A + 2B$ (simple, ρ_0); $A + (\tau - 1)B$ (double, ρ_1); and $A - \tau B$ (double, ρ_2). The oscillation patterns for ρ_1 are shown schematically in Figure 0.2.

The case ρ_2 is a prototype for ρ_l on $\mathbf{D}_n$ when n and l are coprime but $l \neq 1$. We can obtain ρ_2 by composing the standard ρ_1 with the map $\zeta \to \zeta^2$. This has the effect of reordering the cells from 01234 to 02413 (replacing the pentagon with the pentacle). If the patterns for ρ_1 are relabeled in this order, the patterns for ρ_2 result. Note that the $\mathbf{Z}_2(\kappa)$ solutions for ρ_1 and ρ_2 yield the same pattern. This happens because the map $\gamma \mapsto \gamma^2$ defines an automorphism of $\mathbf{D}_n$ which preserves the orbits of $\mathbf{Z}_2(\kappa)$ and interchanges the two representations ρ_1 and ρ_2.

Alexander and Auchmuty [1986] have also considered Hopf bifurcation in rings of oscillators. They work in a different context, seeking solutions in which all cells have *identical* waveforms $x_p(t)$, with possible phase lags. This ansatz rules out solutions other than those with isotropy subgroup $\tilde{\mathbf{Z}}_n$ from those found previously. In their corollary to Theorem 2 they prove (in a "global" context) the existence of this $\tilde{\mathbf{Z}}_n$ branch. Further, when $n = 4$ (or more generally $n = 4k$) they find a torus of periodic solutions of the form

$$\begin{aligned} x_0(t) &= p(t) \\ x_1(t) &= p(t + \chi) \\ x_2(t) &= p(t + \pi) \\ x_3(t) &= p(t + \pi + \chi) \end{aligned} \tag{4.10}$$

where χ is an arbitrary phase shift. (When $\chi = \pi$ this is our $\mathbf{Z}_2(\kappa)$ solution.) However, the existence of these solutions requires restrictions both on the equations for each individual cell on $\mathbb{R}^p$ (they must be odd) and on the couplings between oscillators. See Fiedler [1987].

Exercises

4.1. Analyze possible oscillation patterns for a ring of six cells. Pay special attention to ρ_2 and ρ_3, whose subscripts divide 6.

4.2. Develop a similar analysis for interactions that are not nearest-neighbor but still have $\mathbf{D}_n$ symmetry. Show that the linearization L has the form

$$\begin{bmatrix} A & B_1 & B_2 & & \cdots & & B_3 & B_2 & B_1 \\ B_1 & A & B_1 & B_2 & B_3 & \cdots & & B_3 & B_2 \\ B_2 & B_1 & A & B_1 & B_2 & & \cdots & & B_3 \\ \cdots & \cdots & \cdots & \cdots & \cdots & \cdots & \cdots & \cdots & \cdots \\ \cdots & & B_3 & B_2 & B_1 & A & B_1 & B_2 & \\ B_2 & & \cdots & B_3 & B_2 & B_1 & A & B_1 & \\ B_1 & B_2 & & \cdots & B_3 & B_2 & B_1 & A & \end{bmatrix}$$

and that the eigenvalues of L are those of L_j, where

$$L_j = A + (\zeta^j + \zeta^{-j})B_1 + (\zeta^{2j} + \zeta^{-2j})B_2 + \cdots.$$

If L_j has purely imaginary eigenvalues show that (generically) oscillatons of pattern determined by the representation ρ_j of $\mathbf{D}_n$ occur, just as in the nearest-neighbor case.

4.3. Consider a system of four cells with complete symmetry under all permutations (equivariance under the symmetric group $\mathbf{S}_4$). Show that

$$L = \begin{bmatrix} A & B & B & B \\ B & A & B & B \\ B & B & A & B \\ B & B & B & A \end{bmatrix}.$$

Determine its eigenvalues and the corresponding representations of $\mathbf{S}_4$. Hence classify the $\mathbf{S}_4$ oscillation patterns.

(*Remark*: It is possible to perform a similar analysis for a system of identical cells coupled according to some (possibly directed) graph, in terms of the action of the automorphism group of that graph.)

4.4. Let $\mathbf{D}_n$ act irreducibly on $\mathbb{R}^2$. Complexify to get an action of $\mathbf{D}_n$ on $\mathbb{C}^2$ and consider the induced action of $\mathbf{Z}_n$ on $\mathbb{C}^2 = \mathbb{C} \oplus \mathbb{C}$. This, in suitable coordinates, must be given by

$$\xi(z_1, z_2) = (\zeta^p z_1, \zeta^q z_2), \tag{4.11}$$

where ξ generates $\mathbf{Z}_n$ and $\zeta = e^{2\pi i/n}$. Show that $q = -p$. Conversely, if $\mathbf{D}_n$ acts on $\mathbb{C}^2$ in such a way that $\mathbf{Z}_n$ acts by (4.11) with $q = -p$, show that there is a real subspace $\mathbb{R}^2$ (invariant under the flip κ) on which $\mathbf{D}_n$ acts via ρ_p.

4.5. Verify the assertion in the text that for eigenvalues of $A + l_j B$ crossing the imaginary axis, the relevant representation of $\mathbf{D}_n$ is ρ_j. (*Hint*: First consider the action of $\mathbf{Z}_n$ on a complexification, with eigenvectors

$$[v, \zeta^j v, \ldots \zeta^{(n-1)j} v], \qquad v \in \mathbb{R}^k.$$

Pair complex conjugates to obtain real invariant subspaces for $\mathbf{D}_n$ of dimension 2 and analyze the resulting action.)

4.6. Show by constructing explicit examples that the double eigenvalue case of Theorem 4.2 can occur for all n. (*Comment*: As stated, this is straightforward. To show that the double eigenvalue case can occur as the first bifurcation from a stable steady state is considerably more tricky.)

4.7. Analyze a system of n identical cells with unidirectional coupling, equivariant under $\mathbf{Z}_n$ rather than $\mathbf{D}_n$. Show that there exist various branches of rotating waves, one for each representation of $\mathbf{Z}_n$ that can occur.

§5.[†] Hopf Bifurcation with **O**(3) Symmetry

In this section we summarize, without proofs or calculational details, the information on Hopf bifurcation with spherical symmetry that can be obtained by the methods of Chapter XVI, §§4 and 9. We give for each irreducible representation V_l of the orthogonal group $\mathbf{O}(3)$, a list of isotropy subgroups $\Sigma \subset \mathbf{O}(3) \times \mathbf{S}^1$ for which our methods prove the existence of a branch of periodic solutions. We do not consider whether other periodic branches or other kinds of dynamics may occur; neither do we compute stabilities. Specifically, we classify those Σ for which $\dim \operatorname{Fix}(\Sigma) = 2$, when V is any irreducible representation of $\mathbf{O}(3)$.

We begin by recalling the results of Chapter XIII on representations of $\mathbf{O}(3)$. There are two distinct irreducible representations in each odd dimension $2l + 1$, defined in terms of the space V_l of spherical harmonics of degree l. This consists of homogeneous polynomials of degree l in x, y, z satisfying the Laplace equation. We decompose $\mathbf{O}(3)$ as $\mathbf{Z}_2^c \oplus \mathbf{SO}(3)$, where $\mathbf{Z}_2^c = \{\pm I\}$, and let $\gamma \in \mathbf{SO}(3)$ act on V_l in the standard way:

$$\gamma p(x, y, z) = p(\gamma(x, y, z))$$

for $p \in V_l$. If $-I \in \mathbf{Z}_2^c$ acts as the identity on V_l, we have the *plus* representation of $\mathbf{O}(3)$; if as minus the identity on V_l, we have the *minus* representation. The

Table 5.1. Isotropy Subgroups of $\mathbf{O}(3) \times \mathbf{S}^1$ on $V_l \oplus V_l$, Having Two-Dimensional Fixed-Point Subspaces

J (See note [3])	Type of K	Twist $\theta(H)$	Value of l Plus Representation	Minus Representation
$\mathbf{O}(2)$	II	$\mathbb{1}$	Even l	
$\mathbf{O}(2)$	I	$\mathbf{Z}_2$		Even l
$\mathbf{O}(2)$	II	$\mathbf{Z}_2$	Odd l	
$\mathbf{O}(2)$	III	$\mathbf{Z}_2$		Odd l
$\mathbf{SO}(2)$	II	$\mathbf{S}^1$ $[k = 1, \ldots, l]$ (See note [1])	All l	
$\mathbf{SO}(2)$	III	$\mathbf{S}^1$ $[k = 1, \ldots, l]$		All l
$\mathbb{I}$	II	$\mathbb{1}$	6, 10, 12, 16, 18, 20, 22, 24, 26, 28, 32, 34, 38, 44; 21, 25, 27, 31, 33, 35, 37, 39, 41, 43, 47, 49, 53, 59	
$\mathbb{I}$	I	$\mathbf{Z}_2$		6, 10, 12, 16, 18, 20, 22, 24, 26, 28, 32, 34, 38, 44; 21, 25, 27, 31, 33, 35, 37, 39, 41, 43, 47, 49, 53, 59
$\mathbb{O}$	II	$\mathbb{1}$	4, 6, 8, 10, 14; 9, 13, 15, 17, 19, 23	
$\mathbb{O}$	I	$\mathbf{Z}_2$		4, 6, 8, 10, 14; 9, 13, 15, 17, 19, 23
$\mathbb{O}$	II	$\mathbf{Z}_2$	6, 10, 12, 14, 16, 20; 3, 7, 9, 11, 13, 17	
$\mathbb{O}$	III	$\mathbf{Z}_2$		6, 10, 12, 14, 16, 20; 3, 7, 9, 11, 13, 17
$\mathbb{T}$	II	$\mathbf{Z}_3$	2, 4, 6; 5, 7, 9	
$\mathbb{T}$	I	$\mathbf{Z}_6$		2, 4, 6; 5, 7, 9
$\mathbf{D}_n$	I	$\mathbf{Z}_2$		$l/2 < n \leq l$
$\mathbf{D}_n$ (See note [2])	II	$\mathbf{Z}_2$	$l < n \leq 2l$	

Notes.

[1] For $\mathbf{S}^1$ twists, θ: $\mathbf{SO}(2) \to \mathbf{S}^1$ is given by $\theta(\psi) = k\psi$ and $k = 1, \ldots, l$ occur.

[2] For $\mathbf{D}_n$, II is $\mathbf{Z}_2^c \oplus \mathbf{Z}_n$.

[3] Here $H^\theta = \mathbf{Z}_2^c \oplus J$ is the isotropy subgroup, and $K = \ker \theta$.

natural representation, occurring in most applications, has sign $(-1)^l$. This is the action induced on V_l by the standard action of $\mathbf{O}(3)$ on $\mathbb{R}^3$.

Recall that $\mathbf{O}(3)$ has three types of subgroup:

(I) Subgroups of $\mathbf{SO}(3)$,
(II) Subgroups containing $\mathbf{Z}_2^c$,
(III) Subgroups intersecting $\mathbf{Z}_2^c$ trivially but not contained in $\mathbf{SO}(3)$.

These are classified up to conjugacy as follows:

Class I: $\mathbf{SO}(3)$, $\mathbf{O}(2)$, $\mathbf{SO}(2)$, $\mathbb{I}$, $\mathbb{O}$, $\mathbb{T}$, $\mathbf{D}_n$, $\mathbf{Z}_n$;
Class II: $\mathbf{Z}_2^c \oplus$ Type I;
Class III: $\mathbf{O}(2)^-$, $\mathbb{O}^-$, $\mathbf{D}_n^z$ $(n \geq 2)$, $\mathbf{D}_{2n}^d$ $(n \geq 2)$, $\mathbf{Z}_{2n}^-$ $(n \geq 1)$.

See Chapter XIII for notation and precise definitions.

Let H^θ be an isotropy subgroup of $\mathbf{O}(3) \times \mathbf{S}^1$ with two-dimensional fixed-point subspace. Here $H \subset \mathbf{O}(3)$ and $\theta: H \to \mathbf{S}^1$ is a twist, as in XVI, §8. It turns out that H is always of type II, so that $H = \mathbf{Z}_2^c \oplus J$ for $J \subset \mathbf{SO}(3)$. In almost all cases, H^θ is determined up to conjugacy by the type (I, II, or III) of $K = \ker \theta$. The results are summarized in Table 5.1. We have corrected a minor error in the original list of Golubitsky and Stewart [1985].

For concreteness, we summarize the classification for the "natural" $(-1)^l$ representation for $l \leq 6$ in Table 5.2. This is just a reformulation of parts of Table 5.1 and may easily be extended to all l by combining the even l entries in the "plus" column of Table 5.1 with the odd l entries in the "minus" column. (For this reason the even and odd entries are separated in Table 5.1.) However, the remarks of Chapter XV, §5, should be borne in mind: in typical applications (e.g., to the spherical Bénard problem) large values of l are likely to apply only in extremely small parameter ranges, and more complicated dynamics via mode interactions can be expected.

One amusing entry in Table 5.1 is the $\mathbf{Z}_3$-twisted tetrahedral group, occurring when $l = 2, 4, 5, 6, 7, 9$. This represents an oscillation having tetrahedral symmetry (cubic when λ is even since $(-I, \pi) \in \mathbf{O}(3) \times \mathbf{S}^1$ acts trivially) except for phase lags of $\pm 2\pi/3$ distributed in a certain way. Notice that the tetrahedral group does not occur as an isotropy subgroup (with one-dimensional fixed-point subspace) in static $\mathbf{O}(3)$ bifurcation; see Chapter XIII, §9.

The calculation, which is group-theoretic in nature, is described in some detail in Golubitsky and Stewart [1985], §15. (An error in §14, I, of that paper led to the inclusion of too many $\mathbf{Z}_2$-twisted dihedral subgroups. The "type IIA $\mathbf{D}_n$s" in their Table 14.1 are conjugate to subgroups of the "type IIB $\mathbf{D}_n$s" and must be omitted.) The main tools are the trace formula of Theorem XVI, 9.1, and an analysis of which twists θ are possible. Define the *twist type* of H to be the image $\theta(H) \subset \mathbf{S}^1$. For instance, H has twist type $\mathbf{Z}_3$ if $\theta(H) = \mathbf{Z}_3$, and so on. The possible twist types for H can be determined group-theoretically from the structure of H and are as in Table 5.3. With the exception of the $\mathbf{Z}_2^c \oplus \mathbf{Z}_n$ entry, only the twist types $\mathbb{1}$, $\mathbf{Z}_2$, $\mathbf{Z}_3$, and $\mathbf{Z}_6$ occur. These may be dealt with by using the trace formula, Theorem XVI, 9.1. A maximality

Table 5.2. Low-Dimensional Isotropy Subgroups of $\mathbf{O}(3) \times \mathbf{S}^1$ on $V_l \oplus V_l$ with the Natural $(-1)^l$ Representation, Having Two-Dimensional Fixed-Point Subspaces

l	Twist $\theta(H)$	J	Number of Branches Given by Equivariant Hopf Theorem
1	$\mathbf{Z}_2$	$\mathbf{O}(2)$	2
	$\mathbf{S}^1$ $[k = 1]$	$\mathbf{SO}(2)$	
2	$\mathbb{1}$	$\mathbf{O}(2)$	5
	$\mathbf{Z}_2$	$\mathbf{D}_4$	
	$\mathbf{Z}_3$	$\mathbb{T}$	
	$\mathbf{S}^1$ $[k = 1, 2]$	$\mathbf{SO}(2)$	
3	$\mathbf{Z}_2$	$\mathbf{O}(2)$, $\mathbb{O}$, $\mathbf{D}_2$, $\mathbf{D}_3$	7
	$\mathbf{S}^1$ $[1 \leq k \leq 3]$	$\mathbf{SO}(2)$	
4	$\mathbb{1}$	$\mathbf{O}(2)$, $\mathbb{O}$	9
	$\mathbf{Z}_2$	$\mathbf{D}_6$, $\mathbf{D}_8$	
	$\mathbf{Z}_3$	$\mathbb{T}$	
	$\mathbf{S}^1$ $[1 \leq k \leq 4]$	$\mathbf{SO}(2)$	
5	$\mathbf{Z}_2$	$\mathbf{O}(2)$, $\mathbf{D}_3$, $\mathbf{D}_4$, $\mathbf{D}_5$	10
	$\mathbf{Z}_6$	$\mathbb{T}$	
	$\mathbf{S}^1$ $[1 \leq k \leq 5]$	$\mathbf{SO}(2)$	
6	$\mathbb{1}$	$\mathbf{O}(2)$, $\mathbb{I}$, $\mathbb{O}$	14
	$\mathbf{Z}_2$	$\mathbb{O}$, $\mathbf{D}_8$, $\mathbf{D}_{10}$, $\mathbf{D}_{12}$	
	$\mathbf{Z}_3$	$\mathbb{T}$	
	$\mathbf{S}^1$ $[1 \leq k \leq 6]$	$\mathbf{SO}(2)$	

Note. Here $H = \mathbf{Z}_2^c \oplus J$ twisted by θ.

Table 5.3. Possible Twist Types for Closed Type III Subgroups of $\mathbf{O}(3)$

H	Twist Types
$\mathbf{O}(3)$	$\mathbb{1}$, $\mathbf{Z}_2$
$\mathbf{Z}_2^c \oplus \mathbf{O}(2)$	$\mathbb{1}$, $\mathbf{Z}_2$
$\mathbf{Z}_2^c \oplus \mathbf{SO}(2)$	$\mathbb{1}$, $\mathbf{S}^1$, $\mathbf{Z}_2$
$\mathbf{Z}_2^c \oplus \mathbb{I}$	$\mathbb{1}$, $\mathbf{Z}_2$
$\mathbf{Z}_2^c \oplus \mathbb{O}$	$\mathbb{1}$, $\mathbf{Z}_2$
$\mathbf{Z}_2^c \oplus \mathbb{T}$	$\mathbb{1}$, $\mathbf{Z}_2$, $\mathbf{Z}_3$, $\mathbf{Z}_6$
$\mathbf{Z}_2^c \oplus \mathbf{D}_n$	$\mathbb{1}$, $\mathbf{Z}_2$
$\mathbf{Z}_2^c \oplus \mathbf{Z}_n$	$\mathbb{1}$, $\mathbf{Z}_d$ ($d \mid 2n$ [n odd], $d \mid n$ [n even])

argument, based on the fact that $\mathbf{Z}_n \subset \mathbf{S}^1$, shows that twisted $\mathbf{Z}_2^c \oplus \mathbf{Z}_n$s do not occur. The remainder of the proof consists of a case-by-case calculation of $\dim \mathrm{Fix}_{V_l \oplus V_l}(H^\theta)$ and will not be given here.

The stability of the periodic solutions listed in Table 14.2 for $l = 2$ is analyzed by Iooss and Rossi [1987]. They find that four of the five solutions can be asymptotically stable for suitable parameter ranges, the exception being one of the rotating waves ($J = \mathbf{SO}(2)$ and $k = 1$). They also find emanating from this bifurcation quasiperiodic motions, which appear to correspond to an isotropy subgroup whose fixed-point subspace is of dimension higher than 2.

Exercises

5.1. Let V_2 be the five-dimensional representation of $\mathbf{O}(3)$ and consider $\mathbf{O}(3) \times \mathbf{S}^1$ acting on $V_2 \oplus V_2$. Show that $V_2 \oplus V_2$ may be identified with the space of 3×3 symmetric trace 0 matrices A with complex entries. Show that the action of $\mathbf{O}(3) \times \mathbf{S}^1$ can be put in the form

$$\gamma \cdot A = \gamma^{-1} A \gamma \qquad (\gamma \in \mathbf{O}(3))$$

$$\theta \cdot A = e^{i\theta} A \qquad (\theta \in \mathbf{S}^1).$$

5.2. Show that for any integers k_j, l_j the function $V_2 \oplus V_2 \to \mathbb{C}$

$$\mathrm{tr}(A^{k_1} \bar{A}^{l_1} \ldots A^{k_r} \bar{A}^{l_r})$$

is $\mathbf{O}(3)$-invariant. Show that under the action of θ this function is multiplied by $e^{(k_1 + \cdots + k_r - l_1 - \cdots - l_r)i\theta}$ and hence that suitable products of such functions are also $\mathbf{S}^1$-invariant. In particular, show that $\mathbf{O}(3) \times \mathbf{S}^1$ has the following $\mathbb{C}$-valued invariants:

Order 2: $\mathrm{tr}\, A\bar{A}$;
Order 4: $\mathrm{tr}(A^2\bar{A}^2)$, $\mathrm{tr}(A\bar{A})^2$, $(\mathrm{tr}\, A^2)(\mathrm{tr}\, \bar{A}^2)$, $(\mathrm{tr}\, A\bar{A})^2$;
Order 6: $(\mathrm{tr}\, A^3)(\mathrm{tr}\, \bar{A}^3)$, $(\mathrm{tr}\, A^2\bar{A})(\mathrm{tr}\, A\bar{A}^2)$, plus products of lower order invariants.

(*Remark*: The real and imaginary parts are then $\mathbb{R}$-valued invariants. In fact a Hilbert basis for $\mathbf{O}(3) \times \mathbf{S}^1$ acting on $V_2 \oplus V_2$ can be formed from functions of this type. The main result needed to complete the analysis is Formula (2.2) of Spencer and Rivlin [1959], which gives a Hilbert basis for $\mathbf{O}(3)$ acting on $V_2 \oplus V_2$.)

§6.† Hopf Bifurcation on the Hexagonal Lattice

In Case Study 4 we described the branching of steady-state solutions for the simplest bifurcation problems set on a (planar) hexagonal lattice. We indicated there how such steady states correspond to (small-amplitude) solutions of the Boussinesq equations, which, in turn, represent time-independent, spatially (doubly) periodic states for the Bénard problem in the infinite plane. In this section we extend the analysis to describe typical small-amplitude time-periodic solutions that may be obtained by Hopf bifurcation in differential equations posed on the hexagonal lattice. These results are due to Roberts, Swift, and Wagner [1986], and details may be found in their paper. Such periodic states are *not* of importance for the classical Bénard problem, since the first eigenvalue (in λ) of the Boussinesq equations is always real. However, in doubly diffusive convection in a horizontal planar layer, the first eigenvalue may be purely imaginary, and these results are then relevant. See Turner [1974], Huppert and Moore [1976], and Nagata and Thomas [1986]. We note here only that thermal diffusion is one of the diffusive processes, whereas the other may, for example, be salt or magnetic diffusion.

Roberts, Swift, and Wagner [1986] prove that generically at least eleven branches of periodic solutions are spawned from the simplest Hopf bifurcation on the hexagonal lattice. These periodic solutions fall into three classes:

(i) Four standing waves corresponding to branches of steady-state solutions on the hexagonal lattice. These correspond to fluid motions with associated patterns that are both time-independent and fixed in space.
(ii) Three traveling waves for which the spatial pattern of the corresponding fluid motion is time-independent but moves with constant speed in a fixed direction in the plane.
(iii) Four periodic solutions whose corresponding fluid motions have spatial patterns that *vary* periodically in time.

From a physical point of view, the type (iii) solutions are the most remarkable and the most interesting.

In the remainder of this section we sketch how these three classes of solutions may be found. First we clarify what we mean by bifurcation on the hexagonal lattice. Recall that if $k_1, k_2 \in \mathbb{R}^2$ are two unit vectors satisfying $k_1 \cdot k_2 = -\frac{1}{2}$, then the hexagonal lattice is

$$L = \{n_1 k_1 + n_2 k_2 \colon n_1, n_2 \in \mathbb{Z}\}.$$

Although we consider differential equations defined on $\mathbb{R}^2$ that are invariant under the full Euclidean group of rotations, reflections, and translations, we seek only solutions that are doubly periodic with respect to the lattice L. This restriction implies that any resulting bifurcation problem commutes with the group

$$\Gamma = \mathbf{T}^2 \dotplus \mathbf{D}_6$$

(where the $\dotplus$ indicates a semidirect product).

In Case Study 4, §1(a), we argued that the simplest steady-state bifurcation problems consistent with these assumptions reduce (via Liapunov–Schmidt) to Γ-equivariant bifurcation problems

$$h \colon V \times \mathbb{R} \to V$$

where V is a six-dimensional vector space generated by three independent roll states

$$e^{ik_1 \cdot X}, e^{ik_2 \cdot X}, \quad \text{and} \quad e^{i(k_1+k_2) \cdot X}.$$

Each exponential contributes two real dimensions to V, namely its real and imaginary parts. In Case Study 4, §1(b), we showed that Γ acts absolutely irreducibly on V.

Similarly, the simplest Hopf bifurcations (from Γ-invariant states) must, by our general theory, reduce to Γ-equivariant bifurcation problems

$$g: (V \oplus V) \times \mathbb{R} \to V \oplus V.$$

By Theorem XVI, 4.1, we may find periodic solutions on the hexagonal lattice by classifying those isotropy subgroups of $\Gamma \times \mathbf{S}^1$ acting on $V \oplus V$ that have two-dimensional fixed-point subspaces. It is this group-theoretic problem which is solved in Roberts, Swift, and Wagner [1986]; there are eleven such isotropy subgroups up to conjugacy.

Four of them—the type (i) solutions—may be found directly by recalling another issue raised in Case Study 4. The steady-state solutions on the hexagonal lattice depend crucially on whether the midplane reflection is a symmetry of the original differential equation. If it is, we find rolls, hexagons, regular triangles, and patchwork quilts; if not, then only rolls and hexagons occur generically.

When the midplane reflection acts on V, it does so as $-Id$. But in Hopf bifurcation the phase shift $\pi \in \mathbf{S}^1$ always acts on $V \oplus V$ as $-Id$. Hence the action of the midplane reflection is effectively identified with a phase shift. Thus, Theorem XVI, 9.1(a), implies the existence of four branches of time-periodic standing waves, having the spatial symmetries of rolls, hexagons, regular triangles, and patchwork quilts.

We now turn to the other seven types of solution. First we describe the way $(\mathbf{T}^2 \dotplus \mathbf{D}_6) \times \mathbf{S}^1$ acts on $V \oplus V$. Identify V with $\mathbb{C}^3$ and $V \oplus V$ with $\mathbb{C}^6$, and denote the complex coordinates on $\mathbb{C}^6$ by $(z, w) = (z_1, z_2, z_3, w_1, w_2, w_3)$. The action of $\mathbf{T}^2 \times \mathbf{S}^1$ is:

(a) $[\theta_1, \theta_2] \cdot (z, w) = (e^{-i\theta_1} z_1, e^{i(\theta_1+\theta_2)} z_2, e^{-i\theta_2} z_3, e^{i\theta_1} w_1, e^{-i(\theta_1+\theta_2)} w_2, e^{i\theta_2} w_3)$

for $[\theta_1, \theta_2] \in \mathbf{T}^2$; (6.1)

(b) $\psi \cdot (z, w) = e^{i\psi}(z, w)$ for $\psi \in \mathbf{S}^1$.

Now $\mathbf{D}_6$ is generated by $\mathbf{D}_3$ and an element r of order 2. Choose r so that

$$r(z, w) = (w, z). \tag{6.2}$$

The group $\mathbf{D}_3$ is isomorphic to the permutation group $\mathbf{S}_3$ on three symbols and acts on (z, w) by permuting the zs and ws separately. For example, the cyclic permutation (123) acts on $\mathbb{C}^6$ by

$$(123) \cdot (z, w) = (z_2, z_3, z_1, w_2, w_3, w_1). \tag{6.3}$$

Table 6.1 lists the eleven isotropy subgroups and their (two-dimensional)

Table 6.1. Isotropy Subgroups with Two-Dimensional Fixed-Point Subspaces for $(\mathbf{T}^2 \dot{+} \mathbf{D}_6) \times \mathbf{S}^1$ Acting on $\mathbb{C}^6$

Spatial Symmetries in $\mathbf{T}^2 \dot{+} \mathbf{D}_6$	Spatiotemporal Symmetries	Fixed-Point Subspace	Name
(1) $[0, \theta_2]$ (23) r	$\{[\pi, 0], \pi\}$	$z_2 = z_3 = 0$ $w = z$	Standing rolls
(2) $\mathbf{D}_6$	—	$z_1 = z_2 = z_3$ $w = z$	Standing hexagons
(3) $\mathbf{D}_3$	$\{r, \pi\}$	$z_1 = z_2 = z_3$ $w = -z$	Standing regular triangles
(4) (23) r	$\{[0, \pi], \pi\}$	$z_1 = 0$ $z_2 = z_3$ $w = z$	Standing patchwork quilt
(5) $[0, \theta_2]$ (23)	$\{[\psi, 0], \psi\}$	$z_2 = z_3 = 0$ $w = 0$	Traveling rolls
(6) (13)	$\{[\psi, \psi], \psi\}$	$z_2 = 0$ $z_1 = z_3$ $w = 0$	Traveling patchwork quilt (1)
(7) $r \circ (13)$	$\{[\psi, -\psi], \psi\}$	$z_2 = z_3 = 0$ $w = z$	Traveling patchwork quilt (2)
(8) $\mathbf{D}_3$	$\{[2\pi/3, 2\pi/3], 2\pi/3]$	$z_1 = z_2 = z_3$ $w = 0$	Oscillating triangles
(9) $r \circ [0, \pi]$	$\{(13) \circ [\pi/2, \pi/2], \pi/2]\}$	$z_1 = w_1 = z_3 = -w_3$ $z_2 = w_2 = 0$	Wavy rolls (1)
(10) —	$\{r \circ (231), \pi/3\}$	$z_2 = \omega z_1$ $z_3 = \omega^2 z_1$ $w = -z$ $(\omega = e^{2\pi i/3})$	Wavy rolls (2)
(11) r	$\{(312), 2\pi/3\}$	$z_2 = \omega z_1$ $z_3 = \omega^2 z_1$ $w = z$	Twisted patchwork quilt

fixed-point subspaces. The first column gives generators for the (spatial) symmetries in $\mathbf{T}^2 \dot{+} \mathbf{D}_6$, the second indicates generators for the (mixed spatio-temporal) symmetries in $(\mathbf{T}^2 \dot{+} \mathbf{D}_6) \times \mathbf{S}^1$ having a nonzero $\mathbf{S}^1$ component. The first four entries are the standing waves, the next three the traveling waves; the final four are the solutions with time-dependent patterns.

Roberts, Swift, and Wagner [1986] prove:

(i) No two of the isotropy subgroups listed in Table 6.1 are conjugate in $(\mathbf{T}^2 \dot{+} \mathbf{D}_6) \times \mathbf{S}^1$.

(ii) Every isotropy subgroup of $(\mathbf{T}^2 \dot{+} \mathbf{D}_6) \times \mathbf{S}^1$ having a two-dimensional fixed-point subspace is conjugate to one of these eleven subgroups.

(iii) Knowing the reduced bifurcation equation g up to order 3 is sufficient (generically) to determine the direction of branching of all eleven solutions.
(iv) Generically the asymptotic stability of seven of these solution branches is determined at third order in g. In certain instances, however, some fifth order terms in g are needed to determine whether the standing hexagons or the standing regular triangles are stable. Similarly, fifth order terms are sometimes needed to determine whether wavy rolls (2) or traveling patchwork quilts are asymptotically stable.
(v) Only the traveling patchwork quilts (1) are (generically) always unstable. Any of the other ten branches may, depending on the third and fifth order terms in g, be asymptotically stable.

It is worth emphasizing here that these results do not preclude the generic existence of periodic solutions with submaximal isotropy, nor for that matter the occurrence of dynamics more complicated than periodic in $V \oplus V$. Indeed there is no claim that all periodic states are obtained through Hopf bifurcation.

We end this section by illustrating the spatial patterns associated with the eleven types of periodic solution. To do so we identify (z, w) with the time-dependent real-valued function of $x \in \mathbb{R}^2$ given by

$$u(x, t) = \mathrm{Re}\left\{\sum_{j=1}^{3} (z_j e^{i(t-k_j \cdot x)} + w_j e^{i(t+k_j \cdot x)})\right\} \tag{6.4}$$

where k_1 and k_2 are the vectors in $\mathbb{R}^2$ that generate the hexagonal lattice, and $k_3 = k_1 + k_2$.

For each of the eleven periodic solutions we choose (z, w) in the corresponding fixed-point subspace and graph the level surfaces of $u(x, t)$ as a function of time. Of course, these level surfaces are independent of time for the standing solutions and move in a fixed direction (in $\mathbb{R}^2$) for the traveling solutions. The patterns for the standing waves are the same as the steady solutions (see Figure 2.1 of Case Study 4). Regarding the traveling waves, rolls travel perpendicular to their level surfaces, whereas patchwork quilt can travel parallel to either symmetry axis. The most interesting pictures are those associated with the remaining four type (iii) solutions. See Figures 6.1–6.4, taken from Roberts, Swift, and Wagner [1986].

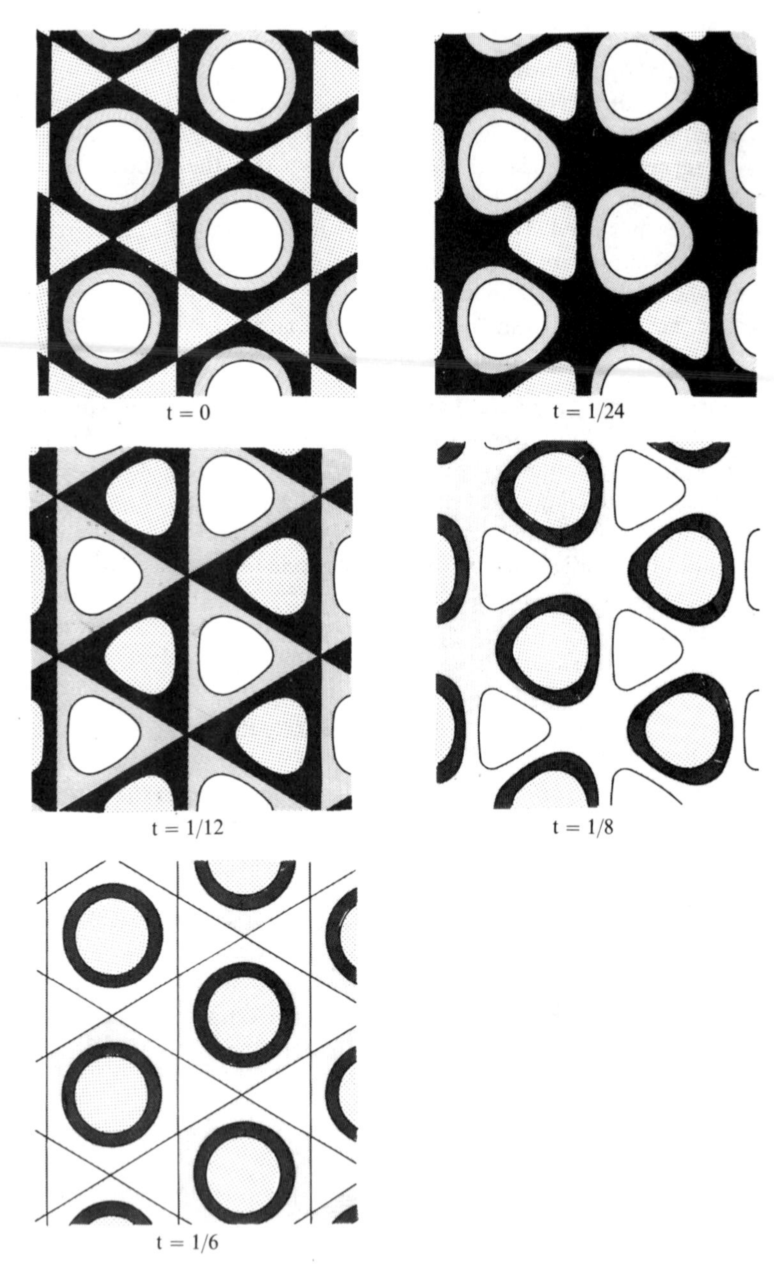

Figure 6.1. Hopf bifurcation on the hexagonal lattice: oscillating triangles.

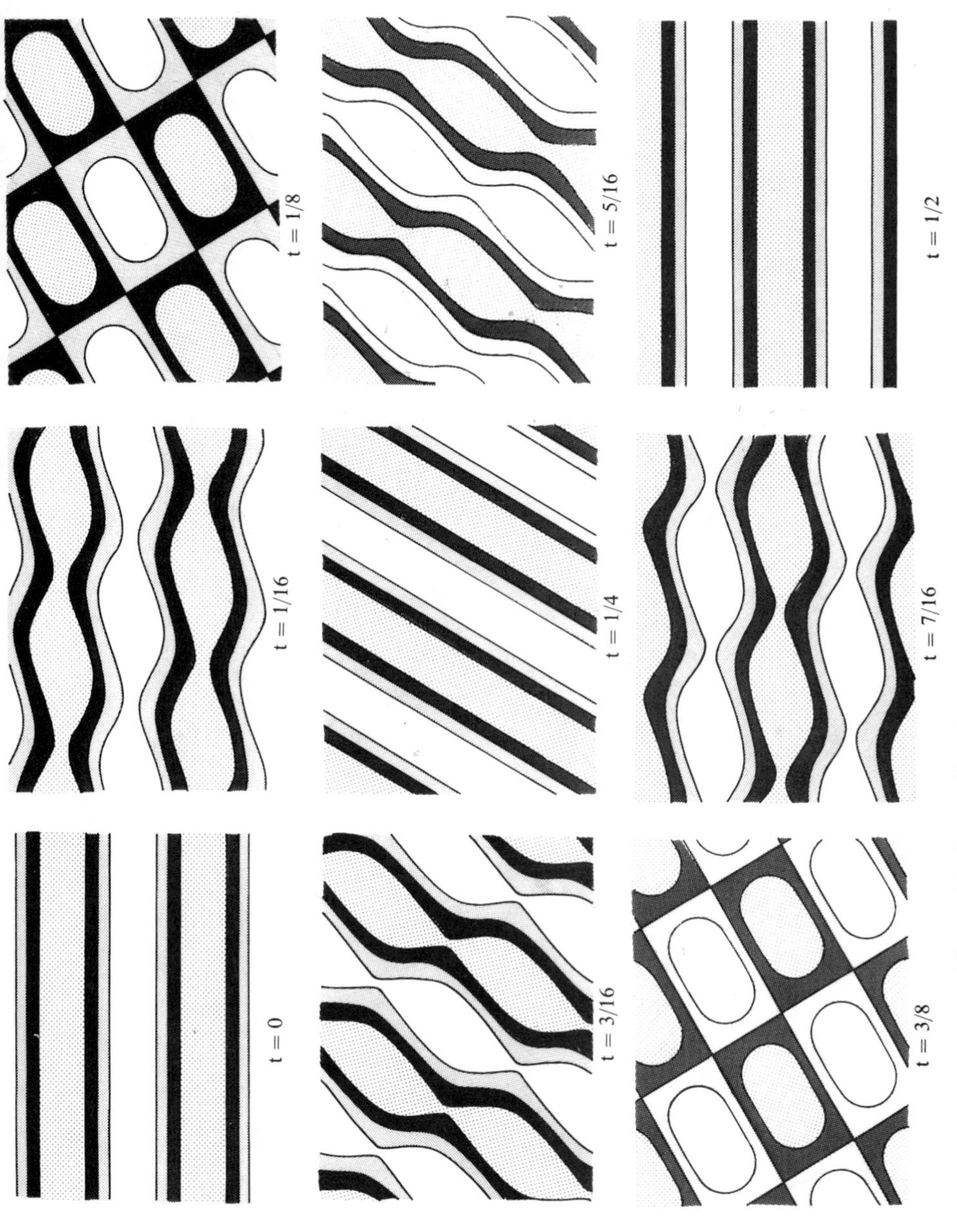

Figure 6.2. Hopf bifurcation on the hexagonal lattice: wavy rolls (1).

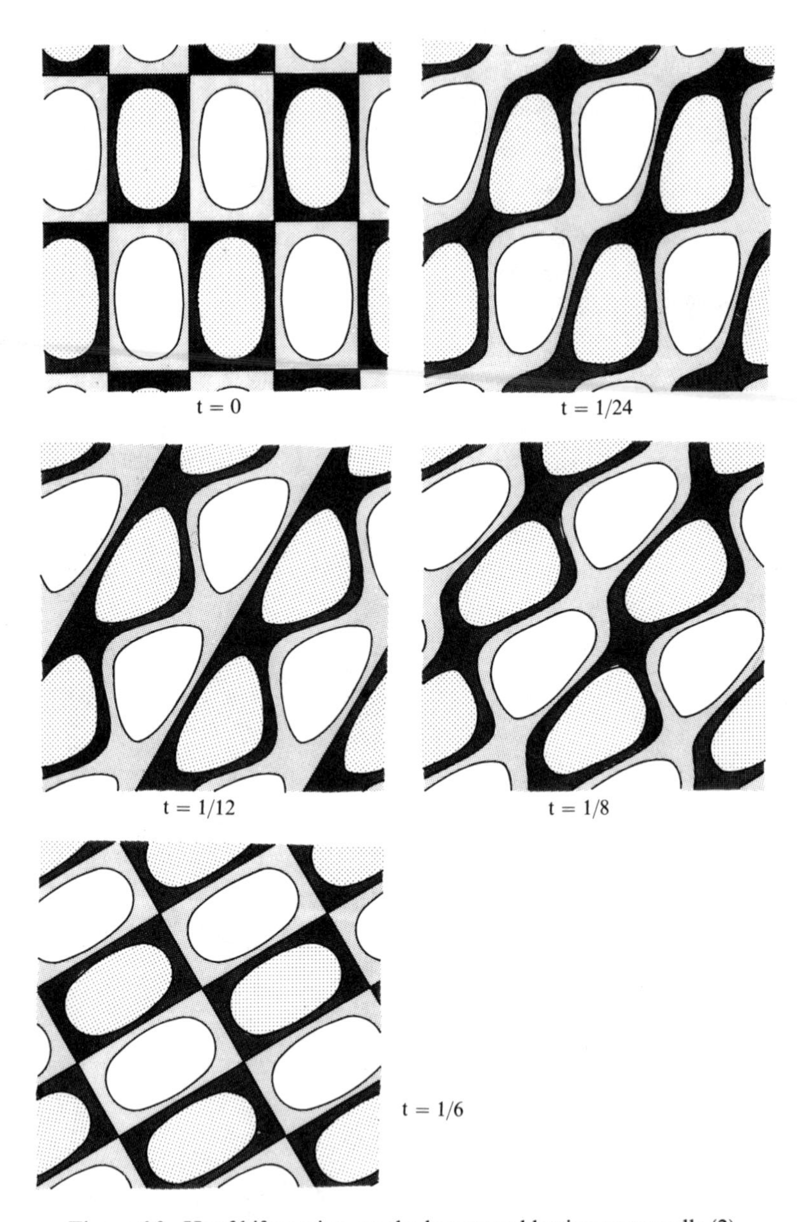

Figure 6.3. Hopf bifurcation on the hexagonal lattice: wavy rolls (2).

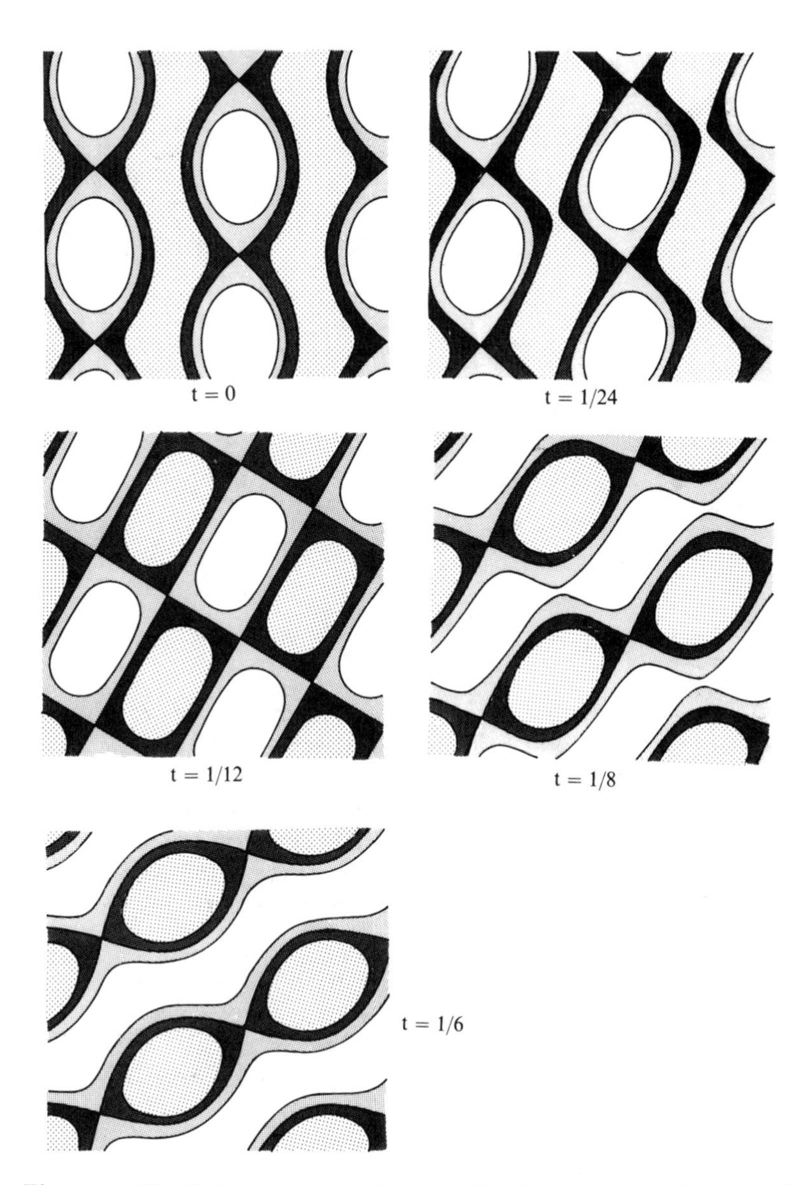

Figure 6.4. Hopf bifurcation on the hexagonal lattice: twisted patchwork quilt.

CHAPTER XIX

Mode Interactions

§0. Introduction

In this chapter we introduce the idea of mode interaction. In order to do this, we review some by now elementary material concerning steady-state and Hopf bifurcation without symmetry.

Perhaps the single most important problem in classical bifurcation theory is to identify the ways in which an equilibrium of a system of ODEs can lose stability as a parameter is varied. Roughly speaking, this loss of stability occurs by one of two methods: static bifurcation or Hopf bifurcation.

More precisely, let

$$\dot{x} + f(x, \lambda) = 0 \tag{0.1}$$

be a system of ODEs where $x \in \mathbb{R}^n$ and $\lambda \in \mathbb{R}$. Assume that $x = 0$ is a "trivial" steady state, so

$$f(0, \lambda) \equiv 0. \tag{0.2}$$

Assume further that the equilibrium $x = 0$ is asymptotically stable for $\lambda < 0$; that is, all eigenvalues of $(df)_{0,\lambda}$ have positive real part when $\lambda < 0$; and that $x = 0$ loses stability at $\lambda = 0$; that is, some eigenvalue of $(df)_{0,0}$ lies on the imaginary axis. Generically, there are two possibilities:

(a) $(df)_{0,0}$ has a simple zero eigenvalue and no other imaginary eigenvalues;
(b) $(df)_{0,0}$ has a simple eigenvalues $\pm\omega i$ ($\omega \neq 0$) and no other imaginary eigenvalues. (0.3)

We refer to (0.3(a)) as *steady-state* bifurcation and to (0.3(b)) as *Hopf* bifurcation. Thus we use these terms to describe degeneracies in the linear terms of the vector field f.

As we saw in Volume I, generic restrictions on the quadratic terms in f show that the simplest steady-state bifurcation is the transcritical case of simple bifurcation with normal form

$$\pm x^2 - \lambda x = 0. \tag{0.4}$$

Generic restrictions on the quadratic and cubic terms in f show that the simplest Hopf bifurcation is that with normal form

$$\pm x^3 - \lambda x = 0; \tag{0.5}$$

see Theorem XIII, 2.1.

We now pose the same question—how does a steady state lose stability?—for multiparameter systems. To simplify the discussion, we assume that f depends on just one additional parameter α and that

$$f(0, \lambda, \alpha) \equiv 0. \tag{0.6}$$

In the multiparameter system

$$\dot{x} + f(x, \lambda, \alpha) = 0 \tag{0.7}$$

assumptions that are generic for the one-parameter system (0.1) may fail generically, but (presumably) only at isolated values of α.

We saw earlier that there are two types of genericity assumption on (0.1): assumptions on the eigenvalue structure of $(df)_{0,\lambda}$ and assumptions on the nonlinear terms of f. Until now our discussion of degeneracies has focused almost exclusively on degeneracies in the *nonlinear* terms of f. We have shown that the ideas of determinacy and unfoldings yield quasiglobal information about the number and arrangement of solutions to (0.7).

In this chapter and the next we indicate how degeneracies associated with multiple eigenvalues on the imaginary axis lead to a similarly rich bifurcation structure. Indeed, the analysis of the full dynamics associated with such degeneracies is the basis of the modern local theory of dynamical systems. This program originated in the work of Smale, Arnold, Bogdanov, and Takens and is described in detail in Guckenheimer and Holmes [1983]. Our purpose here is to indicate which parts of this program can be understood by using singularity theory techniques. Our analysis is based on the Birkhoff normal form method described in XVI, §5.

In many fields of application the eigenvectors (of the linearized equation $\dot{x} + (df)_0 = 0$) corresponding to simple eigenvalues are called *modes*. A mode whose eigenvalues lie on the imaginary axis is said to be *critical*. Generically we expect, in a one-parameter system, to have only one critical mode. However, as discussed previously, we expect multiple critical modes in systems with more than one parameter (that is, α as well as λ). Our purpose here is to understand the types of secondary bifurcation that may arise in nonlinear systems near a parameter value at which there are multiple critical modes.

These secondary solutions are thought of as resulting from a nonlinear interaction of several critical modes. For short, this process is called *mode interaction.*

In the discussion of mode jumping in the buckling of a rectangular plate (see Case Study 3 and Chapter X) we analyzed one case of mode interaction, with $\mathbf{Z}_2 \oplus \mathbf{Z}_2$ symmetry. In that example a "mixed mode" was produced from the interaction of two "pure" modes.

Since there are two types of critical mode (steady-state and Hopf) there are three types of mode interaction in two-parameter systems:

(a) Steady-state/steady-state,
(b) Hopf/steady-state, (0.8)
(c) Hopf/Hopf.

When discussing these we assume, for simplicity, that all eigenvalues of $(df)_{0,0}$ are on the imaginary axis. That is, we assume that a center manifold reduction has already been performed; see Carr [1981] or Guckenheimer and Holmes [1983]. In consequence $n = 2, 3, 4$ for the cases (0.8(a, b, c)), respectively.

We now discuss each type briefly. Steady-state/steady-state interactions lead to a 2×2 system (0.7) in which both eigenvalues of $(df)_{0,0,0}$ are zero. In the absence of symmetry, generically df is nonzero and after a linear change of coordinates takes the form

$$(df)_{0,0,0} = \begin{bmatrix} 0 & 1 \\ 0 & 0 \end{bmatrix}. \tag{0.9}$$

Vector fields of the form

$$\dot{x} = \begin{bmatrix} 0 & 1 \\ 0 & 0 \end{bmatrix} x + \cdots \tag{0.10}$$

have been studied in detail by Takens [1974] and Bogdanov [1975]. In the unfolding of (0.10), the dynamics include homoclinic and periodic orbits. We refer the interested reader to Guckenheimer and Holmes [1983], p. 364.

Our methods do not apply to the analysis of this particular mode interaction. A steady-state analysis via Liapunov–Schmidt reduction leads to a bifurcation problem in one state variable of a standard type. On the other hand, an analysis of the periodic solutions in this problem using Hopf bifurcation techniques does not work, for the following reason. Some small perturbations of (0.9), namely

$$\begin{bmatrix} 0 & 1 \\ -\alpha^2 & 0 \end{bmatrix},$$

have purely imaginary eigenvalues at $\pm\alpha i$. Since the analysis must be carried out near $\alpha = 0$, these eigenvalues cannot be scaled (uniformly) to $\pm i$. (It is the uniformity, not the scaling, that is the problem: near $\alpha = 0$ the period of solutions tends to ∞.) As a result, the standard methods for analyzing Hopf bifurcation do not apply. The preferred method for studying (0.10) is through Birkhoff normal forms. Again, see Guckenheimer and Holmes [1983].

We end the discussion of steady-state/steady-state mode interactions with two remarks. In Chapter IX we studied the simplest steady-state singularities with two state variables (and without symmetry), that is, where $(df)_{0,0} = 0$. These singularities, called *hilltop bifurcation*, were shown to have codimension 3 and hence are not expected to appear in (0.7), which depends only on one auxiliary parameter α.

Finally, when symmetries are present, interaction of steady-state modes leading to $(df)_{0,0,0} \equiv 0$ is possible, even when the system depends on only one auxiliary parameter. We analyzed such an example with $\mathbf{Z}_2 \oplus \mathbf{Z}_2$ symmetry in Chapter X, showing that the simplest such singularities have topological codimension one. Such a singularity can, therefore, appear generically in (0.7). We discuss another example, this time with $\mathbf{O}(2)$ symmetry, in Chapter XX.

The main focus of this chapter is on Hopf/steady-state and Hopf/Hopf interactions without symmetry. In the Hopf/steady-state case, the eigenvalues of $(df)_{0,0,0}$ are $(0, \pm\omega i)$, where $\omega \neq 0$, and may be rescaled to $0, \pm i$. As we show in §1, using Birkhoff normal forms, this leads to a bifurcation problem of the form

$$g(y, \lambda, \alpha) = 0 \tag{0.11}$$

where $y \in \mathbb{R}^2$ and g commutes with the action of $\mathbf{Z}_2$ given by $(y_1, y_2) \mapsto (y_1, -y_2)$. This $\mathbf{Z}_2$ symmetry is all that remains of the $\mathbf{S}^1$ phase-shift symmetry of Hopf bifurcation. Here $y_2 = 0$ corresponds to the pure steady-state mode, and $y_1 = 0$ corresponds to the pure Hopf mode.

The simplest such singularities were studied by Langford [1979], who showed that in the unfoldings of these singularities, invariant 2-tori can appear via secondary bifurcation. These 2-tori can be detected by changes in stability along branches of periodic solutions. More recently, Dangelmayr and Armbruster [1983] have classified the $\mathbf{Z}_2$-singularities (0.11). These results are interpreted for Hopf/steady-state mode interactions in Armbruster, Dangelmayr, and Güttinger [1985]. We present part of these results, by giving the $\mathbf{Z}_2$-classification up to topological codimension two, in §2. In §3 we present the bifurcation diagrams for both the nondegenerate and certain degenerate cases in this mode interaction.

The Hopf/Hopf interaction has been studied by Takens [1974]. See also Langford and Iooss [1980] and Guckenheimer and Holmes [1983], p. 397. The main dynamic feature of this interaction is the occurrence of an invariant 3-torus. The system itself is four-dimensional, and $(df)_{0,0,0}$ is assumed to have eigenvalues $\pm\omega_1 i$, $\pm\omega_2 i$. Here we run into the issue of resonance. The important point is that when the system is nonresonant, the Birkhoff normal form commutes with the 2-torus $\mathbf{T}^2$, but when the system is resonant, it commutes only with the circle group $\mathbf{S}^1$. Compare Theorem XVI, 4.1.

Under the assumption of nonresonance these normal form results lead to amplitude/phase equations. It is the amplitude equations that are used to analyze these singularities. Remarkably, the amplitude equations have $\mathbf{Z}_2 \oplus \mathbf{Z}_2$ symmetry, allowing us to interpret Takens's results by using the results of Chapter X. We do this in §4.

The dynamical systems community usually refers to the mode interactions described here as *codimension two* degeneracies; see Guckenheimer and Holmes [1983], Chapter VII. In our language, which presumes a distinguished bifurcation parameter, these degeneracies have codimension one.

§1. Hopf/Steady-State Interaction

Consider the system of ODEs

$$\dot{u} + f(u, \lambda) = 0 \tag{1.1}$$

where $u = (u_0, u_1, u_2) \in \mathbb{R}^3$ and

$$\begin{aligned} &\text{(a)} \quad f(0,0) = 0 \\ &\text{(b)} \quad (df)_{0,0} = L \equiv \begin{bmatrix} 0 & 0 & 0 \\ 0 & 0 & -1 \\ 0 & 1 & 0 \end{bmatrix}. \end{aligned} \tag{1.2}$$

In effect, we assume that $u = 0$ is an equilibrium of (1.1) when $\lambda = 0$ and that $(df)_{0,0}$ has eigenvalues $0, \pm\omega i$. Then, by performing a linear change of coordinates and rescaling time in (1.1), we may assume $(df)_{0,0} = L$, as in (1.2).

Periodic solutions to (1.1) may be found by using either a Liapunov–Schmidt reduction, as observed by Langford [1979] and Shearer [1981], or Poincaré–Birkhoff normal form, as discussed in Guckenheimer and Holmes [1983] and references therein. Since our discussion of other mode interactions will require normal form theory, we use that approach here.

By Theorem XVI, 5.3, the Birkhoff normal form of (1.1) may be chosen to commute with $\mathbf{S}^1$, where $\mathbf{S}^1$ acts on the (u_1, u_2)-plane as rotations while leaving u_0 fixed. By Lemma VIII, 2.5, the normal form equations are

$$\dot{z} + p\begin{bmatrix} 1 \\ 0 \\ 0 \end{bmatrix} + q\begin{bmatrix} 0 \\ u_1 \\ u_2 \end{bmatrix} + r\begin{bmatrix} 0 \\ -u_2 \\ u_1 \end{bmatrix} = 0 \tag{1.3}$$

where p, q, r are functions of u_0, $u_1^2 + u_2^2$, λ. By (1.2)

$$p(0) = p_{u_0}(0) = q(0) = 0, \qquad r(0) = 1. \tag{1.4}$$

Of course, we can assume only that (1.3) is valid up to any finite order. We do not address here the issues raised by the remaining high order "tail."

As with standard Hopf bifurcation, the system (1.3) is best analyzed by changing to polar coordinates and deriving amplitude/phase equations. Thus we let

$$\begin{aligned} &\text{(a)} \quad x = u_0 \\ &\text{(b)} \quad u_1 + iu_2 = ye^{i\theta}. \end{aligned} \tag{1.5}$$

Now (1.3) becomes

$$\begin{aligned} &\text{(a)} \quad \dot{x} + p(x, y^2, \lambda) = 0 \\ &\text{(b)} \quad \dot{y} + q(x, y^2, \lambda)y = 0 \\ &\text{(c)} \quad \dot{\theta} + r(x, y^2, \lambda) = 0. \end{aligned} \tag{1.6}$$

Since $r(0) = 1$, the dynamics of (1.6) is determined by the system of two equations

$$\begin{aligned} &\text{(a)} \quad \dot{x} + p(x, y^2, \lambda) = 0 \\ &\text{(b)} \quad \dot{y} + q(x, y^2, \lambda)y = 0. \end{aligned} \tag{1.7}$$

As a system, the function in (1.7) has the form

$$g(x, y, \lambda) = (p(x, y^2, \lambda), q(x, y^2, \lambda)y) \tag{1.8}$$

where, using (1.4),

$$p(0) = p_x(0) = q(0) = 0. \tag{1.9}$$

Now (1.8 and 1.9) give the general bifurcation problem on $\mathbb{R}^2$ with $\mathbf{Z}_2$ symmetry, where the reflection acts as

$$(x, y) \mapsto (x, -y).$$

This symmetry is all that is left of the $\mathbf{S}^1$ symmetry in the Birkhoff normal form (1.6) and corresponds to phase shift by half a period.

Zeros (x, y, λ) of (1.8) correspond to steady-state solutions to (1.6) if $y = 0$ and to periodic solutions if $y \neq 0$. These steady states and periodic solutions of (1.6) are asymptotically stable precisely when the corresponding zeros of (1.8) are asymptotically stable. This fact is clear geometrically: compare Lemma XVII, 4.1.

We write the equations for these steady-state and periodic solutions more explicitly:

$$\begin{aligned} &\text{(a)} \quad y = 0, \qquad p(x, 0, \lambda) = 0 \qquad \text{[steady-state]} \\ &\text{(b)} \quad y \neq 0, \qquad p(x, y^2, \lambda) = q(x, y^2, \lambda) = 0 \qquad \text{[periodic]}. \end{aligned} \tag{1.10}$$

In the next section we describe the singularity theory analysis of bifurcation problems with this $\mathbf{Z}_2$ symmetry.

§2. Bifurcation Problems with $\mathbf{Z}_2$ Symmetry

In this section we classify bifurcation problems in two state variables with $\mathbf{Z}_2$ symmetry $(x, y) \mapsto (x, -y)$. The main result is Table 2.1, which lists normal forms with topological $\mathbf{Z}_2$-codimension ≤ 2. Solutions of the recognition problem, and universal unfoldings, for all these singularities are in Tables 2.2

Table 2.1. $\mathbf{Z}_2$-Bifurcation Problems with Topological $\mathbf{Z}_2$-Codimension ≤ 2

Normal Form $[p, q]$	$\mathbf{Z}_2$-Codimension	Topological $\mathbf{Z}_2$-Codimension
(1) $[\varepsilon_1 u^2 + \varepsilon_2 v + \varepsilon_3 \lambda, \varepsilon_4 u]$	1	1
(2) $[\varepsilon_1 u^3 + \varepsilon_2 v + \varepsilon_3 \lambda, \varepsilon_4 u]$	2	2
(3) $[\varepsilon_1 u^2 + \varepsilon_2 v^2 + \varepsilon_3 \lambda, \varepsilon_4 u]$	2	2
(4) $[\varepsilon_1 u^2 + \varepsilon_2 v + m\lambda^2, \varepsilon_3 u + \varepsilon_4 \lambda]$ $m \neq 0, -\varepsilon_1$	3	2
(5) $[\varepsilon_1 u^2 + \varepsilon_2 v + \varepsilon_3 \lambda, mu^2 + \varepsilon_4 v]$ $m \neq 0, \varepsilon_1\varepsilon_2\varepsilon_4$	3	2
Each $\varepsilon_j = \pm 1$		

Table 2.2. Determinacy Conditions for $\mathbf{Z}_2$-Bifurcation Problems Listed in Table 2.1; the defining conditions always include $p = p_u = q = 0$, all derivatives are evaluated at the origin

Normal Form $[p, q]$	Defining Conditions	Nondegeneracy Conditions	
(1)	—	$\varepsilon_1 = \operatorname{sgn} p_{uu}$	$\varepsilon_2 = \operatorname{sgn} p_v$
		$\varepsilon_3 = \operatorname{sgn} p_\lambda$	$\varepsilon_4 = \operatorname{sgn} q_u$
(2)	$p_{uu} = 0$	$\varepsilon_1 = \operatorname{sgn} p_{uuu}$	$\varepsilon_2 = \operatorname{sgn} p_v$
		$\varepsilon_3 = \operatorname{sgn} p_\lambda$	$\varepsilon_4 = \operatorname{sgn} q_u$
(3)	$p_v = 0$	$\varepsilon_1 = \operatorname{sgn} p_{uu}$	$\varepsilon_3 = \operatorname{sgn} p_\lambda$
		$\varepsilon_4 = \operatorname{sgn} q_u$	$\varepsilon_2 = \operatorname{sgn}(p_{vv} q_u^2 - 2p_{uv} q_u + p_{uu} q_v^2)$
(4)	$p_\lambda = 0$	$\varepsilon_1 = \operatorname{sgn} p_{uu}$	$\varepsilon_2 = \operatorname{sgn} p_v$
	$m = \varepsilon_1 \dfrac{[p_{uu} p_{\lambda\lambda} - p_{u\lambda}^2] q_u^2}{[q_\lambda p_{uu} - q_u p_{u\lambda}]^2}$	$\varepsilon_3 = \operatorname{sgn} q_u$ $m \neq 0, -\varepsilon_1$	$\varepsilon_4 = \varepsilon_1 \operatorname{sgn}(q_\lambda p_{uu} - q_u p_{u\lambda})$
(5)	$q_u = 0$	$\varepsilon_1 = \operatorname{sgn} p_{uu}$	$\varepsilon_2 = \operatorname{sgn} p_v$
	$m = \varepsilon_3 \dfrac{\lvert p_v \rvert (q_{uu} p_\lambda - q_\lambda p_{uu})}{\lvert p_{uu} \rvert \lvert q_v p_\lambda - q_\lambda p_v \rvert}$	$\varepsilon_3 = \operatorname{sgn} p_\lambda$ $m \neq 0, \varepsilon_1\varepsilon_2\varepsilon_4$	$\varepsilon_4 = \varepsilon_3 \operatorname{sgn}(q_v p_\lambda - q_\lambda p_v)$

Table 2.3. Universal $\mathbf{Z}_2$-Unfoldings of $\mathbf{Z}_2$-Bifurcation Problems Listed in Table 2.1; here α, β are near 0 and μ is near m

(1)	$[\varepsilon_1 u^2 + \varepsilon_2 v + \varepsilon_3 \lambda, \varepsilon_4 u + \alpha]$
(2)	$[\varepsilon_1 u^3 + \varepsilon_2 v + \varepsilon_3 \lambda + \beta u, \varepsilon_4 u + \alpha]$
(3)	$[\varepsilon_1 u^2 + \varepsilon_2 v^2 + \varepsilon_3 \lambda + \beta v, \varepsilon_4 u + \alpha]$
(4)	$[\varepsilon_1 u^2 + \varepsilon_2 v + \mu\lambda^2 + \beta, \varepsilon_3 u + \varepsilon_4 \lambda + \alpha]$
(5)	$[\varepsilon_1 u^2 + \varepsilon_2 v + \varepsilon_3 \lambda, \mu u^2 + \beta u + \varepsilon_4 v + \alpha]$

and 2.3, respectively. The verification of these results is a (now somewhat routine, though extensive) exercise in singularity theory; nonetheless it is included for completeness. The reader may very well prefer just to look at the list and then continue rapidly to §3, where the consequences for Hopf/steady-state mode interactions are discussed.

The singularity theory mechanics for this classification were set up in Golubitsky and Schaeffer [1979b]; the actual classification, up to topological $\mathbf{Z}_2$-codimension five, was completed by Dangelmayr and Armbruster [1983].

We recall from (1.7) that the general form of a bifurcation problem $g(x, y, \lambda)$ with this $\mathbf{Z}_2$ symmetry is

$$[p, q] = (p(u, v, \lambda), q(u, v, \lambda)y) \tag{2.1}$$

where $u = x$, $v = y^2$ is a Hilbert basis for $\mathscr{E}(\mathbf{Z}_2)$. As just remarked, the details of the $\mathbf{Z}_2$-classification are shown in Tables 2.1–2.3.

In the remainder of this section we establish the results in Tables 2.1–2.3. We begin by solving the recognition problem for the normal forms in Table 2.1. This is the most difficult step. Our method for solving the recognition problem for a normal form g involves two main steps:

(a) Identify the higher order terms for g by computing

$$\operatorname{Itr} \mathscr{K}(g, \Gamma). \tag{2.2}$$

Here we use Theorem XIV, 7.4, to show that terms in (2.2) are among the higher order terms of g.

(b) Use explicit $\mathbf{Z}_2$-changes of coordinates to transform intermediate terms into the given normal form. In this step we derive formulas for the modal parameters and the nondegeneracy conditions.

To make sense of (2.2) we must recall how the restricted tangent space $RT(g, \mathbf{Z}_2)$ is computed, and we must also determine which submodules of $\vec{\mathscr{E}}(\mathbf{Z}_2)$ are intrinsic. In Chapter XIV, Table 3.1, we observed that $RT([p, q], \mathbf{Z}_2)$ is the submodule of $\vec{\mathscr{E}}(\mathbf{Z}_2)$ generated by

$$\begin{aligned} &\text{(a)} \quad [p, 0], [vq, 0], [0, p], [0, q], \\ &\text{(b)} \quad [vp_v, vq_v] \quad \text{and} \quad u, v, \lambda \text{ times } [p_u, q_u]. \end{aligned} \tag{2.3}$$

Then for $\Gamma = \mathbf{Z}_2$ we have

$$\mathscr{K}(g, \mathbf{Z}_2) = \mathscr{M} \cdot RT(g, \mathbf{Z}_2) + \mathbb{R}\{[vq, 0], [0, p]\} \tag{2.4}$$

since the matrix generators $\left[\begin{smallmatrix} 0 & y \\ 0 & 0 \end{smallmatrix}\right]$ and $\left[\begin{smallmatrix} 0 & 0 \\ y & 0 \end{smallmatrix}\right]$ of $\vec{\mathscr{E}}(\mathbf{Z}_2)$ have degree ≥ 1.

Lemma 2.1.

(i) *All sums and products of the ideals $\mathscr{M}$, $\langle \lambda \rangle$, and $\langle v \rangle$ in $\mathscr{E}(\mathbf{Z}_2)$ are intrinsic.*

(ii) *For each intrinsic ideal $\mathscr{I} \subset \mathscr{E}(\mathbf{Z}_2)$, the submodules*

$$[\mathscr{I}, \mathscr{I}] \quad \text{and} \quad [\mathscr{I}v, \mathscr{I}] \tag{2.5}$$

are intrinsic.

(iii) *The submodule* $[\mathscr{I}_1, \mathscr{I}_2]$ *is intrinsic if*

$$\begin{aligned} &\text{(a)} \quad \mathscr{I}_1 \text{ and } \mathscr{I}_2 \text{ are intrinsic ideals,} \\ &\text{(b)} \quad \mathscr{I}_1 \subset \mathscr{I}_2 \\ &\text{(c)} \quad \langle v \rangle \mathscr{I}_2 \subset \mathscr{I}_1. \end{aligned} \tag{2.6}$$

PROOF.
(i) Recall that an ideal $\mathscr{I} \subset \mathscr{E}(\mathbf{Z}_2)$ is intrinsic if for each $g \in \mathscr{I}$ and each $\mathbf{Z}_2$ change of coordinates $\Phi \in \mathscr{E}(\mathbf{Z}_2)$ we have $g(\Phi(x, y, \lambda), \lambda) \in \mathscr{I}$. The ideals $\mathscr{M} = \{g \in \mathscr{E}(\mathbf{Z}_2) | g(0) = 0\}$ and $\langle \lambda \rangle$ are invariant under all changes of coordinates and are thus intrinsic.

To show that $\langle v \rangle$ is intrinsic, recall that $\mathbf{Z}_2$-changes of coordinates Φ have the form

$$\Phi(x, y, \lambda) = (A(u, v, \lambda), B(u, v, \lambda)y) \tag{2.7}$$

where $A(0) = 0$, $A_u(0) > 0$, $B(0) > 0$. Since $v = y^2$ we have

$$\begin{aligned} &\text{(a)} \quad u \circ \Phi = A(u, v, \lambda) \\ &\text{(b)} \quad v \circ \Phi = B(u, v, \lambda)^2 v. \end{aligned} \tag{2.8}$$

Now (2.8(b)) shows that $\langle v \rangle$ is intrinsic.

(ii) Definition XIV, 6.1, states that a submodule $\mathscr{J}$ of $\vec{\mathscr{E}}(\mathbf{Z}_2)$ is intrinsic if every germ strongly $\mathbf{Z}_2$-equivalent to a germ in $\mathscr{J}$ is also in $\mathscr{J}$. Now every $\mathbf{Z}_2$-equivalence has the form

$$g \mapsto Sg(\Phi)$$

where Φ is a $\mathbf{Z}_2$ change of coordinates and $S \in \vec{\mathscr{E}}(\mathbf{Z}_2)$. Thus to show $\mathscr{J}$ is intrinsic we show that it is invariant under the transformations

$$\begin{aligned} &\text{(a)} \quad g \mapsto g \circ \Phi \\ &\text{(b)} \quad g \mapsto Sg. \end{aligned} \tag{2.9}$$

Using the notation $g = [p, q]$ and $\Phi = [A, B]$ we calculate

$$g \circ \Phi = [p(\Phi), q(\Phi)B]. \tag{2.10}$$

Next, recall from Chapter XIV, Table 3.1, that the general matrix $S \in \vec{\mathscr{E}}(\mathbf{Z}_2)$ has the form

$$S = C\begin{bmatrix} 1 & 0 \\ 0 & 0 \end{bmatrix} + D\begin{bmatrix} 0 & y \\ 0 & 0 \end{bmatrix} + E\begin{bmatrix} 0 & 0 \\ y & 0 \end{bmatrix} + F\begin{bmatrix} 0 & 0 \\ 0 & 1 \end{bmatrix}, \tag{2.11}$$

where C, D, E, and F are $\mathbf{Z}_2$-invariant functions with $C(0) > 0$ and $F(0) > 0$. Now calculate

$$S[p, q] = [Cp + Dvq, Ep + Fq]. \tag{2.12}$$

From (2.10 and 2.12) the ideals in (2.5) are intrinsic.

(iii) Let $\mathscr{I} = [\mathscr{I}_1, \mathscr{I}_2]$ be an intrinsic submodule of $\vec{\mathscr{E}}(\mathbf{Z}_2)$. By (2.10), $\mathscr{I}_1$ and $\mathscr{I}_2$ are intrinsic ideals in $\mathscr{E}(\mathbf{Z}_2)$ since $B(0) \neq 0$. Since D in (2.12) is arbitrary we have $\langle v \rangle \mathscr{I}_2 \subset \mathscr{I}_1$, and since E in (2.12) is also arbitrary, $\mathscr{I}_1 \subset \mathscr{I}_2$. We have also proved that

$$\mathscr{I} = [\mathscr{I}_1, \mathscr{I}_1] + [\langle v \rangle \mathscr{I}_2, \mathscr{I}_2],$$

a sum of submodules of the form (2.5). □

We can now calculate $\operatorname{Itr} \mathscr{K}(g, \mathbf{Z}_2)$ for the normal forms in Table 2.1. The results are listed in Table 2.4.

The second step in solving the recognition problem is to put the intermediate order terms in normal form by using explicit $\mathbf{Z}_2$-equivalences. The data in Table 2.4 let us enumerate the intermediate terms for each normal form in Table 2.1. The results are shown in Table 2.5. Note that we do not include terms set to zero by the defining conditions.

We start by writing down the general $\mathbf{Z}_2$-equivalence. This formula follows directly from (2.10 and 2.12) and includes the change of coordinates $\lambda \mapsto \Lambda(\lambda)$, where $\Lambda(0) = 0$, $\Lambda'(0) > 0$. Namely,

Table 2.4. Algebraic Data for Normal Forms of Table 2.1

Normal Form	$RT(g, \mathbf{Z}_2)$	$\operatorname{Itr} \mathscr{K}(g, \mathbf{Z}_2)$
(1)	$[\mathscr{M}^2 + \mathscr{M}\langle v, \lambda \rangle, \mathscr{M}]$	$[\mathscr{M}^3 + \mathscr{M}\langle v, \lambda \rangle, \mathscr{M}^2]$
(2)	$[\mathscr{M}^3 + \mathscr{M}\langle v, \lambda \rangle, \mathscr{M}]$	$[\mathscr{M}^4 + \mathscr{M}\langle v, \lambda \rangle, \mathscr{M}^2]$
(3)	$[\mathscr{M}^2 + \langle \lambda \rangle, \mathscr{M}]$	$[\mathscr{M}^3 + \mathscr{M}\langle \lambda \rangle, \mathscr{M}^2 + \langle \lambda \rangle]$
(4)	$[\mathscr{M}^3 + \langle v \rangle, \mathscr{M}^2 + \langle v \rangle]$ $+ \mathbb{R}\{[\varepsilon_1 u^2 + m\lambda^2, 0], [0, \varepsilon_3 u + \varepsilon_4 \lambda],$ $[2\varepsilon_1 u^2, \varepsilon_3 u], [2\varepsilon_1 u\lambda, \varepsilon_3 \lambda]\}$	$[\mathscr{M}^3 + \mathscr{M}\langle v \rangle, \mathscr{M}^2 + \langle v \rangle]$
(5)	$[\mathscr{M}^3 + \mathscr{M}\langle v, \lambda \rangle, \mathscr{M}^3 + \mathscr{M}\langle v, \lambda \rangle]$ $+ \mathbb{R}\{[\varepsilon_1 u^2 + \varepsilon_2 v + \varepsilon_3 \lambda, 0],$ $[0, \varepsilon_1 u^2 + \varepsilon_2 v + \varepsilon_3 \lambda],$ $[0, mu^2 + \varepsilon_4 v], [\varepsilon_1 u^2, mu^2],$ $[\varepsilon_2 v, \varepsilon_4 v]\}$	$[\mathscr{M}^3 + \mathscr{M}\langle v, \lambda \rangle, \mathscr{M}^3 + \mathscr{M}\langle v, \lambda \rangle]$

Table 2.5. Intermediate Order Terms for Normal Forms Listed in Table 2.1

Normal Form	Terms in $p(u, v, \lambda)$	Terms in $q(u, v, \lambda)$
(1)	v, λ, u^2	u, v, λ
(2)	v, λ, u^3	u, v, λ
(3)	λ, u^2, uv, v^2	u, v
(4)	v, u^2, $u\lambda$, λ^2	u, λ
(5)	v, λ, u^2	v, λ, u^2

$$[p,q] \mapsto [Cp(A,B^2v,\Lambda) + Dvq(A,B^2v,\Lambda)B, Ep(A,B^2v,\Lambda) + Fq(A,B^2v,\Lambda)B]. \tag{2.13}$$

Since $B(0) > 0$ we may simplify (2.13) by relabeling the invariant functions DB, FB, and B^2 by D, F, and B, respectively, obtaining $[p,q] \mapsto [P,Q]$ where

$$\begin{aligned} &\text{(a)} \quad P = Cp(A,Bv,\Lambda) + Dq(A,Bv,\Lambda)v \\ &\text{(b)} \quad Q = Ep(A,Bv,\Lambda) + Fq(A,Bv,\Lambda). \end{aligned} \tag{2.14}$$

This general $\mathbf{Z}_2$-equivalence must satisfy

$$\begin{aligned} &\text{(a)} \quad A(0) = 0, \\ &\text{(b)} \quad A_u(0), B(0), C(0), F(0), \Lambda'(0) > 0. \end{aligned} \tag{2.15}$$

Our method for putting the intermediate order terms into normal form is to compute the derivatives of P, Q with respect to the terms listed in Table 2.5 and to solve the resulting equations to obtain the normal form.

A bifurcation problem $[p,q]$ with $\mathbf{Z}_2$ symmetry always satisfies

$$p(0) = p_u(0) = q(0) = 0,$$

and the transformed bifurcation problem $[P,Q]$ in (2.14) also satisfies

$$P = P_u = Q = 0$$

at the origin. (From now on all derivatives are evaluated at 0, but for simplicity we omit the (0).) From (2.14) we have

$$\begin{aligned} &\text{(a)} \quad P_v = CBp_v \\ &\text{(b)} \quad P_\lambda = C\Lambda' p_\lambda \\ &\text{(c)} \quad P_{uu} = CA_u^2 p_{uu} \\ &\text{(d)} \quad Q_u = FA_u q_u \\ &\text{(e)} \quad Q_v = EBp_v + FA_v q_u + FBq_v \\ &\text{(f)} \quad Q_\lambda = EP_\lambda \Lambda' + FA_\lambda q_u + F\Lambda' q_\lambda. \end{aligned} \tag{2.16}$$

Using (2.16) we can check that each of the defining conditions in Table 2.2 is an invariant of $\mathbf{Z}_2$-equivalence. We now discuss the explicit solution of each of the recognition problems for the normal forms of Table 2.1.

Normal form (1): Using (2.16(a–d)) we may set

$$|P_v| = |P_\lambda| = \frac{1}{2}|P_{uu}| = |Q_u| = 1$$

by choosing

$$B, C, A_u, \quad \text{and} \quad F$$

suitably. Since (2.15(b)) implies that these coefficients are positive, we can scale the relevant coefficients of P and Q only to ± 1. Since F and q_u are assumed nonzero, we may choose A_v and A_λ so that in the transformed equations $Q_v = Q_\lambda = 0$. Moreover, since all other terms are (probably of) higher order, we have verified the normal form for singularity (1) in Table (2.2).

Normal form (2) ($p_{uu} = 0$): To complete the solution of the recognition problem for (2) we must compute

$$P_{uuu} = 3A_u[A_u C_u + A_{uu} C] p_{uu} + CA_u^3 p_{uuu}. \tag{2.17}$$

Thus when $p_{uu} = 0$ we have

$$P_{uuu} = CA_u^3 p_{uuu}.$$

It is now a simple matter to scale the coefficients P_v, P_λ, $\frac{1}{6}P_{uuu}$ and Q_u to have absolute value 1, just as in normal form (1).

Normal form (3) ($p_v = 0$): Here Table 2.4 shows that p_{uv} and p_{vv} are intermediate order terms. We compute

$$\begin{aligned} &\text{(a)} \quad P_{uv}|_{p_v=0} = A_u A_v C p_{uu} + A_u BC p_{uv} + A_u D q_u \\ &\text{(b)} \quad P_{vv}|_{p_v=0} = C[A_v^2 p_{uu} + A_v B p_{uv} + B^2 p_{vv}] + 2D[A_v q_u + B q_v]. \end{aligned} \tag{2.18}$$

Then we set $Q_v = 0$ by choosing A_v. This is possible since q_u is assumed nonzero; see (2.16(e)). Since $p_v = 0$ we may set $P_{uv} = 0$ by solving

$$A_v C p_{uu} + BC p_{uv} + A_u D q_u = 0 \tag{2.19}$$

for D. On substituting for A_v and A_λ in (2.18(b)) we arrive at the formula

$$P_{vv} = B^2 C[q_u p_{vv} - q_v p_{uu}]/q_u. \tag{2.20}$$

Thus we can scale $\frac{1}{2}P_{vv}$ to have absolute value 1, provided $q_u p_{vv} - q_v p_{uu} \neq 0$.

Normal form (4) ($p_\lambda = 0$): The additional derivatives to be calculated are:

$$\begin{aligned} &\text{(a)} \quad P_{u\lambda}|_{p_\lambda=0} = CA_u(A_\lambda p_{uu} + \Lambda' p_{u\lambda}), \\ &\text{(b)} \quad P_{\lambda\lambda}|_{p_\lambda=0} = C[A_\lambda^2 p_{uu} + 2A_\lambda \Lambda' p_{u\lambda} + (\Lambda')^2 p_{\lambda\lambda}]. \end{aligned} \tag{2.21}$$

We set $P_{u\lambda} = 0$ by solving for A_λ; see (2.21(a)). This is possible since q_u, F, and p_{uu} are assumed nonzero. In particular,

$$A_\lambda = -\Lambda' p_{u\lambda}/p_{uu}. \tag{2.22}$$

Next we scale P_v, $\frac{1}{2}P_{uu}$, and Q_u to have absolute value 1, by specifying B, C, and F. See (2.16(a, c, d)).

The intermediate terms in Table 2.5(4) that remain are Q_λ and $P_{\lambda\lambda}$. Substitute (2.22) into (2.16(f)) and (2.21(b)) to get:

$$\begin{aligned} &\text{(a)} \quad Q_\lambda = \Lambda'[p_{uu} q_\lambda - p_{u\lambda} q_u]/A_u |q_u| p_{uu}, \\ &\text{(b)} \quad P_{\lambda\lambda} = 2(\Lambda')^2[p_{uu} p_{\lambda\lambda} - p_{u\lambda}^2]/A_u^2 p_{uu} |p_{uu}|. \end{aligned} \tag{2.23}$$

By choosing Λ'/A_u we can scale $|Q_\lambda|$ to 1, assuming that $p_{uu} q_\lambda - p_{u\lambda} q_u \neq 0$. This leads to the expression for the modal parameter:

$$
\begin{aligned}
m &= \tfrac{1}{2}P_{\lambda\lambda} \\
&= \operatorname{sgn}(p_{uu})[p_{uu}p_{\lambda\lambda} - p_{u\lambda}^2][q_u/(q_\lambda p_{uu} - q_u p_{u\lambda})]^2.
\end{aligned} \tag{2.24}
$$

Normal form (5) ($q_u = 0$): The only additional derivative we must compute is

$$
Q_{uu}|_{q_u=0} = A_u^2[Ep_{uu} + Fq_{uu}]. \tag{2.25}
$$

First, scale $|P_v| = |P_\lambda| = \frac{1}{2}|P_{uu}| = 1$ by choosing B, Λ', and A_u^2 in (2.16(a, b, c)), respectively. The terms that remain are:

$$
\begin{aligned}
&\text{(a)}\quad Q_v = (Ep_v + Fq_v)/C|p_v| \\
&\text{(b)}\quad Q_\lambda = (Ep_\lambda + Fq_\lambda)/C|p_\lambda| \\
&\text{(c)}\quad Q_{uu} = 2(Ep_{uu} + Fq_{uu})/C|p_{uu}|.
\end{aligned} \tag{2.26}
$$

Equations (2.26) are obtained from (2.16(e, f)) and (2.25(c)) by substituting the values for B, Λ', and A_u^2. Next, set $Q_\lambda = 0$ by choosing

$$
E = -Fq_\lambda/p_\lambda. \tag{2.27}
$$

Now substitute (2.27) into (2.26) to obtain

$$
\begin{aligned}
&\text{(a)}\quad Q_v = F(q_v p_\lambda - q_\lambda p_v)/p_\lambda|p_v|C, \\
&\text{(b)}\quad Q_{uu} = 2F(q_{uu}p_\lambda - q_\lambda p_{uu})/p_\lambda|p_{uu}|C.
\end{aligned} \tag{2.28}
$$

By choosing F/C we can scale $|Q_v|$ to 1, provided $q_v p_\lambda - q_\lambda p_v \neq 0$, which we assume. Then

$$
m = \frac{1}{2}Q_{uu} = \frac{q_{uu}p_\lambda - q_\lambda p_{uu}}{|q_v p_\lambda - q_\lambda p_v|}\frac{|p_v|}{|p_{uu}|}\operatorname{sgn}(p_\lambda). \tag{2.29}
$$

This completes the classification.

Exercises

2.1. Calculate $\operatorname{Itr}\mathscr{M}\cdot RT(g, \mathbf{Z}_2)$ for normal forms (3)–(5) and observe the advantage in calculation obtained by using the result $\operatorname{Itr}\mathscr{K}(g, \mathbf{Z}_2) \subset \mathscr{P}(g, \mathbf{Z}_2)$.

2.2. Define

$$
G(x, y, \lambda, c) = (\delta_1 x^2 + \delta_2 y^2 + \delta_3 \lambda x, \rho xy + \delta_4 \lambda y) \tag{2.30}
$$

where $\delta_j = \pm 1$, $j = 1, \ldots, 4$, and $\rho \neq 0$, $\delta_1\delta_3\delta_4$, or $2\delta_1\delta_3\delta_4$. Show that G is equivalent to normal form (4) where

$$
\begin{aligned}
&\text{(a)}\quad \varepsilon_1 = \delta_1 \\
&\text{(b)}\quad \varepsilon_2 = \delta_2 \\
&\text{(c)}\quad \varepsilon_3 = \operatorname{sgn}(\rho) \\
&\text{(d)}\quad \varepsilon_4 = \operatorname{sgn}(2\delta_4 - \delta_1\delta_3\rho) \\
&\text{(e)}\quad m = -\varepsilon_1\rho^2/(\rho - 2\delta_1\delta_3\delta_4)^2.
\end{aligned} \tag{2.31}
$$

In particular, G may be used as an alternative normal form in case (4) when $\varepsilon_1 m < 0$.

§3. Bifurcation Diagrams with $\mathbf{Z}_2$ Symmetry

As we said in §1, the main reason for studying the $\mathbf{Z}_2$-normal form in §2 is to understand how steady-state and Hopf bifurcations interact in low codimension. In particular, we recover here Langford's [1979] observation that such interactions can lead to quasiperiodic motion on a 2-torus. Our discussion is based on work of Armbruster, Dangelmayr, and Güttinger [1985], but the details are organized differently.

The normal forms in Table 2.1 can be classified according to the types of degenerate steady-state or Hopf bifurcation involved in the mode interaction. Normal form (1) is the least degenerate case, in which a generic steady-state bifurcation (limit point) interacts with a generic Hopf bifurcation. This singularity corresponds to the "saddle-node" Hopf bifurcation; see Guckenheimer [1984] and Guckenheimer and Holmes [1983]. Algebraically there are two types of codimension one steady-state bifurcation (hysteresis points $\pm x^3 \pm \lambda$ and bifurcation/isola points $\pm x^2 \pm \lambda^2$), and two types of codimension one degenerate Hopf bifurcation (namely $\pm x^5 \pm \lambda x$, $\pm x^3 \pm \lambda^2 x$; see (VIII, 5.6)). Thus it is no surprise that there are four types of (topological) codimension two Hopf/steady-state interactions, one corresponding to each of the codimension one single mode degeneracies.

Normal form (2) of Table 2.1 represents the confluence of a hysteresis point and a generic Hopf bifurcation. This singularity is discussed in detail by Langford [1986]. Normal form (4) represents the interaction of either a simple steady-state bifurcation ($\varepsilon_2 m < 0$) or an isola center ($\varepsilon_2 m > 0$) with a generic Hopf bifurcation. The interaction of the simple steady-state bifurcation with a generic Hopf bifurcation is the only one of these mode interactions that has a "trivial" steady state for all values of λ. For this reason, it was the first to be analyzed (Langford [1979]). This particular normal form, however, was studied in the context of two interacting steady-state modes, one with $\mathbf{Z}_2$ symmetry, by Keener [1976] and Golubitsky and Schaeffer [1979b, 1981]. Later Boivin [1981] described, in detail, all the persistent bifurcation diagrams, including the possibilities for invariant 2-tori.

Normal forms (3) and (5) represent the interaction of the generic limit point bifurcation with codimension 1 degenerate Hopf bifurcations. These singularities, along with the isola-center/Hopf interaction, were first studied in Armbruster, Dangelmayr, and Güttinger [1985].

We now present the specific normal forms whose bifurcation diagrams we shall discuss. The normal forms in Table 2.1 each involve a choice of four signs $\varepsilon_1, \ldots, \varepsilon_4$. However, by allowing certain (non-$\mathbf{Z}_2$) equivalences we can substantially reduce the number of sign choices. We allow

Table 3.1. Normal Forms of Unfoldings of Hopf/Steady-State Mode Interactions

(1)	$(x^2 + \varepsilon_2 y^2 - \lambda, \varepsilon_4(x - \alpha)y)$	
(2)	$(x^3 + y^2 - \lambda - \beta x, \varepsilon_4(x - \alpha)y)$	
(3)	$(x^2 + \varepsilon_2(y^4 - \beta y^2) - \lambda, \varepsilon_4(x - \alpha)y)$	
(4a)	$(x^2 + \varepsilon_2 y^2 - \lambda x - \alpha, (\rho x + \varepsilon_4 \lambda - 2\beta)y)$	$\rho \neq 0, -\varepsilon_4, -2\varepsilon_4$
(4b)	$(x^2 + \varepsilon_2 y^2 + \mu\lambda^2 - \beta, (\varepsilon_3 x - \lambda + \alpha)y)$	$\mu > 0$
(5)	$(x^2 + \varepsilon_2 y^2 - \lambda, (\mu x^2 - \beta x + \varepsilon_2 y^2 - \alpha)y)$	$\mu \neq 0, 1, 2$

Table 3.2. Solutions and Stability of $g \equiv [p, q] = 0$

Isotropy Subgroup	Equations	Eigenvalues
$\mathbf{Z}_2$ (steady-state)	$y = p(x, 0, \lambda) = 0$	$p_u(x, 0, \lambda)$ $q(x, 0, \lambda)$
$\mathbb{1}$ (periodic)	$p = q = 0$	$\operatorname{tr} dg = p_u + 2vq_v$ $\det dg = 2(p_u q_v - p_v q_u)v$

$$\begin{aligned} &\text{(a)} \quad \lambda \mapsto -\lambda \\ &\text{(b)} \quad g \mapsto -g \\ &\text{(c)} \quad (x, y) \mapsto (-x, -y). \end{aligned} \tag{3.1}$$

The effect of (3.1(a)) is to interchange left and right in the bifurcation diagrams, that is, to interchange super- and subcriticality. The effect of (3.1(b, c)) is to change the signs of all the eigenvalues of dg.

In Table 3.1 we list the universal unfoldings of the bifurcation problems to be studied in this section, obtained by making specific choices of signs. Table 3.1 differs from Table 2.1 in one further way. We have divided normal form (4) of Table 2.1 into (4a and 4b) in Table 3.1. The form (4a) models the interaction of a simple steady-state bifurcation with a generic Hopf bifurcation (see Exercise 2.2); the form (4b) models the interaction of an isola center with a generic Hopf bifurcation.

Recall that for each $\mathbf{Z}_2$-bifurcation problem $g \equiv [p, q]$ there are two types of solution to $g = 0$, as described in (1.10): the "steady-state" solutions having $\mathbf{Z}_2$ symmetry, and the "periodic" solutions having trivial isotropy. In Table 3.2 we recall (1.10) and indicate how to determine the eigenvalues of dg for the two types of solution. This result is obtained by direct calculation.

For each normal form in Table 3.1 the data in Table 3.2 are evaluated to produce Table 3.3.

There is one theoretical issue that must be faced: the asymptotic stability of periodic solutions is *not* preserved by $\mathbf{Z}_2$-equivalence. This issue was discussed in XIV, §5, for general Γ-equivalence; here we describe what happens for $\mathbf{Z}_2$-equivalence. We know that $\mathbf{Z}_2$-equivalence preserves the zeros of g:

Table 3.3.

$\mathbf{Z}_2$		Signs of Eigenvalues	$\mathbb{1}$	sgn(tr)	sgn(det)
(1)	$y = 0$	x	$x = \alpha$	x	$-\varepsilon_2\varepsilon_4$
	$\lambda = x^2$	$\varepsilon_4(x - \alpha)$	$\lambda = \alpha^2 + \varepsilon_2 y^2$		
(2)	$y = 0$	$3x^2 - \beta$	$x = \alpha$	$3x^2 - \beta$	$-\varepsilon_4$
	$\lambda = x^3 - \beta x$	$\varepsilon_4(x - \alpha)$	$\lambda = \alpha^3 - \beta\alpha + y^2$		
(3)	$y = 0$	$2x$	$x = \alpha$	$2x$	$\varepsilon_2\varepsilon_4(\beta - 2y^2)$
	$\lambda = x^2$	$\varepsilon_4(x - \alpha)$	$\lambda = \alpha^2 + \varepsilon_2(y^4 - \beta y^2)$		
(4a)	$y = 0$	$2x - \lambda$	$(1 + \rho\varepsilon_4)x^2 - 2\beta\varepsilon_4 x$ $+ \varepsilon_2 y^2 = \alpha$	$2x - \lambda$	$\varepsilon_2\rho$
	$x^2 - \lambda x = \alpha$	$\rho x + \varepsilon_4\lambda - 2\beta$	$\lambda = \varepsilon_4(2\beta - \rho x)$		
(4b)	$y = 0$	x	$\lambda = \varepsilon_3 x + \alpha$	x	$-\varepsilon_2\varepsilon_3$
$\mu > 0$	$x^2 + \mu\lambda^2 = \beta$	$\varepsilon_3 x + \alpha - \lambda$	$x^2 + \varepsilon_2 y^2 + \mu\lambda^2 = \beta$		
(5)	$y = 0$	x	$\lambda = x^2 + \varepsilon_2 y^2$	$x + \varepsilon_2 y^2$	$\varepsilon_2[\beta + 2(1 - \mu)x]$
	$\lambda = x^2$	$\mu x^2 - \beta x + \alpha$	$\mu(x - \beta/2\mu)^2 + \varepsilon_2 y^2$ $= \alpha + \beta^2/4\mu$		

it was defined for that purpose. We also know that $\mathbf{Z}_2$-equivalence preserves the signs of eigenvalues of $\mathbf{Z}_2$-symmetric solutions. (The $\mathbf{Z}_2$ isotropy forces dg to be diagonal in this case; see Theorem XIV, 5.3.)

In addition, the sign of $\det dg$ is preserved by $\mathbf{Z}_2$-equivalence. Thus, for a normal form g, if a solution with trivial isotropy is unstable because $\det dg < 0$, such a solution is unstable for any problem $\mathbf{Z}_2$-equivalent to g. Finally, if in the universal unfolding of g a branch with trivial isotropy bifurcates from one with $\mathbf{Z}_2$ isotropy, then (generically) this bifurcation has a one-dimensional kernel for dg, hence is a pitchfork. Therefore, exchange of stability (determined by the eigenvalue of dg that goes through zero) holds near this bifurcation.

However, if $\det dg > 0$, then the stability of solutions with trivial isotropy need not be preserved by $\mathbf{Z}_2$-equivalence since the sign of $\operatorname{tr} dg$ may change. Using normal form (1) we give an explicit example that illustrates this difficulty.

The bifurcation diagrams for normal form (1) are shown in Figure 3.1. The discussion focuses on the case $\varepsilon_2 = 1$, $\varepsilon_4 = -1$, when

$$g(x, y, \lambda) = (x^2 + y^2 - \lambda, -(x - \alpha)y). \tag{3.2}$$

When $\alpha = 0$ the bifurcation diagram is as in Figure 3.2.

This property may, however, be perturbed by the addition of higher order terms. For example, consider

$$h = (x^2 + y^2 - \lambda, -(x + y^2 - \alpha)y). \tag{3.3}$$

It is trivial to check—using Table 2.2(1)—that (3.3) is $\mathbf{Z}_2$-equivalent to (3.2). The bifurcation data for (3.3) are similar to those for (3.2). Both have $\mathbf{Z}_2$-solutions

$$y = 0, \qquad \lambda = x^2$$

and the eigenvalues along this branch are x and $\alpha - x$. The periodic solutions

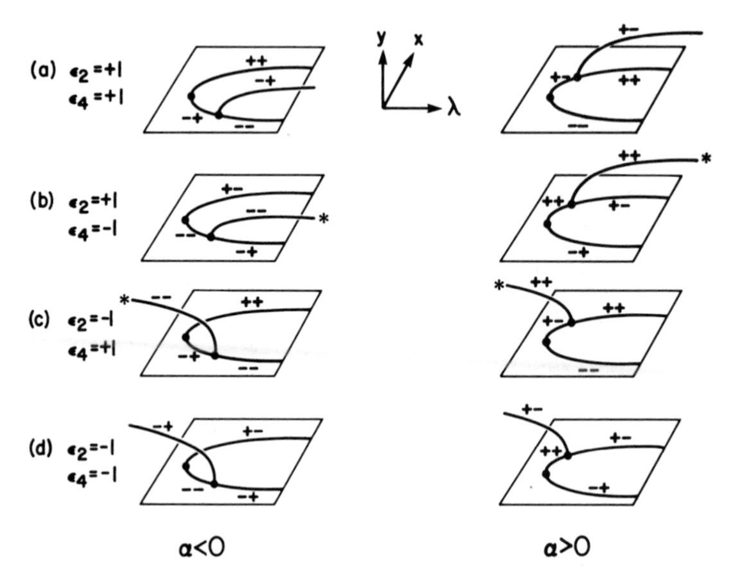

Figure 3.1. Normal form (1) $(x^2 + \varepsilon_2 y^2 - \lambda, \varepsilon_4(x - \alpha)y) = 0$. * indicates a possible bifurcation to an invariant 2-torus along the branch.

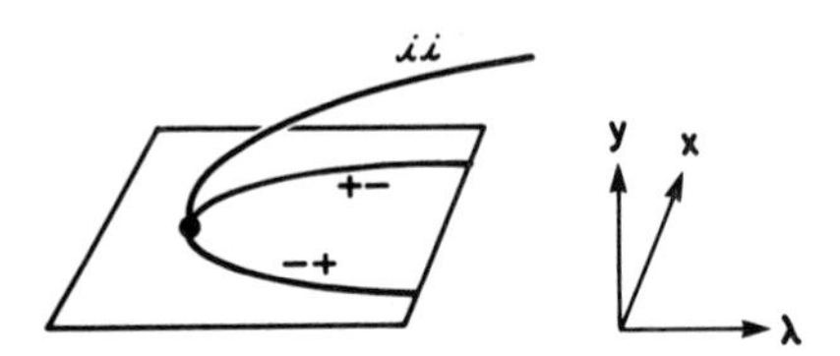

Figure 3.2. Bifurcation diagram of (3.2) when $\alpha = 0$. The stability assignment 'ii' indicates that eigenvalues along this branch of periodic solutions are purely imaginary.

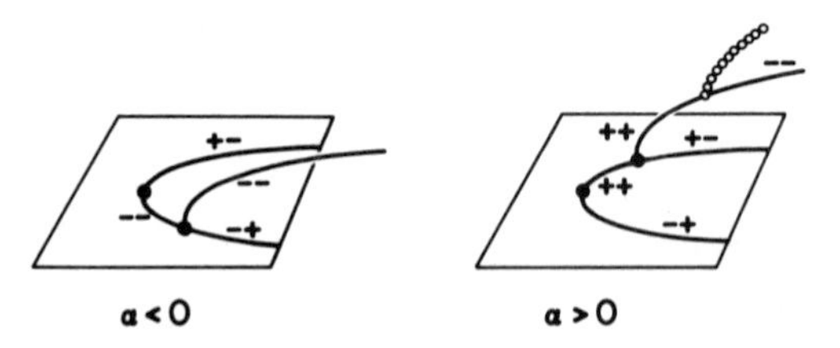

Figure 3.3. Bifurcation diagrams for $h = (x^2 + y^2 - \lambda, -x(x + y^2 - \alpha)y)$. Tertiary branch of 2-tori indicated by circles.

for (3.3) are given by

$$\begin{aligned} &\text{(a)} \quad x = \alpha - y^2 \\ &\text{(b)} \quad \lambda = x^2 + y^2. \end{aligned} \tag{3.4}$$

The eigenvalues along this branch are determined by

$$\begin{aligned} &\text{(a)} \quad \det dh = 2(1 - 2x)y^2 \\ &\text{(b)} \quad \operatorname{tr} dh = 2(x - y^2). \end{aligned} \tag{3.5}$$

As expected, near the origin, $\det dh > 0$ so the stability of solutions is determined by the sign of $\operatorname{tr} dh$. By (3.5(b)), $\operatorname{tr} dh = 0$ when $x = y^2$. Substituting this into (3.4) we find that when

$$\lambda = \frac{\alpha}{2}\left(1 + \frac{\alpha}{2}\right), \qquad x = \frac{\alpha}{2}, \qquad y^2 = \frac{\alpha^2}{2} \tag{3.6}$$

there is a change in the stability of the periodic solutions: eigenvalues of dh cross the imaginary axis with nonzero speed. Thus a Hopf bifurcation takes place and nonstationary periodic solutions appear in the amplitude equations of the Hopf/steady-state interaction. Such a periodic solution corresponds to an invariant 2-torus in the original Birkhoff normal form equation of §1.

To summarize: when $\alpha > 0$ there is a change of stability along the branch of solutions with trivial isotropy in (3.4). This change of stability occurs at values of x, y, λ that approach the origin as $\alpha \to 0$, hence is a *local* phenomenon. Moreover, this change of stability corresponds to a Hopf bifurcation in (3.4). The bifurcation diagram is shown in Figure 3.3, with the branch of tori included.

Compare with Figure 3.1(b). Whether this branch of tori is supercritical (and stable) requires a calculation using the Hopf theorem (VIII, 4.1). We have not performed the calculation here, but numerical evidence indicates that the branch is supercritical and asymptotically stable.

Remark. In Figure 3.2, and in the following figures, we indicate by $*$ any branch of solutions with trivial isotropy along which a torus bifurcation may occur. By appropriate choice of higher order terms, it is possible to arrange for multiple torus bifurcations. Typically, however, we expect at most one such bifurcation.

Finally, suppose that a branch of solutions with trivial isotropy connects branches with $\mathbf{Z}_2$ symmetry. Suppose further that exchange of stability forces the eigenvalues at one end of the branch to be $++$ and at the other end to be $--$. If these eigenvalues are never zero along the branch, then a Hopf bifurcation must occur. This observation is the essence of Langford's [1979] result that Hopf/steady-state interactions can led to invariant 2-tori. See also Chapter X, Figure 4.5.

In Figures 3.4 to 3.17 we draw the persistent diagrams for normal forms (2)–(5) in Table 3.1. For normal form (4a) we consider only the case where $\varepsilon_4 = -1$, leaving $\varepsilon_4 = 1$ as an exercise.

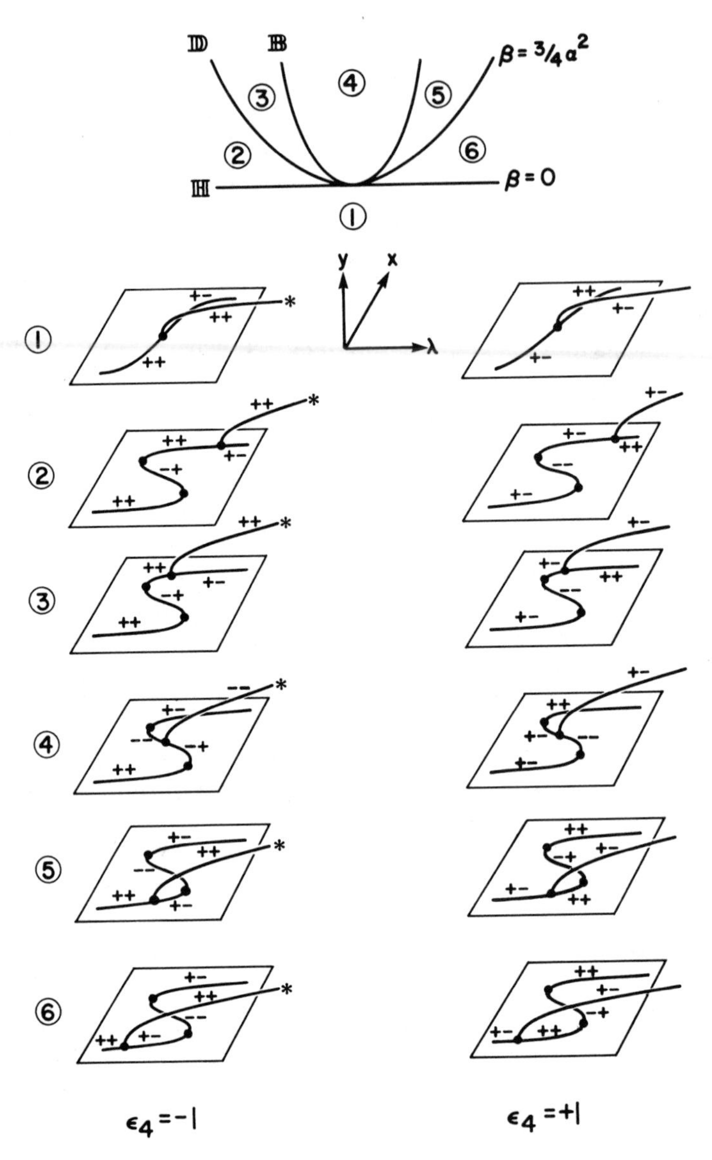

Figure 3.4. Normal form (2): $(x^3 - \beta x + y^2 - \lambda, \varepsilon_4(x - \alpha)y)$.

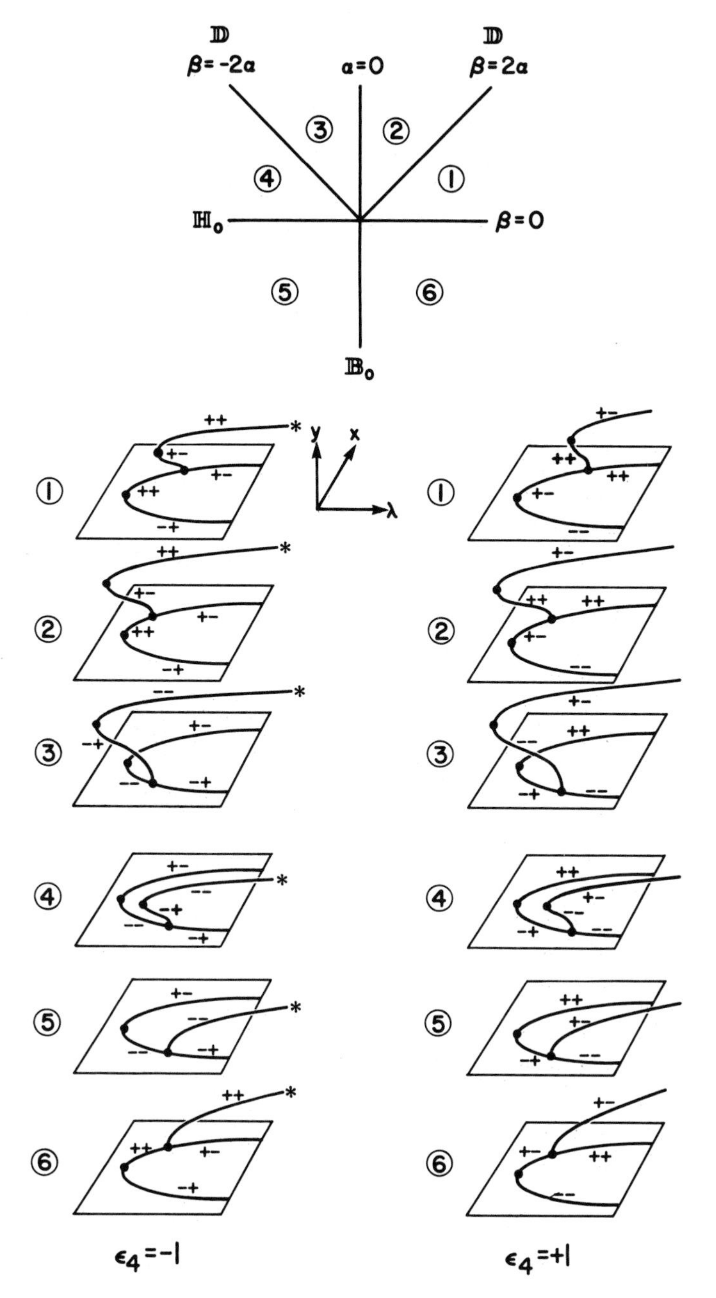

Figure 3.5. Normal form (3) ($\varepsilon_2 = 1$): $(x^2 + y^4 - \beta y^2 - \lambda, \varepsilon_4(x - \alpha)y)$.

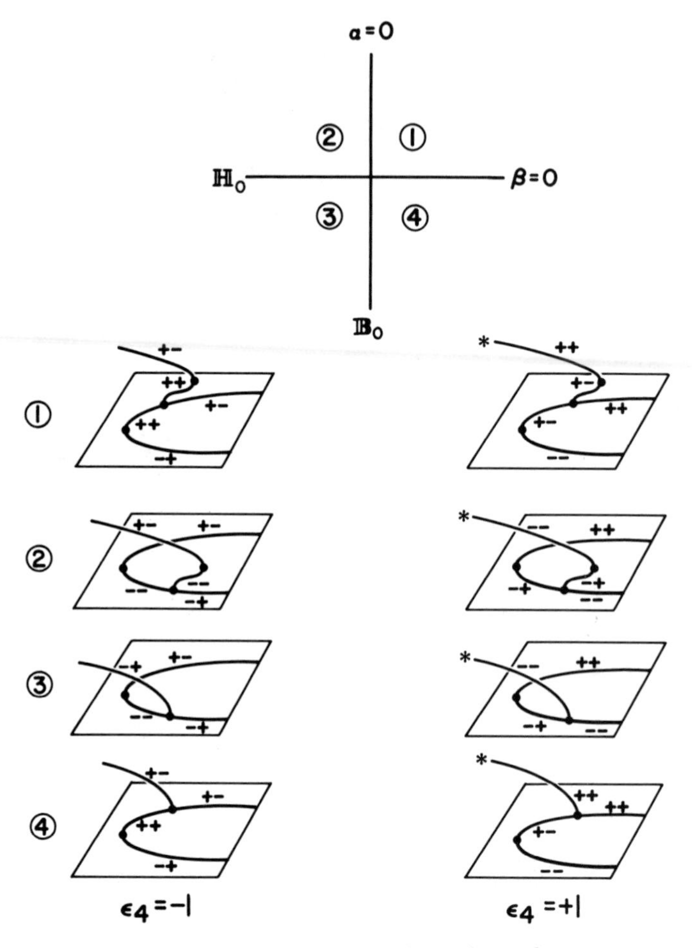

Figure 3.6. Normal form (3) ($\varepsilon_2 = -1$): $(x^2 - (y^4 - \beta y^2) - \lambda, \varepsilon_4(x - \alpha)y)$.

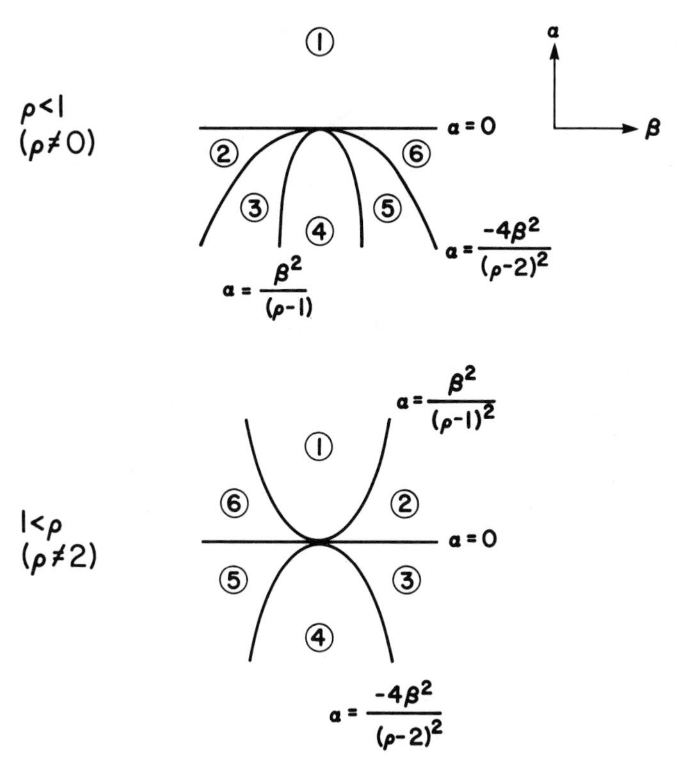

Figure 3.7. Normal form (4a): transition varieties.

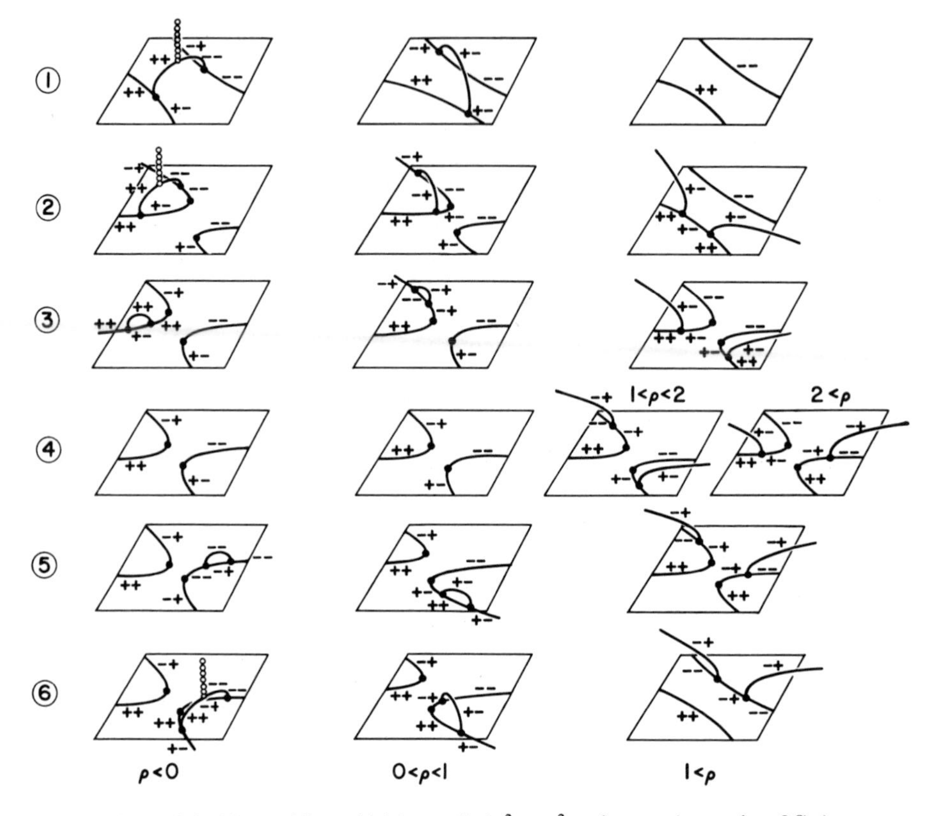

Figure 3.8. Normal form (4a) ($\varepsilon_2 = 1$): $(x^2 + y^2 - \lambda x - \alpha, (\rho x - \lambda - 2\beta)y)$.

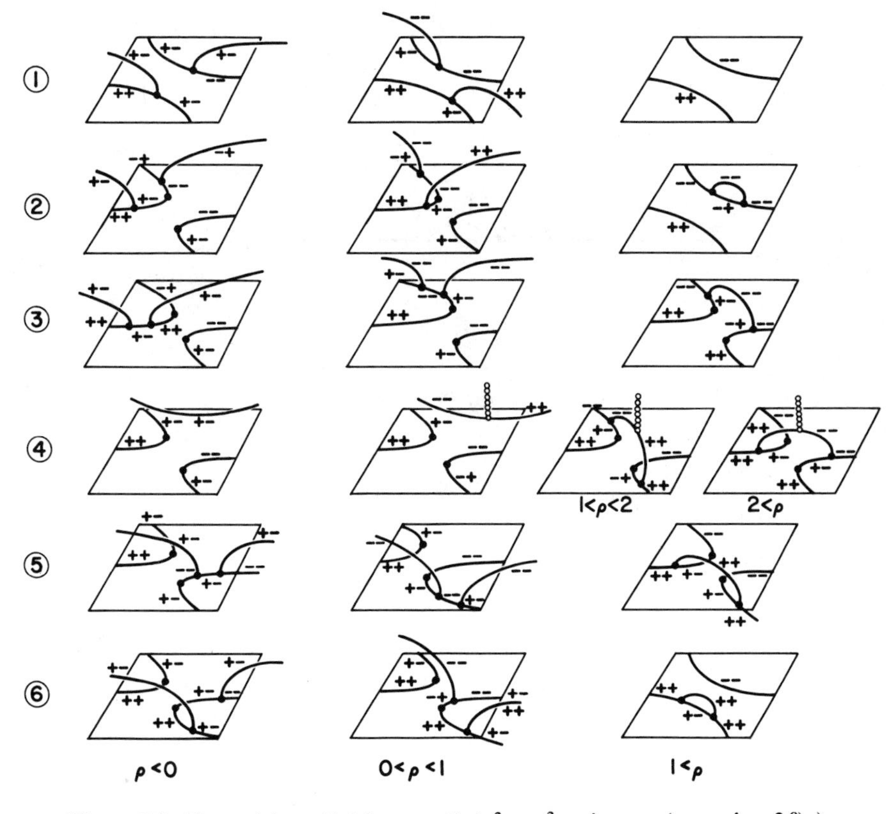

Figure 3.9. Normal form (4a) ($\varepsilon_2 = -1$): $(x^2 - y^2 - \lambda x - \alpha, (\rho x - \lambda - 2\beta)y)$.

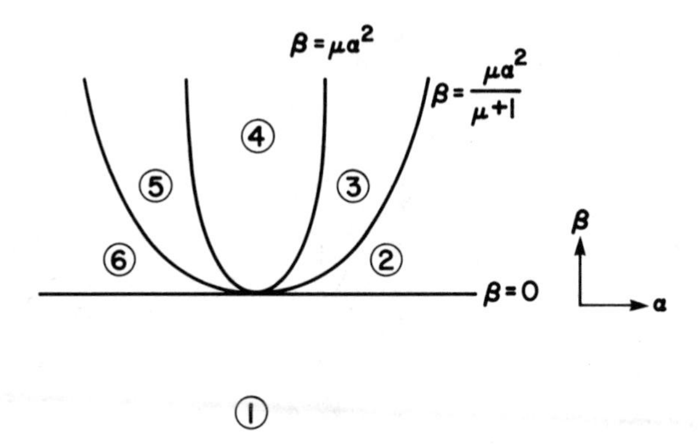

Figure 3.10. Normal form (4b): transition variety.

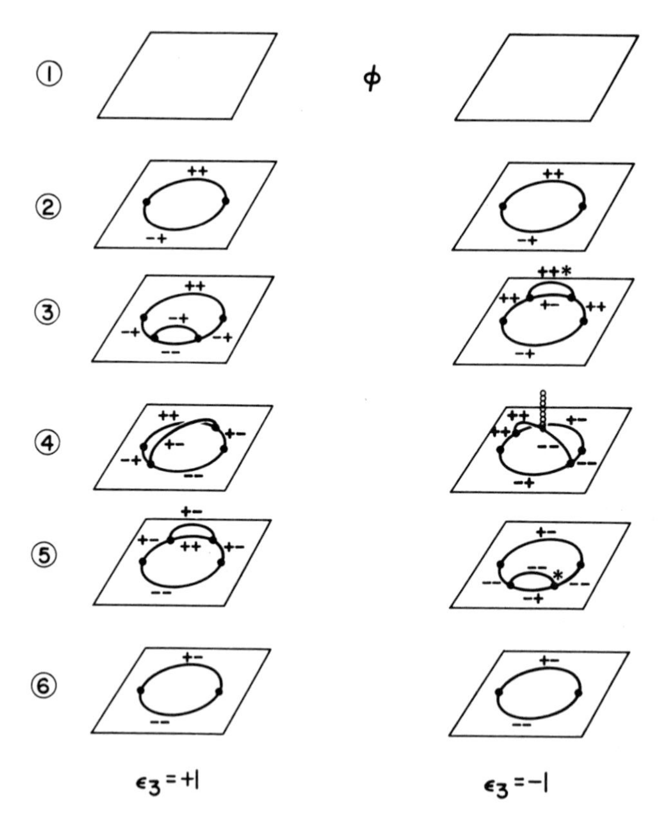

Figure 3.11. Normal form (4b) ($\varepsilon_2 = 1$): $(x^2 + y^2 + \mu\lambda - \beta, (\varepsilon_3 x - \lambda + \alpha)y)$.

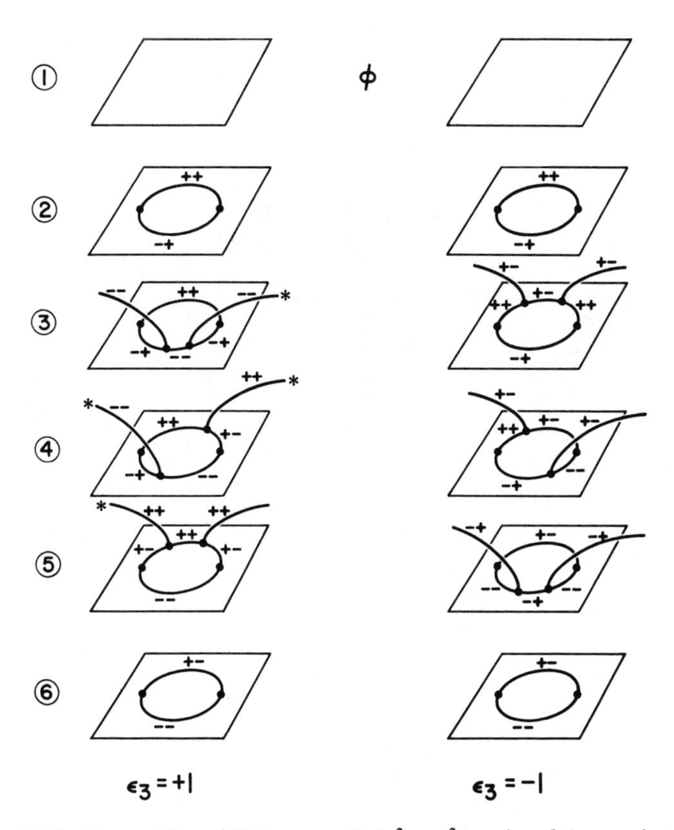

Figure 3.12. Normal form (4b) ($\varepsilon_2 = -1$): $(x^2 - y^2 + \mu\lambda - \beta, (\varepsilon_3 x - \lambda + \alpha)y)$.

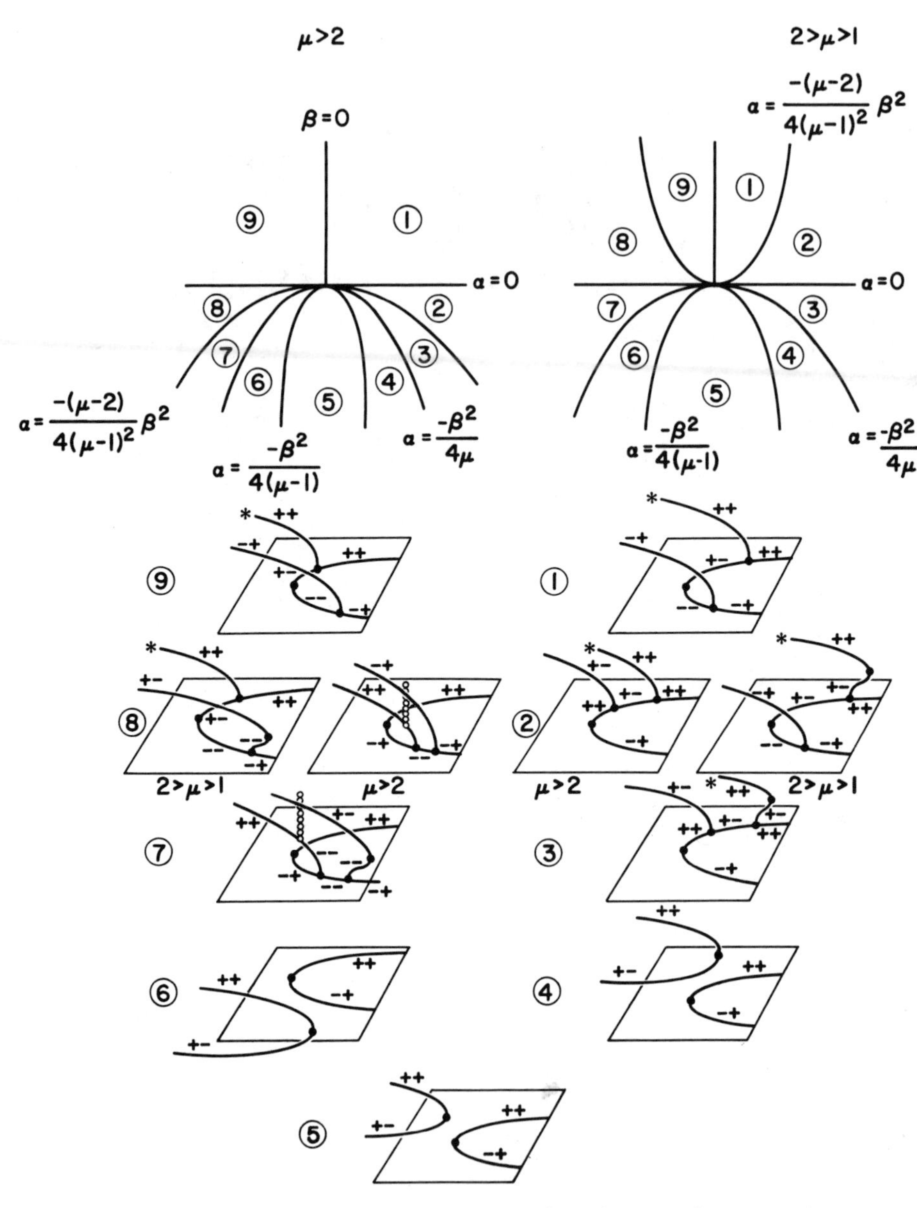

Figure 3.13. Normal form (5) ($\varepsilon_2 = -1, \mu > 1$): $(x^2 - y^2 - \lambda, (\mu x^2 - \beta x + y^2 - \alpha)y)$ $0 > \mu$.

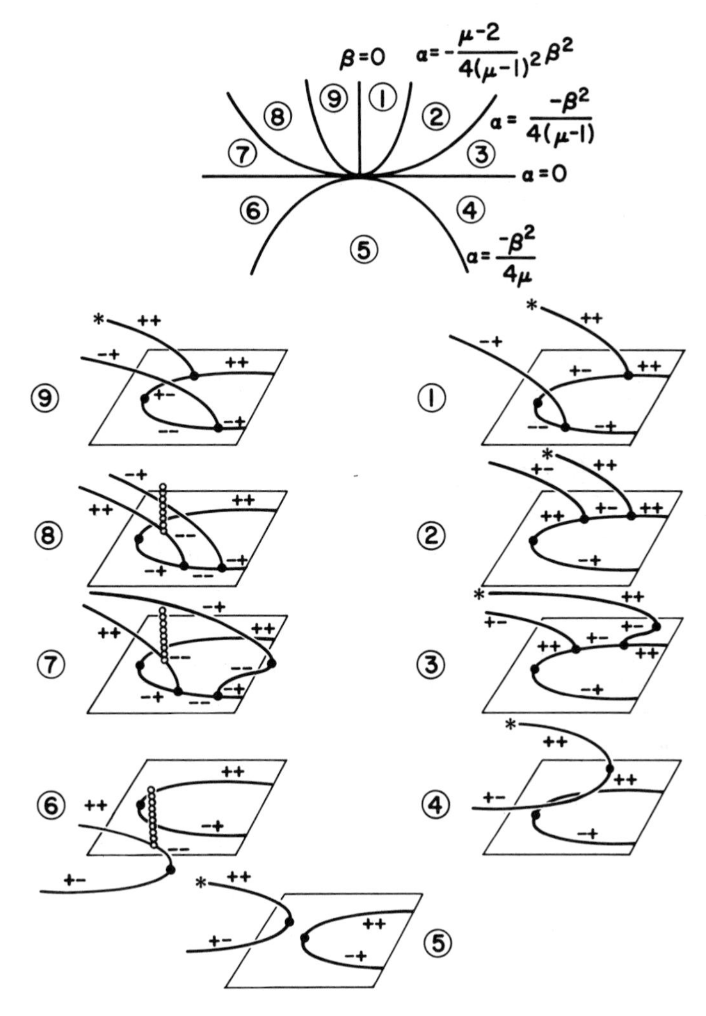

Figure 3.14. Normal form (5) ($\varepsilon_2 = -1, 1 > \mu > 0$): $(x^2 - y^2 - \lambda, (\mu x^2 - \beta x - y^2 - \alpha)y)$ $1 > \mu > 0$.

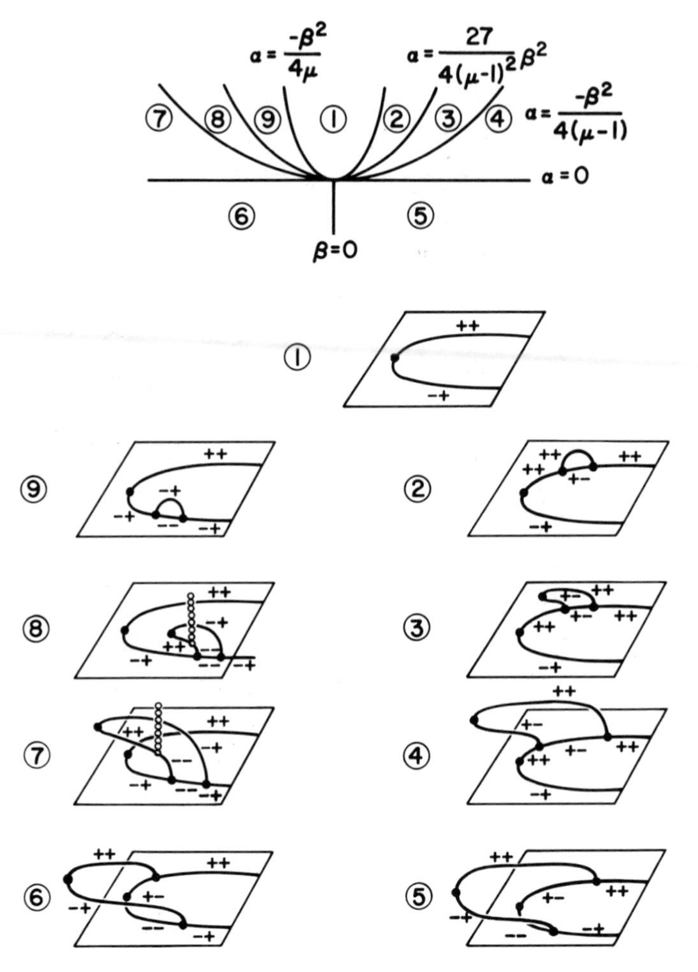

Figure 3.15. Normal form (5) ($\varepsilon_2 = -1, 0 > \mu$): $(x^2 + y^2 - \lambda, (\mu x^2 - \beta x - y^2 - \alpha)y)$ $0 > \mu$.

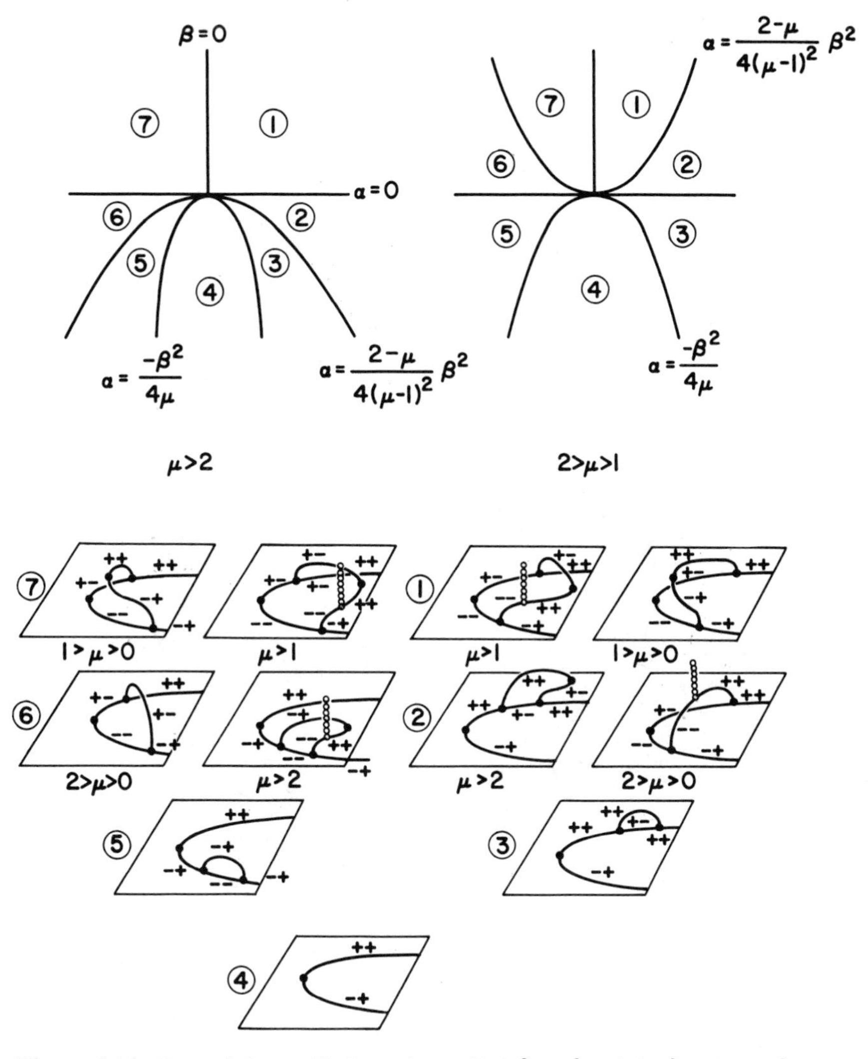

Figure 3.16. Normal form (5) ($\varepsilon_2 = 1, \mu > 0$): $(x^2 + y^2 - \lambda, (\mu x^2 - \beta x + y^2 - \alpha)y)$ $\mu > 0$.

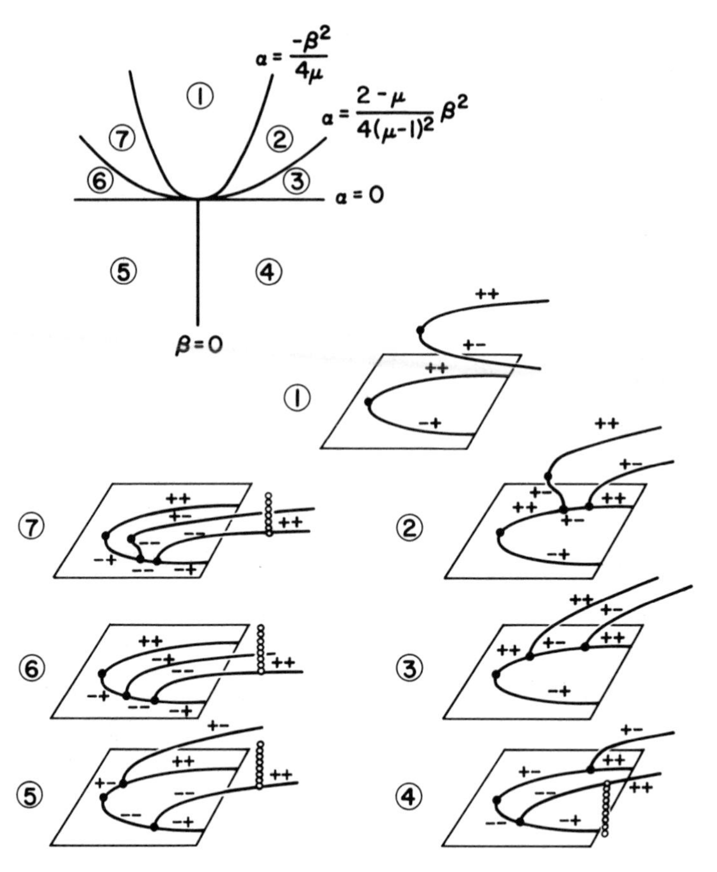

Figure 3.17. Normal form (5) ($\varepsilon_2 = 1, 0 > \mu$): $(x^2 + y^2 - \lambda, (\mu x^2 - \beta x + y^2 - \alpha)y)$ $0 > \mu$.

§4. Hopf/Hopf Interaction

In this section, we consider a system of ODEs

$$\dot{x} + f(x, \lambda) = 0 \tag{4.1}$$

where $x \in \mathbb{R}^4$, $f(0,0) = 0$, and

$$(df)_{0,0} = L \equiv \left[\begin{array}{cc|cc} 0 & \omega_1 & \multicolumn{2}{c}{0} \\ -\omega_1 & 0 & & \\ \hline \multicolumn{2}{c|}{0} & 0 & \omega_2 \\ & & -\omega_2 & 0 \end{array}\right]. \tag{4.2}$$

This is the simplest vector field whose linear part has two distinct pairs of complex conjugate purely imaginary eigenvalues $\pm\omega_1 i$, $\pm\omega_2 i$. We say that

L is *resonant* if there exist nonzero integers n_1 and n_2 such that

$$n_1\omega_1 + n_2\omega_2 = 0, \tag{4.3}$$

and *nonresonant* otherwise.

Our goal here is limited. We assume that (4.1) is in Birkhoff normal form and study the dynamics associated with this normal form. Since the linear part L has two frequencies, it is not surprising that two-frequency tori appear in the unfoldings of Hopf/Hopf interactions. What is more surprising is that three-frequency tori are produced by this interaction. Our goal is to understand this point. To make the discussion easier we assume nonresonance; the problem of resonance is discussed briefly at the end.

In the nonresonant case, the Birkhoff normal form commutes with the 2-torus $\mathbf{T}^2$, which acts on $\mathbb{R}^4 \equiv \mathbb{C}^2$ by

$$(\psi_1, \psi_2)\cdot(z_1, z_2) = (e^{i\psi_1}z_1, e^{i\psi_2}z_2) \tag{4.4}$$

where $(\psi_1, \psi_2) \in \mathbf{T}^2$ and $(z_1, z_2) \in \mathbb{C}^2$. Compare Theorem XVI, 5.1.

Theorem 4.1. *Assume that the vector field f in* (4.1) *has the linear part L in* (4.2), *which is nonresonant as in* (4.3). *Assume that f is in Birkhoff normal form, that is, f commutes with the action of $\mathbf{T}^2$ in* (4.4). *Then f has the form*

$$f(z_1, z_2, \lambda) = (P_1 z_1, P_2 z_2) \tag{4.5}$$

where P_1 and P_2 are complex-valued functions of $|z_1|^2$, $|z_2|^2$, and λ, satisfying

$$P_j(0) = \omega_j i, \qquad j = 1, 2. \tag{4.6}$$

Proof. The proof is left to the reader as Exercise 4.1. □

Using (4.5) we can write (4.1) in terms of amplitude/phase equations. Set

$$z_j = r_j e^{i\theta_j}, \qquad j = 1, 2. \tag{4.7}$$

Then (4.1) becomes

$$\left.\begin{aligned} &\text{(a)} \quad \dot{r}_j + p_j r_j = 0 \\ &\text{(b)} \quad \dot{\theta}_j + q_j = 0 \end{aligned}\right\} \quad j = 1, 2 \tag{4.8}$$

where $P_j = p_j + iq_j$, $p_j(0) = 0$, $q_j(0) = \omega_j$ and p_j, q_j are functions of r_1^2, r_2^2, and λ.

Since $q_j(0) \neq 0$, the dynamics of the amplitude/phase equations (4.8) is (essentially) determined by the amplitude equations (4.8(a)). Moreover, if we define

$$g(r_1, r_2) = (p_1(r_1^2, r_2^2)r_1, p_2(r_1^2, r_2^2)r_2), \tag{4.9}$$

then the form of (4.9) is that of the general mapping commuting with the standard action of $\mathbf{Z}_2 \oplus \mathbf{Z}_2$ in the plane. See Chapter X, specifically Lemma X, 1.1. All that remains of the $\mathbf{T}^2$ phase-shift symmetries in the amplitude/phase equations are the half-period phase shifts.

Table 4.1. Types of Solution to (4.1) Characterized by Isotropy

Isotropy	Equations ($g = 0$)	Type of Solution
$\mathbf{Z}_2 \oplus \mathbf{Z}_2$	$r_1 = r_2 = 0$	Steady state
$\mathbf{Z}_2(-1, 1)$	$r_1 = 0 = p_2$	Periodic solution of period near $2\pi/\omega_2$
$\mathbf{Z}_2(1, -1)$	$p_1 = 0 = r_2$	Periodic solution of period near $2\pi/\omega_1$
$\mathbb{1}$	$p_1 = p_2 = 0$	Invariant 2-torus

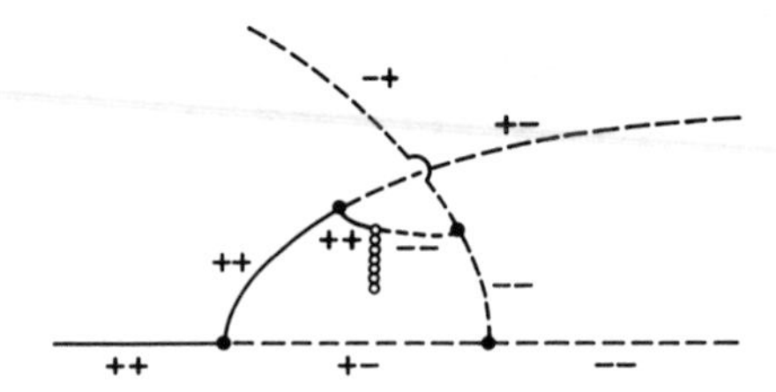

Figure 4.1. How $\mathbf{Z}_2 \oplus \mathbf{Z}_2$ steady-state mode interactions can force Hopf bifurcation.

We now assume that f in (4.1) and hence g in (4.9) depend on a bifurcation parameter λ. Steady-state solutions of the amplitude equations $g = 0$ correspond to steady states, periodic solutions, and invariant 2-tori of the original ODE (4.1). Specifically, there are four types of solution corresponding to four isotropy subgroups (see X, 1.4), and each corresponds to a type of solution in (4.1). This information is recalled and summarized in Table 4.1.

The classification of nondegenerate $\mathbf{Z}_2 \oplus \mathbf{Z}_2$ bifurcation problems in Proposition X, 2.3, and Theorem X, 2.4, lets us draw conclusions here about the dynamics in (4.1). Specifically, the presistent bifurcation diagrams of Chapter X, Figure 4.3, show that periodic solutions of periods near $2\pi/\omega_1$ and $2\pi/\omega_2$ may be expected as primary bifurcations off the steady state. (These occur through standard Hopf bifurcation.) Moreover, we expect invariant 2-tori to be produced in this mode interaction, corresponding to the mixed mode solutions of Chapter X, Figure 4.3. Indeed, these tori may be asymptotically stable.

Finally, as noted in Chapter X, Figure 4.5, reproduced here as Figure 4.1, a "Hopf bifurcation" may be forced to occur in the amplitude equations, along the branch of mixed mode solutions with trivial isotropy. This Hopf bifurcation is forced by exchange of stability and implies the existence of a branch of invariant 3-tori for (4.1). This point is one of the most important made by Takens [1974] in his study of Hopf/Hopf interaction. See Iooss and Langford [1980] and §7.5 of Guckenheimer and Holmes [1983] for further details of the dynamics of (4.1).

For the remainder of this section we briefly discuss the way the analysis of (4.1) changes when resonance occurs. Suppose that

$$n_1\omega_1 + n_2\omega_2 = 0 \tag{4.10}$$

where we may assume n_1 and n_2 are coprime. The remark after Theorem XVI, 5.1, now implies that the Birkhoff normal form of (4.1) commutes only with $\mathbf{S}^1$, under the action

$$\psi \cdot (z_1, z_2) = (e^{in_1\psi} z_1, e^{in_2\psi} z_2). \tag{4.11}$$

We now state the normal form.

Theorem 4.2. *Assume that f in* (4.1) *has linear part L in* (4.2) *which is resonant and satisfies* (4.10). *Assume that f is in Birkhoff normal form, that is, commutes with the action* (4.11) *of* $\mathbf{S}^1$. *Then f has the form*

$$(P_1 z_1 + Q\bar{z}_1^{n_2 - 1} z_2^{n_1}, P_2 z_2 + Q_2 z_1^{n_2} \bar{z}_2^{n_1 - 1}) \tag{4.12}$$

where P_j and Q_j are complex-valued functions of $|z_1|^2$, $|z_2|^2$, $\mathrm{Re}(z_1^{n_2}\bar{z}_2^{n_1})$, *and* $\mathrm{Im}(z_1^{n_2}\bar{z}_2^{n_1})$.

Proof. The proof is left to the reader as Exercise 4.2. □

We say that the resonance (4.10) is *weak* if the form of (4.12) is the same as the form of (4.5) *to cubic order* and *strong* otherwise. Recall (Proposition X, 2.3) that terms of order greater than 3 may be ignored in the singularity theory analysis of (4.9), at least when (4.9) is nondegenerate. Thus up to order 3, the cases of weak resonance and no resonance are identical in form and admit the same analysis described earlier for the nonresonant case.

These remarks do *not* constitute a proof that weakly resonant systems behave similarly to nonresonant systems. However, they do: see Arnold [1983].

The cases of strong resonance are easily enumerated: we need $n_1 + n_2 \leq 4$. We may assume that $n_1 \geq n_2$ and hence that

$$n_1 : n_2 = 1:1, 2:1, 3:1, \quad \text{and} \quad 4:1$$

for strong resonance. These cases are discussed in Arnold [1983].

CHAPTER XX

Mode Interactions with **O**(2) Symmetry

§0. Introduction

In the final chapter of this volume we extend the ideas of the previous chapter to the **O**(2)-symmetric case by studying mode interactions in **O**(2)-symmetric systems.

Before describing the results, however, let us consider general mode interactions in a Γ-equivariant system

$$\dot{y} + g(y, \lambda, \alpha) = 0,$$

where $y \in \mathbb{R}^n$ and g commutes with Γ. We have allowed for the presence of extra parameters α; here it suffices to take $\alpha \in \mathbb{R}$.

First, assume α is not present. The generic (in λ) linear degeneracies are single "modes," either steady-state or Hopf. Steady-state modes occur when $(dg)_{0,0,0}$ has a zero eigenvalue; Hopf modes when it has an imaginary eigenvalue. For a steady-state mode the corresponding eigenspace is generically Γ-irreducible; for a Hopf mode it is generically Γ-simple.

When an extra parameter α is present, however, we may expect, generically, to find linear degeneracies at isolated values of α. These are the symmetric analogues of double zero eigenvalues, simultaneous zero and imaginary eigenvalues, and double imaginary eigenvalues, in a nonsymmetric system. The values of α are generically isolated since such a degeneracy can be destroyed by an arbitrarily small change in α. (Consider perturbations by projections onto individual components W_0 or W_1, much as in the proof of Propositions XIII, 3.4, and XVI, 1.4.) These degeneracies are typically of three kinds:

(a) The 0-eigenspace of $(dg)_{0,0,0}$ decomposes as the direct sum of two Γ-irreducible subspaces $W_0 \oplus W_1$.

(b) A zero eigenvalue and a purely imaginary one occur simultaneously. The problem can be posed on the direct sum of the two corresponding eigenspaces. The zero eigenspace is generically Γ-irreducible, the imaginary eigenspace Γ-simple.
(c) There may be a degeneracy in the imaginary eigenspace, in which case it splits as $W_1 \oplus W_2$ where each W_j is Γ-simple.

In all three cases, as α is varied near 0 the bifurcation at 0 "splits" into two separate bifurcations, one along the W_0 mode, the other along the W_1 mode. Secondary bifurcations, or other features of the interactions of these modes, may appear.

A general theory for the study of such interactions for arbitrary Γ is not currently known. However, when $\Gamma = \mathbf{O}(2)$ we can use Birkhoff normal form and amplitude/phase equations, much as in Chapter XIX. The remainder of this chapter develops these methods and illustrates the results that may be obtained. (We do not address the issues raised if the system is not in Birkhoff normal form to all orders.) The three previous cases (a, b, c) are described in §§1, 2, 3, respectively.

Thus §1 concerns the interaction of two $\mathbf{O}(2)$-symmetric steady-state modes. After Liapunov–Schmidt reduction we can assume that the space is of dimension two, three, or four. The dimensions less than four correspond to $\mathbf{O}(2)$ acting trivially, or nearly so, on at least one mode, and we therefore concentrate on dimension four as being the most interesting case. Here the space can be written as $V_l \oplus V_m$ where $\mathbf{O}(2)$ acts by l-fold rotations on V_l and m-fold rotations on V_m. That is, we must consider nonstandard actions. We mention one key feature of the analysis: the possibility of generating periodic solutions by such an interaction.

In §2 we consider Hopf/steady-state mode interaction with $\mathbf{O}(2)$ symmetry. This situation arises in the analysis of Taylor–Couette flow in Case Study 6, and our main aim is to derive the results needed there. We obtain numerous branches of solutions, together with conditions on their directions of branching and stabilities.

Finally in §3 we study Hopf/Hopf interactions with $\mathbf{O}(2)$ symmetry. Here there is a multitude of cases. The dimension of the space can be assumed to be 4, 6, or 8, depending on whether the eigenvalues are simple or double. The four-dimensional case is effectively that studied in Chapter XIX since in this case $\mathbf{O}(2)$ acts trivially or via its factor group $\mathbf{Z}_2 = \mathbf{O}(2)/\mathbf{SO}(2)$. For this reason only the six- and eight-dimensional cases are studied, in two subsections. In addition, the $\mathbf{O}(2)$ action may be nonstandard, requiring further case distinctions.

In the six-dimensional case the full Birkhoff normal form equations decouple into amplitude and phase equations. One interesting feature is that a Hopf bifurcation can be forced along a secondary branch, leading to the occurrence of an invariant 3-torus.

In the eight-dimensional case the equations do not decouple, but the important dynamics occur on six-dimensional fixed-point subspaces, and when restricted to these, the Birkhoff normal form again decouples into amplitude and phase equations. Again there is the possibility of a forced Hopf bifurcation, this time to an invariant 4-torus.

§1.† Steady-State Mode Interaction

Let $g: \mathbb{R}^n \times \mathbb{R} \to \mathbb{R}$ be a bifurcation problem with $\mathbf{O}(2)$ symmetry. Suppose that g has a branch of $\mathbf{O}(2)$-invariant steady states which, for simplicity, we assume is the origin, so

$$g(0, \lambda) \equiv 0. \tag{1.1}$$

We also assume that g has a steady-state bifurcation at $\lambda = 0$, so

$$V = \ker(dg)_{0,0} \tag{1.2}$$

is nonzero. Generically, the representation of $\mathbf{O}(2)$ on V is irreducible. Thus either

$$\dim V = 1 \quad \text{or} \quad \dim V = 2, \tag{1.3}$$

and if $\dim V = 1$ then $\mathbf{O}(2)$ acts either trivially, or $\mathbf{SO}(2)$ acts trivially and $\mathbf{O}(2) \sim \mathbf{SO}(2)$ acts as $-Id$. If $\dim V = 2$ then we recall from XVII, 1(d), that the $\mathbf{O}(2)$ actions are indexed by an integer $m \geq 1$ as follows. Identify V with $\mathbb{C}$. Then

$$\begin{aligned} \theta \cdot z &= e^{mi\theta} z \qquad (\theta \in \mathbf{SO}(2)) \\ \kappa \cdot z &= \bar{z} \qquad (\kappa = \text{flip}). \end{aligned} \tag{1.4}$$

This action is denoted by ρ_m. The standard action is ρ_1.

In PDEs such actions of $\mathbf{O}(2)$ arise naturally when the equation is posed on a bounded domain with periodic boundary conditions. The $\mathbf{O}(2)$ symmetry is then obtained by spatial translation and reflection. See the Bénard problem in Case Study 4, and the Taylor–Couette system in Case Study 6. In such systems the kernel V often has the form

$$V_m = \mathbb{R}\{\cos(mx)U_0, \sin(mx)U_0\} \tag{1.5}$$

where U_0 is a fixed vector in some Banach space. The action of $\mathbf{O}(2)$ on V_m is given by

$$\begin{aligned} &\text{(a)} \quad x \mapsto x + \theta \\ &\text{(b)} \quad x \mapsto -x \end{aligned} \tag{1.6}$$

and is isomorphic to ρ_m.

In XVII, §1(d), we also remarked that both the invariants and the equivariants for the actions ρ_m of $\mathbf{O}(2)$ are the same for different values of m. As a result, generic steady-state bifurcation is much the same for all the actions ρ_m. Indeed the matrices in $\rho_m(\mathbf{O}(2))$ are the same as those in $\rho_1(\mathbf{O}(2))$: they are just repeated m times. In algebraic language, the kernel of ρ_m is

$$\mathbf{Z}_m = \{2k\pi/m: 0 \leq k \leq m - 1\}$$

and $\mathbf{O}(2)/\mathbf{Z}_m \cong \mathbf{O}(2)$. The action ρ_m factors through the standard action:

$$\rho_m(\gamma) \cdot x = \rho_1(v(\gamma)) \cdot x$$

where v: $\mathbf{O}(2) \to \mathbf{O}(2)$ is the natural quotient map with kernel $\mathbf{Z}_m$. The general $\mathbf{O}(2)$-equivariant mapping on $\mathbb{C}$ is

$$g(z, \lambda) = p(|z|^2, \lambda)z$$

so these bifurcations are easily studied.

In this section we consider the linear degeneracy where two steady-state modes interact. Thus we assume (1.1 and 1.2) where

$$V = V_l \oplus V_m. \tag{1.7}$$

Now we *cannot* assume that l and m are both 1: the exact values of l and m are important. However, we can again factor out the kernel of the action of $\mathbf{O}(2)$ on V in (1.7), which is $\mathbf{Z}_l \cap \mathbf{Z}_m = \mathbf{Z}_d$ where $d = \gcd(l, m)$. This lets us assume that l and m are coprime in the abstract analysis. (However, when interpreting the results in a specific application, the kernel $\mathbf{Z}_d$ must be restored.)

This $\mathbf{O}(2)$ mode interaction, with $(l, m) = (1, 2)$, occurs in the work of Jones and Proctor [1987] on Bénard convection in cylindrical geometry.

For the rest of this section, we assume that

$$g: V \times \mathbb{R} \to V; \tag{1.8}$$

that is, a Liapunov–Schmidt reduction has already been performed.

Remarks 1.1.

(a) The case $l = m = 1$ is markedly different from all others. From (1.8) we know that $L = (dg)_{0,0}$ has all eigenvalues zero and commutes with $\mathbf{O}(2)$. When $l \neq m$ commutativity implies

$$L = \begin{bmatrix} aI & 0 \\ 0 & dI \end{bmatrix}$$

and zero eigenvalues imply $L = 0$. However, when $l = m$ commutativity implies

$$L = \begin{bmatrix} aI & bI \\ cI & dI \end{bmatrix}$$

and zero eigenvalues imply $d = -a$ and $ad - bc = 0$. Thus L is generically nonzero and nilpotent, with Jordan canonical form

$$L = \begin{bmatrix} 0 & 1 \\ 0 & 0 \end{bmatrix}.$$

This case, the **O**(2) *symmetric Takens–Bogdanov bifurcation*, leads to homoclinic orbits and other interesting dynamics. See Guckenheimer [1986] and Dangelmayr and Knobloch [1987a].

(b) Often in PDE models the modes are indexed by m, the *wave number*, and mode interactions with **O**(2) symmetry occur for the simultaneous instability of a steady state to two consecutive modes. That is, $m = l + 1$. In order to reduce the number of different cases under discussion, we shall usually assume that

$$m = l + 1, \qquad l \geq 2.$$

In particular, we shall exclude the cases $(l, m) = (1, 1)$ and $(1, 2)$.

The remainder of this section is divided into three subsections:

(a) Invariant theory and the isotropy lattice,
(b) Bifurcation diagrams,
(c) Stability and dynamics.

(a) Invariant Theory and the Isotropy Lattice

Identify $V = V_l \oplus V_m$ with $\mathbb{C}^2$, using coordinates $z = (z_1, z_2)$. By (1.5) the action of **O**(2) on $\mathbb{C}^2$ is

$$\begin{aligned} &\text{(a)} \quad \theta \cdot (z_1, z_2) = (e^{li\theta} z_1, e^{mi\theta} z_2) \\ &\text{(b)} \quad \kappa \cdot (z_1, z_2) = (\bar{z}_1, \bar{z}_2). \end{aligned} \tag{1.9}$$

The first result follows routine lines and is left as an exercise.

Theorem 1.2.

(a) *A Hilbert basis for the* **O**(2)*-invariants on* $\mathbb{C}^2$ *is*

$$u = z_1 \bar{z}_1, \qquad v = z_2 \bar{z}_2, \qquad w = z_1^m \bar{z}_2^l + \bar{z}_1^m z_2^l.$$

(b) *The* **O**(2)*-equivariants on* $\mathbb{C}^2$ *are generated by*

$$\begin{bmatrix} z_1 \\ 0 \end{bmatrix}, \begin{bmatrix} \bar{z}_1^{m-1} z_2^l \\ 0 \end{bmatrix}, \begin{bmatrix} 0 \\ z_2 \end{bmatrix}, \begin{bmatrix} 0 \\ z_1^m \bar{z}_2^{l-1} \end{bmatrix}.$$

Thus g in (1.8) has the form

$$g(z, \lambda) = (p_1 z_1 + q_1 \bar{z}_1^{m-1} z_2^l, p_2 z_2 + q_2 z_1^m \bar{z}_2^{l-1}) \tag{1.10}$$

where p_1, p_2, q_1, q_2 are functions of u, v, w, λ, and

$$p_1(0) = p_2(0) = 0.$$

Our main interest is in the zeros of g. The first step is to find the isotropy lattice. Since we have excluded the case $l = m$ we may assume $m > l \geq 1$. The reader should check that the lattice is the one shown in Figure 1.1.

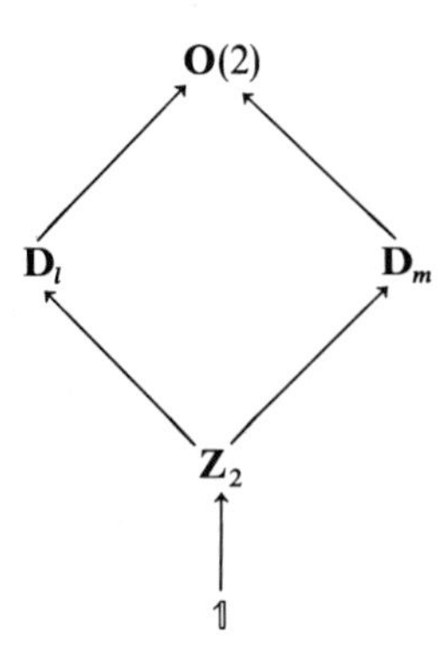

Figure 1.1. Isotropy lattice for $\mathbf{O}(2)$ acting on $\mathbb{C}^2$ when $m > l \geq 2$.

Table 1.1. Isotropy Data for $\mathbf{O}(2)$ Acting on $\mathbb{C}^2$, $m > l \geq 3$

Isotropy Σ	$\mathrm{Fix}(\Sigma)$	$g \vert \mathrm{Fix}(\Sigma) = 0$	Signs of Eigenvalues of dg
$\mathbf{O}(2)$	$\{0\}$	—	p_1 [twice]
			p_2 [twice]
$\mathbf{D}_l$	$\mathbb{R}\{(1,0)\}$	$p_1(x^2,0,0,\lambda) = 0$	p_2 [twice]
		$z_1 = x \in \mathbb{R}$	$p_{1,u}$
		$z_2 = 0$	0
$\mathbf{D}_m$	$\mathbb{R}\{(0,1)\}$	$p_2(0,y^2,0,\lambda) = 0$	p_1 [twice]
		$z_1 = 0$	$p_{2,v}$
		$z_2 = y \in \mathbb{R}$	0
$\mathbf{Z}_2$	$\mathbb{R} \oplus \mathbb{R}$	$h_1 = p_1 + q_2 x^{m-2} y^l = 0$	$mp_1 + lp_2$
	$z = (x, y)$	$h_2 = p_2 + q_2 x^m y^{l-2} = 0$	0
			Eigenvalues of $\dfrac{\partial(h_1, h_2)}{\partial(x, y)}$
$\mathbb{1}$	$\mathbb{C}^2$	$p_1 = p_2 = 0$	—
		$q_1 = q_2 = 0$	

Here $\mathbf{Z}_2 = \{1, \kappa\}$ where κ acts as in (1.9(b)). The fixed-point subspaces and the restrictions of $g = 0$ to these subspaces are given in Table 1.1. We also include the signs of the eigenvalues of dg. However, the prescription for finding these eigenvalues on the $\mathbf{D}_l$ solution is valid only when $l \geq 3$. When $l = 2$ commutativity of dg with $\mathbf{D}_2$ does not force the eigenvalue to be double.

By Table 1.1, generically steady-state solutions with trivial isotropy do not occur, since generically at points of mode interaction $q_1(0)$ and $q_2(0)$ are nonzero. Thus, when seeking steady-state solutions, we can restrict attention to the two-dimensional fixed-point subspace $\mathbb{R}^2$ of $\mathbf{Z}_2$. Dangelmayr [1986] points out that generically, interesting dynamics (such as periodic solutions) may be expected to occur off this subspace. This point is discussed in more detail in subsection (c).

We call the solutions with $\mathbf{D}_l$ or $\mathbf{D}_m$ isotropy *pure modes*, and those with only $\mathbf{Z}_2$ isotropy *mixed modes*, by analogy with mode interactions with $\mathbf{Z}_2 \oplus \mathbf{Z}_2$ symmetry in Chapter X.

(b) Bifurcation Diagrams

We now restrict attention to interaction between consecutive modes, so $m = l + 1$, $l \geq 2$. We also make the generic hypothesis that $q_1(0)q_2(0) \neq 0$, which implies that all solutions to the universal $\mathbf{O}(2)$-unfolding of $g = 0$ lie in the two-dimensional fixed-point subspace $\mathrm{Fix}(\mathbf{Z}_2) \times \mathbb{R}$. With $x = \mathrm{Re}(z_1)$, $y = \mathrm{Re}(z_2)$ we have

$$g|\mathrm{Fix}(\mathbf{Z}_2) \times \mathbb{R} = (p_1 x + q_1 x^l y^l, p_2 y + q_2 x^{l+1} y^{l-1}), \tag{1.11}$$

where p_j, q_j are functions of $u = x^2$, $v = y^2$, $w = 2x^l y^{l+1}$, and λ. The singularity theory for (1.11) is given in Armbruster and Dangelmayr [1986].

The smallest term in (1.11) involving l is homogeneous of degree $2l$. To third order $g|\mathrm{Fix}(\mathbf{Z}_2) \times \mathbb{R}$ has the form

$$k(x, y, \lambda) = (Ax^3 + Bxy^2 + \alpha\lambda x, Cx^2 y + Dy^3 + \beta\lambda y). \tag{1.12}$$

Thus k is the general third order $\mathbf{Z}_2 \oplus \mathbf{Z}_2$-equivariant bifurcation problem on $\mathbb{R}^2$; see Lemma X, 1.1. Moreover, we have shown that nondegenerate $\mathbf{Z}_2 \oplus \mathbf{Z}_2$-bifurcation problems (Definition X, 2.2) are 3-*determined*, that is, $\mathbf{Z}_2$-equivalent to their truncations at degree 3. Henceforth we assume that k is nondegenerate, a hypothesis that is valid generically.

Thus we may think of $g|\mathrm{Fix}(\mathbf{Z}_2) \times \mathbb{R}$ as breaking the $\mathbf{Z}_2 \oplus \mathbf{Z}_2$ symmetry of k at high order, and we might expect the bifurcation diagrams of the universal $\mathbf{O}(2)$-unfolding of g to be small "symmetry-breaking" perturbations of those in the universal $\mathbf{Z}_2 \oplus \mathbf{Z}_2$-unfolding of k. (Recall Chapter X, Figures 4.1–4.3.) To a large extent this view is correct; however, there are some important exceptions which are most easily understood by using group theory. As l becomes large we might expect the diagrams for $g|\mathrm{Fix}(\mathbf{Z}_2) \times \mathbb{R} = 0$ to become closer to those for $k = 0$ since the symmetry-breaking terms begin at degree $2l$. Indeed, there are differences between $l = 2$ and $l = 3$ (where the first symmetry-breaking term is of degree ≤ 6) and $l \geq 4$ (where it has degree ≥ 8).

We begin the group-theoretic comparison by asking how much of the apparent $\mathbf{Z}_2 \oplus \mathbf{Z}_2$ symmetry on $\mathrm{Fix}(\mathbf{Z}_2)$ actually comes from $\mathbf{O}(2)$. (The occurrence of symmetry-breaking terms shows that the full $\mathbf{Z}_2 \oplus \mathbf{Z}_2$ symmetry cannot be present.) There are two points:

(a) Rotation through a half-period π gives a symmetry on $\mathrm{Fix}(\mathbf{Z}_2)$. (1.13)

(b) There is a symmetry on a one-dimensional subspace of $\mathrm{Fix}(\mathbf{Z}_2)$ given by rotation either through a quarter period $\pi/2$ or through an eighth period $\pi/4$, depending on the parity of l.

It is (1.13(b)) that is perhaps unexpected.

We discuss these points in turn, assuming for definiteness that l is odd. Similar observations hold when l is even. Direct computation establishes (1.13(a)):

$$\pi \cdot (z_1, z_2) = (e^{li\pi} z_1, e^{mi\pi} z_2) = (-z_1, z_2).$$

Thus the action of π restricted to $\mathrm{Fix}(\mathbf{Z}_2)$ is

$$\pi \cdot (x, y) = (-x, y). \tag{1.14}$$

Abstractly, this follows from Lemma XIII, 10.2: the normalizer $N(\Sigma)$ of Σ leaves $\mathrm{Fix}(\Sigma)$ invariant (and indeed is the largest subgroup of Γ with this property). Here

$$N(\mathbf{Z}_2) = \{1, \kappa, \pi, \kappa\pi\}.$$

Thus the only elements in $\mathbf{O}(2)$ that act nontrivially on $\mathrm{Fix}(\mathbf{Z}_2)$ are π and $\kappa\pi$, and these have the same action. (Or again apply Lemma XIII, 10.2: it is $N(\Sigma)/\Sigma$ that acts nontrivially.)

Observation (1.13(b)) also follows by direct calculation:

$$\frac{\pi}{2} \cdot (z_1, z_2) = (e^{l\pi i/2} z_1, e^{(l+1)\pi i/2} z_2).$$

Thus if $l + 1 \equiv 2 \pmod 4$, that is, $\frac{1}{2}(l + 1)$ is odd, then

$$\frac{\pi}{2} \cdot (0, y) = (0, -y), \tag{1.15}$$

whereas if $l + 1 \equiv 0 \pmod 4$, that is, $\frac{1}{2}(l + 1)$ is even, then

$$\frac{\pi}{8} \cdot (0, y) = (0, -y). \tag{1.16}$$

Abstractly, these symmetries on subspaces derive from the notion of *hidden symmetry* of Golubitsky, Marsden, and Schaeffer [1984]. The idea is as follows. Let $\gamma \in \Gamma \sim N(\Sigma)$ and suppose that

$$W = \gamma \cdot \mathrm{Fix}(\Sigma) \cap \mathrm{Fix}(\Sigma) \neq \{0\}.$$

Now $W \neq \mathrm{Fix}(\Sigma)$ since $\gamma \notin N(\Sigma)$. By definition

$$\gamma\colon W \to W$$

and γ is a symmetry on the subspace W of $\mathrm{Fix}(\Sigma)$. Any Γ-equivariant map g satisfies

$$g(\gamma w) = \gamma g(w), \qquad (w \in W).$$

Since $g(w), \gamma g(w) \in \mathrm{Fix}(\Sigma)$, these hidden symmetries γ place extra restrictions on $g|\mathrm{Fix}(\Sigma)$, in addition to those imposed by the "overt" symmetries in $N(\Sigma)$. Symmetries on subspaces were first noticed by Hunt [1981, 1982] when discussing $\mathbf{O}(2)$ mode interactions in the buckling of a cylindrical shell. See also the Exercises in XII, §4.

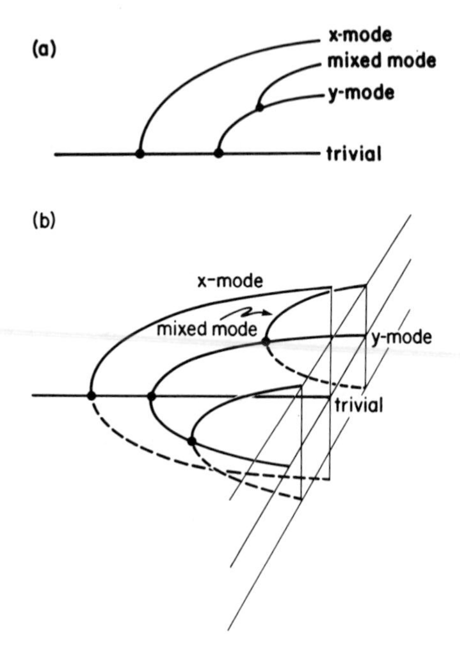

Figure 1.2. (a) Schematic $\mathbf{Z}_2 \oplus \mathbf{Z}_2$ bifurcation diagram. (b) The same diagram, drawn in (x, y, λ) space and distinguishing points in the same group orbit.

We have shown that (1.15, 1.16) are symmetries of $g|\mathrm{Fix}(\mathbf{Z}_2)$. If (1.16) could be extended to a symmetry $(x, y) \to (x, -y)$ then we would have full $\mathbf{Z}_2 \oplus \mathbf{Z}_2$ symmetry.

In fact $\mathbf{O}(2)$ symmetry imposes more restrictions on $g|\mathrm{Fix}(\Sigma)$ than those obtained from (1.14, 1.15, 1.16). For example, when $l = 3$ the term $(x^2y, 0)$ commutes with (1.16) but does not appear in $g|\mathrm{Fix}(\mathbf{Z}_2)$. The abstract question, of determining the precise restrictions placed on $g|\mathrm{Fix}(\Sigma)$ by Γ-equivariance of g, is far from understood.

How should the breaking of $\mathbf{Z}_2 \oplus \mathbf{Z}_2$ symmetry to (1.15, 1.16) affect the bifurcation diagrams for the universal $\mathbf{O}(2)$-unfolding of $g|\mathrm{Fix}(\mathbf{Z}_2)$? For argument's sake, recall (Figure 1.2(a)) a typical bifurcation diagram in the universal unfolding of a nondegenerate $\mathbf{Z}_2 \oplus \mathbf{Z}_2$ bifurcation problem. (See Figure X, 4.3(1), $\sigma > 0$.) As explained in XIII, §5(a), Figure 1.2(a) is a *schematic* bifurcation diagram, each point representing a $\mathbf{Z}_2 \oplus \mathbf{Z}_2$ group orbit of solutions. In particular, x-mode solutions $(x, 0, \lambda)$ with $x \neq 0$ actually represent two solutions $(\pm x, 0, \lambda)$ and similarly for y-mode solutions. Mixed mode solutions (x, y, λ) where $xy \neq 0$ represent four solutions $(\pm x, \pm y, \lambda)$. These solutions are all drawn in (x, y, λ)-space in Figure 1.2(b).

Even when the $\mathbf{Z}_2 \oplus \mathbf{Z}_2$ symmetry is broken, we expect—because of (1.14, 1.15, 1.16)—the pure mode solutions to occur in pairs. Thus there is just

Figure 1.3. Schematic bifurcation diagram when symmetry-breaking terms are included.

one orbit for each pure mode solution. However, now the four mixed mode solutions also occur in pairs $(\pm x, y, \lambda)$ since the symmetry (1.15, 1.16) acts trivially on these solutions. Thus we expect *two* orbits of mixed mode solutions. On the other hand, the $\mathbf{Z}_2 \oplus \mathbf{Z}_2$ symmetry-breaking terms that are added to g are of "high" order. So it is reasonable to expect the solution branches in Figure 1.2(b) not to disappear. This is precisely what happens. Thus the schematic bifurcation diagram that we expect to find when symmetry-breaking terms are added is Figure 1.3. For further details, see Buzano and Russo [1987] and Dangelmayr [1986].

(c) Stability and Dynamics

We begin our analysis of the dynamics of the 4×4 system of ODEs

$$\dot{z} + g(z, \lambda) = 0 \tag{1.17}$$

by discussing the asymptotic stability of the steady-state solutions described in subsection (b). Again we focus on the similarities and differences between the $\mathbf{O}(2)$ and $\mathbf{Z}_2 \oplus \mathbf{Z}_2$ mode interactions.

Table 1.1 showed how to compute the stabilities for each solution type when $l \geq 3$. These results, obtained in the standard way by using the isotropy subgroups, show that for the trivial and pure mode solutions stability assignments for the two types of mode interaction are identical, in the following sense: For the trivial solution, the two eigenvalues of $dg|\mathrm{Fix}(\mathbf{Z}_2)$ are each double as eigenvalues of dg. Thus the stability of the trivial solution is the same as the stability of $dg|\mathrm{Fix}(\mathbf{Z}_2)$, and this stability is determined by the same coefficients in both cases. Similarly for pure mode solutions, the two eigenvalues of $dg|\mathrm{Fix}(\mathbf{Z}_2)$ determine the stability as follows: (when $l \geq 3$) one of these two eigenvalues is doubled (forced by $\mathbf{D}_l$ isotropy) and the fourth is forced to zero by $\mathbf{O}(2)$ symmetry. (When $l = 2$ the eigenvalue need not be doubled and a third nonzero eigenvalue must be controlled.)

The eigenvalues for the mixed mode solution behave somewhat differently. Indeed, there are two eigenvalues given by $dg|\mathrm{Fix}(\mathbf{Z}_2)$ and one eigenvalue forced to zero by $\mathbf{O}(2)$ symmetry. Now the fourth eigenvalue, which by Table 1.1 is $(l + 1)p_1 + lp_2$, is independent of the two eigenvalues of $dg|\mathrm{Fix}(\mathbf{Z}_2)$. This eigenvalue is also the source of a very interesting bifurcation, as we shall see.

The eigenvalues of $dg|\text{Fix}(\mathbf{Z}_2)$ are determined by those of the 2×2 matrix $\partial(h_1, h_2)/\partial(x, y)$ in Table 1.1 and correspond to the eigenvalues along mixed mode solutions in the $\mathbf{Z}_2 \oplus \mathbf{Z}_2$ interaction. As in Chapter X, Figure 4.3, these eigenvalues may cross the imaginary axis away from zero, causing a Hopf bifurcation to a periodic solution. We have seen this phenomenon for amplitude equations in several different contexts. It produces 3-tori in degenerate Hopf bifurcation with **O**(2) symmetry (XVII, §7), 2-tori in steady-state/Hopf mode interactions without symmetry (XIX, §3), and 3-tori in Hopf/Hopf interactions without symmetry (XIX, §4). Periodic solutions can also be generated by this mechanism applied to the mixed mode solutions for steady-state **O**(2) mode interactions.

Dangelmayr [1986] points out that the "extra" eigenvalue $mp_1 + lp_2$ can also change sign, and that when it does two periodic *rotating wave* solutions are formed. Moreover, when $l \geq 2$ these rotating waves can be asymptotically stable. We reiterate that the crossing through zero of a single eigenvalue does *not* produce a new steady state, but rather a periodic solution. The existence of the zero eigenvalue forced by symmetry is crucial to this kind of behavior. More precisely, this phenomenon requires a whole group orbit of mixed mode solutions (in this case a circle's worth) to bifurcate simultaneously.

Dangelmayr's analysis proceeds as follows. Try to write the system (1.17) in phase/amplitude form by setting

$$z_1 = re^{i\varphi}, \qquad z_2 = se^{i\psi}. \tag{1.18}$$

From (1.10) we obtain a system in which the amplitude equations depend explicitly on a mixed phase variable θ:

$$\begin{aligned} &\text{(a)} \quad \dot{r} + rp_1 + r^{m-1}s^l q_1 \cos\theta = 0 \\ &\text{(b)} \quad \dot{s} + sp_2 + r^m s^{l-1} q_2 \cos\theta = 0 \\ &\text{(c)} \quad \dot{\theta} - (mq_1 s^2 + lq_2 r^2) r^{m-2} s^{l-2} \sin\theta = 0 \end{aligned} \tag{1.19}$$

where

$$\theta = m\varphi - l\psi. \tag{1.20}$$

These equations hold for general l and m, but our discussion of them will again be limited to the case $m = l + 1$. Effectively we can use the **SO**(2) symmetry to eliminate *one* phase, but not both, from the system of ODEs. More precisely,

$$r^2 q_2 \dot{\varphi} + s^2 q_1 \dot{\psi} \equiv 0 \tag{1.21}$$

is an integral for the 4×4 system (1.17).

There are three types of steady state for the equation (1.19(c)):

(1) $r = 0$, or $s = 0$, or both. These correspond to the pure mode and trivial solutions of (1.17).
(2) $\sin\theta = 0$ and $rs \neq 0$. These steady states correspond to steady mixed mode solutions of (1.17). (Note that $\cos\theta = \pm 1$.)

(3) $mq_1 s^2 + lq_2 r^2 = 0$. These are steady-state solutions of (1.19) that correspond to periodic solutions of (1.17).

To understand this last statement, recall (1.20), which implies that

$$\dot{\theta} = m\dot{\varphi} - l\dot{\psi}. \tag{1.22}$$

If $\dot{\theta} = 0$ we can couple (1.21) and (1.22) to form the linear system

$$\begin{bmatrix} m & -l \\ r^2 q_2 & s^2 q_1 \end{bmatrix} \begin{bmatrix} \dot{\varphi} \\ \dot{\psi} \end{bmatrix} = 0. \tag{1.23}$$

As long as the determinant $D = ms^2 q_1 + lr^2 q_2 \neq 0$, then (1.23) implies that $\dot{\varphi} = \dot{\psi} = 0$; that is, the steady state of (1.19) is a true steady state of (1.17). Now D is nonzero for the pure and mixed mode solutions but vanishes for type (3) solutions. Thus for type (3) solutions, $\dot{\varphi}$ and $\dot{\psi}$ are nonzero. From $\dot{\theta} = 0$ we see that

$$\dot{\varphi}/\dot{\psi} = l/m. \tag{1.24}$$

Therefore, the steady state of (1.19) corresponds to a periodic solution $(z_1(t), z_2(t))$ of (1.17), where $z_1(t)$ winds l times around the origin and $z_2(t)$ winds m times around.

The planes $\theta = 0, \pi$ are invariant under the flow of the (r, s, θ) system. If no type (3) solution occurs then it may be shown that $\theta \to 0$ or $\theta \to \pi$ as $t \to \pm\infty$. Hence no other interesting dynamics occur.

Next, we claim that each of these periodic solutions is a rotating wave with isotropy $\widetilde{\mathbf{SO}}(2) \subset \mathbf{SO}(2) \times \mathbf{S}^1$. To see why, recall from (1.9(a)) that $\mathbf{SO}(2) \subset \mathbf{O}(2)$ acts on $\mathbb{C}^2$ by an l-fold rotation on the first component and an m-fold rotation on the second. Thus temporal evolution of the system coincides with spatial rotation of the solution trajectory. This verifies the claim. The $\mathbf{O}(2)$ symmetry, in particular κ in (1.9(b)), forces two such trajectories to occur in the system (1.17).

To show the complexities that can arise, we present in Figure 1.4 a single persistent bifurcation diagram that occurs in the case $(l, m) = (2, 3)$. It is taken from Dangelmayr [1986], who gives a detailed discussion, including cases that we have not studied here.

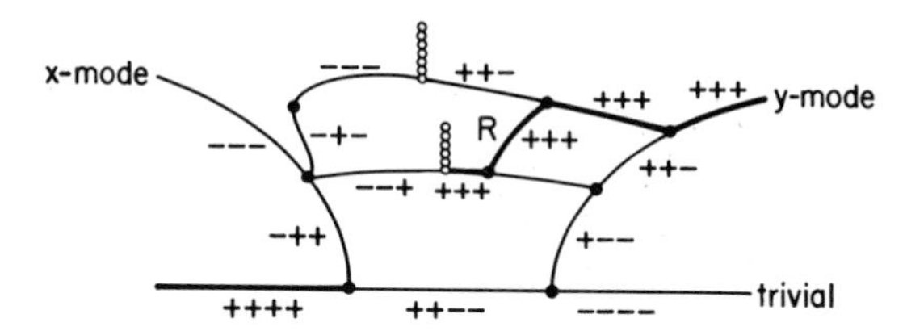

Figure 1.4. Some of the complexities of the case $(l, m) = (2, 3)$. R = rotating wave.

§2. Hopf/Steady-State Mode Interaction

Let

$$\dot{y} + G(y, \lambda, \alpha) = 0 \tag{2.1}$$

be an **O**(2)-equivariant system of ODEs that depends on two parameters λ, α. We assume that $y = 0$ is a steady state, so

$$G(0, \lambda, \alpha) \equiv 0.$$

As in XIX, §1, generically in two-parameter systems there can exist isolated points at which steady-state and Hopf bifurcations occur simultaneously. Specifically, assume that at $\lambda = \alpha = 0$ the linearization $(dG)_{0,0,0}$ has eigenvalues 0 and $\pm\omega i$ $(\omega \neq 0)$.

Because of the **O**(2) symmetry, generically each such eigenvalue is simple or double. In this section we assume all three eigenvalues are double; this will be the situation encountered in Case Study 6 in the discussion of the Taylor–Couette experiment. The analysis of the other cases is similar; see Exercises 2.1 and 2.2. Assuming all eigenvalues double amounts to assuming that both the steady-state bifurcation and the Hopf bifurcation break the **O**(2) symmetry.

Our main aim is to determine the various steady-state and periodic solutions of (2.1). To accomplish this we assume that a Liapunov–Schmidt reduction has been performed, yielding a mapping

$$g\colon \mathbb{R}^6 \times \mathbb{R}^2 \times \mathbb{R} \to \mathbb{R}^6$$

depending on parameters λ, α, τ, where τ is a period-scaling parameter introduced by the reduction procedure.

To simplify notation, identify $\mathbb{R}^6$ with $\mathbb{C}^3 = \{(z_0, z_1, z_2)\}$ where $z_0 = 0$ corresponds to the four-dimensional eigenspace for eigenvalues $\pm\omega i$, and $z_1 = z_2 = 0$ corresponds to the two-dimensional eigenspace for the eigenvalue 0. We choose the coordinates (z_1, z_2) as we did for **O**(2)-symmetric Hopf bifurcation in (XVII, 1.5). The action of **O**(2) on $\mathbb{C}^3$ is then generated by

$$\begin{aligned} &\text{(a)} \quad \varphi\cdot(z_0, z_1, z_2) = (e^{mi\varphi}z_0, e^{li\varphi}z_1, e^{-li\varphi}z_2) \\ &\text{(b)} \quad \kappa\cdot(z_0, z_1, z_2) = (\bar{z}_0, z_2, z_1). \end{aligned} \tag{2.2}$$

See (XVII, 1.6). When analyzing **O**(2)-symmetric Hopf bifurcation we were able to assume, without loss of generality, that $l = 1$. Here the simultaneous action of **O**(2) on z_0 prevents this. However, we can still factor out the kernel of the **O**(2)-action on $\mathbb{C}^3$ or, equivalently, assume that l and m are relatively prime. (As usual, the kernel must be restored when interpreting the results in any particular application.)

The reduced bifurcation equation $g = 0$ has $\mathbf{S}^1$ phase-shift symmetry as well as **O**(2) symmetry. Rescale time so that $\theta \in \mathbf{S}^1$ acts by

$$\theta\cdot(z_0, z_1, z_2) = (z_0, e^{i\theta}z_1, e^{i\theta}z_2). \tag{2.3}$$

Later we study:

(a) The invariant theory of $\mathbf{O}(2) \times \mathbf{S}^1$ acting on $\mathbb{C}^3$ by (2.2, 2.3),
(b) The isotropy lattice,
(c) The branching equations and stabilities of some of the possible solutions,
(d) Bifurcation to tori,
(e) Sample bifurcation diagrams.
The results in (c–e) are only sketched, and many details are left to the reader.

(a) Invariant Theory

In this subsection we prove the following:

Theorem 2.1. *Assume that m in* (2.2) *is odd. Then*
(a) *A Hilbert basis for the* $\mathbf{O}(2) \times \mathbf{S}^1$*-invariants is*

$$\rho = |z_0|^2, \qquad N = |z_1|^2 + |z_2|^2, \qquad \Delta = \delta^2, \qquad \Phi = \operatorname{Re} A, \qquad \Psi = \delta \operatorname{Im} A,$$

where $\delta = |z_2|^2 - |z_1|^2$ *and* $A = \bar{z}_0^{2l}(z_1 \bar{z}_2)^m$.
(b) *The* $\mathbf{O}(2) \times \mathbf{S}^1$*-equivariants are generated over the invariants by the twelve mappings* $(z_0, z_1, z_2) \mapsto$:

$$\begin{aligned}
V^1 &= (z_0, 0, 0),\ i\delta V^1,\\
V^2 &= (\bar{z}_0^{2l-1}(z_1 \bar{z}_2)^m, 0, 0),\ i\delta V^2,\\
V^3 &= (0, z_1, z_2),\ iV^3,\\
\delta V^4 &= \delta(0, z_1, -z_2),\ i\delta V^4,\\
V^5 &= (0, z_0^{2l}\bar{z}_1^{m-1} z_2^m, \bar{z}_0^{2l} z_1^m \bar{z}_2^{m-1}),\ iV^5,\\
\delta V^6 &= \delta(0, z_0^{2l}\bar{z}_1^{m-1} z_2^m, -\bar{z}_0^{2l} z_1^m \bar{z}_2^{m-1}),\ i\delta V^6.
\end{aligned}$$

The (by now routine) details of the proof are not needed later and the reader may prefer to skip directly to subsection (b). We begin with the following:

Lemma 2.2. *Define an action of* $\mathbf{O}(2)$ *on* $\mathbb{C}^2 \times \mathbb{R}$ *by*

$$\text{(a)} \quad \varphi \cdot (w, v, \delta) = (e^{mi\varphi} w, e^{ni\varphi} v, \delta)$$

$$\text{(b)} \quad \kappa \cdot (w, v, \delta) = (\bar{w}, \bar{v}, -\delta)$$

where m and n are coprime. Then
(a) *A Hilbert basis for the* $\mathbf{O}(2)$*-invariants is*

$$w\bar{w},\ v\bar{v},\ \delta^2,\ \operatorname{Re} C,\ \delta \operatorname{Im} C$$

where $C = w^n \bar{v}^m$.
(b) *The* $\mathbf{O}(2)$*-equivariants* $\mathbb{C}^2 \times \mathbb{R} \to \mathbb{C}$ *are generated by*

$$w,\ i\delta w,\ \overline{w}^{n-1}v^m,\ i\delta\overline{w}^{n-1}v^m$$

where $\mathbf{O}(2)$ *acts on the range* $\mathbb{C}$ *by* $\varphi\cdot z = e^{mi\varphi}z$, $\kappa\cdot z = \overline{z}$.

PROOF. This is a routine computation along standard lines and is left as an exercise. □

We are now ready for the following:

PROOF OF THEOREM 2.1.
(a) We follow the approach outlined in XVI, §9. By Lemma XVI, 9.2, we can determine a Hilbert basis for the $\mathbb{C}$-valued $\mathbf{S}^1$-invariants for (2.3). They are

$$z_0,\ \overline{z}_0,\ N,\ \delta,\ v,\ \overline{v}, \tag{2.4}$$

where $N = |z_1|^2 + |z_2|^2$, $\delta = |z_1|^2 - |z_2|^2$, $v = z_1\overline{z}_2$. The relation

$$4v\overline{v} = (N^2 - \delta^2) \tag{2.5}$$

holds. There is an action of $\mathbf{O}(2)$ on (2.4) induced by (2.2). The existence of this action is guaranteed by Lemma XVI, 9.1, but it may also be established directly. In particular,

$$\begin{aligned} &\text{(a)}\quad \varphi\cdot(z_0, N, \delta, v) = (e^{mi\varphi}z_0, N, \delta, e^{2li\varphi}v)\\ &\text{(b)}\quad \kappa\cdot(z_0, N, \delta, v) = (\overline{z}_0, N, -\delta, \overline{v}). \end{aligned} \tag{2.6}$$

By Lemma 2.2(a),

$$z_0\overline{z}_0,\ v\overline{v},\ N,\ \operatorname{Re} A,\ \delta\operatorname{Im} A,\ \delta^2 \tag{2.7}$$

is a Hilbert basis for the $\mathbf{O}(2)\times\mathbf{S}^1$-invariants, where $A = z_0^{2l}\overline{v}^m$. But $v\overline{v}$ is redundant by (2.5).
(b) Let $g = (Z_0, Z_1, Z_2)$ be an $\mathbf{O}(2)\times\mathbf{S}^1$-equivariant mapping $\mathbb{C}^3\to\mathbb{C}^3$. Commutativity with κ, as in (2.2(b)), implies

$$\begin{aligned} &\text{(a)}\quad \overline{Z}_0(z_0, z_1, z_2) = Z_0(\overline{z}_0, z_2, z_1)\\ &\text{(b)}\quad Z_2(z_0, z_1, z_2) = Z_1(\overline{z}_0, z_2, z_1). \end{aligned} \tag{2.8}$$

By (2.8(b)) we need only derive forms for Z_0 and Z_1. We consider Z_0 first.
Commutativity of g with $\mathbf{SO}(2)\times\mathbf{S}^1$ implies

$$\begin{aligned} &\text{(a)}\quad Z_0(z_0, e^{i\theta}z_1, e^{i\theta}z_2) = Z_0(z_0, z_1, z_2)\\ &\text{(b)}\quad Z_0(e^{mi\psi}z_0, e^{li\psi}z_1, e^{-li\psi}z_2) = e^{mi\psi}Z_0(z_0, z_1, z_2). \end{aligned} \tag{2.9}$$

Now (2.9(a)) states that Z_0 is a $\mathbb{C}$-valued $\mathbf{S}^1$-invariant, so by (2.4)

$$Z_0 = W_1(z_0, N, \delta, v).$$

Now (2.6) implies that

$$\begin{aligned} &\text{(a)}\quad W_1(e^{mi\psi}z_0, N, \delta, e^{2li\psi}v) = e^{mi\psi}W_1(z_0, N, \delta, v)\\ &\text{(b)}\quad W_1(z_0, N, -\delta, v) = \overline{W_1}(z_0, N, \delta, v). \end{aligned} \tag{2.10}$$

Since m is odd, Lemma 2.2(b) shows that V^1, $i\delta V^1$, V^2, and $i\delta V^2$ generate the Z_0-coordinate of the $\mathbf{O}(2) \times \mathbf{S}^1$-equivariants.

Finally we consider Z_1. The restrictions imposed by (2.2(a) and 2.3) are

$$\begin{aligned} &\text{(a)} \quad Z_1(e^{mi\psi}z_0, e^{li\psi}z_1, e^{-li\psi}z_2) = e^{li\psi}Z_1(z_0, z_1, z_2) \\ &\text{(b)} \quad Z_1(z_0, e^{i\theta}z_1, e^{i\theta}z_2) = e^{i\theta}Z_1(z_0, z_1, z_2). \end{aligned} \tag{2.11}$$

Write

$$Z_1 = \sum A_{abcdef} z_0^a \bar{z}_0^b z_1^c \bar{z}_1^d z_2^e \bar{z}_2^f. \tag{2.12}$$

By (2.11) $A_{abcdef} = 0$ unless

$$\begin{aligned} &\text{(a)} \quad m(a-b) + l(c-d-e+f-1) = 0 \\ &\text{(b)} \quad c-d+e-f-1 = 0. \end{aligned} \tag{2.13}$$

Substitute (2.13(b)) into (2.13(a)) to obtain

$$m(a-b) = 2l(e-f).$$

Since $2l$ and m are coprime,

$$\begin{aligned} &\text{(a)} \quad a-b = 2ls \\ &\text{(b)} \quad e-f = ms \\ &\text{(c)} \quad c-d = 1-ms \end{aligned} \tag{2.14}$$

for some integer s. Modulo $z_0\bar{z}_0$, $z_1\bar{z}_1 = \frac{1}{2}(N-\delta)$, and $z_2\bar{z}_2 = \frac{1}{2}(N+\delta)$, we may assume

$$a = 0 \quad \text{or} \quad b = 0$$

and

$$c = 0 \quad \text{or} \quad d = 0$$

and

$$e = 0 \quad \text{or} \quad f = 0.$$

Moreover, (2.14) implies that $a-b$ and $e-f$ have the same sign and (except when $s = 0$ or $m = s = 1$) $c-d$ has the opposite sign. Hence either $a = e = d = 0$ or $b = f = c = 0$.

This shows that

$$\begin{aligned} &z_0^{2ls}\bar{z}_1^{ms-1}z_2^{ms} \qquad (s > 1), \\ &\bar{z}_0^{2ls}z_1^{ms+1}\bar{z}_2^{ms} \qquad (s \geq 0) \end{aligned} \tag{2.15}$$

is a set of generators for the $\mathbf{O}(2) \times \mathbf{S}^1$-equivariants with complex coefficients depending on $z_0\bar{z}_0$, N, and δ. Multiplying each of these generators by $1, i, \delta, \delta i$ yields a set of generators with real coefficients depending on the $\mathbf{O}(2) \times \mathbf{S}^1$-invariants $w\bar{w}$, N, and δ^2.

There are identities

$$(\bar{z}_0^{2l} z_1^m \bar{z}_2^m)^s = \begin{cases} 2\,\mathrm{Re}(A)(z_0^{2l} z_1^m \bar{z}_2^m)^{s-1} - |z_0|^{4l} |z_1 z_2|^{2m} (z_0^{2l} z_1^m \bar{z}_2^m)^{s-2} & (s \geq 2) \\ 2\,\mathrm{Re}(A) - z_0^{2l} \bar{z}_1^m z_2^m \quad (s = 1). \end{cases} \tag{2.16}$$

Since $\mathrm{Re}\, A$ is an $\mathbf{O}(2) \times \mathbf{S}^1$-invariant and $|z_0|^{2l} |z_1 z_2|^m$ depends on $z_0 \bar{z}_0$, N, and δ, we can use induction on s to eliminate all of the last set of generators except for z_1. Similarly,

$$z_0^{2ls} z_1^{ms-1} \bar{z}_2^{ms} = A^{s-1} z_0^{2l} \bar{z}_1^{m-1} z_2^m$$

whence induction using a formula similar to (2.16) reduces the first set of generators to just $z_0^{2l} \bar{z}_1^{m-1} z_2^m$. Apply (2.8(b)) to complete the proof. □

(b) The Isotropy Lattice

We next find the orbits, isotropy subgroups, and fixed-point subspaces of $\mathbf{O}(2) \times \mathbf{S}^1$ acting on $\mathbb{C}^3$ by (1.2 and 1.3). In order to avoid a multiplicity of cases depending on l and m we henceforth assume that $l = m = 1$. This is the situation that obtains in Case Study 6 and was studied in Golubitsky and Stewart [1986a] in less convenient coordinates.

The space $\mathbb{C}^3$ decomposes into irreducibles $\mathbb{C} \oplus \mathbb{C}^2$ corresponding to the individual modes. We have already discused the orbits and isotropy subgroups for $\mathbf{O}(2) \times \mathbf{S}^1$ acting on $\mathbb{C}^2$ when discussing $\mathbf{O}(2)$ Hopf bifurcation in XVII, §1(c). There is a simple iterative procedure that lets us read off the orbits of $\mathbf{O}(2) \times \mathbf{S}^1$ acting on $\mathbb{C}^3$, which we formulate in general terms as follows:

Proposition 2.3. *Let the group Γ act on the sum $U \oplus V$ of two Γ-invariant subspaces. Let $\{u_\alpha : \alpha \in A\}$ be a list of orbit representatives for the action of Γ on U, and let $\Sigma_\alpha \subset \Gamma$ be the isotropy subgroup of u_α. Let $\{v_{\alpha\beta} : \beta \in B_\alpha\}$ be a list of orbit representatives of Σ_α acting on V, and let $T_{\alpha\beta}$ be the isotropy subgroup of $v_{\alpha\beta}$. Then*

$$\{(u_\alpha, v_{\alpha\beta}) : \alpha \in A, \beta \in B_\alpha\} \tag{2.17}$$

is a list of orbit representatives for Γ acting on $U \oplus V$, and $T_{\alpha\beta}$ is the isotropy subgroup of $(u_\alpha, v_{\alpha\beta})$ in Γ.

Remark. If the u_α and $v_{\alpha\beta}$ are chosen without redundancy, then there is no redundancy in the list of orbit representatives $(u_\alpha, v_{\alpha\beta})$. But there may be redundancies in the list $T_{\alpha\beta}$ of isotropy subgroups since distinct orbits can have the same isotropy subgroup.

PROOF. Let $(u, v) \in U \oplus V$. There exists $\gamma \in \Gamma$ such that $\gamma u = u_\alpha$ for some α, hence

$$\gamma(u, v) = (\gamma u, \gamma v) = (u_\alpha, \gamma v).$$

By construction there exists $\delta \in \Sigma_\alpha$ such that $\delta\gamma v = v_{\alpha\beta}$ for some β, hence

$$\delta\gamma(u, v) = (\delta\gamma u, \delta\gamma v) = (u_\alpha, v_{\alpha\beta}).$$

Thus (2.17) is a list of orbit representatives.

Next we show that $T_{\alpha\beta}$ is the isotropy subgroup of (u_α, v_β). By construction $T_{\alpha\beta}$ fixes $(u_\alpha, v_{\alpha\beta})$. Conversely, suppose that σ fixes $(u_\alpha, v_{\alpha\beta})$. Then $\sigma u_\alpha = u_\alpha$, so $\sigma \in \Sigma_\alpha$, and then $\sigma v_{\alpha\beta} = v_{\alpha\beta}$, so $\sigma \in T_{\alpha\beta}$. □

We apply this procedure with $U = \mathbb{C}^2$, $V = \mathbb{C}$, $\Gamma = \mathbf{O}(2) \times \mathbf{S}^1$. By Chapter XVII, Table 1.2, the u_α and Σ_α are as listed in Table 2.1. Proposition 2.3 yields Table 2.2, and this leads directly to the orbit data for $\mathbf{O}(2) \times \mathbf{S}^1$ acting on $\mathbb{C}^2 \oplus \mathbb{C}$ of Table 2.3.

The isotropy lattice is shown in Figure 2.1. All inclusions are clear except possibly that between $\mathbf{Z}_2(\kappa\pi, \pi)$ and $\mathbf{Z}_2(\kappa) \times \mathbf{S}^1$. Recall that the lattice consists of *conjugacy classes*. The group $\mathbf{Z}_2(\kappa\pi, \pi)$ is conjugate in $\mathbf{O}(2) \times \mathbf{S}^1$ to $\mathbf{Z}_2(\kappa, \pi)$, which is contained in $\mathbf{Z}_2(\kappa) \times \mathbf{S}^1$. When checking noninclusion, conjugacy must also be taken into account, but this is easy.

Table 2.1. Isotropy Subgroups for $\mathbf{O}(2) \times \mathbf{S}^1$ Acting on $\mathbb{C}^2$; the group $\mathbf{Z}_2(\kappa) = \{0, \kappa\}$ was called $\mathbf{Z}_2$ in Chapter XVII, Table 1.2

u_α		Σ_α
$(0, 0)$		$\mathbf{O}(2) \times \mathbf{S}^1$
$(a, 0)$	$a > 0$	$\widetilde{\mathbf{SO}}(2) = \{(\varphi, -\varphi)\colon \quad \varphi \in \mathbf{SO}(2)\}$
(a, a)	$a > 0$	$\mathbf{Z}_2(\kappa) \oplus \mathbf{Z}_2^c = \{(0, 0), (\kappa, 0), (\pi, \pi), (\kappa\pi, \pi)\}$
(a, b)	$a > b > 0$	$\mathbf{Z}_2^c = \{(0, 0), (\pi, \pi)\}$

Table 2.2. Data for the Application of Proposition 2.3

Σ_α	$v_{\alpha\beta} \in \mathbb{C}$		$T_{\alpha\beta}$
$\mathbf{O}(2) \times \mathbf{S}^1$	0		$\mathbf{O}(2) \times \mathbf{S}^1$
	$x > 0$		$\mathbf{Z}_2(\kappa) \times \mathbf{S}^1$
$\widetilde{\mathbf{SO}}(2)$	0		$\widetilde{\mathbf{SO}}(2)$
	$x > 0$		$\mathbb{1}$
$\mathbf{Z}_2(\kappa) \oplus \mathbf{Z}_2^c$	0		$\mathbf{Z}_2(\kappa) \oplus \mathbf{Z}_2^c$
	$x > 0$		$\mathbf{Z}_2(\kappa)$
	$iy,\ y > 0$		$\mathbf{Z}_2(\kappa\pi, \pi)$
	$w,\ \mathrm{Re}\, w > 0,$	$\mathrm{Im}\, w > 0$	$\mathbb{1}$
$\mathbf{Z}_2^c$	0		$\mathbf{Z}_2^c$
	$w \neq 0$		$\mathbb{1}$

Table 2.3. Orbit Data for $\mathbf{O}(2) \times \mathbf{S}^1$ Acting on $\mathbb{C}^3$

Orbit Representative	Isotropy	$\mathrm{Fix}(\Sigma)$	$\dim \mathrm{Fix}(\Sigma)$
$(0,0,0)$	$\mathbf{O}(2) \times \mathbf{S}^1$	$(0,0,0)$	0
$(x,0,0)$	$\mathbf{Z}_2(\kappa) \times \mathbf{S}^1$	$(x,0,0)$	1
$(0,a,0)$	$\widetilde{\mathbf{SO}}(2)$	$(0,z,0)$	2
$(0,a,a)$	$\mathbf{Z}_2(\kappa) \oplus \mathbf{Z}_2^c$	$(0,z,z)$	2
(x,a,a)	$\mathbf{Z}_2(\kappa)$	(x,z,z)	3
(iy,a,a)	$\mathbf{Z}_2(\kappa\pi,\pi)$	(iy,z,z)	3
$(0,a,b)$	$\mathbf{Z}_2^c$	$(0,z_1,z_2)$	4
$(x,a,0)$			
(w,a,a)	$\mathbb{1}$	$\mathbb{C}^3$	6
(w,a,b)			
$a > b > 0$			
$x, y > 0$			
$w \neq 0$			

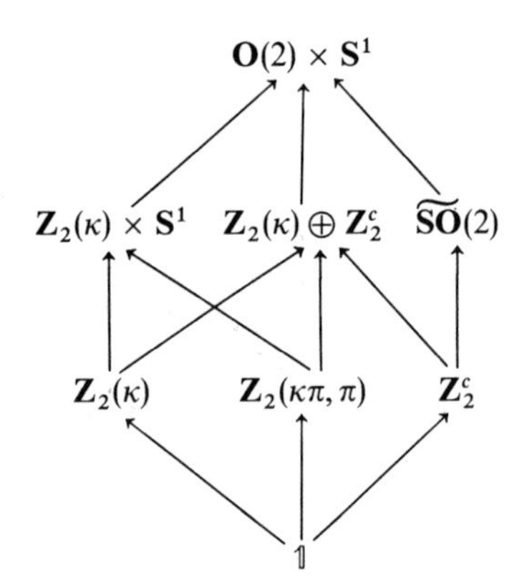

Figure 2.1. Lattice of isotropy subgroups of $\mathbf{O}(2) \times \mathbf{S}^1$ acting on $\mathbb{C}^3$.

(c) Branching Equations and Stability

By Theorem 2.1 the $\mathbf{O}(2) \times \mathbf{S}^1$-symmetric reduced bifurcation problem g has the form

$$\begin{aligned} g(z_0, z_1, z_2, \lambda, \alpha, \tau) &= (c^1 + i\delta c^2)\begin{bmatrix} z_0 \\ 0 \\ 0 \end{bmatrix} + (c^3 + i\delta c^4)\begin{bmatrix} \bar{z}_0 z_1 \bar{z}_2 \\ 0 \\ 0 \end{bmatrix} \\ &+ (p^1 + iq^1)\begin{bmatrix} 0 \\ z_1 \\ z_2 \end{bmatrix} + (p^2 + iq^2)\delta\begin{bmatrix} 0 \\ z_1 \\ -z_2 \end{bmatrix} \\ &+ (p^3 + iq^3)\begin{bmatrix} 0 \\ z_0^2 z_2 \\ \bar{z}_0^2 z_1 \end{bmatrix} + (p^4 + iq^4)\delta\begin{bmatrix} 0 \\ z_0^2 z_2 \\ -\bar{z}_0^2 z_1 \end{bmatrix}, \end{aligned} \tag{2.18}$$

where c^j, p^j, q^j are functions of invariants ρ, N, Δ, Φ, and Ψ, and parameters λ, α, τ. By general properties of the Liapunov–Schmidt reduction, the linear terms satisfy

$$\begin{aligned} &\text{(a)} \quad c^1(0) = 0, \qquad p^1(0) = 0, \qquad q^1(0) = 0 \\ &\text{(b)} \quad p^1_\tau(0) = 0, \qquad q^1_\tau(0) = \omega \neq 0. \end{aligned} \tag{2.19}$$

For reference we recall the action (2.2, 2.3) of $\mathbf{O}(2) \times \mathbf{S}^1$:

$$\begin{aligned} &\text{(a)} \quad \varphi \cdot (z_0, z_1, z_2) = (e^{i\varphi} z_0, e^{i\varphi} z_1, e^{-i\varphi} z_2) \qquad (\varphi \in \mathbf{SO}(2)) \\ &\text{(b)} \quad \kappa \cdot (z_0, z_1, z_2) = (\bar{z}_0, z_2, z_1) \\ &\text{(c)} \quad \theta \cdot (z_0, z_1, z_2) = (z_0, e^{i\theta} z_1, e^{i\theta} z_2) \qquad (\theta \in \mathbf{S}^1). \end{aligned} \tag{2.20}$$

In this section we compute:

(a) The branching equations of $g|\mathrm{Fix}(\Sigma) \times \mathbb{R}^3$ for each of the isotropy subgroups in Figure 2.1,
(b) The signs of the eigenvalues of dg for most of these types of solution.

We begin by listing in Table 2.4 the branching equations for solutions with specified isotropy. To perform these calculations we consider g evaluated at the orbit representatives listed in Table 2.3. We also use $q^1_\tau(0) \neq 0$ to solve the equations involving q^1 for τ as a function of the remaining variables. The implicitly defined function τ is of order at least 2 in the z-variables.

Next we expand each branching equation using terms up to third order in the Taylor expansion of g, defined in (2.18). See Table 2.5. Assuming that the nondegeneracy conditions of Table 2.6 are satisfied, higher order terms of g are not required to identify solutions. Note also that in Table 2.6 we list nondegeneracy conditions which preclude the existence of solutions with isotropy $\mathbf{Z}_2^c$ and with orbit representatives $(x, a, 0)$ and (z_0, a, a) with isotropy $\mathbb{1}$.

Table 2.4. Branching Equations

Isotropy	Orbit Representative	Branching Equations
$\mathbf{O}(2) \times \mathbf{S}^1$	$(0,0,0)$	—
$\mathbf{Z}_2(\kappa) \times \mathbf{S}^1$	$(x,0,0)$	$c^1 = 0$
$\widetilde{\mathbf{SO}}(2)$	$(0,a,0)$	$p^1 - p^2 a^2 = 0$
$\mathbf{Z}_2(\kappa) \oplus \mathbf{Z}_2^c$	$(0,a,a)$	$p^1 = 0$
$\mathbf{Z}_2(\kappa)$	(x,a,a)	$c^1 + c^3 a^2 = 0$
		$p^1 + p^3 x^2 = 0$
$\mathbf{Z}_2(\kappa\pi,\pi)$	(iy,a,a)	$c^1 - c^3 a^2 = 0$
		$p^1 - p^3 y^2 = 0$
$\mathbf{Z}_2(\pi,\pi)$	$(0,a,b)$	$p^1 = 0$
		$p^2 = 0$
		$q^2 = 0$
$\mathbb{1}$	$(x,a,0)$	$c^1 = c^2 = 0$
		$p^1 - a^2 p^2 = 0$
		$p^3 - a^2 p^4 = 0$
		$q^3 - a^2 q^4 = 0$
	(z_0,a,a)	$c^1 = c^3 = 0$
		$p^1 + p^3 \operatorname{Re}(z_0^2) - q^3 \operatorname{Im}(z_0^2) = 0$
	(z_0,a,b)	$(c^1 + i\delta c^2) z_0 + (c^3 + i\delta c^4)\bar{z}_0 ab = 0$
		$P^1 a + P^2 \delta a + P^3 z_0^2 b + P^4 \delta z_0^2 b = 0$
		$P^1 b - P^2 \delta b + P^3 \bar{z}_0^2 a - P^4 \delta \bar{z}_0^2 a = 0$
		where $P^j = p^j + iq^j$.

Table 2.5. Branching Equations Truncated at Third Order

Isotropy	Truncated Branching Equations
$\mathbf{Z}_2(\kappa) \times \mathbf{S}^1$	$c_\mu^1(0)\cdot\mu + c_\rho^1(0)x^2 = 0$
$\widetilde{\mathbf{SO}}(2)$	$p_\mu^1(0)\cdot\mu + (p_N^1(0) - p^2(0))a^2 = 0$
$\mathbf{Z}_2(\kappa) \oplus \mathbf{Z}_2^c$	$p_\mu^1(0)\cdot\mu + 2p_N^1(0)a^2 = 0$
$\mathbf{Z}_2(\kappa)$	$c_\rho^1(0)x^2 + (2c_N^1(0) + c^3(0))a^2 = -c_\mu^1(0)\cdot\mu$
	$(p_\rho^1(0) + p^3(0))x^2 + 2p_N^1(0)a^2 = -p_\mu^1(0)\cdot\mu$
$\mathbf{Z}_2(\kappa\pi,\pi)$	$c_\rho^1(0)y^2 + (2c_N^1(0) - c^3(0))a^2 = -c_\mu^1(0)\cdot\mu$
	$(p_\rho^1(0) - p^3(0))y^2 + 2p_N^1(0)a^2 = -p_\mu^1(0)\cdot\mu$

Table 2.6. Nondegeneracy Conditions for Existence of Solutions

Nondegeneracy Condition*	Intepretation
(a) $c_\rho^1 \neq 0$	Existence of $\mathbf{Z}_2(\kappa) \times \mathbf{S}^1$ branch
(b) $c_\mu^1 \neq 0$	Existence of $\mathbf{Z}_2(\kappa) \times \mathbf{S}^1$ branch
(c) $p_N^1 - p^2 \neq 0$	Existence of $\widetilde{\mathbf{SO}}(2)$ branch
(d) $p_\mu^1 \neq 0$	Existence of $\widetilde{\mathbf{SO}}(2)$ branch
(e) $p_1^N \neq 0$	Existence of $\mathbf{Z}_2(\kappa) \oplus \mathbf{Z}_2^c$ branch
(f) $2c_\rho^1 p_N^1 - (p_\rho^1 + p^3)(2c_N^1 + c^3)$	Existence of $\mathbf{Z}_2(\kappa)$ branch
(g) $2c_\rho^1 p_N^1 - (p_\rho^1 - p^3)(2c_N^1 - c^3)$	Existence of $\mathbf{Z}_2(\kappa\pi,\pi)$ branch
(h) $p^2 \neq 0$ or $q^2 \neq 0$	Nonexistence of $\mathbf{Z}_2^c$ solutions
(i) $c^2 \neq 0$, $p^3 \neq 0$, or $q^3 \neq 0$	Nonexistence of solutions
(j) $c^3 \neq 0$	with isotropy $\mathbb{1}$

* All derivatives evaluated at the origin.

We believe that solutions with orbit representatives (z_0, a, b) do not occur generically but have not proved this assumption.

In Table 2.5 μ is the vector of parameters, $\mu = (\lambda, \alpha)$. Thus

$$c_\mu^1 = (c_\lambda^1, c_\alpha^1) \quad \text{and} \quad p_\mu^1 = (p_\lambda^1, p_\alpha^1).$$

Note that to third order the implicitly defined function τ does *not* enter into the calculations.

We now discuss computing the stabilities of the solutions in Table 2.5. We have chosen to study stationary and periodic solutions of (2.1) using Liapunov–Schmidt reduction, so asymptotic stability of such solutions may not follow directly from knowledge of the eigenvalues of dg. However, Chossat and Golubitsky [1987a] show that to third order the reduced bifurcation equations g with $\tau = 0$ are identical to the center manifold equations in Birkhoff normal form. Therefore, if the signs of the eigenvalues of dg are determined by the third order terms of g, then these signs do determine the stability of the given solutions. We now show that the nondegeneracy conditions of Tables 2.6 and 2.9 (later) imply that stability is determined at third order.

As we know, the form of dg at a solution to $g = 0$ is restricted in two ways: dg must commute with the isotropy subgroup Σ of the solution, and certain vectors are forced by the group action to be null vectors for dg. Since dg commutes with Σ it leaves the isotypic components of the action of Σ invariant; see Theorem XII, 3.5. The null vectors forced by the action of $\mathbf{SO}(2) \times \mathbf{S}^1$ are obtained by differentiating (2.20(a,c)) with respect to φ and θ, respectively. They are:

$$\begin{aligned} &\text{(a)} \quad (iz_0, iz_1, -iz_2) \\ &\text{(b)} \quad (0, iz_1, iz_2). \end{aligned} \tag{2.21}$$

Properties of dg that are determined by these group-theoretic arguments are summarized in Table 2.7. We always take $V_0 = \mathrm{Fix}(\Sigma)$ in the isotypic decomposition.

Table 2.7 shows that $\dim V_j$ is either 1 or 2 for the primary solutions. Moreover, when $\dim V_j = 2$ either a null vector of dg is in V_j, or else $dg|V_j$ commutes with an irreducible action of a circle group. Therefore, $\mathrm{tr}(dg|V_j)$ gives either the remaining nonzero eigenvalue of $dg|V_j$, or else the real part of a complex conjugate pair of eigenvalues (because 2×2 matrices that commute with rotations are scalar multiples of rotations). The eigenvalues of dg along primary branches are listed in Table 2.8.

The entries in Table 2.8 are more easily computed by using the complex notation of the normal form. Recall that the 2×2 Jacobian of a mapping $f\colon \mathbb{C} \to \mathbb{C}$ is given by

$$(df)\zeta = f_z\zeta + f_{\bar{z}}\bar{\zeta}$$

where $\zeta \in \mathbb{C}$. Therefore,

$$\mathrm{tr}(df) = 2\,\mathrm{Re}(f_z).$$

Table 2.7. Isotypic Decomposition by Isotropy Subgroups of $\mathbf{O}(2) \times \mathbf{S}^1$

Isotropy	Isotypic Decomposition		Null Vectors
$\mathbf{O}(2) \times \mathbf{S}^1$	$\mathbb{C}^3$		—
$\mathbf{Z}_2(\kappa) \times \mathbf{S}^1$	$V_0 = \mathbb{R}\{(1,0,0)\}$,	$V_1 = \mathbb{R}\{(i,0,0)\}$	$(i,0,0)$
	$V_2 = \mathbb{C}\{(0,1,1)\}$,	$V_3 = \mathbb{C}\{(0,1,-1)\}$	
$\widetilde{\mathbf{SO}}(2)$	$V_0 = \mathbb{C}\{(0,1,0)\}$		$(0,i,0)$
	$V_1 = \mathbb{C}\{(1,0,0)\}$,	$V_2 = \mathbb{C}\{(0,0,1)\}$	
$\mathbf{Z}_2(\kappa) \oplus \mathbf{Z}_2^c$	$V_0 = \mathbb{C}\{(0,1,1)\}$,	$V_1 = \mathbb{R}\{(1,0,0)\}$	$(0,i,i)$
	$V_2 = \mathbb{R}\{(i,0,0)\}$,	$V_3 = \mathbb{C}\{(0,1,-1)\}$	$(0,i,-i)$
$\mathbf{Z}_2(\kappa)$	$V_0 = \{(\eta,\xi,\xi): \eta \in \mathbb{R}, \xi \in \mathbb{C}\}$		$(0,i,i)$
	$V_1 = \{(i\eta,\xi,-\xi): \eta \in \mathbb{R}, \xi \in \mathbb{C}\}$		$(ix, ia, -ia)$
$\mathbf{Z}_2(\kappa\pi, \pi)$	$V_0 = \{(i\eta,\xi,\xi): \eta \in \mathbb{R}, \xi \in \mathbb{C}\}$		$(0,i,i)$
	$V_1 = \{(\eta,\xi,-\xi): \eta \in \mathbb{R}, \xi \in \mathbb{C}\}$		$(-y, ia, -ia)$

Table 2.8. Eigenvalues along Primary Branches

Isotropy Subgroup	Signs of Eigenvalues	
$\mathbf{O}(2) \times \mathbf{S}^1$		p_1 [4 times]
		c_1 [twice]
$\mathbf{Z}_2(\kappa) \times \mathbf{S}^1$	$dg\|V_0 \to$	c_ρ^1
	$dg\|V_1 \to$	0
	$dg\|V_2 \to$	$p^1 + x^2 p^3$ [*]
	$dg\|V_3 \to$	$p^1 - x^2 p^3$ [*]
$\widetilde{\mathbf{SO}}(2)$	$dg\|V_0 \to$	$0,\ p_N^1 - p^2 + a^2(2p_\Delta^1 - p_N^2) - 2a^4 p_\Delta^2$
	$dg\|V_1 \to$	c^1 [*]
	$dg\|V_2 \to$	p^2 [*]
$\mathbf{Z}_2(\kappa) \oplus \mathbf{Z}_2^c$	$dg\|V_0 \to$	$0,\ p_N^1$
	$dg\|V_1 \to$	$c^1 + a^2 c^3$
	$dg\|V_2 \to$	$c^1 - a^2 c^3$
	$dg\|V_3 \to$	$0,\ -p^2$

[*] Indicates the real part of a complex conjugate pair.

The Jacobian of g is

$$(dg)(\zeta_0,\zeta_1,\zeta_2) = g_{z_0}\zeta_0 + g_{\bar{z}_0}\bar{\zeta}_0 + g_{z_1}\zeta_1 + g_{\bar{z}_1}\bar{\zeta}_1 + g_{z_2}\zeta_2 + g_{\bar{z}_2}\bar{\zeta}_2 \tag{2.22}$$

where $g = (g^0, g^1, g^2)$ in coordinates and $g_{z_j} = (g_{z_j}^0, g_{z_j}^1, g_{z_j}^2)$.

For example, along the $\widetilde{\mathbf{SO}}(2)$ branch

$$(dg|V_0)(0,\zeta,0) = g_{z_1}^1\zeta + g_{\bar{z}_1}^1\bar{\zeta}$$

whence

$$\operatorname{tr} dg|V_0 = 2\operatorname{Re} g_{z_1}^1. \tag{2.23}$$

Similarly

$$\begin{aligned} &\text{(a)} \quad \operatorname{tr} dg|V_1 = 2 \operatorname{Re} g^0_{z_0} \\ &\text{(b)} \quad \operatorname{tr} dg|V_2 = 2 \operatorname{Re} g^2_{z_2}. \end{aligned} \tag{2.24}$$

Next we compute $g^j_{z_j}$ evaluated at the orbit representatives of the $\widetilde{\mathbf{SO}}(2)$ branch $(0, a, 0)$, obtaining

$$\begin{aligned} &\text{(a)} \quad \operatorname{Re} g^0_{z_0} = c^1 \\ &\text{(b)} \quad \operatorname{Re} g^1_{z_1} = p^1 + p^1_{z_1} a + (p^2 \delta z_1)_{z_1} \\ &\text{(c)} \quad \operatorname{Re} g^2_{z_2} = p^1 + p^2 a^2. \end{aligned} \tag{2.25}$$

The entry in Table 2.8 for $dg|V_1$ is (2.25(a)). By Table 2.4, along the $\widetilde{\mathbf{SO}}(2)$ branch $p^1 = p^2 a^2$, so (2.25(c)) implies that $\operatorname{sgn} \operatorname{Re} g^2_{z_2} = \operatorname{sgn} p^2$, yielding the $dg|V_2$ entry. Finally, expanding the z_1-derivatives in (2.25(b)) yields the eigenvalues of $dg|V_0$.

From Table 2.8 we can explicitly determine the contribution to the eigenvalues made by the third order terms of g. The results are shown in Table 2.9.

Table 2.9. Eigenvalues of dg Using Only Third Order Terms of g

Isotropy	Eigenvalues of dg to Third Order in g
$\mathbf{O}(2) \times \mathbf{S}^1$	$p^1_\mu \cdot \mu$ [4 times]
	$c^1_\mu \cdot \mu$ [twice]
$\mathbf{Z}_2(\kappa) \times \mathbf{S}^1$	c^1_ρ
	$p^1_\mu \cdot \mu + (p^1_\rho + p^3)x^2$ [*]
	$p^1_\mu \cdot \mu + (p^1_\rho - p^3)x^2$ [*]
	0
$\widetilde{\mathbf{SO}}(2)$	$p^1_N - p^2$
	$c^1_\mu \cdot \mu + c^1_N a^2$ [*]
	p^2
	0
$\mathbf{Z}_2(\kappa) \oplus \mathbf{Z}^c_2$	p^1_N
	$c^1_\mu \cdot \mu + (2c^1_N + c^3)a^2$
	$c^1_\mu \cdot \mu + (2c^1_N - c^3)a^2$
	$-p_2$
	0 [twice]
$\mathbf{Z}_2(\kappa)$	eigenvalues of $M_0 \equiv \begin{bmatrix} c^1_\rho x^2 & (2c^1_N + c^3)ax \\ (p^1_\rho + p^3)ax & 2p^1_N a^2 \end{bmatrix}$ [*]
	eigenvalues of $M_1 \equiv \begin{bmatrix} p^3 x^2 + 2p^2 a^2 & -q^3 x^2 \\ -2(c^2 - q^2)a^2 + q^3 x^2 & p^3 x^2 + c^3 a^2 \end{bmatrix}$ [*]
	0 [twice]
$\mathbf{Z}_2(\kappa\pi, \pi)$	eigenvalues of $N_0 \equiv \begin{bmatrix} c^1_\rho y^2 & (2c^1_N - c^3)ay \\ (p^1_\rho - p^3)ay & 2p^1_N a^2 \end{bmatrix}$ [*]
	eigenvalues of $N_1 \equiv \begin{bmatrix} p^3 y^2 - 2p^2 a^2 & -q^3 y^2 \\ 2(c^2 - q^2)a^2 + q^3 y^2 & p^3 y^2 + c^3 a^2 \end{bmatrix}$ [*]
	0 [twice]

[*] Indicates possible secondary or tertiary bifurcation.

Table 2.10. Nondegeneracy Conditions in Addition to Those in Table 2.6 Needed to Determine Stability of Solutions at Third Order in g

Nondegeneracy Condition	Interpretation
$p_\rho^1 + p^3 \neq 0$	Eigenvalues along $\mathbf{Z}_2(\kappa) \times \mathbf{S}^1$ branch
$p_\rho^1 - p^3 \neq 0$	Eigenvalues along $\mathbf{Z}_2(\kappa) \times \mathbf{S}^1$ branch
$c_N^1 \neq 0$	Eigenvalues along $\widetilde{\mathbf{SO}}(2)$ branch
$2c_N^1 + c^3 \neq 0$	Eigenvalues along $\mathbf{Z}_2(\kappa) \oplus \mathbf{Z}_2^c$ branch
$2c_N^1 - c^3 \neq 0$	Eigenvalues along $\mathbf{Z}_2(\kappa) \oplus \mathbf{Z}_2^c$ branch

Assuming the nondegeneracy conditions of Tables 2.6 and 2.9 we see that the signs of the eigenvalues along the primary branches are determined by the third order terms in g.

Table 2.9 includes a prescription for finding the eigenvalues of dg along the secondary $\mathbf{Z}_2(\kappa)$ and $\mathbf{Z}_2(\kappa\pi, \pi)$ branches. Here the calculations for general g are too tedious to be carried out by hand. However, the derivation using the general third order truncation of g is tractable, and the remainder of this subsection justifies the entries in Table 2.9.

We explain the computations for the $\mathbf{Z}_2(\kappa\pi, \pi)$ branch: those for $\mathbf{Z}_2(\kappa)$ are similar. By Table 2.7 the isotypic components V_0, V_1 for $\mathbf{Z}_2(\kappa\pi, \pi)$ are both three-dimensional, and a null vector of dg lies in each isotypic component. By linear algebra, the remaining two eigenvalues for each of $dg|V_j$ must be the eigenvalues of some 2×2 matrix. We claim these are the matrices N_0 and N_1 in Table 2.9.

Use (2.22) to write $dg|V_0$ explicitly as

$$(dg)(i\eta, \zeta, \zeta) = i\eta(g_{z_0} - g_{\bar{z}_0}) + \zeta(g_{z_1} + g_{z_2}) + \bar{\zeta}(g_{\bar{z}_1} + g_{\bar{z}_2}).$$

Since $(0, i, i)$ is a null vector for dg we have

$$g_{z_1} + g_{z_2} = g_{\bar{z}_1} + g_{\bar{z}_2}$$

along the $\mathbf{Z}_2(\kappa\pi, \pi)$ branch, so

$$(dg)(i\eta, \zeta, \zeta) = i\eta(g_{z_0} + g_{\bar{z}_0}) + (\zeta + \bar{\zeta})(g_{z_1} + g_{z_2}). \tag{2.26}$$

Next we write $dg|V_0$ in terms of the real coordinates (η, ζ', ζ'') where $\zeta = \zeta' + i\zeta''$. Formula (2.26) becomes

$$\begin{bmatrix} \operatorname{Re}(g^0_{z_0} - g^0_{\bar{z}_0}) & 2\operatorname{Im}(g^0_{z_1} + g^0_{z_2}) & 0 \\ -\operatorname{Im}(g^1_{z_0} - g^1_{\bar{z}_0}) & 2\operatorname{Re}(g^1_{z_1} + g^1_{z_2}) & 0 \\ * & * & 0 \end{bmatrix} \begin{bmatrix} \eta \\ \zeta' \\ \zeta'' \end{bmatrix}.$$

Thus the remaining two eigenvalues of $dg|V_0$ are those of the 2×2 matrix

$$N_0 = \begin{bmatrix} \operatorname{Re}(g^0_{z_0} - g^0_{\bar{z}_0}) & 2\operatorname{Im}(g^0_{z_1} + g^0_{z_2}) \\ -\operatorname{Im}(g^1_{z_0} - g^1_{\bar{z}_0}) & 2\operatorname{Re}(g^1_{z_1} + g^1_{z_2}) \end{bmatrix}. \tag{2.27}$$

It is now straightforward to verify the entry in Table 2.9.

Computing the eigenvalues of $dg|V_1$ requires slightly more sophisticated linear algebra. First write $dg|V_1$ explicitly as:

$$(dg)(\eta, \zeta, -\zeta) = \eta(g_{z_0} + g_{\bar{z}_0}) + \zeta(g_{z_1} - g_{z_2}) + \bar{\zeta}(g_{\bar{z}_1} - g_{\bar{z}_2}).$$

Suppose that $dg|V_1$ is written in real coordinates (η, ζ', ζ'') where $\zeta = \zeta' + i\zeta''$, as the 3×3 matrix

$$Q = \begin{bmatrix} A_1 & B_1 & C_1 \\ A_2 & B_2 & C_2 \\ A_3 & B_3 & C_3 \end{bmatrix}.$$

In these coordinates the null vector of $dg|V_1$ is $(-y, 0, a)$; see Table 2.7. Setting

$$T = \begin{bmatrix} -y & 0 & 0 \\ 0 & 1 & 0 \\ a & 0 & 1 \end{bmatrix}$$

and computing $T^{-1}QT$ yields a 3×3 matrix whose null vector is $(1, 0, 0)$. Thus

$$T^{-1}QT = \begin{bmatrix} 0 & * & * \\ 0 & B_2 & C_2 \\ 0 & B_3 + ay^{-1}B_1 & C_3 + ay^{-1}C_1 \end{bmatrix}.$$

Thus the relevant 2×2 matrix is

$$N_2 = \begin{bmatrix} B_2 & C_2 \\ B_3 + ay^{-1}B_1 & C_3 + ay^{-1}C_1 \end{bmatrix}. \tag{2.28}$$

Setting

$$\begin{aligned} &\text{(a)} \quad Q_j = g^j_{z_1} - g^j_{z_2} \\ &\text{(b)} \quad \tilde{Q}_j = g^j_{\bar{z}_1} - g^j_{\bar{z}_2} \end{aligned} \tag{2.29}$$

we see that

$$\begin{aligned} &\text{(a)} \quad B_1 = \operatorname{Re}(Q_0 + \tilde{Q}_0) \\ &\text{(b)} \quad B_2 = \operatorname{Re}(Q_1 + \tilde{Q}_1) \\ &\text{(c)} \quad B_3 = \operatorname{Im}(Q_1 + \tilde{Q}_1) \\ &\text{(d)} \quad C_1 = -\operatorname{Im}(Q_0 - \tilde{Q}_0) \\ &\text{(e)} \quad C_2 = -\operatorname{Im}(Q_1 - \tilde{Q}_1) \\ &\text{(f)} \quad C_3 = \operatorname{Re}(Q_1 - \tilde{Q}_1). \end{aligned} \tag{2.30}$$

After a longish calculation using the third order truncation of g and (2.28–2.30) we obtain the entry for N_1 in Table 2.9.

(d) Bifurcation to Tori

In this section we briefly discuss the eigenvalue of dg that may cross the imaginary axis along a primary or secondary branch of solutions to $g = 0$. Here we refer to the entries in Table 2.9 marked [*].

The eigenvalues marked [*] along the branch of steady-state $\mathbf{Z}_2(\kappa) \times \mathbf{S}^1$ solutions produce possible Hopf bifurcations to periodic solutions. When

$$p_\mu^1 \cdot \mu + p_\rho^1 + p^3$$

changes sign there is a bifurcation to periodic solutions with $\mathbf{Z}_2(\kappa)$ symmetry, and when

$$p_\mu^1 \cdot \mu + p_\rho^1 - p^3$$

changes sign there is a bifurcation to periodic solutions with $\mathbf{Z}_2(\kappa\pi, \pi)$ symmetry.

When the eigenvalue

$$c_\mu^1 \cdot \mu + c_N^1 a^2$$

crosses zero a Hopf bifurcation from the rotating wave periodic solutions $\widetilde{\mathbf{SO}}(2)$ occurs. The resulting bifurcation leads to quasiperiodic motion on an invariant 2-torus, which we call a *modulated rotating wave*; see Exercise 2.2. Remarkably, the asymptotic stability of these new quasiperiodic solutions can be determined from the third order truncation of g. They are asymptotically stable precisely when

$$p_N^1 - p^2 > 0, \qquad p^2 > 0,$$

and

$$H = c_\rho^1 - \frac{p_\rho^1 c_N^1}{p_N^1 - p^2} - \frac{c^3}{2}\operatorname{Re}\left[\frac{p^3 - iq^3}{p^2 + i(c^2 - q^2)}\right] > 0. \tag{2.31}$$

This formula is derived in Crawford, Golubitsky, and Langford [1987].

Next we consider the $\mathbf{Z}_2(\kappa)$ branch. We claim that eigenvalues of M_0 can cross the imaginary axis, yielding a Hopf bifurcation, and that eigenvalues of M_1 can cross either through 0 or through complex conjugate purely imaginary eigenvalues.

By the nondegeneracy condition (f) in Table 2.6, $\det M_0 \neq 0$. Thus if eigenvalues of M_0 cross the imaginary axis they do so away from zero. This can happen when $c_\rho^1(0)$ and $p_N^1(0)$ have opposite signs. If we consider the ODE $\dot{x} + g(x) = 0$ restricted to $V_0 = \mathrm{Fix}(\mathbf{Z}_2(\kappa))$ we find a Hopf bifurcation to an invariant 2-torus. The $\mathbf{O}(2)$ symmetry forces motion on an invariant 3-torus which is foliated by invariant 2-tori. Higher order terms are needed to determine the stability of these two-frequency motions.

An eigenvalue of M_1 can cross through 0 along a branch of periodic solutions with $\mathbf{Z}_2(\kappa)$ symmetry. This crossing can be detected from the third order truncation of g by solving $\det M_1 = 0$. When such a bifurcation occurs a new branch of invariant 2-tori appears. Roughly speaking, the periodic solutions with isotropy conjugate to $\mathbf{Z}_2(\kappa)$ foliate an invariant 2-torus. When bifurcation occurs, motion on a new invariant torus appears, having one frequency near that of the primary $\mathbf{Z}_2(\kappa)$ solution, and one frequency near zero, corresponding to a slow drift along the $\mathbf{O}(2)$ group orbit. See Chossat and Golubitsky [1987b].

Finally, it is possible for complex conjugate eigenvalues of M_1 to cross the imaginary axis. When this happens, motion on an invariant 3-torus is created. This motion has one frequency near that of the primary $\mathbf{Z}_2(\kappa)$ solution, one frequency near the magnitude of the imaginary eigenvalues concerned, and one frequency near zero corresponding to drift along the group orbit. Again see Chossat and Golubitsky [1987b]. Similar bifurcations to tori are possible along the branch of periodic solutions with $\mathbf{Z}_2(\kappa\pi, \pi)$ symmetry.

(e) Sample Bifurcation Diagrams and Summary

We have seen that in this Hopf/steady-state mode interaction with $\mathbf{O}(2)$ symmetry, three primary bifurcating branches, three secondary branches, and possibly up to six tertiary branches of solutions can occur. Moreover, the existence and much of the determination of asymptotic stability of these solutions may be found from the normal form equations truncated at order 3. Thus, when an $\mathbf{O}(2)$-invariant steady-state loses stability simultaneously to a symmetry-breaking steady-state mode and a symmetry-breaking Hopf mode, complicated dynamics occur.

The three primary branches—one of steady-state solutions, and two of periodic solutions, respectively rotating and standing waves—are to be expected. In addition, generically we find three secondary branches: two corresponding to periodic solutions with less symmetry, and one corresponding to 2-frequency motion bifurcating from the branch of rotating waves. We call these solutions *modulated rotating waves.* Finally, depending on the exact signs occurring in the nondegeneracy conditions of Tables 2.6 and 2.10, up to six tertiary branches of tori may bifurcate from the secondary branches of periodic solutions. The determination of the direction of branching of these tori is beyond the scope of the present work and surely involves terms of at least degree 5 in the normal form equations.

It is, however, still a nontrivial task to enumerate all of the different bifurcation diagrams. Different choices of sign in Tables 2.6 and 2.10 indicate that a conservative lower bound for the number of possibilities is 1000. We prefer to address the following question: *Suppose that we know the order* 3 *truncated reduced bifurcation equations* g *exactly, and suppose that the nondegeneracy conditions are satisfied. How do we draw the associated bifurcation diagram*? This is the situation that pertains when we discuss bifurcation from Couette flow in Case Study 6.

Since there is no distinguished bifurcation parameter in this problem we propose to draw the bifurcation diagram as follows. In the two-parameter (λ, α)-space we draw the curves along which zero or purely imaginary eigen-

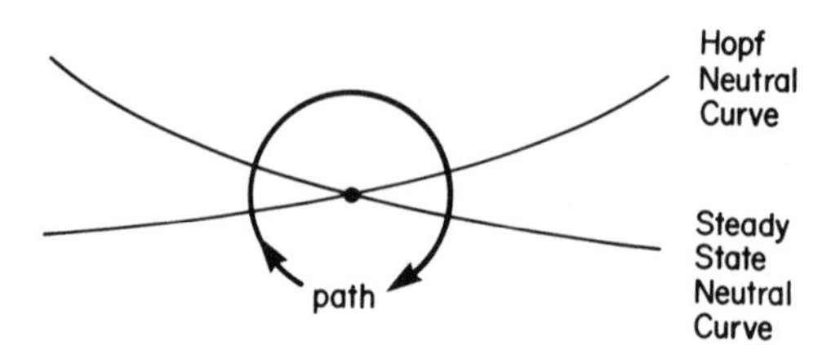

Figure 2.2. Clockwise path around bicritical point.

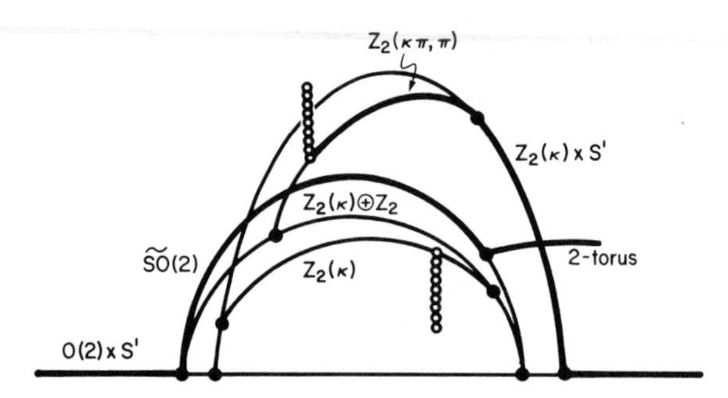

Figure 2.3. Bifurcation diagram corresponding to path in Figure 2.2.

values of dG (where G is defined in (2.1)) occur along the $\mathbf{O}(2)$-invariant trivial solution. In a Hopf/steady-state mode interaction these *neutral stability curves* must intersect, as in Figure 2.2.

We then consider the circular path in parameter space shown on the figure. We follow the path around the bicritical point (the point of mode interaction where the curves cross), and, using the explicit truncated g, we find which transitions in state are possible. One such scenario is shown in Figure 2.3. Along this path, starting in the south, we travel west and arrive at a Hopf bifurcation to a stable rotating wave and an unstable standing wave. Halfway around the circle the rotating wave loses stability to a stable 2-torus of modulated rotating waves. If we start in the south and move eastward we first go to a steady-state bifurcation to a stable noninvariant steady-state, which then bifurcates to a stable periodic solution with $\mathbf{Z}_2(\kappa\pi, \pi)$ symmetry. This periodic solution loses stability to motion on a 2-torus whose direction of branching and stability we cannot determine. Other unstable solutions exist.

Exercises

2.1. Show that the ring of $\mathbf{O}(2) \times \mathbf{S}^1$-invariants is not a polynomial ring by establishing the relation

$$\Psi^2 = \Delta\left[\rho^{2l}\left(\frac{N^2 - \Delta}{4}\right)^m - \Phi^2\right].$$

2.2. Using the actions (2.2, 2.3) of $\mathbf{O}(2) \times \mathbf{S}^1$ on $\mathbb{C} \oplus \mathbb{C}^2$, and Proposition 2.3, show that the isotropy data in Table 2.3 change as follows:

	Orbit	Isotropy	Fix(Σ)	dim Fix(Σ)
0	$(0,0,0)$	$\mathbf{O}(2) \times \mathbf{S}^1$	$(0,0,0)$	0
1	$(x,0,0)$	$\mathbf{D}_m \times \mathbf{S}^1$	$(x,0,0)$	1
2	$(0,a,0)$	$\widetilde{\mathbf{SO}}^l(2) = \{(\varphi, -l\varphi)\}$	$(0,z,0)$	2
3	$(0,a,a)$	$\tilde{\mathbf{D}}_{2l} = \langle \kappa, (\pi/l, \pi) \rangle$	$(0,z,z)$	2
4	(x,a,a)	$\tilde{\mathbf{D}}_h^- = \langle \kappa, (2\pi/h, -2\pi/h) \rangle$ $h = \operatorname{hcf}(m, 2l)$	(x,z,z)	3
5	$(x,a,-a)$	$\tilde{\mathbf{D}}_h^+ = \langle \kappa, (2\pi/h, 2\pi/h) \rangle$ $h = \operatorname{hcf}(m, 2l)$	$(x,z,-z)$	3
6	$(0,a,b)$	$\langle (\pi/l, \pi) \rangle$	$(0,z_1,z_2)$	4
7	$(x,a,0)$	$\langle (2\pi/m, -2l\pi/m) \rangle$	$(z_0,z_1,0)$	4
8	(w,a,a)	$\langle (\pi, l\pi) \rangle$ [m even]	$\mathbb{C} \oplus \mathbb{C}^2$	6
	(w,a,b)	$\mathbb{1}$ [m odd]		

Here, to avoid duplication, the following must be noted:

When $l = 1$ and m is even, **3** is the same as **4**.
When $l = m$ and m is even, **6** is the same as **8**.

§3.† Hopf/Hopf Mode Interaction

As we discussed in §0, two-parameter $\mathbf{O}(2)$-equivariant systems of ODEs

$$\dot{x} + g(x, \lambda, \alpha) = 0 \qquad (x \in \mathbb{R}^n) \tag{3.1}$$

may be expected to have isolated points at which distinct Hopf modes interact. In this section we discuss the simplest forms that such interactions may take and some of the associated dynamics. The number of possibilities, even in the simplest cases, is quite large, and consequently our discussion will be far from complete. To keep the length within bounds, we omit details of all calculations. A more complete discussion may be found in Chossat, Golubitsky, and Keyfitz [1986]. Examples of Hopf/Hopf mode interaction with symmetry occur in certain models of flame propagation; see Matkowsky and Olagunju [1982], Keyfitz et al. [1986], and Booty et al. [1986].

For ease of notation we assume that the mode interaction occurs at $(x, \lambda, \alpha) = (0,0,0)$. In the simplest such interactions, $(dg)_{0,0,0}$ has two pairs of purely imaginary eigenvalues $\pm\omega_0 i$, $\pm\omega_1 i$, whose (generalized) eigenspaces W_0 and W_1 are $\mathbf{O}(2)$-simple. The $\mathbf{O}(2)$ symmetry forces these eigenspaces to have dimension 2 or 4, depending on whether the corresponding eigenvalue is simple or double. If both eigenspaces are two-dimensional then we are (essentially) in the case of Hopf/Hopf mode interaction without symmetry discussed in XIX, §4.

In this section we assume that $\dim W_1 = 4$ and consider in two subsections the cases

$$\text{(a)} \quad \dim W_0 = 2,$$

$$\text{(b)} \quad \dim W_0 = 4.$$

Again, for simplicity we assume that $\mathbb{R}^n = W_0 \oplus W_1$, or equivalently that a center manifold reduction has already been performed. Thus we call (a) the six-dimensional case and (b) the eight-dimensional case.

We wish to study the local dynamics occurring in **O**(2)-symmetric perturbations of the Birkhoff normal form for (3.1). In this section, however, we assume that (3.1) is in Birkhoff normal form. The simplest Birkhoff normal forms occur when $\omega_0 i$ and $\omega_1 i$ are nonresonant, that is, when ω_1/ω_0 is irrational. Under this assumption, Theorem XVI, 5.9, implies that the normal form of (3.1) commutes with the 2-torus $\mathbf{T}^2$ as well as with **O**(2). This 2-torus is the closure of the one-parameter subgroup generated by the linearization $L = (dg)_{0,0,0}$.

In the six-dimensional case we show that the Birkhoff normal form equations decouple into amplitude and phase equations and, as in other instances of such a splitting, the dynamics of the amplitude equations more or less determine those of the normal form equations. This decoupling is not surprising since the normal form equations commute with the 3-torus $\mathbf{SO}(2) \times \mathbf{T}^2$. Thus the normal form equations are determined on group orbits (amplitudes constant) by the vector field at a single point on the group orbit.

In the eight-dimensional case, the full normal form equations do not decouple into amplitude and phase form. However, generically the interesting dynamics occur on six-dimensional fixed-point subspaces, and on these subspaces the normal form equations do decouple.

(a) The Six-Dimensional Case

By a linear change of coordinates we may assume that

$$(dg)_{0,0,0} = \left[\begin{array}{cc|cc} 0 & -\omega_0 & \multicolumn{2}{c}{0} \\ \omega_0 & 0 & & \\ \hline \multicolumn{2}{c|}{0} & 0 & -\omega_1 I \\ & & \omega_1 I & 0 \end{array}\right], \tag{3.2}$$

where I is the 2×2 identity matrix. We begin by describing the group action of $\mathbf{O}(2) \times \mathbf{T}^2$. Since $\dim W_0 = 2$ the action of **SO**(2) on W_0 is trivial and **O**(2)/**SO**(2) acts by $\pm I$ on W_0. The analyses of these two cases are identical, and we shall assume that **O**(2) acts trivially on $W_0 \cong \mathbb{C}$.

The action of **SO**(2) on $W_1 \cong \mathbb{C}^2$ has the form

$$\psi \cdot (z_1, z_2) = (e^{ki\psi} z_1, e^{-ki\psi} z_2)$$

for some positive integer k. Dividing by the kernel we may assume that $k = 1$. Thus we may assume that the action of $\mathbf{O}(2)$ on $\mathbb{C} \oplus \mathbb{C}^2$ is generated by

$$\begin{aligned} &\text{(a)} \quad \psi\cdot(z_0, z_1, z_2) = (z_0, e^{i\psi}z_1, e^{-i\psi}z_2) \qquad (\psi \in \mathbf{SO}(2)) \\ &\text{(b)} \quad \kappa\cdot(z_0, z_1, z_2) = (z_0, z_2, z_1). \end{aligned} \tag{3.3}$$

Compare with (XVII, 1.3).

From (3.2) we may take the action of $\mathbf{T}^2$ to be

$$(\theta_0, \theta_1)\cdot(z_0, z_1, z_2) = (e^{i\theta_0}z_0, e^{i\theta_1}z_1, e^{i\theta_1}z_2). \tag{3.4}$$

The Birkhoff normal form for vector fields commuting with the action (3.2, 3.3) of $\mathbf{O}(2) \times \mathbf{T}^2$ may now be derived as follows:

Theorem 3.1. *Vector fields commuting with* $\mathbf{O}(2) \times \mathbf{T}^2$ *have the form*

$$(p_0 + iq_0)\begin{bmatrix} z_0 \\ 0 \\ 0 \end{bmatrix} + (p_1 + iq_1)\begin{bmatrix} 0 \\ z_1 \\ z_2 \end{bmatrix} + (p_2 + iq_2)\delta\begin{bmatrix} 0 \\ z_1 \\ -z_2 \end{bmatrix} \tag{3.5}$$

where p_j, q_j *are functions of* ρ, N, Δ *and parameters, and*

$$\rho = |z_0|^2, \qquad N = |z_1|^2 + |z_2|^2, \qquad \Delta = \delta^2, \qquad \delta = |z_2|^2 - |z_1|^2.$$

PROOF. See Chossat, Golubitsky, and Keyfitz [1986], Proposition 2.3. □

From (3.2) we have

$$\begin{aligned} &\text{(a)} \quad p_0(0) = 0, \qquad p_1(0) = 0 \\ &\text{(b)} \quad q_0(0) = \omega_0, \qquad q_1(0) = \omega_1. \end{aligned} \tag{3.6}$$

The Birkhoff normal form (3.5) decomposes into amplitude and phase equations. With $r_j = |z_j|$, the amplitude equations are:

$$\begin{aligned} &\text{(a)} \quad \dot{r}_0 + p_0 r_0 = 0 \\ &\text{(b)} \quad \dot{r}_1 + (p_1 + \delta p_2)r_1 = 0 \\ &\text{(c)} \quad \dot{r}_2 + (p_1 - \delta p_2)r_2 = 0, \end{aligned} \tag{3.7}$$

where p_0, p_1 are functions of $\rho = r_0^2$, $N = r_1^2 + r_2^2$, and $\Delta = \delta^2$ where $\delta = r_2^2 - r_1^2$.

Zeros (r_0, r_1, r_2) of (3.7) correspond to periodic orbits, invariant 2-tori, and invariant 3-tori for the Birkhoff normal form, depending on whether two, one, or none of the r_j are zero.

The amplitude equations (3.7) define a mapping $f\colon \mathbb{R}^3 \to \mathbb{R}^3$ where

$$f(r_0, r_1, r_2) = (p_0 r_0, (p_1 + \delta p_2)r_1, (p_1 - \delta p_2)r_2). \tag{3.8}$$

The mappings (3.8) are precisely those that commute with $\mathbf{Z}_2 \times \mathbf{D}_4$ acting on $\mathbb{R}^3$ as follows. The group $\mathbf{Z}_2$ is generated by κ_0, and $\mathbf{D}_4$ is generated by three elements κ_1, κ_2, and κ, all of order 2, acting by

$$\begin{aligned}
&\text{(a)} \quad \kappa_0(r_0, r_1, r_2) = (-r_0, r_1, r_2)\\
&\text{(b)} \quad \kappa_1(r_0, r_1, r_2) = (r_0, -r_1, -r_2)\\
&\text{(c)} \quad \kappa_2(r_0, r_1, r_2) = (r_0, r_1, -r_2)\\
&\text{(d)} \quad \kappa(r_0, r_1, r_2) = (r_0, r_2, r_1).
\end{aligned} \tag{3.9}$$

This is no surprise: there are corresponding reductions for ordinary Hopf bifurcation to a $\mathbf{Z}_2$-equivariant bifurcation equation and for $\mathbf{O}(2)$ Hopf bifurcation to a $\mathbf{D}_4$-equivariant equation; see Chapter VIII and XVII, §4.

We use the $\mathbf{Z}_2 \times \mathbf{D}_4$-symmetry to detect different types of solution by isotropy. The conjugacy classes of isotropy subgroups are given in Table 3.1, together with their fixed-point subspaces and the conditions for $f|\mathrm{Fix}(\Sigma)$ to vanish. The isotropy lattice is shown in Figure 3.1. Its structure is reflected in the bifurcation diagrams, which we draw later.

A zero of the amplitude equations corresponds to an equilibrium orbit of the six-dimensional system. The (orbital) asymptotic stability of the latter

Table 3.1. Orbit Data for $\mathbf{Z}_2 \times \mathbf{D}_4$

	Σ	$\mathrm{Fix}(\Sigma)$	$f\|\mathrm{Fix}(\Sigma) = 0$
0	$\mathbf{Z}_2 \times \mathbf{D}_4$	0	—
1	$\mathbf{D}_4$	$(r_0, 0, 0)$	$p_0 = 0$
2	$\mathbf{Z}_2 \times \langle\kappa_2\rangle$	$(0, r_1, 0)$	$p_1 - r_1^2 p_2 = 0$
3	$\mathbf{Z}_2 \times \langle\kappa\rangle$	$(0, r_1, r_1)$	$p_1 = 0$
4	$\langle\kappa_2\rangle$	$(r_0, r_1, 0)$	$p_0 = p_1 - r_1^2 p_2 = 0$
5	$\mathbf{Z}_2$	$(0, r_1, r_2)$	$p_1 = p_2 = 0$
6	$\langle\kappa\rangle$	(r_0, r_1, r_1)	$p_0 = p_1 = 0$
7	$\mathbb{1}$	(r_0, r_1, r_2)	$p_0 = p_1 = p_2 = 0$

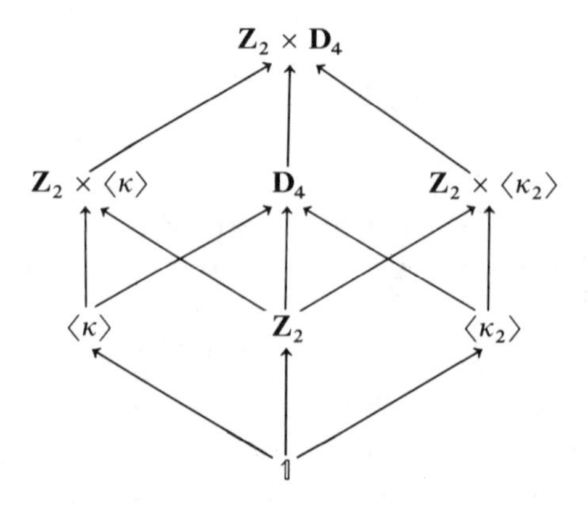

Figure 3.1. Isotropy lattice for $\mathbf{Z}_2 \times \mathbf{D}_4$.

Table 3.2. Form of Eigenvalues of df

Type	df	Eigenvalues
0	$\begin{bmatrix} A & 0 & 0 \\ 0 & B & 0 \\ 0 & 0 & B \end{bmatrix}$	A, B, B
1	$\begin{bmatrix} A & 0 & 0 \\ 0 & B & 0 \\ 0 & 0 & B \end{bmatrix}$	A, B, B
2	$\begin{bmatrix} A & 0 & 0 \\ 0 & B & 0 \\ 0 & 0 & C \end{bmatrix}$	A, B, C
3	$\begin{bmatrix} A & 0 & 0 \\ 0 & B & C \\ 0 & C & B \end{bmatrix}$	$A, B + C, B - C$
4	$\begin{bmatrix} A & D & 0 \\ E & B & 0 \\ 0 & 0 & C \end{bmatrix}$	C, eigenvalues of $N_1 = \begin{bmatrix} A & D \\ E & B \end{bmatrix}$
6	$\begin{bmatrix} A & D & D \\ E & B & C \\ E & C & B \end{bmatrix}$	$B - C$, eigenvalues of $N_2 = \begin{bmatrix} A & E \\ 2D & B + C \end{bmatrix}$

corresponds to asymptotic stability of the former. Suppose that $z(t) = (z_0(t), z_1(t), z_2(t)) \in \mathbb{C}^3$ is a trajectory of the invariant six-dimensional vector field. Then $r(t) = (|z_0(t)|, |z_1(t)|, |z_2(t)|)$ is the corresponding trajectory of the amplitude equations. Note that $r(t)$ limits on a point $r \in \mathbb{R}^3$ precisely when $z(t)$ limits on the group orbit corresponding to r.

The asymptotic stability of a zero of the amplitude equations may be computed, using linearized stability, from the eigenvalues of the 3×3 Jacobian df. The form of df, as restricted by isotropy, and its eigenvalues, are given in Table 3.2 for most solution types. Since generically (when $p_2 \neq 0$) solutions of types **5** and **7** do not occur, we have not included them.

Each of the eigenvalues of df can now be computed along primary branches (solution types **0**–**3**) by evaluating the relevant partial derivatives of f. Along secondary branches (types **4** and **6**) the best we can do in general is to find the trace and determinant of the specified 2×2 matrix. These results are summarized in Table 3.3.

It is now possible, in theory, to determine the bifurcation diagrams associated to the amplitude equations using the $F|\text{Fix}(\Sigma)$ column of Table 3.3. We note here only that generically ($p_2 \neq 0$) solutions of types **5** and **7** do not occur and that it is not possible for solutions on branches **2** and **3** and branches **4** and **6** to be simultaneously asymptotically stable.

Some aspects of the Birkhoff normal form equations $\tilde{f}$ truncated at order 3

Table 3.3. Computation of Eigenvalues of df

Solution Type	Signs of Eigenvalues
0	p_0
	p_1 [twice]
1	$p_{0,\rho}$
	p_1 [twice]
2	p_0
	$p_{1,N} - p_2 + r_1^2(2p_{1,\Delta} - p_{2,N} - 2r_1p_{2,\Delta})$
	p_2
3	p_0
	$p_{1,N}$
	$-p_2$
4	p_2
	$\operatorname{tr} N_1 = \rho p_{0,\rho} + r_1^2[p_{1,N} - p_2 + r_1^2(2p_{1,\Delta} - p_{2,N} - 2r_1^2p_{2,\Delta})]$
	$\det N_1 = p_{0,\rho}(p_{1,N} - p_2) - p_{0,N}p_{1,\rho} + r_1^2[p_{0,\rho}(2p_{1,\Delta} - p_{2,N} - r_1^2p_{2,\Delta}) + p_{0,N}p_{2,\Delta} - 2p_{0,\Delta}p_{1,\rho} + 2r_1^2p_{0,\Delta}p_{2,\rho}]$
6	$-p_2$
	$\operatorname{tr} N_2 = \rho p_{0,\rho} + 2r_1^2p_{1,N}$
	$\det N_2 = p_{0,\rho}p_{1,N} - p_{0,N}p_{1,\rho}$

are studied in Chossat, Golubitsky, and Keyfitz [1986]. We mention four of their observations.

Assume that $\tilde{f}$ has the form

$$\begin{aligned} &\text{(a)} \quad p_0 = \alpha_0(\beta - \lambda + a_0\rho + b_0N) \\ &\text{(b)} \quad p_1 = \alpha_1(-\lambda + a_1\rho + b_1N) \\ &\text{(c)} \quad p_2 = \alpha_1 p_2 \end{aligned} \tag{3.10}$$

where $\alpha_0, \alpha_1, a_0, a_1, b_1, p_2$ are constant.

(1) As for a single mode bifurcation in the presence of **O**(2) symmetry, the branches **2** and **3** emerge together, and generically one of them is stable only if both are supercritical. Then precisely one of them is stable at the bifurcation point if and only if $\beta > 0$, that is, if this pair bifurcates first.
(2) The same principle governs the bifurcation of the secondary branches **4** and **6** from **1**. If this bifurcation occurs, then both branches emerge together and precisely one gains stability (by an exchange of stabilities) only when **1** is supercritical. Then either **1** is stable before the bifurcation and both **4** and **6** are supercritical, or **1** is unstable before the bifurcation and both **4** and **6** are subcritical.
(3) The secondary branches may be finite (Fig. 3.2) or infinite (Fig. 3.3).
(4) A tertiary Hopf bifurcation may occur on a secondary branch that is finite and forms a transition between a subcritical and a supercritical primary branch. This Hopf bifurcation can be detected by a change of stability along the branch. If either end is stable, under the conditions in **2** then there is an arrangement of fifth order terms that guarantees a stable

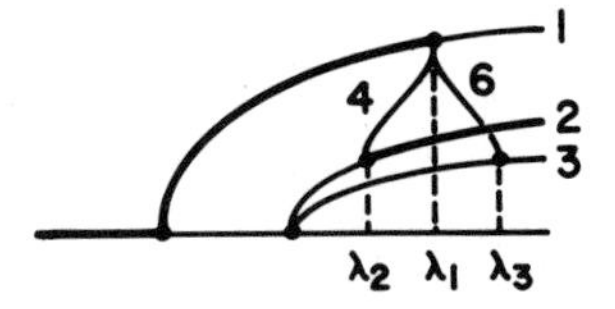

Figure 3.2. Finite secondary branches.

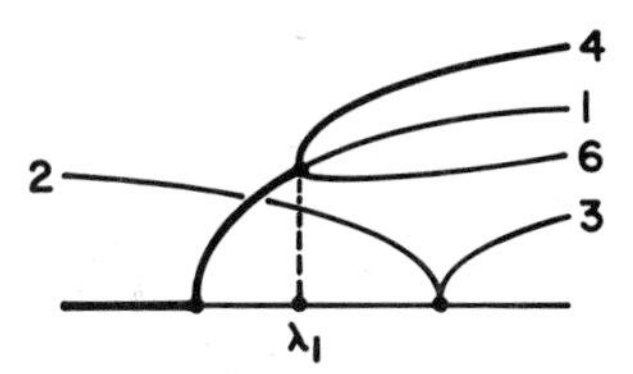

Figure 3.3. Infinite secondary branches.

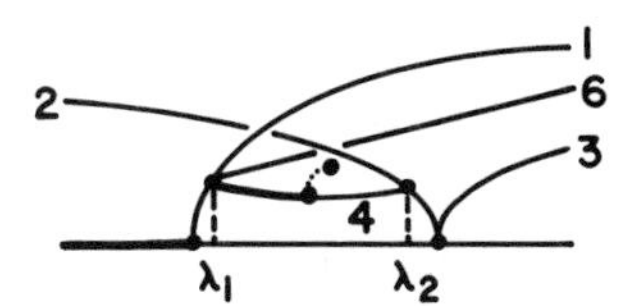

Figure 3.4. Tertiary Hopf bifurcation in amplitude equations.

tertiary branch. See Figure 3.4. This tertiary Hopf bifurcation corresponds to an invariant 3-torus in the dynamics of the Birkhoff normal form equations.

(b) The Eight-Dimensional Case

In the eight-dimensional case we may assume

$$(dg)_{0,0,0} = \left[\begin{array}{cc|cc} 0 & -\omega_0 I & & \\ \omega_0 I & 0 & \multicolumn{2}{c}{0} \\ \hline & & 0 & -\omega_1 I \\ \multicolumn{2}{c|}{0} & \omega_1 I & 0 \end{array}\right], \tag{3.11}$$

and we write $W_0 \oplus W_1 = \mathbb{C}^2 \oplus \mathbb{C}^2$ with coordinates (z_1, z_2, z_3, z_4). The action of $\mathbf{O}(2) \times \mathbf{T}^2$ on $\mathbb{C}^2 \oplus \mathbb{C}^2$ is

$$\begin{aligned} &\text{(a)} \quad (\theta_1, \theta_2) \cdot (z_1, z_2, z_3, z_4) = (e^{i\theta_1} z_1, e^{i\theta_1} z_2, e^{i\theta_2} z_3, e^{i\theta_2} z_4) \\ &\text{(b)} \quad \kappa \cdot (z_1, z_2, z_3, z_4) = (z_2, z_1, z_4, z_3) \\ &\text{(c)} \quad \psi \cdot (z_1, z_2, z_3, z_4) = (e^{li\psi} z_1, e^{-li\psi} z_2, e^{mi\psi} z_3, e^{-mi\psi} z_4), \end{aligned} \tag{3.12}$$

for $(\theta_1, \theta_2) \in \mathbf{T}^2$ and $\psi \in \mathbf{SO}(2)$. Note that ψ acts as rotation by $l\psi$ on the first $\mathbb{C}^2$ and rotation by $m\psi$ on the second. Because two wave numbers l and m exist, we cannot scale them both to 1 as we did in the six-dimensional case, but as usual, by factoring out the kernel, we may assume that l and m are coprime.

The form of $\mathbf{O}(2) \times \mathbf{T}^2$-equivariant vector fields can be computed by first finding the general form for $\mathbf{SO}(2) \times \mathbf{T}^2$-equivariance and then determining the restrictions imposed by the flip κ.

Theorem 3.2. *Let g be a nonresonant* **SO**(2)*-invariant vector field with linear part* (3.11). *Then g has Birkhoff normal form*

$$\begin{bmatrix} (p_1 + iq_1)z_1 + (r_1 + is_1)\bar{z}_1^{m-1} z_2^m (z_3 \bar{z}_4)^l \\ (p_2 + iq_2)z_2 + (r_2 + is_2) z_1^m \bar{z}_2^{m-1} (\bar{z}_3 z_4)^l \\ (p_3 + iq_3)z_3 + (r_3 + is_3)(z_1 \bar{z}_2)^m \bar{z}_3^{l-1} z_4^l \\ (p_4 + iq_4)z_4 + (r_4 + is_4)(\bar{z}_1 z_2)^m z_3^l \bar{z}_4^{l-1} \end{bmatrix}. \tag{3.13}$$

Here p_j, q_j, r_j, s_j are functions of $\rho_i = |z_i|^2$, $\mathrm{Re}\,\alpha$, $\mathrm{Im}\,\alpha$, where

$$\alpha = (z_1 \bar{z}_2)^m (\bar{z}_3 z_4)^l.$$

If further g is **O**(2)*-equivariant, then* (3.13) *must also satisfy*

$$\begin{aligned} g_2(z) &= g_1(\kappa \cdot z) \\ g_4(z) &= g_3(\kappa \cdot z). \end{aligned} \tag{3.14}$$

PROOF. See Chossat, Golubitsky, and Keyfitz [1986]. □

We next discuss the isotropy subgroups of $\mathbf{O}(2) \times \mathbf{T}^2$ acting on $\mathbb{C}^2 \oplus \mathbb{C}^2$ and use these to classify some of the dynamics occuring in (3.11). We must consider three types of subgroup of $\mathbf{O}(2) \times \mathbf{T}^2$, as follows. Each element of $\mathbf{O}(2) \times \mathbf{T}^2$ has the form

$$\text{(a)} \quad (\psi, \theta_1, \theta_2) \in \mathbf{SO}(2) \times \mathbf{T}^2$$

or (3.15)

$$\text{(b)} \quad (\kappa\psi, \theta_1, \theta_2)$$

where κ is the flip in $\mathbf{O}(2)$. We define

$$\begin{aligned} &\text{(a)} \quad \mathbf{Z}(\psi, \theta_1, \theta_2) = \langle(\psi, \theta_1, \theta_2)\rangle \subset \mathbf{SO}(2) \times \mathbf{T}^2 \\ &\text{(b)} \quad \mathbf{Z}_\kappa(\psi, \theta_1, \theta_2) = \langle(\kappa\psi, \theta_1, \theta_2)\rangle \\ &\text{(c)} \quad \mathbf{S}(k, l, m) = \{(k\theta, l\theta, m\theta) \colon \theta \in \mathbf{S}^1\}. \end{aligned} \tag{3.16}$$

Here $\langle\ \rangle$ means "group generated by." Because of its special role we define

$$\mathbf{Z}_2^\kappa = \mathbf{Z}_\kappa(0, 0, 0) = \langle\kappa\rangle.$$

Table 3.4. Isotropy Subgroups of $\mathbf{O}(2) \times \mathbf{T}^2$ on $\mathbb{C}^2 \oplus \mathbb{C}^2$

	Isotropy Subgroup	Fixed-Point Subspace	Dimension	Solution Type
0	$\mathbf{O}(2) \times \mathbf{T}^2$	$\{0\}$	0	
1	$\mathbf{S}(0,0,1) \times \mathbf{S}(1,-l,0)$	$z_2 = z_3 = z_4 = 0$	2	l-rotating wave
2	$\mathbf{S}(0,0,1) \times \mathbf{Z}_2^\kappa \times \mathbf{Z}(\pi/l,\pi,0)$	$z_1 = z_2, \quad z_3 = z_4 = 0$	2	l-standing wave
3	$\mathbf{S}(0,1,0) \times \mathbf{Z}_2^\kappa \times \mathbf{Z}(\pi/m,0,\pi)$	$z_1 = z_2 = 0, \quad z_3 = z_4$	2	m-standing wave
4	$\mathbf{S}(0,1,0) \times \mathbf{S}(1,0,m)$	$z_1 = z_2 = z_3 = 0$	2	m-rotating wave
5	$\mathbf{S}(0,0,1) \times \mathbf{Z}(\pi/l,\pi,0)$	$z_3 = z_4 = 0$	4	2-torus
6	$\mathbf{S}(0,1,0) \times \mathbf{Z}(\pi/m,0,\pi)$	$z_1 = z_2 = 0$	4	2-torus
7	$\mathbf{S}(1,l,m)$	$z_1 = z_3 = 0$	4	2-torus
8	$\mathbf{S}(1,l,-m)$	$z_1 = z_4 = 0$	4	2-torus
9	$\mathbf{Z}_2^\kappa \times \mathbf{Z}(\pi, l\pi, m\pi)$	$z_1 = z_2, \quad z_3 = z_4$	4	2-torus
10	$\mathbf{Z}_k(0,\pi,0) \times \mathbf{Z}(\pi, l\pi, m\pi)$ (m odd)	$z_1 = -z_2, \quad z_3 = z_4$	4	2-torus
11	$\mathbf{Z}_k(0,0,\pi) \times \mathbf{Z}(\pi, l\pi, m\pi)$ (l odd, m even)	$z_1 = z_2, \quad z_3 = -z_4$	4	2-torus
12	$\mathbf{Z}(\pi/l, \pi, m\pi/l) \quad (l \neq 1)$	$z_3 = 0$	6	3-torus
13	$\mathbf{Z}(\pi/m, l\pi/m, \pi) \quad (m \neq 1)$	$z_1 = 0$	6	3-torus
14	$\mathbf{Z}(\pi, l\pi, m\pi)$	$\mathbb{C}^2 \oplus \mathbb{C}^2$	8	

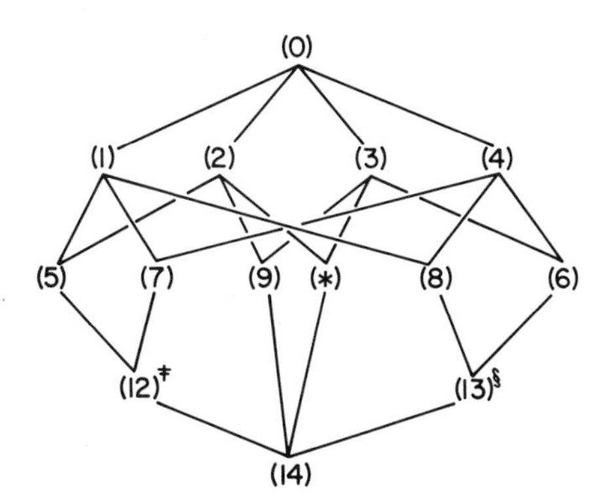

Figure 3.5. Isotropy lattice of $\mathbf{O}(2) \times \mathbf{T}^2$ acting on $\mathbb{C}^2 \oplus \mathbb{C}^2$. * = (10) if m odd, (11) if m even. ‡ absent if $l = 1$; (14) is contained directly in (5) and (7). § absent if $m = 1$; (14) is contained directly in (6) and (8).

The conjugacy classes of isotropy subgroups of $\mathbf{O}(2) \times \mathbf{T}^2$ and their fixed-point subspaces are given in Table 3.4. Figure 3.5 shows the isotropy lattice. It depends on whether l and m are even, odd, or 1.

It is now possible to determine the restricted vector fields $g|\mathrm{Fix}(\Sigma)$ for each isotropy subgroup Σ, and to show that for each of these restrictions the associated system of ODEs decouples into amplitude and phase equations. As usual, zeros of the amplitude equations correspond to invariant tori in the normal form equations. Table 3.4 shows the dimensions of these tori (periodic solutions, 2-torus, or 3-torus), which depend only on the isotropy subgroup. Hopf bifurcation along a branch of invariant 3-tori seems possible and leads to the existence of invariant 4-tori.

CASE STUDY 6

The Taylor–Couette System

With William F. Langford

§0. Introduction

The Taylor–Couette apparatus consists of a fluid contained between two coaxial, independently rotating cylinders. The experiments involve setting the speeds of rotation of the inner and the outer cylinders and observing the nature of the fluid flow between them. What astonishes observers of the experiment is the beautiful patterns that develop in the fluid flow. In this respect both theoreticians and experimentalists are united in an attempt to explain how these patterns form and why there are so many different types. Since the Taylor–Couette experiment is one of the simplest fluid flow experiments, it is a natural place to try to connect theory with experiment.

In this case study we attempt to relate fluid flow pattern formation to the existence of (approximate) symmetry. We study only those patterns found near an invariant steady state and compare our theoretical and numerical predictions with experimental findings.

This case study is divided into three sections. In the first we introduce the flow patterns and set the background for a bifurcation analysis of the Navier–Stokes equations based on $\mathbf{O}(2)$ symmetry. The analysis and the numerical results concerning the Navier–Stokes equations are presented in §2. There we show that mode interactions with $\mathbf{O}(2)$ symmetry occur in this model, and we indicate how to find the third order truncation of the bifurcation equations of steady-state/Hopf mode interaction with $\mathbf{O}(2)$ symmetry. Using the theoretical results of Chapter XX, we make a prediction concerning the existence of asymptotically stable solutions in the Taylor–Couette system, and then we describe experiments that confirm this prediction. Finally in §3 we briefly discuss finite length effects.

Some of the exposition and many of the results appearing in this case study are taken from the papers of Langford et al. [1988], Golubitsky & Langford [1987] and Tagg, Hirst & Swinney [1988]. We also acknowledge the excellent works of Chossat, Iooss, and Demay, whose theoretical and numerical studies on the Couette–Taylor system have complemented our own work. Finally we thank Randy Tagg and Harry Swinney for their help in relating our theoretical work to the experiments on the Taylor–Couette system.

§1. Detailed Overview

We have three major goals in this section:

(i) To describe the most elementary fluid flow patterns observed in experiments on the Taylor–Couette apparatus,
(ii) To explain how symmetries enter the theoretical analysis of the Taylor–Couette system and how the states described in (i) can be derived by using the techniques of bifurcation theory with symmetry,
(iii) To show how genericity, elementary group theory, and the consideration of the experimentally observed states permit an educated guess that mode interactions as studied in Chapter XX will appear in the analysis of (ii).

The section itself is divided into six parts. In subsection (a) we introduce the basic flow patterns and discuss both their dynamic nature and their symmetries. In (b) we show how $\mathbf{O}(2) \times \mathbf{SO}(2)$ symmetry is introduced into the mathematical model of the Taylor–Couette system, the Navier–Stokes equations, through the assumption of periodic boundary conditions. Part (c) is devoted to showing how, in a natural way, we can expect the study of bifurcation with $\mathbf{O}(2) \times \mathbf{SO}(2)$ symmetry to reduce to the study of bifurcation with $\mathbf{O}(2)$ symmetry. The general idea for performing the linear theory calculations necessary for this bifurcation analysis is discussed in (d). In that subsection we describe both the experiments and the theoretical analysis contained in the seminal paper of Taylor [1923]. In (e) we indicate how steady-state/Hopf mode interactions arise in the model with periodic boundary conditions model. In the final subsection (f) we illustrate conclusions that can be drawn for the PDE model, based on the nonlinear theory of mode interactions developed in Chapter XX.

We focus on three experimentally controllable parameters in the Taylor–Couette system:

Ω_1 = the angular velocity of the inner cylinder,

Ω_2 = the angular velocity of the outer cylinder,

η = the ratio of the radius of the inner cylinder to the radius of the outer cylinder.

The first two parameters are easily varied in an experiment, and the third parameter *can* be varied—but only at the expense of building a new apparatus.

It is most natural to use cylindrical geometry when describing the Taylor–Couette system. The *axial direction* refers to the direction along the axis of the cylinder; the *azimuthal direction* refers to the direction around the cylinder; the *radial direction* needs no definition.

(a) Basic Experimental Facts

We begin our story with a description of selected experimental observations. When the outer cylinder is held fixed and the inner cylinder is rotated slowly at speed Ω_1, the fluid particles seem to rotate in circles with a speed that is a function of their distance from the outer cylinder. This state of motion of the fluid is called *Couette flow*; see Figure 1.1(a). As Ω_1 is increased a critical speed is reached where Couette flow loses stability to a spatially structured state known as *Taylor vortices*; see Figure 1.1(b). This state is time-independent in the sense that the velocity field of the fluid flow does not depend on time. Of course, the fluid particles themselves move in time. Note that vortices normally occur in pairs when the orientation of the fluid motion is considered.

In the experiments the flow between the cylinders is visualized by adding small platelike particles—called *kalliroscope*—to the fluid, which align themselves in the direction of the flow and reflect light. The straight dark lines in the Taylor vortex pattern consist of points where the fluid flow is horizontal. Observe that the Taylor vortex pattern appears to be spatially periodic in the axial direction, at least near the midplane of a moderately long cylinder. As noted, the spatial period is measured by the extent of a vortex pair.

The *spiral vortices* of Figure 1.1(c) are another fluid flow pattern that is observed as a direct transition from Couette flow. To find this state the inner cylinder is rotated at a fixed velocity $\Omega_1 < 0$ and the outer cylinder is counter-rotated at a velocity $\Omega_2 > 0$. When Ω_1 is small, Couette flow is still the observed state. As Ω_1 is increased, however, there is a critical speed at which Couette flow loses stability to the spiral vortex pattern.

Unlike Taylor vortices, the spiral vortex state is time-dependent. To understand this point, look at Figure 1.1(c) and imagine the spiral vortices spiraling up the cylinder. At a fixed level on the outer cylinder the velocity field must change in time as the spiral wave crosses that level. Like Taylor vortices the spiral vortices appear to be spatially periodic in the axial direction, again if one focuses on the center part of a moderately long cylinder. Moreover, this spatial periodicity suggests that the spiral vortices should be time-periodic (the period being the length of time it takes the spiral wave to move up one vortex pair along the cylinder) and a traveling wave (since the time evolution of spiral vortices may be obtained by rotating the apparatus). These points are confirmed by the experiments.

A number of other fluid states have been observed in the Taylor–Couette experiment. This is amply illustrated by the findings of Andereck, Liu, and Swinney [1986] and reproduced in Figure 1.2. Some states that arise as

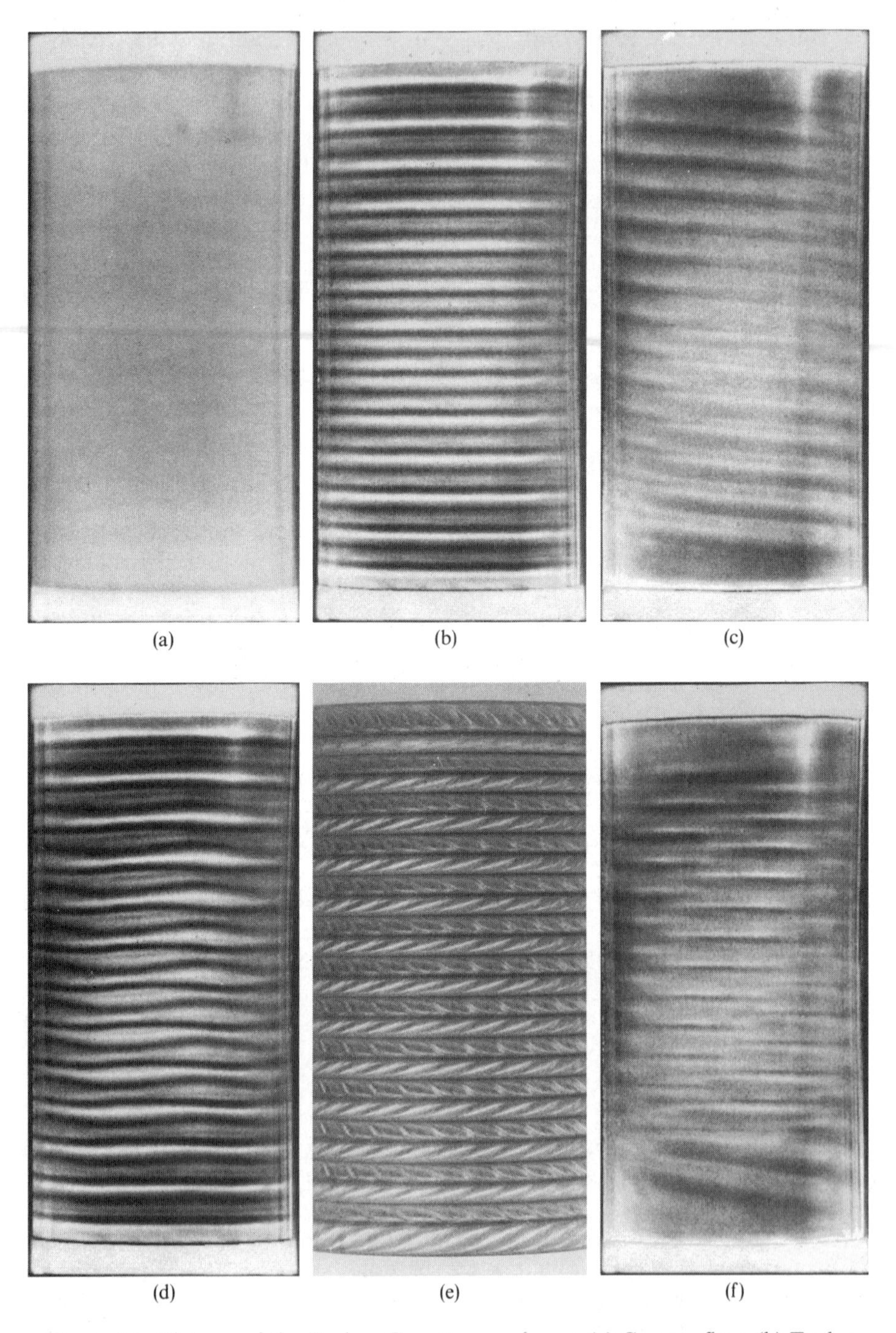

Figure 1.1. Pictures of the Taylor–Couette experiment: (a) Couette flow; (b) Taylor vortices; (c) spiral vortices; (d) wavy vortices; (e) twisted vortices; (f) interpenetrating spirals. (Courtesy of Harry Swinney and Randy Tagg.)

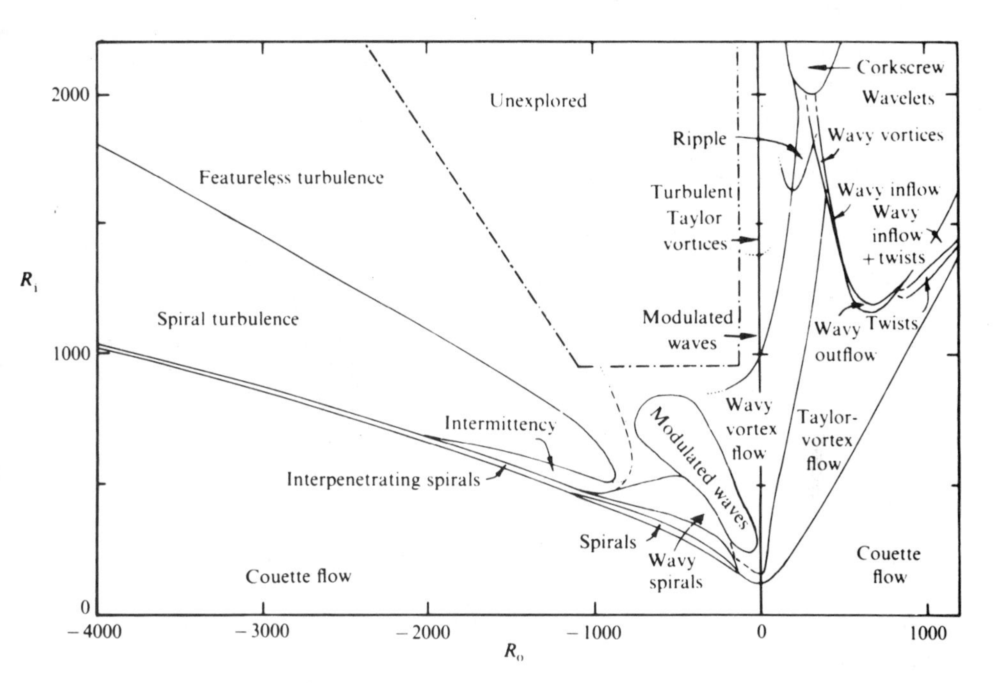

Figure 1.2. Experimentally determined flow regimes in the Taylor–Couette system. (Adapted from Andereck, Liu, and Swinney [1986].)

secondary transitions are picured in Figure 1.1(d)–(f). Again observe the striking spatially periodic patterns of *wavy vortices*, *twisted vortices*, and *interpenetrating spirals*.

(b) The Infinite Cylinder Approximation

Time evolution of the fluid flow vector field is governed by the three-dimensional Navier–Stokes equations. The Taylor–Couette experiment is modeled by augmenting these equations with appropriately chosen boundary conditions.

The accepted choice for boundary conditions on the inner and outer cylinder are "no slip" boundary conditions, in which the fluid is assumed to move along with the cylinder itself. The choice of boundary conditions on the ends of the cylinders is more problematic. There are two distinct choices to be made, depending on whether one is primarily interested in modeling the experimental apparatus or the experimentally observed states. One assumes physically realistic no slip boundary conditions in which the flow field vanishes on the fixed ends of the cylinder. The other assumes periodic boundary conditions in the axial direction, this choice being motivated by the periodicity apparent in the observed states.

The goals of analyses differ depending on which choice of axial boundary conditions is made. With the choice of realistic boundary conditions one hopes to study an exact mathematical model of the experiment and to match up steady and time-periodic solutions of the equations with the experimentally observed states. The difficulty with this choice, however, is that direct mathematical analysis is not possible—for example, there is no simple analytic formula for Couette flow, since even in Couette flow the motion of the fluid near the ends of the cylinder is complicated. With present-day computers it is possible, of course, to integrate the three-dimensional Navier–Stokes equations and to compare the computed solutions with experiment. This approach has been followed by several authors; see, for example, Marcus [1984], Cliffe and Spence [1985], and Dinar and Keller [1986]. Two drawbacks of this approach are that it is quite expensive to compute a large number of cases and the method does not provide a mathematical answer to the question, Why do the observed patterns actually appear?

The choice of periodic boundary conditions is motivated by two considerations. First and foremost, many of the observed states appear to be spatially periodic in the axial direction, and the class of solutions to these equations should provide insight into the nature of the patterns themselves. Second, the flow of the fluid far from the axial boundaries is unaffected (at least approximately) by boundary effects. We call this model the *infinite cylinder approximation* since any solution with periodic boundary conditions can be extended periodically to a solution with infinite extent in the axial direction.

As we shall see in §2, one virtue of the infinite cylinder approximation is the existence of an exact formula for Couette flow. Once a trivial solution exists, it is possible to perform a bifurcation analysis by searching for critical speeds of the inner and outer cylinders in which the trivial solution Couette flow loses stability and then computing the relevant higher order terms to determine stable nontrivial states. Such computations also require computer assistance, but of a much simpler sort than required by the exact equations. For example, a standard desktop PC is sufficiently powerful.

The goal of the infinite cylinder analysis is to provide an understanding of pattern formation. Since the model is at best approximate one must make a careful comparison of theoretical results with experiments to verify that the approximate model is reasonable. We use this approach in §§ 1 and 2.

(c) Bifurcation with $\mathbf{O}(2) \times \mathbf{SO}(2)$ Symmetry

Since the Navier–Stokes equations are invariant under the Euclidean group $\mathbf{E}_3$ of all translations, rotations and reflections of space, the group of symmetries of a given model is a subgroup of $\mathbf{E}_3$ determined by the shape of the domain and the boundary conditions. In the Taylor–Couette system, the simplest symmetries are obtained by rotating the whole apparatus about the common axis of the cylinders. These rotations lead to an $\mathbf{SO}(2)$ *azimuthal* symmetry. (Note that reflecting the apparatus through a plane containing the cylinder axis is not a symmetry of the system as this reflection changes the signs of the rotation speeds Ω_j.)

In the infinite cylinder approximation, translations along the axis of the cylinder yield symmetries of the equations, as do reflections of the cylinder through a plane perpendicular to its axis. As we have seen before, the assumption of periodic boundary conditions allows us to identify these translations with the action of a circle group. Thus the *axial* symmetries, induced by periodic boundary conditions, may be identified with the group $\mathbf{O}(2)$. Since the axial $\mathbf{O}(2)$ and the azimuthal $\mathbf{SO}(2)$ symmetries commute, the spatial symmetries of the infinite cylinder model form the group $\Gamma = \mathbf{O}(2) \times \mathbf{SO}(2)$.

We now discuss how we may use both the experimentally observed Taylor and spiral vortex states and genericity to make informed guesses about the types of bifurcation that should be expected from Couette flow in the infinite cylinder approximation. We claim that in order to find Taylor and spiral vortices the typical steady-state and Hopf bifurcations with symmetry group Γ should have the same structure as those same bifurcations with $\mathbf{O}(2)$ symmetry. We begin by considering steady-state bifurcation and let V be the eigenspace corresponding to the zero eigenvalue of the Navier–Stokes equations linearized about Couette flow. Recall from Theorem XII, 3.1, that in generic steady-state bifurcation we expect the action of Γ on V to be absolutely irreducible; that is, the only matrices commuting with the group are scalar multiples of the identity. Since the $\mathbf{SO}(2)$ part of Γ commutes with the whole of Γ, it must act trivially on V. Factoring out this trivial action we see that V is an absolutely irreducible representation of $\mathbf{O}(2)$ and is hence either one- or two-dimensional. One-dimensional spaces V correspond to solutions of the full nonlinear equations that are invariant under translation. Since Taylor vortices are not invariant under translation in the axial direction, double zero eigenvalues are to be expected in an analysis of the infinite cylinder approximation.

Similar reasoning applies in our abstract discussion of Hopf bifurcation. Here we let W be the real part of the eigenspaces associated with the complex conjugate pair of purely imaginary eigenvalues of the linearized equations. Recall Proposition XVI, 1.4, which states that typically either W is the direct sum of two identical absolutely irreducible representations or W is a non–absolutely irreducible representation of Γ. In the first case, as discussed, $\mathbf{SO}(2)$ acts trivially. Since $\mathbf{SO}(2)$ acts nontrivially on spiral vortices we do not expect this case to occur in our analysis. Next note that the irreducible representations of Γ are at most four-dimensional. (To verify this point recall that $\mathbf{SO}(2) \times \mathbf{SO}(2)$ is abelian and hence its irreducible representations are at most two-dimensional. See Theorem XII, 7.1. The two-element group $\Gamma/(\mathbf{SO}(2) \times \mathbf{SO}(2))$ can act nontrivially on such a two-dimensional space to give a four-dimensional irreducible representation of Γ.) Thus the purely imaginary eigenvalues of the linearized equations are (generically) either simple or double.

In fact, the spiral vortex state is consistent only with double eigenvalues.

To verify this point, begin by noting that no element of $\Gamma \sim \mathbf{SO}(2) \times \mathbf{SO}(2)$ can act trivially on W since any such element acts nontrivially on spiral vortices by transforming an upward spiraling vortex to a downward spiraling one. Thus, the kernel of the representation ρ of Γ on W must be contained in $\mathbf{SO}(2) \times \mathbf{SO}(2)$. Moreover, general principles imply that $\ker \rho$ is a normal subgroup of Γ. Now suppose that $\dim W = 2$. Then $\dim \ker \rho \geq 1$ since $\dim \Gamma = 2$ and the dimension of the orthogonal matrices on W is 1. Now normal subgroups of Γ with dimension at least 1 that are contained in $\mathbf{SO}(2) \times \mathbf{SO}(2)$ must either equal $\mathbf{SO}(2) \times \{0\}$ or contain the azimuthal symmetry group $\{0\} \times \mathbf{SO}(2)$ (see Exercise 1.1). As noted, the existence of spiral vortices implies that $\{0\} \times \mathbf{SO}(2)$ cannot act trivially on W. Similarly, axial translations in $\mathbf{SO}(2) \times \{0\}$ act nontrivially on spiral vortices and hence on W. Therefore, $\dim W = 4$.

Once we know that the eigenvalues of the linearization are double and that the representation ρ of Γ on W is irreducible, it follows that ρ is essentially the same representation as that of $\mathbf{O}(2) \times \mathbf{S}^1$, which appears in Hopf bifurcation with $\mathbf{O}(2)$ symmetry. (Here we have identified $\mathbf{S}^1$ with the azimuthal symmetries $\mathbf{SO}(2)$.) See Exercise 1.2(a).

We end this subsection by discussing two consequences of identifying the $\mathbf{S}^1$ phase shift symmetry with the azimuthal $\mathbf{SO}(2)$ symmetry.

(i) Every periodic solution is a "rotating wave" in the sense that time evolution of the solution can be obtained by azimuthal rotations of the cylinder.
(ii) Analysis of this Hopf bifurcation by a center manifold reduction leads to a vector field which is in Poincaré–Birkhoff normal form.

The identification of $\mathbf{S}^1$ with $\mathbf{SO}(2)$ means precisely that the time evolution of a periodic solution emanating from this bifurcation may be obtained merely by rotating the apparatus, thus verifying (i). This is *not* the same identification that would be obtained by noting that axial translation of the spiral vortex solution may also be obtained by rotation of the cylinders (due to the helical structure of spiral vortices). The existence of these two different symmetries in the isotropy subgroup of spiral vortices is related to the (sometimes confusing) fact that there are three circle groups of symmetries in the Taylor–Couette system: axial, azimuthal, and phase shift.

Another source of confusion is related to the fact that $\mathbf{O}(2)$ Hopf bifurcation typically spawns two types of periodic solution: rotating waves and standing waves. In the Taylor–Couette experiment the $\mathbf{O}(2)$ rotating wave is a time-periodic solution whose time evolution is obtained by axial translation. The spiral vortices have this property. The $\mathbf{O}(2)$ standing waves correspond to a pattern which does *not* travel in the axial direction. Such patterns have been called *ribbons* by Demay and Iooss [1984]. Note that ribbons must be invariant under an axial flip, and hence the flow field must be horizontal along a straight line—just like Taylor vortices. Nevertheless, even though ribbons are $\mathbf{O}(2)$ axial standing waves, they are $\mathbf{SO}(2)$ azimuthal rotating waves, as indicated by (i). Chossat, Demay, and Iooss [1987] have shown that under certain circumstances ribbons are asymptotically stable. Preliminary experimental evidence of Tagg (personal communication) indicates that ribbons have been observed, though not at the parameter values suggested by the analysis of Chossat, Demay, and Iooss.

(d) Basic Facts from the Numerical Calculations

The choice of periodic boundary conditions in the axial direction introduces a new parameter into the problem: the *axial wave number* k. Since the spatial axial period $(2\pi/k)$ is not specified by the choice of boundary conditions, it must be chosen in the analysis. Early experiments provide a rationale for making such a choice, which we now describe.

The first comprehensive set of experiments was performed by Taylor [1923]. In these experiments, Taylor fixed the speed Ω_2 of the outer cylinder and then varied the speed Ω_1 of the inner cylinder quasistatically. In this way, he could search for instability of Couette flow; his results are reproduced in Figure 1.3. Thus the experimental data consist of points (Ω_2, Ω_1^*) where the outer cylinder is fixed at velocity Ω_2 and Couette flow loses stability at $\Omega_1 = \Omega_1^*$.

In his 1923 paper Taylor also made theoretical predictions for the value of Ω_1^*. He used the infinite cylinder approximation to compute the instability of Couette flow to axisymmetric, time-independent disturbances (Taylor vortices). More precisely, Taylor fixed an axial wave number k and performed a linear bifurcation analysis to find the critical speed $\Omega_1^*(k)$ at which Couette flow loses stability. He then observed (at least numerically) that, as a function of k, $\Omega_1^*(k)$ had a unique global minimum and set Ω_1^* to be that minimum value. Taylor's predictions are also presented in Figure 1.3.

For small values of Ω_2 the agreement between theory and experiment is remarkably good, whereas for values of $\Omega_2 \ll 0$ there is a discrepancy between the experimental and theoretical values, the experimental value being consis-

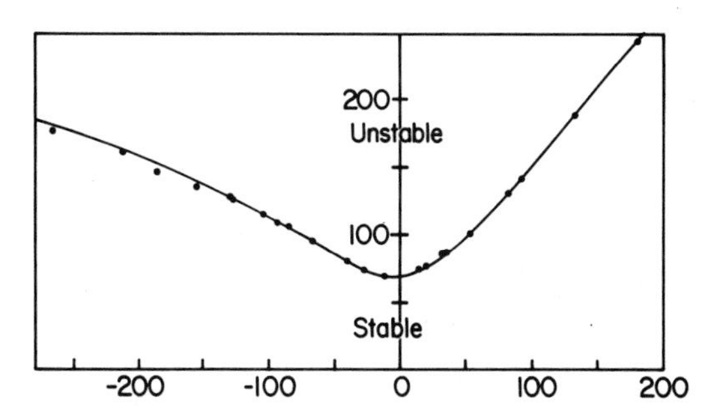

Figure 1.3. Comparison between observed and calculated speeds at which instability of Couette flow first appears when $\eta = 0.88$. (Adapted from Taylor [1923].)

tently smaller than the theoretical value. The later experiments of Snyder [1968] and the yet later and more comprehensive experiments of Andereck, Liu, and Swinney [1986] summarized in Figure 1.2 explain this discrepancy. In these circumstances the primary instability of Couette flow is caused by time-dependent disturbances (spiral vortices) rather than by time-independent disturbances.

This observation suggests that in order to determine Ω_1^* it is necessary to keep track of those eigenvalues of the Navier–Stokes equations that cross the imaginary axis at nonzero values. This was considered by Kreuger, Gross, and DiPrima [1966]. We use group-theoretic notions to help explain how their calculations are organized. We can interpret physically the choice of axial period as the axial length of one pair of Taylor vortices or one pair of spiral vortices. Hence we may assume (after rescaling by the axial period) that the axial $\mathbf{O}(2)$ symmetries act by the standard representation of $\mathbf{O}(2)$ on the eigenspace corresponding to the purely imaginary eigenvalues. The action of the azimuthal $\mathbf{SO}(2)$ symmetries, however, may be by *any* irreducible representation. Irreducible representations of $\mathbf{SO}(2)$ are indexed by a nonnegative integer. In the Taylor–Couette experiment this integer is called the *azimuthal wave number* and denoted by m. For the spiral vortex state m is the number of vortex pairs that intertwine to make up the spiral vortices. Taylor vortices correspond to $m = 0$.

Kreuger, Gross, and DiPrima [1966] approached the problem of time-independent instabilities as follows: For each m they computed numerically the curve of critical values $\Omega_{1,m}^*(k)$ at which Couette flow loses stability to a disturbance with azimuthal wave number m. Then they found a global minimum $\Omega_{1,m}^*$ for each (small) m and observed that except for the first few ms the values of $\Omega_{1,m}^*$ increase with m. Hence they could choose

$$\Omega_1^* \equiv \min \Omega_{1,m}^* \leq \Omega_{1,0}^*.$$

Most importantly, they found that when the speed of counterrotation $\Omega_2 \ll 0$, then the minimum Ω_1^* can occur for $m = 1$, 2 or higher.

(e) Mode Interactions

Both the experiments and the numerical calculations suggest that there should exist critical values of Ω_2, the speed of the outer cylinder, at which Couette flow loses stability simultaneously to Taylor vortices and to spiral vortices. That is, in the infinite cylinder approximation, there should be values of Ω_1 and Ω_2 at which steady-state/Hopf mode interactions with $\mathbf{O}(2)$ symmetry occur. This point was first explored by DiPrima and Grannick [1971] and more recently by Langford et al. [1988]. In §2 we describe the numerical results and compare them with experiments. There is one theoretical issue, however, that must be discussed when trying to make the existence of this mode interaction precise.

When determining the existence of steady state and Hopf bifurcations from

Couette flow in the infinite cylinder approximation one chooses an axial wave number k as described previously. At a point of mode interaction, however, there is no reason why the wave numbers chosen for these two types of bifurcation should be equal. In fact, both the calculations and the experiments lead to different but approximately equal values of k. In any theoretical analysis of the infinite cylinder approximation, however, the two values of k must be equal in order to perform a rigorous reduction to a finite-dimensional system (using either a Liapunov–Schmidt reduction or a center manifold reduction). This equality is needed to construct an appropriate function space on which the reduction may be performed.

There are at least two different methods for dealing with this difficulty, neither of which is totally satisfactory from a theoretical viewpoint. One is to assume that the cylinder has a definite axial length l, ignore boundary effects, and choose only axial wave numbers k for which l/k is an integer. This is the method used by Chossat, Demay, and Iooss [1987]. Another method is to let the instability to Taylor vortices determine the axial wave number k and fix this k to be the axial wave number of the instability to spiral vortices. This is the method used in DiPrima and Sijbrand [1982] and in Golubitsky and Langford [1987]. The two methods yield results that are very much in agreement and, as we shall see, are also in good agreement with experiments.

As remarked in subsection (d), there are values of Ω_2 at which the first instability of Couette flow is by a Hopf bifurcation to spiral vortices with azimuthal wave number m, with $m = 1, 2$ or higher. Therefore, we may expect that there are values of Ω_2 at which Hopf bifurcations to spiral vortices with wave numbers m and $m + 1$ occur simultaneously. This point will be verified in §2. See Chossat, Demay, and Iooss [1987].

Recall our discussion in subsection (c), where we showed that in steady-state and Hopf bifurcation we expect only $\mathbf{O}(2)$ symmetry to be present rather than $\mathbf{O}(2) \times \mathbf{SO}(2)$ symmetry. This turns out to be the case for Hopf/steady-state mode interactions as well, for the same reasons. See Exercise 1.2(b). Moreover, the consequences of this observation (basically that phase-shift symmetry and azimuthal rotation symmetry can be identified) are the same for this mode interaction as for Hopf bifurcation alone. Center manifold vector fields are always in normal form, and periodic solutions are always azimuthal rotating waves.

In Hopf/Hopf mode interactions the identification of phase-shift symmetry and azimuthal rotation symmetry does not happen automatically. In fact, such an identification is unlikely to occur. This makes the rigorous analysis of Hopf/Hopf mode interactions more difficult.

(f) Higher Order Terms

The preceding discussion explains why and how steady-state/Hopf mode interaction with $\mathbf{O}(2)$ symmetry occurs in the Taylor–Couette system. As we saw in XX, §2, there is a rich variety of secondary and tertiary branches of solutions that can be expected as a product of this mode interaction. Besides the primary branches corresponding to Taylor vortices, spiral vortices, and ribbons, there must exist three secondary branches of solutions. One of these solutions corresponds to a time-periodic pattern that is invariant under a flip in the axial direction and may be identified with twisted vortices. The second corresponds to a time-periodic solution that, when flippped in the axial direction, is half a period out of phase; this solution may be identified with wavy vortices. The last of the secondary branches corresponds to a two-frequency motion obtained by modulating the time-periodic spiral vortices. We refer to these states as *modulated spiral vortices.*

The tertiary states all correspond to invariant 2- and 3-tori, and their existence depends on the precise values of the third order terms in the reduced bifurcation equations.

Precisely which of the secondary states are asymptotically stable in the infinite cylinder approximation and which of the tertiary bifurcations actually occur can be determined only by computing the third order truncation of the reduced bifurcation equations. This reduction is discussed in the next section. We anticipate the results described there by stating that the third order truncation implies that Taylor vortices, spiral vortices, and wavy vortices are stable in a small neighborhood of the bicritical point. In addition, we find that spiral vortices and either Taylor or wavy vortices may be stable simultaneously for fixed values of Ω_1, Ω_2. A sample bifurcation diagram for $\eta = 0.80$ is shown in Figure 1.4. The prediction that multiple stable states exist and that wavy vortices are stable has been confirmed experimentally; see Tagg, Hirst, and Swinney [1988].

The experimental results of Andereck, Liu, and Swinney [1986], Figure 1.2, show that near the bicritical point, the interpenetrating spirals state seems to be observed. This state does not occur in the steady-state/Hopf mode interaction model, but it does appear as a result of Hopf/Hopf mode interaction; see Chossat [1986]. The reason that interpenetrating spirals are observed near the steady-state/Hopf mode interaction is something of an accident. The radius ratio $\eta = 0.883$ of the cylinders in the experiment is such that the steady-state/Hopf bicritical point occurs at a value of Ω_2 which is almost

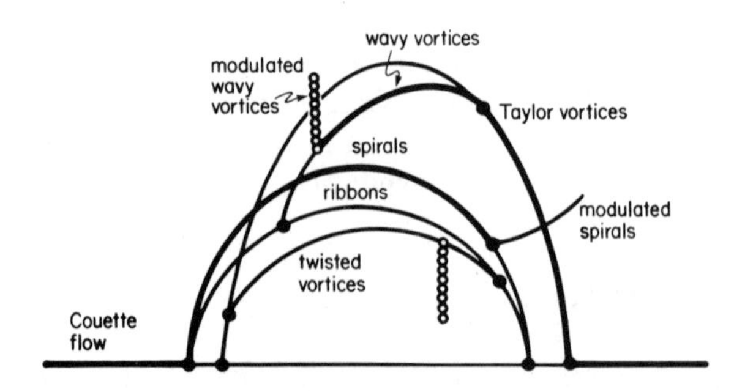

Figure 1.4. Bifurcation diagram around bicritical point for $\eta = 0.80$.

within experimental error of the value of Ω_2 where the bicritical point for Hopf/Hopf mode interaction occurs. This issue is discussed in §2. At smaller radius ratios the two bicritical points separate.

We make one last remark about higher order terms in the various bifurcation analyses. As we mentioned in subsection (c) there are values of the Ω_j at which ribbons are shown to be stable, and there are values at which spiral vortices are shown to be stable. It thus comes as no surprise that there are $\mathbf{O}(2)$-symmetric Hopf bifurcations from Couette flow where the cubic term in the bifurcation analysis, whose sign determines which of spiral vortices and ribbons are stable, is zero. That is, a degenerate Hopf bifurcation with $\mathbf{O}(2)$ symmetry appears in the Taylor–Couette system. As we saw in XVII, §6, the product of this degeneracy is the existence of a branch of invariant 2-tori on which the motion is a linear flow. To determine whether or not these solutions are asymptotically stable requires the computation of certain terms up to order 7 in the bifurcation equations. This computation has been completed by Laure and Demay [1987]; the results indicate that near this degeneracy the 2-torus is always unstable. Consequently, there should exist values of Ω_j at which ribbons and spiral vortices are simultaneously stable.

Exercises

1.1. A subgroup Σ of Γ is *normal* if $\gamma\Sigma\gamma^{-1} = \Sigma$, $\forall\gamma \in \Gamma$.

(a) Let $\rho\colon \Gamma \to GL(V)$ be a representation of Γ on V. Define

$$\ker\rho = \{\gamma \in \Gamma\colon \rho(\gamma) = I_V\}.$$

Show that $\ker\rho$ is a closed normal subgroup of Γ.

(b) Let Σ be a normal subgroup of $\Gamma = \mathbf{O}(2) \times \mathbf{SO}(2)$. Assume that $\Sigma \subset \mathbf{SO}(2) \times \mathbf{SO}(2)$ and that $\dim\Sigma = 1$. Show that either

$$\Sigma = \mathbf{SO}(2) \times \{0\} \quad \text{or} \quad \Sigma \supset \{0\} \times \mathbf{SO}(2).$$

(*Hint*: Suppose that $\Sigma \neq \mathbf{SO}(2) \times \{0\}$. Then the connected component of the identity of Σ contains a subgroup of the form $\{(m\theta, \theta)\}$. Use the fact that Σ is normal to show that $m = 0$.)

1.2. Let $\mathbf{O}(2) \times \mathbf{S}^1$ act on $W \cong \mathbb{C}^2$ as in the study of Hopf bifurcation with $\mathbf{O}(2)$ symmetry in (XVII, 1.3). Namely:

$$\begin{aligned} &\text{(a)} \quad \theta(z_1, z_2) = (e^{i\theta}z_1, e^{i\theta}z_2) \qquad (\theta \in S^1)\\ &\text{(b)} \quad \phi(z_1, z_2) = (e^{-i\phi}z_1, e^{i\phi}z_2) \qquad (\phi \in \mathbf{SO}(2))\\ &\text{(c)} \quad \kappa(z_1, z_2) = (z_2, z_1). \end{aligned} \tag{1.1}$$

Denote this representation of $\mathbf{O}(2) \times \mathbf{S}^1$ by ρ. Let γ be an orthogonal matrix on W that commutes with $\mathbf{O}(2) \times \mathbf{S}^1$.

(a) Show that $\gamma = \rho(\theta)$ for some $\theta \in S^1$.

(*Comment*: Since the azimuthal symmetries $\mathbf{SO}(2)$ commute with $\mathbf{O}(2) \times \mathbf{S}^1$ it follows that $\mathbf{S}^1$ can be identified $\mathbf{SO}(2)$ by the map $\theta \to l\theta$ for some positive integer l.)

(b) Extend the action of $\mathbf{O}(2) \times \mathbf{S}^1$ to $\mathbb{C}^3$ by letting:

$$\theta z_0 = z_0, \qquad \phi z_0 = e^{i\phi} z_0 \quad \text{and} \quad \kappa z_0 = \bar{z}_0. \tag{1.2}$$

Assume now that γ is an orthogonal matrix on $\mathbb{C}^3$ which commutes with $\mathbf{O}(2) \times \mathbf{S}^1$. Show again that $\gamma = \rho(\theta)$ for some $\theta \in S^1$.

§2. The Bifurcation Theory Analysis

This section is divided into six parts. We begin in (a) by presenting the infinite cylinder approximation PDE model discussed in §1(b), that is, the Navier–Stokes equations with periodic boundary conditions. We then give an explicit formula for the Couette flow solution. In part (b) we linearize these equations about Couette flow and present numerical results concerning the eigenvalues of the linearized equations. In doing so we show that mode interactions actually do occur in the infinite cylinder approximation. The form of the eigenfunctions for the linearized equations is presented in part (c). Once these eigenfunctions are determined explicitly it is easy to show that the symmetries of the Taylor–Couette system act in the way suggested by the discussion in §1. In (d) we present experimental findings of Tagg, Hirst, and Swinney (see Langford et al. [1988]), which illustrate the agreement between the linear analysis and the experiments). In (e) we briefly describe the results of a Liapunov–Schmidt reduction determining the third order truncation of the bifurcation equations. The relationship between the results of the nonlinear theory and experiment are discussed in the last subsection (f).

(a) The Navier–Stokes Equations and Couette Flow

The analysis begins with the Navier–Stokes equations

$$\begin{aligned} \mathbf{u}_t &= \nu \nabla^2 \mathbf{u} - (\mathbf{u} \cdot \nabla)\mathbf{u} - \frac{1}{\rho}\nabla p \\ \nabla \cdot \mathbf{u} &= 0, \end{aligned} \tag{2.1}$$

where $\mathbf{u}(t, x)$ = velocity vector at time t and $x \in \mathbb{R}^3$, p = pressure, ρ = mass density, ν = kinematic viscosity. (2.2)

Let us introduce the additional notation

$$\begin{aligned} r_1 &= \text{radius of inner cylinder} \\ r_2 &= \text{radius of outer cylinder} \\ d &= \text{gap width} = r_2 - r_1 \\ \eta &= \text{radius ratio} = r_1/r_2 \end{aligned}$$

$$\begin{aligned}
\Omega_1 &= \text{angular velocity of inner cylinder} \\
\Omega_2 &= \text{angular velocity of outer cylinder} \\
R_1 &= r_1\Omega_1 d/\nu = \text{Reynolds number of inner cylinder} \\
R_2 &= r_2\Omega_2 d/\nu = \text{Reynolds number of outer cylinder} \\
\mu &= \Omega_1/\Omega_2.
\end{aligned} \tag{2.3}$$

Since we are interested in the case in which the cylinders counterrotate, we assume $R_1 > 0$ and $R_2 < 0$. It is convenient to bring (2.1) into nondimensional form by rescaling lengths by d, velocities by the inner cylinder velocity $r_1\Omega_1$, and time t by the quantity d^2/ν. Furthermore, given the cylindrical symmetry of the apparatus, it is natural to express (2.1) in cylindrical coordinates (r, θ, z). We let u, v, w denote the corresponding components of $\mathbf{u}$ in cylindrical coordinates. Then (2.1) in the new variables, with partial derivatives denoted by subscripts, is

$$\begin{aligned}
u_t &= \nabla^2 u - \frac{2}{r^2}v_\theta - \frac{u}{r^2} - p_r - R_1\left[uu_r + \frac{v}{r}u_\theta + wu_z - \frac{v^2}{r}\right] \\
v_t &= \nabla^2 v + \frac{2}{r^2}u_\theta - \frac{v}{r^2} + \frac{1}{r}p_\theta - R_1\left[uv_r + \frac{v}{r}v_\theta + wv_z + \frac{uv}{r}\right] \\
w_t &= \nabla^2 w - p_z - R_1\left[uw_r + \frac{v}{r}w_\theta + ww_z\right] \\
\nabla\cdot\mathbf{u} &= u_r + \frac{1}{r}u + \frac{1}{r}v_\theta + w_z = 0.
\end{aligned} \tag{2.4}$$

The "no slip" boundary conditions at the cylinder walls take the form

$$\begin{aligned}
&\text{(a)}\quad (u, v, w) = (0, 1, 0) \quad \text{at} \quad r = \eta/(1-\eta) \\
&\text{(b)}\quad (u, v, w) = (0, \mu/\eta, 0) \quad \text{at} \quad r = 1/(1-\eta).
\end{aligned} \tag{2.5}$$

In the infinite cylinder approximation we assume that the flow $\mathbf{u}$ is periodic in the z direction, with period $2\pi/k$, where k is the axial wave number to be determined numerically as §1(d). We solve (2.4) over one period in z, with periodic end conditions, as well as azimuthal periodicity

$$\begin{aligned}
&\text{(a)}\quad \mathbf{u}(r, \theta, z) = \mathbf{u}(r, \theta, z + 2\pi/k) \\
&\text{(b)}\quad \mathbf{u}(r, \theta, z) = \mathbf{u}(r, \theta + 2\pi, z).
\end{aligned} \tag{2.6}$$

It is well known that the problem (2.4)–(2.6) has an exact time-independent solution, known as Couette flow, given by

$$\begin{aligned}
&\mathbf{u}_c = (0, v_c(r), 0), \qquad p = p_c(r), \qquad v_c(r) = Ar + B/r, \\
&A = -\frac{\eta^2 - \mu}{\eta(1+\eta)}, \qquad B = \frac{\eta(1-\mu)}{(1-\eta)(1-\eta^2)}, \qquad p_c(r) = R_1\int\frac{v_c(r)^2}{r}\,dr.
\end{aligned} \tag{2.7}$$

Since Couette flow is independent of z and θ, it has the symmetry of the full group $\mathbf{O}(2) \times \mathbf{SO}(2)$ described in §1. We are interested in the bifurcation from Couette flow to solutions with greater spatial structure, i.e., to solutions that break symmetry.

(b) Sample Results from the Linear Eigenvalue Problem

In this subsection we translate Couette flow to the trivial solution, write down explicitly the eigenvalue problem that must be solved to verify the stability of Couette flow, and present some sample numerical calculations showing that mode interactions occur in bifurcation from Couette flow. We begin by substituting

$$\mathbf{u} = \mathbf{u}_c + \hat{\mathbf{u}}, \qquad p = p_c + \hat{p}, \tag{2.8}$$

and then dropping the "hats." The resulting system is

$$\begin{aligned}
u_t &= \nabla^2 u - \frac{u}{r^2} - \frac{2}{r^2} v_\theta - p_r - C(r) u_\theta + 2C(r) v \\
&\quad - R_1 \left[u u_r + \frac{v}{r} u_\theta + w u_z - \frac{v^2}{r} \right] \\
v_t &= \nabla^2 v - \frac{v}{r^2} + \frac{2}{r^2} u_\theta - \frac{1}{r} p_\theta - C(r) v_\theta + 2Du \\
&\quad - R_1 \left[u v_r + \frac{v}{r} v_\theta + w v_z + \frac{uv}{r} \right] \\
w_t &= \nabla^2 w - p_z - C(r) w_\theta - R_1 \left[u w_r + \frac{v}{r} w_\theta + w w_z \right] \\
& u_r + \frac{1}{r} u + \frac{1}{r} v_\theta + w_z = 0.
\end{aligned} \tag{2.9}$$

The coefficients $C(r)$ and D appearing in (2.9) are defined by

$$\begin{aligned}
D &= -R_1 A = \frac{\eta R_1 - R_2}{1 + \eta}, \\
C(r) &= R_1(A + B/r^2) = -D + \frac{\eta(R_1 - \eta R_2)}{(1-\eta)(1-\eta^2) r^2}.
\end{aligned} \tag{2.10}$$

The boundary conditions for (2.9) states that the velocity is zero on the cylinder walls and that the periodicity conditions (2.6) are satisfied.

The analysis of bifurcation from Couette flow begins with a linear stability analysis of the trivial solution of (2.9). (The validity of the principle of linearized stability for the Navier–Stokes equations has been examined by several authors; see Sattinger [1973] and references therein.) The linearization of (2.9) is found by dropping the bilinear terms (in square brackets). Then the stability of the trivial solution is determined by the eigenvalues λ of the eigenvalue problem

$$\begin{aligned}
\lambda u &= \nabla^2 u - \frac{u}{r^2} - \frac{2}{r^2} v_\theta - p_r - C(r) u_\theta + 2C(r) v \\
\lambda v &= \nabla^2 v - \frac{v}{r^2} + \frac{2}{r^2} u_\theta - \frac{1}{r} p_\theta - C(r) v_\theta + 2Du \\
\lambda w &= \nabla^2 w - p_z - C(r) w_\theta \\
\nabla \cdot u &= 0.
\end{aligned} \tag{2.11}$$

This eigenvalue problem has been studied by many authors. The axisymmetric case was investigated by Taylor [1923] and Chandrasekhar [1961]. The non-axisymmetric case was studied by Krueger, Gross, and DiPrima [1966]; Demay and Iooss [1984]; and Langford et al. [1988], among others; see the survey of DiPrima and Swinney [1981]. The results that we describe here are taken from Langford et al. [1988]. The reader is referred to this work for tables of numerical values and graphs of the neutral stability curves, from which Figures 2.1 and 2.2 have been taken, as well as for details of the numerical procedure used to solve the eigenvalue problem (2.11).

These linear investigations show that, for each value of the radius ratio η, there are smooth curves in the (R_1, R_2) plane, along which (2.11) has critical eigenvalues, either $\lambda = 0$ (real) or $\lambda = \pm i\omega$ (purely imaginary); see Figure 2.1. Here m is the azimuthal wave number, as defined in §1(e); we note that $\lambda = 0$ if and only if $m = 0$. In every case the eigenvalues have been found to have multiplicity 2, as predicted by symmetry arguments in §1.

It is worth observing in Figure 2.1(b) just how close the intersection of the $m = 0$ and $m = 1$ neutral stability curves is with the intersection of the $m = 1$ and $m = 2$ neutral stability curves. The actual values for R_2 are -128.9 and

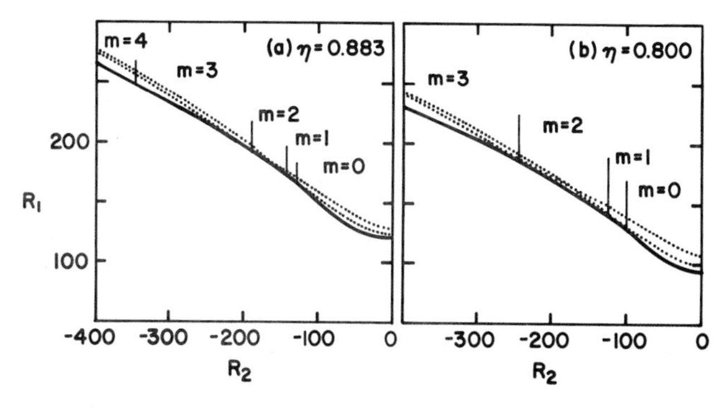

Figure 2.1. Set of neutral stability curves for counterrotating cylinders: (a) $\eta = 0.883$; (b) $\eta = 0.80$. (From Langford et al. [1988].)

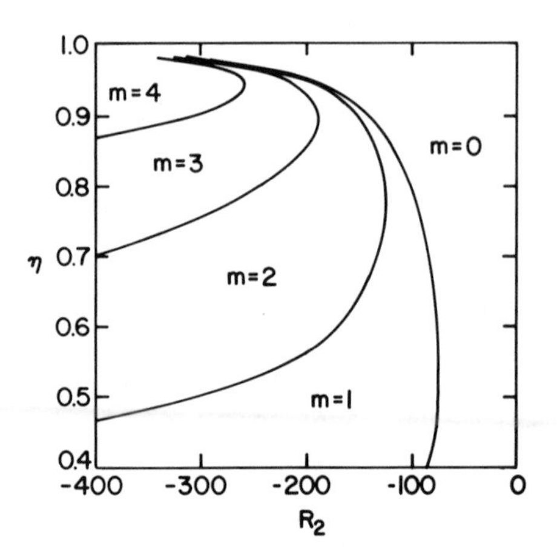

Figure 2.2. Location of bicritical points in R_2 as η varies, $0.45 < \eta < 0.98$. If (R_2, η) lies in a region marked $m = k$, then as R_1 increases Couette flow first loses stability to a disturbance with axial wave number k. (From Langford et al. [1988].)

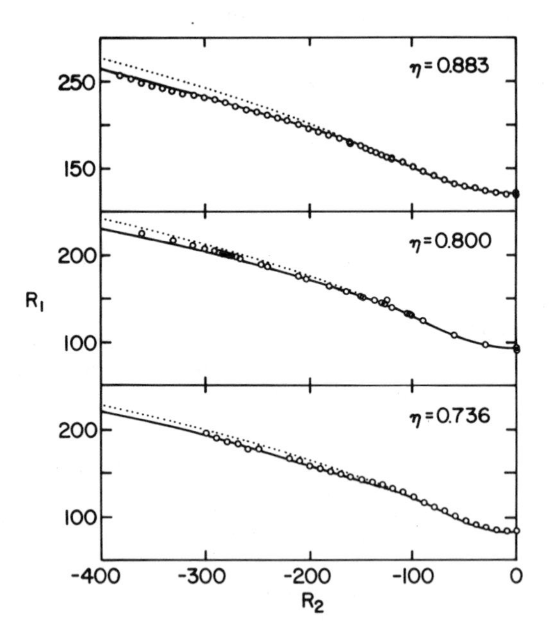

Figure 2.3. Experimentally determined critical values of R_1 for $\eta = 0.883$, 0.80, and 0.76. The solid curves are the theoretical envelopes of the lowest critical R_1 for all possible m. The dashed curves are the theoretical values of R_1 for $m = 0$ only. (From Langford et al. [1988].)

-142.8, respectively. Due to long relaxation times, this difference cannot be resolved in the results of experiments reported by Andereck, Liu, and Swinney [1986], and shown in Figure 1.2. As we mentioned in §1(d), this near coalescence of the bicritical points seems to explain why interpenetrating spirals are observed near the bicritical point of steady-state/Hopf mode interaction. One also sees that these bicritical points separate for smaller radius ratios η. For example, when $\eta = 0.80$ the value of R_2 at which the Hopf/Hopf bicritical point occurs is about 25% higher than the value of R_2 at which the steady-state/Hopf bicritical point occurs. For $\eta = 0.883$ this difference is only 11%. It should, therefore, be easier to see the differences between the two mode interactions by using smaller radius ratios in the experiments. In fact, in subsection (e), we show that there are additional complications predicted by the nonlinear analysis for experiments with radius ratio $\eta = 0.883$.

In this case study we are concerned with the bicritical points, as in Figure 2.1, where (2.11) simultaneously has eigenvalues $\lambda = 0$ and $\lambda = \pm i\omega$ (i.e., $m = 0$ and $m \neq 0$ in Figure 2.1). The analysis shows that below the envelope of the curves in Figure 2.1 all the eigenvalues have negative real part; i.e., Couette flow is linearly stable to small disturbances. As R_1 increases in Figure 2.1, Couette flow loses stability successively to an increasing number of modes (eigenfunctions of (2.11)). Therefore, intersection points above the lower envelope in Figure 2.1 are of less interest than those on the envelope, where Couette flow first loses stability. Here we focus on the intersection of the $m = 0$ and $m = 1$ curves in Figure 2.1. For a similar study of the intersections of m and $m + 1$ curves with $m \geq 1$, see Chossat et al. [1987].

The bicritical points in Figure 2.1 move as the radius ratio η varies. However, the numerical results in Langford et al. [1988] show that their ordering in R_2 remains monotone. See Figure 2.2. In Figure 2.3 we present experimentally determined values of R_1 at which Couette flow loses stability, and compare these values with the linear theory.

(c) Eigenfunctions and Symmetry

Let us now assume that the parameters have been chosen, as determined numerically, so that the eigenvalue problem (2.11) has eigenvalues 0 and $\pm i\omega$, each of which is double, and no others on the imaginary axis. As in Langford et al. [1988], the corresponding six eigenfunctions can be written

$$\Phi_0 = e^{ikz}\begin{bmatrix} U_0(r) \\ V_0(r) \\ iW_0(r) \end{bmatrix}, \quad \Phi_1 = e^{i(\theta+kz)}\begin{bmatrix} U_1(r) \\ V_1(r) \\ iW_1(r) \end{bmatrix}, \quad \Phi_2 = e^{i(\theta-kz)}\begin{bmatrix} U_1(r) \\ V_1(r) \\ -iW_1(r) \end{bmatrix}, \tag{2.12}$$

together with the three complex conjugates $\bar{\Phi}_j$. The exponential dependence on z and θ is obtained by elementary separation of variables, then the r-dependent vectors are computed numerically as the solution of a two-point boundary value problem. A factor i is included in the third components (equivalent to a phase shift of $\pi/4$) so that U_0, V_0, and W_0 are real functions. In every case, U_1, V_1, and W_1 are complex. Note that in (2.12) we have assumed that the azimuthal wave number m is 1.

Let us now consider the time-dependent linearized equations, that is, (2.9) with the bilinear terms dropped. Again, by separation of variables, this has solutions

$$\phi_0(t, x) = \Phi_0(x), \qquad \phi_1(t, x) = e^{-i\omega t}\Phi_1(x)_1, \qquad \phi_2(t, x) = e^{-i\omega t}\Phi_2(x) \quad (2.13)$$

where Φ_js are defined in (2.12), together with the three complex conjugates $\bar{\phi}_j$. Here $i\omega$ is the numerically computed purely imaginary eigenvalue, and $\omega > 0$. The general real solution of the linearized differential equation is then given by

$$\sum_{j=0}^{2} [z_j\phi_j(t, x) + \bar{z}_j\bar{\phi}_j(t, x)] \qquad \text{where } z_j \in \mathbb{C}. \quad (2.14)$$

The sum in (2.14) spans a six-dimensional linear space of smooth functions, satisfying the boundary conditions of (2.9), and periodic in t with period $2\pi/\omega$. We refer to this space as the *six-dimensional kernel*; it is the space on which the Liapunov–Schmidt reduction of the nonlinear equations will be performed. Details of this reduction may be found in Golubitsky and Langford [1987].

Using (2.12)–(2.14) we can compute the way the symmetries of the Taylor–Couette apparatus act on $\mathbb{C}^3$. In this calculation we set

$$\begin{aligned} &\text{(a)} \quad Z = kz \\ &\text{(b)} \quad \theta = \theta - \omega t. \end{aligned} \quad (2.15)$$

The group $\mathbf{O}(2)$ of axial translations and flips is generated by

$$\begin{aligned} &\text{(a)} \quad S_\phi = (Z, \theta) = (Z + \phi, \theta) \\ &\text{(b)} \quad \kappa(Z, \theta) = (-Z, \theta) \end{aligned} \quad (2.16)$$

where, in addition, κ changes the sign of the third component of the velocity field. It follows that S_ϕ and κ induce actions on $\mathbb{C}^3$ by

$$\begin{aligned} &\text{(a)} \quad S_\phi(z_0, z_1, z_2) = (e^{i\phi}z_0, e^{i\phi}z_1, e^{-i\phi}z_2) \\ &\text{(b)} \quad \kappa(z_0, z_1, z_2) = (\bar{z}_0, z_2, z_1). \end{aligned} \quad (2.17)$$

The group $\mathbf{SO}(2)$ of azimuthal rotations acts by

$$T_\theta(Z, \theta) = (z, \theta + \theta) \quad (2.18)$$

and induces the action on $\mathbb{C}^3$

$$T_\theta(z_0, z_1, z_2) = (z_0, e^{i\theta}z_1, e^{i\theta}z_2). \quad (2.19)$$

Thus the reduced bifurcation equations obtained by a Liapunov–Schmidt reduction must commute with the action (2.17, 2.19) of $\mathbf{O}(2) \times \mathbf{SO}(2)$ on $\mathbb{C}^3$;

Table 2.1. Comparison of Experimental and Theoretical Results at the Steady-State/Hopf Bicritical Point for Radius Ratio $\eta = 0.80$

	Theoretical	Experimental	Difference
R_2	-99.2	-100.7	-1.5%
R_1	129.5	131.7	1.1%
k_0	3.57	3.59	0.6%
k_1	3.55	3.39	-1.7%
ω_1/Ω_1	0.345	0.339	-1.8%

see Sattinger [1983] and VII, §3. Moreover, this action is the one promised in our general discussions in §1.

The $\mathbf{S}^1$ phase shift symmetry $t \to t - t_0$ can be computed from (2.13, 2.14). After rescaling time by ω we find that phase shifts act by

$$R_{t_0}(z_0, z_1, z_2) = (z_0, e^{it_0}z_1, e^{it_0}z_2). \tag{2.20}$$

Comparing (2.19) and (2.20) shows that indeed azimuthal rotations and phase-shift symmetries can be identified, as discussed in §1.

(d) Comparison of the Linear Calculations with Experiment

Experimental results of Tagg, and Swinney are recorded in Figure 2.3. In that figure a circle at (R_2, R_1) indicates a value of R_1 at which Couette flow is observed to lose stability when the outer cylinder is counterrotated at the speed R_2. The dark curves show the numerical values of instability of Couette flow, taken from Figure 2.1. (The dotted line indicates the points of instability of Couette flow to Taylor vortices, also taken from Figure 2.1.)

When the radius ratio $\eta = 0.8$ a precise determination of the steady-state/Hopf bicritical point was made, both numerically and experimentally. The results are presented in Table 2.1 (along with the axial periods and the time frequency of the Hopf mode). Observe that the agreement is to within 2%, thus suggesting that the infinite cylinder approximation is a reasonable one for a cylinder of moderate length. See Langford et al. [1988].

(e) The Liapunov–Schmidt Reduction

We are now in a position to apply the general results on mode interactions with $\mathbf{O}(2)$ symmetry discussed in XX, §2. We present some of the results of the nonlinear analysis in this subsection. Because of the $\mathbf{O}(2) \times \mathbf{SO}(2)$ symmetry present in this steady-state/Hopf mode interaction, we know the form that the equations g obtained by Liapunov–Schmidt reduction must take. (See (2.18) of Chapter XX.) We recall this form now.

Let

$$\begin{aligned}
&\text{(a)} \quad \rho = |z_0|^2 \\
&\text{(b)} \quad N = |z_1|^2 + |z_2|^2 \\
&\text{(c)} \quad \delta = |z_2|^2 - |z_1|^2 \\
&\text{(d)} \quad \Delta = \delta^2 \\
&\text{(e)} \quad A = z_0^2 \bar{z}_1 z_2 \\
&\text{(f)} \quad \Phi = \operatorname{Re} A \\
&\text{(g)} \quad \Psi = \delta \operatorname{Im} A.
\end{aligned} \tag{2.21}$$

Let $g\colon \mathbb{C}^3 \times \mathbb{R}^4 \to \mathbb{C}^3$ be the $\mathbf{O}(2) \times \mathbf{SO}(2)$-equivariant mapping obtained by Liapunov–Schmidt reduction on the infinite cylinder approximation mode at a bicritical point of steady-state/Hopf mode interaction. The parameters listed in $\mathbb{R}^4$ include the system parameters R_1, R_2, and η and the perturbed period parameter τ. The details of this reduction may be found in Golubitsky and Langford [1987]. Then g has the form

$$\begin{aligned}
g(z, \mu) = {} & (c^1 + i\delta c^2)\begin{bmatrix} z_0 \\ 0 \\ 0 \end{bmatrix} + (c^3 + i\delta c^4)\begin{bmatrix} \bar{z}_0 z_1 \bar{z}_2 \\ 0 \\ 0 \end{bmatrix} + (p^1 + iq^1)\begin{bmatrix} 0 \\ z_1 \\ z_2 \end{bmatrix} \\
& + (p^2 + iq^2)\delta\begin{bmatrix} 0 \\ z_1 \\ -z_2 \end{bmatrix} + (p^3 + iq^3)\begin{bmatrix} 0 \\ z_0^2 z_2 \\ \bar{z}_0^2 z_1 \end{bmatrix} + (p^4 + iq^4)\delta\begin{bmatrix} 0 \\ z_0^2 z_2 \\ -\bar{z}_0^2 z_1 \end{bmatrix}
\end{aligned} \tag{2.22}$$

where c^j, p^j, and q^j are functions of $\rho, N, \Delta, \Phi, \Psi$ and the vector of parameters $\mu = (R_1, R_2, \eta, \tau)$. In the Liapunov–Schmidt reduction all linear terms must vanish, so

$$c^1(0) = p^1(0) = q^1(0) = 0. \tag{2.23}$$

Further, as noted in (XX, 2.19(b))

$$c_\tau^1(0) = p_\tau^1(0) = 0, \; q_\tau^1(0) \neq 0. \tag{2.24}$$

In our analysis we will work explicitly with g truncated at third order in z and linear order in the parameters. Thus we assume that c^j, p^j, q^j have the form

$$\begin{aligned}
&\text{(a)} \quad c^1 = c_\mu^1(0) \cdot \mu + c_\rho^1(0)\rho + c_N^1(0)N \\
&\text{(b)} \quad c^2 = c^2(0) \\
&\text{(c)} \quad c^3 = c^3(0) \\
&\text{(d)} \quad c^4 = 0
\end{aligned}$$

$$\begin{aligned}
&\text{(e)} \quad p^1 = p^1_\mu(0)\cdot\mu + p^1_\rho(0)\rho + p^1_N(0)N \\
&\text{(f)} \quad q^1 = q^1_\mu(0)\cdot\mu + q^1_\rho(0)\rho + \rho^1_N(0)N \\
&\text{(g)} \quad p^2 = p^2(0) \\
&\text{(h)} \quad q^2 = q^2(0) \\
&\text{(i)} \quad p^3 = p^3(0) \\
&\text{(j)} \quad q^3 = q^3(0) \\
&\text{(k)} \quad p^4 = 0 \\
&\text{(l)} \quad q^4 = 0.
\end{aligned} \tag{2.25}$$

As shown in XX, §2, the existence and stability of solutions corresponding to Taylor vortices, spiral vortices, ribbons, wavy vortices, and twisted vortices are determined at third order, assuming certain nondegeneracy conditions. In addition the existence of certain branches of tori is determined by the third order truncation of g. In Table 2.2 we give the values of the derivatives (2.25) for some representative radius ratios. See Golubitsky and Langford [1987] for more details.

Table 2.2. Third Order Truncation

η	0.736	0.800	0.883
$c^1_{R_1}(0)$	0.537	0.457	0.342
$c^1_{R_2}(0)$	0.294	0.248	0.182
$c^1_\rho(0)$	−39.2	−13.4	1.06
$c^1_N(0)$	−164.	−84.4	−27.2
$c^2(0)$	−168.	−97.9	−37.9
$c^3(0)$	−169.	−99.2	−42.8
$p^1_{R_1}(0)$	0.488	0.429	0.330
$p^1_{R_2}(0)$	0.233	0.210	0.166
$p^1_\rho(0)$	−123.	−64.1	−21.1
$p^1_N(0)$	−101.	−49.8	−13.6
$p^2(0)$	−35.0	−21.9	−10.2
$q^2(0)$	−104.	−61.9	−24.5
$p^3(0)$	−94.3	−56.5	−24.4
$q^3(0)$	−51.2	−32.0	−13.2

(f) Predictions from the Nonlinear Theory

In Figure 2.4 we present the bifurcation diagrams of the third order truncated g for $\eta = 0.883$; recall Figure 1.4 where the bifurcation diagram for $\eta = 0.800$ is given. These bifurcation diagrams are presented, following the pattern for

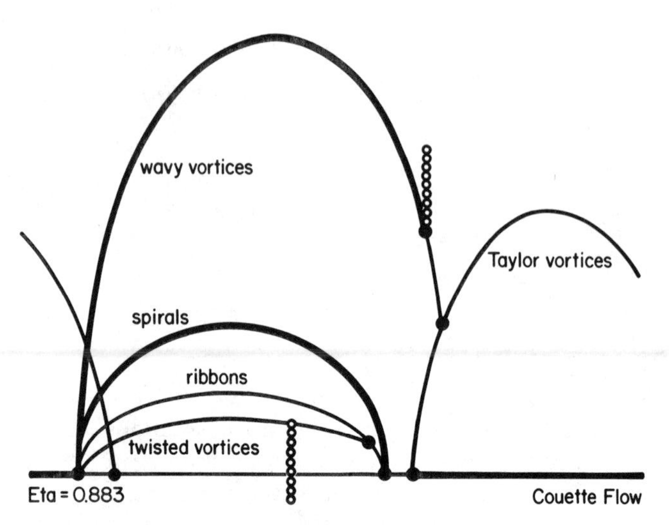

Figure 2.4. Bifurcation diagram around bicritical point for $\eta = 0.883$.

bifurcation diagrams around bicritical points described in XX, §2(e). In particular, we follow the path shown in Chapter XX, Figure 2.2 in a clockwise direction. The light lines in that figure indicate the neutral stability curves for $m = 0$ and $m = 1$.

Note the tori shown in these diagrams. At third order we can determine that the modulated spiral vortices are unstable (see XX, (2.31)); we have not attempted to determine the stability and direction of branching of the other tori, and these branches are shown as vertical for that reason.

There is one major prediction that can be made from the calculations on the nonlinear equations. At $\eta = 0.8$, the steady Taylor vortices state should lose stability to an asymptotically stable time-periodic ($m = 1$) wavy vortices state and this wavy vortices state should lose stability via a torus bifurcation to a two-frequency motion on an invariant 2-torus. If that 2-torus is asymptotically stable, then we would expect it to be observed in experiments; if it is not stable, then we would expect a jump bifurcation from wavy vortices to spiral vortices. In recent experiments Tagg, Hirst, and Swinney [1988] have found the wavy vortices state, thus confirming this prediction of theory, and a jump from wavy vortices directly to spiral vortices.

Another observation that can be made on the basis of the calculations from the nonlinear equations is that between $\eta = 0.80$ and $\eta = 0.883$ the coefficient $c_\rho^1(0)$ changes sign. This implies that the direction of branching and the stability of Taylor vortices change; thus the transition from Couette flow to Taylor vortices should not continuous when $\eta = 0.883$, and some hysteresis effects should be expected. The codimension three degeneracy which occurs when $c_\rho^1(0) = 0$ has not been yet been fully explored, either theoretically or experimentally. On the theoretical side a first analysis is given by Signoret and

Iooss [1987]. Other codimension three points are also shown to exist by the calculations; see Golubitsky and Langford [1987].

§3. Finite Length Effects

In this section we briefly discuss the relaxation of the infinite cylinder hypothesis. We suppose the outer cylinder and the end plates are at rest, and we consider only the first bifurcation from (approximately) Couette flow to steady Taylor vortices. In a finite-length apparatus the number of Taylor vortices is of course finite. This observation, despite its triviality, raises a difficult question: How does the number of cells, an integer, depend on the length L of the apparatus, a real parameter? Necessarily, there must be some discontinuous behavior.

Before this question was investigated carefully, it was generally felt that the number of cells $N(L)$ was simply a step function, and indeed, this is true of the idealized problem with periodic boundary conditions. However, Benjamin [1978] challenged this view as regards physically realistic boundary conditions. He argued on grounds of genericity and then substantiated his claims with experiment. More complete experiments were later performed by Mullin [1982]. In this case study we focus on the latter, specifically on the competition between four and six vortices.

Let us insert here the parenthetical remark that if the end plates are fixed, then under usual experimental conditions the number of Taylor vortices is always *even*. (In particular, this is true if the spin-up is gradual. We will call the *primary* flow the flow which develops if spin-up is gradual.) To understand this phenomenon, recall that the circulation of the bifurcating flows is driven by centrifugal force, which is approximately independent of the axial variable z away from the end faces but falls off near the ends where the velocity vanishes. Thus there exists a preference for inward flow along the end faces, and this can only happen along *both* faces if the number of cells is even.

The relevant part of Mullin's data is presented in Figure 3.1; in this figure R, the Reynolds number, is the nondimensionalized speed of rotation of the inner cylinder, and W, the width of the gap, is chosen as the unit of length. Spin-up is performed gradually with the height L held constant, say $L = L^*$. If $L^* > L_2$, $L_1 < L^* < L_2$, or $L^* < L_1$ the primary flow has six, six, or four cells, respectively; however, when $L_1 < L^* < L_2$ the evolution of the flow as R is increased is not smooth, but suffers a jump as R passes the middle of the three intersections of the line $\{L = L^*\}$ with the cusped curve C in the figure. In other words, the anticipated jump in $N(L)$ is broadened into a more subtle discontinuity which is spread out over the interval (L_1, L_2).

More generally, the curve C divides the plane into two regions such that there are one or two stable, experimentally observable flows according to whether (R, L) lies in region I or region II, respectively. We may refer to the

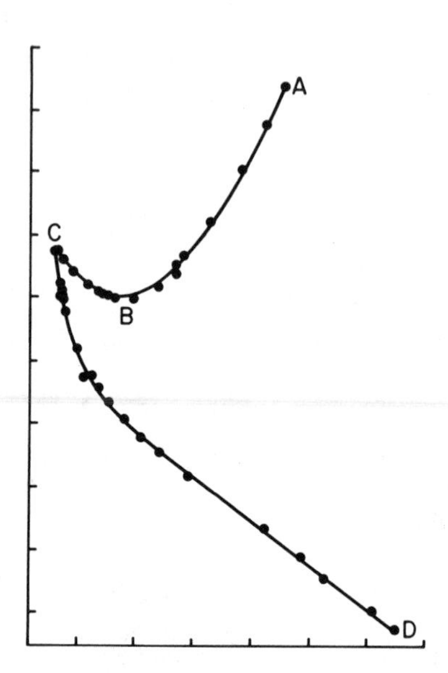

Figure 3.1. Experimental results on the competition between four and six vortices. (After Mullin [1982].)

two stable states in region II as four-cell or six-cell, although their cellular structure is only partially developed at the moderate values of the Reynolds number occurring in the figure. One or the other of the two flows in region II loses stability across C, giving rise to the possibility of jumps when, as here, (R, L) crosses the boundary moving from region II to region I. In general if (R, L) is varied quasistatically along a path which begins in region I, passes through region II, and reenters region I, a jump will occur on reentry if and only if the point H lies between the two points of C where the path crosses the boundary (between in the sense of distance along Γ).

Schaeffer [1980] discussed an analytical model which sheds insight on these experimental results. His model is a perturbed bifurcation problem containing two state variables (corresponding to the amplitudes of the competing four-cell and six-cell modes), the bifurcation parameter R, and two auxiliary parameters (the height L and a homotopy parameter τ). A plausibility argument deriving this model from the full PDE is also presented. The homotopy parameter τ, which varies between 0 and 1, determines the boundary conditions on the end plates as follows: For $\tau = 0$, periodic boundary conditions are imposed; for $\tau = 1$, "no slip" conditions are imposed; and for $0 < \tau < 1$, boundary conditions interpolating between these extremes are imposed. Schaeffer solved the equations in the model for small, nonzero τ and showed that the resulting solutions possess the same qualitative behavior as shown in Figure 3.2 provided the coefficients in the equations satisfied certain inequalities (in more technical language, provided the modal parameters lay in

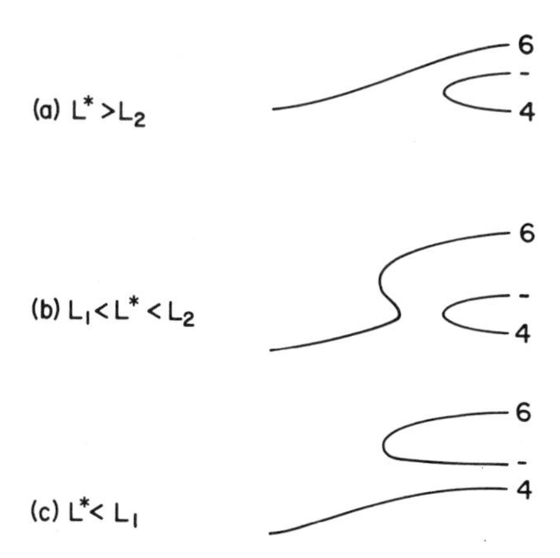

Figure 3.2. Qualitative behavior of the competition between four and six vortices in the model of Schaeffer [1980].

a certain region). The calculations of Hall [1982] later showed that these inequalities are indeed satisfied.

For more details about this work, we refer the reader to Schaeffer [1980] and to the later papers to be discussed. One reason for not discussing it here is that it raises issues lying outside the scope of this volume. Specifically, in deriving the bifurcation equations with two state variables from the PDE, Schaeffer made an ad hoc assumption, based on genericity, about how choosing τ nonzero would affect the equations. The relation of this assumption to the general theory is unclear at present; indeed, this assumption is related to the challenging open problem of developing a generic theory for breaking of symmetry in the equations. (see Golubitsky and Schaeffer [1983]).

We conclude this section by mentioning several more recent papers on finite length effects. On the experimental side, Mullin, Pfister, and Lorenzen [1982] report on laser-velocimetry measurements of the flow speed, especially near the point H in Figure 3.1 where one expects a hysteresis point. The quantitative data from these experiments confirm the presence of a hysteresis point. (These data also provide further tests for the calculations to be reported later, but comparisons have not yet been made.)

Dinar and Keller [1986] solve the steady-state Navier–Stokes equations numerically for various values of R and L. In effect, they repeat Mullin's experiments numerically. Besides the intrinsic challenge of such computations, they also allow one to follow unstable solution branches, an impossibility for physical experiments. In particular, Dinar and Keller's results confirm that in the full bifurcation diagrams, the stable solutions observed experimentally are connected with unstable branches, as shown in Figure 3.2. Roughly speaking, their calculations agree with the experimental results to better than 5%.

Finite cylinder effects that we have not discussed in this section include anomalous modes (Bolstad and Keller [1987], Cliffe and Mullin [1985]), the special circumstances surrounding the two-, four-cell competition in Benjamin's original experiment (Tavener and Cliffe [1987]), and the bifurcation problem in which the end plates both *rotate* (Tavener, Mullin, and Cliffe [1987]).

Bibliography

Abraham, R. and Marsden, J.E. [1978]. *Foundations of Mechanics.* (Second Edition) Benjamin/Cummings, New York.
Abraham, R., Marsden, J.E., and Ratiu, T. [1983]. *Manifolds, Tensor Analysis and Applications.* Addison-Wesley, Reading, Mass.
Abud, M., Anastaze, G., Eckert, P., and Ruegg, H. [1985]. Minima of Higgs potentials corresponding to non-maximal isotropy subgroups. *Ann. Physics* **162**, 155–191.
Adams, J.F. [1969]. *Lectures on Lie Groups.* Benjamin/Cummings, New York.
Alexander, J.C. [1986]. Patterns at primary Hopf bifurcations of a plexus of identical oscillators. *SIAM J. Appl. Math.* **46**, 199–221.
Alexander, J. and Auchmuty, J.F.G. [1979]. Global branching of waves. *Manus. Math.* **27**, 159–166.
Alexander, J.C. and Auchmuty, G. [1986]. Global bifurcations of phase-locked oscillators. *Arch. Rational Mech. Anal.* **93**, No. 3, 253–270.
Andereck, C.D., Liu, S.S., and Swinney, H.L. [1986]. Flow regimes in a circular Couette system with independently rotating cylinders. *J. Fluid Mech.* **164**, 155–183.
Armbruster, D. and Dangelmayr, G. [1983a]. Topological singularities and phonon focusing. *Z. Phys.* B, **52**, 87–94.
Armbruster, D. and Dangelmayr, G. [1983b]. Singularities in phonon focussing. *Z. Kristallographie* **162**, 236–237.
Armbruster, D. and Dangelmayr, G. [1984]. Structurally stable bifurcations in optical bistability. In: Proc. NATO ASI on Nonequilibrium Cooperative Phenomena in Physics and Related Fields (M. Velarde, Ed.) Plenum, 137–143.
Armbruster, D. and Dangelmayr, G. [1985]. Structurally stable transitions in optical tristability. *Nuovo Cimento* **85B**, 125–141.
Armbruster, D. and Dangelmayr, G. [1986]. Corank two bifurcations for the Brusselator with non-flux boundary conditions. *Dynam. & Stab. Sys.* **1**, 187–200.
Armbruster, D. and Dangelmayr, G. [1987]. Coupled stationary bifurcations in non-flux boundary value problems. *Math. Proc. Camb. Phil. Soc.* **101**, 167–192.
Armbruster, D., Dangelmayr, G., and Güttinger, W. [1984]. Nonlinear phonon focussing. In: *Phonon Scattering and Condensed Matter* (W. Eisenmenger, et al., Eds.) Plenum, 75–77.
Armbruster, D., Dangelmayr, G., and Güttinger, W. [1985]. Imperfection sensitivity

of interacting Hopf and steady-state bifurcations and their classification. *Physica* **16D**, 99–123.

Armbruster, D., Guckenheimer, J., and Holmes, P. [1987a]. Heteroclinic cycles and modulated travelling waves in systems with **O**(2) symmetry. *Physica* D. To appear.

Armbruster, D., Guckenheimer, J., and Holmes, P. [1987b]. Kuramoto–Sivashinsky dynamics on the center-unstable manifold. Cornell University. Preprint.

Arnold, V.I. [1978]. Critical points of functions on a manifold with boundary, the simple Lie groups B_k, C_k, and F_4, and singularities of evolutes. *Russian Math. Surveys* **33**: 5, 99–116.

Arnold, V.I. [1983]. *Geometrical Methods in the Theory of Ordinary Differential Equations*, Grundlehren **250**, Springer-Verlag, New York.

Ascher, E. [1962]. Role of particular maximal subgroups in continuous phase transitions. *Phys. Lett.* **20**, No. 4, 352–354.

Atkinson, F.V., Langford, W.F. and Mingarelli, A.B. [1987]. *Oscillation, Bifurcation and Chaos. CMS Conf. Proc.* **8**, AMS, Providence.

Auchmuty, J.F.G. [1979]. Bifurcating waves. In: *Bifurcation Theory and Applications in Scientific Disciplines* (O. Gurel and O.E. Rössler, Eds.) *Ann. New York Acad. Sci.* **316**, 263–278.

Auchmuty, J.F.G. [1984]. Bifurcation analysis of reaction–diffusion equations V. Rotating waves on a disc. In: *Partial Differential Equations and Dynamical Systems* (W.E. Fitzgibbon, Ed.) Res. Notes in Math **101**, Pitman, San Francisco, 35–63.

Aulbach, B. [1984]. *Continuous and Discrete Dynamics Near Manifolds of Equilibria.* Lecture Notes in Mathematics **1058**, Springer-Verlag, Berlin.

Bajaj, A.K. [1982]. Bifurcating periodic solutions in rotationally symmetric systems. *SIAM J. Appl. Math.* **42**, No. 5, 1078–1098.

Bajaj, A.K. and Sethna, P.R. [1982]. Bifurcations in three-dimensional motions of articulated tubes. *Trans. ASME*, **49**, 606–618.

Bajaj, A.K. and Sethna, P.R. [1984]. Flow induced bifurcations to three-dimensional oscillatory motions in continuous tubes. *SIAM J. Appl. Math.* **44**, No. 2, 270–286.

Ball, J.M. and Schaeffer, D.G. [1983]. Bifurcation and stability of homogeneous equilibrium configurations of an elastic body under dead-load tractions. *Math. Proc. Camb. Phil. Soc.* **94**, 315–339.

Bauer, L., Keller, H.B., and Reiss, E.L. [1975]. Multiple eigenvalues lead to secondary bifurcation. *SIAM J. Appl. Math.* **17**, 101–122.

Bay, J.S. and Hemami, H. [1987]. Modeling of a neural pattern generator with coupled nonlinear oscillators. *IEEE Trans. Biomed. Eng.* **34**, 297–306.

Benjamin, T.B. [1978]. Bifurcation phenomena in steady flows of a viscous fluid. *Proc. R. Soc. Lond. Ser.* A, **359**, 1–26, 27–43.

Bierstone, E. [1977a]. Generic equivariant maps in real and complex singularities. In: *Oslo 1976* (Holm, Ed.) Sijthoff and Noordhoff.

Bierstone, E. [1977b]. General position of equivariant maps. *Trans. A.M.S.* **234**, 447–466.

Bierstone, E. [1980]. *The Structure of Orbit Spaces and the Singularities of Equivariant Mappings.* I.M.P.A., Rio de Janeiro.

Bogdanov, R.I. [1975]. Versal deformations of a singular point of a vector field on the plane in the case of two zero eigenvalues. *Func. Anal. Appl.* **9** (2), 144–145.

Boivin, J.F. [1981]. Catastrophe theory and bifurcation. M.S. Thesis. McGill University.

Bolstad, J.H. and Keller, H.B. [1987]. Computation of anomalous modes in the Taylor experiment. *J. Comp. Phys.* **69**, 230–251.

Booty, M.R., Margolis, S.B., and Matkowsky, B.J. [1986]. Interaction of pulsating and spinning waves in condensed phase combustion. *SIAM J. Appl. Math.* **45**, 801–843.

Bourbaki, N. [1960]. *Groupes et Algebras de Lie*, Ch. I, *Act. Sc. et Ind.* **1285**, Ch. IV, V, VI [1968] Hermann, Paris.

Bredon, G.E. [1972]. *Introduction to Compact Transformation Groups.* Pure & Appl. Math. **46**, Academic Press, New York.

Bröcker, T. [1975]. *Differential Germs and Catastrophes.* (Transl. by L. Lander.) London Math. Soc., Lect. Notes Ser. **17**, Cambridge University Press, Cambridge.
Bruce, J.W. [1984]. Functions on discriminants. *J. London. Math. Soc.* **30**, 551–567.
Bruce, J.W., du Plessis, A.A., and Wall. C.T.C. [1987]. Determinacy and unipotency. *Invent. Math.* **88**, 521–554.
Busse, F.H. [1962]. *Das Stabilitätsverhalten der Zellarkonvektion bei endlicher Amplitude.* Thesis, University of Munich. (Engl. transl. by S.H. Davis, Rand Rep. LT-69-19, Rand Corporation, Santa Monica, Calif.)
Busse, F.H. [1975]. Pattern of convection in spherical shells. *J. Fluid Mech.* **72**, 65–85.
Busse, F.H. [1978]. Nonlinear properties of thermal convection. *Rep. Prog. Phys.* **41**, 1929–1967.
Busse, F.H. and Riahi, N. [1982]. Pattern of convection in spherical shells, II. *J. Fluid Mech.* **123**, 283–291.
Buzano, E., Geymonat, G., and Poston, T. [1985]. Post-buckling behavior of a nonlinearly hyperelastic thin rod with cross-section invariant under the dihedral group $\mathbf{D}_n$. *Arch. Rational Mech. Anal.* **89**, No. 4, 307–388.
Buzano, E. and Golubitsky, M. [1983]. Bifurcation on the hexagonal lattice and the planar Bénard problem. *Phil. Trans. R. Soc. Lond.* **A308**, 617–667.
Buzano, E. and Russo, A. [1987]. Bifurcation problems with $\mathbf{O}(2) \oplus \mathbf{Z}_2$ symmetry and the buckling of a cylindrical shell. *Annali di Matematica Pura ed Applicata* (IV) **146**, 217–262.
Carr, J. [1981]. *Applications of Centre Manifold Theory.* Applied Math. Sci. **35**. Springer-Verlag, New York.
Chandrasekhar, S. [1961]. *Hydrodynamic and Hydromagnetic Stability.* Oxford University Press, Oxford.
Chatelin, F. [1983]. *Spectral Approximations of Linear Operators.* Academic Press, New York.
Chillingworth, D.R.J. [1986]. Bifurcation from an orbit of symmetry. In: *Singularities and Dynamical Systems* (S. Pnevmatikos, Ed.) North-Holland, 285–294.
Chillingworth, D.R.J. [1987]. The ubiquitous astroid. In: Güttinger and Dangelmayr [1987].
Chillingworth, D.R.J., Marsden, J.E., and Wan, Y-H. [1982]. Symmetry and bifurcation in three-dimensional elasticity I. *Arch. Rational Mech. Anal.* **80**, 295–331; II, ibid. **83**, 363–395.
Chossat, P. [1979]. Bifurcation and stability of convective flows in a rotating or not rotating spherical shell. *SIAM J. Appl. Math.* **37**, 624–647.
Chossat, P. [1982]. *Le Problème de Bénard dans une Couche Spherique.* These d'État, Université de Nice.
Chossat, P. [1983]. Solutions avec symétrie diédrale dans les problèmes de bifurcation invariants par symétrie sphérique. *C.R. Acad. Sci. Paris, Serie I*, **297**, 639–642.
Chossat, P. [1984]. Bifurcations en présence de symétrie dans les problèmes classiques de l'hydrodynamique. *J. Mech. Theor. et Appl., Numéro special*, 157–192.
Chossat, P. [1985a]. Interaction d'ondes rotatives dans le problème de Couette–Taylor. *C.R. Acad. Sci. Paris* **300**, I, No. 8, 251–255.
Chossat, P. [1985b]. Bifurcation d'ondes rotatives superposées. *C.R. Acad. Paris* **300**, I, No. 7, 209.
Chossat, P. [1986]. Bifurcation secondaire de solutions quasi-periodiques dans un problème de bifurcation de Hopf invariant par symétrie $\mathbf{O}(2)$. *C.R. Acad. Sci. Paris* **302**, 539–541.
Chossat, P., Demay, Y., and Iooss, G. [1984]. Recent results about secondary bifurcations in the Couette–Taylor problem. In: *Proceedings of the International Conference on Nonlinear Mechanics, Shanghai* (Chien Wei-Zang, Ed.) Science Press, Beijing, 1001–1004.
Chossat, P., Demay, Y., and Iooss, G. [1987]. Interactions des modes azimutaux dans le problème de Couette–Taylor. *Arch. Rational Mech. Anal.* **99**, No. 3, 213–248.

Chossat, P. and Golubitsky, M. [1987a]. Hopf bifurcation in the presence of symmetry, center manifold and Liapunov–Schmidt reduction. In: Atkinson et. al. [1987], 343–352.

Chossat, P. and Golubitsky, M. [1987b]. Iterates of maps with symmetry. *SIAM J. Math. Anal.* To appear.

Chossat, P. and Golubitsky, M. [1988]. Symmetry increasing bifurcation of chaotic attractors. *Physica* D, submitted.

Chossat, P., Golubitsky, M., and Keyfitz, B.L. [1987]. Hopf–Hopf mode interactions with **O**(2) symmetry. *Dyn. Stab. Sys.* **1**, 255–292.

Chossat, P. and Iooss, G. [1985]. Primary and secondary bifurcations in the Couette–Taylor problem. *Japan J. Appl. Math.* **2**, No. 1, 37–68.

Chossat, P. and Lauterbach, R. [1987]. The instability of axisymmetric solutions in problems with spherical symmetry. *SIAM J. Math. Anal.* To appear.

Chow, S.N. and Hale, J. [1982]. *Methods of Bifurcation Theory.* Grundlehren **251**. Springer-Verlag, New York.

Chow, S.N. and Lauterbach, R. [1986]. A bifurcation theorem for critical points of variational problems. IMA Preprint #179, University of Minnesota.

Cicogna, G. [1981]. Symmetry breakdown from bifurcations. *Lettere al Nuovo Cimento* **31**, 600–602.

Cicogna, C. and Gaeta, G. [1985]. Periodic solutions from quaternionic bifurcation. *Lettere al Nuovo Cimento* **44**, 65.

Cicogna, C. and Gaeta, G. [1986a]. Spontaneous linearization and periodic solutions in Hopf and symmetric bifurcations. *Phys. Lett.* A, **116**, 303–306.

Cicogna, C. and Gaeta, G. [1986b]. Quaternionic bifurcation and **SU**(2) symmetry. University of Pisa. Preprint.

Cicogna, C. and Gaeta, G. [1987a]. Quaternionic-like bifurcation in the absence of symmetry. *J. Phys.* A, **20**, 79–89.

Cicogna, C. and Gaeta, G. [1987b]. Hopf-type bifurcation in the presence of multiple critical eigenvalues. *J. Phys.* A, **20**, L425–L427.

Cliffe, K.A. [1984]. Numerical calculations of the primary flow exchange process in the Taylor problem. *J. Fluid Mech.* To appear.

Cliffe, K.A., Jepson, A.D., and Spence, A. [1985]. The numerical solution of bifurcation problems with symmetry with application to the finite Taylor problem. *Numerical Methods for Fluid Dynamics II* (K.W. Morton and M.J. Baines, Eds.) Oxford University Press, 155–176.

Cliffe, K.A. and Mullin, T. [1985]. A numerical and experimental study of anomalous modes in the Taylor experiment. *J. Fluid Mech.* **153**, 243–258.

Cliffe, K.A. and Mullin, T. [1986]. A numerical and experimental study of the Taylor problem with asymmetric end conditions. *Proceedings of the 6th International Symposium on Finite Element Methods in Flow Problems*, Antibes, June 1986, 377–381.

Cliffe, K.A. and Spence, A. [1983]. The calculation of high order singularities in the finite Taylor problem. In: *Numerical Methods for Bifurcation Problems, Dortmund*, (T. Kupper, H.D. Mittelmann and H. Weber, Eds.) Birkhäuser, Basel, 129–144.

Cliffe, K.A. and Spence, A. [1985]. Numerical calculation of bifurcations in the finite Taylor problem. *Numerical Methods for Fluid Dynamics II* (K.W. Morton and M.J. Baines, Eds.) Oxford University Press, 177–197.

Coullet, P. and Repaux, D. [1987]. Models of pattern formation from a singularity theory point of view. In: *Instabilities and Nonequilibrium Structures* (E. Tirapegui and D. Villanoel, Eds.). Reidel, Dordrecht, 179–195.

Cowan, J.D. [1982]. Spontaneous symmetry breaking in large scale nervous activity. *Inter. J. Quantum Chem.* **22**, 1059–1082.

Cowan, J.D. [1986]. Brain mechanisms underlying visual hallucinations. University of Chicago. Preprint.

Coxeter, H.S.M. [1963]. *Regular Polytopes.* Macmillan, New York.

Crandall, M. and Rabinowitz, P. [1971]. Bifurcation from simple eigenvalues. *J. Func. Anal.* **8**, 321–340.
Crandall, M. and Rabinowitz, P. [1973]. Bifurcation, perturbation of simple eigenvalues, and linearized stability. *Arch. Rational Mech. Anal.* **52**, 161–180.
Crawford, J.D., Golubitsky, M., and Langford, W.F. [1987]. Modulated rotating waves in **O**(2) mode interactions. *Dyn. Stab. Sys.*, submitted.
Crawford, J.D. and Knobloch, E. [1987]. Classification and unfolding of degenerate Hopf bifurcations with **O**(2)-symmetry: no distinguished parameter. *Physica* D. To appear.
Cushman, R. and Sanders, J.A. [1986]. Nilpotent normal forms and representation theory of $sl(2, R)$. In: Golubitsky and Guckenheimer [1986], 31–52.
Damon, J. [1983]. Private communication.
Damon, J. [1984]. *The unfolding and determinacy theorems for subgroups of $\mathscr{A}$ and $\mathscr{K}$. Memoirs, A.M.S.* **306**, Providence.
Damon, J. [1986]. On a theorem of Mather and the local structure of nonlinear Fredholm maps. *Proc. Sympos. Pure Math.* **45**, Part I, 339–352.
Damon, J. [1987]. Time dependent nonlinear oscillations with many periodic solutions. *SIAM J. Math. Anal.* **18**, No. 5, 1294–1316.
Dancer, E.N. [1980a]. On the existence of bifurcating solutions in the presence of symmetries. *Proc. Royal Soc. Edin.* **85A**, 321–336.
Dancer, E.N. [1980b]. An implicit function theorem with symmetries and its application to nonlinear eigenvalue problems. *Bull. Austral. Math. Soc.* **21**, 404–437.
Dancer, E.N. [1982a]. Symmetries, degree, homotopy indices and asymptotically homogeneous problems. *Nonlinear Analysis* **6**, 667–686.
Dancer, E.N. [1982b]. The G-invariant implicit function theorem in infinite dimensions. *Proc. Roy. Soc. Edinburgh* **92A**, 13–30.
Dancer, E.N. [1983a]. Bifurcation under continuous groups of symmetry. In: *Systems of Nonlinear Partial Differential Equations.* (J.M. Ball, Ed.) Reidel, 343–350.
Dancer, E.N. [1983b]. Breaking of symmetries for forced equations. *Math. Ann.* **262**, 473–486.
Dancer, E.N. [1984]. Perturbation of zeros in the presence of symmetries. *J. Austral. Math. Soc.* **36A**, 106–125.
Dancer, E.N. [1986]. The G-invariant implicit function theorem in infinite dimensions Part II. *Proc. Roy. Soc. Edinburgh* **102A**, 211–220.
Dangelmayr, G. [1986]. Steady-state mode interactions in the presence of **O**(2) symmetry. *Dyn. Stab. Sys.* **1**, 159–185.
Dangelmayr, G. [1987a]. Degenerate bifurcations near a double eigenvalue in the Brusselator. *J. Austral. Math. Soc. Ser.* B, **28**, 486–535.
Dangelmayr, G. [1987b]. Wave patterns in coupled stationary bifurcations with **O**(2) symmetry. In: Kupper et al. [1987], 27–37.
Dangelmayr, G. and Armbruster, D. [1983]. Classification of $\mathbf{Z}_2$-equivariant imperfect bifurcations with corank 2. *Proc. London Math. Soc.* (3) **46**, 517–546.
Dangelmayr, G. and Armbruster, D. [1986]. Steady state mode interactions in the presence of **O**(2) symmetry and in non-flux boundary conditions. In: Golubitsky and Guckenheimer [1986], 53–68.
Dangelmayr, G. and Guckenheimer, J. [1987]. On a four parameter family of planar vector fields. *Arch. Rational Mech. Anal.* **97**, No. 4, 321–352.
Dangelmayr, G. and Knobloch, E. [1986]. Interactions between steady and travelling waves and steady states in magnetoconvection. *Phys. Lett.* **117**, 394–398.
Dangelmayr, G. and Knobloch, E. [1987a]. The Takens–Bogdanov bifurcation with **O**(2) symmetry. *Phil. Trans. R. Soc. London* **A322**, 243–279.
Dangelmayr, G. and Knobloch, E. [1987b]. On the Hopf bifurcation with broken **O**(2) symmetry. In: Güttinger and Dangelmayr [1987], 387–393.
Dangelmayr, G. and Stewart, I. [1984]. Classification and unfoldings of sequential

bifurcations. *SIAM J. Math. Anal.* **15**, 423–445.
Dangelmayr, G. and Stewart, I. [1985]. Sequential bifurcations in continuous stirred tank chemical reactors. *SIAM J. Appl. Math.* **45**, 895–918.
Deane, A.E., Knobloch, E., and Toomre, J. [1987]. Travelling waves and chaos in thermosolutal convection. *Phys. Rev.* **A36**, 2862–2869.
Deane, A.E., Knobloch, E., and Toomre, J. [1988]. Large aspect ratio thermosolutal convection. *Phys. Rev. Lett.* **A37**, 1817–1820.
DeKleine, H.A., Kennedy, E., and MacDonald, N. [1982]. A study of coupled oscillators. Preprint.
Demay, Y. and Iooss, G. [1984]. Calcul des solutions bifurquées pour le problème de Couette–Taylor avec les deux cylindres en rotation. *J. de Mech. Theor. et Appl.*, Numéro special, 193–216.
Dieudonné, [1960]. *Foundations of Modern Analysis.* Academic Press, New York.
Dinar, N. and Keller, H.B. [1986]. Computations of Taylor vortex flows using multigrid methods. To appear.
DiPrima, R.C. and Grannick, R.N. [1971]. A nonlinear investigation of the stability of flow between counter-rotating cylinders. In: *Instability of Continuous Systems* (H. Leipholz, Ed.) Springer-Verlag, Berlin, 55–60.
DiPrima, R.C. and Sijbrand, J. [1982]. Interactions of axisymmetric and non-axisymmetric disturbances in the flow between concentric rotating cylinders: bifurcations near multiple eigenvalues. In: *Stabilitiy in the Mechanics of Continua* (F.H. Schroeder, Ed.) Springer-Verlag, Berlin, 383–386.
DiPrima, R.C. and Swinney, H.L. [1981]. Instabilities and transition in flow between concentric rotating cylinders. In: *Hydrodynamic Instabilities and the Transition to Turbulence* (H.L. Swinney and J.P. Gollub, Eds.) Topics in Applied Physics **45**, Springer-Verlag, Berlin, 139–180.
Dubrovin, B.A., Fomenko, A.T., and Novikov, S.P. [1984]. *Modern Geometry—Methods and Applications*, Vol. 1. Springer-Verlag, New York.
Duff, G.F.D. and Naylor, D. [1966]. *Differential Equations of Applied Mathematics.* Wiley, New York.
Elphick, C. [1987]. Some remarks on the normal forms of Hamiltonian and conservative systems. University of Nice. Preprint.
Elphick, C., Tirapegui, E., Brachet, M.E., Coullet, P., and Iooss, G. [1987]. A simple global characterization for normal forms of singular vector fields. *Physica* **29D**, 95–127.
Erneux, T. and Herschkowitz-Kaufman, M. [1977]. Rotating waves as asymptotic solutions of a model chemical reaction. *J. Chem. Phys.* **66**, 248–253.
Erneux, T. and Matkowsky, B.J. [1984]. Quasi-periodic waves along a pulsating propagating front in a reaction–diffusion system. *SIAM J. Appl. Math.* **44**, 536–544.
Farr, W. and Aris, R. [1986]. "Yet who would have thought the old man to have so much blood in him?" Reflections on the multiplicity of steady states of the stirred tank reactor. *Chem. Eng. Sci.* **41**, 1385–1402.
Farr, W.W. and Aris, R. [1987]. Degenerate Hopf bifurcations in the CSTR with reactions $A \to B \to C$. In: Atkinson et. al. [1987], 397–418.
Fiedler, B. [1986]. Global Hopf bifurcation of two parameter flows. *Arch. Rational Mech. Anal.* **94**, 59–81.
Fiedler, B. [1987]. *Global Bifurcation of Periodic Solutions with Symmetry.* Habilitationsschrift, Universität Heidelberg.
Fiedler, B. and Kunkel, P. [1987]. A quick multiparameter test for periodic solutions. In Kupper et al. [1987].
Field, M. [1976]. Transversality in G-manifolds. *Trans. A.M.S.* **231**, No. 2, 429–450.
Field, M. [1977]. Stratifications of equivariant varieties. *Bull. Aust. Math. Soc.* **16**, 279–295.
Field, M. [1980]. Equivariant dynamical systems. *Trans. A.M.S.* **259**, No. 1, 185–205.

Field, M. [1983]. Isotropy and stability of equivariant diffeomorphisms. *Proc. London Math. Soc.* **46**, No. 3, 563–576.

Field, M. [1986a]. Unfolding equivariant diffeomorphisms: theory and practice. University of Sydney. Preprint.

Field, M. [1986b]. Equivariant dynamics. In Golubitsky and Guckenheimer [1986], 69–95.

Field, M. [1987]. Symmetry breaking for the Weyl groups $W(D_k)$. University of Sydney. Preprint.

Field, M. and Richardson, R.W. [1987]. Symmetry breaking and the maximal isotropy subgroup conjecture for finite reflection groups. *Arch. Rational Mech. Anal.* To appear.

Gaffney, T. [1986]. Some new results in the classification theory of bifurcation problems. In: Golubitsky and Guckenheimer [1986], 97–116.

Geiger, C., Güttinger, W., and Haug, P. [1985]. Bifurcations in particle physics and crystal growth. *Inter. Symp. on Synergetics*, Schloss Eliman, Springer-Verlag.

Gils, S.A. van [1984]. *Some Studies in Dynamical Systems Theory*, Thesis, Vrije Universitet, Amsterdam.

Gils, S.A. van and Mallet-Paret, J. [1984]. Hopf bifurcation and symmetry: standing and travelling waves on a circle. *Proc. Royal Soc. Edinburgh* **104A**, 279–307.

Gils, S.A. van and Valkering, T. [1986]. Hopf bifurcation and symmetry: standing and travelling waves in a circular chain. *Japan J. Appl. Math.* **3**, 207–222.

Glaeser, G. [1963]. Fonctions composées différentiables. *Ann. Math.* **77**, 193–209.

Golubitsky, M. [1978]. An introduction to catastrophe theory and its applications. *SIAM Rev.* **20**, 352–387.

Golubitsky, M. [1983]. The Bénard problem, symmetry and the lattice of isotropy subgroups. In: *Bifurcation Theory, Mechanics and Physics* (C.P. Bruter et. al., Eds.) Reidel, Dordrecht, 225–256.

Golubitsky, M. and Guckenheimer, J. [1986]. *Multiparameter Bifurcation Theory.* Contemporary Mathematics **56**, A.M.S., Providence.

Golubitsky, M. and Guillemin, V. [1973]. *Stable Mappings and Their Singularities.* Graduate Texts in Math. **14**, Springer-Verlag, New York.

Golubitsky, M. and Langford, W.F. [1981]. Classification and unfoldings of degenerate Hopf bifurcations. *J. Diff. Eqns.* **41**, 375–415.

Golubitsky, M. and Langford, W.F. [1987]. Pattern formation and bistability in flow between counterrotating cylinders. *Physica* D. Submitted.

Golubitsky, M., Marsden, J.E., and Schaeffer, D.G. [1984]. Bifurcation problems with hidden symmetries. In: *Partial Differential Equations and Dynamical Systems* (W.E. Fitzgibbon III, Ed.) Research Notes in Math. **101**, Pitman, San Francisco, 181–210.

Golubitsky, M. and Roberts, M. [1987]. A classification of degenerate Hopf bifurcations with **O**(2) symmetry. *J. Diff. Eqns.* **69**, 216–264.

Golubitsky, M. and Schaeffer, D. [1979a]. A theory for imperfect bifurcation theory via singularity theory. *Commun. Pure and Appl. Math.* **32**, 21–98.

Golubitsky, M. and Schaeffer, D. [1979b]. Imperfect bifurcation in the presence of symmetry. *Commun. Math. Phys.* **67**, 205–232.

Golubitsky, M. and Schaeffer, D. [1982]. Bifurcation with **O**(3) symmetry including applications to the Bénard problem. *Commun. Pure Appl. Math.* **35**, 81–111.

Golubitsky, M. and Schaeffer, D. [1983]. A discussion of symmetry and symmetry breaking. *Proc. Symp. Pure Math.* **40**, Part I, 499–515.

Golubitsky, M. and Schaeffer, D.G. [1985]. *Singularities and Groups in Bifurcation Theory*, Vol. 1. Appl. Math. Sci. **51**, Springer-Verlag, New York.

Golubitsky, M. and Stewart, I.N. [1985]. Hopf bifurcation in the presence of symmetry. *Arch. Rational Mech. Anal.* **87**, No. 2, 107–165.

Golubitsky, M. and Stewart, I.N. [1986a]. Symmetry and stability in Taylor–Couette flow. *SIAM J. Math. Anal.* **17**, No. 2, 249–288.

Golubitsky, M. and Stewart, I.N. [1986b]. Hopf bifurcation with dihedral group symmetry: coupled nonlinear oscillators. In: Golubitsky and Guckenheimer [1986], 131–173.

Golubitsky, M. and Stewart, I.N. [1987]. Generic bifurcation of Hamiltonian systems with symmetry. *Physica* **24D**, 391–405 (with Appendix by J. Marsden).

Golubitsky, M., Swift, J.W., and Knobloch, E. [1984]. Symmetries and pattern selection in Rayleigh–Bénard convection. *Physica* **10D**, 249–276.

Guckenheimer, J. [1984]. Multiple bifurcation problems of codimension two. *SIAM J. Math. Anal.* **15**, No. 1, 1–49.

Guckenheimer, J. [1986]. A codimension two bifurcation with circular symmetry. In: Golubitsky and Guckenheimer [1986], 175–184.

Guckenheimer, J. and Holmes, P. [1983]. *Nonlinear Oscillations, Dynamical Systems, and Bifurcations of Vector Fields.* Appl. Math. Sci. **42**. Springer-Verlag, New York.

Guckenheimer, J. and Holmes, P. [1987]. Structurally stable heteroclinic cycles. *Math. Proc. Camb. Phil. Soc.* To appear.

Güttinger, W. [1984]. Bifurcation geometry in physics. *Proc. NATO Adv. Study Inst.*, Santa Fe.

Güttinger, W. and Dangelmayr, G. [1987]. *The Physics of Structure Formation.* Springer-Verlag, Heidelberg.

Hall, P. [1982]. Centrifugal instabilities of circumferential flows in finite cylinders: the wide gap problem. *Proc. Roy. Soc. Lond.* **A384**, 359–379.

Halmos, P.R. [1974]. *Finite-Dimensional Vector Spaces.* UTM, Springer-Verlag, New York.

Hassard, B., Kazarinoff, N., and Y.-H. Wan [1981]. *Theory and Applications of Hopf Bifurcation.* London Math. Soc. Lect. Note Ser. **41**. Cambridge University Press, Cambridge.

Hassard, B. and Wan, Y.-H. [1978]. Bifurcation formulae derived from the center manifold theory. *J. Math. Anal. Appl.* **63**, 297–312.

Healey, T.J. [1988a]. Symmetry and equivariance in nonlinear elastostatics I: differential field equations. *Arch. Rational Mech. Anal.* To appear.

Healey, T.J. [1988b]. A group theoretic approach to computational bifurcation problems with symmetry. *Comput. Meth. Appl. Mech. & Eng.* To appear.

Healey, T.J. [1988c]. Global bifurcation and continuation in the presence of symmetry with an application in solid mechanics. *SIAM J. Math. Anal.* **19**. To appear.

Hirsch, M. and Smale, S. [1974]. *Differential Equations, Dynamical Systems, and Linear Algebra.* Academic Press, New York.

Hochschild, G. [1965]. *The Structure of Lie Groups.* Holden-Day, San Francisco.

Hoffman, K. and Kunze, R. [1971]. *Linear Algebra.* (Second Edition). Prentice-Hall, Englewood Cliffs, N.J.

Howard, L.N. [1979]. Nonlinear oscillations. In: *Oscillations in Biology* (F.R. Hoppensteadt, Ed.) *AMS Lecture Notes in Applied Mathematics* **17**, 1–69.

Hummel, A. [1979]. *Bifurcation and Periodic Points.* Thesis, University of Groningen.

Hunt, G.W. [1981]. An algorithm for the nonlinear analysis of compound bifurcation. *Phil. Trans. R. Soc. Lond.* **A300**, 447–471.

Hunt, G.W. [1982]. Symmetries of elastic buckling. *Eng. Struct.* **4**, 21–28.

Huppert, H.E. and Moore, D.R. [1976]. Nonlinear double-diffusive convection. *J. Fluid Mech.* **78**, 821–854.

Ihrig, E. and Golubitsky, M. [1984]. Pattern selection with **O**(3) symmetry. *Physica* **12D**, 1–33.

Iooss, G. [1986]. Secondary bifurcations of Taylor vortices into wavy inflow and outflow boundaries. *J. Fluid Mech.* **173**, 273–288.

Iooss, G. [1987]. Reduction of the dynamics of a bifurcation problem using normal forms and symmetries. In: *Instabilities and Nonequilibrium Structures* (E. Tirapegui

and D. Villanoel, Eds.). Reidel, Dordrecht, 3–40.
Iooss, G., Coullet, P. and Demay, Y. [1986]. Large scale modulations in the Taylor–Couette problem with counterrotating cylinders. Preprint No. 89, Université de Nice.
Iooss, G. and Joseph, D.D. [1981]. *Elementary Stability and Bifurcation Theory.* Springer-Verlag, New York.
Iooss, G. and Langford, W.F. [1980]. Conjectures on the routes to turbulence. *Annals of the New York Academy of Sciences*, **357**, 489–505.
Iooss, G. and Rossi, M. [1987]. Hopf bifurcation in the presence of spherical symmetry: analytical results. *SIAM J. Math. Anal.* To appear.
Ize, J. Massabo, I., Pejsachowicz, J., and Vignoli, A. [1985]. Global results on continuation and bifurcation of equivariant maps.
Janeczko, S. [1983]. On *G*-versal Lagrangian submanifolds. *Bull. Acad. Polon. Sci.* **31**, 183–190.
Jepson, A.D., Spence, A., and Cliffe, K.A. [1987]. The numerical solution of nonlinear equations having several parameters, III: equations with $\mathbf{Z}_2$ symmetry. *SIAM J. Num. Anal.* To appear.
Jones, C.A. and Proctor, M.R.E. [1987]. Strong spatial resonance and travelling waves in Bénard convection. Preprint.
Jones, D.F. and Treloar, L.R.G. [1975]. The properties of rubber in pure homogeneous strain. *J. Phys. D (Appl. Phys.)* **8**, 1285–1304.
Keener, J. [1976]. Secondary bifurcation in non-linear diffusion reaction equations. *Stud. Appl. Math.* **55**, 187–211.
Keyfitz, B.L. [1984]. Classification of one state variable singularities with distinguished parameter up to codimension seven. *Dyn. Stab. Sys.* **1**, 1–41.
Keyfitz, B.L., Golubitsky, M., Gorman, M., and Chossat, P. [1986]. The use of symmetry and bifurcation techniques in studying flame stability. *Lectures in Applied Mathematics* **24**, AMS, Providence, 293–315.
Kirchgässner, K. [1979]. Exotische lösungen des Bénardschen problems. *Math. Meth. Appl. Sci.* **1**, 453–467.
Kirillov, A.A. [1976]. *Elements of the Theory of Representations*, Grundlehren **220**, Springer-Verlag, Berlin.
Knightly, G.H. and Sather, D. [1980]. Buckled states of a sphere under uniform external pressure. *Arch. Rational Mech. Anal.* **72**, 315–380.
Knobloch, E. [1980a]. Oscillatory convection in binary mixtures. *Phys. Rev.* A, **34**, No. 2, 1538–1549.
Knobloch, E. [1986b]. On the degenerate Hopf bifurcation with **O**(2) symmetry. In: Golubitsky and Guckenheimer [1986], 193–202.
Knobloch, E. [1986c]. On convection in a horizontal magnetic field with periodic boundary conditions. *Geophys. Astophys. Fluid Dyn.* **36**, 161–177.
Knobloch, E. [1986d]. Normal form coefficients for the nonresonant double Hopf bifurcation. *Phys. Lett.* A, **116**, No. 8, 365–369.
Knobloch, E. and Weiss, J.B. [1987]. Chaotic advection by modulated travelling waves. *Phys. Rev.* A, **36**, 1522–1524.
Kopell, N. [1983]. Forced and coupled oscillators in biological applications. Proc. Int. Congr. Math. Warsaw.
Koschmieder, E.L. [1974]. Bénard convection. *Adv. Chem. Phys.* **26**, 177–188.
Krasnosel'skii, M.A. [1964]. *Topological Methods in the Theory of Nonlinear Equations.* Pergamon Press, Oxford.
Kreuger, E.R., Gross, A., and DiPrima, R.C. [1966]. On the relative importance of Taylor-vortex and nonaxisymmetric modes in flow between rotating cylinders. *J. Fluid Mech.* **24**, 521–538.
Kupper, T., Seydel, R., and Troger, H. [1987]. *Bifurcation, Analysis, Algorithms, Applications.* Birkhäuser, Basel.

Labouriau, I.S. [1985]. Degenerate Hopf bifurcation and nerve impulse. *SIAM J. Math. Anal.* **16**, No. 6, 1121–1133.

Langford, W.F. [1979]. Periodic and steady-state mode interactions lead to tori. *SIAM J. Appl. Math.* **37**, 22–48.

Langford, W.F. [1986]. Hopf bifurcation at a hysteresis point. In: *Differential Equations: Qualitative Theory*, Colloq. Math. Soc. Janos Bolyai V **47**, North-Holland, Amsterdam.

Langford, W.F. and Iooss, G. [1980]. Interactions of Hopf and pitchfork bifurcaons. In: *Bifurcation Problems and Their Numerical Solution* (H.D. Mittelmann and H. Weber, Eds.) International Series on Numerical Mathematics **54**, Birkhäuser-Verlag, Basel, 103–134.

Langford, W.F., Tagg, R., Kostelich, E., Swinney, H.L., and Golubitsky, M. [1987]. Primary instability and bicriticality in flow between counterrotating cylinders. *Phys. Fluids* **31**, 776.

Laure, P. and Demay, Y. [1987]. Symbolic computation and equation on the center manifold: application to the Couette–Taylor problem. University of Nice. Preprint No. 120.

Lauterbach, R. [1986]. An example of symmetry breaking with submaximal isotropy. In: Golubitsky and Guckenheimer [1986], 217–222.

Lindtner, E., Steindl, A., and Troger, H. [1987]. Stabilitätsverlust der gestreckten lage eines raumlichen doppelpendels mit elastischer endlegerung unter einer folgelast. *ZAMM* **67**, T105–T107.

Marcus, P.S. [1984]. Simulation of Taylor–Couette flow. Part I: Numerical methods and comparison with experiment. Part II: Numerical results for wavy-vortex flow with one travelling wave. *J. Fluid Mech.* **146**, 45–113.

Marsden, J.E. and McCracken, M. [1976]. *The Hopf Bifurcation and Its Applications.* Appl. Math. Sci. **19**. Springer-Verlag, New York.

Martinet, J. [1976]. Déploiements versels des applications différentiables et classification des applications stables. In: *Singularités d'Applications Différentiables, Plans-sur-Bex 1975* (O. Burlet and F. Ronga, Eds.) Lec. Notes Math **535**. Springer-Verlag, Berlin, 1–44.

Martinet, J. [1982]. *Singularities of Smooth Functions and Maps* (Transl. by C. Simon.) London Math. Soc. Lect. Notes Ser. **58**, Cambridge University Press, Cambridge.

Mather, J.N. [1968]. Stability of C^∞ mappings, III. Finitely determined map germs. *Publ. Math. I.H.E.S.* **35**, 127–156.

Mather, J.N. [1977]. Differential invariants. *Topology* **16**, 145–155.

Matkowsky, B.J. and Olagunju, D.O. [1982]. Travelling waves along the front of a pulsating flame. *SIAM J. Appl. Math.* **42**, 486–501.

Meer, J.C. van der [1982]. Nonsemisimple 1:1 resonance at an equilibrium. *Cel. Mech.* **27**, 131–149.

Meer, J.C. van der [1985]. *The Hamiltonian Hopf Bifurcation*, Lecture Notes in Mathematics **1160**. Springer-Verlag, Berlin.

Meer, J.C. van der [1987]. Hamiltonian Hopf bifurcation with symmetry. Preprint, University of Houston.

Melbourne, I. [1986]. A singularity theory analysis of bifurcation problems with octahedral symmetry. *Dyn. Stab. Sys.* **1**, 293–321.

Melbourne, I. [1987a]. The recognition problem for equivariant singularities. *Nonlinearity* **1**, 215–240.

Melbourne, I. [1987b]. Bifurcation problems with octahedral symmetry. Thesis, University of Warwick.

Michel, L. [1972]. Nonlinear group action: Smooth actions of compact Lie groups on manifolds. In: *Statistical Mechanics and Field Theory* (R.N. Sen and C. Weil, Eds.) Israel University Press, Jerusalem, 133–150.

Michel, L. [1980]. Symmetry defects and broken symmetry configurations. Hidden symmetry. *Rev. Mod. Phys.* **52**, 617–651.

Miller, W. [1972]. *Symmetry Groups and Their Applications.* Academic Press, New York.

Montaldi, J.A., Roberts, R.M., and Stewart, I.N. [1987]. Nonlinear normal modes of symmetric Hamiltonian systems. In: Güttinger and Dangelmayr [1987], 354–371.

Montaldi, J., Roberts, M., and Stewart, I. [1988]. Periodic solutions near equilibria of symmetric Hamiltonian systems. *Phil. Trans. Roy. Soc. London* **A325**, 237–293.

Montgomery, D. and Zippin, L. [1955]. *Topological Transformation Groups.* Interscience Publishers, New York.

Mullin, T. [1982]. Cellular mutations in Taylor flow. *J. Fluid Mech.* **121**, 207–218.

Mullin, T. and Cliffe, K.A. [1986]. Symmetry breaking and the onset of time dependence. *Proc. Malvern Seminar on Dynamical Systems,* (S. Sarker, Ed.).

Mullin, T., Cliffe, K.A., and Pfister, G. [1987]. Chaotic phenomena in Taylor–Couette flow at moderately low Reynolds numbers. *Phys. Rev. Lett.* **58**, 2212–2215.

Mullin, T., Pfister, G., and Lorenzen, A. [1982]. New observations on hysteresis effects in Taylor–Couette flow. *Phys. Fluids* **25**, 1134–1136.

Nagata, W. [1986a]. Unfoldings of degenerate Hopf bifurcations with **O**(2) symmetry. *Dyn. Stab. Sys.* **1**, 125–158.

Nagata, W. [1986b]. Symmetric Hopf bifurcations in magnetoconvection. In: Golubitsky and Guckenheimer [1986], 237–266.

Nagata, W. and Thomas, J.W. [1986]. Bifurcation in doubly-diffusive systems, II: time periodic solutions. *SIAM J. Math. Anal.* **17**, 289–311.

Neveling, M., Lang, D., Haug, P., Güttinger, W., and Dangelmayr, G. [1987]. Interactions of stationary modes in systems with two and three spatial degrees of freedom. In: Güttinger and Dangelmayr [1987], 153–165.

Othmer, H.G. and Scriven, L.E. [1971]. Instability and dynamic pattern in cellular networks. *J. Theor. Biol.* **32**, 507–537.

Palis, J.J. and de Melo, W. [1982]. *Geometric Theory of Dynamical Systems.* Springer-Verlag, New York.

Poénaru, V. [1976]. *Singularités C^∞ en Présence de Symétrie.* Lecture Notes in Mathematics **510**, Springer-Verlag, Berlin.

Pospiech, C. [1986]. Global bifurcation with symmetry breaking. University of Heidelberg. Preprint.

Rand, D. [1982]. Dynamics and symmetry: predictions for modulated waves in rotating fluids. *Arch. Rational Mech. Anal.* **79**, 1–38.

Renardy, M. [1982]. Bifurcation from rotating waves. *Arch. Rational Mech. Anal.* **75**, 49–84.

Richtmyer, R.D. [1978]. *Principles of Advanced Mathematical Physics.* Springer-Verlag, New York.

Rivlin, R.S. [1948]. Large elastic deformations of isotropic materials, II. Some uniqueness theorems for pure, homogeneous, deformations. *Phil. Trans R. Soc. Lond.* **A240**, 491–508.

Rivlin, R.S. [1974]. Stability of pure homogeneous deformations of an elastic cube under dead loading. *Quart. Appl. Math.* **32**, 265–271.

Roberts, M. [1985a]. Equivariant Milnor numbers and invariant Morse approximations. *J. Lond. Math. Soc.* **31**, 487–500.

Roberts, M. [1985b]. On the genericity of some properties of equivariant map germs. *J. Lond. Math. Soc.* **32**, 177–192.

Roberts, M. [1986]. Characterisations of finitely determined equivariant map germs. *Math. Ann.* **275**, 583–597.

Roberts, M., Swift, J.W., and Wagner, D. [1986]. The Hopf bifurcation on a hexagonal lattice. In: Golubitsky and Guckenheimer [1986], 283–318.

Ruelle, D. [1973]. Bifurcations in the presence of a symmetry group. *Arch. Rational Mech. Anal.* **51**, 136–152.

Sattinger, D.H. [1973]. *Topics in Stability and Bifurcation Theory.* Lecture Notes in Mathematics **309**. Springer-Verlag, Berlin.

Sattinger, D.H. [1978]. Group representation theory, bifurcation theory and pattern formation. *J. Func. Anal.* **28**, 58–101.

Sattinger, D.H. [1979]. *Group Theoretic Methods in Bifurcation Theory*. Lecture Notes in Mathematics **762**. Springer-Verlag, Berlin.

Sattinger, D.H. [1980]. Spontaneous symmetry-breaking in bifurcation problems. In: *Symmetries in Science*, Plenum, New York, 365–383.

Sattinger, D.H. [1983]. *Branching in the Presence of Symmetry. CBMS-NSF Conference Notes* **40**. SIAM, Philadelphia.

Sawyers, K.N. [1976]. Stability of an elastic cube under dead loading: two equal forces. *Int. J. Nonlinear Mech.* **11**, 11–23.

Schaeffer, D.G. [1980]. Qualitative analysis of a model for boundary effects in the Taylor problem. *Math. Proc. Camb. Phil. Soc.* **87**, 307–337.

Schaeffer, D.G. [1983]. Topics in bifurcation theory. In: *Systems of Nonlinear Partial Differential Equations* (J. Ball, Ed.) Reidel, Dordrecht, 219–262.

Schaeffer, D. and Golubitsky, M. [1979]. Boundary conditions and mode jumping in the bucking of a rectangular plate. *Commun. Math. Phys.* **69**, 209–236.

Schaeffer, D. and Golubitsky, M. [1981]. Bifurcation analysis near a double eigenvalue of a model chemical reaction. *Arch. Rational Mech. Anal.* **75**, 315–347.

Schecter, S. [1976]. Bifurcation with symmetry. In: Marsden and McCracken [1976], 224–249.

Schluter, A., Lortz, D., and Busse, F. [1965]. On the stability of steady finite amplitude convection. *J. Fluid Mech.* **23**, 129–144.

Schwarz, G. [1975]. Smooth functions invariant under the action of a compact Lie group. *Topology* **14**, 63–68.

Shearer, M. [1981]. Coincident bifurcation of equilibrium and periodic solutions of evolution equations. *J. Math. Anal. Appl.* **84**, 113–132.

Siegel, C.L. and Moser, J. [1956]. *Lectures on Celestial Mechanics*. Springer-Verlag, New York.

Signoret, F. and Iooss, G. [1987]. Une singularité de codimension 3 dans le problème de Couette–Taylor. Preprint No. **142**, Université de Nice.

Silber, M. and Knobloch, E. [1987]. Pattern selection in ferrofluids. *Physica* D. To appear.

Simonelli, F. and Gollub, J.P. [1987]. The masking of symmetry by degeneracy in the dynamics of interacting modes. Harvard University. Preprint.

Smale, S. [1974]. A mathematical model of two cells via Turing's equation. In: *Some Mathematical Questions in Biology V* (J.D. Cowan, Ed.) *AMS Lecture Notes on Mathematics in the Life Sciences* **6**, 15–26.

Smoller, J. and Wasserman, A.G. [1985]. Symmetry-breaking for semilinear elliptic equations. In: *Ordinary and Partial Differential Equations* (B.D. Sleeman and R.J. Jarvis, Eds.) Lecture Notes in Mathematics **1151**, Springer-Verlag, New York, 325–334.

Smoller, J. and Wasserman, A.G. [1986a]. Symmetry-breaking for positive solutions of semilinear elliptic equations. *Arch. Rational Mech. Anal.* **95**, 217–225.

Smoller, J. and Wasserman, A.G. [1986b]. Symmetry-breaking for positive solutions of semilinear elliptic equations with general boundary conditions. University of Michigan. Preprint.

Snyder, H.A. [1968]. Wave number selection at finite amplitude in rotating Couette flow. *J. Fluid Mech.* **35**, 273–298.

Snyder, H.A. [1968]. Stability of rotating Couette flow, I. Asymmetric waveforms. *Phys. Fluids* **11**, 728–734, 1599–1605.

Spencer, A.J.M. and Rivlin, R.S. [1959]. Symmetric 3×3 matrices. *Arch. Rational Mech. Anal.* **2**, 435–446.

Steindl, A. and Troger, H. [1987a]. Bifurcations of the equilibrium of a spherical double pendulum at a multiple eigenvalue. *Int. Ser. of Num. Math.* **79**, Springer-Verlag, 227–287.

Steindl, A. and Troger, H. [1987b]. Bifurcations and oscillations in mechanical systems with symmetries. *Proceedings of the 11th International Conference on Nonlinear Oscillations*. Budapest.

Stewart, I.N. [1987a]. Bifurcations with symmetry. In: *New Directions in Dynamical Systems* (T. Bedford and J.W. Swift, Eds.) Cambridge University Press, 235–283.

Stewart, I.N. [1987b]. Stability of periodic solutions in symmetric Hopf bifurcation. *Dyn. Stab. Sys.* To appear.

Swift, J.W. [1984a]. Bifurcation and Symmetry in Convection. Thesis, University of California, Berkeley.

Swift, J.W. [1984b]. Convection in a rotating fluid layer. *Contemp. Math.* **28**, 435–448.

Swift, J.W. [1986]. Hopf bifurcation with the symmetry of the square. *Nonlinearity*. To appear.

Tagg, R., Hirst, D., and Swinney, H.L. [1988]. Critical dynamics near the spiral-Taylor vortex codimension two point. Preprint.

Takens, F. [1974]. Singularities of vector fields. *Publ. Math. I.H.E.S.* **43**, 47–100.

Taliaferro, S.D. [1987]. Stability of bifurcating solutions in the presence of symmetry. *Nonlinear Analysis. Th., Meth. & Appl.* **8**, No. 10, 821–839.

Tavener, S. and Cliffe, A. [1987]. Primary flow exchange mechanisms in Taylor–Couette flow applying nonflux boundary conditions. *Proc. Roy. Soc. Lond.* Submitted.

Tavener, S., Mullin, T., and Cliffe, A. [1987]. Bifurcation in Taylor–Couette flow with rotating ends. *J. Fluid Mech.* Submitted.

Taylor, G.I. [1923]. Stability of a viscous liquid contained between two rotating cylinders. *Phil. Trans. Roy. Soc. London* **A223**, 289–343.

Tsotsis, T.T., [1981]. Nonuniform steady states in systems of interacting catalyst particles: the case of negligible interparticle mass transfer coefficient. *Chem. Eng. Commun.* **11**, 27–58.

Turing, A. [1952]. The chemical basis of morphogenesis. *Phil. Trans. Roy. Soc.* **B237**, 37–72.

Turner, J.S. [1974]. Double diffusive phenomena. *Ann. Rev. Fluid Mech.* **6**, 37–56.

Vanderbauwhede, A. [1980]. *Local Bifurcation and Symmetry*. Habilitation Thesis, Rijksuniversiteit Gent.

Vanderbauwhede, A. [1982]. *Local Bifurcation and Symmetry*. Res. Notes Math. **75**. Pitman, Boston.

Vanderbauwhede, A. [1984a]. Hopf bifurcation in symmetric systems I, II. University of Gent. Preprints.

Vanderbauwhede, A. [1984b]. Bifurcation of periodic solutions in symmetric systems: an example. University of Gent. Preprint.

Vanderbauwhede, A. [1985]. Bifurcation of periodic solutions in a rotationally symmetric oscillation system. *J. Reine u. Angew. Math.* **360**, 1–18.

Vanderbauwhede, A. [1987a]. Center manifolds, normal forms and elementary bifurcations. University of Gent. Preprint.

Vanderbauwhede, A. [1987b]. Symmetry-breaking at positive solutions. *Z. Angew. Math. Phys.* **38**, 315–326.

Wall, C.T.C. [1985a]. Equivariant jets. *Math. Ann.* **272**, 41–65.

Wall, C.T.C. [1985b]. Infinite determinacy of equivariant map germs. *Math. Ann.* **272**, 67–82.

Wall, C.T.C. [1985c]. Survey of recent results on singularities of equivariant maps. University of Liverpool. Preprint.

Wan, Y-H. and Marsden, J.E. [1983]. Symmetry and bifurcation in three-dimensional elasticity III. *Arch. Rational Mech. Anal.* **84**, 203–233.

Werner, B. and Spence, A. [1984]. The computation of symmetry-breaking bifurcation points. *SIAM J. Numer. Anal.* **21**, 388–399.

Weyl, H. [1946]. *The Classical Groups*, Princeton University Press, Princeton.

Whitney, H. [1943]. Differentiable even functions. *Duke Math. J.* **10**, 159–160.

Whittaker, E.T. and Watson, G.N. [1948]. *A Course on Mathematical Analysis.* Cambridge University Press, Macmillan, New York.

Wigner, E.P. [1959]. *Group Theory.* Academic Press, New York.

Wilkinson, J.H. [1965]. *The Algebraic Eigenvalue Problem.* Oxford University Press, Oxford.

Winfree, A.T. [1980]. *The Geometry of Biological Time.* Springer-Verlag, New York.

Winters, K.H., Cliffe, K.H., and Jackson, C. [1987]. The prediction of instabilities using bifurcation theory. In Numerical Methods for Transient and Coupled Problems (R.W. Lewis et al., Eds.) Wiley, New York, 179–198.

Winters, K.H. and Cliffe, K.A. [1986]. The onset of convection in a bounded fluid with a free surface. *J. Comp. Phys.* Submitted.

Wolf, J.A. [1967]. *Spaces of Constant Curvature.* McGraw-Hill, New York.

Young, R.E. [1974]. Finite amplitude thermal convection in a spherical shell. *J. Fluid Mech.* **63**, 695–722.

Zeeman, E.C. [1981]. Bifurcation, catastrophe and turbulence. In: *New Directions in Applied Mathematics* (P.J. Hilton and G.S. Young, Eds.) Springer-Verlag, New York, 110–153.

Index

T

Applied Mathematical Sciences

cont. from page ii

44. Pazy: **Semigroups of Linear Operators and Applications to Partial Differential Equations.**
45. Glashoff/Gustafson: **Linear Optimization and Approximation: An Introduction to the Theoretical Analysis and Numerical Treatment of Semi-Infinite Programs.**
46. Wilcox: **Scattering Theory for Diffraction Gratings.**
47. Hale et al.: **An Introduction to Infinite Dimensional Dynamical Systems—Geometric Theory.**
48. Murray: **Asymptotic Analysis.**
49. Ladyzhenskaya: **The Boundary-Value Problems of Mathematical Physics.**
50. Wilcox: **Sound Propagation in Stratified Fluids.**
51. Golubitsky/Schaeffer: **Bifurcation and Groups in Bifurcation Theory, Vol. I.**
52. Chipot: **Variational Inequalities and Flow in Porous Media.**
53. Majda: **Compressible Fluid Flow and Systems of Conservation Laws in Several Space Variables.**
54. Wasow: **Linear Turning Point Theory.**
55. Yosida: **Operational Calculus: A Theory of Hyperfunctions.**
56. Chang/Howes: **Nonlinear Singular Perturbation Phenomena: Theory and Applications.**
57. Reinhardt: **Analysis of Approximation Methods for Differential and Integral Equations.**
58. Dwoyer/Hussaini/Voigt (eds.): **Theoretical Approaches to Turbulence.**
59. Sanders/Verhulst: **Averaging Methods in Nonlinear Dynamical Systems.**
60. Ghil/Childress: **Topics in Geophysical Dynamics: Atmospheric Dynamics, Dynamo Theory and Climate Dynamics.**
61. Sattinger/Weaver: **Lie Groups and Algebras with Applications to Physics, Geometry, and Mechanics.**
62. LaSalle: **The Stability and Control of Discrete Processes.**
63. Grasman: **Asymptotic Methods of Relaxation Oscillations and Applications.**
64. Hsu: **Cell-to-Cell Mapping: A Method of Global Analysis for Nonlinear Systems.**
65. Rand/Armbruster: **Perturbation Methods, Bifurcation Theory and Computer Algebra.**
66. Hlaváček/Haslinger/Nečas/Lovíšek: **Solution of Variational Inequalities in Mechanics.**
67. Cercignani: **The Boltzmann Equation and Its Applications.**
68. Temam: **Infinite Dimensional Dynamical System in Mechanics and Physics.**
69. Golubitsky/Stewart/Schaeffer: **Singularities and Groups in Bifurcation Theory, Vol. II.**